Results and Problems in Cell Differentiation

Volume 67

Series editors
Jacek Z. Kubiak, Rennes, France
Malgorzata Kloc, Houston, TX, USA

More information about this series at http://www.springer.com/series/400

Malgorzata Kloc

Editor

The Golgi Apparatus and Centriole

Functions, Interactions and Role in Disease

 Springer

Editor
Malgorzata Kloc
Department of Surgery
Houston Methodist Hospital
Houston, Texas, USA

ISSN 0080-1844   ISSN 1861-0412   (electronic)
Results and Problems in Cell Differentiation
ISBN 978-3-030-23172-9   ISBN 978-3-030-23173-6   (eBook)
https://doi.org/10.1007/978-3-030-23173-6

This Springer imprint is published by the registered company Springer Nature Switzerland AG.
The registered company address is: Gewerbestrasse 11, 6330 Cham, Switzerland

Preface

This volume reviews the current knowledge on the structure, composition, and functions of the Golgi and centriole/centrosome and the functional partnership and codependence of these two organelles, their roles in the establishment of cell and organ geometry and morphogenesis, and how the disruptions of their structure and positioning lead to the various diseases.

The first part of this volume describes structural diversity and evolution of centriole, the role of acetylated proteins and cytoskeletal remodeling proteins (formins) in the centriole and Golgi biology, and the role of intracellular transport and RhoA and Rab GTPase signaling in the formation of Golgi and Golgi/centriole complex.

The second part is devoted to the description of mechanisms involved in the positioning of Golgi and centriole in resting and directionally moving cells, the significance of their positioning, and the methods for studying the Golgi dynamics in the semi-intact cell system.

The third part describes how centrosome coordinates divisions during *Drosophila* early embryogenesis and focuses on the role of the centriole and Golgi in the establishment of cell geometry, organ branching, tubulogenesis, neurogenesis, and differentiation of neurons and hypothesizes how the Golgi may communicate with the cell periphery.

The fourth part summarizes our current knowledge on the role of Golgi and centriole in stress response and various diseases and describes how the changes in the Golgi/centriole structure/number may lead to development or/and progression of cancer.

We believe that this volume besides being highly informative and scientifically inspiring will shed new light on the mechanisms and role of the Golgi/centriole functional partnership during development and in health and disease.

Houston, TX   Malgorzata Kloc

Abstract

This book reviews the most recent knowledge on the evolution, structure, functions, codependence, and interactions of centriole and Golgi apparatus; what roles they play in the establishment of cell and organ geometry and development; and how their disruption leads to cancer and other diseases.

The book covers the following subjects: the evolution of centriole structure and the role of intracellular transport and centriole in the formation of the Golgi ribbon; the role of small GTPases and acetylated proteins in the Golgi and centriole/centrosome structure and function; the mechanisms and methods to study the dynamics and the role of positioning of Golgi/centriole in different cell types and how they communicate with cell periphery; the role of centriole/Golgi in embryo development, and in the establishment of geometry and polarity of cells and organs; and how the inherited or acquired defects in centriole or Golgi lead to cancer and other diseases.

This book should give the readers a new and often unrecognized perspective on the roles of the centriole and Golgi complex, structural and functional codependence and partnership between these two organelles, and their importance for various aspects of cell and organ functions.

Keywords Golgi · Centriole · Centrosome · Polarity · Geometry · Morphogenesis · Evolution

Contents

Part I
Golgi and Centriole Structure, Assembly and Regulation

Chapter 1
The Evolution of Centriole Structure: Heterochrony, Neoteny, and Hypermorphosis

Tomer Avidor-Reiss and Katerina Turner

Abstract Centrioles are subcellular organelles that were present in the last eukaryotic common ancestor, where the centriole's ancestral role was to form cilia. Centrioles have maintained a remarkably conserved structure in eukaryotes that have cilia, while groups that lack cilia have lost their centrioles, highlighting the structure–function relationship that exists between the centriole and the cilium. In contrast, animal sperm cells, a ciliated cell, exhibit remarkable structural diversity in the centriole. Understanding how this structural diversity evolved may provide insight into centriole assembly and function, as well as their unique role in sperm. Here, we apply concepts used in the study of the evolution of animal morphology to gain insight into the evolution of centriole structure. We propose that centrioles with an atypical structure form because of changes in the timing of centriole assembly events, which can be described as centriolar "heterochrony." Atypical centrioles of insects and mammals appear to have evolved through different types of heterochrony. Here, we discuss two particular types of heterochrony: neoteny and hypermorphosis. The centriole assembly of insect sperm cells exhibits the retention of "juvenile" centriole structure, which can be described as centriolar "neoteny." Mammalian sperm cells have an extended centriole assembly program through the addition of novel steps such as centrosome reduction and centriole remodeling to form atypical centrioles, a form of centriole "hypermorphosis." Overall, centriole heterochrony appears to be a common mechanism for the development of the atypical centriole during the evolution of centriole assembly of various animals' sperm.

1.1 Introduction

Centrioles are present in most eukaryotic cell types and are essential for the development and physiology of humans and many animals. Because centrioles are so essential for life, they have been studied using multiple approaches in many in vitro

T. Avidor-Reiss (✉) · K. Turner
Department of Biological Sciences, University of Toledo, Toledo, OH, USA
e-mail: Tomer.AvidorReiss@utoledo.edu; Katerina.Turner@utoledo.edu

© Springer Nature Switzerland AG 2019
M. Kloc (ed.), *The Golgi Apparatus and Centriole*, Results and Problems in Cell
Differentiation 67, https://doi.org/10.1007/978-3-030-23173-6_1

and in vivo systems. Over the years, it has become evident that, while centriole structure and function are highly conserved, centrioles exhibit distinct and sometimes dramatic differences (Jana et al. 2018; Riparbelli et al. 2010). Many studies focus on the more universal aspects of centrioles to draw conclusions that are applicable across species because conservation suggests a similar underlying mechanism (Jana et al. 2016; Winey and O'Toole 2014; Sluder 2016). However, differences between centrioles are also significant for several reasons. First, some differences provide a unique opportunity to overcome a difficulty in investigating a process (e.g., the presence of the giant centriole cartwheel in some species was instrumental in elucidating its detailed structure) (Guichard et al. 2012). Second, understanding the differences can provide conceptual insight that would otherwise be hidden. For example, the observation that in some species centrioles with one symmetry can nucleate a centriole with a different symmetry, suggests that the preexisting centriole does not act as the template for centriole organization (Phillips 1967). Third, differences are commonly present and are essential for animal or tissue-specific function; impacting them can result in devastating pathologies. Fourth, differences provide a basis for tissue-specific therapeutics with minimal systemic side effects. Last, there are evolutionary reasons for differences—they are beneficial. For these reasons, in this chapter, we will focus on the diversity in centriole structure and how differently shaped centrioles evolved.

Here, to gain insight into centriole structural diversity, we take the approach best described by Theodosius Dobzhansky: "Nothing in Biology Makes Sense Except in the Light of Evolution" (Dobzhansky 1973). We will apply concepts from the study of animal development such as heterochrony, neoteny, and hypermorphosis to study the evolution of the centriole. We focus on sperm because, due to the postcopulatory sexual selection, it underwent rapid evolution, during which time the typical structure of the centriole changed in many species (Lupold and Pitnick 2018; Mordhorst et al. 2016). This chapter starts with background on heterochrony and centriole structure and function. We continue with describing two types of centriole changes: a neotenic change in insect proximal centrioles and a hypermorphotic change in mammalian distal centrioles. We will then discuss potential molecular mechanisms that may be essential to this evolutionary change. Finally, we propose that applying the concept of heterochrony, which was originally intended to explain organismal evolution, to organelle evolution is beneficial to understanding the molecular basis of heterochrony.

1.2   Heterochrony, Neoteny, and Hypermorphosis

A comparative biology approach is routinely used in the study of sperm where the sperm centriole mainly acts as a tool to determine the phylogenetic relationship between groups of animals (see for example Dias et al. 2015). Here, we borrow concepts from evolutionary developmental biology that are generally used to describe animal development, to explain changes in centriole assembly and structure. One general concept we focus on is heterochrony, a term originally coined by the nineteenth century German biologist Ernst Haeckel in the context of the theory of recapitulation. The modern premise of heterochrony, as explained by the twentieth century evolutionary biologist Stephen Jay Gould, is that the development of an organism (ontogeny) and the evolution of an organism (phylogeny) are related; and changes in the timing and the rate of developmental processes explain evolutionary change (Gould 1977; McNamara and McKinney 2005). For example, the developmental process for the formation of vertebrae may be happening quicker or slower resulting in a relatively longer or shorter spine in similar species (Keyte and Smith 2014). At its core, heterochrony provides an explanation for the differences observed in various species in terms of evolutionary change and timing of development. Other ideas that we do not discuss here are that the evolutionary change can be mediated by changing the location of a process (i.e., Heterotopy).

Heterochrony can be divided into two broad categories of changes (Smith 2002): (1) changes that result in a juvenile or simple shape in comparison to the ancestral shape and (2) changes that result in a more complex shape in comparison to the ancestral shape. Here we focus on a specific example for each category, known as neoteny and hypermorphosis. Neoteny is a decrease in the rate of development or a maturation arrest at an early stage. Hypermorphosis is an acceleration or extension of a preexisting process to accommodate additional steps.

The concept of neoteny has already been "borrowed" to describe a cellular process; the term "cellular neoteny" was used to describe the differentiation program that generates various neuronal and neuroendocrine cells. It was suggested that these cell types might represent different stages of differentiation by cells "arresting" along a linear development pathway, whose endpoint is a cholinergic sympathetic neuron (Anderson 1989). Here, we apply this concept to the subcellular level, which in our case is the alteration of the timing of centriole assembly events. We create distinct analogies between "animal" and "centriole," "development of an animal" and "assembly of a centriole," and "evolution of an animal" and "evolution of a centriole." The centrioles of sperm cells are particularly suitable for this analysis because postmating sexual selection drove the rapid evolution of sperm, during which time centriole structure changed in many species (Mordhorst et al. 2016; Lupold and Pitnick 2018).

1.3 The Centriole and Cilium Structure–Function Relationship Restricts Centriole Diversity

Centrioles are barrel-shaped structures made of nine triplet microtubule blades that form a wall surrounding the centriole lumen (Fig. 1.1a–ii). Each blade is made up of three connected microtubules (named A, closest to the lumen, B, and C, furthest from the lumen) and therefore is referred to as triplet microtubules. Centrioles have two essential functions inside the cell (Bornens 2012). The centrioles form centrosomes, which are large microtubule-organizing centers in the cell; the resulting organized microtubules mediate cell division and intracellular transport. Centrioles

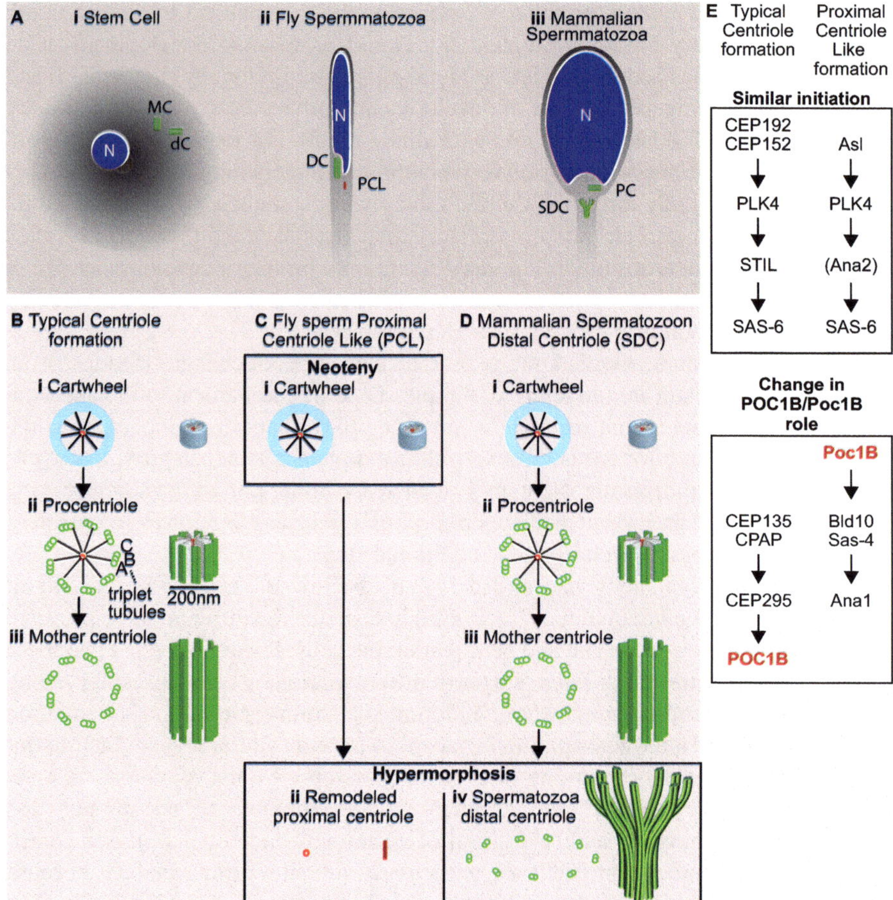

Fig. 1.1 Model of centriole development in various animal groups. The centrioles are depicted via cross section at the centriole base and side view. (**a**) A model depicting the two centrioles in a stem cell (i), fly spermatozoon (ii), and non-rodent mammal spermatozoon (iii). *N* nucleus, *MC* mother centriole, *dC* daughter centriole, *DC* distal centriole, *PCL* proximal centriole like, *SDC* spermatozoon distal centriole, *PC* proximal centriole. (**b–d**) Models depicting the mechanism of a typical centriole formation in a stem cell (**b**), of an atypical centriole in fly sperm (**c**), and of an atypical centriole in mammalian sperm (**d**). (**b**) A typical centriole forms from a cartwheel made of a central tubule with spokes surrounded by an amorphous wall (i). Then, the procentriole develops a wall of nine singlet tubules, which grows to doublet tubules, and then triplet tubules (ii). Next, the procentriole elongates and loses its cartwheel (iii). (**c**) The neotenic sperm centriole of flies (the PCL) initially resembles the cartwheel stage and is made of a central tubule with spokes and an amorphous wall (i). Then, the neotenic centriole is remodeled, losing its amorphous wall (ii) in a hypermorphic step. (**d**) The hypermorphic sperm centriole of non-rodent mammals starts its formation like a typical centriole with a cartwheel (i), procentriole (ii), and a mature centriole (iii). Finally, the centriole is remodeled by splaying the microtubules in a hypermorphic step (iv). (**e**) The molecular pathway of human typical centriole formation (left column) and PCL formation (right column). Genes in the same row are orthologues to each other in humans and flies, except for Poc1B that changes position in the pathway. The figure shows that the same molecular pathway initiates the typical centriole and fly PCL, but Poc1B gains an earlier essential function in the formation of the PCL as compared to the human typical centriole pathway

are also responsible for the formation of cilia, which are hair-like organelles that are essential for cell motility as well as cell–cell communication. The centriole also provides a stable anchor for the cilium and centrosome after their formation when they perform their respective functions. A typical animal cell has two centrioles (Fig. 1.1a-i). These centrioles are different from each other in their age, structure, composition, and function. The older centriole (aka mother centriole) is structurally and compositionally mature, and it is functionally competent to form a centrosome or a cilium. The younger centriole (aka daughter centriole) is immature; thus, it is unable to build a centrosome or a cilium.

Animal centrioles form centrosomes, and most animal cells require two centrosomes for normal mitosis (Nigg and Raff 2009; Bornens 2012). The centrosome nucleates and anchors asters of microtubules and determines the location of the mitotic spindle pole (Tang and Marshall 2012). When present, centrosomes are the dominant microtubule-organizing center in the cell. When centrosomes are normally absent, as in the oocyte, a self-assembly mechanism can mediate mitosis (Petry 2016). However, when centrosomes are abnormally absent, there is an increased rate of chromosome missegregation during mitosis (Poulton et al. 2014). An abnormal number of centrosomes can lead to mono- or multipolar spindles, which often results in cell death (Prosser and Pelletier 2017). An exception to this outcome occurs in cancer cells, which overcome the centrosome's dominance by clustering the centrosomes in a bipolar spindle (Leber et al. 2010). However, asymmetric clustering of centrosomes can also cause chromosome missegregation (Cosenza et al. 2017). Altogether, mature centrosomes, and the centrioles within them, are microtubule organization centers whose *precise number* is essential for normal animal development.

Centriole number control is achieved through a two-part process: first, by regulating the number of newly assembled centrioles in the cell and, second, by precisely segregating centriole pairs, each made up of one old and one new, during cell division (Firat-Karalar and Stearns 2014). New centrioles are assembled in association with a preexisting (mature) centriole that serves as a platform to restrict centriole formation to one centriole per preexisting centriole per cell cycle. Many proteins that are key to centriole assembly have been identified, but the precise mechanism that assures that only a single new centriole forms near an old centriole is still under intensive investigation. However, it appears that centriole microtubules do not have an essential role in centriole duplication (Avidor-Reiss 2018). Altogether, having precisely two centrioles in a cell is essential for cellular function, animal viability, and reproductive success; the control of centriole formation requires a preexisting centriole, but centriolar microtubules are dispensable for the assembly of new centrioles or for centrosome function.

The ancestral role of cilia in eukaryotes is to produce cellular motility. This motility is generated by molecular machines known as dynein arms, which contain dynein motor proteins (Viswanadha et al. 2017). The dynein arms are permanently attached to each of the microtubule blades on one side and are transiently binding to a nearby microtubule blade to exert the force that produces motility. This force results in one microtubule blade sliding relative to the other microtubule blade. Each microtubule blade is made of two connected microtubules (called A and B) and are therefore referred to as doublet microtubules. There are nine doublets arranged in a circle, such that each of the nine microtubule doublets can slide against another doublet. This ninefold arrangement is conserved in animal evolution and found across many groups. These microtubules form the cilium skeleton that is named the axoneme, and they are the cilium's most fundamental structural element. More details on cilium motility can be found in Downing and Sui (2007).

In addition to cell motility, cilia function as a cell receiver or antenna in cell signaling (Malicki and Johnson 2017). In many of these cases, the cilia are immotile and the dynein arms are missing. In order to be an efficient signaling device, the cilium is compartmentalized from the rest of the cell by a cilium gate and the cilium transport machinery allows entry of specific ciliary cargo. The cilium gate (aka transition zone in general or annulus in sperm cells) and the cilium transport machinery (aka intraflagellar transport) are built around and travel along the axoneme microtubules. The cilium gate connects the microtubule doublets and the ciliary membrane to form a barrier between both the cilioplasm and cytoplasm, and the cilium membrane and cell membrane. More details on cilium gate and cilium transport machinery can be found in Malicki and Avidor-Reiss (2014). The critical point to our discussion is that cilia mediate signals utilizing an axoneme made of microtubule doublets organize in ninefold symmetry.

During cilium formation, the centriolar microtubules extend to form the cilium microtubules. Therefore, the centriole's microtubules dictate the symmetry of the axoneme microtubules, which are critical to the cilium's motility and signaling function. Because the centriole's structure has such an important role in axoneme structure, it makes sense that centriole structure is highly conserved throughout evolution.

Centriole assembly is conserved in protists, invertebrates, and vertebrates (Azimzadeh 2014). The new centriole initially forms as a cartwheel structure surrounded by electron dense material at the base of the preexisting centriole, near to the wall (Fig. 1.1b-i,ii). Next, microtubules are built around the cartwheel to create the procentriole. First, the A microtubules are formed and later the B and C microtubules. The completed procentriole structure is 200 nm long and 200 nm wide, including the wall made of nine microtubule triplets and a centriole lumen filled by the cartwheel. The formation of the cartwheel and procentriole usually happens in the early S phase of the cell cycle and is very rapid. The next step in centriole formation is the elongation of the centriole, which starts in the G2 phase of the cell cycle. In this stage, the microtubules of the centriole elongate to about 400–500 nm in length. The cartwheel does not elongate and is restricted to the base of the centriole. Finally, the cartwheel is eliminated from the centriole base and the distal lumen is formed, which has a distinct structure composed of rings and columns (Fig. 1.1b-iii). Altogether, centriole formation is a step-by-step process in which a cartwheel forms, then develops to become a procentriole, and further matures into a centriole.

1.4 Centriole Neoteny in Sperm Cells

During development, certain traits can be advantageous to a young animal, but those same traits become a detriment when the animal reaches maturity so they are replaced by adult features. Neoteny describes the inverse; it is a biological phenomenon where an adult animal retains juvenile features, presumably because those features remain advantageous (Gould 1977). The classic example of neoteny in an organism is the *Ambystoma mexicanum*, or axolotl, a species of salamander. Most salamanders start their life as a larva; at this stage, they live in water, and have external gills, and a caudal fin. They then develop into an adult form that lives on land and breathes air. However, unlike other salamanders, the adult axolotl retains some larval characteristics as it matures, it continues living in water, and has external gills, and a caudal fin (Rosenkilde and Ussing 1996).

Identifying neoteny in nature is useful because it provides insight into the type of evolutionary changes that led to the morphology of an animal and is likely linked to developmental genes. Here we propose that the term neoteny has a broader application and can be applied to subcellular structures that retain immature features in an otherwise mature subcellular system. We hypothesize that these structures may also exhibit neoteny by arresting early in certain specialized cells. We propose that the centriole found in insect sperm cells is a neotenic subcellular structure.

In most animals, round spermatids (haploid cells that differentiate to form spermatozoa) have two mature centrioles, named the distal centriole and the proximal centriole (Avidor-Reiss et al. 2015). However, insect spermatids for a long time were thought to have only one centriole, the distal centriole, which has the typical barrel-shaped structure with a microtubule wall. Recently, an early form of the procentriole was identified in the insect spermatid near the distal centriole (Khire et al. 2016; Blachon et al. 2014; Gottardo et al. 2015; Dallai et al. 2017; Fishman et al. 2017) (Fig. 1.1a-ii). This structurally immature form of sperm centriole was named the proximal centriole-like structure or PCL and may represent an example of subcellular neoteny; the structure maintains juvenile traits while the sperm itself matures from spermatid to spermatozoon. During spermatid differentiation, both the distal centriole and the PCL undergo remodeling that further modifies their structure (Fig. 1.1c). Both centrioles are deposited in the egg after fertilization and both function in zygotes like mature centrioles, which include nucleating new centrioles.

When neoteny is exhibited, it is thought that the halt of development is evolutionarily beneficial. In the case of humans, neoteny may provide more time to increase brain size after birth and more time to develop social skills (Skulachev et al. 2017; Bufill et al. 2011). The reason for sperm centriole neoteny is not yet clear, but it may be an advantage for sperm to have an immature centriole when competing with other sperm trying to fertilize the egg. The smaller size of the centriole does not deform the neck of the sperm, thus improving motility. Neoteny, like other evolutionary changes in development, is mainly thought to be a result of mutations in the regulation of genes that control development, but the precise mutations are not known. Similarly, the centriole neoteny that forms the PCL may be due to mutations in genes that control the development of centrioles in the sperm. One potential gene to mediate PCL neoteny is the gene *poc1* (see Sect. 1.6).

1.5 Centriole Hypermorphosis in Sperm Cells

Adult animals exhibit certain traits that are characteristic of their maturation. Hypermorphosis is a biological phenomenon where development is extended, for example, by the addition of new developmental stages at the end of the ancestral development sequence. The common example of hypermorphosis is the enlargement of a body part relative to the rest of the body, such as the large antlers of reindeer or the large upper canine teeth of saber-toothed tigers. Interestingly, it was proposed that hypermorphosis may be a mechanism for the evolution of male weaponry (Kelly and Adams 2010). Similar to animal development, subcellular structures can also have developmental programs that reach a "mature" state, which then could be extended. Here, we propose that the centrioles found in mammalian sperm cells exhibit hypermorphosis.

In most animals, a spermatozoon has two centrioles, each with typical mature centrioles morphology (Avidor-Reiss et al. 2015). However, most mammalian spermatozoon only has one typical centriole, the proximal centriole. Recently, a distinctly shaped centriole was identified in the spermatozoon of non-rodent mammals (Fishman et al. 2018; Avidor-Reiss and Fishman 2018) (Fig. 1.1d-iv). This shape results from the remodeling of the distal centriole during spermatid differentiation. Both centrioles, the typical centriole and the atypical centriole, are deposited in the egg after fertilization, and both function in the zygote like mature centrioles, which includes forming centrosomes and nucleating new centrioles. A more moderate form of distal centriole remodeling is observed in insects (Khire et al. 2016; Dallai et al. 2018; Fishman et al. 2017). We propose that the alteration of the distal centriole's structure is due to the addition of new developmental stages after the end of normal centriole maturation when the sperm is maturing from spermatid to spermatozoon and, therefore, is an example of centriolar hypermorphosis (Fig. 1.1c).

Sperm centriolar hypermorphosis can take several forms in various animal groups. Compared to other mammals, the rodent spermatozoon's distal centriole is further modified, resulting in the apparent degeneration of the DC. Furthermore, the rodent spermatozoon's proximal sperm centriole is also degenerated after it is fully formed (Simerly et al. 2016). Similarly, in insects, the neotenic proximal centriole, the PCL, undergoes further remodeling after its neotenic formation is finished, suggesting that the PCL is a product of two heterochronic processes: neoteny and hypermorphosis. Currently, it is unclear if the two processes evolved together, or one after the other.

1.6 The Genetic and Molecular Control of Centriole Heterochrony

In the last two decades, some progress has been made in understanding the molecular changes underlying heterochrony, but the complexity of studying whole animal development presents a major barrier to that progress (Keyte and Smith 2014). Centriole assembly is much simpler than animal development and may provide some insight into the understanding of the molecular basis of heterochrony. It would also be interesting to compare the molecular basis of heterochrony at a subcellular level and at the whole animal level to determine if there are general rules that affect developmental timing. Here, we suggest that the appearance of a neotenic centriole in flies is linked to a change in the essential function of the gene Protein of Centriole 1 (*poc1*), based on the comparison of the molecular pathways that form the PCL and the centriole.

Poc1 is a family of proteins that is evolutionarily conserved and is found throughout the eukaryotic tree of life suggesting it was present in the ancestral eukaryote that had a centriole (Hodges et al. 2010). Poc1 family members are found only in eukaryotes that have centrioles, pointing to its specific role in centriole biology. However, Poc1 members are absent in some eukaryotes, such as nematodes, indicating it is not one of the core essential centriole proteins. In vertebrates, the Poc1 family is made of two genes (POC1A and POC1B), in invertebrates the Poc1 family is made of one gene, *poc1*. In flies, the *poc1* gene codes for two splice isoforms: Poc1A, which localizes to the typical centriole (the DC), and Poc1B, which localizes to the atypical centriole (the PCL) (Khire et al. 2016). Depletion of Poc1 proteins in human cells and fly sperm results in short centrioles that are unstable, hinting that Poc1 is essential after the initial formation of the procentriole (Keller et al. 2009; Pearson et al. 2009; Blachon et al. 2009). In fly sperm, Poc1 depletion also results in an abnormal looking PCL (Khire et al. 2015).

The placement of Poc1 proteins in the molecular pathway of centriole assembly was studied based on whether Poc1 was required or dispensable for the localization of other centriolar proteins to the centriole. In the centriole of human cells, the last steps in centriole assembly are Centrosomal Protein 135 (CEP135), which recruits Centrosomal Protein 295 (CEP295), which then recruits Protein of Centriole 1B (POC1B) (Chang et al. 2016) (Fig. 1.1e). In the fly PCL, the order of recruitment seems to be reversed; the fly ortholog gene of human POC1B (Poc1B) is essential for the recruitment of the fly CEP295 protein ortholog Anastral spindle 1 (Ana1) and the fly CEP135 protein ortholog Bald 10 (Bld10) (Fig. 1.1e) (Blachon et al. 2009). Together, these studies suggest that Poc1B gained a new essential early function in the centriole formation pathway in flies that is not observed in human typical centrioles. This new essential function may allow the cartwheel to be a stable structure and become the PCL, instead of being an intermediate structure that normally continues to develop into a stable centriole. To test this hypothesis, it would be critical to determine this new essential function more precisely.

One insight into the origin of Poc1's essential function in the early centriole is its localization during early centriole formation. Poc1 is recruited to the procentriole and localizes to the cartwheel in *Tetrahymena thermophila*, a ciliated protozoan, although Poc1 does not appear to have an essential function at that stage (Pearson et al. 2009). Therefore, one possible scenario is that Poc1 was recruited to the cartwheel by an ancestral mechanism, and the nonessential function of Poc1 evolved to an essential function in the fly. The Poc1 recruitment mechanism and the molecular change that made Poc1 essential are currently unknown. Altogether, small perturbations in proteins already functioning in the centriole (possibly through the generation new splice isoforms) may be the mechanism of centriole heterochrony.

1.7 Conclusions

Heterochrony, neoteny, and hypermorphosis are useful concepts for the study of the evolution of centrioles and other subcellular structures. Here, using these terms enables us to describe the different types of changes that occur in the centriole assembly pathway resulting in the formation of an atypical centriole shape. This creates a conceptual framework to study the evolution of the centriole. The future challenge is to understand the genetic and molecular basis of centriole heterochrony. The molecular pathway that assembles centrioles is extensively studied in a variety of eukaryotes that are amenable for genetic analysis, including vertebrates, invertebrates, and protists. Therefore, in the future we should be able to draw the ancestral pathway of centriole assembly and the step-by-step evolutionary changes that produce a variety of diverse centriole forms.

Acknowledgements We would like to thank Lilli Fishman for her assistance in preparing the manuscript. This work was supported by grant R03 HD087429 and R21 HD092700 from Eunice Kennedy Shriver National Institute of Child Health & Human Development (NICHD).

Conflict of Interest The authors declare that they do not have any conflicts of interest.

References

Anderson DJ (1989) Cellular 'neoteny': a possible developmental basis for chromaffin cell plasticity. Trends Genet 5(6):174–178

Avidor-Reiss T (2018) Rapid evolution of sperm produces diverse centriole structures that reveal the most rudimentary structure needed for function. Cells 7(7):67

Avidor-Reiss T, Fishman EL (2018) It takes two (centrioles) to Tango. Reproduction. https://doi.org/10.1530/REP-18-0350

Avidor-Reiss T, Khire A, Fishman EL, Jo KH (2015) Atypical centrioles during sexual reproduction. Front Cell Dev Biol 3:21. https://doi.org/10.3389/fcell.2015.00021

Azimzadeh J (2014) Exploring the evolutionary history of centrosomes. Philos Trans R Soc Lond B Biol Sci 369(1650). https://doi.org/10.1098/rstb.2013.0453

Blachon S, Cai X, Roberts KA, Yang K, Polyanovsky A, Church A, Avidor-Reiss T (2009) A proximal centriole-like structure is present in Drosophila spermatids and can serve as a model to study centriole duplication. Genetics 182(1):133–144. https://doi.org/10.1534/genetics.109. 101709

Blachon S, Khire A, Avidor-Reiss T (2014) The origin of the second centriole in the zygote of *Drosophila melanogaster*. Genetics 197(1):199–205. https://doi.org/10.1534/genetics.113.160523

Bornens M (2012) The centrosome in cells and organisms. Science 335(6067):422–426. https://doi. org/10.1126/science.1209037

Bufill E, Agusti J, Blesa R (2011) Human neoteny revisited: the case of synaptic plasticity. Am J Hum Biol 23(6):729–739. https://doi.org/10.1002/ajhb.21225

Chang CW, Hsu WB, Tsai JJ, Tang CJ, Tang TK (2016) CEP295 interacts with microtubules and is required for centriole elongation. J Cell Sci 129(13):2501–2513. https://doi.org/10.1242/jcs. 186338

Cosenza MR, Cazzola A, Rossberg A, Schieber NL, Konotop G, Bausch E, Slynko A, Holland-Letz T, Raab MS, Dubash T, Glimm H, Poppelreuther S, Herold-Mende C, Schwab Y, Kramer A (2017) Asymmetric centriole numbers at spindle poles cause chromosome missegregation in cancer. Cell Rep 20(8):1906–1920. https://doi.org/10.1016/j.celrep.2017.08.005

Dallai R, Mercati D, Lino-Neto J, Dias G, Lupetti P (2017) Evidence of a procentriole during spermiogenesis in the coccinellid insect *Adalia decempunctata* (L): an ultrastructural study. Arthropod Struct Dev 46(6):815–823. https://doi.org/10.1016/j.asd.2017.10.004

Dallai R, Mercati D, Lino-Neto J, Dias G, Folly C, Lupetti P (2018) The peculiar structure of the flagellar axoneme in Coccinellidae (Insecta-Coleoptera). Arthropod Struct Dev. https://doi.org/ 10.1016/j.asd.2018.11.004

Dias G, Lino-Neto J, Dallai R (2015) The sperm ultrastructure of *Stictoleptura cordigera* (Fussli, 1775) (Insecta, Coleoptera, Cerambycidae). Tissue Cell 47(1):73–77. https://doi.org/10.1016/j. tice.2014.11.007

Dobzhansky T (1973) Nothing in biology makes sense except in the light of evolution. Am Biol Teacher 35(3):125–129. https://doi.org/10.2307/4444260

Downing KH, Sui H (2007) Structural insights into microtubule doublet interactions in axonemes. Curr Opin Struct Biol 17(2):253–259. https://doi.org/10.1016/j.sbi.2007.03.013

Firat-Karalar EN, Stearns T (2014) The centriole duplication cycle. Philos Trans R Soc Lond B Biol Sci 369(1650). https://doi.org/10.1098/rstb.2013.0460

Fishman EL, Jo K, Ha A, Royfman R, Zinn A, Krishnamurthy M, Avidor-Reiss T (2017) Atypical centrioles are present in Tribolium sperm. Open Biol 7(3). https://doi.org/10.1098/rsob.160334

Fishman EL, Jo K, Nguyen QPH, Kong D, Royfman R, Cekic AR, Khanal S, Miller AL, Simerly C, Schatten G, Loncarek J, Mennella V, Avidor-Reiss T (2018) A novel atypical sperm centriole is functional during human fertilization. Nat Commun 9(1):2210. https://doi.org/10.1038/s41467-018-04678-8

Gottardo M, Callaini G, Riparbelli MG (2015) Structural characterization of procentrioles in Drosophila spermatids. Cytoskeleton (Hoboken) 72(11):576–584. https://doi.org/10.1002/cm. 21260

Gould SJ (1977) Ontogeny and phylogeny. Harvard University Press, Cambridge

Guichard P, Desfosses A, Maheshwari A, Hachet V, Dietrich C, Brune A, Ishikawa T, Sachse C, Gonczy P (2012) Cartwheel architecture of Trichonympha basal body. Science 337(6094):553. https://doi.org/10.1126/science.1222789

Hodges ME, Scheumann N, Wickstead B, Langdale JA, Gull K (2010) Reconstructing the evolutionary history of the centriole from protein components. J Cell Sci 123(Pt 9):1407–1413. https:// doi.org/10.1242/jcs.064873

Jana SC, Bettencourt-Dias M, Durand B, Megraw TL (2016) *Drosophila melanogaster* as a model for basal body research. Cilia 5:22. https://doi.org/10.1186/s13630-016-0041-5

Jana SC, Mendonca S, Machado P, Werner S, Rocha J, Pereira A, Maiato H, Bettencourt-Dias M (2018) Differential regulation of transition zone and centriole proteins contributes to ciliary base diversity. Nat Cell Biol 20(8):928–941. https://doi.org/10.1038/s41556-018-0132-1

Keller LC, Geimer S, Romijn E, Yates J III, Zamora I, Marshall WF (2009) Molecular architecture of the centriole proteome: the conserved WD40 domain protein POC1 is required for centriole duplication and length control. Mol Biol Cell 20(4):1150–1166. https://doi.org/10.1091/mbc. E08-06-0619

Kelly CD, Adams DC (2010) Sexual selection, ontogenetic acceleration, and hypermorphosis generates male trimorphism in Wellington Tree Weta. Evol Biol 37(4):200–209. https://doi. org/10.1007/s11692-010-9096-1

Keyte AL, Smith KK (2014) Heterochrony and developmental timing mechanisms: changing ontogenies in evolution. Semin Cell Dev Biol 34:99–107. https://doi.org/10.1016/j.semcdb. 2014.06.015

Khire A, Vizuet AA, Davila E, Avidor-Reiss T (2015) Asterless reduction during spermiogenesis is regulated by Plk4 and is essential for zygote development in Drosophila. Curr Biol 25 (22):2956–2963. https://doi.org/10.1016/j.cub.2015.09.045

Khire A, Jo KH, Kong D, Akhshi T, Blachon S, Cekic AR, Hynek S, Ha A, Loncarek J, Mennella V, Avidor-Reiss T (2016) Centriole remodeling during spermiogenesis in Drosophila. Curr Biol 26(23):3183–3189. https://doi.org/10.1016/j.cub.2016.07.006

Leber B, Maier B, Fuchs F, Chi J, Riffel P, Anderhub S, Wagner L, Ho AD, Salisbury JL, Boutros M, Kramer A (2010) Proteins required for centrosome clustering in cancer cells. Sci Transl Med 2(33):33ra38. https://doi.org/10.1126/scitranslmed.3000915

Lupold S, Pitnick S (2018) Sperm form and function: what do we know about the role of sexual selection? Reproduction 155(5):R229–R243. https://doi.org/10.1530/REP-17-0536

Malicki J, Avidor-Reiss T (2014) From the cytoplasm into the cilium: bon voyage. Organogenesis 10(1):138–157. https://doi.org/10.4161/org.29055

Malicki JJ, Johnson CA (2017) The cilium: cellular antenna and central processing unit. Trends Cell Biol 27(2):126–140. https://doi.org/10.1016/j.tcb.2016.08.002

McNamara KJ, McKinney ML (2005) Heterochrony, disparity, and macroevolution. Paleobiology 31(S2):17–26

Mordhorst BR, Wilson ML, Conant GC (2016) Some assembly required: evolutionary and systems perspectives on the mammalian reproductive system. Cell Tissue Res 363(1):267–278. https:// doi.org/10.1007/s00441-015-2257-x

Nigg EA, Raff JW (2009) Centrioles, centrosomes, and cilia in health and disease. Cell 139 (4):663–678. https://doi.org/10.1016/j.cell.2009.10.036. S0092-8674(09)01362-2 [pii]

Pearson CG, Osborn DP, Giddings TH Jr, Beales PL, Winey M (2009) Basal body stability and ciliogenesis requires the conserved component Poc1. J Cell Biol 187(6):905–920. https://doi. org/10.1083/jcb.200908019

Petry S (2016) Mechanisms of mitotic spindle assembly. Annu Rev Biochem 85:659–683. https:// doi.org/10.1146/annurev-biochem-060815-014528

Phillips DM (1967) Giant centriole formation in Sciara. J Cell Biol 33(1):73–92

Poulton JS, Cuningham JC, Peifer M (2014) Acentrosomal Drosophila epithelial cells exhibit abnormal cell division, leading to cell death and compensatory proliferation. Dev Cell 30 (6):731–745. https://doi.org/10.1016/j.devcel.2014.08.007

Prosser SL, Pelletier L (2017) Mitotic spindle assembly in animal cells: a fine balancing act. Nat Rev Mol Cell Biol 18(3):187–201. https://doi.org/10.1038/nrm.2016.162

Riparbelli MG, Dallai R, Callaini G (2010) The insect centriole: a land of discovery. Tissue Cell 42 (2):69–80. https://doi.org/10.1016/j.tice.2010.01.002

Rosenkilde P, Ussing AP (1996) What mechanisms control neoteny and regulate induced metamorphosis in urodeles? Int J Dev Biol 40(4):665–673

Simerly C, Castro C, Hartnett C, Lin CC, Sukhwani M, Orwig K, Schatten G (2016) Post-testicular sperm maturation: centriole pairs, found in upper epididymis, are destroyed prior to sperm's release at ejaculation. Sci Rep 6:31816. https://doi.org/10.1038/srep31816

Skulachev VP, Holtze S, Vyssokikh MY, Bakeeva LE, Skulachev MV, Markov AV, Hildebrandt TB, Sadovnichii VA (2017) Neoteny, prolongation of youth: from naked mole rats to "naked apes" (humans). Physiol Rev 97(2):699–720. https://doi.org/10.1152/physrev.00040.2015

Sluder G (2016) Using sea urchin gametes and zygotes to investigate centrosome duplication. Cilia 5(1):20. https://doi.org/10.1186/s13630-016-0043-3

Smith KK (2002) Sequence heterochrony and the evolution of development. J Morphol 252 (1):82–97

Tang N, Marshall WF (2012) Centrosome positioning in vertebrate development. J Cell Sci 125 (Pt 21):4951–4961. https://doi.org/10.1242/jcs.038083

Viswanadha R, Sale WS, Porter ME (2017) Ciliary motility: regulation of axonemal dynein motors. Cold Spring Harb Perspect Biol 9(8). https://doi.org/10.1101/cshperspect.a018325

Winey M, O'Toole E (2014) Centriole structure. Philos Trans R Soc Lond B Biol Sci 369(1650). https://doi.org/10.1098/rstb.2013.0457

Chapter 2
The Role of Protein Acetylation in Centrosome Biology

Delowar Hossain and William Y. Tsang

Abstract Acetylation is among the most prevalent posttranslational modifications in cells and regulates a number of physiological processes such as gene transcription, cell metabolism, and cell signaling. Although initially discovered on nuclear histones, many non-nuclear proteins have subsequently been found to be acetylated as well. The centrosome is the major microtubule-organizing center in most metazoans. Recent proteomic data indicate that a number of proteins in this subcellular compartment are acetylated. This review gives an overview of our current knowledge on protein acetylation at the centrosome and its functional relevance in organelle biology.

2.1 Acetylation: An Overview

Posttranslational modifications (PTMs) are biochemical modifications in which new functional groups are added to proteins. Similar to ubiquitination and phosphorylation, acetylation is one of the most common PTMs in cells and has been intensively studied over the years (Verdin and Ott 2015). Acetylation was first discovered on histones in the 1960s (Phillips 1963). About 20 years later, the first cytoplasmic acetylated protein, tubulin, was described (L'Hernault and Rosenbaum 1983). Since then, numerous nuclear and non-nuclear proteins have been reported to be acetylated. Acetylation entails the transfer of an acetyl group from acetyl CoA to the amino group of a protein. While acetylation reactions can occur non-enzymatically

D. Hossain
Institut de recherches cliniques de Montreal, Montreal, Quebec, Canada

Division of Experimental Medicine, McGill University, Montreal, Quebec, Canada
e-mail: mohammaddelowar.hossain@ircm.qc.ca

W. Y. Tsang (✉)
Institut de recherches cliniques de Montreal, Montreal, Quebec, Canada

Division of Experimental Medicine, McGill University, Montreal, Quebec, Canada

Department of Pathology and Cell Biology, Universite de Montreal, Montreal, Quebec, Canada
e-mail: william.tsang@ircm.qc.ca

© Springer Nature Switzerland AG 2019

M. Kloc (ed.), *The Golgi Apparatus and Centriole*, Results and Problems in Cell Differentiation 67, https://doi.org/10.1007/978-3-030-23173-6_2

under certain conditions (Wagner and Payne 2013), most are catalyzed by enzymes. There are two different classes of enzymes responsible for acetylation: N-terminal acetyltransferases (NATs) and lysine acetyltransferases (KATs). NATs catalyze the transfer of an acetyl group to the amino terminus of proteins (Aksnes et al. 2015), whereas KATs transfer an acetyl group to the epsilon-amino group of lysine residues of proteins (Allis et al. 2007). The former reaction is considered to be irreversible, while the latter can be reversed by lysine deacetylases (KDACs) or histone deacetylases (HDACs), enzymes that remove an acetyl group from the amino terminus of proteins (Taunton et al. 1996). At the molecular level, addition of an acetyl group causes the lysine residue to become bulkier and lose its positive charge, changing the conformation of the protein which in turn can alter its subcellular localization, stability, activity, and/or ability to interact with DNA or binding partners. In general, acetylation serves as a mechanism to regulate various physiological processes such as gene transcription, cell metabolism, and cell signaling (Choudhary et al. 2014). Defects in protein acetylation can lead to severe human diseases such as cancer, diabetes, neurodegenerative diseases, and lung disorders (Drazic et al. 2016).

The acetylation status of many cellular proteins is controlled by the action of KATs and KDACs. To date, at least 17 eukaryotic KATs have been identified, and these can be grouped into three families (GNAT or general control non-depressible 5-related N-terminal acetyltransferase, MYST or Morf, Ybf2, Sas2 and Tip60, and p300/CBP or CREB-binding protein) based on their catalytic domains (Drazic et al. 2016). KATs are predominantly nuclear, although some exhibit localization to the cytoplasm and/or mitochondria. The 18 KDACs found in eukaryotes are phylogenetically divided into four classes, requiring either Zn^{2+} or NAD^+ for catalytic activity (Drazic et al. 2016). Unlike KATs, there are several more KDACs reported outside the nucleus, especially in the cytoplasm and mitochondria. Given that as many as several thousand cellular proteins are believed to be acetylated (Choudhary et al. 2009; Lundby et al. 2012; Weinert et al. 2013), a given KAT or KDAC, on average, might potentially act on hundreds of substrates.

2.2 Centrosome Biology

The cytoskeleton of a eukaryotic cell is composed of microtubules, actin filaments, and intermediate filaments. Centrosomes are the major microtubule-organizing centers in most metazoans, including cnidarians, ctenophores, insects, macrostomidans, mollusks, nematodes, sponges, and vertebrates (Azimzadeh 2014), and influence all microtubule-related processes such as cell division, migration, adhesion, and cilia formation (Conduit et al. 2015). In the G1 phase, a single centrosome is composed of two centrioles, a mother and a daughter, surrounded by a pericentriolar matrix (PCM) that enables microtubule nucleation. This centrosome duplicates once in the

S phase, increasing in number from one to two. After duplication, the two centrosomes undergo maturation wherein the PCMs enlarge and substantially increase in size. At the onset of mitosis, the two centrosomes separate, migrating to the opposite end of the cell and establishing the spindle poles. When the cell finally divides, one centrosome is distributed to each incipient daughter cell. Upon cell cycle exit, the mother centriole of a centrosome nucleates the formation of a cilium, a cellular antenna possessing motility and/or sensory function (Hossain and Tsang 2018).

Centrosomes are membraneless organelles made up of hundreds of proteins. Some are core components of centrioles and the PCM; others are associated with the organelle in a transient fashion. Because centrosomes are highly dynamic structures whose number, size, and composition change in the cell cycle, it is not surprising that the localization, stability, and function of many centrosome-associated proteins are regulated by PTMs. While there are ample examples of centrosomal proteins that undergo phosphorylation and ubiquitination (Hossain and Tsang 2018; Fry et al. 2000), less is known about acetylation of centrosomal proteins other than tubulin. The topic of tubulin acetylation has been reviewed elsewhere (Wloga et al. 2017a, b). Here, we will provide an overview on KATs and KDACs localization to the centrosome, centrosomal proteins that are thought to be acetylated, and the importance of KATs, KDACs, and centrosomal protein acetylation in organelle biology.

2.3   KATs and KDACs at the Centrosome

Although some KATs and KDACs are located in the cytoplasm, their precise localization within the cytoplasmic compartment has not been documented in detail until recently. Endogenous HDAC1 is enriched at the centrosome in the M phase (Sakai et al. 2002), whereas endogenous HDAC6 can be found on centrosomes of proliferating and non-proliferating cells (Ran et al. 2015). In order to examine the localization of all known KDACs, Ling et al. expressed individual enzymes in human cells and showed that recombinant HDAC1, HDAC4, HDAC10, HDAC11, Sirtuin1 (SIRT1), and SIRT2 are present on unduplicated and duplicated centrosomes in the G1 and S/G2 phases, respectively (Ling et al. 2012). In contrast, HDAC5 and HDAC6 are associated with unduplicated centrosomes in the G1 phase only, while the remaining KDACs do not localize to the centrosome at all (Ling et al. 2012). With regard to KATs, KAT2A and KAT2B are demonstrated to be centrosomal (Fournier et al. 2016). KAT2A, in particular, is targeted to the centrosome in the late G1 and early S phases (Fournier et al. 2016). Thus, it appears that multiple KATs and KDACs are localized to the centrosome at specific times in the cell cycle, where they might function to acetylate or deacetylate certain substrates and regulate certain aspects of organelle function.

2.4   Acetylated Proteins at the Centrosome

A number of large-scale proteomic analyses performed within the past 10–15 years reveal that lysine acetylation is a highly prevalent PTM found in many organisms, including eukaryotes (humans, mice, rats, flies, and yeast) and prokaryotes (bacteria) (Kim et al. 2006; Choudhary et al. 2009; Zhao et al. 2010; Beli et al. 2012; Lundby et al. 2012; Weinert et al. 2013; Chen et al. 2012; Fournier et al. 2016). These results also show that acetylation does not occur only on nuclear histones or chromatin-associated proteins. Rather, the majority of acetylation events are associated with non-nuclear proteins. Among the non-nuclear proteins that are found to be acetylated, some are components of the centrosome (Table 2.1). These components localize to centrioles, PCM, cilia, and centriolar satellites (electron dense granules surrounding the centrosome), suggesting that acetylation of centrosomal proteins is a common phenomenon that can occur at different locations within the organelle. So far, only acetylations of Centrin, Polo-like kinase 2 (Plk2), Septin7, Plk4, Centrosomal protein of 76 kDa (Cep76), and Breast cancer gene 1 (BRCA1) have been confirmed (Ling et al. 2012; Fournier et al. 2016; Barbelanne et al. 2016), and the acetylation status of the remaining centrosomal proteins would require further validation. For most of the acetylated centrosomal proteins, it is currently unknown which KAT(s) is/are responsible for their acetylation. Plk4 is one exception, as this protein interacts with, and is a bona fide substrate of, KAT2A and KAT2B (Fournier et al. 2016). BRCA1 is also known to be acetylated by KAT2A and KAT2B, and can be deacetylated by SIRT1 (Lahusen et al. 2018). Plk2 might be acetylated by KAT2A and KAT2B as well, but this remains to be validated (Fournier et al. 2016).

2.5   Role of KATs, KDACs, and Centrosomal Protein Acetylation

In light of the observations that several KDACs are localized to unduplicated and duplicated centrosomes, further experiments were conducted to assess if these enzymes might regulate centrosome amplification, a condition that can be triggered by prolonged S phase arrest with a DNA synthesis inhibitor hydroxyurea. Indeed, there is a correlation between the localization of KDACs to the centrosome and the ability of KDACs to suppress centrosome amplification. In particular, overexpression of HDAC1, HDAC5, or SIRT1 markedly inhibits centrosome amplification, and conversely, depletion of HDAC1, HDAC5, or SIRT1 in S phase-arrested cells enhances amplification (Ling et al. 2012). These three KDACs appear to function in a mechanistically distinct manner. Besides centrosome amplification, HDAC5 and SIRT1, but not HDAC1, also delay/suppress centrosome duplication in cycling cells (Ling et al. 2012). Moreover, the deacetylase activity of HDAC1 and SIRT1, but not HDAC5, is required to prevent centrosome amplification (Ling et al. 2012).

Table 2.1 List of acetylated centrosomal proteins identified by mass spectrometry

Centrosomal protein	Function	Acetylation validated?	References
AKAP9	Centrosome maturation	No	Lundby et al. (2012)
ALMS1	Ciliogenesis	No	Beli et al. (2012), Fournier et al. (2016)
ASPM	Mitotic spindle regulation	No	Lundby et al. (2012)
BRCA1	Centrosome re-duplication	Yes	Lahusen et al. (2018)
CCDC11	Ciliogenesis	No	Zhao et al. (2010)
Centrin2	Centrosome duplication	Yes	Lundby et al. (2012), Ling et al. (2012)
Cep41	Ciliogenesis	No	Lundby et al. (2012), Weinert et al. (2013)
Cep68	Centrosome separation	No	Zhao et al. (2010), Lundby et al. (2012)
Cep76	Centriole re-duplication	Yes	Barbelanne et al. (2016), Zhao et al. (2010)
Cep152	Centrosome duplication	No	Zhao et al. (2010)
Cep170	Microtubule dynamics	No	Chen et al. (2012)
Cep192	Centrosome duplication	No	Lundby et al. (2012)
Cep250	Centrosome separation	No	Choudhary et al. (2009), Beli et al. (2012)
Cep290	Ciliogenesis	No	Zhao et al. (2010), Lundby et al. (2012)
Cep350	Ciliogenesis	No	Zhao et al. (2010)
CSPP1	Ciliogenesis	No	Lundby et al. (2012)
DNAH3	Ciliary beating	No	Zhao et al. (2010)
DNAH7	Ciliary beating	No	Zhao et al. (2010)
GCP2	Centrosome maturation	No	Choudhary et al. (2009)
IFT20	Ciliogenesis	No	Lundby et al. (2012)
IFT74	Ciliogenesis	No	Zhao et al. (2010)
KIF11	Centrosome separation	No	Fournier et al. (2016)
Kif3a	Ciliogenesis	No	Zhao et al. (2010)
NEK9	Centrosome separation	No	Choudhary et al. (2009)
NINL	Microtubule nucleation	No	Fournier et al. (2016)
NPM1	Centrosome duplication	No	Zhao et al. (2010), Kim et al. (2006), Beli et al. (2012)
ODF2	Ciliogenesis	No	Zhao et al. (2010)
PCM1	Centriolar satellite formation	No	Choudhary et al. (2009)
Plk2	Centrosome duplication	No	Ling et al. (2012), Fournier et al. (2016)
Plk4	Centrosome duplication	Yes	Fournier et al. (2016)
Rootletin	Centrosome separation	No	Zhao et al. (2010)

(continued)

Table 2.1 (continued)

Centrosomal protein	Function	Acetylation validated?	References
Septin7	Ciliogenesis	Yes	Lundby et al. (2012), Ling et al. (2012)
TTBK2	Ciliogenesis	No	Fournier et al. (2016)

AKAP9 A-kinase anchoring protein 9, *ALMS1* Alstrom syndrome protein 1, *ASPM* abnormal spindle microtubule assembly, *BRCA1* breast cancer gene 1, *CCDC11* coiled-coil domain containing 11, *Cep41* centrosomal protein of 41 kDa, *Cep68* centrosomal protein of 68 kDa, *Cep76* centrosomal protein of 76 kDa, *Cep152* centrosomal protein of 152 kDa, *Cep170* centrosomal protein of 170 kDa, *Cep192* centrosomal protein of 192 kDa, *Cep250* centrosomal protein of 250 kDa, *Cep290* centrosomal protein of 290 kDa, *Cep350* centrosomal protein of 350 kDa, *CSPP1* centrosome and spindle pole associated protein 1, *DNAH3* dynein axonemal heavy chain 3, *DNAH7* dynein axonemal heavy chain 7, *GCP2* gamma-tubulin complex component 2, *IFT20* intraflagellar transport 20, *IFT74* intraflagellar transport 74, *KIF11* kinesin family member 11, *Kif3a* kinesin family member 3a, *NEK9* never in mitosis gene A related kinase 9, *NINL* ninein-like protein, *NPM1* nucleophosmin 1, *ODF2* outer dense fiber of sperm tails 2, *PCM1* pericentriolar material 1, *Plk2* polo-like kinase 2, *Plk4* polo-like kinase 4, *TTBK2* tau tubulin kinase 2

Besides the aforementioned KDACs, the KAT KAT2A plays a role in the regulation of centrosome duplication. First, KAT2A localizes to the centrosome in the late G1 and early S phases when organelle duplication takes place (Fournier et al. 2016). Second, KAT2A acetylates Plk4, a kinase essential for centriole duplication (Habedanck et al. 2005), impairing its activity and ability to promote duplication (Fournier et al. 2016). Third, depletion of KAT2A or ectopic expression of a catalytic dead mutant of KAT2A induces centrosome amplification (Fournier et al. 2016). These data collectively suggest that one critical function of KAT2A is to prevent centrosome amplification through Plk4 acetylation (Fig. 2.1).

The function of Cep76, an important regulator of centriole number, is also controlled by acetylation. Depletion of Cep76 leads to the accumulation of extra centrioles and centriolar intermediates, whereas its overexpression suppresses centrosome amplification in S phase-arrested cells (Tsang et al. 2009). Cep76 is acetylated at lysine 279 (K279), and overexpression of an acetyl-mimetic mutant K279Q, but not an acetyl-resistant mutant K279R, is unable to suppress centrosome amplification in S phase-arrested cells (Barbelanne et al. 2016). Likewise, overexpression of wild-type Cep76 cannot prevent centrosome amplification in G2 phase-arrested cells (Barbelanne et al. 2016), presumably because this protein is highly acetylated in this cell cycle phase. Thus, although it is not known which KAT (s) and KDAC(s) act on Cep76, acetylation of this protein seems to hamper its ability to suppress centriole amplification (Fig. 2.1).

BRCA1 is a well-known tumor suppressor involved in DNA damage repair, transcription, and ubiquitination (Baer and Ludwig 2002; Deng 2006). Although initially discovered as a nuclear protein, BRCA1 is also located at the centrosome where it serves to regulate organelle copy number and microtubule nucleation (Xu et al. 1999; Sankaran et al. 2005). Loss or inactivation of BRCA1 results in a defective G2/M checkpoint and centrosome amplification (Xu et al. 1999). A recent

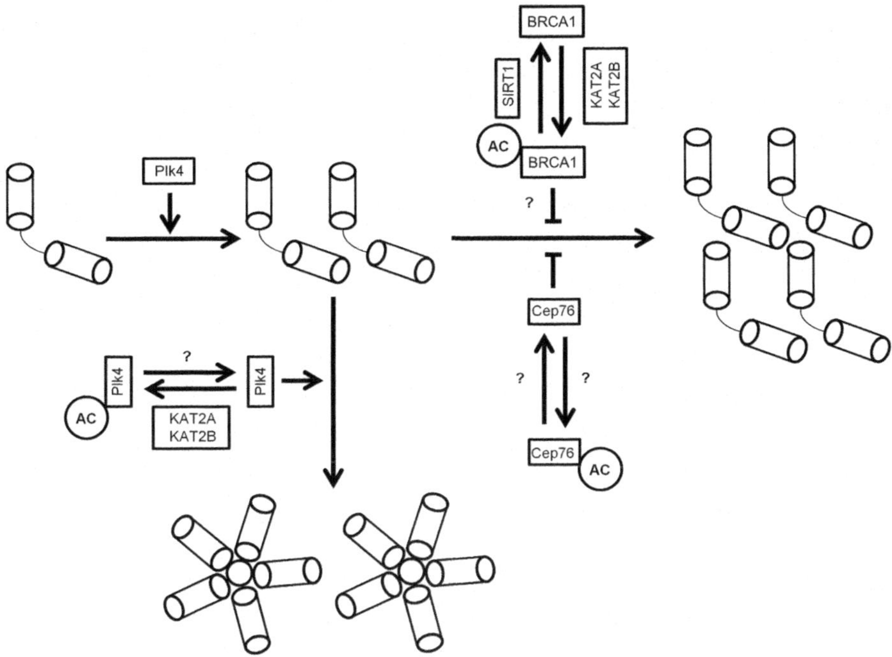

Fig. 2.1 Acetylation of centrosomal components in organelle biology. Centrosome duplication requires Plk4. Excessive Plk4 activity leads to centrosome amplification characterized by the generation of multiple new centrioles surrounding the existing mother and daughter centrioles. Plk4 activity is regulated by KAT2A and KAT2B and an unknown deacetylase. Although the enzymes responsible for acetylating and deacetylating Cep76 have not been identified, acetylation of Cep76 is unable to prevent centrosome re-duplication, leading to centrosome amplification. Acetylation of BRCA1 by KAT2A or KAT2B activates BRCA1, whereas deacetylation by SIRT1 inhibits BRCA1 activity. Whether or not acetylation of BRCA1 can regulate centrosome amplification is currently unknown. AC denotes an acetyl group

paper demonstrated that KAT2A and KAT2B acetylate BRCA1, leading to its functional activation at the intra-S checkpoint after DNA damage (Lahusen et al. 2018). Despite these observations, it is not understood if BRCA1 acetylation occurs at the centrosome, and whether or not this acetylation event can prevent centrosome amplification (Fig. 2.1).

2.6   Conclusions

Several lines of evidence implicate acetylation in the regulation of centriole and centrosome number. So far, one KAT, several KDACs, and acetylation of a small number of centrosomal proteins is shown to play a role in controlling organelle duplication/amplification, and data from proteomic studies indicate that additional centrosomal proteins may be involved and their activities might be regulated by acetylation. In addition, it is likely that acetylation of centrosomal proteins regulates other aspects of organelle function such as centrosome separation and cilium formation. Moving forward, it would be important to validate additional acetylated proteins at the centrosome, identify the enzymes responsible for modulating their acetylation levels, and determine the biological relevance of these acetylation and deacetylation events. In light of the frequent deregulation of KATs or KDACs in human diseases, a mechanistic understanding of the role of protein acetylation at the centrosome may form the basis for therapeutic intervention in the future.

Acknowledgements W.Y.T. was a Canadian Institutes of Health Research New Investigator and a Fonds de recherche Santé Junior 2 Research Scholar. This work was supported by grants from the Natural Sciences and Engineering Research Council of Canada and Cancer Research Society to W.Y.T.

References

Aksnes H, Hole K, Arnesen T (2015) Molecular, cellular, and physiological significance of N-terminal acetylation. Int Rev Cell Mol Biol 316:267–305

Allis CD, Berger SL, Cote J, Dent S, Jenuwien T, Kouzarides T, Pillus L, Reinberg D, Shi Y, Shiekhattar R, Shilatifard A, Workman J, Zhang Y (2007) New nomenclature for chromatin-modifying enzymes. Cell 131(4):633–636

Azimzadeh J (2014) Exploring the evolutionary history of centrosomes. Philos Trans R Soc Lond B Biol Sci 369(1650)

Baer R, Ludwig T (2002) The BRCA1/BARD1 heterodimer, a tumor suppressor complex with ubiquitin E3 ligase activity. Curr Opin Genet Dev 12(1):86–91

Barbelanne M, Chiu A, Qian J, Tsang WY (2016) Opposing post-translational modifications regulate Cep76 function to suppress centriole amplification. Oncogene 35(41):5377–5387

Beli P, Lukashchuk N, Wagner SA, Weinert BT, Olsen JV, Baskcomb L, Mann M, Jackson SP, Choudhary C (2012) Proteomic investigations reveal a role for RNA processing factor THRAP3 in the DNA damage response. Mol Cell 46(2):212–225

Chen Y, Zhao W, Yang JS, Cheng Z, Luo H, Lu Z, Tan M, Gu W, Zhao Y (2012) Quantitative acetylome analysis reveals the roles of SIRT1 in regulating diverse substrates and cellular pathways. Mol Cell Proteomics 11(10):1048–1062

Choudhary C, Kumar C, Gnad F, Nielsen ML, Rehman M, Walther TC, Olsen JV, Mann M (2009) Lysine acetylation targets protein complexes and co-regulates major cellular functions. Science 325(5942):834–840

Choudhary C, Weinert BT, Nishida Y, Verdin E, Mann M (2014) The growing landscape of lysine acetylation links metabolism and cell signalling. Nat Rev Mol Cell Biol 15(8):536–550

Conduit PT, Wainman A, Raff JW (2015) Centrosome function and assembly in animal cells. Nat Rev Mol Cell Biol 16(10):611–624

Deng CX (2006) BRCA1: cell cycle checkpoint, genetic instability, DNA damage response and cancer evolution. Nucleic Acids Res 34(5):1416–1426

Drazic A, Myklebust LM, Ree R, Arnesen T (2016) The world of protein acetylation. Biochim Biophys Acta 1864(10):1372–1401

Fournier M, Orpinell M, Grauffel C, Scheer E, Garnier JM, Ye T, Chavant V, Joint M, Esashi F, Dejaegere A, Gonczy P, Tora L (2016) KAT2A/KAT2B-targeted acetylome reveals a role for PLK4 acetylation in preventing centrosome amplification. Nat Commun 7:13227

Fry AM, Mayor T, Nigg EA (2000) Regulating centrosomes by protein phosphorylation. Curr Top Dev Biol 49:291–312

Habedanck R, Stierhof YD, Wilkinson CJ, Nigg EA (2005) The Polo kinase Plk4 functions in centriole duplication. Nat Cell Biol 7(11):1140–1146

Hossain D, Tsang WY (2018) The role of ubiquitination in the regulation of primary cilia assembly and disassembly. Semin Cell Dev Biol. (in press)

Kim SC, Sprung R, Chen Y, Xu Y, Ball H, Pei J, Cheng T, Kho Y, Xiao H, Xiao L, Grishin NV, White M, Yang XJ, Zhao Y (2006) Substrate and functional diversity of lysine acetylation revealed by a proteomics survey. Mol Cell 23(4):607–618

Lahusen TJ, Kim SJ, Miao K, Huang Z, Xu X, Deng CX (2018) BRCA1 function in the intra-S checkpoint is activated by acetylation via a pCAF/SIRT1 axis. Oncogene 37(17):2343–2350

L'Hernault SW, Rosenbaum JL (1983) Chlamydomonas alpha-tubulin is posttranslationally modified in the flagella during flagellar assembly. J Cell Biol 97(1):258–263

Ling H, Peng L, Seto E, Fukasawa K (2012) Suppression of centrosome duplication and amplification by deacetylases. Cell Cycle 11(20):3779–3791

Lundby A, Lage K, Weinert BT, Bekker-Jensen DB, Secher A, Skovgaard T, Kelstrup CD, Dmytriyev A, Choudhary C, Lundby C, Olsen JV (2012) Proteomic analysis of lysine acetylation sites in rat tissues reveals organ specificity and subcellular patterns. Cell Rep 2 (2):419–431

Phillips DM (1963) The presence of acetyl groups of histones. Biochem J 87:258–263

Ran J, Yang Y, Li D, Liu M, Zhou J (2015) Deacetylation of alpha-tubulin and cortactin is required for HDAC6 to trigger ciliary disassembly. Sci Rep 5:12917

Sakai H, Urano T, Ookata K, Kim MH, Hirai Y, Saito M, Nojima Y, Ishikawa F (2002) MBD3 and HDAC1, two components of the NuRD complex, are localized at Aurora-A-positive centrosomes in M phase. J Biol Chem 277(50):48714–48723

Sankaran S, Starita LM, Groen AC, Ko MJ, Parvin JD (2005) Centrosomal microtubule nucleation activity is inhibited by BRCA1-dependent ubiquitination. Mol Cell Biol 25(19):8656–8668

Taunton J, Hassig CA, Schreiber SL (1996) A mammalian histone deacetylase related to the yeast transcriptional regulator Rpd3p. Science 272(5260):408–411

Tsang WY, Spektor A, Vijayakumar S, Bista BR, Li J, Sanchez I, Duensing S, Dynlacht BD (2009) Cep76, a centrosomal protein that specifically restrains centriole reduplication. Dev Cell 16 (5):649–660

Verdin E, Ott M (2015) 50 years of protein acetylation: from gene regulation to epigenetics, metabolism and beyond. Nat Rev Mol Cell Biol 16(4):258–264

Wagner GR, Payne RM (2013) Widespread and enzyme-independent Nepsilon-acetylation and Nepsilon-succinylation of proteins in the chemical conditions of the mitochondrial matrix. J Biol Chem 288(40):29036–29045

Weinert BT, Scholz C, Wagner SA, Iesmantavicius V, Su D, Daniel JA, Choudhary C (2013) Lysine succinylation is a frequently occurring modification in prokaryotes and eukaryotes and extensively overlaps with acetylation. Cell Rep 4(4):842–851

Wloga D, Joachimiak E, Fabczak H (2017a) Tubulin post-translational modifications and microtubule dynamics. Int J Mol Sci 18(10)

Wloga D, Joachimiak E, Louka P, Gaertig J (2017b) Posttranslational modifications of tubulin and cilia. Cold Spring Harb Perspect Biol 9(6)

Xu X, Weaver Z, Linke SP, Li C, Gotay J, Wang XW, Harris CC, Ried T, Deng CX (1999) Centrosome amplification and a defective G2-M cell cycle checkpoint induce genetic instability in BRCA1 exon 11 isoform-deficient cells. Mol Cell 3(3):389–395

Zhao S, Xu W, Jiang W, Yu W, Lin Y, Zhang T, Yao J, Zhou L, Zeng Y, Li H, Li Y, Shi J, An W, Hancock SM, He F, Qin L, Chin J, Yang P, Chen X, Lei Q, Xiong Y, Guan KL (2010) Regulation of cellular metabolism by protein lysine acetylation. Science 327(5968):1000–1004

Chapter 3
Formins, Golgi, and the Centriole

John Copeland

Abstract Formin homology proteins (formins) are a highly conserved family of cytoskeletal remodeling proteins that are involved in a diverse array of cellular functions. Formins are best known for their ability to regulate actin dynamics, but the same functional domains also govern stability and organization of microtubules. It is thought that this dual activity allows them to coordinate the activity of these two major cytoskeletal networks and thereby influence cellular architecture. Golgi ribbon assembly is dependent upon cooperative interactions between actin filaments and cytoplasmic microtubules originating both at the Golgi itself and from the centrosome. Similarly, centrosome assembly, centriole duplication, and centrosome positioning are also reliant on a dialogue between both cytoskeletal networks. As presented in this chapter, a growing body of evidence suggests that multiple formin proteins play essential roles in these central cellular processes.

3.1 Formin Homology Proteins

Formins are a highly conserved family of cytoskeletal remodeling proteins found in all eukaryotes from budding yeast to plants to vertebrates. Their name derives from the founding member of the family—Formin1 (Fmn1), the product of the mouse *limb deformity* gene (Woychik et al. 1990). There are 15 family members encoded by the human genome identified by two regions of homology designated formin homology 1 (FH1) and formin homology 2 (FH2) (Castrillon and Wasserman 1994). These 15 proteins are divided into two major groups: the Diaphanous Related Formins (DRFs) and the non-Diaphanous-Related Formins (Fig. 3.1). The DRFs include DIAPH 1, 2, and 3 (mDia1, 3, & 2); DAAM1 and 2, FMNL1, 2, and 3; and FHOD 1 and 3 and are identified by the regulatory GTPase-binding domain (GBD) and Diaphanous Inhibitory Domain (DID) in the N-terminal half of the protein and by the Diaphanous Autoregulatory Domain (DAD) C-terminal to FH2. The non-DRFs

J. Copeland (✉)
Faculty of Medicine, Department of Cellular and Molecular Medicine, University of Ottawa, Ottawa, ON, Canada
e-mail: john.copeland@uottawa.ca

© Springer Nature Switzerland AG 2019
M. Kloc (ed.), *The Golgi Apparatus and Centriole*, Results and Problems in Cell Differentiation 67, https://doi.org/10.1007/978-3-030-23173-6_3

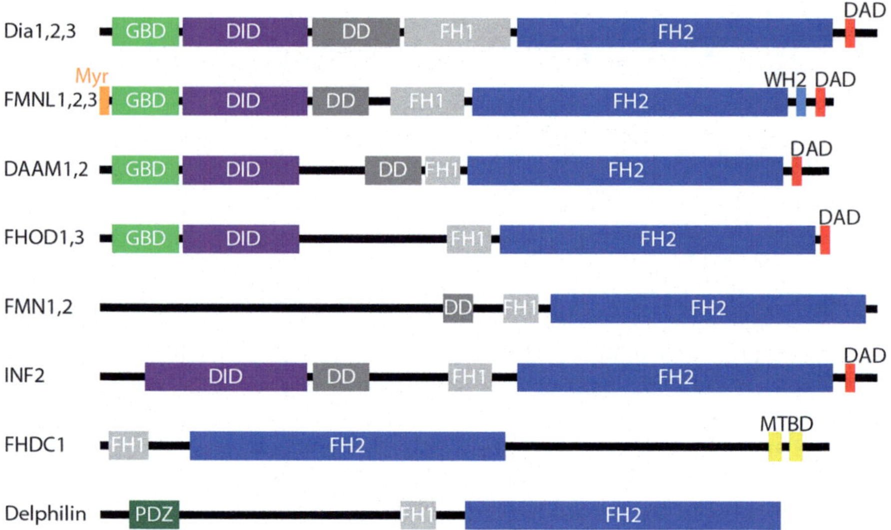

Fig. 3.1 Mammalian formins. The 15 mammalian formin proteins can be grouped into subcategories based on their domain architecture. *GBD* GTPase binding domain, *DID* diaphanous inhibitory domain, *DD* N-terminal dimerization domain, *FH1* formin homology 1, *FH2* formin homology 2, *WH2* WASP homology 2, *DAD* diaphanous autoregulatory domain, *Myr* N-myristoylation motif, *MTBD* microtubule binding domain, *PDZ* PSD-95, *Dlg1*, ZO-1 domain

include FMN1, FMN2, FHDC1 (aka INF1), INF2, and Delphilin each of which lack some or all of these regulatory motifs (Hegsted et al. 2017; Higgs 2005; Schonichen and Geyer 2010). Formins promote the formation of long unbranched actin filaments via the action of the FH1 and FH2 cytoskeletal regulatory unit (Courtemanche 2018). FH2 forms a head-to-tail dimer and FH2 dimerization is essential for the nucleation and elongation of actin filaments (Copeland et al. 2004; Lu et al. 2007; Moseley et al. 2004; Shimada et al. 2004; Xu et al. 2004). Two actin-binding motifs have been defined in FH2. The first contains an absolutely conserved isoleucine residue (I704 in mDia2, I649 in FMNL3, and I180 in FHDC1) that is essential for F-actin barbed-end binding and FH2-dependent actin regulation (Bartolini et al. 2008; Copeland et al. 2016; Harris et al. 2006; 2010; Lu et al. 2007; Peladeau et al. 2016; Xu et al. 2004). The second contains an absolutely conserved lysine residue (K853 in mDia2 and K800 in FMNL3), but it is unclear if this residue is absolutely essential for FH2 activity (Copeland et al. 2004; Ishizaki et al. 2001; Scott et al. 2011; Thompson et al. 2013). The FH2 dimer overcomes the rate-limiting step in nucleation of actin polymerization by stabilizing formation of a G-actin dimer or trimer (Pring et al. 2003), although the exact mechanism is still not clearly defined (Courtemanche 2018; Thompson et al. 2013). Once polymerization has initiated, the FH2 dimer acts as a "leaky capper" that stays bound to the elongating barbed-end while still allowing the incorporation of G-actin subunits (Higashida et al. 2004; Kovar and Pollard 2004; Mizuno et al. 2011; Pring et al. 2003; Romero et al. 2004; Zigmond et al. 2003).

Barbed-end binding by FH2 also protects the actin filament from capping protein (Harris et al. 2004; Romero et al. 2004).

In vitro, the isolated FH2 domain nucleates polymerization from free G-actin; however, in the cell nearly all G-actin is bound by the small actin-binding protein profilin. Utilization of profilin-actin by FH2 requires the action of two accessory domains, FH1 and the FH2 C-terminal tail. FH1 consists of multiple poly-proline repeats and is a profilin ligand. It recruits profilin–actin complexes to the growing barbed-end and feeds them to FH2 as the filament elongates. A similar role is also proposed for the FH2 C-terminal tail (Gould et al. 2011; Heimsath and Higgs 2012; Thompson et al. 2013). The FH2 C-terminal tail contains a WH2-like motif that functions similarly to the poly-proline repeats of the FH1 domain. In addition, the C-terminal WH2-like motif in FMNL2 and FMNL3 is also required for F-actin bundling (Heimsath and Higgs 2012; Vaillant et al. 2008) and the same motif participates in INF2-induced F-actin severing (Gurel et al. 2014; Ramabhadran et al. 2012).

Formin regulation is best understood in the DRF subfamily. DRFs contain an N-terminal Diaphanous Inhibitory Domain (DID) that binds directly to the C-terminal Diaphanous Autoregulatory Domain (DAD) to inhibit DRF function. In the autoinhibited conformation, the interaction of DID with DAD obscures the actin-binding surface of the FH2 domain. Activation occurs when the DID/DAD interaction is disrupted by binding of the appropriate Rho GTPase to the N-terminal GBD. This releases the DID/DAD autoinhibitory interaction and unmasks the FH2 domain. This mechanism has been defined functionally both in vitro and in vivo and the molecular details have been confirmed through studies on the full-length mDia1 protein (Alberts 2001; Copeland et al. 2004, 2007; Li and Higgs 2003; Maiti et al. 2012; Nezami et al. 2010; Watanabe et al. 1999). The non-DRFs FMN1, FMN2, and INF2 are also regulated by autoinhibition (Kobielak et al. 2004; Quinlan et al. 2007; Ramabhadran et al. 2013). Exceptions to the formin autoregulatory rule are the novel formins FHDC1 and Delphilin. Both of these proteins are constitutively active in vivo and in vitro (Silkworth et al. 2018; Young et al. 2008). Surprisingly, studies of DRF autoregulation have also shown that the binding of Rho-GTP to the GBD is not sufficient to fully relieve autoinhibition (Li and Higgs 2003, 2005; Maiti et al. 2012; Nezami et al. 2006; Otomo et al. 2005; Rose et al. 2005), suggesting that additional factors must be at work to fully activate these proteins. Consistent with this hypothesis, autoinhibition of mDia2 and FHOD1 is also directly regulated by ROCK-induced phosphorylation (Gasteier et al. 2003; Staus et al. 2011; Takeya et al. 2008; Truong et al. 2013) and other formins are also thought to be regulated by additional kinases (Cheng et al. 2011; Iskratsch et al. 2010; Wang et al. 2004; Zaidel-Bar 2018). Additional mechanisms regulating formin activity include subcellular targeting (Copeland et al. 2007; Gorelik et al. 2011; Seth et al. 2006), lipid modification (Block et al. 2012; Han et al. 2009; Moriya et al. 2012; Peladeau et al. 2016), and mechanical tension (Chan et al. 2010; Courtemanche et al. 2013; Higashida et al. 2013; Jegou et al. 2013; Kozlov and Bershadsky 2004; Zimmermann and Kovar 2019).

Formins promote the formation of a variety of actin-based cellular structures including stress fibers, filopodia, and lamellipodia (Block et al. 2012; Hotulainen and

Lappalainen 2006; Moseley et al. 2007; Schulze et al. 2014; Takeya et al. 2008; Watanabe et al. 1999). They accomplish this either through their direct regulation of actin dynamics or in cooperation with other remodeling factors. In some cases, the interaction of formins with other cytoskeletal remodeling proteins is antagonistic. For example, in HeLa and MTLn3 adenocarcinoma cells, EGF-induced stimulation of lamellipodia formation requires activation of the Arp2/3 complex and concomitant inhibition of DRF activity (Beli et al. 2008; Sarmiento et al. 2008). In contrast, FMNL2 and Arp2/3 complex cooperate directly to form lamellipodia in B16-F1 melanoma cells; mDia2 and the WAVE complex cooperate in lamellipodia formation in HeLa cells and mDia2 and Arp2/3 cooperate to assemble filopodia-like structures in vitro (Block et al. 2012; Lee et al. 2010; Yang et al. 2007). More recently, it has been shown that formin activity can be potentiated via their interaction with the microtubule-binding proteins APC (Okada et al. 2010) and Clip170 (Henty-Ridilla et al. 2016). Clip-170 is a "plus-tip" binding protein that is able to recruit mDia1, and other formins, to the plus end of the microtubules. Once bound, Clip170 stimulates mDia1-induced F-actin nucleation and the Clip170/mDia1 dimer tracks the growing barbed end of the actin filament, while the pointed end stays bound to the microtubule plus tip. This type of interaction is thought to mediate coordinated regulation of actin and microtubule dynamics and likely accounts for the ability of some formins to induce co-alignment of F-actin with the microtubule network (Ishizaki et al. 2001; Thurston et al. 2012).

Formins also play a direct role in regulating microtubule dynamics either through FH1+FH2 or through additional accessory domains (Bartolini and Gundersen 2010; Bartolini et al. 2012; Copeland et al. 2016; Fernandez-Barrera and Alonso 2018; Gaillard et al. 2011; Roth-Johnson et al. 2014; Thurston et al. 2012; Young et al. 2008). The FH2 domain of mDia1, mDia2, and INF2, exon 2 of Fmn1 and the FHDC1 microtubule-binding domain (MTBD) all bind microtubules directly and induce microtubule stabilization. Stable microtubules undergo extensive posttranslational modifications including removal of the C-terminal tyrosine residue of α-tubulin (detyrosination) and acetylation of α-tubulin on lysine 40. The precise function of these modifications is still unclear, but they likely affect interactions with microtubule-associated proteins, vesicle trafficking, and microtubule stability (Gadadhar et al. 2017; Nieuwenhuis and Brummelkamp 2019). Stabilization and detyrosination have been suggested to result from actin-independent effects of formin activity, although the precise mechanism is not well-understood (Bartolini et al. 2008, 2012, 2016). In contrast, formin-induced MT-acetylation is actin-dependent (Thurston et al. 2012) and follows from induction of α-TAT (tubulin acetyl-transferase) expression via activation of the MRTF/SRF signal transduction pathway, an actin-regulated transcriptional response (Fernandez-Barrera et al. 2018). Cytoplasmic Golgi-derived microtubules (GDMT) are enriched for acetylated tubulin as are the microtubules of the ciliary axoneme. Microtubule stabilization and detyrosination are associated with establishing centrosome positioning and cell polarity. Golgi ribbon assembly, centrosome polarity, and centriole assembly are also each subject to regulation by the coordinated effects of actin and MT dynamics and are regulated in part by formin-dependent regulation of these cytoskeletal networks.

3.2 Formins and the Golgi Ribbon

The Golgi apparatus is the central director of cellular trafficking with cargo passed from the endoplasmic reticulum through to the stacked *cis*, *medial*, and *trans* cisternae of the Golgi before being distributed to its final cellular destination. In vertebrates, adjacent cisternae units are tethered together to form a perinuclear Golgi ribbon. This organization is an evolutionary innovation of the vertebrate lineage (Nakamura et al. 2012) and thus this discussion will be limited to studies from vertebrate systems. Golgi ribbon assembly is dependent upon the coordinated action of the actin cytoskeleton and microtubule network (Thyberg and Moskalewski 1985; Valderrama et al. 1998). Golgi-derived microtubules (GDMT) link cisternae units into ribbons which are held in their perinuclear position by attachments to the centrosome-derived microtubule network (Miller et al. 2009). Treatment with nocodazole, or other microtubule disrupting drugs, results in dispersion of the Golgi into ministacks which are still able to support trafficking (Thyberg and Moskalewski 1985). Actin filaments play a complex role in Golgi assembly and function (Egea et al. 2013). Treatment with either actin-disrupting or actin-stabilizing drugs causes Golgi compaction; however, at the ultrastructural level actin-stabilizing drugs cause the compacted cisternae to swell, while actin-disrupting drugs induce fragmentation (Valderrama et al. 1998). Given the ability of formins to regulate directly both actin and microtubule dynamics, it was suspected that they would play an important role in both Golgi assembly and subcellular positioning. In support of this hypothesis, inhibition of formin activity using the pan-formin inhibitor smiFH2 induces Golgi dispersion and inhibits expression of the Golgi resident protein Giantin (Isogai et al. 2015). Although smiFH2 inhibits FH2-dependent regulation of actin dynamics, Golgi dispersion in smiFH2-treated cells is associated with the loss of the Golgi-derived microtubule network, thus pointing to the complex mechanisms underlying Golgi ribbon assembly. In addition, smiFH2 effects on Golgi dispersion are cell-type specific with U2OS and HCT116 cells affected, but mouse embryonic fibroblasts (MEFs) are not (Isogai et al. 2015). This may reflect both differing profiles of formin expression in different cell types and cell-type-specific mechanisms regulating Golgi assembly. Similarly, treatment of U251 and other glioblastoma cell lines with smiFH2 or the mDia agonists IMM01 or IMM02 did not disrupt Golgi architecture, but did affect Golgi repolarization in migrating cells (Arden et al. 2015). Specifically, formin inhibition blocked Golgi reorientation, while mDia activation promoted it. Surprisingly, both treatments reduced cell motility, likely attributable to interference with the cell's internal guidance system. A similar effect has been observed with the inhibition of the Diaphanous-Related Formin DAAM1. DAAM1 is a component of the planar cell polarity (PCP) pathway and a number of studies have linked its activity with organelle reorientation in migrating cells (Ang et al. 2010; Guillabert-Gourgues et al. 2016; Ju et al. 2010; LaMonica et al. 2009). Overexpression of the DAAM1 N-terminus in f9 teratocarcinoma cells interferes with planar polarity and prevents Golgi reorientation (LaMonica et al. 2009). The same effect was observed following depletion of

DAAM1 expression in endothelial cells; this blocked Golgi repolarization through inhibition of PCP signaling (Guillabert-Gourgues et al. 2016). It is not clear if these events are direct effects on the intrinsic Golgi assembly machinery, or if they are secondary to overall cell polarity and positioning of the centrosome. In contrast, direct connections have been made between specific formins and assembly of the Golgi ribbon.

3.3　FMNL1, FMNL2, and FMNL3 and Golgi Assembly

The FMNL proteins are a subgroup of DRFs. As with other DRFs, FMNL1, 2, and 3 are autoregulated via the DID/DAD interaction and activated via their GBD. FMNL1 has been reported to be an effector for the small GTPases Rac1, Cdc42, and RhoA (Kuhn and Geyer 2014; Wang et al. 2015); FMNL2 is purported to act downstream of Cdc42, Rac1, and RhoC (Block et al. 2012; Grobe et al. 2018; Kage et al. 2017; Kitzing et al. 2010; Kuhn et al. 2015; Woodham et al. 2017), while FMNL3 may act downstream of Cdc42 and RhoC (Kuhn and Geyer 2014; Vega et al. 2011). Despite their similarities, the three proteins have differing effects on actin dynamics. All three proteins are able to bind and bundle F-actin (Esue et al. 2008; Vaillant et al. 2008); FMNL1 is a weak nucleator of actin polymerization, FMNL2 cannot nucleate, and FMNL3 is a good nucleator (Harris et al. 2006; Heimsath and Higgs 2012). FMNL protein subcellular localization is also regulated by N-myristoylation where a myristoyl group is coupled to an N-terminal glycine residue (Han et al. 2009; Moriya et al. 2012; Peladeau et al. 2016). This lipidation modification targets the modified protein to cellular membranes where they may impact on trafficking and membrane dynamics (Farazi et al. 2001). In separate studies, all three proteins have been associated with regulating distinct aspects of Golgi assembly (Colon-Franco et al. 2011; Kage et al. 2017), although the role of N-myristoylation in this process is not clear.

FMNL1 expression is largely restricted to white blood cells, but it is also expressed in a variety of metastatic cancer cell lines. It was found that HeLa cells express FMNL1 at levels comparable to Jurkat cells, an immortalized T lymphocyte cell line that represents a sub-type of leukocytes involved in cell-mediated immunity. In both Jurkat and HeLa cells, the endogenous FMNL1 is associated with the Golgi and co-localizes with both the *cis*-Golgi marker GM130 and the *trans*-Golgi marker Golgin 97 (Colon-Franco et al. 2011). Depletion of FMNL1 expression has the surprising effect of increasing overall F-actin levels with an overabundance of stress fibers in HeLa cells and increased cortical actin in Jurkat cells. This unexpected increase in F-actin accumulation is associated with disruption of normal Golgi ribbon assembly in HeLa cells. In this case, the normally compact Golgi ribbon is dispersed, but still maintains some degree of its asymmetric perinuclear positioning suggesting that FMNL1 is required for normal Golgi assembly. Loss of FMNL1 also disrupted specific aspects of Golgi-dependent trafficking. The secretory marker VSV-G still trafficked normally to the plasma membrane, but the lysosomal markers mannose-6-phosphate receptor and cathepsin D were mis-sorted and the lysosomes themselves were misshapen. These defects were thought to arise from an overabundance of F-actin at the Golgi itself and could be partially rescued by treatment with the actin-disrupting drug latrunculin B (Colon-Franco et al. 2011). Curiously, only the γ isoform, but not α or β, of FMNL1 was able to localize to the Golgi and rescue the knockdown phenotype. The FMNL1 isoforms originate from alternative splicing and only differ at their C-terminal tails; it is not clear how the unique C-terminal sequence of FMNL1γ would affect its activity or subcellular targeting. Rescue was dependent on barbed-end binding, but neither rescue nor Golgi localization was dependent upon membrane targeting by N-terminal myristoylation of FMNL1γ. On this basis, it was suggested that the unique C-terminal tail of the γ isoform targets FMNL1 to the Golgi where it maintains Golgi integrity by preventing excessive F-actin assembly via barbed-end capping and F-actin severing. Although reexpression of FMNL1γ is able to rescue the FMNL1 knockdown phenotype, the subcellular localization of the transgene does not precisely match the localization of the endogenous protein. It remains to be seen how this isoform is specifically targeted to the Golgi ribbon. In addition, it was noted that no effect on Golgi assembly was observed in these cells following depletion of FMNL2, FMNL3, mDia1, or mDia2 (Colon-Franco et al. 2011).

As with FMNL1, FMNL2 and FMNL3 also play a cell-type-specific role in Golgi assembly (Gardberg et al. 2010; Kage et al. 2017). FMNL2 and FMNL3, but not FMNL1, are expressed in B16-F1 mouse melanoma cells and NIH 3T3 fibroblasts where these proteins normally accumulate at the leading edge of migrating cells (Block et al. 2012; Peladeau et al. 2016). All three FMNL proteins are posited to be Cdc42 effectors (Block et al. 2012; Kage et al. 2017; Kuhn et al. 2015; Kuhn and Geyer 2014) and Cdc42 is associated with normal Golgi assembly and polarity (Chi et al. 2013; Farhan and Hsu 2016; Long and Simpson 2017). Expression of a constitutively active mutant of Cdc42 (Q61L) was sufficient to relocate the endogenous FMNL2 and FMNL3 proteins to the Golgi and was also able to recruit GFP-tagged versions of FMNL1α as well as FMNL2 and 3 to this organelle. Unlike FMNL1 in HeLa cells, FMNL2 and 3 localization to the Golgi was dependent on N-myristoylation and inhibited by the presence of an N-terminal GFP tag. Super-resolution imaging revealed that FMNL2 and 3 are most closely associated with the trans-Golgi network. Accordingly, loss of FMNL2 and 3 through CRISPR-mediated knockout or siRNA-mediated knockdown induced Golgi dispersion and disrupted secretory and lysosomal trafficking. The defects in Golgi assembly and trafficking were associated with loss of F-actin assembly. This suggests that despite the similarities between the three proteins, FMNL2/3 are playing decidedly different roles in Golgi assembly. In HeLa cells, FMNL1γ, but not FMNL2 or 3 (Colon-Franco et al. 2011), is required to maintain normal ribbon assembly by limiting F-actin assembly and neither the FMNL1α nor FMNL1β isoforms are able to localize to the Golgi (Colon-Franco et al. 2011). In contrast, in mouse fibroblasts or mouse melanoma cells, FMNL2 and 3 are needed to ensure that the necessary actin filaments are maintained to hold the Golgi ribbon together. In these cells, exogenous FMNL1α is also able to be recruited to the Golgi by Cdc42 and for all three FMNL proteins Golgi localization is dependent on their N-terminal myristoylation (Kage et al. 2017). Thus, both the mode of localization and their function at the Golgi seem quite distinct between these cell types. It is not clear if these differences arise from species (human versus mouse) or cell type (epithelial versus fibroblast)-specific mechanisms.

3.4 Golgi Dispersion and mDia1

mDia1 (human DIAPH1) is a RhoA effector and one of the best studied of the vertebrate formins (Courtemanche 2018; Ishizaki et al. 2001; Lin and Windhorst 2016; Mizuno and Watanabe 2012; Watanabe et al. 1997, 1999). As with the FMNL proteins, manipulation of mDia1 activity also disrupts Golgi morphology. mDia1 is a potent nucleator of actin polymerization and is able to induce microtubule stabilization, acetylation, and detyrosination (Bartolini and Gundersen 2010; Palazzo et al. 2001; Thurston et al. 2012). Association of mDia1 activity with the Golgi ribbon was first noted in HeLa cells where treatment with the Golgi disrupting drug brefeldin A interfered with mDia1-induced actin stress fiber bundling (Ishizaki et al. 2001). Golgi polarization in migrating NIH 3T3 fibroblasts could also be disrupted by overexpression of a constitutively active mDia1 derivative, apparently through its effects on microtubule organization (Magdalena et al. 2003). A direct role for mDia1 in regulating Golgi assembly has also been described in the HeLa JW cell line (Paran et al. 2006; Zilberman et al. 2011). In these cells, mDia1 induces dispersion of the Golgi ribbon into ministacks downstream of RhoA activation, following either LPA stimulation or expression of active RhoA mutants. GTP-loaded RhoA recruits mDia1 to the Golgi and shRNA-mediated depletion of mDia1 results in Golgi compaction and blocks RhoA-induced Golgi dispersion. Dispersion is both actin and microtubule dependent and can be rescued by treatment with latrunculin B. In accordance with its role downstream of RhoA, expression of a constitutively active (CA) mDia1 derivative also induces Golgi dispersion and associated with the formation of F-actin patches around the dispersed Golgi remnants. CA-mDia1 expression blocks Golgi reassembly following washout of the drug in nocodazole-treated cells. Concomitant with dispersion it was noted that RhoA activation also induced enhanced formation of Rab6 positive vesicles, a component of the secretory machinery exiting from the trans-Golgi network. This was also mDia1 dependent and apparently a separate effect from the dispersion of the Golgi into ministacks. The endogenous role of mDia1 in maintaining Golgi architecture in resting cells remains unclear, but it clearly participates in balancing Golgi-specific F-actin assembly.

3.5 Golgi and the "Inverted" Formins

The so-called inverted formins, FHDC1 (aka INF1) and INF2, are outliers within the formin family (Hegsted et al. 2017). Despite the name, the two "inverted" formins are not closely related; however, they both play a part in regulating Golgi dynamics. A role for INF2 in vesicular trafficking was first noted in HepG2 hepatoma cells (Madrid et al. 2010). Here INF2 cooperates with the transmembrane protein MAL2 to facilitate vesicle movement from basolateral to apical membranes during transcytotic trafficking. Subsequently, it was shown that a specific INF2 isoform also played a direct role in Golgi ribbon assembly in HeLa, U2OS, and Jurkat cells. There are two INF2 isoforms generated by alternative splicing, the farnesylated CAAX isoform predominant in NIH 3T3 cells and the non-CAAX isoform predominant in U2OS, Jurkat, and HeLa cells (Chhabra et al. 2009; Madrid et al. 2010; Ramabhadran et al. 2011). INF2-CAAX co-localizes with the endoplasmic reticulum, while the non-CAAX isoform is found in a perinuclear actin meshwork overlapping the Golgi ribbon. Knockdown of INF2 in U2OS cells disrupts normal Golgi morphology resulting in either dispersion or ribbon compaction. Along with Golgi dispersion, INF2 depletion also causes loss of discrete patches of F-actin surrounding the Golgi. The dispersed Golgi can be partially rescued by latrunculin B treatment suggesting dispersion results from an overabundance of F-actin at the Golgi, similar to the effect of FMNL1 depletion in HeLa cells (Colon-Franco et al. 2011).

FHDC1, like the Golgi ribbon, is a vertebrate evolutionary innovation with no obvious orthologue in lower organisms (Hegsted et al. 2017; Young et al. 2008). A true "inverted" formin, FHDC1 has its functional FH1 + FH2 domains in the N-terminus, and, unique to the formin family, it has a discrete C-terminal microtubule-binding domain (MTBD) (Young et al. 2008). In cycling cells, the endogenous FHDC1 protein preferentially accumulates on acetylated microtubules that are part of the Golgi-derived microtubule (GDMT) network (Copeland et al. 2016). FHDC1 overexpression causes the Golgi to disperse into functional ministacks still capable of supporting normal Golgi trafficking. Dispersion is dependent upon both the FH2 and MTBD of FHDC1 and can be partially rescued by latrunculin B treatment. FHDC1-induced dispersion is not affected by depletion of centrosome-derived microtubules in cells treated with the PLK4 inhibitor centrinone, but is blocked by knockdown of GM130 expression (Copeland et al. 2016; Wong et al. 2015). Loss of GM130 prevents assembly of the GDMT network and points to this network as the site of FHDC1 action. FHDC1 depletion also results in dramatic defects in Golgi assembly consistent with loss of GDMT. On this basis, it was proposed that FHDC1 acts as an actin and microtubule bridging factor that promotes stability of the GDMT as they are handed off from *cis*- to *trans*-Golgi. The defects that arise from the loss of FHDC1 expression apparently originate from defects in GDMT; however, the actin-dependent FHDC1 overexpression phenotype underlines the importance of maintaining a fine balance of F-actin assembly at the Golgi. Even though the effects of FHDC1 on Golgi assembly were entirely dependent on the GDMT, FHDC1 was also found to play a separate and distinct role in the control of centriole function and maturation (Copeland et al. 2018). Indeed, despite the well-known interconnected relationship between the Golgi and the centrosome (Bisel et al. 2008; Hurtado et al. 2011; Kodani et al. 2009; Rios 2014; Sutterlin and Colanzi 2010; Vinogradova et al. 2012), few direct links have been made that connect formin effects on Golgi assembly to effects on the centrosome itself. Nonetheless, the ability of formins to regulate cytoskeletal dynamics has also been found to have an important impact on centrosome positioning and cell polarity.

3.6 Diaphanous-Related Formins and Centrosome Positioning

The centrosome is the primary microtubule-organizing center (MTOC) in metazoan cells (Ito and Bettencourt-Dias 2018; Petry and Vale 2015; Werner et al. 2017; Wu and Akhmanova 2017). The relative positioning of the centrosome and Golgi as well as the assembly of their respective microtubule networks is a central determiner of overall cell polarity (Agircan et al. 2014; Rios 2014; Wu and Akhmanova 2017). More recently, it has been shown that the centrosome may also act as an actin-organizing center, although this activity is formin-independent (Farina et al. 2016). Still, a number of studies suggest that formins play a role in establishing the subcellular position of the centrosome (Andres-Delgado et al. 2012, 2013; Ang et al. 2010; Ercan-Sencicek et al. 2015; Gomez et al. 2007; Kutscheidt et al. 2014; LaMonica et al. 2009; Shinohara et al. 2012). In fibroblasts, the centrosome is repositioned to face the leading edge as a result of FHOD1-dependent rearward nuclear movement (Kutscheidt et al. 2014). In other cell types, the Diaphanous-Related Formins DAAM1, mDia1, and mDia3 act more directly to establish centrosome polarity.

DAAM1 is a Diaphanous-Related Formin that is broadly expressed across many tissues and cancer cell lines (Ang et al. 2010). DAAM1 connects the Planar Cell Polarity (PCP) pathway to small GTPase signaling (Aspenstrom et al. 2006; Guillabert-Gourgues et al. 2016; Habas et al. 2001; Zhu et al. 2012). The DAAM1 C-terminus interacts with the PCP protein DVL3 and its N-terminus binds and activates RhoA. Depletion of DAAM1 in U2OS or Cos-7 cells prevents centrosome and Golgi reorientation in cells responding to a scratch wound (Ang et al. 2010). This results in less persistent directional cell migration and is associated with altered morphology of cellular protrusions at the leading edge. Overexpression of DAAM1 in U2OS cells enhanced formation of stress fibers containing myosin IIB. Myosin IIB, but not IIA, is thought to define front-back cell polarity through effects on Golgi and centrosome reorientation (Vicente-Manzanares et al. 2007, 2011). A similar association of DAAM1 with myosin IIB is proposed to regulate cell polarity in endothelial cells. Following Wnt activation, DAAM1 forms a trimeric complex with DVL3 and Kif26b, a kinesin-11 family member, to govern front-back polarity in these cells (Guillabert-Gourgues et al. 2016). As in U2OS cells, knockdown of DAAM1 expression in endothelial cells blocks centrosome and Golgi reorientation during cell migration. This is accompanied by mis-localization of myosin IIB to the rear of the cell as well as defects in microtubule acetylation and stabilization. Similar results were obtained with Kif26b depletion and overexpression of Kif26b was able to rescue defective cell polarity and myosin IIB distribution in both DAAM1 and Kif26b knockdown cells. Together these data point to DAAM1 effects on cell polarity being mediated by myosin IIB and thereby affecting centrosome positioning (Guillabert-Gourgues et al. 2016).

The effects of DAAM on centrosome polarity are not restricted to migrating cells; it also plays a role in basal body positioning downstream of DVL during the formation of multiciliated epithelium in *Xenopus*. In this case, DAAM1 is required to form a subapical actin layer that connects adjacent basal bodies. This subapical actin layer is linked to the basal body via the interaction of DAAM1 with NPHP4 and Inturned, a PDZ-containing CPLANE protein that connects to the basal body (Adler and Wallingford 2017; Zeng et al. 2010). Loss of any member of the trimeric complex dislocates the actin network and inhibits coordinated beating of the motile apical cilia (Yasunaga et al. 2015).

mDia1 also plays an important cell-type-specific role in centrosome positioning. A homozygous nonsense mutation in the human *Diaph1* gene is associated with microcephaly, a neurodevelopmental disorder that arises from defective centrosome function in neurons and neuronal precursors (Ercan-Sencicek et al. 2015). Cell-type-specific defects in centrosome movement were also found in mDia1/mDia3 double knockout mice. Loss of mDia1/3 inhibited migration of neuronal precursors destined to become cortical and olfactory inhibitory interneurons, but did not affect neuronal precursors that will make up the cortical excitatory layer (Shinohara et al. 2012). Neuroblasts isolated from the double knockout mice also displayed defective migration in vitro. These cells migrate by first extending a cellular process followed by the retraction of the nucleus and cell body. The saltatory nuclear translocation is preceded by forward motion of the centrosome into a central cellular swelling that opens up as the leading cellular process extends forward. In wild-type cells, mDia1-EGFP accumulates at this swelling prior to movement of the centrosome and both centrosome and nuclear translocation are defective in the mDia1/3 double knockout neuroblasts. The defects do not arise from abnormal cell polarity as the cells extend their leading process normally. Instead, impaired centrosome migration is associated with loss of mDia-dependent anterograde movement of F-actin into the leading process which likely acts to pull the centrosome forward. It is not clear how actin filaments connect to the centrosome to provide this pulling force or what role mDia might play in connecting F-actin to the microtubule-organizing center.

3.7 Formins and Centrosome Polarization at the Immune Synapse

The immune synapse (IS) is a specialized structure that forms at the junction between T cells and antigen presenting cell (APC). IS assembly is induced by T-cell receptor (TCR) activation after binding its ligand on the APC. IS formation is accompanied by TCR clustering and requires reorganization of the T-cell microtubule network, reorientation of the centrosome to face the synapse, and increased actin polymerization. Re-localization of the centrosome requires formation of an asymmetric stable network of detyrosinated microtubules directed toward the synapse and marks a shift away from the acetylated microtubules predominant in resting T-cells (Andres-Delgado et al. 2013). Formins regulate microtubule stabilization, acetylation, and detyrosination (Bartolini et al. 2008, 2012, 2016; Fernandez-Barrera and Alonso 2018; Palazzo et al. 2001; Thurston et al. 2012) suggesting a possible role for these proteins in immune synapse formation. Indeed, *mDia1* expression is induced by TCR activation (Vicente-Manzanares et al. 2003) and *mDia1* knockout mice are reported to have multiple defects in T-cell function (Eisenmann et al. 2007; Peng et al. 2007; Shi et al. 2009). In addition to mDia1, FMNL1 and INF2 also play an important role in T-cell function and all three proteins play a role in immune synapse formation (Andres-Delgado et al. 2012, 2013; Gomez et al. 2007). In Jurkat cells, the three formins were found distributed along the microtubule network as well as on the actin filaments assembled at the IS. A pool of mDia1 and FMNL1 also accumulated at the centrosome (Gomez et al. 2007). Depletion of any three of these formins did not affect F-actin accumulation at the IS, but did inhibit microtubule detyrosination and centrosome reorientation. Surprisingly, the knockdown phenotype of each of the three formins could be rescued by reexpression of the FH2 domain of any one of them; for example, reexpression of the INF2 FH2 domain was able to rescue FMNL1, mDia1, or INF2 depletion (Andres-Delgado et al. 2012). This rescue was not actin-dependent and was likely mediated by FH2-induced microtubule stabilization. In contrast, the effects of mDia1/3 on centrosome positioning in neuroblasts were entirely actin-dependent (Shinohara et al. 2012) suggesting very different mechanisms are at work in these two distinct cell types. Similar to the situation with INF2 and Golgi ribbon assembly, it is thought that INF2 likely acts in direct cooperation with other formins to affect centrosome positioning during IS formation (Andres-Delgado et al. 2013). How these formins are activated downstream of TCR engagement is not clear. Nor is it clear how the activity of the isolated FH2 domain is targeted to enable the specific reorientation of the centrosome towards the IS.

3.8 FHDC1 and the Centriole Cycle

As noted above, FHDC1 acts at the Golgi-derived microtubule network to facilitate normal Golgi ribbon assembly. Separate from its effects on Golgi, FHDC1 also plays a role in the centriole cycle and ciliogenesis (Fig. 3.2). FHDC1 overexpression in NIH 3T3 fibroblasts has dramatic effects on cilia assembly, either blocking cilia formation or inducing dramatic cilia elongation (Copeland et al. 2018). In both ciliated and nonciliated cells, the centriole maturation cycle is clearly disrupted: more than two centrioles are present per cell; they are widely separated and multiple centrioles are positive for the mother centriole markers Cep170 and cenexin. This effect was unique to FHDC1 and was not observed following overexpression of constitutively active derivatives of 12 other vertebrate formins (FMN2 and Delphilin were not tested). As with the effects of FHDC1 on Golgi assembly, the FHDC1 MTBD and FH2 were required to affect ciliogenesis. Many of the effects of

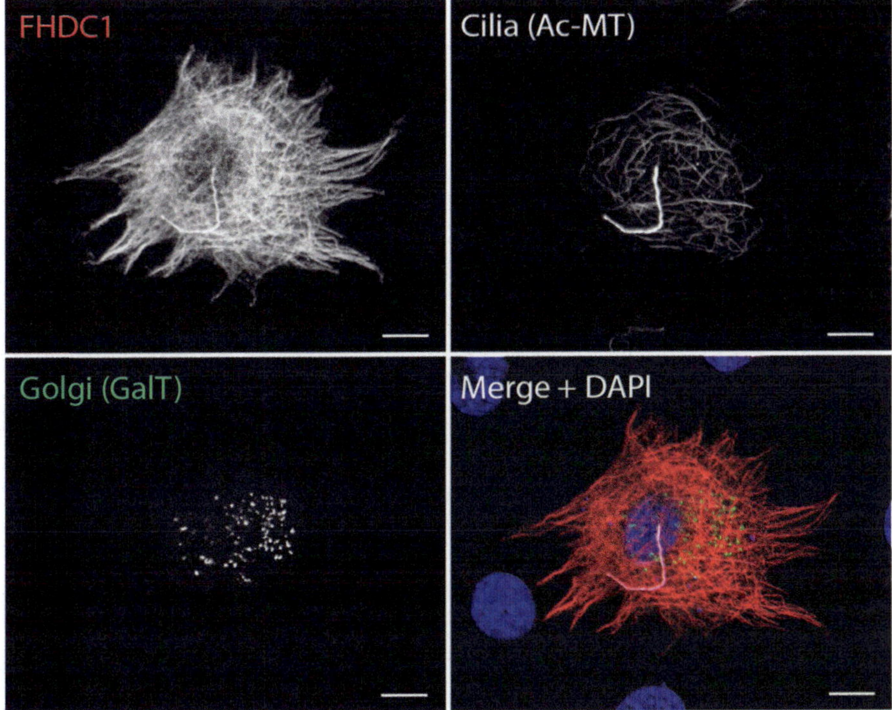

Fig. 3.2 FHDC1 overexpression acts at distinct microtubule networks to induce Golgi dispersion and disrupt regulation of ciliogenesis. FHDC1-mCherry (red) overexpression in NIH 3T3 fibroblasts induces cilia elongation (Acetylated-tubulin, Ac-MT, white) by disrupting the junction between the subdistal appendages and cytoplasmic microtubules which provides a signaling platform for PLK1. FHDC1 expression induces Golgi dispersion (GalT-GFP reporter, green) through a separate effect by disrupting the connection between the Golgi ribbon and the Golgi-derived microtubules. Scale bar = 10 μm

overexpression were actin-dependent and both cilia length and centriole separation could be rescued by treatment with latrunculin B. Conversely, FHDC1 depletion inhibited ciliogenesis and inhibited centrosome assembly. The subdistal appendage protein Cep170 connects the mother centriole to the cytoplasmic microtubule network (Guarguaglini et al. 2005) and Cep170 was identified as an FHDC1-binding protein. The endogenous FHDC1 protein accumulates on a knot of microtubules converging on the basal body and led to the proposal that FHDC1 promotes the Cep170-dependent connection of the centriole/basal body to microtubules. The subdistal appendages serve as a signaling platform for the polo-like kinase PLK1 (Guarguaglini et al. 2005), and it is likely that the effects of FHDC1 overexpression or depletion are mediated by interfering with assembly of this platform which is required for normal PLK1 signaling. PLK1 regulates certain aspects of the centriole maturation cycle (Kong et al. 2014) and the FHDC1 overexpression phenotype could be rescued by co-expression of a constitutively active PLK1 derivative. Despite the interconnections between the Golgi ribbon and centrosome (Agircan et al. 2014; Kodani et al. 2009; Rios 2014), the effects of FHDC1 on these two organelles are discrete and separate effects targeting distinct microtubule populations (Copeland et al. 2016, 2018). It is not clear what the endogenous role of FHDC1-dependent regulation of actin dynamics might be in either process. It is clear, however, that the effects of FHDC1 are cell-type specific and likely contribute to the diversity in centriole structure observed across cell types (Loncarek and Bettencourt-Dias 2018). A similar role for formins at the centrioles of other cell types has not yet been described.

3.9 Conclusions

Numerous studies have connected formin activity to maintenance, function, and organization of the Golgi ribbon and centrosome. Despite these efforts, no unifying picture comes through of a single factor playing the same role in all cell types or of different factors filling the same role across cell types. Indeed, the same protein may follow different rules as the cell type varies; e.g., FMNL1γ is recruited to the Golgi by its unique C-terminal tail in HeLa cells, but is targeted to the Golgi by N-myristoylation in B16-F1 melanoma cells. In part, this may reflect the varied structure and function of the Golgi ribbon as determined by the function of the cell type itself. It is also worth noting that assembly of a Golgi ribbon is unique to the vertebrate lineage and that the diversity of formin function may be coevolving with it. One common theme that has emerged from these studies is that formins play an important role in maintaining a balance of F-actin assembly at the Golgi ribbon and this can be easily upset by formin depletion or overexpression. How this balanced activity is specifically maintained at the Golgi remains to be determined. Similar questions are also left unanswered at the centriole. While many formins seem to be involved in centrosome reorientation, the role and mechanism clearly vary between cell types. A direct connection to centriole function has only been established for

FHDC1 and it is likely that its effects will also be restricted to only certain cell types. To date no systematic study has been carried out to characterize the role of formins in Golgi ribbon assembly, centriole replication, or centrosome formation. The studies discussed in this chapter highlight how obtaining a comprehensive picture will require a focused effort across diverse cell types.

Acknowledgements   Thanks to S.J. Copeland for preparation of figures. Work in the Copeland lab is supported by NSERC Discovery Grant 05921/RGPIN/2016.

References

Adler PN, Wallingford JB (2017) From planar cell polarity to ciliogenesis and back: the curious tale of the PPE and CPLANE proteins. Trends Cell Biol 27:379–390

Agircan FG, Schiebel E, Mardin BR (2014) Separate to operate: control of centrosome positioning and separation. Philos Trans R Soc Lond Ser B Biol Sci 369

Alberts AS (2001) Identification of a carboxyl-terminal diaphanous-related formin homology protein autoregulatory domain. J Biol Chem 276:2824–2830

Andres-Delgado L, Anton OM, Bartolini F, Ruiz-Saenz A, Correas I, Gundersen GG, Alonso MA (2012) INF2 promotes the formation of detyrosinated microtubules necessary for centrosome reorientation in T cells. J Cell Biol 198:1025–1037

Andres-Delgado L, Anton OM, Alonso MA (2013) Centrosome polarization in T cells: a task for formins. Front Immunol 4:191

Ang SF, Zhao ZS, Lim L, Manser E (2010) DAAM1 is a formin required for centrosome re-orientation during cell migration. PLoS One 5

Arden JD, Lavik KI, Rubinic KA, Chiaia N, Khuder SA, Howard MJ, Nestor-Kalinoski AL, Alberts AS, Eisenmann KM (2015) Small-molecule agonists of mammalian Diaphanous-related (mDia) formins reveal an effective glioblastoma anti-invasion strategy. Mol Biol Cell 26:3704–3718

Aspenstrom P, Richnau N, Johansson AS (2006) The diaphanous-related formin DAAM1 collaborates with the Rho GTPases RhoA and Cdc42, CIP4 and Src in regulating cell morphogenesis and actin dynamics. Exp Cell Res 312:2180–2194

Bartolini F, Gundersen GG (2010) Formins and microtubules. Biochim Biophys Acta 1803:164–173

Bartolini F, Moseley JB, Schmoranzer J, Cassimeris L, Goode BL, Gundersen GG (2008) The formin mDia2 stabilizes microtubules independently of its actin nucleation activity. J Cell Biol 181:523–536

Bartolini F, Ramalingam N, Gundersen GG (2012) Actin-capping protein promotes microtubule stability by antagonizing the actin activity of mDia1. Mol Biol Cell 23:4032–4040

Bartolini F, Andres-Delgado L, Qu X, Nik S, Ramalingam N, Kremer L, Alonso MA, Gundersen GG (2016) An mDia1-INF2 formin activation cascade facilitated by IQGAP1 regulates stable microtubules in migrating cells. Mol Biol Cell 27:1797–1808

Beli P, Mascheroni D, Xu D, Innocenti M (2008) WAVE and Arp2/3 jointly inhibit filopodium formation by entering into a complex with mDia2. Nat Cell Biol 10:849–857

Bisel B, Wang Y, Wei JH, Xiang Y, Tang D, Miron-Mendoza M, Yoshimura S, Nakamura N, Seemann J (2008) ERK regulates Golgi and centrosome orientation towards the leading edge through GRASP65. J Cell Biol 182:837–843

Block J, Breitsprecher D, Kuhn S, Winterhoff M, Kage F, Geffers R, Duwe P, Rohn JL, Baum B, Brakebusch C, Geyer M, Stradal TE, Faix J, Rottner K (2012) FMNL2 drives actin-based protrusion and migration downstream of Cdc42. Curr Biol 22:1005–1012

Castrillon DH, Wasserman SA (1994) Diaphanous is required for cytokinesis in Drosophila and shares domains of similarity with the products of the limb deformity gene. Development 120:3367–3377

Chan MW, Chaudary F, Lee W, Copeland JW, McCulloch CA (2010) Force-induced myofibroblast differentiation through collagen receptors is dependent on mammalian diaphanous (mDia). J Biol Chem 285:9273–9281

Cheng L, Zhang J, Ahmad S, Rozier L, Yu H, Deng H, Mao Y (2011) Aurora B regulates formin mDia3 in achieving metaphase chromosome alignment. Dev Cell 20:342–352

Chhabra ES, Ramabhadran V, Gerber SA, Higgs HN (2009) INF2 is an endoplasmic reticulum-associated formin protein. J Cell Sci 122:1430–1440

Chi X, Wang S, Huang Y, Stamnes M, Chen JL (2013) Roles of rho GTPases in intracellular transport and cellular transformation. Int J Mol Sci 14:7089–7108

Colon-Franco JM, Gomez TS, Billadeau DD (2011) Dynamic remodeling of the actin cytoskeleton by FMNL1gamma is required for structural maintenance of the Golgi complex. J Cell Sci 124:3118–3126

Copeland JW, Copeland SJ, Treisman R (2004) Homo-oligomerization is essential for F-actin assembly by the formin family FH2 domain. J Biol Chem 279:50250–50256

Copeland SJ, Green BJ, Burchat S, Papalia GA, Banner D, Copeland JW (2007) The diaphanous inhibitory domain/diaphanous autoregulatory domain interaction is able to mediate heterodimerization between mDia1 and mDia2. J Biol Chem 282:30120–30130

Copeland SJ, Thurston SF, Copeland JW (2016) Actin- and microtubule-dependent regulation of Golgi morphology by FHDC1. Mol Biol Cell 27:260–276

Copeland SJ, McRae A, Guarguaglini G, Trinkle-Mulcahy L, Copeland JW (2018) Actin-dependent regulation of cilia length by the inverted formin FHDC1. Mol Biol Cell 29:1611–1627

Courtemanche N (2018) Mechanisms of formin-mediated actin assembly and dynamics. Biophys Rev 10:1553–1569

Courtemanche N, Lee JY, Pollard TD, Greene EC (2013) Tension modulates actin filament polymerization mediated by formin and profilin. Proc Natl Acad Sci U S A 110:9752–9757

Egea G, Serra-Peinado C, Salcedo-Sicilia L, Gutierrez-Martinez E (2013) Actin acting at the Golgi. Histochem Cell Biol 140:347–360

Eisenmann KM, West RA, Hildebrand D, Kitchen SM, Peng J, Sigler R, Zhang J, Siminovitch KA, Alberts AS (2007) T cell responses in mammalian diaphanous-related formin mDia1 knock-out mice. J Biol Chem 282:25152–25158

Ercan-Sencicek AG, Jambi S, Franjic D, Nishimura S, Li M, El-Fishawy P, Morgan TM, Sanders SJ, Bilguvar K, Suri M, Johnson MH, Gupta AR, Yuksel Z, Mane S, Grigorenko E, Picciotto M, Alberts AS, Gunel M, Sestan N, State MW (2015) Homozygous loss of DIAPH1 is a novel cause of microcephaly in humans. Eur J Hum Genet 23:165–172

Esue O, Harris ES, Higgs HN, Wirtz D (2008) The filamentous actin cross-linking/bundling activity of mammalian formins. J Mol Biol 384:324–334

Farazi TA, Waksman G, Gordon JI (2001) The biology and enzymology of protein N-myristoylation. J Biol Chem 276:39501–39504

Farhan H, Hsu VW (2016) Cdc42 and cellular polarity: emerging roles at the Golgi. Trends Cell Biol 26:241–248

Farina F, Gaillard J, Guerin C, Coute Y, Sillibourne J, Blanchoin L, Thery M (2016) The centrosome is an actin-organizing centre. Nat Cell Biol 18:65–75

Fernandez-Barrera J, Alonso MA (2018) Coordination of microtubule acetylation and the actin cytoskeleton by formins. Cell Mol Life Sci 75:3181–3191

Fernandez-Barrera J, Bernabe-Rubio M, Casares-Arias J, Rangel L, Fernandez-Martin L, Correas I, Alonso MA (2018) The actin-MRTF-SRF transcriptional circuit controls tubulin acetylation via alpha-TAT1 gene expression. J Cell Biol 217:929–944

Gadadhar S, Bodakuntla S, Natarajan K, Janke C (2017) The tubulin code at a glance. J Cell Sci 130:1347–1353

Gaillard J, Ramabhadran V, Neumanne E, Gurel P, Blanchoin L, Vantard M, Higgs HN (2011) Differential interactions of the formins INF2, mDia1, and mDia2 with microtubules. Mol Biol Cell 22:4575–4587

Gardberg M, Talvinen K, Kaipio K, Iljin K, Kampf C, Uhlen M, Carpen O (2010) Characterization of Diaphanous-related formin FMNL2 in human tissues. BMC Cell Biol 11:55

Gasteier JE, Madrid R, Krautkramer E, Schroder S, Muranyi W, Benichou S, Fackler OT (2003) Activation of the Rac-binding partner FHOD1 induces actin stress fibers via a ROCK-dependent mechanism. J Biol Chem 278:38902–38912

Gomez TS, Kumar K, Medeiros RB, Shimizu Y, Leibson PJ, Billadeau DD (2007) Formins regulate the actin-related protein 2/3 complex-independent polarization of the centrosome to the immunological synapse. Immunity 26:177–190

Gorelik R, Yang C, Kameswaran V, Dominguez R, Svitkina T (2011) Mechanisms of plasma membrane targeting of formin mDia2 through its amino terminal domains. Mol Biol Cell 22:189–201

Gould CJ, Maiti S, Michelot A, Graziano BR, Blanchoin L, Goode BL (2011) The formin DAD domain plays dual roles in autoinhibition and actin nucleation. Curr Biol 21:384–390

Grobe H, Wustenhagen A, Baarlink C, Grosse R, Grikscheit K (2018) A Rac1-FMNL2 signaling module affects cell-cell contact formation independent of Cdc42 and membrane protrusions. PLoS One 13:e0194716

Guarguaglini G, Duncan PI, Stierhof YD, Holmstrom T, Duensing S, Nigg EA (2005) The forkhead-associated domain protein Cep170 interacts with Polo-like kinase 1 and serves as a marker for mature centrioles. Mol Biol Cell 16:1095–1107

Guillabert-Gourgues A, Jaspard-Vinassa B, Bats ML, Sewduth RN, Franzl N, Peghaire C, Jeanningros S, Moreau C, Roux E, Larrieu-Lahargue F, Dufourcq P, Couffinhal T, Duplaa C (2016) Kif26b controls endothelial cell polarity through the Dishevelled/Daam1-dependent planar cell polarity-signaling pathway. Mol Biol Cell 27:941–953

Gurel PS, Ge P, Grintsevich EE, Shu R, Blanchoin L, Zhou ZH, Reisler E, Higgs HN (2014) INF2-mediated severing through actin filament encirclement and disruption. Curr Biol 24:156–164

Habas R, Kato Y, He X (2001) Wnt/Frizzled activation of Rho regulates vertebrate gastrulation and requires a novel Formin homology protein Daam1. Cell 107:843–854

Han Y, Eppinger E, Schuster IG, Weigand LU, Liang X, Kremmer E, Peschel C, Krackhardt AM (2009) Formin-like 1 (FMNL1) is regulated by N-terminal myristoylation and induces polarized membrane blebbing. J Biol Chem 284:33409–33417

Harris ES, Li F, Higgs HN (2004) The mouse formin, FRLalpha, slows actin filament barbed end elongation, competes with capping protein, accelerates polymerization from monomers, and severs filaments. J Biol Chem 279:20076–20087

Harris ES, Rouiller I, Hanein D, Higgs HN (2006) Mechanistic differences in actin bundling activity of two mammalian formins, FRL1 and mDia2. J Biol Chem 281:14383–14392

Harris ES, Gauvin TJ, Heimsath EG, Higgs HN (2010) Assembly of filopodia by the formin FRL2 (FMNL3). Cytoskeleton (Hoboken) 67:755–772

Hegsted A, Yingling CV, Pruyne D (2017) Inverted formins: a subfamily of atypical formins. Cytoskeleton 74:405–419

Heimsath EG Jr, Higgs HN (2012) The C terminus of formin FMNL3 accelerates actin polymerization and contains a WH2 domain-like sequence that binds both monomers and filament barbed ends. J Biol Chem 287:3087–3098

Henty-Ridilla JL, Rankova A, Eskin JA, Kenny K, Goode BL (2016) Accelerated actin filament polymerization from microtubule plus ends. Science 352:1004–1009

Higashida C, Miyoshi T, Fujita A, Oceguera-Yanez F, Monypenny J, Andou Y, Narumiya S, Watanabe N (2004) Actin polymerization-driven molecular movement of mDia1 in living cells. Science 303:2007–2010

Higashida C, Kiuchi T, Akiba Y, Mizuno H, Maruoka M, Narumiya S, Mizuno K, Watanabe N (2013) F- and G-actin homeostasis regulates mechanosensitive actin nucleation by formins. Nat Cell Biol 15:395–405

Higgs HN (2005) Formin proteins: a domain-based approach. Trends Biochem Sci 30:342–353

Hotulainen P, Lappalainen P (2006) Stress fibers are generated by two distinct actin assembly mechanisms in motile cells. J Cell Biol 173:383–394

Hurtado L, Caballero C, Gavilan MP, Cardenas J, Bornens M, Rios RM (2011) Disconnecting the Golgi ribbon from the centrosome prevents directional cell migration and ciliogenesis. J Cell Biol 193:917–933

Ishizaki T, Morishima Y, Okamoto M, Furuyashiki T, Kato T, Narumiya S (2001) Coordination of microtubules and the actin cytoskeleton by the Rho effector mDia1. Nat Cell Biol 3:8–14

Iskratsch T, Lange S, Dwyer J, Kho AL, dos Remedios C, Ehler E (2010) Formin follows function: a muscle-specific isoform of FHOD3 is regulated by CK2 phosphorylation and promotes myofibril maintenance. J Cell Biol 191:1159–1172

Isogai T, van der Kammen R, Innocenti M (2015) SMIFH2 has effects on Formins and p53 that perturb the cell cytoskeleton. Sci Rep 5:9802

Ito D, Bettencourt-Dias M (2018) Centrosome remodelling in evolution. Cells 7

Jegou A, Carlier MF, Romet-Lemonne G (2013) Formin mDia1 senses and generates mechanical forces on actin filaments. Nat Commun 4:1883

Ju R, Cirone P, Lin S, Griesbach H, Slusarski DC, Crews CM (2010) Activation of the planar cell polarity formin DAAM1 leads to inhibition of endothelial cell proliferation, migration, and angiogenesis. Proc Natl Acad Sci U S A 107:6906–6911

Kage F, Steffen A, Ellinger A, Ranftler C, Gehre C, Brakebusch C, Pavelka M, Stradal T, Rottner K (2017) FMNL2 and -3 regulate Golgi architecture and anterograde transport downstream of Cdc42. Sci Rep 7:9791

Kitzing TM, Wang Y, Pertz O, Copeland JW, Grosse R (2010) Formin-like 2 drives amoeboid invasive cell motility downstream of RhoC. Oncogene 29:2441–2448

Kobielak A, Pasolli HA, Fuchs E (2004) Mammalian formin-1 participates in adherens junctions and polymerization of linear actin cables. Nat Cell Biol 6:21–30

Kodani A, Kristensen I, Huang L, Sutterlin C (2009) GM130-dependent control of Cdc42 activity at the Golgi regulates centrosome organization. Mol Biol Cell 20:1192–1200

Kong D, Farmer V, Shukla A, James J, Gruskin R, Kiriyama S, Loncarek J (2014) Centriole maturation requires regulated Plk1 activity during two consecutive cell cycles. J Cell Biol 206:855–865

Kovar DR, Pollard TD (2004) Insertional assembly of actin filament barbed ends in association with formins produces piconewton forces. Proc Natl Acad Sci U S A 101:14725–14730

Kozlov MM, Bershadsky AD (2004) Processive capping by formin suggests a force-driven mechanism of actin polymerization. J Cell Biol 167:1011–1017

Kuhn S, Geyer M (2014) Formins as effector proteins of Rho GTPases. Small GTPases 5:e29513

Kuhn S, Erdmann C, Kage F, Block J, Schwenkmezger L, Steffen A, Rottner K, Geyer M (2015) The structure of FMNL2-Cdc42 yields insights into the mechanism of lamellipodia and filopodia formation. Nat Commun 6:7088

Kutscheidt S, Zhu R, Antoku S, Luxton GW, Stagljar I, Fackler OT, Gundersen GG (2014) FHOD1 interaction with nesprin-2G mediates TAN line formation and nuclear movement. Nat Cell Biol 16:708–715

LaMonica K, Bass M, Grabel L (2009) The planar cell polarity pathway directs parietal endoderm migration. Dev Biol 330:44–53

Lee K, Gallop JL, Rambani K, Kirschner MW (2010) Self-assembly of filopodia-like structures on supported lipid bilayers. Science 329:1341–1345

Li F, Higgs HN (2003) The mouse Formin mDia1 is a potent actin nucleation factor regulated by autoinhibition. Curr Biol 13:1335–1340

Li F, Higgs HN (2005) Dissecting requirements for auto-inhibition of actin nucleation by the formin, mDia1. J Biol Chem 280:6986–6992

Lin YN, Windhorst S (2016) Diaphanous-related formin 1 as a target for tumor therapy. Biochem Soc Trans 44:1289–1293

Loncarek J, Bettencourt-Dias M (2018) Building the right centriole for each cell type. J Cell Biol 217:823–835

Long M, Simpson JC (2017) Rho GTPases operating at the Golgi complex: implications for membrane traffic and cancer biology. Tissue Cell 49:163–169

Lu J, Meng W, Poy F, Maiti S, Goode BL, Eck MJ (2007) Structure of the FH2 domain of Daam1: implications for formin regulation of actin assembly. J Mol Biol 369:1258–1269

Madrid R, Aranda JF, Rodriguez-Fraticelli AE, Ventimiglia L, Andres-Delgado L, Shehata M, Fanayan S, Shahheydari H, Gomez S, Jimenez A, Martin-Belmonte F, Byrne JA, Alonso MA (2010) The formin INF2 regulates basolateral-to-apical transcytosis and lumen formation in association with Cdc42 and MAL2. Dev Cell 18:814–827

Magdalena J, Millard TH, Machesky LM (2003) Microtubule involvement in NIH 3T3 Golgi and MTOC polarity establishment. J Cell Sci 116:743–756

Maiti S, Michelot A, Gould C, Blanchoin L, Sokolova O, Goode BL (2012) Structure and activity of full-length formin mDia1. Cytoskeleton (Hoboken) 69:393–405

Miller PM, Folkmann AW, Maia AR, Efimova N, Efimov A, Kaverina I (2009) Golgi-derived CLASP-dependent microtubules control Golgi organization and polarized trafficking in motile cells. Nat Cell Biol 11:1069–1080

Mizuno H, Watanabe N (2012) mDia1 and formins: screw cap of the actin filament. Biophysics (Nagoya-shi) 8:95–102

Mizuno H, Higashida C, Yuan Y, Ishizaki T, Narumiya S, Watanabe N (2011) Rotational movement of the formin mDia1 along the double helical strand of an actin filament. Science 331:80–83

Moriya K, Yamamoto T, Takamitsu E, Matsunaga Y, Kimoto M, Fukushige D, Kimoto C, Suzuki T, Utsumi T (2012) Protein N-myristoylation is required for cellular morphological changes induced by two formin family proteins, FMNL2 and FMNL3. Biosci Biotechnol Biochem 76:1201–1209

Moseley JB, Sagot I, Manning AL, Xu Y, Eck MJ, Pellman D, Goode BL (2004) A conserved mechanism for Bni1- and mDia1-induced actin assembly and dual regulation of Bni1 by Bud6 and profilin. Mol Biol Cell 15:896–907

Moseley JB, Bartolini F, Okada K, Wen Y, Gundersen GG, Goode BL (2007) Regulated binding of adenomatous polyposis coli protein to actin. J Biol Chem 282:12661–12668

Nakamura N, Wei JH, Seemann J (2012) Modular organization of the mammalian Golgi apparatus. Curr Opin Cell Biol 24:467–474

Nezami AG, Poy F, Eck MJ (2006) Structure of the autoinhibitory switch in formin mDia1. Structure 14:257–263

Nezami A, Poy F, Toms A, Zheng W, Eck MJ (2010) Crystal structure of a complex between amino and carboxy terminal fragments of mDia1: insights into autoinhibition of diaphanous-related formins. PLoS One 5

Nieuwenhuis J, Brummelkamp TR (2019) The tubulin detyrosination cycle: function and enzymes. Trends Cell Biol 29:80–92

Okada K, Bartolini F, Deaconescu AM, Moseley JB, Dogic Z, Grigorieff N, Gundersen GG, Goode BL (2010) Adenomatous polyposis coli protein nucleates actin assembly and synergizes with the formin mDia1. J Cell Biol 189:1087–1096

Otomo T, Otomo C, Tomchick DR, Machius M, Rosen MK (2005) Structural basis of Rho GTPase-mediated activation of the formin mDia1. Mol Cell 18:273–281

Palazzo AF, Cook TA, Alberts AS, Gundersen GG (2001) mDia mediates Rho-regulated formation and orientation of stable microtubules. Nat Cell Biol 3:723–729

Paran Y, Lavelin I, Naffar-Abu-Amara S, Winograd-Katz S, Liron Y, Geiger B, Kam Z (2006) Development and application of automatic high-resolution light microscopy for cell-based screens. Methods Enzymol 414:228–247

Peladeau C, Heibein A, Maltez MT, Copeland SJ, Copeland JW (2016) A specific FMNL2 isoform is up-regulated in invasive cells. BMC Cell Biol 17:32

Peng J, Kitchen SM, West RA, Sigler R, Eisenmann KM, Alberts AS (2007) Myeloproliferative defects following targeting of the Drf1 gene encoding the mammalian diaphanous related formin mDia1. Cancer Res 67:7565–7571

Petry S, Vale RD (2015) Microtubule nucleation at the centrosome and beyond. Nat Cell Biol 17:1089–1093

Pring M, Evangelista M, Boone C, Yang C, Zigmond SH (2003) Mechanism of formin-induced nucleation of actin filaments. Biochemistry 42:486–496

Quinlan ME, Hilgert S, Bedrossian A, Mullins RD, Kerkhoff E (2007) Regulatory interactions between two actin nucleators, Spire and Cappuccino. J Cell Biol 179:117–128

Ramabhadran V, Korobova F, Rahme GJ, Higgs HN (2011) Splice variant-specific cellular function of the formin INF2 in maintenance of Golgi architecture. Mol Biol Cell 22:4822–4833

Ramabhadran V, Gurel PS, Higgs HN (2012) Mutations to the formin homology 2 domain of INF2 protein have unexpected effects on actin polymerization and severing. J Biol Chem 287:34234–34245

Ramabhadran V, Hatch AL, Higgs HN (2013) Actin monomers activate inverted formin 2 by competing with its autoinhibitory interaction. J Biol Chem 288:26847–26855

Rios RM (2014) The centrosome-Golgi apparatus nexus. Philos Trans R Soc Lond Ser B Biol Sci 369

Romero S, Le Clainche C, Didry D, Egile C, Pantaloni D, Carlier MF (2004) Formin is a processive motor that requires profilin to accelerate actin assembly and associated ATP hydrolysis. Cell 119:419–429

Rose R, Weyand M, Lammers M, Ishizaki T, Ahmadian MR, Wittinghofer A (2005) Structural and mechanistic insights into the interaction between Rho and mammalian Dia. Nature 435:513–518

Roth-Johnson EA, Vizcarra CL, Bois JS, Quinlan ME (2014) Interaction between microtubules and the Drosophila formin Cappuccino and its effect on actin assembly. J Biol Chem 289:4395–4404

Sarmiento C, Wang W, Dovas A, Yamaguchi H, Sidani M, El-Sibai M, Desmarais V, Holman HA, Kitchen S, Backer JM, Alberts A, Condeelis J (2008) WASP family members and formin proteins coordinate regulation of cell protrusions in carcinoma cells. J Cell Biol 180:1245–1260

Schonichen A, Geyer M (2010) Fifteen formins for an actin filament: a molecular view on the regulation of human formins. Biochim Biophys Acta 1803:152–163

Schulze N, Graessl M, Blancke Soares A, Geyer M, Dehmelt L, Nalbant P (2014) FHOD1 regulates stress fiber organization by controlling transversal arc and dorsal fiber dynamics. J Cell Sci 127:1379–1393

Scott BJ, Neidt EM, Kovar DR (2011) The functionally distinct fission yeast formins have specific actin-assembly properties. Mol Biol Cell 22:3826–3839

Seth A, Otomo C, Rosen MK (2006) Autoinhibition regulates cellular localization and actin assembly activity of the diaphanous-related formins FRLalpha and mDia1. J Cell Biol 174:701–713

Shi Y, Zhang J, Mullin M, Dong B, Alberts AS, Siminovitch KA (2009) The mDia1 formin is required for neutrophil polarization, migration, and activation of the LARG/RhoA/ROCK signaling axis during chemotaxis. J Immunol 182:3837–3845

Shimada A, Nyitrai M, Vetter IR, Kuhlmann D, Bugyi B, Narumiya S, Geeves MA, Wittinghofer A (2004) The core FH2 domain of diaphanous-related formins is an elongated actin binding protein that inhibits polymerization. Mol Cell 13:511–522

Shinohara R, Thumkeo D, Kamijo H, Kaneko N, Sawamoto K, Watanabe K, Takebayashi H, Kiyonari H, Ishizaki T, Furuyashiki T, Narumiya S (2012) A role for mDia, a Rho-regulated actin nucleator, in tangential migration of interneuron precursors. Nat Neurosci 15:373–380, S371–372

Silkworth WT, Kunes KL, Nickel GC, Phillips ML, Quinlan ME, Vizcarra CL (2018) The neuron-specific formin Delphilin nucleates nonmuscle actin but does not enhance elongation. Mol Biol Cell 29:610–621

Staus DP, Taylor JM, Mack CP (2011) Enhancement of mDia2 activity by Rho-kinase-dependent phosphorylation of the diaphanous autoregulatory domain. Biochem J 439:57–65

Sutterlin C, Colanzi A (2010) The Golgi and the centrosome: building a functional partnership. J Cell Biol 188:621–628

Takeya R, Taniguchi K, Narumiya S, Sumimoto H (2008) The mammalian formin FHOD1 is activated through phosphorylation by ROCK and mediates thrombin-induced stress fibre formation in endothelial cells. EMBO J 27:618–628

Thompson ME, Heimsath EG, Gauvin TJ, Higgs HN, Kull FJ (2013) FMNL3 FH2-actin structure gives insight into formin-mediated actin nucleation and elongation. Nat Struct Mol Biol 20:111–118

Thurston SF, Kulacz WA, Shaikh S, Lee JM, Copeland JW (2012) The ability to induce microtubule acetylation is a general feature of formin proteins. PLoS One 7:e48041

Thyberg J, Moskalewski S (1985) Microtubules and the organization of the Golgi complex. Exp Cell Res 159:1–16

Truong D, Brabant D, Bashkurov M, Wan LC, Braun V, Heo WD, Meyer T, Pelletier L, Copeland J, Brumell JH (2013) Formin-mediated actin polymerization promotes Salmonella invasion. Cell Microbiol 15:2051–2063

Vaillant DC, Copeland SJ, Davis C, Thurston SF, Abdennur N, Copeland JW (2008) Interaction of the N- and C-terminal autoregulatory domains of FRL2 does not inhibit FRL2 activity. J Biol Chem 283:33750–33762

Valderrama F, Babia T, Ayala I, Kok JW, Renau-Piqueras J, Egea G (1998) Actin microfilaments are essential for the cytological positioning and morphology of the Golgi complex. Eur J Cell Biol 76:9–17

Vega FM, Fruhwirth G, Ng T, Ridley AJ (2011) RhoA and RhoC have distinct roles in migration and invasion by acting through different targets. J Cell Biol 193:655–665

Vicente-Manzanares M, Rey M, Perez-Martinez M, Yanez-Mo M, Sancho D, Cabrero JR, Barreiro O, de la Fuente H, Itoh K, Sanchez-Madrid F (2003) The RhoA effector mDia is induced during T cell activation and regulates actin polymerization and cell migration in T lymphocytes. J Immunol 171:1023–1034

Vicente-Manzanares M, Zareno J, Whitmore L, Choi CK, Horwitz AF (2007) Regulation of protrusion, adhesion dynamics, and polarity by myosins IIA and IIB in migrating cells. J Cell Biol 176:573–580

Vicente-Manzanares M, Newell-Litwa K, Bachir AI, Whitmore LA, Horwitz AR (2011) Myosin IIA/IIB restrict adhesive and protrusive signaling to generate front-back polarity in migrating cells. J Cell Biol 193:381–396

Vinogradova T, Paul R, Grimaldi AD, Loncarek J, Miller PM, Yampolsky D, Magidson V, Khodjakov A, Mogilner A, Kaverina I (2012) Concerted effort of centrosomal and Golgi-derived microtubules is required for proper Golgi complex assembly but not for maintenance. Mol Biol Cell 23:820–833

Wang Y, El-Zaru MR, Surks HK, Mendelsohn ME (2004) Formin homology domain protein (FHOD1) is a cyclic GMP-dependent protein kinase I-binding protein and substrate in vascular smooth muscle cells. J Biol Chem 279:24420–24426

Wang F, Zhang L, Duan X, Zhang GL, Wang ZB, Wang Q, Xiong B, Sun SC (2015) RhoA-mediated FMNL1 regulates GM130 for actin assembly and phosphorylates MAPK for spindle formation in mouse oocyte meiosis. Cell Cycle 14:2835–2843

Watanabe N, Madaule P, Reid T, Ishizaki T, Watanabe G, Kakizuka A, Saito Y, Nakao K, Jockusch BM, Narumiya S (1997) p140mDia, a mammalian homolog of Drosophila diaphanous, is a target protein for Rho small GTPase and is a ligand for profilin. EMBO J 16:3044–3056

Watanabe N, Kato T, Fujita A, Ishizaki T, Narumiya S (1999) Cooperation between mDia1 and ROCK in Rho-induced actin reorganization. Nat Cell Biol 1:136–143

Werner S, Pimenta-Marques A, Bettencourt-Dias M (2017) Maintaining centrosomes and cilia. J Cell Sci 130:3789–3800

Wong YL, Anzola JV, Davis RL, Yoon M, Motamedi A, Kroll A, Seo CP, Hsia JE, Kim SK, Mitchell JW, Mitchell BJ, Desai A, Gahman TC, Shiau AK, Oegema K (2015) Cell biology. Reversible centriole depletion with an inhibitor of Polo-like kinase 4. Science 348:1155–1160

Woodham EF, Paul NR, Tyrrell B, Spence HJ, Swaminathan K, Scribner MR, Giampazolias E, Hedley A, Clark W, Kage F, Marston DJ, Hahn KM, Tait SW, Larue L, Brakebusch CH, Insall RH, Machesky LM (2017) Coordination by Cdc42 of actin, contractility, and adhesion for melanoblast movement in mouse skin. Curr Biol 27:624–637

Woychik RP, Maas RL, Zeller R, Vogt TF, Leder P (1990) 'Formins': proteins deduced from the alternative transcripts of the limb deformity gene. Nature 346:850–853

Wu J, Akhmanova A (2017) Microtubule-organizing centers. Annu Rev Cell Dev Biol 33:51–75

Xu Y, Moseley JB, Sagot I, Poy F, Pellman D, Goode BL, Eck MJ (2004) Crystal structures of a Formin Homology-2 domain reveal a tethered dimer architecture. Cell 116:711–723

Yang C, Czech L, Gerboth S, Kojima S, Scita G, Svitkina T (2007) Novel roles of formin mDia2 in lamellipodia and filopodia formation in motile cells. PLoS Biol 5:e317

Yasunaga T, Hoff S, Schell C, Helmstadter M, Kretz O, Kuechlin S, Yakulov TA, Engel C, Muller B, Bensch R, Ronneberger O, Huber TB, Lienkamp SS, Walz G (2015) The polarity protein Inturned links NPHP4 to Daam1 to control the subapical actin network in multiciliated cells. J Cell Biol 211:963–973

Young KG, Thurston SF, Copeland S, Smallwood C, Copeland JW (2008) INF1 is a novel microtubule-associated formin. Mol Biol Cell 19:5168–5180

Zaidel-Bar R (2018) Cell cycle pacemaker keeps adhesion in step with division. J Cell Biol 217:2981–2982

Zeng H, Hoover AN, Liu A (2010) PCP effector gene Inturned is an important regulator of cilia formation and embryonic development in mammals. Dev Biol 339:418–428

Zhu Y, Tian Y, Du J, Hu Z, Yang L, Liu J, Gu L (2012) Dvl2-dependent activation of Daam1 and RhoA regulates Wnt5a-induced breast cancer cell migration. PLoS One 7:e37823

Zigmond SH, Evangelista M, Boone C, Yang C, Dar AC, Sicheri F, Forkey J, Pring M (2003) Formin leaky cap allows elongation in the presence of tight capping proteins. Curr Biol 13:1820–1823

Zilberman Y, Alieva NO, Miserey-Lenkei S, Lichtenstein A, Kam Z, Sabanay H, Bershadsky A (2011) Involvement of the Rho-mDia1 pathway in the regulation of Golgi complex architecture and dynamics. Mol Biol Cell 22:2900–2911

Zimmermann D, Kovar DR (2019) Feeling the force: formin's role in mechanotransduction. Curr Opin Cell Biol 56:130–140

Chapter 4
Role of Intracellular Transport in the Centriole-Dependent Formation of Golgi Ribbon

Alexander A. Mironov, Ivan D. Dimov, and Galina V. Beznoussenko

Abstract The intracellular transport is the most confusing issue in the field of cell biology. The Golgi complex (GC) is the central station along the secretory pathway. It contains Golgi glycosylation enzymes, which are responsible for protein and lipid glycosylation, and in many cells, it is organized into a ribbon. Position and structure of the GC depend on the position and function of the centriole. Here, we analyze published data related to the role of centriole and intracellular transport (ICT) for the formation of Golgi ribbon and specifically stress the importance of the delivery of membranes containing cargo and membrane proteins to the cell centre where centriole/centrosome is localized. Additionally, we re-examined the formation of Golgi ribbon from the point of view of different models of ICT.

Abbreviations

Apo B	Apolipoprotein B
ArfGAP	Arf GTPase activating protein
Bet3	Trafficking protein particle complex subunit BET3
CMC	*Cis*-most cisterna
CMPM	Compartment (cisterna) maturation progression model
COP	Coatomer
DM	Diffusion model
EGC	ER-Golgi carrier
EGT	ER-Golgi transport
ER	Endoplasmic reticulum
ERES	ER exit site
GC	Golgi complex

A. A. Mironov (✉) · G. V. Beznoussenko
The FIRC Institute of Molecular Oncology, Milan, Italy
e-mail: alexandre.mironov@ifom.eu; galina.beznusenko@ifom.eu

I. D. Dimov
Department of Anatomy, Saint Petersburg State Paediatric Medical University, Saint Petersburg, Russia

© Springer Nature Switzerland AG 2019
M. Kloc (ed.), *The Golgi Apparatus and Centriole*, Results and Problems in Cell Differentiation 67, https://doi.org/10.1007/978-3-030-23173-6_4

GFP	Green fluorescent protein
GMAP210	Golgi microtubule-associated protein 210 KDa
GTP	Guanosine-5′-triphosphate
ICC	Inter-cisternal connections
ICT	Intracellular transport
IGT	Intra-Golgi transport
KARM	Kiss-and-run model
KIFC3	Kinesin family member C3
Man	Mannosidase
MT	Microtubule
NSF	*N*-ethylmaleimide-sensitive factor
PM	Plasma membrane
Rab	Ras-related in the brain
Sar1	Secretion-associated RAS superfamily-related gene
Sec	Secretory clone
SNAP	Synaptosomal-associated protein
SNARE	Soluble NSF attachment receptor
TGN	*Trans*-Golgi network
TMC	*Trans*-most cisterna
TRIP11	Thyroid receptor-interacting protein 11
VLDL	Very low-density lipoprotein
VM	Vesicular model

4.1 Centriole and the Golgi Complex

After being associated with proteins of peri-centriolar matrix, the centriole becomes the functional microtubule-organizing centre/centrosome (Sukhorukov and Meyer-Hermann 2015; Carvajal-Gonzalez et al. 2016; Wu and Akhmanova 2017). Centrosomes are important for Golgi ribbon organization and the position of the Golgi complex (GC). These aspects are discussed below. However, here, we do not describe the structure and molecular organization of centrosome and microtubule aster because these subjects are covered by other chapters of this volume. Also, we do not discuss here the mitotic Golgi fragmentation.

In the majority of animal cell types, the microtubule (MT) component of the cytoskeleton includes a star-shaped array (aster), where MTs are anchored to the centrosome. Within the centrosome, MTs are nucleated to form tubulin polymers. The opposite (plus) ends of MTs often reach cell periphery where they interact physically with actin filaments. Centrosomes are usually located in the geometric centre of the cell close to the nucleus and play a significant role in organization and function of the GC (Carvajal-Gonzalez et al. 2016).

In nonpolarized mammalian and many other animal cells, the centrioles localize near the cell geometric centre. The GC surrounds centrosome forming a three-dimensional ribbon (Figs. 4.1a–d and 4.2a, b). The position of the centrosome and the GC depends on cell polarity and its migration. During cell migration in vivo or in vitro, in wound healing/scratch assays, the centrosome and the GC reorient towards the leading edge of the migrating cells. There, the formation of actin filament network is observed just near the plasma membrane (PM) of the leading edge. In fully polarized epithelial cells (in cell culture and in situ), the centrosome is shifted towards the apical PM. There, centrioles may become the basal bodies to form and build the cilia (ciliogenesis; Carvajal-Gonzalez et al. 2016). In these cells, the nucleus is localized more basally, whereas the GC resides between the apical PM and the nucleus. Cytochalasin D, an inhibitor of actin polymerization, interferes with basal body migration and ciliary development in epithelial cells (Boisvieux-Ulrich et al. 1990; Carvajal-Gonzalez et al. 2016). In cytotoxic T-lymphocytes, centrosome and the GC move towards the immune synapse (Kloc et al. 2014; Carvajal-Gonzalez et al. 2016).

4.2 The Golgi Ribbon

In plants and yeast, the MT pattern is different than in mammalian cells, and thus, these cells have no Golgi ribbon. In contrast, if a protist cell has a MT-sorganizing centre, then the GC is usually organized into a Golgi ribbon (Soksolova and Mironov 2008; Mironov et al. 2016). In plants and yeast and in some insect cells and protists, the GC is fragmented and distributed within the entire cytoplasm (peripheral type of Golgi fragmentation). In protist cells with the centrally located centrosome, the GC is usually organized into a ribbon (Sokolova and Mironov 2008; Beznoussenko et al. 2016). In some insect cells, the mini ribbons composed of two mini stacks are observed. Tubules interconnect these neighbouring stacks, and this pair of mini stacks seems to function as a ribbon (reviewed by Kondylis and Rabouille 2003). However, in differentiated insect cells, the GC usually forms a ribbon (Conti et al. 2010).

The GC can exhibit the following forms: (1) tubular (in microsporidia), (2) isolated disks with perforations (in *Saccharomyces cerevisiae*), and (3) disks with perforations near their edges organized in a stack containing different compartments (in other cells). Golgi cisternae form stacks. Stacks can be isolated, or they can form ribbon, and these two forms can differ in the number of disks and vesicles. Not only disks but also tubules are among Golgi elements. In addition, the GC can be situated in the centre, close to the centriole (1) or at the periphery (2). The number of disks in the stack can be high (>8) or low (<8). The number of tubules can be high or low. The number of 52-nm vesicles also could be high, low or they can be absent. The Golgi stacks can contain the perforated *cis*-most cisterna (CMC) and the perforated *trans*-most cisterna (TMC). Their attachment depends on the functional status of the GC (see below). These variations represent functional, experimental and pathological phenotypes of GC. When the GC transports cargo proteins, it is composed of the perforated CMC, three or six (rarely more) medial cisternae in stack, TMC and the *trans*-cisterna of the endoplasmic

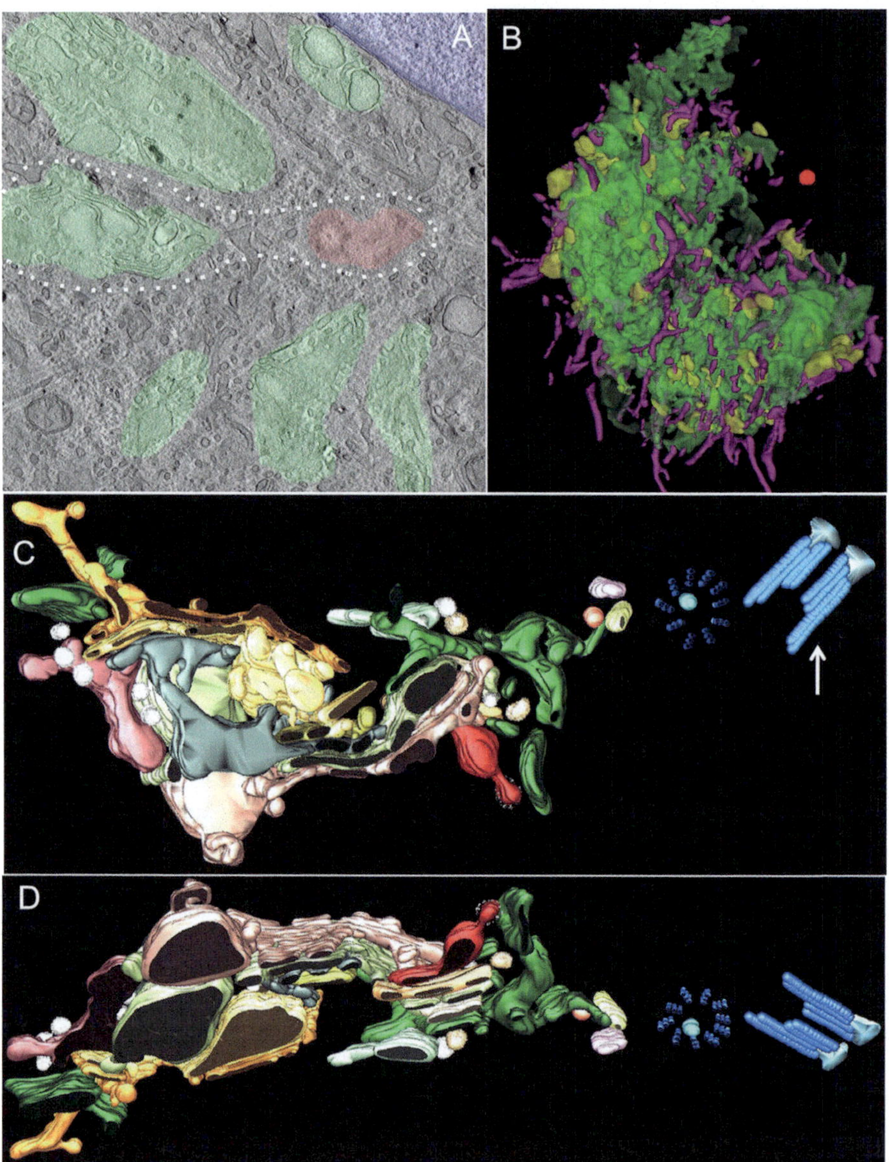

Fig. 4.1 Centriole and the Golgi. (**a**) The tomographic section of Golgi complex in the human fibroblasts. Centriole is coloured in red. The nucleus is pictured in blue. Golgi stacks are green. Dished line indicates the Golgi stack, three-dimensional (3D) reconstruction of which is shown in plate (**c** and **d**; view from two opposite sides). The white arrow shows the centriole. (**b**) 3DED reconstruction of the Golgi ribbon, which forms a cap around centriole (the position of it is indicated with red dot). Golgi stacks are pictured in green, the ER cisternae in magenta and ERES are yellow. Scale bars: 500 nm (**a**); 2 μm (**b**); 200 nm (**c, d**)

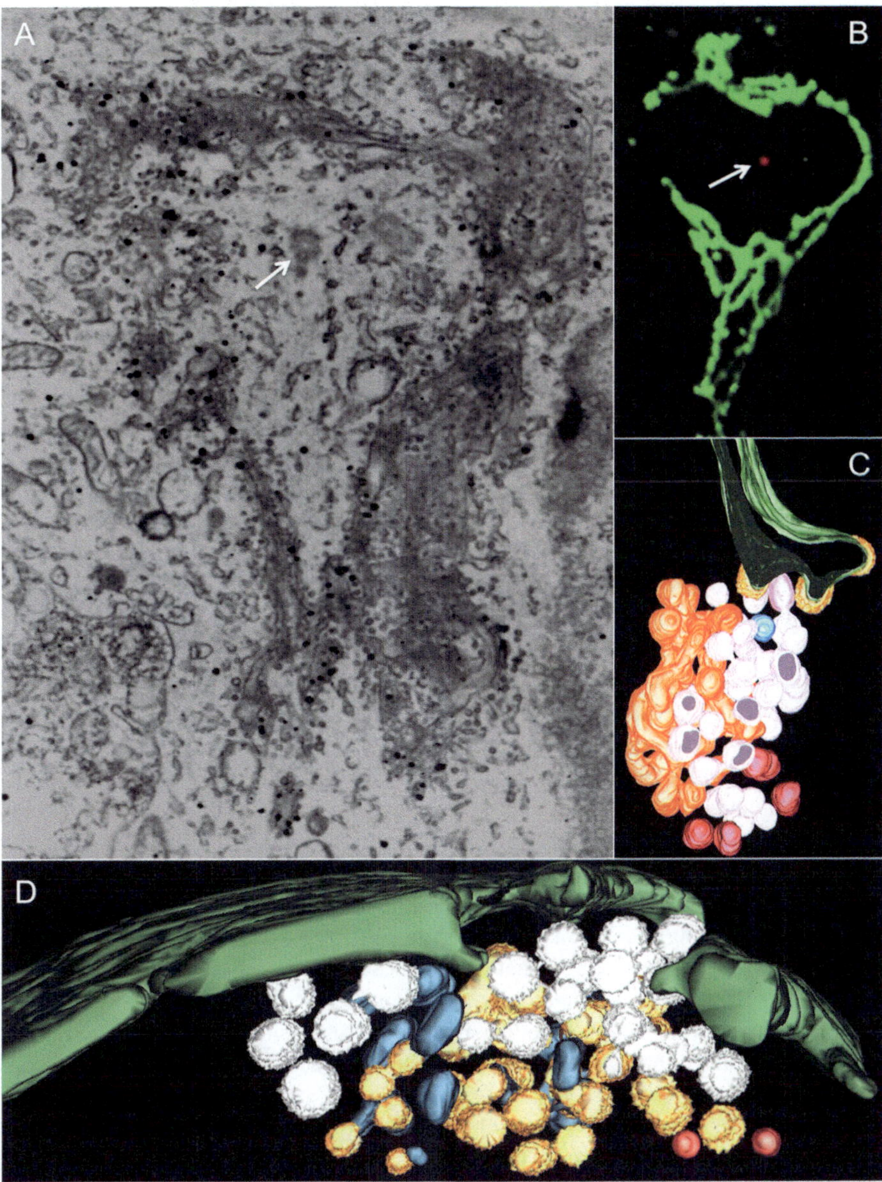

Fig. 4.2 Centriole and ER-Golgi compartments. (**a**, **b**) Correlative light (**b**) electron (**a**) microscopy of the Golgi ribbon. In (**a**, **b**) the Golgi is labelled for GM130 [gold labelling in (**a**) and green colour in (**b**)]. White arrows in (**a**, **b**) show position of centriole, which is indicated with red dot. (**c**, **d**) 3D reconstruction of ERES after chemical fixation. Separated 50-nm spheres are coloured in magenta. The ER is green. COPII coat is yellow. Tubular networks are orange. Varicose tubules are white. Views from opposite sides. Scale bars: 800 nm (**a**); 2.4 μm (**b**); 250 nm (**c**, **d**)

reticulum (ER) attached to TMC. In the transporting GC, the number of inter-cisternal connections (ICC) is high, whereas the number of 52-nm coatomer I (COPI)-dependent vesicles is low. Stacks are connected with each other by tubules, perforated cisternae or solid cisternae to form the Golgi ribbon. The ribbon exhibits a rather complicated three-dimensional structure (Fig. 4.1). When the GC does not transport cargos, it contains more 52-nm vesicles and less ICCs (Trucco et al. 2004), Golgi stacks are localized near and around the centrosome and many of the Golgi stacks are not connected with each other. Therefore, the Golgi ribbon is less developed.

The main function of the GC is lipid and protein glycosylation and protein sorting. The GC cooperates with the trans-Golgi network (TGN) localized nearby. The details of Golgi function and organization are described in Mironov et al. (2016). Of interest, the GC can also nucleate MTs (reviewed by Polishchuk and Mironov 2004; Mironov and Pavelka 2008). Proteins, which regulate behaviour of the microtubule minus end, also affect Golgi organization and morphology (Yang et al. 2017). The Golgi ribbon is also a site for localization and activation of several proteins involved in cell signalling (Gosavi et al. 2018). Several centrosomal proteins stabilize microtubules associated with the GC (Hoppeler-Lebel et al. 2007).

In mammals, dozens or even hundreds of Golgi stacks are laterally linked together to form an interconnected, ribbon-like single organelle, which is located in the perinuclear area. Golgi ribbon is formed from the mini stacks localized near the centrosome. MTs growing at a high density from the centrosome 'push' the Golgi membranes out, forcing them to stay away from the cell centre. The connectivity between individual stacks has been shown in fixed cells using scanning electron microscopy and observation of the complementary fractures of the GC (Inoue 1992) and in living cells using fluorescence restoration after photobleaching of Golgi resident proteins tagged with the green fluorescent protein (GFP). For instance, several Golgi resident proteins exhibit rapid diffusion exchange between stacks (Cole et al. 1996). Golgi stacks are connected (1) by tubules or noncompact zones or (2) by single cisterna. In human fibroblasts, the neighbouring stacks are connected by a single cisterna that sometimes extends out of the stacks and is transformed into a tubule connecting corresponding cisterna of the neighbouring stacks (Mironov et al. 2016). Of interest, cisternal distensions containing aggregates of procollagen are situated at the peripheral (oriented to the centriole) side of the GC (Mironov et al. 2016).

The development of the Golgi ribbon varies in different cells. For instance, in fibroblasts and in spinal neurons, the ribbon is well developed. In contrast, in HeLa cells the GC is fragmented into several pieces (reviewed by Polishchuk and Mironov 2004; Mironov and Pavelka 2008; Mironov et al. 2016). Golgi ribbon is not found in some oocytes (Motta et al. 1994), highly differentiated uroepithelial cells (Kreft et al. 2010), myotubes (Rahkila et al. 1996) and in mammalian cells completely deprived of microtubules (Polishchuk et al. 1999; Trucco et al. 2004; Mironov et al. 2016). Golgi fragmentation has been also observed in some mammalian neurons (Horton et al. 2005; Hanus and Ehlers 2008). Golgi stacks are connected with each other by tubular-reticular noncompact zones and membranous tubules. The formation and positioning of the Golgi ribbon depend on the organization of MT aster and the position of the centrosome. In polarized epithelial cells, the MTs are primarily oriented with their plus ends located near the GC and their minus ends in the apical cytoplasm. The ribbon is necessary for the delivery of post-Golgi carriers towards the leading edge of the migrating cells (Mironov and Beznoussenko 2011; Mironov et al. 2016).

The MT motors such as dynein are involved in the central delivery (centralization) of ER-to-Golgi carriers and Golgi mini stacks, and thus, the motors are important for the formation of the Golgi ribbon (reviewed by Mironov and Beznoussenko 2011). When the delivery of ER-Golgi carrier (EGC) to the cell centre (but not the exit of cargo from the ER) was blocked, the GC appeared as an assembly of the peripheral fragments. The peripherally located Golgi fragments (mini stacks) are fully functional (Trucco et al. 2004) and can reform the ribbon after restoration of centralization. For instance, upon nocodazole washout, when the already fragmented GC is placed in conditions where the MT aster with a united centrosome is restored, the Golgi fragments immediately undergo centralization, and the ribbon of interconnected stacks is rebuilt (Ho et al. 1989). When the GC is fragmented due to the absence of ER-to-Golgi transport, the relative number of coatomer I (COPI)-dependent vesicles increases due to the relative increase of cisternal rims in comparison to the Golgi ribbon. In contrast, the centrally located GC fragments usually are devoid of CMC and TMC and are surrounded by many 52-nm vesicles (Marra et al. 2007; Mironov et al. 2016). Fragmentation of the Golgi ribbon results in the modulation of many signalling pathways (Makhoul et al. 2018).

4.3   Ribbon Factors

Mechanisms responsible for the generation of the Golgi ribbon include Golgi centralization system, which is composed of centrioles/centrosome, Golgins, SNAREs (soluble NSF [*N*-ethylmaleimide-sensitive factor] attachment receptors) and other molecular machines responsible for membrane fusion and several other less studied factors. Interpretation of all these findings is highly affected by the model of intracellular transport (ICT), which had been used for data analysis. In most mammals, the Golgi ribbon depends on the centralization of Golgi membranes and membranes containing cargos. There are several factors involved in Golgi centralization: (1) centriole/centrosome and microtubule aster and (2) Golgi membrane-associated minus end-directed microtubule motors, mainly dynein, which move the Golgi stacks inwards and cause concentration of the Golgi stacks around the centrosome. There are also factors inhibiting centralization: (1) kinesin and (2) GMAP-210/TRIP11 (Golgi microtubule-associated protein 210 or thyroid receptor-interacting protein 11), a protein participating in attachment of the GC to MTs, which prevents complete centralization of the GC. Thus, the formation of Golgi ribbon occurs only when there are (1) functional MTOC and dynamic MT, (2) central delivery of membranes from the ER exit sites, and (3) functional coatomers I and II. The balance between the activity of the minus-end motor dynein and the plus-end motor kinesin determines the final positioning of GC. For instance, KIFC3 (kinesin family member C3) plays a significant role in the MT-dependent centralization and positioning of the GC in some polarized epithelial cells. Also, acetylated MTs are enriched at the GC, their deacetylation resulting in Golgi fragmentation (Mironov and Beznoussenko 2011). A mechanism for functional coupling of COPII-coated membranes to the microtubule cytoskeleton has been described. This mechanism regulates the delivery or EGCs to the cell centre (Watson et al. 2005).

Centralization can be blocked by several methods: (1) depolymerization of MTs by specific drugs (e.g., nocodazole; Polishchuk et al. 1999; Trucco et al. 2004) or impairment of their dynamics (for instance, treatment of cells with low concentrations of nocodazole; Minin 1997); (2) stabilization of MTs, for instance, by Taxol; and (3) impairment of the function of dynein and proteins regulating MT function (Mironov and Beznoussenko 2011). The blockage of MT polymerization induces transformation of the Golgi ribbon into many peripheral fragments. The inactivation of kinesins, proteins that move towards the plus end of MTs, in nonpolarized cells results in the collapse of the GC and the ER around the centrosome (Feiguin et al. 1994) and, strangely, microinjection of an anti-kinesin antibody inhibits centralization of ER-to-Golgi carriers (Lippincott-Schwartz et al. 1995). Also, after depolymerization of actin, the GC shifts to the cell centre (Valderrama et al. 1998).

There are several proteins and other factors involved in the formation of Golgi ribbon. We call them the gluing factors. Among them Golgins and SNAREs/Ca^{2+} play central role. However, these gluing factors function only when the membranes filled with cargos are delivered to the GC (Marra et al. 2007; Mironov and Beznoussenko 2011). Gluing makes fusion of Golgi cisternae and consecutive formation of Golgi ribbon easier. Understanding of the cargo role explains many other seemingly strange findings. For instance, silencing, depletion or inhibition of ßCOP (Styers et al. 2008; Razi et al. 2009) and proteins cooperating with COPI such as protein number 24 (Mitrovic et al. 2008; Koegler et al. 2010), COPII (Cutrona et al. 2013), Rab 1 and 2 (Venditti et al. 2012; Romero et al. 2013) or trafficking protein particle complex subunit 3 (Venditti et al. 2012) leads to Golgi fragmentation because the delivery of cargo to the GC was blocked. The cells lacking COPII coat do not form peripheral mini stacks when treated with nocodazole (Cutrona et al. 2013). The role of cargo domain for the formation of Golgi ribbon is shown in Fig. 4.3. After arrival of the EGC, its membranes are integrated between the neighbouring Golgi stacks and after fusion with them function as a bridge between the stacks.

Thus, the formation of Golgi ribbon depends on the delivery of EGCs containing cargos towards the GC. Also, the role of IGT in the formation of Golgi ribbon is important. In order to understand how the Golgi ribbon cargo participates in its formation, it is necessary to know mechanisms of EGT and IGT. Currently, there are four main models of ICT: (1) the vesicular model (VM), (2) the compartment maturation progression model (CMPM), (3) the diffusion model (DM) and (4) the kiss-and-run model (KARM). VM is based on the assumption that COPI-dependent vesicles transport cargo proteins and membrane from one more proximal compartment to another more distal compartment. CMPM poses that newly formed *cis*-Golgi

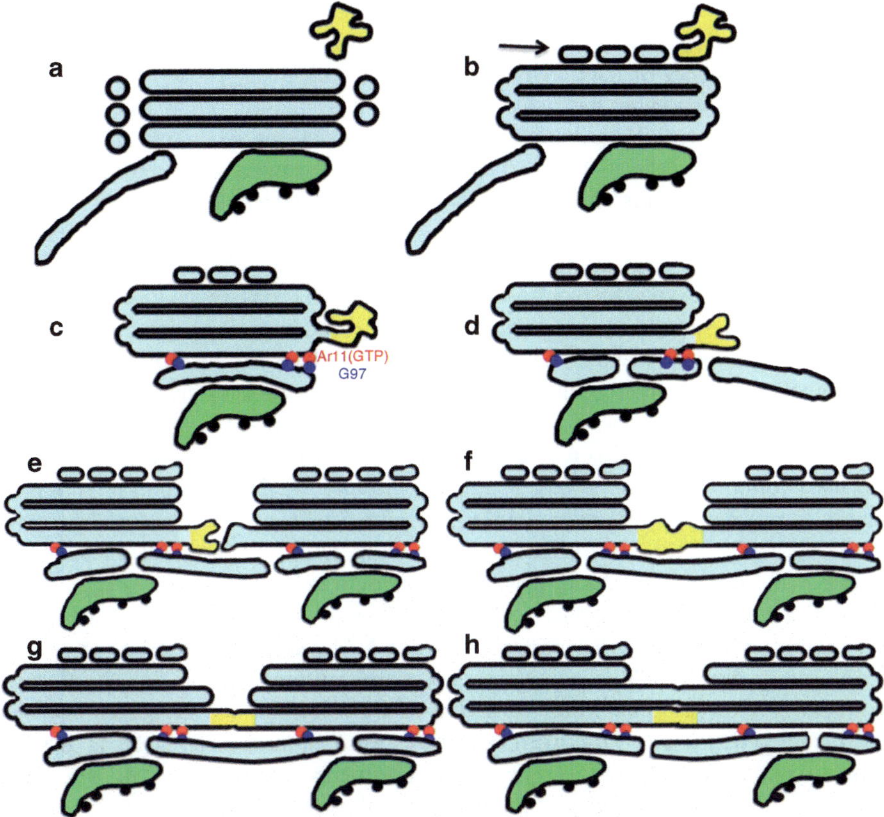

Fig. 4.3 Possible scheme of the Golgi ribbon formation. Role of cargo for gluing of Golgi stacks. (**a**) Resting Golgi stack without CMC and TMC and ER-Golgi carrier (EGC; yellow). The trans ER is coloured in green. (**b**) Attachment of EGC to the Golgi stack with the formation of perforated CMC (black arrow) and inter-cisternal connections. (**c**) Cargo domain is transported (yellow) across the Golgi stack. Simultaneously there is the Golgin97 and Arl1-dependent attachment of TMC to the medial Golgi cisternae. (**d**) The cargo (yellow) reaches the *trans* side of the stack. (**e**, **f**) Fusion of cargo domain (yellow) with cisternae of another stack. (**g**, **h**) Consecutive steps of the formation of the common cisterna for both stacks and attachment of TMC. After its arrival at the GC, ER-Golgi carrier serves as a bridge between adjacent Golgi cisternae and fuses with them forming Golgi ribbon

cisternae undergo progression through the Golgi stack with simultaneous recycling of resident Golgi proteins with retrograde vesicles. DM claims that GC represents a single membrane compartment and cargo moves along it using simple diffusion. KARM exists in symmetrical and asymmetrical variants and suggests that there are repetitive fusion and subsequent fission between two Golgi cisternae. Fusion/fission could occur in the same site (the symmetrical variant) or in different sites (asymmetrical variant). In the latter case, fusion would be between the leading edge of the distal compartment and the rare edge of the proximal one, whereas fission would be somewhere within the distal compartment.

There could be several strategies. The simplest solution is to propose the different model for each cargo transport at different steps of ICT and then find corresponding molecular machinery for such transport. We tried to use this approach combining the diffusion mechanism with the cisterna (compartment) maturation progression model (Beznoussenko et al. 2014). Another strategy is to insist that the dogma is correct and propose explanations for the data, which contradict the main aspects of the dogma. For instance, proponents of VM suggested that in our paper (Mironov et al. 2003), we should use anti-Sec13 antibody (although we used anti-Sec31 antibody, we also confirmed our data with the help of anti-Sec13 antibody, but these data were not shown) because Sec13 has only one isoform (Sec31 has two isoforms). Thus, according to the authors, using anti-Sec31 antibody, we could not detect mega-vesicle because these mega-vesicles could be covered by another isoform of Sec31, which was undetectable by our antibodies. However, we used polyclonal antibody, and we also did not observe mega-vesicle at the electron microscopy level. Trying to convince readers that VM is correct, it is also possible to suggest that the antibodies we used cannot detect some mega-vesicles (Gorur et al. 2017). Another way is to claim (in our experiments with silencing of both Sar1A and B isoforms where we could not find COPII-coated buds and COPII vesicles): 'It is as yet unclear how possible mechanisms such as kiss and run could retain sufficient selectivity to prevent non-selective transport. It remains possible, however, that the small remaining amount of COPII proteins in the RNAi experiments described here is sufficient to direct COPII-dependent selectivity' (https://stephenslab.wordpress. com/2013/02/27/comments-on-silencing-of-mammalian-sar1-isoforms-reveals-copii-independent-protein-sorting-and-transport/). Another strategy is to find other explanations for the cornerstone experiments supporting VM. The third approach is to find a model that would explain all the available data.

4.4 ER-Golgi Transport

VM of the ER-Golgi transport (EGT) poses that COPII-dependent vesicles are involved in EGT as the carriers. According to CMPM of EGT, immature EGCs are formed by protrusion from the ER, and ER resident proteins are eliminated from EGC by retrograde COPI-dependent vesicles. Within the framework of DM, EGT occurs by diffusion along constant connections between the ER and the GC. ER-Golgi connections were shown several times (reviewed by Mironov et al. 1997). For example, Sesso et al. (1994) demonstrated ER-Golgi connections. High-resolution three-dimensional analysis revealed connection between this partial cisterna, positioned between medial Golgi cisternae and the ER. This cisterna contains pores and membrane buds with typical necks (Ladinsky et al. 1999).

Finally, KARM assumes that EGT is realized by fusion–fission mechanism: initially the membrane protrusion filled with a cargo is formed, and then this protrusion fuses with the tubule emanating from the GC and in particular from CMC. After this fusion, the fission occurs near the neck connecting the protrusion and the ER. The tubule delivers dynein to the immature EGC. This motor moves EGC towards the GC. The arrival of this carrier at the GC generates flux of Ca^{2+} from ER exit sites (Micaroni et al. 2010) and stimulates fusion of the carrier with the medial GC. The hug-and-kiss model, proposed by Kurokawa et al. (2013), actually represents KARM.

These models compete with each other for the role of the paradigm within this field. There are many unresolved questions. For instance, within the framework of VM and CMPM, it is not possible to explain the role of cargo delivery in the formation of Golgi ribbon. It is not clear why Golgi ribbon undergoes fragmentation when delivery of cargo is blocked. Also, the formation of the centrally fragmented and the peripherally fragmented GC cannot be explained within the framework of VM, DM or CMPM. The important role of cargo in the formation of Golgi ribbon necessitates a description of mechanisms of EGT and IGT. Several evidences against VM of EGT were published. For instance, EGT occurs even in the absence of COPII (reviewed by Mironov 2014). For example, Bard et al. (2006) showed that after elimination of COPII subunits, EGT still occurs although at the slightly lower level (54% of the control). Caco-2 cells lacking Sec13 (this protein has only one isoform in humans) could grow although their polarization was defective. These cells form a multilayered sheet of cells (Townley et al. 2012). Their ability to divide suggests that the delivery of membrane to the PM necessary for membrane duplication during cell division was normal. Thus, in the absence of Sec13, Caco-2 cells can transport membrane proteins. Also, Sec13 depletion leads to a defect in deposition of large ECM components such as collagen (Townley et al. 2008). However, there is the lack of detectable defects in the transport or secretion of small, soluble, freely diffusible proteins or transmembrane proteins in Sec13-suppressed HeLa cells (Townley et al. 2008) and Caco-2 cells (Townley et al. 2012).

The absence of COPII does not affect transport of soluble and membrane cargos although exit of procollagen from the ER is inhibited (Cutrona et al. 2013). Not only the absence of COPII coat but also slowdown (after depletion of Sedlin; Venditti et al. 2012) or acceleration of COPII turnover (the Sec23A-M702V mutation, Kim et al. 2012) inhibited procollagen exit from the ER. We did not find membrane buds on the ER in *Saccharomyces cerevisiae* (Beznoussenko et al. 2016). Other researchers also did not demonstrate the presence of these buds (reviewed by Beznoussenko et al. 2016). In *Microsporidia* parasite, the ER and Golgi remnants also have no such buds and separated vesicles (Beznoussenko et al. 2007). Using electron cryomicroscopy and pure liposomes, it was demonstrated that the size of COPII-dependent vesicles should be about 75–80 nm (Antonny et al. 2003; Fig. 3a in Lee et al. 2005; Bacia et al. 2011). Moreover, within the ER exit sites, only 50-nm vesicles were found using a quick-freezing technique while the 70–80 nm vesicles were absent (Mironov et al. 2003). Also, Zeuschner et al. (2006) found only free 50-nm diameter vesicles. Finally, Bannykh et al. (1996) also did not find the 70–80-nm separated vesicles within the ER exit sites (ERES). Typical images of this structure are shown in Fig. 4.2c, d.

Additionally, we found the depletion of the anterograde cargos in the 50-nm vesicles localized within the ERES (Mironov et al. 2003). Also, nobody demonstrated how COPII vesicles or aggregates of COPII vesicles (as it was proposed by Bannykh et al. 1996) move towards the GC. Of interest, ERES described in cell cultures were found in situ only in the tissues with high mitotic index such as bone marrow, basal cells of epidermis, enterocytes in intestinal crypts (Rhodin 1974; Pavelka and Roth 2005), fibroblasts of the rat food pad (Marchi and Leblond 1983), odontoblasts (Weinstock and Leblond 1974; Leblond 1989) and in cancer cells (Ghadialli 1982) as well as in professional secretors such as the acinar pancreatic cells or ß-cells ofs Langerhans islets (Jamieson and Palade 1968; Sesso et al. 1994). In neurons, ERES are present only near the GC (Peters et al. 1991). Here, in order to explain all these contradictions, we try to find another interpretation of data presented in the cornerstone papers, supporting different models of EGT.

The main support for VM of EGT is the paper by Kaiser and Schekman (1990). The authors generated the temperature-sensitive mutants of proteins that are responsible for EGT (COPII subunits Sec23 and Sec13) and proteins involved into membrane fusion (Sec17 [known also as αSNAP] and Sec18 [known also as NSF]). When cells were heated up to 37°C and the membrane fusion was blocked, they found an accumulation of 50-nm vesicles (60 vesicles per 1 μm^3 of cytosol, whereas at normal temperature, there were 13 vesicles per 1 μm^3). When they simultaneously inhibited both Sec17 and Sec23, a significant accumulation of 50-nm vesicles was not observed (Fig. 3 by Kaiser and Schekman 1990).

Their interpretations of these results were within the framework of VM, which poses that COPII-dependent vesicles execute EGT, whereas IGT is performed by COPI-dependent vesicles. When the fusion of COPII-vesicles generated by Sec23 or Sec13 is blocked, COPII vesicles have to accumulate, whereas when not only fusion of vesicles with the distal compartments but also their generation is inhibited, the accumulation of the vesicles should not be observed. However, their interpretation contains contradiction because during ICT, not only COPII vesicles were generated but also COPI vesicles should appear, and their number has to be higher than that of COPII vesicles because their volume is smaller. According to Bacia et al. (2011) and Lee et al. (2005), the size of COPII vesicle should be 70–80 nm, whereas COPI vesicle in *Saccharomyces cerevisiae* has a diameter of 50 nm; see Beznoussenko et al. (2016). Therefore, when function of Sec17 was blocked and putative COPII vesicles were not formed, the GC still existed, and COPI-dependent vesicles should be generated from it.

Here, we try to interpret all existing data within the framework of KARM. Initially, we have to emphasize that the authors blocked the proteins responsible for ER-GC transport by heating cells to 37°C and examined cells 60 min after the heating. According to KARM, Sec23 and Sec13 are important for ER-GC transport although COPII vesicles are very rare and do not function as EGCs. According to KARM, when the authors inhibited Sec17, there should be accumulation of COPI vesicles because COPII vesicles have no transport role. Indeed, the authors observed accumulation of 50-nm vesicles. In *Saccharomyces cerevisiae*, 50 nm is the diameter of COPI-dependent vesicles (Beznoussenko et al. 2016). Also, when both Sec17 and Sec23 were inhibited, presumably, there should be accumulation of COPI vesicle but only if the GC remain separated from the ER. However, in *Saccharomyces cerevisiae*, when the delivery of cargo at the GC is blocked, the GC disappears in 4 min (Morin-Ganet et al. 2000), and the generation of COPI vesicles is stopped. In the absence of cargo delivery, the redistribution of the GC into ER under the action of brefeldin A requires membrane fusion between the GC and the post-Golgi compartments (Fukunaga et al. 1998; Kweon et al. 2004), whereas in *Saccharomyces cerevisiae*, the GC is often connected with the ER (Beznoussenko et al. 2016 and reference therein) and is not dependent on membrane fusion. Indeed, in agreement with the prediction of KARM, Kaiser and Schekman (1990) found only small accumulation of vesicles (from normal 13 vesicles per 1 μm^3 up to 20 vesicles per 1 μm^3). Thus, KARM can explain the results, which were presented in the cornerstone papers and interpreted in favour of VM of EGT.

The second cornerstone paper in favour of VM is that by Barlowe et al. (1994). They isolated COPII-coated vesicles after incubation of yeast microsomes with purified component of COPII in the presence of GTP. Importantly, after incubation of microsomes with COPII subunits and GTP, the authors obtained mixture of tubules and vesicles. Then the authors filtered this membrane fraction through the gel with small pores. The authors concluded that these vesicles are formed by COPII and these vesicles contain cargo proteins. In order to find another explanation of these observations, we should stress that Barlowe et al. (1994) presented EM images of round profiles partially coated with COPI-like coat (Fig. 8E by Barlowe et al. 1994) and images of cryosections of round profiles with diameter of 66–73 nm and labelled for Sec23 (Fig. 8G in Barlowe et al. 1994) and Sec13 (see Fig. 8H in Barlowe et al. 1994) but not for Sar1 (see Fig. 8E in Barlowe et al. 1994). In Fig. 8E, the diameter of vesicles partially coated with COPI-like coat is very uniform and varied from 38 to 55 nm with the average of 51 nm. These characteristics are typical for COPI vesicles (Marsh et al. 2001). Also, the structure of the vesicle coats in the Fig. 8E is similar to that observed on COPI-coated buds localized within ERES (Bannykh et al. 1996) and differs from the structure of COPII coat demonstrated with electron cryo-microscopy (see Bacia et al. 2011). Vesicles shown in Figs. 8B–D and 8F–H by Barlowe et al. (1994) exhibited much higher heterogeneity than 51-nm vesicles in Fig. 8E. These parameters are typical for membranes coated with COPII coat (Lee et al. 2005). Importantly, the same research group using electron cryo-microscopy and pure liposomes demonstrated that the size of COPII-dependent vesicles should be about 75–80 nm (Fig. 3A in Lee et al. 2005; Bacia et al. 2011).

We believe that the authors presented two types of vesicles, COPI-dependent (Fig. 7 and 8E) and COPII-dependent (in Fig. 8B–H). When the primarily isolated COPII-coated tubules were passed through small pores, they could undergo fragmentation into 66-nm COPII-coated vesicles. These vesicles do not contain Sar1 because COPII-coated buds and tubules are devoid of Sar1p (Mironov et al. 2003). Careful analysis of the experimental procedures used by Barlowe et al. (1994) revealed that microsomes contain residual amount of COPI attached to the microsomal membranes. However, in their experiments on isolation of vesicles, the authors do not use the microsomes washed with 2.5 M urea in order to eliminate all peripheral membrane proteins. Thus, this residual amount of COPI present on their microsomes could be sufficient to generate a few COPI-coated vesicles. If these

51-nm vesicles were formed with COPI, why are these vesicles coated? The reason is the following: ArfGAP protein, which is necessary for COPI uncoating, is present on the GC but not on the ER (microsomes). Then, it is necessary to explain why TEM sectioning gave images of COPI vesicles, whereas cryosectioning resulted in COPII vesicles. The reason could be the following: After centrifugation of the membranes passed through small pores, the authors obtained a pellet where heavy COPI-coated vesicles were near the bottom, whereas less heavy COPII-coated structures were at the top. This distribution was described by Kweon et al. (2004). The pellets were embedded into Epon or prepared for cryosectioning. The sectioning of Epon-embedded pellet starts from the very bottom, and therefore on sections, mostly 48–50-nm partially COPI-coated round profiles were visible. In contrast, for cryosectioning it is necessary to prepare a pyramid. Therefore, several layers of the pellet, which were situated near the bottom and containing mostly COPI, were eliminated in order to get the precise pyramid. As a result on the cryosections, one would have only vesicles derived from COPII-coated tubules.

Finally, in another cornerstone paper, Bednarek et al. (1995) described so-called COPII-coated buds on the ER of *Saccharomyces cerevisiae* after cell permeabilization and incubation with COPII subunits. However, again in Material and Methods, they quoted Rexach and Schekman (1992) for the description of isolation of nuclei. However, Rexach and Schekman (1992) did not use 2.5 M urea in order to wash the membrane from COPI. Thus, a significant amount of COPI was present on the external nuclear membrane of the ER, which could be used for the generation of COPI-coated buds on the ER.

But the main contradiction of VM of EGT is the observation of large cargo aggregates, which are formed in the lumen of the ER. Pre-chylomicrons [in enterocytes (Sabesin and Frase 1977; Siddiqi et al. 2003)], low-density lipoproteins and very low-density lipoproteins [in hepatocytes (Claude 1970)] and procollagen aggregates (Mironov et al. 1997; Bonfanti et al. 1998) form in the lumen of the smooth ER and then exit the ER. PCI is known to trimerize in the ER, and only the correctly folded PCI trimers are chaperoned by HSP47 to exit the ER en route to the GC (Mironov et al. 2003; Gorur et al. 2017). After formation of a procollagen aggregate, it does not undergo disassembly during EGT and IGT (Patterson et al. 2008). If EGT occurred by 70–80 nm COPII-dependent vesicles (Mironov 2014), the size of chylomicrons in enterocytes (Sabesin and Frase 1977), very low-density lipoprotein (VLDL) in hepatocytes (Claude 1970) and procollagen I aggregates in fibroblasts (Leblond 1989; Bonfanti et al. 1998; Mironov et al. 2003) would be incompatible with the diameter of 70–80 nm COPII vesicles. Indeed, these aggregates are too big to be packed into 70–85 nm COPII vesicles. For instance, non-bent procollagen I trimers and aggregates composed of them (with a length of up to 300 nm), which form inside the lumen of the ER near ERES, are incompatible in size with COPII vesicles (Mironov et al. 1997, 2003; Bonfanti et al. 1998).

In order to solve these contradictions, it has been proposed that large cargos are transported by 'mega-vesicles', or 'mega-carriers', which are formed by unusual combinations of isoforms of COPII subunits (Fromme and Schekman 2005; Venditti et al. 2012; Malhotra et al. 2015; Santos et al. 2016; Gorur et al. 2017; Raote et al.

2017, 2018). According to mega-vesicles model, large cargo aggregates form in the mega-buds coated with COPII. It has been shown that the membrane protein TANGO1 binds PCVII and that TANGO1 binds the COPII-coated proteins, Sec23/Sec24 (Saito et al. 2009). Knockdown of TANGO1 inhibits export of the bulky PCVII (but not of PCI) from the ER (Saito et al. 2009; Nogueira et al. 2014). However, in order to prove that mega-buds and mega-vesicles exist, it is necessary to show (better with correlative light electron microscopy or immune electron microscopy) buds coated with COPII or separated mega-vesicles coated with COPII. If mega-vesicles exist, Sec13 should form a cap over procollagen aggregate, very low-density lipoproteins (VLDL) or chylomicron. Of interest, until now this requirement was not fulfilled. For instance, Santos et al. (2016) did not demonstrate co-localization between Sec13 and lipids or Apo B (lipid and Apo proteins of VLDL). Importantly, Santos et al. (2016) did not discuss the contradiction of their data with the results by Siddiqi et al. (2003).

In the article by Raote et al. (2018), there is not a single picture that directly confirms the scheme of the formation of a mega-buds proposed by the authors, namely, the COPII ring, then more external ring of TANGO1 and a procollagen-positive spots inside these rings. Claude (1970) and Sabesin and Frase (1977) have not observed lipid particles and pre-chylomicrons in the ER buds. It was shown that the aggregates of PC are formed inside the lumen of the ER cisternae. However, there is no coat visible on the distensions of the ER. Thus, at steady state, mega-buds containing procollagen and coated with COPII-like coat were also not detected (Leblond 1989).

In order to prove the existence of transport mega-vesicles filled with procollagen, Gorur et al. (2017) engineered cells to stably overexpress the human pro-α1 (I) collagen. Using the correlative light electron microscopy based on serial sections with a thickness of 70–100 nm, the authors demonstrated the structure filled with procollagen I with the diameter of 900 nm (Fig. 2C/Z7 by Gorur et al. 2017). The thickness of coat over this structure is more than 40 nm, whereas the typical thickness of COPII coat is 12 nm (Bannykh et al. 1996; Bacia et al. 2011). Also, the significant thickness of serial sections indicates that the resolution along Z-axis was 140–200 nm. Therefore, it is not possible to judge whether this structure is connected with the ER or disconnected. Moreover, in a vast majority of papers, the diameter of the procollagen containing EGCs or Golgi cisterna distensions filled with procollagen never exceeds 350 nm (Leblond 1989, Bonfanti et al. 1998; Patterson et al. 2008; Perinetti et al. 2009). Importantly, in their Fig. 2ii, the labelling for Sec31a does not form ring, as it should be in agreement with the hypothesis of COPII-coated mega-vesicles. Of interest, the area of the labelling for procollagen is wider than the labelling for Sec31A, namely, near the border the green intensity is higher than the red one, whereas in the centre the intensities of red and green colours are equal. The authors used super-resolution light microscopy and have to detect the ring- or cap-like labelling for Sec31A, which should surround the collagen aggregate. Moreover, in their Fig. 3A v–x, the thickness of COPII coat is more than 100 nm, whereas under normal conditions, the thickness of COPII coat is only 12 nm (Bannykh et al. 1996). Moreover, the diameter of procollagen aggregate is only 100 nm, although under normal conditions, their diameter is 300 nm (Mironov et al. 2003). Also, there is an empty space between the PCI spot and Sec31A-positive cap (see Fig. 3A viii by Gorur et al. 2017). Such space has never been observed under normal conditions. On the other hand, in Figs. S5B: i, ii, iii (in Gorur et al. 2017), diameter of the vesicles is about 200 nm. Only in Fig. S5B:iv the vesicle has a diameter of 350 nm, but this vesicle is not coated.

Recently, McCaughey et al. (2019) provided direct evidence suggesting in favour of the very minor (if any) role of mega-vesicles for EGT of procollagen. They demonstrated that EGT of procollagen occurs without formation COPII-coated 200–300 nm carriers. These observations contradict to the mega-vesicle hypothesis proposed by the proponents of VM of EGT. However, the authors did not discuss this issue and simply wrote the following: 'In many cases, small GFP–COL puncta were seen in our experiments that co-localize with Sec31A. These show a distinct size distribution from large, static, circular structures negative for the COPII marker. Therefore, our data are entirely consistent with COP II-dependent trafficking of procollagen from the ER via conventionally described ERES, and we do not dispute an absolute requirement for COP II in this process (page 12).' Also, the authors did not quote two important papers by Patterson et al. (2008) and Mironov et al. (2003) where mechanisms of EGT of procollagen I were described and the role of COPII was questioned. Moreover, Patterson et al. (2008) presented data on the procollagen I transport in live cells, whereas we already demonstrated the rarity of the arrival of EGCs with the diameter of 300 nm to the GC. Also, in contrast to McCaughey et al. (2019), we observed rare EGCs filled with procollagen III. To this end, we bleached the whole Golgi area and monitored the rate of entry of PC-III-GFP into the GC from the ER. PC-III-GFP behaved as expected from our previous experiments on PC-I trafficking (Bonfanti et al. 1998; Mironov et al. 2001; Trucco et al. 2004). At 3 min post-bleaching, some PC-III-GFP aggregates (in the form of distinct bright puncta) had already entered the Golgi area (Beznoussenko et al. 2014). In experiments by McCaughey et al. (2019), the PC-containing dots grew inside the Golgi area. In our experiments, these dots acquired their high brightness at the periphery and then moved to the Golgi area.

We suggest that mega-vesicles observed by Gorur et al. (2017) and large procollagen-positive immobile dots observed by McCaughey et al. (2019) could be generated using a similar mechanism. It seems that unusually large mega-vesicles demonstrated by Gorur et al. (2017) and McCaughey et al. (2019, Fig. 2) represent the ER-derived autophagosomes (reticulophagosomes; Fregno et al. 2018; Fregno and Molinari 2018; Forrester et al. 2019). This phenomenon was observed upon overexpression of the secretory heavy chain of immunoglobulin M lacking some domains; the aggregates of this chain are concentrated in the ER protrusions with the diameter of 400 nm or more. These protrusions are not coated with COPII-like coat. After detachment from the ER, these distensions are delivered to the GC and then secreted or fused with lysosomes.

Finally, Oprins et al. (2001) observed significant concentration of the regulation secretory protein during their exit from the ER. Indeed, the concentration of these cargos in the ER is 57.6-fold lower than inside the ERES compartments. The authors tried to apply CMPM in order to explain such high concentration. However, COPI vesicles presumably operating as carriers for retrograde transport at the level of ERES cannot be used because COPI vesicles have very high ratio between the surface and the volume, and this means that these vesicles are not suitable to explain such high level of cargo concentration. This cannot be explained using maturation progression model. Recycling of COPI vesicles would eliminate surface area but not volume of immature EGCs. Similarly, the assumption of constant connections between the ER and the GC cannot explain the rarity of their findings.

Thus, VM, DM and CMPM cannot explain data, which contradict to their logic, whereas KARM can explain cornerstone observations supporting VM, DM and CMPM.

4.5 Intra-Golgi Transport

VM, DM, CMPM and KARM are also the main models describing mechanisms of IGT. The pros and cons of these models have been extensively discussed (Mironov et al. 1997; 1998, 2005, 2013, 2016; Beznoussenko and Mironov 2002; Mironov and Beznoussenko 2008, 2012; Glick and Nakano 2009; Glick and Luini 2011). Attempts to find the compromise between different models of IGT failed (Emr et al. 2009). Therefore, we re-examine cornerstone experiments, on which each model of IGT is based, and try to reinterpret these results within the framework of another model.

VM was proposed by Palade (1975) and substantiated by Rothman et al. (1980), who found that after accumulation of the temperature-sensitive G protein of the vesicular stomatitis virus (VSVG) in the ER at the restrictive (40 °C) temperature and the consecutive release of this temperature block by placing the cells at 32 °C, VSVG moved along the secretory pathway and in 10 min appeared in the clathrin-coated vesicles isolated from the GC. However, a significant part of VSVG found in these vesicles was not fully processed by Golgi glycosylation enzymes. This suggests that VSVG in these vesicles bypassed some compartments of the GC. The authors proposed that these vesicles are transport carriers.

The explanation of this bypass could be the following: Rothman et al. (1980) accumulated a large amount of VSVG in the ER. When a lot of VSVG molecules start to move through the GC, these molecules could quickly reach the *trans*-side of the GC (Patterson et al. 2008), where clathrin-coated vesicles are formed (Bonfanti et al. 1998). This is due to overloading of this pathway. Then, Rothman et al. (1984) demonstrated that after fusion of two cells, where the GCs and the cargos were differently labelled, and formation of heterokaryon, the cargo derived from the GC of one cell can be quickly delivered to the GC derived from another cell and vice versa. The authors concluded that these observations support VM of IGT. However, in 1986, COPI-dependent vesicles were found near the GC (Orci et al. 1986). After this, instead of clathrin-coated vesicles, COPI-dependent vesicles began to be considered as transport carriers (see details in Mironov et al. 1997, 2005, 2013,

2016). Of interest, after the discovery that all COPI-dependent vesicles are on strings and cannot freely diffuse across the cytosol, the results of the heterokaryon experiments were considered as artefacts (Orci et al. 1998).

However, the main problem for VM is the large cargo aggregates incompatible in size with COPI vesicles, which cannot be transported by COPI vesicles (see Mironov et al. 1997). Transport of VLDL particles through the GC of hepatocytes was demonstrated by Taylor et al. (1997). Transport of procollagen I through the GC was proved by Bonfanti et al. (1998). Transport of chylomicrons through the GC was proved by Sabesin and Frase (1977). Similarly, secretory casein submicelles, which are transported through the GC in lactating mammary glands, are larger than COPI vesicles (Clermont et al. 1993). In order to solve this contradiction, Rothman's group proposed that such large cargos are transported according to CMPM, whereas VSVG is transported by vesicles (Orci et al. 2000b). However, it is established now that a vast majority of cargos are absent in COPI vesicles. The list of cargo proteins that are excluded from COPI vesicles was presented by Mironov et al. (2005). Martinez-Menárguez et al. (1999) showed that the concentration of amylase in COPI-dependent vesicles is lower than in Golgi cisterna. Orci et al. (1986), the main proponents of VM, demonstrated that the concentration of the G protein of the vesicular stomatitis virus within 52–56 nm COPI-dependent vesicles was 1.5-fold lower than in Golgi cisterna (see Table 1 in Orci et al. 1986). In order to support VM, Orci et al. (1997) claimed that pro-insulin is transported by COPI vesicles. However, careful analysis of the paper revealed that in Table 3 by Orci et al. (1997), the concentration of insulin in Golgi-associated round profile is threefold lower than in Golgi cisterna. Trying to solve this contradiction, the authors proposed that there are two populations of COPI vesicles, namely, one population, for the anterograde transport, and another population, for the retrograde transport. Further, trying to provide the additional support for VM, Rothman's group (Pellett et al. 2013) transfected one population of cells with fluorescently tagged cargo tagged with one fluorophore; other cells were transfected with Golgi resident proteins tagged with another fluorophore. Next, heterokaryons were generated. These cargos and enzymes were found in small dots visible in the cytosol. A small portion of these particles contained coatomer. Pellett et al. (2013) measured diameters of the above-mentioned spots in cytoplasm and isolated COPI vesicles out of cytosol using super-resolution light microscopy. Comparing diameters of these particles measured at the level of light microscopy particles with diameter of isolated COPI vesicles, the authors concluded that these particles are dependent COPI vesicles. The authors state that the resolution of their super-resolution method is 80 nm. However, in reality (not in model experiments), STED resolution is about 100 nm (Sesorova et al. 2018). Diameter of COPI is 52 nm (Marsh et al. 2001). It means that the resolution of their method is lower than the size of structures measured. Resolution of super-resolution microscopy depends on the refractory index of the medium, and in vitro this parameter differs from that of cytosol. Also, the method used by Pellett et al. (2013) is very sensitive to the refractive indices of the media (Sesorova et al. 2018). Therefore, under these conditions, there could be significant systematic mistakes. Also, the authors did not take into consideration that COPI-derived vesicles are on strings. This fact was discovered by Orci et al. (1998) and then confirmed by Marsh et al. (2001). Indeed, in mammalian cells, nobody demonstrated 52-nm vesicles coated or non-coated at the distance more than 200 nm (Martinez-Menárguez et al. 1999). Finally, it is known that diffusion of particles with a diameter of more than 50 nm is strongly restricted (Luby-Phelps 1994).

Another problem of the interpretations presented in this chapter is the volume-to-surface ratio of transport carriers observed. According to the authors, during 30 min 25% of membrane protein was transported from the GC of one cell to the GC of another cell. During this time, the authors observed 20,000 such particles. The rate of transport of soluble cargo is identical to that of the membrane cargo (see Fig. 3 in Pellett et al. 2013). The diameter of COPI-dependent vesicle is 52 nm (Marsh et al. 2001). Its internal volume is equal to 0.000045 μm^3, and the surface area is equal to 0.074 μm^2. The Golgi volume is equal to 1500 μm^3 (Mironov and Mironov 1998). If we take into consideration that the ratio between the volume and the surface area of the GC is equal to 140 (Ladinsky et al. 1999), the surface area of the GC would be equal to 210,000 μm^2. If these dots were COPI vesicles, 20,000 such vesicles would transport 1480 μm^2 of surface area (0.7% of the total) and 0.9 μm^3 of Golgi volume (0.06% of the total). These considerations suggest that the authors observed movement of carriers, which are much larger than COPI vesicles.

There are also other problems of VM. There is a significant decrease of the number of COPI vesicles during synchronous intra-Golgi transport (IGT; Rambourg et al. 1993; Fusella et al. 2013). In some organisms, COPI-dependent vesicles are absent [microsporidia *Paranosema grylli* and *Paranosema locustae* (Beznoussenko et al. 2007)] or very few [alga *Ostreococcus tauri*; Henderson et al. (2007)]; parasites *Plasmodium falciparum*; Hohmann-Marriott et al. (2009) and *Trypanosoma cruzi* [see movies and Fig. 2i by Girard-Dias et al. (2012)]. VM cannot explain maturation of Golgi compartments in yeast (Matsuura-Tokita et al. 2006).

There are several observations favouring the DM. In order to be relevant, DM should be based on the structures which are interconnected. Indeed, connections between Golgi cisternae are more abundant in transporting Golgi stacks and after stimulation of cell signalling (Marsh et al. 2001; Trucco et al. 2004; Mironov and Beznoussenko 2012; Mironov et al. 2016). These connections are permeable for albumin (Beznoussenko et al. 2014). Pagano et al. (1989) demonstrated that externally added lipids move from the PM to the GC and then to the ER even in aldehyde-fixed cells and that OsO_4, which freezes the lateral diffusion of lipids, blocks this movement. This lipid movement in aldehyde-fixed cells suggests that there are physical continuities between the GC and the plasma membrane. Furthermore, dicumarol destabilizes Golgi tubules and delays IGT (Mironov et al. 2004), whereas after activation of protein kinase A, when the cisternae of the GC become interconnected, IGT is accelerated (Mavillard et al. 2010). This suggests the important role of the connections. Some lipids (e.g. phosphatidylethanolamine, diacylglycerol) can be easily transported along the secretory pathway at low temperatures when active transport (including the formation of vesicles) and endocytosis are inhibited (Sleight and Pagano 1983; Pagano and Longmuir 1985). In living cells, spots filled with fluorescent cargos can move through the pre-bleached Golgi ribbon gradually losing their intensity (Presley et al. 1997). Finally, Patterson et al. (2008) reported that a cargo, which exits the Golgi area, exhibits exponential kinetics. Such type of kinetics indicates that all compartments within the GC are interconnected. The authors also proposed that large cargos, which diffuses slowly, can exit even from the *cis*-side of the GC. The last explanation is invalid, because PCI always exits from the *trans*-side of the GC (Bonfanti et al. 1998). Another problem of this paper is the following: Patterson et al. (2008) did not examine the GC, which is empty before the restoration of IGT. They examined only the GC which were already filled with cargos. However, the process of Golgi filling with cargos could take a significant time, and under such conditions, the exit kinetic could be different.

The main problem of DM is the existence of protein, lipid and ionic gradients across Golgi stacks and the presence of SNARE complexes within all stages of the secretory pathway. For instance, during IGT, there is a concentration of both soluble cargo (albumin; Beznoussenko et al. 2014) and large cargo aggregates unable to diffuse along ICs (Claude 1970; Sabesin and Frase 1977; Bonfanti et al. 1998; see below). This indicates that the connections are transient. This feature of the connections contradicts to the essence of DM (Mironov and Beznoussenko 2008, 2012; Mironov et al. 2013).

According to CMPM, during IGT, concentration of any cargo is impossible because cargo should not leave the cisterna, which is formed at the *cis*-side of the GC, and then progress through the stack. The second restriction of VM is the following: Golgi resident proteins should not be depleted in COPI-dependent 52-nm vesicles because if the concentration of these proteins in the vesicles were lower than in Golgi cisterna, the recycling of these proteins would be very slow. Indeed, mathematical modelling based on CMPM demonstrated that a significant concentration of the Golgi enzymes in the vesicles is a prerequisite for the good performance of CMPM (Glick et al. 1997). However, depletion of the Golgi resident proteins in COPI vesicles is observed in a vast majority of papers suggesting against CMPM. Indeed, it was shown that concentration of Golgi glycosylation enzymes (Kweon et al. 2004), nucleotide sugar transporters (Fusella et al. 2013), some SNAREs (Orci et al. 2000a, b), several proteins localized within the *cis*- and *trans*-side of the GC (Gilchrist et al. 2006) and Rab escort protein 1 (in *Saccharomyces cerevisiae,* Beznoussenko et al. 2016) is lower than in Golgi cisternae.

In spite of this, Gilchrist et al. (2006) demonstrated that the light membrane fraction of Golgi membranes obtained after incubation of isolated Golgi membranes with cytosol and GTP is depleted of secretory cargo but enriched in Golgi enzymes. Electron microscopic analysis revealed that this fraction is composed of 52-nm vesicles. The authors concluded that COPI-dependent vesicles are retrograde transport carriers for Golgi enzymes. However, careful analysis of their results revealed that when the light fraction was prepared for electron microscopy, it was additionally pelleted onto sucrose cushion (50% (w/w)) at 45,000 rpm. This procedure was not used for biochemical measurement of Golgi enzyme concentration in this fraction. This sucrose-based centrifugation was not used in their previous paper (Lanoix et al. 1999), and the purity of 52-nm vesicles was significantly lower. Therefore, one could propose that in their light fraction, perforated fragments of Golgi cisternae enriched in Golgi enzymes (Kweon et al. 2004) were present, whereas after the additional centrifugation, only 52-nm vesicle remained in the samples prepared for electron microscopy. This could explain why concentration of the enzymes in the light fraction is higher than in the isolated Golgi membranes. Moreover, actually, the paper by Gilchrist et al. (2006) contains information suggesting against the role of COPI vesicles as retrograde carriers because several other resident Golgi proteins have lower concentration in the light fraction than in the isolated GC (see above).

On the other hand, Martinez-Menárguez et al. (1999) demonstrated that in situ, mannosidase II is 1.6-fold more concentrated in COPI-coated peri-Golgiolar round profiles located within 200-nm distance from GC. However, on cryosections it is not possible to distinguish a section of COPI vesicle from a section of a COPI-coated tube. Indeed, it was demonstrated that tangential tubules are coated with COPI (Weidman et al. 1993; Yang et al. 2011). Importantly, a vast majority of 52-nm vesicles within the Golgi area are uncoated (Marsh et al. 2001), whereas sections of tangential tubules could give 52-nm round profiles coated with COPI. We showed that these areas are enriched in Golgi glycosylation enzymes (Kweon et al. 2004). Importantly, other studies performed under the same conditions as those used by Martinez-Menárguez et al. (1999) have reported a depletion of mannosidase II in peri-Golgi round profiles (Cosson et al. 2002; Kweon et al. 2004). Also, in a vast majority of the papers based on different types of immune EM, the density of labelling for the Golgi enzymes upon cisternae is much higher than over nearby round profiles (Orci et al. 2000a, b; Cosson et al. 2002; Kweon et al. 2004; Dunlop et al. 2017; see also Fig. 7 by Velasco et al. 1993).

Recently, Rizzo et al. (2013) prepared the chimeric mannosidase I (ManI), the protein, which is localized at the *cis*-side of the GC. This chimera is able to polymerize after addition of the chemical accelerator. After the polymerization of the chimeric ManI, these polymers moved to the *trans*-side of the GC. The monomeric form of this chimera is depleted in near-Golgi round profiles, whereas after depolymerization of these aggregates, this protein quickly appeared in round profiles. According to the authors, 50% of these round profiles were coated with COPI. These data were interpreted in favour of CMPM. However, Rizzo et al. (2013) did not prove that these ManI aggregates behaved as a cargo because these cargos cannot exit from the GC. At least this is not shown in the paper. Moreover, we think that it is possible to provide the alternative interpretation of these data. Indeed, it is established that ManI is localized within the highly perforated CMC (Marra et al. 2007). There, pores are surrounded by thin tubules with branches. This means that mannosidase I preferred to localize within the highly curved membranes. When the ManI-containing chimera was stimulated to polymerize, this ability to reside within perforated CMC was lost and ManI oligomers were shifted to more solid cisternae at the *trans*-side of the GC. When the ManI polymers were depolymerized, the monomers of ManI chimera cannot reside in the solid Golgi cisterna and need the cisternae composed of tubules such as CMC. This necessity could be caused by the shape of ManI molecule and its molecular interactions. Otherwise normal ManI would not reside within CMC. Sharp appearance of many monomers of ManI could transform the solid Golgi cisterna, which resides near the *trans*-side of the GC, into the tubular network, which would protrude out of Golgi stack. On cryosections, this network could appear as round profiles which were considered by the authors as COPI-dependent vesicles. In order to prove that round profiles represent separated vesicles, the authors show very small in size serial electron microscopic tomography images of only one round profile. However, this round profile exhibits a visible neck connecting it with Golgi cisterna. This neck is visible on frames 45–55 and in Figs. S5j and k by Rizzo et al. (2013). If we took into consideration that the thickness of their tomography slice is 3 nm and the resolution of the presented the images is 10 nm, the obvious conclusion is that this neck represents a membranous structure and that this round profile actually represents a COPI-coated bud. The statement that 50% of round profiles is coated with COPI coat also suggests in favour of our explanation because a vast majority of free vesicles near the GC are uncoated (Marsh et al. 2001). Rizzo et al. (2013) did not use serial cryosections in order to distinguish between the round profile as the projection of cross section of the tubule and the projection of a real vesicle. On random cryosections, this distinction is not possible (Kweon et al. 2004).

Recently, using experimental approach when individual Golgi cisternae are separated and 'land-locked' between mitochondria, Dunlop et al. (2017) provided the additional evidence against CMPM. However, the authors did not exclude the possibility that KARM could also explain their data. Indeed, Golgi cisternae visible within the mitochondria aggregates were rather close to each other and could be temporally connected by tubules (see Fig. 3B by Dunlop et al. 2017). On the other hand, the authors did not observe vesicles on strings, which they described earlier (Orci et al. 1998).

CMPM has several other problems. The full list of CMPM problems was presented in our review (Mironov et al. 2013). For instance, sialyltransferases and fucosyltransferases are present within TMC. However, there are no COPI-coated buds on the trans-most cistern (Ladinsky et al. 1999; Marsh et al. 2001; Mironov et al. 2016). Therefore, it is not clear how the resident proteins undergo recycling from TMC. Even when COPI vesicles were generated at maximal speed, the rate of their generation can support only 10% of vesicles necessary for IGT (Fusella et al. 2013). In some organisms, COPI-dependent vesicles are absent (see above). If we assumed that IGT occurs according to CMPM in *Saccharomyces cerevisiae*, the mechanism of the vectoral delivery of retrograde COPI vesicles would be unclear. Indeed, in mammalian cells, COPI vesicles are on 'strings' and this could explain the vectorality of vesicle movement (Orci et al. 1998). In contrast, in *Saccharomyces cerevisiae*, the different Golgi compartments are localized separately from each other, and hence they are divided by significant space.

One of the main problems of CMPM is the concentration of soluble cargos, regulated secretory cargos and cargo aggregates during IGT (Oprins et al. 2001; Mironov and Arvan 2008; Beznoussenko et al. 2014) and the observation that albumin reached the *trans*-side of the GC faster than VSVG and procollagen I (Beznoussenko et al. 2014). Also, aggregates of cargo proteins inside cisternal distensions of Golgi cisterna move faster than cisterna domains where opposite membranes are connected by protein bridges (Lavieu et al. 2013). Concentration of cisternal distensions filled with PCI was proved by Bonfanti et al. (1998, Fig. 4). Concentration of chylomicrons in cisterna distensions at the *trans*-side of the GC in enterocytes was shown by Sabesin and Frase (1977). Concentration of VLDL in cisternal distensions of the GC in hepatocytes was shown by Claude (1970). Concentration of lipid particles in cisterna distensions at the *trans*-side of the Golgi complex is also demonstrated by Glaumann et al. (1975, Fig. 7b) and Matsuura and Tashiro (1979, Figs. 1, 9, 14). Also, in Figs. 1 and 6 by Dahan et al. (1994), it is visible that the number of Apo E, a marker of lipid particles, increases at the *trans*-side of the Golgi complex in comparison to *cis*-side. Concentration of large cargo aggregates unable to diffuse along ICs not only suggests against CMPM but also suggests against DM.

Although here, we prove that now KARM in the most powerful model of IGT, it has some difficulties. For instance, one of these difficulties is the existence of separated different Golgi compartments in *Saccharomyces cerevisiae*. The observation that different Golgi compartments are rarely connected by tubules could provide the explanation and requirements of KARM (Beznoussenko et al. 2016). On the other hand, in order to be efficient, KARM should be based on the prerequisites that membrane cargo(s), cargo aggregates and SNAREs should be concentrated in cargo domains and these domains have to be separated from the rest of Golgi cisternae by row of pores as the site for fission within the framework of the asymmetric KARM. Indeed, pores separating cisternal distensions from the rest of Golgi cisternae were shown by Claude (1970), in hepatocytes; by Sabesin and Frase (1977) and by Pavelka and Roth (2005, Fig. 101A, p. 205), in enterocytes where the GC transported chylomicrons; by Sesso et al. (1994, Figs. 5, 7a–d, 9, 10), in acinar pancreatic cells; by Ladinsky et al. (2002), after the 20 °C temperature block; by Mironov et al. (2001, Figs. 4C, E, F), in fibroblasts transporting PCI aggregates; and by Pavelka and Roth (2005, Fig. 98B, p. 199), in hepatocytes during IGT of VLDL. Within the framework of KARM, COPI vesicles are important for (1) elimination of excessive membrane curvature (Beznoussenko et al. 2015), (2) extraction of Qb SNAREs and slowing down of IGT (Trucco et al. 2004; Fusella et al. 2013; Beznoussenko et al. 2016) and (3) retention of Golgi enzymes.

Now, post-Golgi transport (PGT) was not considered as being executed by vesicles. Clathrin-dependent vesicles are not used for the delivery of cargo from the GC towards the PM and for recycling of resident proteins (Polishchuk et al. 2000, 2003). On the other hand, the above-mentioned results by Pagano et al. (1989) could be explained on the basis of the proposal that connections between different post-Golgi compartments are constantly formed and then undergo fission. However, at each given moment, there are several connections, which are used for the slow diffusion of the lipid dye in experiments by Pagano et al. (1989). This explanation is within the framework of KARM.

4.6 Conclusions and Perspectives

Centrioles/centrosomes are important for Golgi organization. The formation of Golgi ribbon depends on many molecular machines. However, function of a vast majority of them is examined in vitro. Now, it is necessary to check their role in situ.

Delivery of membrane from the ER towards the GC and through the GC is one of the most important prerequisites for the formation of Golgi ribbon. However, experimental data on ICT are highly controversial. Therefore, now it is important to find the way on how to adapt all these observations to the existing models of ICT. Alternatively, one could explain these experimental results from the point of view of only one model. Here, we demonstrated that the experiments usually considered as the cornerstone of VM, DM and CMPM could be easily explained from the point of view of KARM. For instance, VM, CMPM and DM cannot explain mechanisms of Golgi ribbon formation and the disappearance of the GC in *Saccharomyces cerevisiae* and fragmentation of Golgi ribbon in mammals and other animal cells after blockage of cargo delivery and concentration of cargo during IGT. In contrast, KARM can explain these observations. Our current analysis of the cornerstone experiments within the field of ICT favours the KARM.

References

Antonny B, Gounon P, Schekman R, Orci L (2003) Self-assembly of minimal COPII cages. EMBO Rep 4:419–424

Bacia K, Futai E, Prinz S, Meister A, Daum S, Glatte D, Briggs JA, Schekman R (2011) Multibudded tubules formed by COPII on artificial liposomes. Sci Rep 1:17

Bannykh SI, Rowe T, Balch WE (1996) The organization of endoplasmic reticulum export complexes. J Cell Biol 135:19–35

Bard F, Casano L, Mallabiabarrena A, Wallace E, Saito K, Kitayama H, Guizzunti G, Hu Y, Wendler F, Dasgupta R, Perrimon N, Malhotra V (2006) Functional genomics reveals genes involved in protein secretion and Golgi organization. Nature 439:604–607

Barlowe C, Orci L, Yeung T, Hosobuchi M, Hamamoto S, Salama N, Rexach MF, Ravazzola M, Amherdt M, Schekman R (1994) COPII: a membrane coat formed by Sec proteins that drive vesicle budding from the endoplasmic reticulum. Cell 77:895–907

Bednarek SY, Ravazzola M, Hosobuchi M, Amherdt M, Perrelet A, Schekman R, Orci L (1995) COPI- and COPII-coated vesicles bud directly from the endoplasmic reticulum in yeast. Cell 83:1183–1196

Beznoussenko GV, Mironov AA (2002) Models of intracellular transport and evolution of the Golgi complex. Anat Rec 268:226–238

Beznoussenko GV, Dolgikh VV, Seliverstova EV, Semenov PB, Tokarev YS, Trucco A, Micaroni M, Di Giandomenico D, Auinger P, Senderskiy IV, Skarlato SO, Snigirevskaya ES, Komissarchik YY, Pavelka M, De Matteis MA, Luini A, Sokolova YY, Mironov AA (2007) Analogs of the Golgi complex in microsporidia, structure and avesicular mechanisms of function. J Cell Sci 120:1288–1298

Beznoussenko GV, Parashuraman S, Rizzo R, Polishchuk R, Martella O, Di Giandomenico D, Fusella A, Spaar A, Sallese M, Capestrano MG, Pavelka M, Vos MR, Rikers YG, Helms V, Mironov AA, Luini A (2014) Transport of soluble proteins through the Golgi occurs by diffusion via continuities across cisternae. Elife 3. https://doi.org/10.7554/eLife.02009

Beznoussenko GV, Pilyugin SS, Geerts WJ, Kozlov MM, Burger KN, Luini A, Derganc J, Mironov AA (2015) Trans-membrane area asymmetry controls the shape of cellular organelles. Int J Mol Sci 16:5299–5333

Beznoussenko GV, Ragnini-Wilson A, Wilson C, Mironov AA (2016) Three-dimensional and immune electron microscopic analysis of the secretory pathway in *Saccharomyces cerevisiae*. Histochem Cell Biol 146(5):515–527. https://doi.org/10.1007/s00418-016-1483-y

Boisvieux-Ulrich E, Laine MC, Sandoz D (1990) Cytochalasin D inhibits basal body migration and ciliary elongation in quail oviduct epithelium. Cell Tissue Res 259:443–454

Bonfanti L, Mironov AA Jr, Martínez-Menárguez JA, Martella O, Fusella A, Baldassarre M, Buccione R, Geuze HJ, Mironov AA, Luini A (1998) Procollagen traverses the Golgi stack without leaving the lumen of cisternae: evidence for cisternal maturation. Cell 95:993–1003

Carvajal-Gonzalez JM, Mulero-Navarro S, Mlodzik M (2016) Centriole positioning in epithelial cells and its intimate relationship with planar cell polarity. BioEssays 38:1234–1245

Claude A (1970) Growth and differentiation of cytoplasmic membranes in the course of lipoprotein granule synthesis in the hepatic cell. I. Elaboration of elements of the Golgi complex. J Cell Biol 47:745–766

Clermont Y, Xia L, Rambourg A, Turner JD, Hermo L (1993) Transport of casein submicelles and formation of secretion granules in the Golgi apparatus of epithelial cells of the lactating mammary gland of rat. Anat Rec 235:363–373

Cole NB, Sciaky N, Marotta A, Song J, Lippincott-Schwartz J (1996) Golgi dispersal during microtubule disruption: regeneration of Golgi stacks at peripheral endoplasmic reticulum exit sites. Mol Biol Cell 7:631–650

Comments on: Silencing of mammalian Sar1 isoforms reveals COPII-independent protein sorting and transport. https://stephenslab.wordpress.com/2013/02/27/comments-on-silencing-of-mammalian-sar1-isoforms-reveals-copii-independent-protein-sorting-and-transport/

Conti B, Berti F, Mercati D, Giusti F, Dallai R (2010) The ultrastructure of malpighian tubules and the chemical composition of the cocoon of *Aeolothrips intermedius* Bagnall (Thysanoptera). J Morphol 271:244–254

Cosson P, Amherdt M, Rothman JE, Orci L (2002) A resident Golgi protein is excluded from peri-Golgi vesicles in NRK cells. Proc Natl Acad Sci USA 99:12831–12834

Cutrona MB, Beznoussenko GV, Fusella A, Martella O, Moral P, Mironov AA (2013) Silencing of the mammalian Sar1 isoforms reveals COPII-independent protein sorting and transport. Traffic 14(6):691–708

Dahan S, Ahluwalia JP, Wong L, Posner BI, Bergeron JJ (1994) Concentration of intracellular hepatic apolipoprotein E in Golgi apparatus saccular distensions and endosomes. J Cell Biol 127:1859–1869

Dunlop MH, Ernst AM, Schroeder LK, Toomre DK, Lavieu G, Rothman JE (2017) Land-locked mammalian Golgi reveals cargo transport between stable cisternae. Nat Commun 8:432. https://doi.org/10.1038/s41467-017-00570-z

Emr S, Glick BS, Linstedt AD, Lippincott-Schwartz J, Luini A, Malhotra V, Marsh BJ, Nakano A, Pfeffer SR, Rabouille C, Rothman JE, Warren G, Wieland FT (2009) Journeys through the Golgi—taking stock in a new era. J Cell Biol 187:449–453

Feiguin F, Ferreira A, Kosik KS, Caceres A (1994) Kinesin-mediated organelle translocation revealed by specific cellular manipulations. J Cell Biol 127:1021–1039

Forrester A, De Leonibus C, Grumati P, Fasana E, Piemontese M, Staiano L, Fregno I, Raimondi A, Marazza A, Bruno G, Iavazzo M, Intartaglia D, Seczynska M, van Anken E, Conte I, De Matteis MA, Dikic I, Molinari M, Settembre C (2019) A selective ER-phagy exerts procollagen quality control via a Calnexin-FAM134B complex. EMBO J 38:e99847

Fregno I, Molinari M (2018) Endoplasmic reticulum turnover: ER-phagy and other flavors in selective and non-selective ER clearance. F1000Res 7:454

Fregno I, Fasana E, Bergmann TJ, Raimondi A, Loi M, Soldà T, Galli C, D'Antuono R, Morone D, Danieli A, Paganetti P, van Anken E, Molinari M (2018) ER-to-lysosome-associated degradation of proteasome-resistant ATZ polymers occurs via receptor-mediated vesicular transport. EMBO J 37:e99259

Fromme JC, Schekman R (2005) COPII-coated vesicles: flexible enough for large Cargo? Curr Opin Cell Biol 17:345–352

Fukunaga T, Furuno A, Hatsuzawa K, Tani K, Yamamoto A, Tagaya M (1998) NSF is required for the brefeldin A-promoted disassembly of the Golgi apparatus. FEBS Lett 435:237–240

Fusella A, Micaroni M, Di Giandomenico D, Mironov AA, Beznoussenko GV (2013) Segregation of the Qb-SNAREs GS27 and GS28 into Golgi vesicles regulates intra-Golgi transport. Traffic 14:568–584

Ghadialli FN (1982) Ultrastructural pathology of the cell and matrix. Butterworths, London, 971 p

Gilchrist A, Au CE, Hiding J, Bell AW, Fernandez-Rodriguez J, Lesimple S, Nagaya H, Roy L, Gosline SJ, Hallett M, Paiement J, Kearney RE, Nilsson T, Bergeron JJ (2006) Quantitative proteomics analysis of the secretory pathway. Cell 127:1265–1281

Girard-Dias W, Alcântara CL, Cunha-E-Silva N, de Souza W, Miranda K (2012) On the ultrastructural organization of *Trypanosoma cruzi* using cryopreparation methods and electron tomography. Histochem Cell Biol 138:821–831

Glaumann H, Bergstrand A, Ericsson JL (1975) Studies on the synthesis and intracellular transport of lipoprotein particles in rat liver. J Cell Biol 64:356–377

Glick BS, Luini A (2011) Models for Golgi traffic: a critical assessment. Cold Spring Harb Perspect Biol 3:a005215

Glick BS, Nakano A (2009) Membrane traffic within the Golgi apparatus. Annu Rev Cell Dev Biol 25:113–132

Glick BS, Iston ET, Oster G (1997) A cisternal maturation mechanism can explain the asymmetry of the Golgi stack. FEBS Lett 414:177–181

Gorur A, Yuan L, Kenny SJ, Baba S, Xu K, Schekman R (2017) COPII-coated membranes function as transport carriers of intracellular procollagen I. J Cell Biol. https://doi.org/10.1083/jcb. 201702135

Gosavi P, Houghton FJ, McMillan PJ, Hanssen E, Gleeson PA (2018) The Golgi ribbon in mammalian cells negatively regulates autophagy by modulating mTOR activity. J Cell Sci 131:jcs211987

Hanus C, Ehlers MD (2008) Secretory outposts for the local processing of membrane cargo in neuronal dendrites. Traffic 9:1437–1445

Henderson GP, Gan L, Jensen GJ (2007) 3-D ultrastructure of O. tauri: electron cryotomography of an entire eukaryotic cell. PLoS One 2:e749

Ho WC, Allan VJ, van Meer G, Berger EG, Kreis TE (1989) Reclustering of scattered Golgi elements occurs along microtubules. Eur J Cell Biol 48:250–263

Hohmann-Marriott MF, Sousa AA, Azari AA, Glushakova S, Zhang G, Zimmerberg J, Leapman RD (2009) Nanoscale 3D cellular imaging by axial scanning transmission electron tomography. Nat Methods 6:729–731

Hoppeler-Lebel A, Celati C, Bellett G, Mogensen MM, Klein-Hitpass L, Bornens M, Tassin AM (2007) Centrosomal CAP350 protein stabilises microtubules associated with the Golgi complex. J Cell Sci 120:3299–3308

Horton AC, Racz B, Monson EE, Lin AL, Weinberg RJ, Ehlers MD (2005) Polarized secretory trafficking directs cargo for asymmetric dendrite growth and morphogenesis. Neuron 48:757–771

Inoue T (1992) Complementary scanning electron microscopy, technical notes and applications. Arch Histol Cytol 55:45–51

Jamieson JD, Palade GE (1968) Intracellular transport of secretory proteins in the pancreatic exocrine cell. 3. Dissociation of intracellular transport from protein synthesis. J Cell Biol 39:580–588

Kaiser CA, Schekman R (1990) Distinct sets of SEC genes govern transport vesicle formation and fusion early in the secretory pathway. Cell 61:723–733

Kim SD, Pahuja KB, Ravazzola M, Yoon J, Boyadjiev SA, Hammamoto S, Schekman R, Orci L, Kim J (2012) The SEC23-SEC31 interface plays critical role for export of procollagen from the endoplasmic reticulum. J Biol Chem 287(13):10134–10144

Kloc M, Kubiak JZ, Li XC, Ghobrial RM (2014) The newly found functions of MTOC in immunological response. J Leukoc Biol 95:417–430

Koegler E, Bonnon C, Waldmeier L, Mitrovic S, Halbeisen R, Hauri HP (2010) p28, a novel ERGIC/cis Golgi protein, required for golgi ribbon formation. Traffic 11:70–89

Kondylis V, Rabouille C (2003) A novel role for dp115 in the organization of tER sites in Drosophila. J Cell Biol 162:185–198

Kreft ME, Di Giandomenico D, Beznoussenko GV, Resnik N, Mironov AA, Jezernik K (2010) Golgi apparatus fragmentation as a mechanism responsible for uniform delivery of uroplakins to the apical plasma membrane of uroepithelial cells. Biol Cell 102:593–607

Kurokawa K, Ishii M, Suda Y, Ichihara A, Nakano A (2013) Live cell visualization of Golgi membrane dynamics by super-resolution confocal live imaging microscopy. Methods Cell Biol 118:235–242

Kweon HS, Beznoussenko GV, Micaroni M, Polishchuk RS, Trucco A, Martella O, Di Giandomenico D, Marra P, Fusella A, Di Pentima A, Berger EG, Geerts WJ, Koster AJ, Burger KN, Luini A, Mironov AA (2004) Golgi enzymes are enriched in perforated zones of golgi cisternae but are depleted in COPI vesicles. Mol Biol Cell 15:4710–4724

Ladinsky MS, Mastronarde DN, McIntosh JR, Howell KE, Staehelin LA (1999) Golgi structure in three dimensions, functional insights from the normal rat kidney cell. J Cell Biol 144:1135–1149

Ladinsky MS, Wu CC, McIntosh S, McIntosh JR, Howell KE (2002) Structure of the Golgi and distribution of reporter molecules at 20°C reveals the complexity of the exit compartments. Mol Biol Cell 13:2810–2825

Lanoix J, Ouwendijk J, Lin CC, Stark A, Love HD, Ostermann J, Nilsson T (1999) GTP hydrolysis by arf-1 mediates sorting and concentration of Golgi resident enzymes into functional COPI vesicles. EMBO J 18:4935–4948

Lavieu G, Zheng H, Rothman JE (2013) Stapled Golgi cisternae remain in place as cargo passes through the stack. Elife 2:e00558

Leblond CP (1989) Synthesis and secretion of collagen by cells of connective tissue, bone, and dentin. Anat Rec 224:123–138

Lee MC, Orci L, Hamamoto S, Futai E, Ravazzola M, Schekman R (2005) Sar1p N-terminal helix initiates membrane curvature and completes the fission of a COPII vesicle. Cell 122:605–617

Lippincott-Schwartz J, Cole NB, Marotta A, Conrad PA, Bloom GS (1995) Kinesin is the motor for microtubule-mediated Golgi-to-ER membrane traffic. J Cell Biol 128:293–306

Luby-Phelps K (1994) Physical properties of cytoplasm. Curr Opin Cell Biol 6:3–9

Makhoul C, Gosavi P, Gleeson PA (2018) The Golgi architecture and cell sensing. Biochem Soc Trans 46:1063–1072

Malhotra V, Erlmann P, Nogueira C (2015) Procollagen export from the endoplasmic reticulum. Biochem Soc Trans 43:104–107

Marchi F, Leblond CP (1983) Collagen biogenesis and assembly into fibrils as shown by ultra-structural and 3H-proline radioautographic studies on the fibroblasts of the rat food pad. Am J Anat 168:167–197

Marra P, Salvatore L, Mironov A Jr, Di Campli A, Di Tullio G, Trucco A, Beznoussenko G, Mironov A, De Matteis MA (2007) The biogenesis of the Golgi ribbon, the roles of membrane input from the ER and of GM130. Mol Biol Cell 18:1595–1608

Marsh BJ, Mastronarde DN, Buttle KF, Howell KE, McIntosh JR (2001) Organellar relationships in the Golgi region of pancreatic beta cell line, HIT-T15, visualized by high resolution electron tomography. Proc Natl Acad Sci USA 98:2399–2406

Martínez-Menárguez JA, Geuze HJ, Slot JW, Klumperman J (1999) Vesicular tubular clusters (VTCs) between ER and Golgi mediate concentration of soluble secretory proteins by exclusion from COPI-coated vesicles. Cell 98:81–90

Matsuura S, Tashiro Y (1979) Immunoelectron-microscopic studies of endoplasmic reticulum-Golgi relationships in the intracellular transport process of lipoprotein particles in rat hepato-cytes. J Cell Sci 39:273–290

Matsuura-Tokita K, Takeuchi M, Ichihara A, Mikuriya K, Nakano A (2006) Live imaging of yeast Golgi cisternal maturation. Nature 441:1007–1010

Mavillard F, Hidalgo J, Megias D, Levitsky KL, Velasco A (2010) PKA-mediated Golgi remodeling during cAMP signal transmission. Traffic 11(1):90–109

McCaughey J, Stevenson NL, Cross S, Stephens DJ (2019) ER-to-Golgi trafficking of procollagen in the absence of large carriers. J Cell Biol. https://doi.org/10.1083/jcb.201806035

Micaroni M, Perinetti G, Di Giandomenico D, Bianchi K, Spaar A, Mironov AA (2010) Synchronous intra-Golgi transport induces the release of Ca^{2+} from the Golgi apparatus. Exp Cell Res 316:2071–2086

Minin AA (1997) Dispersal of Golgi apparatus in nocodazole-treated fibroblasts is a kinesin-driven process. J Cell Sci 110:2495–2505

Mironov AA (2014) ER-Golgi transport could occur in the absence of COPII vesicles. Nat Rev Mol Cell Biol. https://doi.org/10.1038/nrm3588-c1

Mironov AA, Arvan P (2008) Origin of the regulated secretory pathway. In: Mironov AA, Pavelka M (eds) Golgi apparatus, Chapter 3.11. Springer, Wien, pp 482–515

Mironov AA, Beznoussenko GV (2008) Intra-Golgi transport. In: Mironov AA, Pavelka M (eds) The Golgi Apparatus. State of the art 110 years after Camillo Golgi's discovery, Chapter 3.2. Springer, Wien, pp 342–357

Mironov AA, Beznoussenko GV (2011) Molecular mechanisms responsible for formation of Golgi ribbon. Histol Histopathol 26:117–133

Mironov AA, Beznoussenko GV (2012) The kiss-and-run model of intra-Golgi transport. Int J Mol Sci 13:6800–6819

Mironov AA Jr, Mironov AA (1998) Estimation of subcellular organelle volume from ultrathin sections through centrioles with a discretized version of vertical rotator. J Microsc 192:29–36

Mironov AA, Pavelka M (2008) The Golgi apparatus as a crossroad in intracellular traffic. In: Mironov AA, Pavelka M (eds) The Golgi apparatus. State of the art 110 years after Camillo Golgi's discovery, Chapter 1.2. Springer, Wien, pp 16–39

Mironov AA, Weidman P, Luini A (1997) Variations on the intracellular transport theme, maturing cisternae and trafficking tubules. J Cell Biol 138:481–484

Mironov AA Jr, Luini A, Mironov AA (1998) A synthetic model of intra-Golgi traffic. FASEB J 12:249–252

Mironov AA, Beznoussenko GV, Nicoziani P, Martella O, Trucco A, Kweon HS, Di Giandomenico D, Polishchuk RS, Fusella A, Lupetti P, Berger EG, Geerts WJ, Koster AJ, Burger KN, Luini A (2001) Small cargo proteins and large aggregates can traverse the Golgi by a common mechanism without leaving the lumen of cisternae. J Cell Biol 155:1225–1238

Mironov AA, Mironov AA Jr, Beznoussenko GV, Trucco A, Lupetti P, Smith JD, Geerts WJ, Koster AJ, Burger KN, Martone ME, Deerinck TJ, Ellisman MH, Luini A (2003) ER-to-Golgi carriers arise through direct en bloc protrusion and multistage maturation of specialized ER exit domains. Dev Cell 5:583–594

Mironov AA, Colanzi A, Polishchuk RS, Beznoussenko GV, Mironov AA Jr, Fusella A, Di Tullio G, Silletta MG, Corda D, De Matteis MA, Luini A (2004) Dicumarol, an inhibitor of ADP-ribosylation of CtBP3/BARS, fragments Golgi non-compact tubular zones and inhibits intra-Golgi transport. Eur J Cell Biol 83:263–279

Mironov AA, Beznoussenko GV, Polishchuk RS, Trucco A (2005) Intra-Golgi transport: a way to a new paradigm? Biochim Biophys Acta 1744:340–350

Mironov AA, Sesorova IV, Beznoussenko GV (2013) Golgi's way: a long path toward the new paradigm of the intra-Golgi transport. Histochem Cell Biol 140:383–393

Mironov AA, Sesorova IS, Seliverstova EV, Beznoussenko GV (2016) Different Golgi ultrastructure across species and tissues: implications under functional and pathological conditions, and an attempt at classification. Tissue Cell. https://doi.org/10.1016/j.tice.2016.12.002

Mitrovic S, Ben-Tekaya H, Koegler E, Gruenberg J, Hauri HP (2008) The cargo receptors Surf4, endoplasmic reticulum-Golgi intermediate compartment (ERGIC)-53, and p25 are required to maintain the architecture of ERGIC and Golgi. Mol Biol Cell 19:1976–1990

Morin-Ganet MN, Rambourg A, Deitz SB, Franzusoff A, Képès F (2000) Morphogenesis and dynamics of the yeast Golgi apparatus. Traffic 1:56–68

Motta PM, Makabe S, Naguro T, Correr S (1994) Oocyte follicle cells association during development of human ovarian follicle. A study by high resolution scanning and transmission electron microscopy. Arch Histol Cytol 57:369–394

Nogueira C, Erlmann P, Villeneuve J, Santos AJ, Martínez-Alonso E, Martínez-Menárguez JÁ, Malhotra V (2014) SLY1 and Syntaxin 18 specify a distinct pathway for procollagen VII export from the endoplasmic reticulum. Elife 3:e02784

Oprins A, Rabouille C, Posthuma G, Klumperman J, Geuze HJ, Slot JW (2001) The ER to Golgi interface is the major concentration site of secretory proteins in the exocrine pancreatic cell. Traffic 2:831–838

Orci L, Glick BS, Rothman JE (1986) A new type of coated vesicular carrier that appears not to contain clathrin: its possible role in protein transport within the Golgi stack. Cell 46:171–184

Orci L, Stamnes M, Ravazzola M, Amherdt M, Perrelet A, Sollner TH, Rothman JE (1997) Bidirectional transport by distinct populations of COPI-coated vesicles. Cell 90:335–349

Orci L, Perrelet A, Rothman JE (1998) Vesicles on strings: morphological evidence for processive transport within the Golgi stack. Proc Natl Acad Sci USA 95:2279–2283

Orci L, Amherdt M, Ravazzola M, Perrelet A, Rothman JE (2000a) Exclusion of Golgi residents from transport vesicles budding from Golgi cisternae in intact cells. J Cell Biol 150:1263–1270

Orci L, Ravazzola M, Volchuk A, Engel T, Gmachl M, Amherdt M, Perrelet A, Sollner TH, Rothman JE (2000b) Anterograde flow of cargo across the golgi stack potentially mediated via bidirectional "percolating" COPI vesicles. Proc Natl Acad Sci USA 97:10400–10405

Pagano RE, Longmuir KJ (1985) Phosphorylation, transbilayer movement, and facilitated intracellular transport of diacylglycerol are involved in the uptake of a fluorescent analog of phosphatidic acid by cultured fibroblasts. J Biol Chem 260:1909–1916

Pagano RE, Sepanski MA, Martin OC (1989) Molecular trapping of a fluorescent ceramide analogue at the Golgi apparatus of fixed cells: interaction with endogenous lipids provides a trans-Golgi marker for both light and electron microscopy. J Cell Biol 109:2067–2079

Palade G (1975) Intracellular aspects of the process of protein synthesis. Science 189:347–359

Patterson GH, Hirschberg K, Polishchuk RS, Gerlich D, Phair RD, Lippincott-Schwartz J (2008) Transport through the Golgi apparatus by rapid partitioning within a two-phase membrane system. Cell 133:1055–1067

Pavelka M, Roth J (2005) Functional ultrastructure. Atlas of tissue biology and pathology. Springer, Wien, 326 p

Pellett PA, Dietrich F, Bewersdorf J, Rothman JE, Lavieu G (2013) Inter-Golgi transport mediated by COPI-containing vesicles carrying small cargoes. Elife 2:e01296

Perinetti G, Müller T, Spaar A, Polishchuk R, Luini A, Egner A (2009) Correlation of 4Pi- and electron microscopy to study transport through single Golgi stacks in living cells with super resolution. Traffic 10:379–391

Peters A, Palay SL, Webster HdeF (1991) Fine structure of the nervous system: neurons and their supporting cells, 3rd edn. Oxford University Press, Oxford, 528 p

Polishchuk RS, Mironov AA (2004) Structural aspects of Golgi function. Cell Mol Life Sci 61:146–158

Polishchuk RS, Polishchuk EV, Mironov AA (1999) Coalescence of Golgi fragments in microtubule-deprived living cells. Eur J Cell Biol 78:170–185

Polishchuk RS, Polishchuk EV, Marra P, Alberti S, Buccione R, Luini A, Mironov AA (2000) Correlative light-electron microscopy reveals the tubular-saccular ultrastructure of carriers operating between Golgi apparatus and plasma membrane. J Cell Biol 148(1):45–58

Polishchuk EV, Di Pentima A, Luini A, Polishchuk RS (2003) Mechanism of constitutive export from the Golgi: bulk flow via the formation, protrusion, and en bloc cleavage of large trans-Golgi network tubular domains. Mol Biol Cell 14:4470–4485

Presley JF, Cole NB, Schroer TA, Hirschberg K, Zaal KJ, Lippincott-Schwartz J (1997) ER-to-Golgi transport visualized in living cells. Nature 389:81–85

Rahkila P, Alakangas A, Väänänen K, Metsikkö K (1996) Transport pathway, maturation, and targeting of the vesicular stomatitis virus glycoprotein in skeletal muscle fibers. J Cell Sci 109:1585–1596

Rambourg A, Clermont Y, Chrétien M, Olivier L (1993) Modulation of the Golgi apparatus in stimulated and nonstimulated prolactin cells of female rats. Anat Rec 235:353–362

Raote I, Ortega Bellido M, Pirozzi M, Zhang C, Melville D, Parashuraman S, Zimmermann T, Malhotra V (2017) TANGO1 assembles into rings around COPII coats at ER exit sites. J Cell Biol 216(4):901–909

Raote I, Ortega-Bellido M, Santos AJ, Foresti O, Zhang C, Garcia-Parajo MF, Campelo F, Malhotra V (2018) TANGO1 builds a machine for collagen export by recruiting and spatially organizing COPII, tethers and membranes. Elife 7:e32723. https://doi.org/10.7554/eLife.32723

Razi M, Chan EY, Tooze SA (2009) Early endosomes and endosomal coatomer are required for autophagy. J Cell Biol 185:305–321

Rexach MF, Schekman RW (1992) Use of sec mutants to define intermediates in protein transport from endoplasmic reticulum. Methods Enzymol 219:267–286

Rhodin JAG (1974) Histology: a text and atlas. Oxford University Press, New York, 803 pp

Rizzo R, Parashuraman S, Mirabelli P, Puri C, Lucocq J, Luini A (2013) The dynamics of engineered resident proteins in the mammalian Golgi complex relies on cisternal maturation. J Cell Biol 201:1027–1036

Romero N, Dumur CI, Martinez H, García IA, Monetta P, Slavin I, Sampieri L, Koritschoner N, Mironov AA, De Matteis MA, Alvarez C (2013) Rab1b overexpression modifies Golgi size and gene expression in HeLa cells and modulates the thyrotrophin response in thyroid cells in culture. Mol Biol Cell 24:617–632

Rothman JE, Bursztyn-Pettegrew H, Fine RE (1980) Transport of the membrane glycoprotein of vesicular stomatitis virus to the cell surface in two stages by clathrin-coated vesicles. J Cell Biol 86:162–171

Rothman JE, Miller RL, Urbani LJ (1984) Intercompartmental transport in the Golgi complex is a dissociative process: facile transfer of membrane protein between two Golgi populations. J Cell Biol 99:260–271

Sabesin SM, Frase S (1977) Electron microscopic studies of the assembly, intracellular transport, and secretion of chylomicrons by rat intestine. J Lipid Res 18:496–511

Saito K, Chen M, Bard F, Chen S, Zhou H, Woodley D, Polischuk R, Schekman R, Malhotra V (2009) TANGO1 facilitates cargo loading at endoplasmic reticulum exit sites. Cell 136(5):891–902

Santos AJ, Nogueira C, Ortega-Bellido M, Malhotra V (2016) TANGO1 and Mia2/cTAGE5 (TALI) cooperate to export bulky pre-chylomicrons/VLDLs from the endoplasmic reticulum. J Cell Biol 213:343–354

Sesorova IS, Beznoussenko GV, Kazakova TE, Sesorov VV, Dimov ID, Mironov AA (2018) New opportunities of light microscopy in cytology and histology. Tsitologia 60(5):319–329

Sesso A, de Faria FP, Iwamura ES, Correa H (1994) A three-dimensional reconstruction study of the rough ER-Golgi interface in serial thin sections of the pancreatic acinar cell of the rat. J Cell Sci 107:517–528

Siddiqi SA, Gorelick FS, Mahan JT, Mansbach CM 2nd (2003) COPII proteins are required for Golgi fusion but not for endoplasmic reticulum budding of the pre-chylomicron transport vesicle. J Cell Sci 116:415–427

Sleight RG, Pagano RE (1983) Rapid appearance of newly synthesized phosphatidylethanolamine at the plasma membrane. J Biol Chem 258:9050–9058

Sokolova YY, Mironov AA (2008) Structure and function of the Golgi organelle in parasitic protists. In: Mironov AA, Pavelka M (eds) Golgi apparatus, Chapter 4.14. Springer, Wien, pp 647–674

Styers ML, O'Connor AK, Grabski R, Cormet-Boyaka E, Sztul E (2008) Depletion of beta-COP reveals a role for COP-I in compartmentalization of secretory compartments and in biosynthetic transport of caveolin-1. Am J Physiol Cell Physiol 294:C1485–C1498

Sukhorukov VM, Meyer-Hermann M (2015) Structural heterogeneity of mitochondria induced by the microtubule cytoskeleton. Sci Rep 5:13924

Taylor RS, Jones SM, Dahl RH, Nordeen MH, Howell KE (1997) Characterization of the Golgi complex cleared of proteins in transit and examination of calcium uptake activities. Mol Biol Cell 8:1911–1931

Townley AK, Feng Y, Schmidt K, Carter DA, Porter R, Verkade P, Stephens DJ (2008) Efficient coupling of Sec23-Sec24 to Sec13-Sec31 drives COPII-dependent collagen secretion and is essential for normal craniofacial development. J Cell Sci 121:3025–3034

Townley AK, Schmidt K, Hodgson L, Stephens DJ (2012) Epithelial organization and cyst lumen expansion require efficient Sec13-Sec31-driven secretion. J Cell Sci 125:673–684

Trucco A, Polishchuk RS, Martella O, Di Pentima A, Fusella A, Di Giandomenico D, San Pietro E, Beznoussenko GV, Polishchuk EV, Baldassarre M, Buccione R, Geerts WJ, Koster AJ, Burger KN, Mironov AA, Luini A (2004) Secretory traffic triggers the formation of tubular continuities across Golgi sub-compartments. Nat Cell Biol 6:1071–1081

Valderrama F, Babià T, Ayala I, Kok JW, Renau-Piqueras J, Egea G (1998) Actin microfilaments are essential for the cytological positioning and morphology of the Golgi complex. Eur J Cell Biol 76:9–17

Velasco A, Hendricks L, Moreman KW, Tulsiani DRP, Touster O, Farquhar MG (1993) Cell-type dependent variations in the subcellular distribution of a-mannosidase I and II. J Cell Biol 122:39–51

Venditti R, Scanu T, Santoro M, Di Tullio G, Spaar A, Gaibisso R, Beznoussenko GV, Mironov AA, Mironov A Jr, Zelante L, Piemontese MR, Notarangelo A, Malhotra V, Vertel BM, Wilson C, De Matteis MA (2012) Sedlin controls the ER export of procollagen by regulating the Sar1 cycle. Science 337:1668–1672

Watson P, Forster R, Palmer KJ, Pepperkok R, Stephens DJ (2005) Coupling of ER exit to microtubules through direct interaction of COPII with dynactin. Nat Cell Biol 7:48–55

Weidman P, Roth R, Heuser J (1993) Golgi membrane dynamics imaged by freeze-etch electron microscopy: views of different membrane coatings involved in tubulation versus vesiculation. Cell 75:123–133

Weinstock M, Leblond CP (1974) Synthesis, migration, and release of precursor collagen by odontoblasts as visualized by radioautography after (3H)proline administration. J Cell Biol 60:92–127

Wu J, Akhmanova A (2017) Microtubule-organizing centers. Annu Rev Cell Dev Biol 33:51–75

Yang JS, Valente C, Polishchuk RS, Turacchio G, Layre E, Branch Moody D, Leslie CC, Gelb MH, Brown WJ, Corda D, Luini A, Hsu VW (2011) COPI acts in both vesicular and tubular transport. Nat Cell Biol. https://doi.org/10.1038/ncb2273

Yang C, Wu J, de Heus C, Grigoriev I, Liv N, Yao Y, Smal I, Meijering E, Klumperman J, Qi RZ, Akhmanova A (2017) EB1 and EB3 regulate microtubule minus end organization and Golgi morphology. J Cell Biol 216:3179–3198

Zeuschner D, Geerts WJ, van Donselaar E, Humbel BM, Slot JW, Koster AJ, Klumperman J (2006) Immuno-electron tomography of ER exit sites reveals the existence of free COPII-coated transport carriers. Nat Cell Biol 8:377–383

Chapter 5
RhoA Pathway and Actin Regulation of the Golgi/Centriole Complex

Malgorzata Kloc, Ahmed Uosef, Jarek Wosik, Jacek Z. Kubiak, and Rafik Mark Ghobrial

Abstract In vertebrate cells, the Golgi apparatus is located in close proximity to the centriole. The architecture of the Golgi/centriole complex depends on a multitude of factors, including the actin filament cytoskeleton. In turn, both the Golgi and centriole act as the actin nucleation centers. Actin organization and polymerization also depend on the small GTPase RhoA pathway. In this chapter, we summarize the most current knowledge on how the genetic, magnetic, or pharmacologic interference with RhoA pathway and actin cytoskeleton directly or indirectly affects architecture, structure, and function of the Golgi/centriole complex.

M. Kloc (✉)
The Houston Methodist Research Institute, Houston, TX, USA

Department of Surgery, The Houston Methodist Hospital, Houston, TX, USA

Department of Genetics, MD Anderson Cancer Center, The University of Texas, Houston, TX, USA
e-mail: mkloc@houstonmethodist.org

A. Uosef · R. M. Ghobrial
The Houston Methodist Research Institute, Houston, TX, USA

Department of Surgery, The Houston Methodist Hospital, Houston, TX, USA

J. Wosik
Department of Electrical and Computer Engineering, University of Houston, Houston, TX, USA

Texas Center for Superconductivity, University of Houston, Houston, TX, USA

J. Z. Kubiak
Laboratory of Epidemiology, Military Institute of Hygiene and Epidemiology (WIHE), Warsaw, Poland

Department of Regenerative Medicine and Cell Biology, Military Institute of Hygiene and Epidemiology (WIHE), Warsaw, Poland

Faculty of Medicine, Cell Cycle Group, Institute of Genetics and Development of Rennes, Univ Rennes, UMR 6290, CNRS, Rennes, France

© Springer Nature Switzerland AG 2019
M. Kloc (ed.), *The Golgi Apparatus and Centriole*, Results and Problems in Cell Differentiation 67, https://doi.org/10.1007/978-3-030-23173-6_5

5.1 Pericentriolar Location of the Golgi Apparatus

In vertebrate cells, the Golgi apparatus consists of stacks of flattened membranous cisternae interconnected by tubular cisternae into a continuous Golgi ribbon. The main function of the Golgi is the posttranslational modification of the proteins delivered from the endoplasmic reticulum (ER) and their subsequent sorting and delivery, within the Golgi-derived secretory vesicles, to the cell surface or to the endosomal–lysosomal system. During interphase, the Golgi apparatus is located near the nucleus, in close proximity to or surrounding the centrioles and their pericentriolar material (PCM; collectively called the centrosome). In preparation for cell division, during the G2 phase of the cell cycle, the Golgi ribbon separates into isolated stacks that migrate to the cell periphery and away from the centrioles that migrate to the opposite poles of the cell to nucleate the microtubules of the spindle (Colanzi et al. 2003; Cervigni et al. 2015; Sütterlin and Colanzi 2010; Rios 2014). In dividing cells, the centrioles and/or centrosomes function as the main microtubule-organizing centers (MTOCs) for the asters and the spindles. Additionally, in ciliated cells, the centrioles are converted into basal bodies and organize cilia (Bettencourt-Dias and Glover 2007).

Although this subject is beyond the scope of this review, the Golgi itself is also able to nucleate microtubules (Miller et al. 2009; Maia et al. 2013; Rios 2014). Interestingly, the Golgi and the centrosome share the molecules needed for microtubule assembly. One such molecule is the scaffolding protein kinase N-associated protein AKAP450, which recruits other proteins required for microtubule formation (EI Din El Homasany et al. 2005). Since Golgi- and centrosome-derived microtubule arrays differ in their geometry, it is believed that they may play complementary and/or different functions (Rios 2014). For example, it was shown that Golgi-derived microtubules controlled the speed of cell migration during wound closure (Hurtado et al. 2011; Rios 2014).

Physical contact between the Golgi and centriole is also required for cell polarization and movement when the Golgi and centriole move in unison to the leading edge of the cell. A striking example of the polarization of the Golgi/centriole complex is the formation of the immunological synapse between interacting immune cells during the immune response (Kloc et al. 2014). Although we still do not fully understand the role of the association between the Golgi and centriole, it is believed that it facilitates polar/directional movement of the Golgi-derived secretory vesicles on the microtubules nucleated at the centrioles. Thus, it defines a polarity of secretion of various molecules (Rios 2014) and the recycling and directional positioning of the receptors in the cell membrane.

In contrast to vertebrate cells, in yeast, invertebrates, and plants, the Golgi stacks are dispersed in the cytoplasm, do not form ribbon, and are not located in the vicinity of the centrioles (daSilva et al. 2004; Kondylis and Rabouille 2009; Preuss et al. 1992). This suggests that pericentriolar localization of the Golgi is not a prerequisite for its most basic functions.

The question that arises is how the peripherally located Golgi stacks are repositioned at the centriole and centrosome after cell division? It has been shown that the movement of the Golgi stacks from the cell periphery toward the cell center and the maintenance of the pericentrosomal location of the Golgi depend on the microtubules and the motor protein dynein (Cole et al. 1996; Corthesy-Theulaz et al. 1992; Harada et al. 1998; Rios 2014; Thyberg and Moskalewski 1999), which is recruited to the Golgi by its resident membrane protein Golgin-160 (Hicks and Machamer 2002; Yadav et al. 2009, 2012; Rios 2014). The anchoring of the Golgi stacks in the vicinity of centriole/centrosome depends on a *cis*-Golgi network-associated protein GMAP-210 (TRIP11) (Infante et al. 1999; Rios et al. 2004). It has been shown that the depletion of GMAP-210 causes dispersion of the Golgi stacks, while the experimental targeting of GAMP-210 to the mitochondria causes their clustering around the centrioles (Yadav et al. 2009, 2012; Rios 2014).

Recently, it was shown that not only does the centriole affect the positioning and function of the Golgi but also the Golgi regulates the positioning of the centriole and centrosome (Bisel et al. 2008; Sütterlin and Colanzi 2010). Bisel et al. (2008) showed that in the rat kidney (NRK) cell wound healing assay, the reorientation of the centrosome and accompanying Golgi toward the leading edge of migrating cell depends on the phosphorylation of the Golgi structural protein GRASP65 by the extracellular signal-regulated kinase (ERK). This study showed that phosphorylation of GRASP65 or treatment with brefeldin A, which both cause disassembly of the Golgi ribbon, is necessary for centrosome orientation and subsequent polarization of migrating cells.

5.2 Golgi/Centriole Complex and Actin

It is well established that the structure of the Golgi depends on centriole- and Golgi-derived microtubules (Kreis 1990 ; Thyberg and Moskalewski 1999; Miller et al. 2009; Maia et al. 2013; Rios 2014). However, in the last decades, numerous studies have shown that actin filaments are also very important for the establishment and maintenance of the Golgi architecture (Colón-Franco et al. 2011; Egea et al. 2006; Guet et al. 2014; Lowe 2011). Surprisingly, it has been shown that both the Golgi and the centriole are able to nucleate actin filaments (Chen et al. 2004; Dubois et al. 2005; Carreno et al. 2004; Guet et al. 2014; Obino et al. 2016).

5.2.1 Golgi and Actin

The Golgi cisternae are stacked parallel to each other and have a very distinct morphology; they are flattened in the center and swollen at the rims, which bud off to form transporting vesicles. The actin filaments associated with the Golgi apparatus maintain the structural integrity of the Golgi stacks, provide mechanical

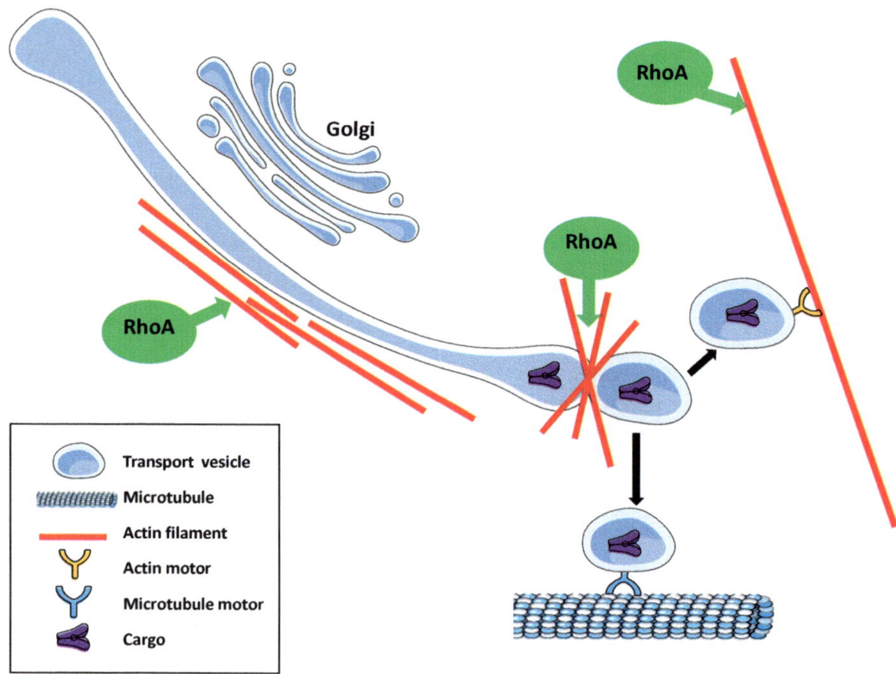

Fig. 5.1 Diagram of actin filament and RhoA role in the scission from the Golgi cisternae and in the transport of the cargo-containing vesicles. The actin filaments associated with the Golgi provide mechanical rigidity and flatness of the cisternae, form a physical barrier preventing the formation of transport vesicles at the center of the cisternae, and facilitate scission of the transport vesicles at the rims of the cisternae. Actin polymerization is regulated by RhoA pathway. The vesicles transport their cargo, such as various receptors, to the cell membrane using actin filament and actin motors, or alternatively, the microtubules and microtubule motors

rigidity and flatness of the cisternae, form a physical barrier preventing the formation of transport vesicles at the center of the cisternae, and facilitate scission of the transport vesicles at the rims of the cisternae (Fig. 5.1). In addition, actin, together with the actin motor proteins, such as myosin II and VI, translocates the vesicles away from the Golgi to their intracellular or extracellular destinations (Fig. 5.1; Brownhill et al. 2009; Buss et al. 2002; di Campli et al. 1999; Egea et al. 2006; Fath 2005; Guet et al. 2014; Miserey-Lenkei et al. 2010). Recently, Capmany et al. (2019) showed that the Golgi-associated myosin MYO1C, which colocalizes with the Golgi-associated actin, is necessary for actin stabilization, Golgi compaction, and anterograde and retrograde vesicular transport. Depolymerization of actin filaments with cytochalasin D, latrunculin B, or mycalolide B induces swelling of the cisternae, while stabilization of actin filaments with jasplakinolide causes fragmentation of the cisternae (Egea et al. 2006). Recently, Guet et al. (2014) used internalized microspheres trapped in optical tweezers to deform Golgi membranes and showed that the alteration in actin dynamics and inhibition of actomyosin (myosin II) decreased the rigidity of Golgi membranes. They also showed that the mechanical force applied to the Golgi membranes disrupts the dynamics of Golgi-associated actin and decreases the formation of Golgi-derived vesicles (Guet et al. 2014). In addition, Makhoul et al. (2019) showed that actin regulation of the Golgi dynamics depends on the interaction between the golgin (trans-Golgi network (TGN), GCC88, protein) and its binding partnerintersectin-1 (ITSN-1).

Another fascinating phenomenon dependent on actin/myosin is the stop-and-go motion of the Golgi described in plant cells (Nebenfuhr and Staehelin 2001). Although the movement of Golgi through the cytoplasm is probably restricted to plant cells, where the Golgi stacks are dispersed in the cytoplasm, Breuer et al. (2017) showed that the Golgi apparatus oscillates (wiggles) in the hypocotyl cells of *Arabidopsis*. The authors believe that because "the wiggling resembles the searching behavior of foraging animals or microbes that has been suggested to optimize food search efficiency," the Golgi may use this motion to more efficiently transport various compounds from the ER or/and distribute Golgi-derived material within the cell (Breuer et al. 2017). Further studies are needed to establish if the fascinating phenomenon of the Golgi oscillation is limited to the plant cells or is a universal phenomenon of all eukaryotic cells.

5.2.2 Centriole and Actin

It has been known for years that the centriole and centrosome interact with actin filaments in interphase and dividing invertebrate and vertebrate cells, eggs, and embryos. For example, it has been shown that actin is associated with the cnidarian sperm centrioles (Kleve and Cark 1980) and that cortical actin plays a role in centrosome separation at the onset of mitosis in *Drosophila* embryos and 3T3-L1 adipocytes (Cao et al. 2010; Wang et al. 2008). In *Drosophila* syncytial blastoderm embryos, centrosomes direct the assembly and position of cortical actin caps and cytokinetic furrows (Stevenson et al. 2001). Ciliated cells in the neural tube have centriole-derived basal bodies that are anchored by actin filaments (Antoniades et al. 2014). However, although the results of proteomic analyses showed the presence of actin and its associated proteins at mammalian and human centrosomes (Andersen et al. 2003; Firat-Karalar et al. 2014; Jakobsen et al. 2011), until recently, there was no direct proof that the centriole and centrosome could nucleate actin filaments. In 2016, Farina and coinvestigators used isolated centrosomes from human Jurkat and HeLa cells to provide evidence that, indeed, the centrosome nucleates actin filaments through the action of the centrosomal Wiskott–Aldrich syndrome protein (WASP) family member, WASH (Zigmond 2000), which stimulates the Arp2/3 complex (Farina et al. 2016). The Arp2/3 protein complex contains seven subunits. Two of these subunits, the actin-related proteins, ARP2 and ARP3, serve as nucleation sites for actin filaments (Veltman and Insall 2010). A question that arises concerns the roles of centrosome-derived actin filaments. Although further studies are needed to discover all potential functions, a recent study by Obino et al. (2016) showed that in the resting lymphocytes, the actin filaments nucleated by centrosomal Arp2/3 anchor the centrosome to the nucleus. Upon lymphocyte activation, which leads to its polarization and formation of the immunological synapse, the level of ARP2/3 at the centrosome decreases. This, in turn, reduces actin nucleation at the centrosome and allows centrosome detachment from the nucleus and its movement together with the accompanying Golgi complex to the immunological synapse (Obino et al. 2016). Ultimately, the positioning of the Golgi at the immunological synapse allows for the polar delivery of cytokines, enzymes, and receptors toward the lymphocyte target (Kloc et al. 2014). Recent studies by Inoue et al. (2019) on the actin present in the lymphocyte centrosomes showed that centrosomal actin physically blocks elongation of the centrosome-derived microtubules.

5.3 The Role of RhoA Pathway in Golgi and Centriole Architecture and Functions

In all eukaryotic cells, the actin cytoskeleton is regulated by the small GTPase RhoA and its downstream effector ROCK1 p160 kinase. RhoA is activated by RhoA-specific guanine exchange factors (GEFs) that activate GTPases by stimulating the release of guanosine diphosphate (GDP) and binding of guanosine triphosphate (GTP; Cherfils and Zeghouf 2013; Fig. 5.2a). RhoA is also reciprocally regulated by the Rho GTPase Rac1 and mTOR pathways (Fig.5.2a; Byrne et al. 2016; Laplante and Sabatini 2009). Recent studies indicate that, surprisingly, ROCK1 kinase is not only the effector of RhoA but also an effector of Rac1 (Fig.5.2a; Soriano-Castell et al. 2017). As previously discussed, the actin cytoskeleton plays a major role in the structure and function of the Golgi/centriole complex. Thus, it should be expected that interference with the RhoA pathway would disrupt actin cytoskeleton and indirectly affect the Golgi/centriole complex. Indeed, studies from our laboratories showed that the inhibition of RhoA GEFs using Rhosin or Y16 inhibitors, the inhibition of ROCK1 using the Y2762 inhibitor, or the genetic deletion of RhoA reorganizes the actin cytoskeleton in mouse macrophages and disrupts the structure of the Golgi, which fragments, disperses, and/or migrates from its perinuclear position to the macrophage tail (Figs. 5.2b and 5.3; Chen et al. 2017, 2018; Liu et al. 2016a, b, c, 2017). Interestingly, actin disorganization does not cause repositioning of the centriole, which remains in the perinuclear position. Thus, the RhoA interference causes uncoupling of the Golgi from the centriole (Fig. 5.3). We also showed that the macrophage-specific deletion of RhoA caused defective recycling, which relies on the Golgi-derived vesicular pathway, of macrophage fractalkine receptor CX3CR1 (Liu et al. 2017). Interestingly, very similar effects on macrophage phenotype, the Golgi and receptors were achieved when the macrophages were exposed to the magnetic field gradient interference (Wosik et al. 2018). In these studies, we have designed a permanent rare-earth magnet setup with defined magnetic field-gradient patterns and investigated the effect of these fields on mouse peritoneal macrophages grown in in vitro culture. We observed that the magnetic

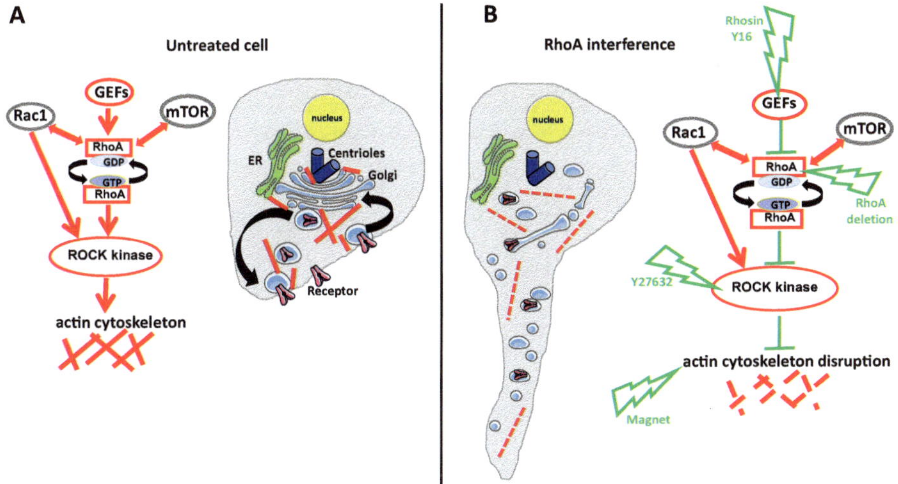

Fig. 5.2 Diagram of the RhoA pathway, actin, centriole, and Golgi. (**a**) In normal, untreated cells, RhoA is activated by RhoA-specific GEFs and is reciprocally regulated by mTOR and the Rac1 pathway. RhoA regulates actin filament polymerization and organization through its downstream effector ROCK kinase. Recently, it has been shown that ROCK is also a direct downstream effector of Rac1 (Soriano-Castell et al. 2017). A properly organized actin cytoskeleton and fully functioning RhoA pathway support perinuclear organization of Golgi and centrioles, the Golgi-derived vesicular transport of various molecules and receptors, and the recycling of the receptors from the membrane back to the Golgi. (**b**) Pharmacologic, genetic, or magnetic interferences with the components of the RhoA pathway, such as GEF inhibition with Rhosin or Y16, Rock inhibition with Y27632, RhoA deletion, or the exposure of macrophages to the magnetic field gradient, elongate macrophage, disrupt actin cytoskeleton, disperse Golgi, and/or cause translocation of its fragments into the macrophage tail. These interferences do not affect perinuclear localization of the centrioles

force elongated macrophages and arranged them in distinctive rows/waves. The location and alignment of magnetic field-elongated macrophages correlated very well with the simulated distribution and orientation of magnetic force lines (Wosik et al. 2018). Also, the exposure of cultured RAW 264 cells (Abelson murine leukemia virus-transformed mouse macrophage) to the magnetic force had a similar effect (Uosef et al. unpublished). We also showed that similar to RhoA interference, the magnetic field exposure elongated macrophages, changed Golgi complex and cation channel receptor TRPM2 distribution, and modified the expression of macrophage molecular markers (Wosik et al. 2018).

All these studies suggested that the effect of RhoA pathway interference on the Golgi/receptor recycling was indirect, resulting from the changes to the actin cytoskeleton. However, there are studies indicating that RhoA and/or its binding partners are associated with the Golgi apparatus and thus are directly involved in the regulation of some aspects of the Golgi structure and functions (Chi et al. 2013; Long and Simpson 2017). Quassollo et al. (2015) showed that in the neuronal cells, RhoA, ROCK, and serine-threonine kinases, LIMK1, and PKD1, are necessary for

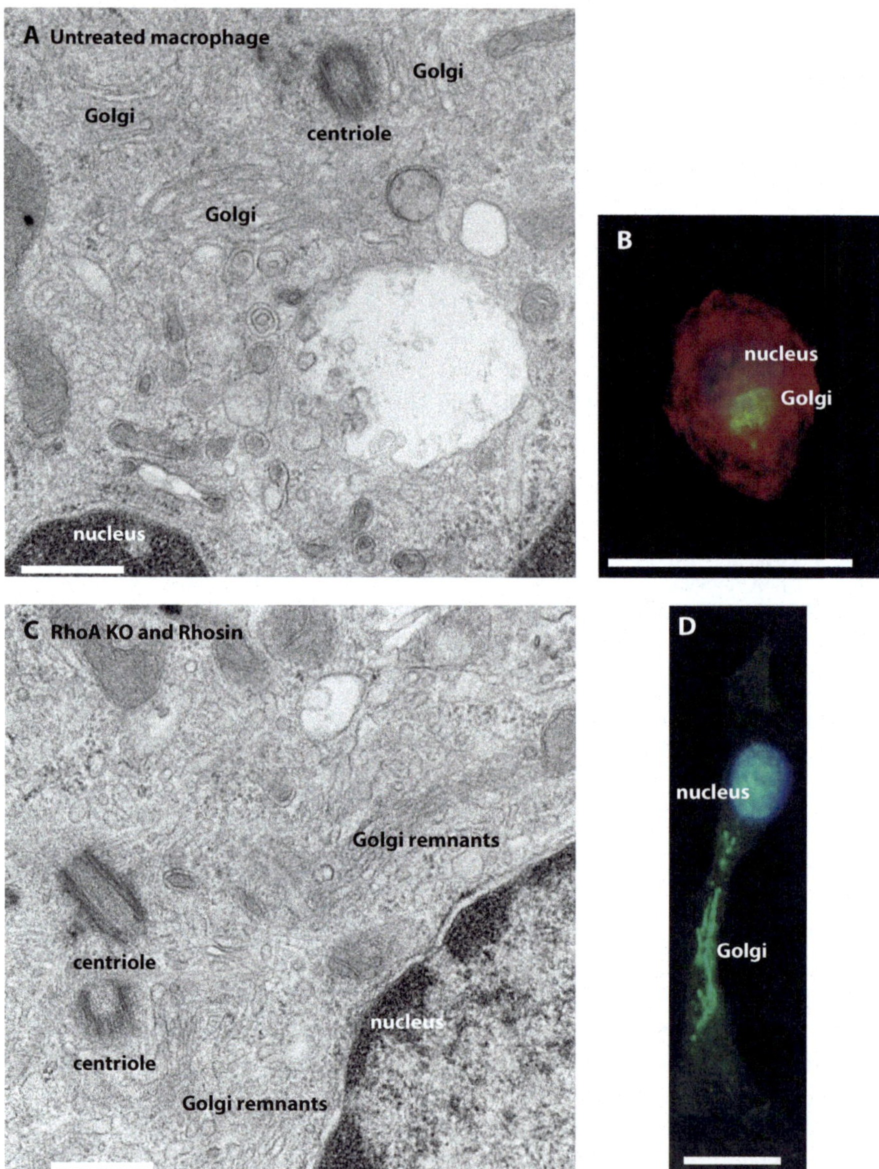

Fig. 5.3 Golgi and centrioles in mouse macrophages. (**a**)Transmission electron microscopy. In untreated macrophages, centrioles are surrounded by the stacks of Golgi cisternae and positioned in the vicinity of the nucleus. Bar is equal to 500 nm. (**b**). Light microscopy. Untreated mouse macrophage immunostained with Golgi marker anti-GM130 antibody and FITC-conjugated secondary antibody (green) and counterstained with Rhodamine phalloidin to visualize actin (red) and DAPI to visualize nucleus (blue). Golgi is visible in the vicinity of the nucleus. Bar is equal to 50 μm. (**c**) Transmission electron microscopy. RhoA deleted macrophages treated with GEFs' inhibitor, Rhosin. The centrioles and remnants of the dispersed Golgi were visible in the vicinity of the nucleus, while the majority of the dispersed Golgi moved away from the nucleus toward the

the scission and fission of the dendritic Golgi outposts (GOPs) needed for the trafficking of synaptic receptors (Fig. 5.1). In another study, Zilberman et al. (2011) showed that the effects of the overexpression of RhoA, which causes dispersion of the Golgi ribbon and defects in the fission of the transporting vesicles, depend on a member of the formin protein, mDia1, localized at the Golgi membranes. Camera et al. (2003, 2008) showed that the Golgi apparatus of neural cells contains RhoA-binding protein Citron-N, which controls actin filament polymerization through the local assembly of ROCK kinase, and actin-binding protein, Profilin IIa. Another study showed a centriolar function of RhoA. Aoki et al. (2009) showed that the RhoA GEFARHGEF10 is localized at the centrosome in G1/S and M phases of the cell cycle. The ARHGEF10 regulates RhoA and controls centrosome duplication through its binding partner—a motor protein KIF3B, which is colocalized with ARHGEF10 at the centrosome.

In summary, these data indicate that the RhoA pathway is both indirectly, through the actin cytoskeleton, and directly, through the Golgi/centriole-associated factors, involved in the regulation of Golgi/centriole structure and functions.

Acknowledgments The authors gratefully acknowledge support from the William and Ella Owens Medical Research Foundation to JW and MK, the William Stamps Farish Fund to MK and RMG, the State of Texas through the Texas Center for Superconductivity at the University of Houston to JW, Polish Ministry of National Defense project "Kościuszko" # 571/2016/DA to JZK and Polish Ministry of National Defense project "Kościuszko" # 5508/2017/DA to JZK. We also thank Kenneth Dunner Jr. for the electron microscopy work and acknowledge CCSG grant NIH P30CA016672 to MDACC High Resolution Electron Microscopy Facility.

References

Andersen JS, Wilkinson CJ, Mayor T, Mortensen P, Nigg EA, Mann M (2003) Proteomic characterization of the human centrosome by protein correlation profiling. Nature 426:570–574

Antoniades I, Stylianou P, Skourides PA (2014) Making the connection: ciliary adhesion complexes anchor basal bodies to the actin cytoskeleton. Dev Cell 28:70–80

Aoki T, Ueda S, Kataoka T, Satoh T (2009) Regulation of mitotic spindle formation by the RhoA guanine nucleotide exchange factor ARHGEF10. BMC Cell Biol 10:56

Fig. 5.3 (continued) macrophage tail (not shown here). Bar is equal to 500 nm. (**d**) Light microscopy. Mouse macrophage treated with GEF inhibitor Rhosin and immunostained with Golgi marker anti-GM130 antibody and FITC-conjugated secondary antibody (green), and counterstained with DAPI to visualize nucleus (blue). Golgi fragments are displaced from the perinuclear position to the macrophage tail. Bar is equal to 18 μm. Panel D is from Fig. 6 in Chen et al. (2017) with the permission of Elsevier

Bettencourt-Dias M, Glover DM (2007) Centrosome biogenesis and function: centrosomics brings new understanding. Nat Rev Mol Cell Biol 8:451–463

Bisel B, Wang Y, Wei J-H, Xiang Y, Tang D, Miron-Mendoza M, Yoshimura S, Nakamura N, Seemann J (2008) ERK regulates Golgi and centrosome orientation towards the leading edge through GRASP65. J Cell Biol 182:837–843

Breuer D, Nowak J, Ivakov A, Somssich M, Persson S, Nikoloski Z (2017) System-wide organization of actin cytoskeleton determines organelle transport in hypocotyl plant cells. Proc Natl Acad Sci U S A 114(28):E5741–E5749

Brownhill K, Wood L, Allan V (2009) Molecular motors and the Golgi complex: staying put and moving through. Semin Cell Dev Biol 20:784–792

Buss F, Luzio JP, Kendrick-Jones J (2002) Myosin VI, an actin motor for membrane traffic and cell migration. Traffic 3:851–858

Byrne KM, Monsefi N, Dawson JC, Degasperi A, Bukowski-Wills J-C, Volinsky N, Dobrzyński M et al (2016) Bistability in the Rac1, PAK, and RhoA signaling network drives actin cytoskeleton dynamics and cell motility switches. Cell Syst 2:38–48

Camera P, da Silva JS, Griffiths G, Giuffrida MG, Ferrara L, Schubert V, Imarisio S, Silengo L, Dotti CG, Di Cunto F (2003) Citron-N is a neuronal Rho-associated protein involved in Golgi organization through actin cytoskeleton regulation. Nat Cell Biol 5:1071–1078

Camera P, Schubert V, Pellegrino M, Berto G, Vercelli A, Muzzi P, Hirsch E, Altruda F, Dotti CG, Di Cunto F (2008) The RhoA-associated protein Citron-N controls dendritic spine maintenance by interacting with spine-associated Golgi compartments. EMBO Rep 9:384–392

Cao J, Crest J, Fasulo B, Sullivan W (2010) Cortical actin dynamics facilitate early-stage centrosome separation. Curr Biol 20:770–776

Capmany A, Yoshimura A, Kerdous R, Caorsi V, Lescure A, Nery ED, Coudrier E, Goud B, Schauer K (2019) MYO1C stabilizes actin and facilitates arrival of transport carriers at the Golgi apparatus. J Cell Sci 132(8):jcs.225029. https://doi.org/10.1242/jcs.225029

Carreno S, Engqvist-Goldstein AE, Zhang CX, McDonald KL, Drubin DG (2004) Actin dynamics coupled to clathrin-coated vesicle formation at the trans-Golgi network. J Cell Biol 165:781–788

Cervigni RI, Bonavita R, Barretta ML, Spano D, Ayala I, Nakamura N, Corda D, Colanzi A (2015) JNK2 controls fragmentation of the Golgi complex and the G2/M transition through phosphorylation of GRASP65. J Cell Sci 128:2249–2260

Chen J-L, Lacomis L, Erdjument-Bromage H, Tempst P, Stamnes M (2004) Cytosol-derived proteins are sufficient for Arp2/3recruitment and ARF/coatomer-dependent actin polymerization on Golgi membranes. FEBS Lett 566:281–286

Chen W, Li XC, Kubiak JZ, Ghobrial RM, Kloc M (2017) Rho-specific Guanine nucleotide exchange factors (Rho-GEFs) inhibition affects macrophage phenotype and disrupts Golgi complex. Int J Biochem Cell Biol 93:12–24

Chen W, Ghobrial RM, Li XC, Kloc M (2018) Inhibition of RhoA and mTORC2/Rictor by Fingolimod (FTY720) induces p21-activated kinase 1, PAK-1 and amplifies podosomes in mouse peritoneal macrophages. Immunobiology 223(11):634–647. https://doi.org/10.1016/j.imbio.2018.07.009

Cherfils J, Zeghouf M (2013) Regulation of small GTPases by GEFs, GAPs, and GDIs. Physiol Rev 93(1):269–309

Chi X, Wang S, Huang Y, Stamnes M, Chen JL (2013) Roles of rho GTPases in intracellular transport and cellular transformation. Int J Mol Sci 14:7089–7108

Colanzi A, Süetterlin C, Malhotra V (2003) Cell-cycle-specific Golgi fragmentation: how and why? Curr Opin Cell Biol 15:462–467

Cole NB, Sciaky N, Marotta A, Song J, Lippincott-Schwartz J (1996) Golgi dispersal during microtubule disruption: regeneration of Golgi stacks at peripheral endoplasmic reticulum exit sites. Mol Biol Cell 7:631–650

Colón-Franco JM, Gomez TS, Billadeau DD (2011) Dynamic remodeling of the actin cytoskeleton by FMNL1γ is required for structural maintenance of the Golgi complex. J Cell Sci 124 (18):3118–3126

Corthesy-Theulaz I, Pauloin A, Pfeffer SR (1992) Cytoplasmic dynein participates in the centrosomal localization of the Golgi complex. J Cell Biol 118:1333–1345

daSilva LL, Snapp EL, Denecke J, Lippincott-Schwartz J, Hawes C, Brandizzi F (2004) Endoplasmic reticulum export sites and Golgi bodies behave as single mobile secretory units in plant cells. Plant Cell 16:1753–1771

di Campli A, Valderrama F, Babia T, De Matteis MA, Luini A, Egea G (1999) Morphological changes in the Golgi complex correlate with actin cytoskeleton rearrangements. Cell Motil Cytoskeleton 43:334–348

Dubois T, Paléotti O, Mironov AA, Fraisier V, Stradal TEB, De Matteis MA, Franco M, Chavrier P (2005) Golgi-localized GAP for Cdc42 functions downstream of ARF1 to control Arp2/3 complex and F-actin dynamics. Nat Cell Biol 7:353–364

Egea G, Lazaro-Dieguez F, Vilella M (2006) Actin dynamics at the Golgi complex in mammalian cells. Curr Opin Cell Biol 18:168–178

EI Din El Homasany BS, Volkov Y, Takahashi M, Ono Y, Keryer G, Delouvée A, Looby E, Long A, Kelleher D (2005) The scaffolding protein CG-NAP/AKAP450 is a critical integrating component of the LFA-1-Induced signaling complex in migratory T cells. J Immunol 175:7811–7818

Farina F, Gaillard J, Guérin C, Couté Y, Sillibourne J, Blanchoin L, Théry M (2016) The centrosome is an actin-organizing centre. Nat Cell Biol 18:65–75

Fath KR (2005) Characterization of myosin-II binding to Golgi stacks in vitro. Cell Motil Cytoskeleton 60:222–235

Firat-Karalar EN, Sante J, Elliott S, Stearns T (2014) Proteomic analysis of mammalian sperm cells identifies new components of the centrosome. J Cell Sci 127:4128–4133

Guet D, Mandal K, Pinot M, Hoffmann J, Abidine Y, Sigaut W, Bardin S, Schauer K, Goud B, Manneville J-B (2014) Mechanical role of actin dynamics in the rheology of the Golgi complex and in Golgi-associated trafficking events. Curr Biol 24:1700–1711

Harada A, Takei Y, Kanai Y, Tanaka Y, Nonaka S, Hirokawa N (1998) Golgi vesiculation and lysosome dispersion in cells lacking cytoplasmic dynein. J Cell Biol 141:51–59

Hicks SW, Machamer CE (2002) The NH2-terminal domain of Golgin-160 contains both Golgi and nuclear targeting information. J Biol Chem 277:35833–35839

Hurtado L, Caballero C, Gavilan MP, Cardenas J, Bornens M, Rios RM (2011) Disconnecting the Golgi ribbon from the centrosome prevents directional cell migration and ciliogenesis. J Cell Biol 193:917–933

Infante C, Ramos-Morales F, Fedriani C, Bornens M, Rios RM (1999) GMAP-210, A cis-Golgi network associated protein, is a minus end microtubule binding protein. J Cell Biol 145:83–98

Inoue D, Obino D, Pineau J, Farina F, Gaillard J, Guerin C, Blanchoin L, Lennon-Duménil AM, Théry M (2019) Actin filaments regulate microtubule growth at the centrosome. EMBO J pii: e99630. https://doi.org/10.15252/embj.201899630

Jakobsen L, Vanselow K, Skogs M, Toyoda Y, Lundberg E, Poser I, Falkenby LG, Bennetzen M, Westendorf J, Nigg EA, Uhlen M, Hyman AA, Andersen JS (2011) Novel asymmetrically localizing components of human centrosomes identified by complementary proteomics methods. EMBO J 30:1520–1535

Kleve MG, Clark WH Jr (1980) Association of actin with sperm centrioles: isolation of centriolar complexes and immunofluorescent localization of actin. J Cell Biol 86:87–95

Kloc M, Kubiak JZ, Li XC, Ghobrial RM (2014) The newly found functions of MTOC in immunological response. J Leukoc Biol 95:417–430

Kondylis V, Rabouille C (2009) The Golgi apparatus: lessons from Drosophila. FEBS Lett 583:3827–3838

Kreis TE (1990) Role of microtubules in the organization of the Golgi apparatus. Cell Motil Cytoskeleton 15:67–70

Laplante M, Sabatini DM (2009) mTOR signaling at glance. J Cell Sci 122:3589–3594

Liu Y, Chen W, Minze LJ, Kubiak JZ, Li XC, Ghobrial RM, Kloc M (2016a) Dissonant response of M0/M2 and M1 bone marrow derived macrophages to RhoA pathway interference. Cell Tissue Res 366:707–720

Liu Y, Minze LJ, Mumma L, Li XC, Ghobrial RM, Kloc M (2016b) Mouse macrophage polarity and ROCK1 activity depend on RhoA and non-apoptotic Caspase 3. Exp Cell Res 341:225–236

Liu Y, Tejpal N, You J, Li XC, Ghobrial RM, Kloc M (2016c) ROCK inhibition impedes macrophage polarity and functions. Cell Immunol 300:54–62

Liu Y, Chen W, Wu C, Minze LJ, Kubiak JZ, Li XC, Kloc M, Ghobrial RM (2017) Macrophage/monocyte-specific deletion of RhoA down-regulates fractalkine receptor and inhibits chronic rejection of mouse cardiac allografts. J Heart Lung Transplant 36:340–354

Long M, Simpson JC (2017) Rho GTPases operating at the Golgi complex: implications for membrane traffic and cancer biology. Tissue Cell 49:163–169

Lowe M (2011) Structural organization of the Golgi apparatus. Curr Opin Cell Biol 23:85–93

Maia AR, Zhu X, Miller P, Gu G, Maiato H, Kaverina I (2013) Modulation of Golgi-associated microtubule nucleation throughout the cell cycle. Cytoskeleton (Hoboken) 70:32–43

Makhoul C, Gosavi P, Duffield R, Delbridge B, Williamson NA, Gleeson PA (2019) Intersectin-1 interacts with the golgin GCC88 to couple the actin network and Golgi architecture. Mol Biol Cell 30(3):370–386. https://doi.org/10.1091/mbc.E18-05-0313

Miller PM, Folkmann AW, Maia AR, Efimova N, Efimov A, Kaverina I (2009) Golgi-derived CLASP dependent microtubules control Golgi organization and polarized trafficking in motile cells. Nat Cell Biol 11:1069–1080

Miserey-Lenkei S, Chalancon G, Bardin S, Formstecher E, Goud B, Echard A (2010) Rab and actomyosin-dependent fission of transport vesicles at the Golgi complex. Nat Cell Biol 12:645–654

Nebenfuhr A, Staehelin LA (2001) Mobile factories: Golgi dynamics in plant cells. Trends Plant Sci 6:160–167

Obino D, Farina F, Malbec O, Sáez PJ, Maurin M, Gaillard J, Dingli F, Loew D, Gautreau A, Yuseff M-I, Blanchoin L, Théry M, Lennon-Duménil A-M (2016) Actin nucleation at the centrosome controls lymphocyte polarity. 2016. Nature Com 7:10969–10983

Preuss D, Mulholland J, Franzusoff A, Segev N, Botstein D (1992) Characterization of the Saccharomyces Golgi complex through the cell cycle by immunoelectron microscopy. Mol Biol Cell 3:789–803

Quassollo G, Wojnacki J, Salas DA, Gastaldi L, Marzolo MP, Conde C, Bisbal M, Couve A, Cáceres A (2015) A RhoA signaling pathway regulates dendritic Golgi outpost formation. Curr Biol 25:971–982

Rios RM (2014) The centrosome–Golgi apparatus nexus. PhilTrans R Soc B 369:20130462

Rios RM, Sanchis A, Tassin AM, Fedriani C, Bornens M (2004) GMAP-210 recruits gamma-tubulin complexes to cis-Golgi membranes and is required for Golgi ribbon formation. Cell 118:323–335

Soriano-Castell D, Chavero A, Rentero C, Bosch M, Vidal-Quadras M, Pol A, Enrich C, Tebar F (2017) ROCK1 is a novel Rac1 effector to regulate tubular endocytic membrane formation during clathrin-independent endocytosis. Sci Rep 7:6866

Stevenson VA, Kramer J, Kuhn J, Theurkauf WE (2001) Centrosomes and the Scrambled protein coordinate microtubule-independent actin reorganization. Nat Cell Biol 3:68–75

Sütterlin C, Colanzi A (2010) The Golgi and the centrosome: building a functional partnership. The Golgi and the centrosome: building a function. J Cell Biol 188:621–628

Thyberg J, Moskalewski S (1999) Role of microtubules in the organization of the Golgi complex. Exp Cell Res 246:263–279

Veltman DM, Insall RH (2010) WASP family proteins: their evolution and its physiological implications. Mol Biol Cell 21:2880–2893

Wang W, Chen L, Ding Y, Jin J, Liao K (2008) Centrosome separation driven by actin-microfilaments during mitosis is mediated by centrosome-associated tyrosine-phosphorylated cortactin. J Cell Sci 121:1334–1343

Wosik J, Chen W, Qin K, Ghobrial RM, Kubiak JZ, Kloc M (2018) Magnetic field changes macrophage phenotype. Biophys J 114:2001–2013

Yadav S, Puri S, Linstedt AD (2009) A primary role for Golgi positioning in directed secretion, cell polarity, wound healing. Mol Biol Cell 20:1728–1736

Yadav S, Puthenveedu MA, Linstedt AD (2012) Golgin160 recruits the dynein motor to position the Golgi apparatus. Dev Cell 23:153–165

Zigmond SH (2000) How wasp regulates actin polymerization. J Cell Biol 150:117–120

Zilberman Y, Alieva NO, Miserey-Lenkei S, Lichtenstein A, Kam Z, Sabanay H, Bershadsky A (2011) Involvement of the Rho/mDia1 pathway in the regulationof Golgi complex architecture and dynamics. Mol Biol Cell 22:2900–2911

Chapter 6
Multiple Roles of Rab GTPases at the Golgi

Cinzia Progida

Abstract The Golgi apparatus is a central sorting station in the cell. It receives newly synthesized molecules from the endoplasmic reticulum and directs them to different subcellular destinations, such as the plasma membrane or the endocytic pathway. Importantly, in the last few years, it has emerged that the maintenance of Golgi structure is connected to the proper regulation of membrane trafficking. Rab proteins are small GTPases that are considered to be the master regulators of the intracellular membrane trafficking. Several of the over 60 human Rabs are involved in the regulation of transport pathways at the Golgi as well as in the maintenance of its architecture. This chapter will summarize the different roles of Rab GTPases at the Golgi, both as regulators of membrane transport, scaffold, and tethering proteins and in preserving the structure and function of this organelle.

6.1 Introduction

The Golgi apparatus is one of the main sorting stations of the cells at the crossroad between the endocytic and the secretory pathway. Newly synthesized proteins, carbohydrates, and lipids are transported from the endoplasmic reticulum (ER) via the Golgi toward their final destination. Some will remain in the Golgi, while others, as newly synthesized lysosomal enzymes, will be transported to the endocytic pathway and others to the plasma membrane (secretory pathway).

The communication between the different intracellular compartments occurs via vesicles or tubules and consists of different steps (Fig. 6.1). The first step is cargo recruitment and membrane budding. Transmembrane cargos or receptors for luminal cargos recruit coat proteins directly or via adaptor proteins, inducing membrane curvature that shapes a nascent bud (Faini et al. 2013). The budding vesicle is then pinched off from the membrane of the donor compartment. The fission process involves dynamin, which forms a constrictive ring around the neck of the bud or actin cytoskeleton (Antonny et al. 2016; Romer et al. 2010). Once the vesicle is

C. Progida (✉)
Department of Biosciences, University of Oslo, Oslo, Norway
e-mail: c.a.m.progida@ibv.uio.no

© Springer Nature Switzerland AG 2019
M. Kloc (ed.), *The Golgi Apparatus and Centriole*, Results and Problems in Cell
Differentiation 67, https://doi.org/10.1007/978-3-030-23173-6_6

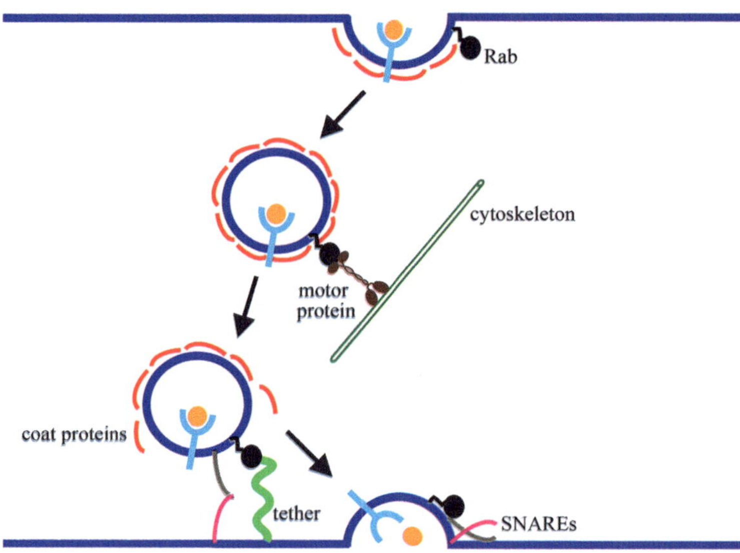

Fig. 6.1 Steps in vesicular transport. Coat proteins induce the formation of a bud containing cargo on the donor membrane. Next, the vesicle is transported to the acceptor compartment along cytoskeletal track by motor proteins. Tethering factors tether the vesicle to the target membrane. Finally, the vesicle-associated SNARE and the SNARE on the acceptor membrane assemble into a trans-SNARE complex, leading to membrane fusion and the delivery of cargo. Rab GTPases are key regulators of all these steps

released in the cytosol, it travels along the cytoskeleton toward its final destination, and coat proteins dissociate (Kirchhausen et al. 2014; Trahey and Hay 2010). Vesicle transport is mediated by different motor proteins that connect, directly or indirectly, through adaptor proteins, the vesicle to actin filaments, or microtubules. The vesicle eventually tethers and fuses to the target membrane, transferring its content to the acceptor compartment (Spang 2016; Witkos and Lowe 2017). Soluble *N*-ethylmaleimide-sensitive factor attachment protein receptor (SNARE) proteins mediate the fusion with the target membrane (Baker and Hughson 2016; Bombardier and Munson 2015).

In order to ensure the correct transport to the right destination, every single step of intracellular transport has to be tightly regulated. The importance of this regulation is demonstrated by the multitude of diseases associated with alterations of any of the abovementioned transport steps including neurological and immune disorders (Bucci et al. 2012; Krzewski and Cullinane 2013; Olkkonen and Ikonen 2006).

6.2 The Multiple Transport Pathways at the Golgi

In the cell, there are two main transport pathways, the biosynthetic (or exocytic/secretory) and the endocytic route, and the Golgi apparatus is the central sorting station at the intersection between them.

The secretory pathway starts at the ER. From there, the newly synthesized proteins and lipids are transported to the Golgi in vesicles that fuse to form the vesicles and tubules of the ER-Golgi intermediate compartment (ERGIC). Golgi-resident proteins are retained in the Golgi, while secretory cargo proteins undergo the posttranslational modification and glycosylation before further transport to different destinations. Secretory cargo proteins are delivered to the cell surface through the secretory pathway, while newly synthesized lysosomal enzymes are delivered to the endosomal pathway. Altogether, this forward transport constitutes the anterograde trafficking pathway (Palmer and Stephens 2004; Viotti 2016). The *trans*-Golgi network (TGN) also receive vesicles from the endosomal pathway for recycling of transport machinery (Progida and Bakke 2016). Furthermore, a Golgi retrograde pathway mediates the transport of ER proteins back to their original location using Coat protein complex I (COPI)-coated vesicles (Borgese 2016; Cottam and Ungar 2012; Spang 2013).

When vesicles arrive in the Golgi proximity, they are captured by tethering factors such as golgins, the Golgi-associated retrograde protein (GARP I) complex at the TGN, or the multisubunit transport protein particles (TRAPP) I/II and conserved oligomeric Golgi (COG) complexes for the ER-Golgi and intra-Golgi traffic (Barrowman et al. 2010; Bonifacino and Hierro 2011; Gillingham and Munro 2016; Miller and Ungar 2012). Golgins are long coiled-coil proteins that contain multiple Rab-binding sites. By protruding from the Golgi membranes, they capture incoming vesicles (Sinka et al. 2008). The distribution of different golgins in different regions of the Golgi has been suggested to contribute to the specificity of membrane recognition and to the ability of capturing vesicles of different origins (Gillingham and Munro 2016; Wong and Munro 2014).

In order to ensure the proper function of all these transport pathways, the Golgi apparatus needs to maintain an ordered structure (Liu and Storrie 2012; Storrie 2005). On the other side, it has been suggested that the balance between membranes arriving or leaving at the *cis*-side and at the *trans*-side is essential for the maintenance of Golgi organization, homeostasis, and morphology (Liu and Storrie 2012, 2015). Thus, increasing evidence indicates that the different transport pathways at the Golgi are connected with its organization and emerging candidates involved in this regulation are Rab proteins (Liu and Storrie 2012, 2015).

6.3 Rab GTPases

Rab proteins are small GTPases consisting of over 60 members in humans (Zhen and Stenmark 2015). Each Rab is characterized by distinct intracellular localization and regulates a specific trafficking route. Rab GTPases interact with a plethora of effector proteins to perform different tasks in the various steps of intracellular transport (Gillingham et al. 2014; Pylypenko et al. 2018). Adaptor proteins, as well as lipid kinases or phosphatases, help Rabs in cargo sorting and membrane remodeling (Christoforidis et al. 1999b; Perrin et al. 2013; Stein et al. 2003). Rabs also interact with motor proteins to mediate vesicular transport along the cytoskeleton (Borg et al.

2014; Horgan and McCaffrey 2011; Jordens et al. 2001; Kjos et al. 2018; Seabra and Coudrier 2004) or with tethering factors and SNAREs to facilitate the fusion of vesicles with the correct target membrane (Cai et al. 2007; Christoforidis et al. 1999a; Epp et al. 2011; Lupashin and Waters 1997; Lurick et al. 2017).

Rab GTPases are characterized by a conserved fold (G-domain) and two switch regions that are subjected to a conformational change upon GTP binding (Goitre et al. 2014; Pylypenko et al. 2018). The nucleotide-dependent conformational change allows these small GTPases to function as molecular switches. In addition, Rab proteins contain five Rab family (RabF) motifs that differentiate them from the other small GTPases and four Rab-subfamily sequence motifs (RabSF) that define subfamilies of Rabs (Pereira-Leal and Seabra 2000, 2001). Both RabF and RabSF motifs are conserved across species (Pereira-Leal and Seabra 2001).

At the C-terminus, Rab proteins contain a hypervariable region, which was originally suggested to be responsible for the targeting of Rabs to specific membranes (Chavrier et al. 1991; Stenmark et al. 1994). However, this hypothesis has later been challenged as membrane targeting seems to be regulated by a more complex mechanism involving different regions of Rab proteins as well as their effectors and other binding partners (Ali et al. 2004; Li et al. 2014).

One or two cysteine prenylation motifs are present at the end of the Rab C-terminus. Posttranslational modifications (geranylgeranylation) at these cysteines permit the association of Rab proteins to membranes (Goitre et al. 2014; Pylypenko et al. 2018; Wu et al. 2009).

6.3.1 Rab Proteins as Molecular Switches

Rab GTPases function as molecular switches by cycling between an active GTP-bound state and an inactive GDP-bound state (Fig. 6.2). This switch regulates the association of Rabs with the membranes and effectors.

Newly synthesized GDP-bound Rab proteins are escorted by a Rab escort protein (REP) to a Rab geranylgeranyl transferase (RabGGT) (Andres et al. 1993; Casey and Seabra 1996). This transferase catalyzes the addition of one or two geranylgeranyl groups to the C-terminal cysteines of the Rabs (Casey and Seabra 1996; Gutkowska and Swiezewska 2012). REP masks the geranylgeranyl group of Rab proteins, thus keeping soluble the prenylated Rabs (Rak et al. 2004).

The GDP dissociation inhibitor (GDI) is another Rab chaperone that binds to Rab proteins in the cytosol. Nevertheless, while REP binds to Rabs regardless of their prenylation status, and therefore is involved in presenting a newly synthesized Rab to the GGT, GDI has higher affinity for Rabs when prenylated (Wu et al. 2007). Consequently, GDI is implicated in the extraction of inactive Rabs from the membranes (Muller and Goody 2017; Pylypenko et al. 2018).

It has been suggested that a GDI displacement factors (GDFs) may assist in targeting and inserting Rab proteins in the appropriate membranes (Collins 2003; Dirac-Svejstrup et al. 1997; Goody et al. 2017). However, what targets different Rabs to their specific membranes is not fully understood yet. Indeed, it has also been

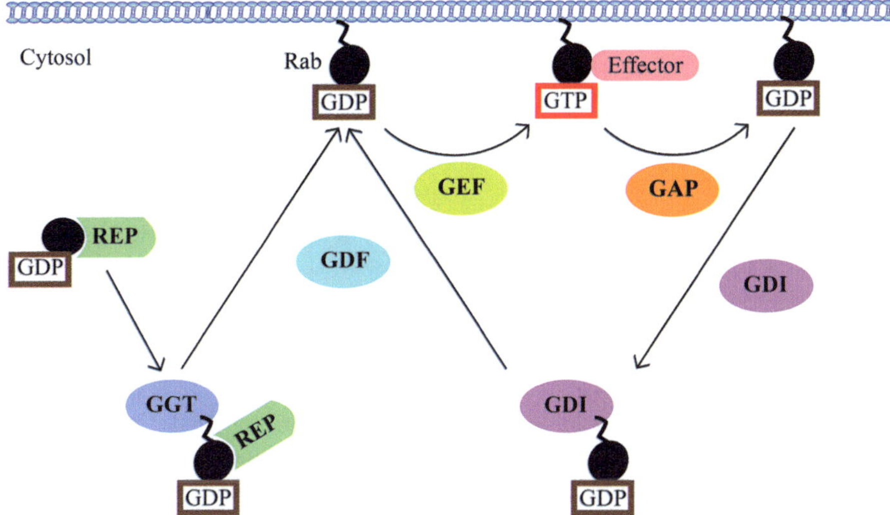

Fig. 6.2 The Rab GTPase cycle. Newly synthesized Rab proteins associate with a Rab escort protein (REP). This complex is recognized by a geranylgeranyl transferase (GGT) that prenylates the Rab C-terminal cysteine motifs. Upon membrane attachment, Rab proteins are activated by guanine nucleotide exchange factors (GEFs) that promotes the exchange of GDP for GTP. The active, membrane-bound Rabs are able to interact with specific effectors to fulfill their functions. Subsequently, specific GTPase activating proteins (GAPs) inactivate the Rabs by promoting GTP hydrolysis. The inactive, GDP-bound Rab is then extracted from the membrane by GDP dissociation inhibitors (GDIs) and can go through another round of activation. This is possible because GDI displacement factors (GDFs) assist Rab proteins in the dissociation from the GDIs and in the reinsertion into the membrane

reported that guanine-nucleotide exchange factors (GEFs) are necessary and sufficient for membrane targeting of Rabs and that GDFs are not required for this process (Blumer et al. 2013; Wu et al. 2010). GEFs stimulate the exchange of GDP for GTP, thereby activating the Rabs. Active Rab GTPases recruit downstream effectors that facilitate the regulation of the trafficking steps (Pylypenko et al. 2018).

Afterward, GTPase-activating proteins (GAPs) inactivate the Rabs by catalyzing the hydrolysis of GTP. The GDP-bound Rab is then a substrate for Rab-GDI. As mentioned before, this protein extracts Rab proteins from the membranes and keeps them inactive in the cytosol until the reactivation for a new round of vesicular transport (Gavriljuk et al. 2013; Muller and Goody 2017; Ullrich et al. 1993).

6.4 Rab GTPases at the Golgi

Several Rabs associate with the Golgi complex and the TGN where they regulate intra-Golgi trafficking as well as different transport pathways to or from the Golgi. In addition, some Rab proteins play also a role in maintaining Golgi architecture. These

Golgi-associated Rabs have been suggested to be the link between Golgi organization and membrane trafficking (Liu and Storrie 2012).

6.4.1 The Multiple Roles of Rab6 at the Golgi

Rab6 is the most studied Golgi-associated Rab. It regulates different trafficking pathways at the Golgi, such as intra- and post-Golgi transport, Golgi-to-ER and endosome-to-Golgi retrograde traffic (Grigoriev et al. 2007; Mallard et al. 2002; Martinez et al. 1994; Mayer et al. 1996; Utskarpen et al. 2006; White et al. 1999). In addition, Rab6 is involved in the maintenance of Golgi integrity and steady-state homeostasis (Starr et al. 2010).

Rab6 has four isoforms in humans. Alternative splicing of the *Rab6A* gene generates Rab6A and Rab6A' that differ in only three amino acids (Echard et al. 2000). Rab6B is encoded by a different gene and is mainly expressed in the brain (Opdam et al. 2000). Rab6C is a retrogene derived from the RAB6A' transcript. It localizes to centrosomes and is involved in cell cycle progression (Young et al. 2010).

Rab6A and Rab6A' are ubiquitously expressed and localize to the *trans*-Golgi and TGN. Some studies report Rab6A to regulate the traffic between the Golgi and the ER, and others indicate the involvement of Rab6A' in the transport from endosomes to the TGN (Del Nery et al. 2006; Echard et al. 2000; Matanis et al. 2002). Despite this evidence, Rab6A and Rab6A' seem to be functionally redundant. Indeed, the retrograde transport of ricin toxin is dependent on both Rab6A and Rab6A', indicating that Rab6A can also regulate the endosome-to-Golgi transport (Utskarpen et al. 2006). Additionally, either the silencing of Rab6A or Rab6A' or the expression of their constitutively active mutants influences Golgi-to-ER transport (Young et al. 2005), demonstrating the overlapping roles of Rab6A or Rab6A' also in the regulation of this transport pathway. For this, and because Rab6A or Rab6A' are quite similar both biochemically and functionally, they are collectively referred to as Rab6.

Rab6B is also present at the Golgi and the TGN (Opdam et al. 2000). Active Rab6 (A or A') or Rab6B interacts with ELKS (glutamic acid/leucine/lysine/serine-rich protein; also known as Rab6IP2 or CAST) on Golgi membranes. It has been earlier suggested that this complex regulates the endosome-to-Golgi transport as expression of either the Rab6-binding domain of Rab6IP2 or of Rab6 dominant-negative mutant or Rab6 depletion inhibits the retrograde transport of the Shiga toxin B subunit (Del Nery et al. 2006; Mallard et al. 2002; Monier et al. 2002). However, more recently it has been demonstrated that Rab6 interacts with ELKS at the cell periphery to mediate exocytosis (Grigoriev et al. 2007, 2011; Kobayashi et al. 2016).

Indeed, Rab6 is also involved in the regulation of the transport from the Golgi to the plasma membrane (Grigoriev et al. 2007, 2011). Within this pathway, this small GTPase has an additional role in the fission of vesicles from the Golgi membranes by binding to the actin motor myosin II and the microtubule motor kinesin family member 20A (KIF20A) (Miserey-Lenkei et al. 2010, 2017). The interaction with

KIF20A is responsible for anchoring Rab6 to the Golgi membranes and for positioning the Rab6-positive vesicles leaving the Golgi along microtubules (Miserey-Lenkei et al. 2017).

Rab6 binds also to another kinesin, KIF1C (Lee et al. 2015). This interaction mediates vesicle transport to the cell surface and influences Golgi organization. Rab6 controls KIF1C's activity by binding to the kinesin motor domain, thus inhibiting the interaction of this motor with microtubules (Lee et al. 2015). However, as another kinesin, KIF5B, has been previously shown to mediate the transport of Rab6-positive exocytotic vesicles to the cell periphery (Grigoriev et al. 2007), it remains to be determined whether these two kinesins work in concert for the transport of Rab6 post-Golgi carriers to the plasma membrane.

In addition to myosin and kinesin motors, Rab6 recruits the dynein motor to Golgi membranes through the interaction with the golgin Bicaudal D2 (BICD2) and p150glued, a subunit of the dynactin complex (Short et al. 2002). Through this binding, Rab6 is involved in the targeting and docking of endosomes with the TGN (Matsuto et al. 2015; Short et al. 2002). Therefore, Rab6 has multiple roles not only in regulating numerous trafficking pathways but also in tethering and fission of vesicles at the Golgi. Furthermore, it is involved in the maintenance of Golgi ribbon organization. Indeed, light microscopy studies show that Rab6 depletion condenses Golgi ribbon (Sun et al. 2007; Young et al. 2005). A more fine analysis by electron microscopy indicates that silencing of Rab6 increases both number and length of Golgi cisternae, as well as cisternal continuity (Ferraro et al. 2014; Micaroni et al. 2013; Storrie et al. 2012).

To regulate and coordinate all its functions at the Golgi, Rab6 not only interacts with several effectors but is also involved in crosstalk with other Rab proteins. It has been suggested that Rab6 and Rab11 sequentially regulate the retrograde transport pathway between recycling endosomes and the Golgi, as they both bind to the Rab6-interacting protein 1 (R6IP1) (Miserey-Lenkei et al. 2007). In addition, Rab6 cooperates with Rab8 to control docking and fusion of exocytotic vesicles (Grigoriev et al. 2011). It has been proposed that Rab6 recruits Rab8 to Golgi-derived vesicles. Subsequently, Rab8 is linked by molecule interacting with CasL protein 3 (MICAL) 3, to the Rab6-interacting cortical factor ELKS. Expression of an inactive MICAL3 mutant accumulates secretory vesicles that dock at the cell cortex but fail to fuse with the plasma membrane. Therefore, MICAL3 may function as linker between these Rabs on the same pathway regulating vesicle docking and fusion (Grigoriev et al. 2011).

Finally, Rab6, together with Rab33b, takes also part in an intra-Golgi Rab cascade. Rab33b is a medial Golgi-localized GTPase that, when active, associates with Ribosomal control protein1 (Ric1) and retrograde Golgi transport homolog 1 (Rgp1) proteins. These proteins form a complex that catalyzes guanine nucleotide exchange of Rab6. Loss of Ric1 or Rgp1 inhibits the Rab6-dependent retrograde Golgi transport, and so the Rab33b-Rab6 cascade has been suggested to regulate a major intra-Golgi retrograde trafficking pathway (Pusapati et al. 2012; Starr et al. 2010).

6.4.2 Rab33

The Rab33 subfamily consists of two members, Rab33a and Rab33b. Rab33a is expressed in the brain and the immune system, while Rab33b is ubiquitously expressed (Zheng et al. 1998). Both Rabs are Golgi-localized, with Rab33b localized to the medial Golgi cisternae (Nakazawa et al. 2012; Zheng et al. 1998). Rab33a is also present on post-Golgi vesicles in growing axons of developing neurons. Therefore, it has been suggested that this small GTPase contributes to the anterograde axonal transport of post-Golgi vesicles, to promote membrane insertion at the growth cones and axon outgrowth (Nakazawa et al. 2012).

Similarly to Rab6, also Rab33b regulates Golgi-to-ER retrograde trafficking and Golgi homeostasis/organization (Starr et al. 2010; Valsdottir et al. 2001). Indeed, overexpression of either Rab6 or Rab33b, or of their constitutively active mutants, relocates Golgi enzymes to the ER (Jiang and Storrie 2005; Martinez et al. 1997; Valsdottir et al. 2001). However, while depletion of Rab33b partially inhibits this redistribution of Golgi enzymes to the ER, Rab6 depletion has barely any effect, suggesting that Rab33b and Rab6 act sequentially in this retrograde Golgi transport pathway (Starr et al. 2010). Indeed, as mentioned earlier, Rab33b and Rab6 take part in a Rab cascade, where Rab33b recruits the GEF for Rab6 to the Golgi membranes for the regulation of subsequent steps in retrograde transport pathways (Pusapati et al. 2012; Starr et al. 2010).

Golgi- and ER-derived vesicles are one of the membrane sources for autophagosomes, the double-membrane organelles responsible for the degradation of cytoplasmic components in eukaryotic cells. As Rab33b specifically interacts with autophagy related 16 like (Atg16L), a factor essential for the initiation of autophagosome formation, it has been proposed that Rab33b has an additional role in the process of autophagosome formation by supplying Golgi-derived membranes (Itoh et al. 2008).

The importance of Rab33b function in Golgi transport pathways and homeostasis is further highlighted by the recent discoveries of mutations in the gene encoding for this small GTPase in patients with Smith-McCort dysplasia or Dyggve-Melchior-Clausen syndrome, two types of skeletal dysplasias characterized by defects in Golgi organization and traffic (Alshammari et al. 2012; Dupuis et al. 2013; Osipovich et al. 2008; Salian et al. 2017).

6.4.3 Golgi-Associated Rabs Regulating Lysosomal Positioning

Two other Rab proteins, Rab34 and Rab36, primarily localize to the Golgi apparatus where they are involved in the regulation of lysosomal positioning. This regulation occurs through the recruitment of the Rab-interacting lysosomal protein (RILP) (Chen et al. 2010; Wang and Hong 2002). Indeed, the expression of wild-type or constitutively active mutant of Rab34 that binds to RILP redistributes lysosomes from the periphery to the Golgi region (Goldenberg et al. 2007; Wang and Hong 2002).

Furthermore, Rab34 regulates the secretory pathway. Depletion of Rab34 or expression of its dominant-negative mutant impairs transports from the Golgi to the plasma membrane, but not ER-Golgi traffic. In more detail, Rab34 acts at the Golgi controlling intra-Golgi transport but not the exit from the TGN (Goldenberg et al. 2007).

6.5 Rab Proteins Mediating the Transport Between Endosomes and Golgi

In addition to the described Rab proteins primarily associated with the Golgi, there are also Rabs that mainly localize to other organelles but regulate the transport in or out from the Golgi. A bidirectional transport between the Golgi and the endosomal pathway ensures a continuous traffic of newly synthesized proteins and their sorting receptors, as well as lipids, bacterial toxins, and other proteins (Hasanagic et al. 2015; Progida and Bakke 2016; Sandvig et al. 2013).

At the TGN, transmembrane sorting receptors, such as mannose 6-phosphate receptors (MPRs) and sortilins (Vps10p-domain receptor family), bind to newly synthesized enzymes for their delivery to the endo-lysosomal pathway. The acidic endosomal environment promotes the dissociation of the ligands from their receptors. In this way, the enzymes continue their journey toward the lysosomes where they will perform their degradative function, while the sorting receptors are recycled back to the TGN to load newly synthesized enzymes and start another transport cycle.

Several molecules mediate this bidirectional transport between Golgi and endosomes, and between them Rab proteins and their effectors are key regulators (Progida and Bakke 2016). The best characterized Rabs regulating these transport pathways are Rab9 and Rab7b.

6.5.1 Rab9

Rab9 was identified almost three decades ago as a small GTPase localized to late endosomes (Chavrier et al. 1990b; Lombardi et al. 1993). Its function was proposed to be the transport of MPR from late endosomes to the TGN (Lombardi et al. 1993). In support of this, the first identified effector of Rab9, a 40-kD protein named p40, was found to promote MPR retrograde trafficking while the dominant-negative mutant of Rab9 to inhibit MPR recycling (Diaz et al. 1997; Riederer et al. 1994). The regulation of MPR recycling mediated by Rab9 also requires Rho-related BTB domain containing 3 (RhoBTB3), an atypical member of the Rho GTPase family and the golgin GCC185. These proteins, indeed, interact with Rab9, and it has been suggested that they aid the docking and tethering of Rab9-positive endosomes to the TGN (Espinosa et al. 2009; Reddy et al. 2006).

By regulating MPR trafficking, Rab9 is also involved in the delivery of enzymes to lysosomes. Indeed, the expression of Rab9-negative mutant decreases the efficiency of enzyme transport to lysosomes and increases their secretion (Riederer et al. 1994). In line with this, Rab9 is important for the proper morphology and localization of late endosomes and lysosomes and its depletion clusters and reduces the size of late endosomes (Ganley et al. 2004). The involvement of Rab9 in the transport from the TGN toward the endo-lysosomal pathway has been more recently confirmed by live imaging studies showing that Rab9-positive vesicles, similarly to MPR-positive vesicles, reach the endosomal pathway at the transition between early and late endosomes (Kucera et al. 2016b).

In this pathway, other effectors are involved, as, for example, RUN and TBC1 domain-containing proteins 1 and 2 (RUTBC1 and RUTBC2). They are GAPs for Rab32, Rab33b, and Rab36, but not for Rab9 (Nottingham et al. 2011, 2012). Even though the role of their interaction with Rab9 has not been fully elucidated, the Rab9-RUTBC1 complex is required for the transport of melanogenic enzymes to melanosomes that are lysosome-related organelles (Mahanty et al. 2016; Marubashi et al. 2016). By regulating the transport of enzymes to lysosomes and melanosomes, Rab9 is involved in the biogenesis of lysosomes and lysosome-related organelles through the interaction with biogenesis of lysosome-related organelle complex 3 (BLOC-3) (Kloer et al. 2010; Kucera et al. 2016b; Lombardi et al. 1993; Mahanty et al. 2016; Riederer et al. 1994).

In addition to the previously mentioned interactors, several other Rab9 effectors have been identified. Tail-interacting protein of 47 kDa (TIP47) was found to bind directly to both Rab9 and MPR to mediate the sorting of this receptor (Carroll et al. 2001; Hanna et al. 2002). However, a subsequent study questioned the role of TIP47 in Rab9-mediated MPR transport, by demonstrating that TIP47 is involved in lipid droplet biogenesis (Bulankina et al. 2009).

The multitude of Rab9 interactors reflects the complexity of the pathways at the intersection between the TGN and the endocytic route regulated by this small GTPase (Kucera et al. 2016a). Furthermore, Rab9 contributes to the formation of autophagosomes by providing membranes derived from the trans-Golgi and late endosomes (Nishida et al. 2009).

Therefore, Rab9 regulates several functions that depend on the transport directed toward the endosomal pathway. The identification of two additional Rab proteins that mediate MPR recycling, Rab7b and Rab29, questioned the sovereignty of Rab9 in the retrograde transport pathway (Progida et al. 2010, 2012; Wang et al. 2014).

6.5.2   Rab7b

A new Rab protein was identified in 2004 and named Rab7b because it shares 65% similarity with Rab7a and similarly localizes to late endosomes and lysosomes (Yang et al. 2004). However, it was later shown that Rab7b is also present at the Golgi and TGN and that it has a different function compared to Rab7a (Progida et al. 2010). While Rab7a mediates the transport from late endosomes to lysosomes, and is necessary for lysosomal degradation (Cantalupo et al. 2001; Ceresa and Bahr 2006), Rab7b mediates endosome-to-Golgi transport (Progida et al. 2010). Indeed, Rab7b depletion delays the retrograde transport to the TGN of the B-subunit of cholera toxin and of the sorting receptors sortilin and MPR (Bucci et al. 2010; Progida et al. 2010, 2012). Furthermore, expression of Rab7b dominant-negative mutant reduces human papillomavirus 16 (HPV16) infection. As HPV16 subunits traffic from late endosomes to the TGN, the reduced infection in presence of Rab7b-negative mutant supports that Rab7b mediates HPV16 subunits traffic in this pathway (Day et al. 2013).

In addition, GTP hydrolysis of Rab7b is required for the correct carrier formation at the TGN. Indeed, expression of Rab7b constitutively active mutant impairs the formation of carriers containing sorting receptors from the TGN (Progida et al. 2012). In line with this, depletion of TBC1D5, a Rab7b GAP, has been recently shown to mimic the effect of Rab7b constitutively active mutant by reducing the number of vesicles containing the sorting receptors MPR and sortilin (Borg Distefano et al. 2018). TBC1D5 is also a GAP for Rab7a and associates with the retromer, a multi-protein complex that recycles transmembrane cargo from endosomes to the TGN (Jia et al. 2016). It has been suggested that this GAP directs the next phase in transport from late endosomes by regulating in time and space Rab7b and Rab7a cargo delivery to the TGN or the endosomal system, respectively (Borg Distefano et al. 2018).

Whereas the expression of Rab7b constitutively active mutant reduces the number of TGN-derived vesicles, the expression of the constitutively active mutant of Rab9 increases carrier formation at the TGN (Kucera et al. 2016b; Progida et al. 2012). Additionally, these two mutants localize to different target compartments: the Golgi for the constitutively active mutant of Rab7b and late endosomes for the constitutively active mutant of Rab9 (Kucera et al. 2016b; Progida et al. 2012). It is therefore likely that Rab7b mediates the endosome-to-Golgi transport and Rab9 primarily regulates the opposite pathway.

6.5.3 Other Rabs in the Transport Between Endosomes and Golgi

Other Rabs have also been reported to mediate the transport pathways between endosome and Golgi; however, the underlying mechanisms are less characterized. Rab29 (also known as Rab7L1) is associated with and maintain the integrity of the TGN. It also regulates the retrograde trafficking of MPRs from endosomes to the TGN (Wang et al. 2014). Rab31 (also known as Rab22b) is involved in the formation and in the transport of tubulo-vesicular carriers from the TGN to endosomes (Ng et al.

2007; Rodriguez-Gabin et al. 2001, 2009). Rab31 and its effector OCRL-1, a phosphatidylinositol 4,5-biphosphate 5-phosphatase that regulates the levels of lipids for vesicular transport at the Golgi, localize to both TGN and endosomes, as well as on MPR-containing carriers that bud from the TGN (Rodriguez-Gabin et al. 2010; Suchy et al. 1995). It has therefore been suggested that Rab31 recruits OCRL-1 to the TGN for the formation and sorting of MPR carriers (Rodriguez-Gabin et al. 2010, 2009). However, further studies are needed to fully understand the intracellular pathway(s) regulated by Rab31 as this small GTPase has also been reported to mediate the transport from early to late endosomes (Chua and Tang 2014; Ng et al. 2009).

Similarly to Rab31, also other Rabs described to function in the transport from the TGN to endosomes, such as Rab11, Rab13, and Rab14, are reported to influence additional intracellular pathways. Rab11 associates with the TGN and TGN-derived vesicles to regulate TGN-to-plasma membrane transport (Chen et al. 1998; de Graaf et al. 2004; Parmar and Duncan 2016; Urbe et al. 1993). However, it has also been reported to regulate the traffic through the recycling endosomes in the pericentriolar recycling compartment (Gidon et al. 2012; Horgan et al. 2010; Ren et al. 1998; Takahashi et al. 2012; Ullrich et al. 1996). As endosomal recycling compartments are frequently concentrated in the perinuclear region in close proximity of the TGN, it has also been suggested that the Rab11-positive recycling endosomes are intermediate compartment for the post-Golgi trafficking and exocytosis (Ang et al. 2004; Lock and Stow 2005).

Likewise Rab11 and Rab13 regulate both membrane trafficking between TGN and recycling endosomes and endocytic recycling to the cell surface (Morimoto et al. 2005; Nokes et al. 2008). Rab14 is localized to biosynthetic (ER, Golgi, and TGN) and endosomal compartments, and it has been suggested to regulate the biosynthetic pathway between the Golgi and endosomal/recycling compartments (Junutula et al. 2004; Kitt et al. 2008; Proikas-Cezanne et al. 2006).

6.6 Rabs Mediating the Transport Between ER and Golgi

Continuous bidirectional membrane traffic between the ER and the Golgi ensures that newly synthesized secretory cargo are transported from the ER along the secretory pathway through the Golgi complex and that transport machineries can be recycled back to the ER. Coat protein complex II (COPII) participates in cargo export from the ER, and coat protein complex I (COPI) in the transport from the Golgi back to the ER, despite a COPI-independent pathway also being described (Girod et al. 1999; Szul and Sztul 2011; Viotti 2016). Several Rab GTPases are also involved in these pathways.

6.6.1 Rab1

The first identified Rab protein, Ypt1, was discovered in yeast over three decades ago (Gallwitz et al. 1983; Schmitt et al. 1986). Ypt1 and its human homolog Rab1 regulate the secretory pathway, in particular the ER-to-Golgi transport (Jedd et al. 1995; Plutner et al. 1991; Segev et al. 1988; Tisdale et al. 1992). Two isoforms of Rab1 exist: Rab1a and Rab1b (Touchot et al. 1989). They are both localized at the ER, early Golgi stack, and ERGIC (Plutner et al. 1991; Saraste et al. 1995), and they have been suggested to be functionally interchangeable (Nuoffer et al. 1994). Indeed, both Rab1a and Rab1b are able to antagonize the effects of the negative mutant of Rab1a that inhibits protein export from the ER and the transport between Golgi compartments (Nuoffer et al. 1994).

Electron microscopy studies show that Rab1a dominant-negative mutant causes an accumulation of cargos in pre-cis-Golgi vesicles and vesicular-tubular clusters, indicating that this small GTPase is essential for the delivery of vesicles to the Golgi (Pind et al. 1994). In agreement with this, active Rab1 interacts with tethering factors at the Golgi, including membrane tethering protein p115 and Golgi matrix protein 130 kD (GM130) (Allan et al. 2000; Weide et al. 2001). Rab1 recruits first p115 to COPII vesicles and, together with SNAREs (soluble N-ethylmaleimide-sensitive factor attachment protein receptors), forms a complex that promotes GM130-mediated targeting to the Golgi apparatus (Allan et al. 2000; Moyer et al. 2001).

Rab1 dominant-negative mutant also disperses the Golgi apparatus to the ER (Nuoffer et al. 1994; Wilson et al. 1994). Similarly, Rab1 depletion causes fusion of Golgi membranes with the ER (Bard et al. 2006). Therefore, Rab1 not only requires for the ER-to-Golgi and intra-Golgi transport but also for the maintenance of the Golgi apparatus. The importance of Rab1 for Golgi biogenesis is highlighted not only by the effect of its depletion or expression of dominant-negative mutant but also by the inactivation caused by the expression of TBC1 domain family member 20 (TBC1D20), a GAP for Rab1 localized to the ER. All cause a block in the ER-to-Golgi transport, and the collapse of the Golgi, supporting the role of Rab1 in the biogenesis and maintenance of Golgi (Haas et al. 2007).

As for Rab6, the role of Rab1 in Golgi organization may be an indirect consequence of its function in membrane trafficking. However, GTP-bound Rab1 binds to Golgin-84, a membrane-anchored golgin that is involved in generating and maintaining the architecture of the Golgi apparatus, suggesting that the role of Rab1 in Golgi maintenance may be dependent on this interaction (Satoh et al. 2003). Also, the Rab1 effector p115 plays a role in maintenance of Golgi structure, and its inhibition induces Golgi vesiculation, similarly to the effect of Rab1-negative mutant (Puthenveedu and Linstedt 2001).

Recently, it has been demonstrated that active Rab1 interacts with WAS protein homolog associated with actin, Golgi membranes, and microtubules (WHAMM), an activator of actin nucleation to promote membrane tubule elongation, and with Golgi phosphoprotein 3 (GOLPH3), a PtdIns(4)P and myosin 18A-binding protein that connects the Golgi to F-actin for efficient tubule and vesicle formation (Dippold et al. 2009; Russo et al. 2016; Sechi et al. 2017). This interaction is important for successful cytokinesis where Rab1, by controlling GOLPH3 localization at the Golgi and at the cleavage site, regulates the secretory vesicle trafficking that is necessary for proper furrow ingression during cytokinesis (Sechi et al. 2017).

6.6.2 Rab2

Rab2 subfamily consists of two members, Rab2a and Rab2b. Even though they are localized to an intermediate compartment between the ER and the Golgi, and regulate the transport from ER to pre-Golgi intermediates, they seem to be involved in the regulation of Golgi morphology (Chavrier et al. 1990a; Ni et al. 2002; Tisdale et al. 1992).

Similarly to Rab1 that mediates the transport from the ER to the Golgi through the interaction with the vesicle tethering factor p115 and the cis-Golgi matrix protein GM130, Rab2 interacts with the coiled-coil protein Golgin-45 which binds in turn to the medial-Golgi matrix protein Golgi reassembly stacking protein of 55 kDa (GRASP55). Golgin-45 together with GRASP55 forms a Rab2 effector complex on the medial-Golgi essential for ER-to-Golgi traffic and for Golgi structure (Short et al. 2001). Indeed, Rab2 depletion causes Golgi fragmentation (Haas et al. 2007). Interestingly, the Golgi fragmentation caused by Rab2b silencing is rescued by the re-expression of Rab2b but not of Rab2a. It has therefore been suggested that these Rabs regulate Golgi morphology by interacting with isoform-specific effectors. One of these effectors is the Golgi-associated Rab2b interactor-like 4 (GARI-L4), a specific Rab2b interactor whose knockdown also induced Golgi fragmentation (Aizawa and Fukuda 2015).

Rab2 has also been reported to bind to other golgins, including Golgi microtubule-associated protein 210 (GMAP-210), GRIP and coiled-coil domain containing 88 kDa and 185 kDa proteins (GCC88 and GCC185), and Golgin-245 (Gillingham et al. 2014; Sato et al. 2015; Sinka et al. 2008). Although many of these golgins bind to Rabs mostly through their coiled-coil regions, the function of these interactions has been long debated. Initially, it was thought that the Rab-binding sites within golgins were responsible for vesicle capture (Sinka et al. 2008). More recently, it has been suggested that Rab2 binding to golgins occurs downstream from vesicle capture and tethering (Sato et al. 2015). The binding of Rab2 to its binding site within the central coiled-coil region of GMAP-210 would then bring the vesicle in closer proximity to the target membrane, leading ultimately to SNARE-mediated fusion (Sato et al. 2015).

Rab2 also interacts with subunits of the homotypic fusion and protein sorting (HOPS) complex. This complex is present on endosomal membranes where it contributes to membrane tethering and fusion. It has been therefore suggested that Rab2, by interacting with HOPS subunits, can regulate the transport of carriers from Golgi to endosomes (Gillingham et al. 2014; Lorincz et al. 2017).

6.7 Rabs Mediating the Transport from the Golgi to the Plasma Membrane

The discovery that Rab proteins regulate intracellular trafficking was initiated over 30 years ago when it was demonstrated that Sec4, the yeast homolog of Rab8, mediates the transport of Golgi-derived vesicles to the plasma membrane (Goud et al. 1988; Salminen and Novick 1987). Similarly, in mammalian cells Rab8 regulates the vesicular transport from TGN to the plasma membrane (Huber et al. 1993).

At the Golgi, Rab8 interacts with optineurin, a protein that links myosin VI to the Golgi complex and plays a central role in Golgi ribbon formation and exocytosis (Sahlender et al. 2005). On exocytotic vesicles, Rab8 is not needed for budding or motility but for docking and fusion. Rab6 promotes the recruitment of Rab8 to exocytotic vesicles and contributes to the interaction of these vesicles with the cortex through direct binding to ELKS. Rab8 also interacts with ELKS at cortical sites through MICAL3, a member of the MICAL family of flavoprotein monooxygenases, promoting docking and fusion of vesicles (Grigoriev et al. 2011).

Rab10 has also been reported to regulate Golgi-to-plasma membrane transport (Schuck et al. 2007; Wang et al. 2010). It localizes to the Golgi during early cell polarization, and it has been suggested to have a role in the transport from the Golgi to the basolateral membrane at early stages of epithelial polarization. Indeed, the expression of its constitutively active mutant inhibits the transport from the Golgi to the plasma membrane and missorts basolateral cargo to the apical membrane (Schuck et al. 2007).

Rab10 regulation of post-Golgi transport is mediated by its interaction with myosin Vb. Inhibition of this interaction or downregulation of myosin Vb expression prevents the fission of Rab10 vesicles from the trans-Golgi and reduces the number of Rab10 transport carriers (Liu et al. 2013b). However, Rab10 has also been reported to function in different trafficking pathways, including endosomal sorting in polarized cells, endo-phagocytic processes, Golgi-to-ER transport, as well as in the regulation of ER dynamics (Babbey et al. 2006; Cardoso et al. 2010; Chua and Tang 2018; Deen et al. 2014; English and Voeltz 2013; Galea et al. 2015; Liu et al. 2018; Wang et al. 2016). How this small GTPase can contribute to the regulation of several different pathways is still unclear even though one possible explanation is that Rab10's roles are dependent on the specificity of the effectors present in different cell types or cell stages.

6.8 Rab Proteins Involved in Golgi Maintenance

Several Rab proteins, including the previously described Rab6, Rab33, Rab1, and Rab2, have a role not only in membrane trafficking but also in the regulation of Golgi organization and maintenance. In a siRNAs screening for human Rabs influencing Golgi morphology, knockdown of Rab1, Rab2, Rab6, or Rab8 caused Golgi fragmentation (Aizawa and Fukuda 2015). In addition also Rab43, Rab18, Rab30, and Rab41 have been reported to influence Golgi organization (Dejgaard et al. 2008; Haas et al. 2007; Kelly et al. 2012; Liu et al. 2013a).

6.8.1   *Rab30*

Rab30 is mainly localized to the Golgi apparatus where it is required for the structural integrity of this organelle (de Leeuw et al. 1998; Kelly et al. 2012). Depletion of Rab30 indeed alters the morphology of the Golgi, by fragmenting the cisternae (Kelly et al. 2012). As Rab30 interacts with *D. melanogaster* orthologs of several human golgin proteins such as Golgin-245, Golgin-97, GM130, GCC88, p115, and Bicaudal D (BicD), it has been suggested that Rab30's ability to regulate Golgi structural integrity is dependent on its association with these golgins (Kelly et al. 2012; Sinka et al. 2008). According to this model, loss of Rab30 prevents the recruitment or the function of the golgins at the Golgi cisternae resulting in the overall loss of Golgi structural integrity.

The function of Rab30 at the Golgi seems to be restricted only to the maintenance of this organelle architecture as perturbation of Rab30 function does not affect anterograde or retrograde trafficking through the Golgi (Kelly et al. 2012). However, it has been more recently shown that Rab30 also interacts with all eight subunits of the exocyst, a complex that mediates tethering of Golgi-derived vesicles at the plasma membrane (Gillingham et al. 2014), but the functional role of this interaction needs to be further investigated.

6.8.2   *Rab41*

Rab41 is also termed Rab6d as it shares 60% similarity with the other members of the Rab6 family (Goud et al. 2018; Pereira-Leal and Seabra 2001). In contrast to Rab6 depletion that increases the number and the continuity of Golgi cisternae, Rab41 depletion affects Golgi organization scattering the Golgi ribbon into punctate elements (Liu et al. 2013a; Storrie et al. 2012). As depletion of Rab6 results in a more compact Golgi ribbon, while depletion of Rab41 disperses the Golgi, it has been suggested that these small GTPases contribute to Golgi ribbon organization by recruiting opposing motors (Liu and Storrie 2015).

Rab6 recruits the effectors myosin II and the plus-end motor kinesin Kif20A, also involved in the regulation of Golgi organization (Liu and Storrie 2015; Majeed et al. 2014). On the other side, Rab41, by interacting with dynactin 6, recruits cytoplasmic dynein to Golgi membranes (Liu et al. 2016). The recruitment of the minus-end motor dynein is important for Golgi organization as it regulates the afferent transport of membrane cargo (Jaarsma and Hoogenraad 2015). Rab41 interacts also with syntaxin 8 and depletion of dynactin 6, and syntaxin 8 results in Golgi fragmentation, a phenotype similar to the one observed upon Rab41 depletion, indicating that these effectors contribute to Rab41 function in the maintenance of Golgi organization (Liu et al. 2016).

Furthermore, Rab41 regulates ER-to-Golgi trafficking, as its silencing or the expression of a dominant-negative mutant partially inhibits ER-to-Golgi transport of Vesicular stomatitis virus G (VSV-G) protein (Liu et al. 2013a).

6.8.3 Other Rabs That Influence Golgi Organization

Similarly to Rab41, other Rabs regulating trafficking between ER and Golgi, such as Rab1, Rab2, Rab18, and Rab43, influence Golgi organization. Indeed, depletion or inactivation of any of these Rab proteins fragments or disperses the Golgi, suggesting that a continuous and regulated anterograde transport is needed to maintain Golgi organization (Dejgaard et al. 2008; Haas et al. 2007; Tisdale et al. 1992).

The dominant-negative (GDP-locked) mutant of Rab43 redistributes Golgi to ER exit sites without blocking the secretory pathway, indicating a function in the anterograde ER-Golgi transport route (Dejgaard et al. 2008). This function has been recently confirmed, as Rab43, by directly interacting with G-protein-coupled receptors (GPCRs), the largest superfamily of cell-surface signaling proteins, regulates the transport of these receptors from the ER to the Golgi (Li et al. 2017).

On the other side, overexpression of Rab18 or its depletion disrupts the Golgi complex and reduces secretion, indicating the involvement of this small GTPase in the retrograde Golgi-to-ER transport pathway (Dejgaard et al. 2008). The regulation of opposite transport pathways for these two Rabs is consistent also with their different intracellular localization: while Rab18 is ER-associated, Rab43 is mostly localized at the cis- and medial- Golgi (Cox et al. 2016; Dejgaard et al. 2008; Gerondopoulos et al. 2014; Jayson et al. 2018; Martin et al. 2005). Nevertheless, both Rab18 and Rab43 are necessary for proper Golgi ribbon organization. Overexpression or downregulation of Rab18, as well as the expression of the dominant-negative mutant of Rab43, disperses the Golgi complex (Dejgaard et al. 2008; Haas et al. 2007).

6.9 Conclusions

Rab GTPases are master regulators of the intracellular membrane trafficking, and, as such, is not surprising that several of them regulate different transport pathways toward, through, and from the Golgi apparatus (Fig. 6.3). What is maybe more remarkable is that a few of them are also involved in other Golgi-associated functions, as, for example, Golgi organization and maintenance. Despite, at least for some Rabs, their role in Golgi organization which can be explained as the result

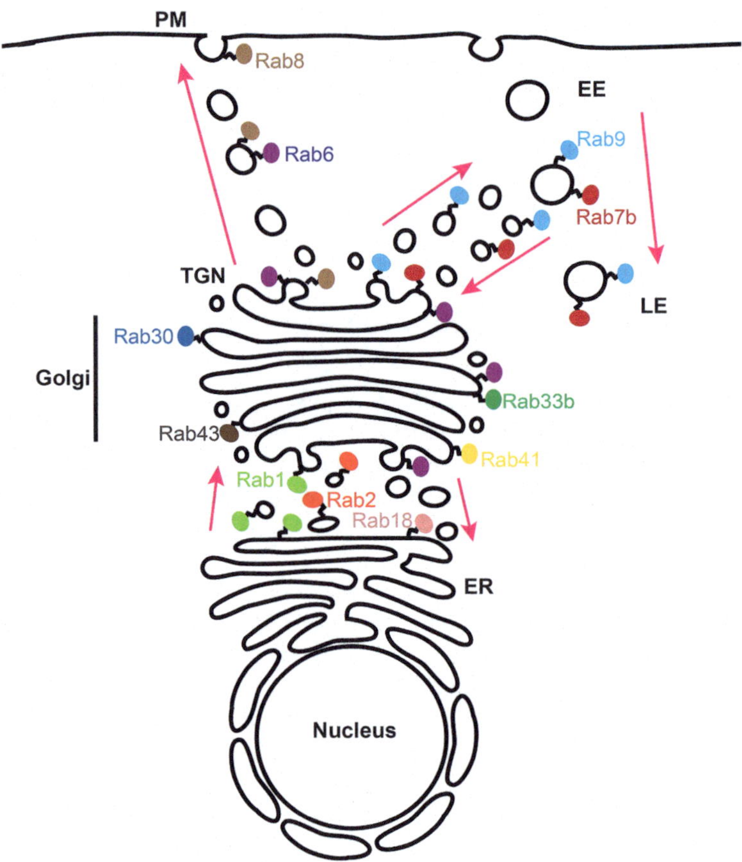

Fig. 6.3 Localization and transport pathways of major Golgi-associated Rab proteins. *ER* endoplasmic reticulum, *TGN* trans Golgi network, *LE* late endosome, *EE* early endosome, *PM* plasma membrane

of an effect on vesicular trafficking, e.g., related to the recycling of transport machineries, for other Rab proteins this remains an open question. It could be a result of interactions with specific Golgi-associated structural proteins as, for example, golgins, or with microtubules and motor proteins (Yadav and Linstedt 2011). According to this hypothesis, the Rab-dependent Golgi ribbon organization would be the result of a balance between the recruitment of minus-end and plus-end directed motors (Liu and Storrie 2015; Majeed et al. 2014). Given the importance of Golgi biogenesis and organization during the cell cycle, further studies are needed to fully elucidate how the Golgi-associated Rab proteins coordinate multiple processes in order to regulate Golgi structure, functions, and dynamics.

Acknowledgments C. Progida is supported by grants from the Research Council of Norway (grant 287560), the Norwegian Cancer Society (grant 198094), the Anders Jahre Foundation, and the S. G. Sønneland Foundation.

References

Aizawa M, Fukuda M (2015) Small GTPase Rab2B and its specific binding protein Golgi-associated Rab2B interactor-like 4 (GARI-L4) regulate Golgi morphology. J Biol Chem 290:22250–22261

Ali BR, Wasmeier C, Lamoreux L, Strom M, Seabra MC (2004) Multiple regions contribute to membrane targeting of Rab GTPases. J Cell Sci 117:6401–6412

Allan BB, Moyer BD, Balch WE (2000) Rab1 recruitment of p115 into a cis-SNARE complex: programming budding COPII vesicles for fusion. Science 289:444–448

Alshammari MJ, Al-Otaibi L, Alkuraya FS (2012) Mutation in RAB33B, which encodes a regulator of retrograde Golgi transport, defines a second Dyggve-Melchior-Clausen locus. J Med Genet 49:455–461

Andres DA, Seabra MC, Brown MS, Armstrong SA, Smeland TE, Cremers FP, Goldstein JL (1993) cDNA cloning of component A of Rab geranylgeranyl transferase and demonstration of its role as a Rab escort protein. Cell 73:1091–1099

Ang AL, Taguchi T, Francis S, Folsch H, Murrells LJ, Pypaert M, Warren G, Mellman I (2004) Recycling endosomes can serve as intermediates during transport from the Golgi to the plasma membrane of MDCK cells. J Cell Biol 167:531–543

Antonny B, Burd C, De Camilli P, Chen E, Daumke O, Faelber K, Ford M, Frolov VA, Frost A, Hinshaw JE, Kirchhausen T, Kozlov MM, Lenz M, Low HH, McMahon H, Merrifield C, Pollard TD, Robinson PJ, Roux A, Schmid S (2016) Membrane fission by dynamin: what we know and what we need to know. EMBO J 35:2270–2284

Babbey CM, Ahktar N, Wang E, Chen CC, Grant BD, Dunn KW (2006) Rab10 regulates membrane transport through early endosomes of polarized Madin-Darby canine kidney cells. Mol Biol Cell 17:3156–3175

Baker RW, Hughson FM (2016) Chaperoning SNARE assembly and disassembly. Nat Rev Mol Cell Biol 17:465–479

Bard F, Casano L, Mallabiabarrena A, Wallace E, Saito K, Kitayama H, Guizzunti G, Hu Y, Wendler F, Dasgupta R, Perrimon N, Malhotra V (2006) Functional genomics reveals genes involved in protein secretion and Golgi organization. Nature 439:604–607

Barrowman J, Bhandari D, Reinisch K, Ferro-Novick S (2010) TRAPP complexes in membrane traffic: convergence through a common Rab. Nat Rev Mol Cell Biol 11:759–763

Blumer J, Rey J, Dehmelt L, Mazel T, Wu YW, Bastiaens P, Goody RS, Itzen A (2013) RabGEFs are a major determinant for specific Rab membrane targeting. J Cell Biol 200:287–300

Bombardier JP, Munson M (2015) Three steps forward, two steps back: mechanistic insights into the assembly and disassembly of the SNARE complex. Curr Opin Chem Biol 29:66–71

Bonifacino JS, Hierro A (2011) Transport according to GARP: receiving retrograde cargo at the trans-Golgi network. Trends Cell Biol 21:159–167

Borg Distefano M, Hofstad Haugen L, Wang Y, Perdreau-Dahl H, Kjos I, Jia D, Morth JP, Neefjes J, Bakke O, Progida C (2018) TBC1D5 controls the GTPase cycle of Rab7b. J Cell Sci 131. https://doi.org/10.1242/jcs.216630

Borg M, Bakke O, Progida C (2014) A novel interaction between Rab7b and actomyosin reveals a dual role in intracellular transport and cell migration. J Cell Sci 127:4927–4939

Borgese N (2016) Getting membrane proteins on and off the shuttle bus between the endoplasmic reticulum and the Golgi complex. J Cell Sci 129:1537–1545

Bucci C, Bakke O, Progida C (2010) Rab7b and receptors trafficking. Commun Integr Biol 3:401–404
Bucci C, Bakke O, Progida C (2012) Charcot-Marie-Tooth disease and intracellular traffic. Prog Neurobiol 99:191–225
Bulankina AV, Deggerich A, Wenzel D, Mutenda K, Wittmann JG, Rudolph MG, Burger KN, Honing S (2009) TIP47 functions in the biogenesis of lipid droplets. J Cell Biol 185:641–655
Cai H, Reinisch K, Ferro-Novick S (2007) Coats, tethers, Rabs, and SNAREs work together to mediate the intracellular destination of a transport vesicle. Dev Cell 12:671–682
Cantalupo G, Alifano P, Roberti V, Bruni CB, Bucci C (2001) Rab-interacting lysosomal protein (RILP): the Rab7 effector required for transport to lysosomes. EMBO J 20:683–693
Cardoso CM, Jordao L, Vieira OV (2010) Rab10 regulates phagosome maturation and its overexpression rescues Mycobacterium-containing phagosomes maturation. Traffic 11:221–235
Carroll KS, Hanna J, Simon I, Krise J, Barbero P, Pfeffer SR (2001) Role of Rab9 GTPase in facilitating receptor recruitment by TIP47. Science 292:1373–1376
Casey PJ, Seabra MC (1996) Protein prenyltransferases. J Biol Chem 271:5289–5292
Ceresa BP, Bahr SJ (2006) rab7 activity affects epidermal growth factor: epidermal growth factor receptor degradation by regulating endocytic trafficking from the late endosome. J Biol Chem 281:1099–1106
Chavrier P, Parton RG, Hauri HP, Simons K, Zerial M (1990a) Localization of low molecular weight GTP binding proteins to exocytic and endocytic compartments. Cell 62:317–329
Chavrier P, Vingron M, Sander C, Simons K, Zerial M (1990b) Molecular cloning of YPT1/SEC4-related cDNAs from an epithelial cell line. Mol Cell Biol 10:6578–6585
Chavrier P, Gorvel JP, Stelzer E, Simons K, Gruenberg J, Zerial M (1991) Hypervariable C-terminal domain of rab proteins acts as a targeting signal. Nature 353:769–772
Chen W, Feng Y, Chen D, Wandinger-Ness A (1998) Rab11 is required for trans-golgi network-to-plasma membrane transport and a preferential target for GDP dissociation inhibitor. Mol Biol Cell 9:3241–3257
Chen L, Hu J, Yun Y, Wang T (2010) Rab36 regulates the spatial distribution of late endosomes and lysosomes through a similar mechanism to Rab34. Mol Membr Biol 27:23–30
Christoforidis S, McBride HM, Burgoyne RD, Zerial M (1999a) The Rab5 effector EEA1 is a core component of endosome docking. Nature 397:621–625
Christoforidis S, Miaczynska M, Ashman K, Wilm M, Zhao L, Yip SC, Waterfield MD, Backer JM, Zerial M (1999b) Phosphatidylinositol-3-OH kinases are Rab5 effectors. Nat Cell Biol 1:249–252
Chua CEL, Tang BL (2014) Engagement of the small GTPase Rab31 protein and its effector, early endosome antigen 1, is important for trafficking of the ligand-bound epidermal growth factor receptor from the early to the late endosome. J Biol Chem 289:12375–12389
Chua CEL, Tang BL (2018) Rab 10-a traffic controller in multiple cellular pathways and locations. J Cell Physiol 233:6483–6494
Collins RN (2003) "Getting it on"–GDI displacement and small GTPase membrane recruitment. Mol Cell 12:1064–1066
Cottam NP, Ungar D (2012) Retrograde vesicle transport in the Golgi. Protoplasma 249:943–955
Cox JV, Kansal R, Whitt MA (2016) Rab43 regulates the sorting of a subset of membrane protein cargo through the medial Golgi. Mol Biol Cell 27:1834–1844
Day PM, Thompson CD, Schowalter RM, Lowy DR, Schiller JT (2013) Identification of a role for the trans-Golgi network in HPV16 pseudovirus infection. J Virol 87:3862–3870
de Graaf P, Zwart WT, van Dijken RA, Deneka M, Schulz TK, Geijsen N, Coffer PJ, Gadella BM, Verkleij AJ, van der Sluijs P, van Bergen en Henegouwen PM (2004) Phosphatidylinositol 4-kinasebeta is critical for functional association of rab11 with the Golgi complex. Mol Biol Cell 15:2038–2047
de Leeuw HP, Koster PM, Calafat J, Janssen H, van Zonneveld AJ, van Mourik JA, Voorberg J (1998) Small GTP-binding proteins in human endothelial cells. Br J Haematol 103:15–19

Deen AJ, Rilla K, Oikari S, Karna R, Bart G, Hayrinen J, Bathina AR, Ropponen A, Makkonen K, Tammi RH, Tammi MI (2014) Rab10-mediated endocytosis of the hyaluronan synthase HAS3 regulates hyaluronan synthesis and cell adhesion to collagen. J Biol Chem 289:8375–8389

Dejgaard SY, Murshid A, Erman A, Kizilay O, Verbich D, Lodge R, Dejgaard K, Ly-Hartig TB, Pepperkok R, Simpson JC, Presley JF (2008) Rab18 and Rab43 have key roles in ER-Golgi trafficking. J Cell Sci 121:2768–2781

Del Nery E, Miserey-Lenkei S, Falguières T, Nizak C, Johannes L, Perez F, Goud B (2006) Rab6A and Rab6A′ GTPases play non-overlapping roles in membrane trafficking. Traffic 7:394–407

Diaz E, Schimmoller F, Pfeffer SR (1997) A novel Rab9 effector required for endosome-to-TGN transport. J Cell Biol 138:283–290

Dippold HC, Ng MM, Farber-Katz SE, Lee SK, Kerr ML, Peterman MC, Sim R, Wiharto PA, Galbraith KA, Madhavarapu S, Fuchs GJ, Meerloo T, Farquhar MG, Zhou H, Field SJ (2009) GOLPH3 bridges phosphatidylinositol-4- phosphate and actomyosin to stretch and shape the Golgi to promote budding. Cell 139:337–351

Dirac-Svejstrup AB, Sumizawa T, Pfeffer SR (1997) Identification of a GDI displacement factor that releases endosomal Rab GTPases from Rab-GDI. EMBO J 16:465–472

Dupuis N, Lebon S, Kumar M, Drunat S, Graul-Neumann LM, Gressens P, El Ghouzzi V (2013) A novel RAB33B mutation in Smith-McCort dysplasia. Hum Mutat 34:283–286

Echard A, Opdam FJ, de Leeuw HJ, Jollivet F, Savelkoul P, Hendriks W, Voorberg J, Goud B, Fransen JA (2000) Alternative splicing of the human Rab6A gene generates two close but functionally different isoforms. Mol Biol Cell 11:3819–3833

English AR, Voeltz GK (2013) Rab10 GTPase regulates ER dynamics and morphology. Nat Cell Biol 15:169–178

Epp N, Rethmeier R, Kramer L, Ungermann C (2011) Membrane dynamics and fusion at late endosomes and vacuoles—Rab regulation, multisubunit tethering complexes and SNAREs. Eur J Cell Biol 90:779–785

Espinosa EJ, Calero M, Sridevi K, Pfeffer SR (2009) RhoBTB3: a Rho GTPase-family ATPase required for endosome to Golgi transport. Cell 137:938–948

Faini M, Beck R, Wieland FT, Briggs JA (2013) Vesicle coats: structure, function, and general principles of assembly. Trends Cell Biol 23:279–288

Ferraro F, Kriston-Vizi J, Metcalf DJ, Martin-Martin B, Freeman J, Burden JJ, Westmoreland D, Dyer CE, Knight AE, Ketteler R, Cutler DF (2014) A two-tier Golgi-based control of organelle size underpins the functional plasticity of endothelial cells. Dev Cell 29:292–304

Galea G, Bexiga MG, Panarella A, O'Neill ED, Simpson JC (2015) A high-content screening microscopy approach to dissect the role of Rab proteins in Golgi-to-ER retrograde trafficking. J Cell Sci 128:2339–2349

Gallwitz D, Donath C, Sander C (1983) A yeast gene encoding a protein homologous to the human c-has/bas proto-oncogene product. Nature 306:704–707

Ganley IG, Carroll K, Bittova L, Pfeffer S (2004) Rab9 GTPase regulates late endosome size and requires effector interaction for its stability. Mol Biol Cell 15:5420–5430

Gavriljuk K, Itzen A, Goody RS, Gerwert K, Kotting C (2013) Membrane extraction of Rab proteins by GDP dissociation inhibitor characterized using attenuated total reflection infrared spectroscopy. Proc Natl Acad Sci USA 110:13380–13385

Gerondopoulos A, Bastos RN, Yoshimura S-I, Anderson R, Carpanini S, Aligianis I, Handley MT, Barr FA (2014) Rab18 and a Rab18 GEF complex are required for normal ER structure. J Cell Biol 205:707–720

Gidon A, Bardin S, Cinquin B, Boulanger J, Waharte F, Heliot L, de la Salle H, Hanau D, Kervrann C, Goud B, Salamero J (2012) A Rab11A/myosin Vb/Rab11-FIP2 complex frames two late recycling steps of langerin from the ERC to the plasma membrane. Traffic 13:815–833

Gillingham AK, Munro S (2016) Finding the Golgi: Golgin coiled-coil proteins show the way. Trends Cell Biol 26:399–408

Gillingham AK, Sinka R, Torres IL, Lilley KS, Munro S (2014) Toward a comprehensive map of the effectors of rab GTPases. Dev Cell 31:358–373

Girod A, Storrie B, Simpson JC, Johannes L, Goud B, Roberts LM, Lord JM, Nilsson T, Pepperkok R (1999) Evidence for a COP-I-independent transport route from the Golgi complex to the endoplasmic reticulum. Nat Cell Biol 1:423–430

Goitre L, Trapani E, Trabalzini L, Retta SF (2014) The Ras superfamily of small GTPases: the unlocked secrets. Methods Mol Biol 1120:1–18

Goldenberg NM, Grinstein S, Silverman M (2007) Golgi-bound Rab34 is a novel member of the secretory pathway. Mol Biol Cell 18:4762–4771

Goody RS, Muller MP, Wu YW (2017) Mechanisms of action of Rab proteins, key regulators of intracellular vesicular transport. Biol Chem 398:565–575

Goud B, Salminen A, Walworth NC, Novick PJ (1988) A GTP-binding protein required for secretion rapidly associates with secretory vesicles and the plasma membrane in yeast. Cell 53:753–768

Goud B, Liu S, Storrie B (2018) Rab proteins as major determinants of the Golgi complex structure. Small GTPases 9:66–75

Grigoriev I, Splinter D, Keijzer N, Wulf PS, Demmers J, Ohtsuka T, Modesti M, Maly IV, Grosveld F, Hoogenraad CC, Akhmanova A (2007) Rab6 regulates transport and targeting of exocytotic carriers. Dev Cell 13:305–314

Grigoriev I, Yu KL, Martinez-Sanchez E, Serra-Marques A, Smal I, Meijering E, Demmers J, Peranen J, Pasterkamp RJ, van der Sluijs P, Hoogenraad CC, Akhmanova A (2011) Rab6, Rab8, and MICAL3 cooperate in controlling docking and fusion of exocytotic carriers. Curr Biol 21:967–974

Gutkowska M, Swiezewska E (2012) Structure, regulation and cellular functions of Rab geranylgeranyl transferase and its cellular partner Rab Escort Protein. Mol Membr Biol 29:243–256

Haas AK, Yoshimura S, Stephens DJ, Preisinger C, Fuchs E, Barr FA (2007) Analysis of GTPase-activating proteins: Rab1 and Rab43 are key Rabs required to maintain a functional Golgi complex in human cells. J Cell Sci 120:2997–3010

Hanna J, Carroll K, Pfeffer SR (2002) Identification of residues in TIP47 essential for Rab9 binding. Proc Natl Acad Sci USA 99:7450–7454

Hasanagic M, Waheed A, Eissenberg JC (2015) Different pathways to the lysosome: sorting out alternatives. Int Rev Cell Mol Biol 320:75–101

Horgan CP, McCaffrey MW (2011) Rab GTPases and microtubule motors. Biochem Soc Trans 39:1202–1206

Horgan CP, Hanscom SR, Jolly RS, Futter CE, McCaffrey MW (2010) Rab11-FIP3 links the Rab11 GTPase and cytoplasmic dynein to mediate transport to the endosomal-recycling compartment. J Cell Sci 123:181–191

Huber LA, Pimplikar S, Parton RG, Virta H, Zerial M, Simons K (1993) Rab8, a small GTPase involved in vesicular traffic between the TGN and the basolateral plasma membrane. J Cell Biol 123:35–45

Itoh T, Fujita N, Kanno E, Yamamoto A, Yoshimori T, Fukuda M (2008) Golgi-resident small GTPase Rab33B interacts with Atg16L and modulates autophagosome formation. Mol Biol Cell 19:2916–2925

Jaarsma D, Hoogenraad CC (2015) Cytoplasmic dynein and its regulatory proteins in Golgi pathology in nervous system disorders. Front Neurosci 9:397

Jayson CBK, Arlt H, Fischer AW, Lai ZW, Farese RV, Walther TC, Barr FA (2018) Rab18 is not necessary for lipid droplet biogenesis or turnover in human mammary carcinoma cells. Mol Biol Cell 29:2045–2054

Jedd G, Richardson C, Litt R, Segev N (1995) The Ypt1 GTPase is essential for the first two steps of the yeast secretory pathway. J Cell Biol 131:583–590

Jia D, Zhang JS, Li F, Wang J, Deng Z, White MA, Osborne DG, Phillips-Krawczak C, Gomez TS, Li H, Singla A, Burstein E, Billadeau DD, Rosen MK (2016) Structural and mechanistic insights into regulation of the retromer coat by TBC1d5. Nat Commun 7:13305

Jiang S, Storrie B (2005) Cisternal rab proteins regulate Golgi apparatus redistribution in response to hypotonic stress. Mol Biol Cell 16:2586–2596

Jordens I, Fernandez-Borja M, Marsman M, Dusseljee S, Janssen L, Calafat J, Janssen H, Wubbolts R, Neefjes J (2001) The Rab7 effector protein RILP controls lysosomal transport by inducing the recruitment of dynein-dynactin motors. Curr Biol 11:1680–1685

Junutula JR, De Maziére AM, Peden AA, Ervin KE, Advani RJ, van Dijk SM, Klumperman J, Scheller RH (2004) Rab14 is involved in membrane trafficking between the Golgi complex and endosomes. Mol Biol Cell 15:2218–2229

Kelly EE, Giordano F, Horgan CP, Jollivet F, Raposo G, McCaffrey MW (2012) Rab30 is required for the morphological integrity of the Golgi apparatus. Biol Cell 104:84–101

Kirchhausen T, Owen D, Harrison SC (2014) Molecular structure, function, and dynamics of clathrin-mediated membrane traffic. Cold Spring Harb Perspect Biol 6:a016725

Kitt KN, Hernández-Deviez D, Ballantyne SD, Spiliotis ET, Casanova JE, Wilson JM (2008) Rab14 regulates apical targeting in polarized epithelial cells. Traffic 9:1218–1231

Kjos I, Vestre K, Guadagno NA, Borg Distefano M, Progida C (2018) Rab and Arf proteins at the crossroad between membrane transport and cytoskeleton dynamics. Biochim Biophys Acta 1865:1397–1409

Kloer DP, Rojas R, Ivan V, Moriyama K, van Vlijmen T, Murthy N, Ghirlando R, van der Sluijs P, Hurley JH, Bonifacino JS (2010) Assembly of the biogenesis of lysosome-related organelles complex-3 (BLOC-3) and its interaction with Rab9. J Biol Chem 285:7794–7804

Kobayashi S, Hida Y, Ishizaki H, Inoue E, Tanaka-Okamoto M, Yamasaki M, Miyazaki T, Fukaya M, Kitajima I, Takai Y, Watanabe M, Ohtsuka T, Manabe T (2016) The active zone protein CAST regulates synaptic vesicle recycling and quantal size in the mouse hippocampus. Eur J Neurosci 44:2272–2284

Krzewski K, Cullinane AR (2013) Evidence for defective Rab GTPase-dependent cargo traffic in immune disorders. Exp Cell Res 319:2360–2367

Kucera A, Bakke O, Progida C (2016a) The multiple roles of Rab9 in the endolysosomal system. Commun Integr Biol 9:e1204498

Kucera A, Borg Distefano M, Berg-Larsen A, Skjeldal F, Repnik U, Bakke O, Progida C (2016b) Spatiotemporal resolution of Rab9 and CI-MPR dynamics in the endocytic pathway. Traffic 17:211–229

Lee PL, Ohlson MB, Pfeffer SR (2015) Rab6 regulation of the kinesin family KIF1C motor domain contributes to Golgi tethering. Elife 4. https://doi.org/10.7554/eLife.06029

Li F, Yi L, Zhao L, Itzen A, Goody RS, Wu YW (2014) The role of the hypervariable C-terminal domain in Rab GTPases membrane targeting. Proc Natl Acad Sci USA 111:2572–2577

Li C, Wei Z, Fan Y, Huang W, Su Y, Li H, Dong Z, Fukuda M, Khater M, Wu G (2017) The GTPase Rab43 controls the anterograde ER-Golgi trafficking and sorting of GPCRs. Cell Rep 21:1089–1101

Liu S, Storrie B (2012) Are Rab proteins the link between Golgi organization and membrane trafficking? Cell Mol Life Sci 69:4093–4106

Liu S, Storrie B (2015) How Rab proteins determine Golgi structure. Int Rev Cell Mol Biol 315:1–22

Liu S, Hunt L, Storrie B (2013a) Rab41 is a novel regulator of Golgi apparatus organization that is needed for ER-to-Golgi trafficking and cell growth. PLoS One 8:e71886

Liu Y, Xu XH, Chen Q, Wang T, Deng CY, Song BL, Du JL, Luo ZG (2013b) Myosin Vb controls biogenesis of post-Golgi Rab10 carriers during axon development. Nat Commun 4:2005

Liu S, Majeed W, Kudlyk T, Lupashin V, Storrie B (2016) Identification of Rab41/6d effectors provides an explanation for the differential effects of Rab41/6d and Rab6a/a′ on Golgi organization. Front Cell Dev Biol 4:13

Liu H, Wang S, Hang W, Gao J, Zhang W, Cheng Z, Yang C, He J, Zhou J, Chen J, Shi A (2018) LET-413/Erbin acts as a RAB-5 effector to promote RAB-10 activation during endocytic recycling. J Cell Biol 217:299–314

Lock JG, Stow JL (2005) Rab11 in recycling endosomes regulates the sorting and basolateral transport of E-cadherin. Mol Biol Cell 16:1744–1755

Lombardi D, Soldati T, Riederer MA, Goda Y, Zerial M, Pfeffer SR (1993) Rab9 functions in transport between late endosomes and the trans Golgi network. EMBO J 12:677–682

Lorincz P, Toth S, Benko P, Lakatos Z, Boda A, Glatz G, Zobel M, Bisi S, Hegedus K, Takats S, Scita G, Juhasz G (2017) Rab2 promotes autophagic and endocytic lysosomal degradation. J Cell Biol 216:1937–1947

Lupashin VV, Waters MG (1997) t-SNARE activation through transient interaction with a rab-like guanosine triphosphatase. Science 276:1255–1258

Lurick A, Gao J, Kuhlee A, Yavavli E, Langemeyer L, Perz A, Raunser S, Ungermann C (2017) Multivalent Rab interactions determine tether-mediated membrane fusion. Mol Biol Cell 28:322–332

Mahanty S, Ravichandran K, Chitirala P, Prabha J, Jani RA, Setty SR (2016) Rab9A is required for delivery of cargo from recycling endosomes to melanosomes. Pigment Cell Melanoma Res 29:43–59

Majeed W, Liu S, Storrie B (2014) Distinct sets of Rab6 effectors contribute to ZW10—and COG-dependent Golgi homeostasis. Traffic 15:630–647

Mallard F, Tang BL, Galli T, Tenza D, Saint-Pol A, Yue X, Antony C, Hong W, Goud B, Johannes L (2002) Early/recycling endosomes-to-TGN transport involves two SNARE complexes and a Rab6 isoform. J Cell Biol 156:653–664

Martin S, Driessen K, Nixon SJ, Zerial M, Parton RG (2005) Regulated Localization of Rab18 to lipid droplets: effects of lipolytic stimulation and inhibition of lipid droplet catabolism. J Biol Chem 280:42325–42335

Martinez O, Schmidt A, Salamero J, Hoflack B, Roa M, Goud B (1994) The small GTP-binding protein rab6 functions in intra-Golgi transport. J Cell Biol 127:1575–1588

Martinez O, Antony C, Pehau-Arnaudet G, Berger EG, Salamero J, Goud B (1997) GTP-bound forms of rab6 induce the redistribution of Golgi proteins into the endoplasmic reticulum. Proc Natl Acad Sci USA 94:1828–1833

Marubashi S, Shimada H, Fukuda M, Ohbayashi N (2016) RUTBC1 functions as a GTPase-activating protein for Rab32/38 and regulates melanogenic enzyme trafficking in melanocytes. J Biol Chem 291:1427–1440

Matanis T, Akhmanova A, Wulf P, Del Nery E, Weide T, Stepanova T, Galjart N, Grosveld F, Goud B, De Zeeuw CI, Barnekow A, Hoogenraad CC (2002) Bicaudal-D regulates COPI-independent Golgi-ER transport by recruiting the dynein-dynactin motor complex. Nat Cell Biol 4:986–992

Matsuto M, Kano F, Murata M (2015) Reconstitution of the targeting of Rab6A to the Golgi apparatus in semi-intact HeLa cells: a role of BICD2 in stabilizing Rab6A on Golgi membranes and a concerted role of Rab6A/BICD2 interactions in Golgi-to-ER retrograde transport. Biochim Biophys Acta 1853:2592–2609

Mayer T, Touchot N, Elazar Z (1996) Transport between cis and medial Golgi cisternae requires the function of the Ras-related protein Rab6. J Biol Chem 271:16097–16103

Micaroni M, Stanley AC, Khromykh T, Venturato J, Wong CX, Lim JP, Marsh BJ, Storrie B, Gleeson PA, Stow JL (2013) Rab6a/a′ are important Golgi regulators of pro-inflammatory TNF secretion in macrophages. PLoS One 8:e57034

Miller VJ, Ungar D (2012) Re'COG'nition at the Golgi. Traffic 13:891–897

Miserey-Lenkei S, Waharte F, Boulet A, Cuif MH, Tenza D, El Marjou A, Raposo G, Salamero J, Heliot L, Goud B, Monier S (2007) Rab6-interacting protein 1 links Rab6 and Rab11 function. Traffic 8:1385–1403

Miserey-Lenkei S, Chalancon G, Bardin S, Formstecher E, Goud B, Echard A (2010) Rab and actomyosin-dependent fission of transport vesicles at the Golgi complex. Nat Cell Biol 12:645–654

Miserey-Lenkei S, Bousquet H, Pylypenko O, Bardin S, Dimitrov A, Bressanelli G, Bonifay R, Fraisier V, Guillou C, Bougeret C, Houdusse A, Echard A, Goud B (2017) Coupling fission and exit of RAB6 vesicles at Golgi hotspots through kinesin-myosin interactions. Nat Commun 8:1254

Monier S, Jollivet F, Janoueix-Lerosey I, Johannes L, Goud B (2002) Characterization of novel Rab6-interacting proteins involved in endosome-to-TGN transport. Traffic 3:289–297

Morimoto S, Nishimura N, Terai T, Manabe S, Yamamoto Y, Shinahara W, Miyake H, Tashiro S, Shimada M, Sasaki T (2005) Rab13 mediates the continuous endocytic recycling of occludin to the cell surface. J Biol Chem 280:2220–2228

Moyer BD, Allan BB, Balch WE (2001) Rab1 interaction with a GM130 effector complex regulates COPII vesicle cis-Golgi tethering. Traffic 2:268–276

Muller MP, Goody RS (2017) Molecular control of Rab activity by GEFs, GAPs and GDI. Small GTPases 9:5–21

Nakazawa H, Sada T, Toriyama M, Tago K, Sugiura T, Fukuda M, Inagaki N (2012) Rab33a mediates anterograde vesicular transport for membrane exocytosis and axon outgrowth. J Neurosci 32:12712–12725

Ng EL, Wang Y, Tang BL (2007) Rab22B's role in trans-Golgi network membrane dynamics. Biochem Biophys Res Commun 361:751–757

Ng EL, Ng JJ, Liang F, Tang BL (2009) Rab22B is expressed in the CNS astroglia lineage and plays a role in epidermal growth factor receptor trafficking in A431 cells. J Cell Physiol 221:716–728

Ni X, Ma Y, Cheng H, Jiang M, Guo L, Ji C, Gu S, Cao Y, Xie Y, Mao Y (2002) Molecular cloning and characterization of a novel human Rab (Rab2B) gene. J Hum Genet 47:548–551

Nishida Y, Arakawa S, Fujitani K, Yamaguchi H, Mizuta T, Kanaseki T, Komatsu M, Otsu K, Tsujimoto Y, Shimizu S (2009) Discovery of Atg5/Atg7-independent alternative macroautophagy. Nature 461:654–658

Nokes RL, Fields IC, Collins RN, Fölsch H (2008) Rab13 regulates membrane trafficking between TGN and recycling endosomes in polarized epithelial cells. J Cell Biol 182:845–853

Nottingham RM, Ganley IG, Barr FA, Lambright DG, Pfeffer SR (2011) RUTBC1 protein, a Rab9A effector that activates GTP hydrolysis by Rab32 and Rab33B proteins. J Biol Chem 286:33213–33222

Nottingham RM, Pusapati GV, Ganley IG, Barr FA, Lambright DG, Pfeffer SR (2012) RUTBC2 protein, a Rab9A effector and GTPase-activating protein for Rab36. J Biol Chem 287:22740–22748

Nuoffer C, Davidson HW, Matteson J, Meinkoth J, Balch WE (1994) A GDP-bound of rab1 inhibits protein export from the endoplasmic reticulum and transport between Golgi compartments. J Cell Biol 125:225–237

Olkkonen VM, Ikonen E (2006) When intracellular logistics fails—genetic defects in membrane trafficking. J Cell Sci 119:5031–5045

Opdam FJ, Echard A, Croes HJ, van den Hurk JA, van de Vorstenbosch RA, Ginsel LA, Goud B, Fransen JA (2000) The small GTPase Rab6B, a novel Rab6 subfamily member, is cell-type specifically expressed and localised to the Golgi apparatus. J Cell Sci 113(Pt 15):2725–2735

Osipovich AB, Jennings JL, Lin Q, Link AJ, Ruley HE (2008) Dyggve-Melchior-Clausen syndrome: chondrodysplasia resulting from defects in intracellular vesicle traffic. Proc Natl Acad Sci USA 105:16171–16176

Palmer KJ, Stephens DJ (2004) Biogenesis of ER-to-Golgi transport carriers: complex roles of COPII in ER export. Trends Cell Biol 14:57–61

Parmar HB, Duncan R (2016) A novel tribasic Golgi export signal directs cargo protein interaction with activated Rab11 and AP-1-dependent Golgi-plasma membrane trafficking. Mol Biol Cell 27:1320–1331

Pereira-Leal JB, Seabra MC (2000) The mammalian Rab family of small GTPases: definition of family and subfamily sequence motifs suggests a mechanism for functional specificity in the Ras superfamily. J Mol Biol 301:1077–1087

Pereira-Leal JB, Seabra MC (2001) Evolution of the Rab family of small GTP-binding proteins. J Mol Biol 313:889–901

Perrin L, Lacas-Gervais S, Gilleron J, Ceppo F, Prodon F, Benmerah A, Tanti JF, Cormont M (2013) Rab4b controls an early endosome sorting event by interacting with the gamma-subunit of the clathrin adaptor complex 1. J Cell Sci 126:4950–4962

Pind SN, Nuoffer C, McCaffery JM, Plutner H, Davidson HW, Farquhar MG, Balch WE (1994) Rab1 and Ca2+ are required for the fusion of carrier vesicles mediating endoplasmic reticulum to Golgi transport. J Cell Biol 125:239–252

Plutner H, Cox AD, Pind S, Khosravi-Far R, Bourne JR, Schwaninger R, Der CJ, Balch WE (1991) Rab1b regulates vesicular transport between the endoplasmic reticulum and successive Golgi compartments. J Cell Biol 115:31–43

Progida C, Bakke O (2016) Bidirectional traffic between the Golgi and the endosomes – machineries and regulation. J Cell Sci 129:3971–3982

Progida C, Cogli L, Piro F, De Luca A, Bakke O, Bucci C (2010) Rab7b controls trafficking from endosomes to the TGN. J Cell Sci 123:1480–1491

Progida C, Nielsen MS, Koster G, Bucci C, Bakke O (2012) Dynamics of Rab7b-dependent transport of sorting receptors. Traffic 13:1273–1285

Proikas-Cezanne T, Gaugel A, Frickey T, Nordheim A (2006) Rab14 is part of the early endosomal clathrin-coated TGN microdomain. FEBS Lett 580:5241–5246

Pusapati GV, Luchetti G, Pfeffer SR (2012) Ric1-Rgp1 complex is a guanine nucleotide exchange factor for the late Golgi Rab6A GTPase and an effector of the medial Golgi Rab33B GTPase. J Biol Chem 287:42129–42137

Puthenveedu MA, Linstedt AD (2001) Evidence that Golgi structure depends on a p115 activity that is independent of the vesicle tether components giantin and GM130. J Cell Biol 155:227–238

Pylypenko O, Hammich H, Yu IM, Houdusse A (2018) Rab GTPases and their interacting protein partners: structural insights into Rab functional diversity. Small GTPases 9:22–48

Rak A, Pylypenko O, Niculae A, Pyatkov K, Goody RS, Alexandrov K (2004) Structure of the Rab7:REP-1 complex: insights into the mechanism of Rab prenylation and choroideremia disease. Cell 117:749–760

Reddy JV, Burguete AS, Sridevi K, Ganley IG, Nottingham RM, Pfeffer SR (2006) A functional role for the GCC185 golgin in mannose 6-phosphate receptor recycling. Mol Biol Cell 17:4353–4363

Ren M, Xu G, Zeng J, De Lemos-Chiarandini C, Adesnik M, Sabatini DD (1998) Hydrolysis of GTP on rab11 is required for the direct delivery of transferrin from the pericentriolar recycling compartment to the cell surface but not from sorting endosomes. Proc Natl Acad Sci USA 95:6187–6192

Riederer MA, Soldati T, Shapiro AD, Lin J, Pfeffer SR (1994) Lysosome biogenesis requires Rab9 function and receptor recycling from endosomes to the trans-Golgi network. J Cell Biol 125:573–582

Rodriguez-Gabin AG, Cammer M, Almazan G, Charron M, Larocca JN (2001) Role of rRAB22b, an oligodendrocyte protein, in regulation of transport of vesicles from trans Golgi to endocytic compartments. J Neurosci Res 66:1149–1160

Rodriguez-Gabin AG, Yin X, Si Q, Larocca JN (2009) Transport of mannose-6-phosphate receptors from the trans-Golgi network to endosomes requires Rab31. Exp Cell Res 315:2215–2230

Rodriguez-Gabin AG, Ortiz E, Demoliner K, Si Q, Almazan G, Larocca JN (2010) Interaction of Rab31 and OCRL-1 in oligodendrocytes: its role in transport of mannose 6-phosphate receptors. J Neurosci Res 88:589–604

Romer W, Pontani LL, Sorre B, Rentero C, Berland L, Chambon V, Lamaze C, Bassereau P, Sykes C, Gaus K, Johannes L (2010) Actin dynamics drive membrane reorganization and scission in clathrin-independent endocytosis. Cell 140:540–553

Russo AJ, Mathiowetz AJ, Hong S, Welch MD, Campellone KG (2016) Rab1 recruits WHAMM during membrane remodeling but limits actin nucleation. Mol Biol Cell 27:967–978

Sahlender DA, Roberts RC, Arden SD, Spudich G, Taylor MJ, Luzio JP, Kendrick-Jones J, Buss F (2005) Optineurin links myosin VI to the Golgi complex and is involved in Golgi organization and exocytosis. J Cell Biol 169:285–295

Salian S, Cho TJ, Phadke SR, Gowrishankar K, Bhavani GS, Shukla A, Jagadeesh S, Kim OH, Nishimura G, Girisha KM (2017) Additional three patients with Smith-McCort dysplasia due to novel RAB33B mutations. Am J Med Genet A 173:588–595

Salminen A, Novick PJ (1987) A ras-like protein is required for a post-Golgi event in yeast secretion. Cell 49:527–538

Sandvig K, Skotland T, van Deurs B, Klokk TI (2013) Retrograde transport of protein toxins through the Golgi apparatus. Histochem Cell Biol 140:317–326

Saraste J, Lahtinen U, Goud B (1995) Localization of the small GTP-binding protein rab1p to early compartments of the secretory pathway. J Cell Sci 108(Pt 4):1541–1552

Sato K, Roboti P, Mironov AA, Lowe M (2015) Coupling of vesicle tethering and Rab binding is required for in vivo functionality of the golgin GMAP-210. Mol Biol Cell 26:537–553

Satoh A, Wang Y, Malsam J, Beard MB, Warren G (2003) Golgin-84 is a rab1 binding partner involved in Golgi structure. Traffic 4:153–161

Schmitt HD, Wagner P, Pfaff E, Gallwitz D (1986) The ras-related YPT1 gene product in yeast: a GTP-binding protein that might be involved in microtubule organization. Cell 47:401–412

Schuck S, Gerl MJ, Ang A, Manninen A, Keller P, Mellman I, Simons K (2007) Rab10 is involved in basolateral transport in polarized Madin-Darby canine kidney cells. Traffic 8:47–60

Seabra MC, Coudrier E (2004) Rab GTPases and myosin motors in organelle motility. Traffic 5:393–399

Sechi S, Frappaolo A, Fraschini R, Capalbo L, Gottardo M, Belloni G, Glover DM, Wainman A, Giansanti MG (2017) Rab1 interacts with GOLPH3 and controls Golgi structure and contractile ring constriction during cytokinesis in *Drosophila melanogaster*. Open Biol 7. https://doi.org/10.1098/rsob.160257

Segev N, Mulholland J, Botstein D (1988) The yeast GTP-binding YPT1 protein and a mammalian counterpart are associated with the secretion machinery. Cell 52:915–924

Short B, Preisinger C, Korner R, Kopajtich R, Byron O, Barr FA (2001) A GRASP55-rab2 effector complex linking Golgi structure to membrane traffic. J Cell Biol 155:877–883

Short B, Preisinger C, Schaletzky J, Kopajtich R, Barr FA (2002) The Rab6 GTPase regulates recruitment of the dynactin complex to Golgi membranes. Curr Biol 12:1792–1795

Sinka R, Gillingham AK, Kondylis V, Munro S (2008) Golgi coiled-coil proteins contain multiple binding sites for Rab family G proteins. J Cell Biol 183:607–615

Spang A (2013) Retrograde traffic from the Golgi to the endoplasmic reticulum. Cold Spring Harb Perspect Biol 5. https://doi.org/10.1101/cshperspect.a013391

Spang A (2016) Membrane tethering complexes in the endosomal system. Front Cell Dev Biol 4:35

Starr T, Sun Y, Wilkins N, Storrie B (2010) Rab33b and Rab6 are functionally overlapping regulators of Golgi homeostasis and trafficking. Traffic 11:626–636

Stein MP, Feng Y, Cooper KL, Welford AM, Wandinger-Ness A (2003) Human VPS34 and p150 are Rab7 interacting partners. Traffic 4:754–771

Stenmark H, Valencia A, Martinez O, Ullrich O, Goud B, Zerial M (1994) Distinct structural elements of rab5 define its functional specificity. EMBO J 13:575–583

Storrie B (2005) Maintenance of Golgi apparatus structure in the face of continuous protein recycling to the endoplasmic reticulum: making ends meet. Int Rev Cytol 244:69–94

Storrie B, Micaroni M, Morgan GP, Jones N, Kamykowski JA, Wilkins N, Pan TH, Marsh BJ (2012) Electron tomography reveals Rab6 is essential to the trafficking of trans-Golgi clathrin and COPI-coated vesicles and the maintenance of Golgi cisternal number. Traffic 13:727–744

Suchy SF, Olivos-Glander IM, Nussbaum RL (1995) Lowe Syndrome, a deficiency of a phosphatidyl-inositol 4,5-bisphosphate 5-phosphatase in the Golgi apparatus. Hum Mol Genet 4:2245–2250

Sun Y, Shestakova A, Hunt L, Sehgal S, Lupashin V, Storrie B (2007) Rab6 regulates both ZW10/RINT-1 and conserved oligomeric Golgi complex-dependent Golgi trafficking and homeostasis. Mol Biol Cell 18:4129–4142

Szul T, Sztul E (2011) COPII and COPI traffic at the ER-Golgi interface. Physiology (Bethesda) 26:348–364

Takahashi S, Kubo K, Waguri S, Yabashi A, Shin HW, Katoh Y, Nakayama K (2012) Rab11 regulates exocytosis of recycling vesicles at the plasma membrane. J Cell Sci 125:4049–4057

Tisdale EJ, Bourne JR, Khosravi-Far R, Der CJ, Balch WE (1992) GTP-binding mutants of rab1 and rab2 are potent inhibitors of vesicular transport from the endoplasmic reticulum to the Golgi complex. J Cell Biol 119:749–761

Touchot N, Zahraoui A, Vielh E, Tavitian A (1989) Biochemical properties of the YPT-related rab1B protein. Comparison with rab1A. FEBS Lett 256:79–84

Trahey M, Hay JC (2010) Transport vesicle uncoating: it's later than you think. F1000 Biol Rep 2:47

Ullrich O, Stenmark H, Alexandrov K, Huber LA, Kaibuchi K, Sasaki T, Takai Y, Zerial M (1993) Rab GDP dissociation inhibitor as a general regulator for the membrane association of rab proteins. J Biol Chem 268:18143–18150

Ullrich O, Reinsch S, Urbe S, Zerial M, Parton RG (1996) Rab11 regulates recycling through the pericentriolar recycling endosome. J Cell Biol 135:913–924

Urbe S, Huber LA, Zerial M, Tooze SA, Parton RG (1993) Rab11, a small GTPase associated with both constitutive and regulated secretory pathways in PC12 cells. FEBS Lett 334:175–182

Utskarpen A, Slagsvold HH, Iversen TG, Walchli S, Sandvig K (2006) Transport of ricin from endosomes to the Golgi apparatus is regulated by Rab6A and Rab6A'. Traffic 7:663–672

Valsdottir R, Hashimoto H, Ashman K, Koda T, Storrie B, Nilsson T (2001) Identification of rabaptin-5, rabex-5, and GM130 as putative effectors of rab33b, a regulator of retrograde traffic between the Golgi apparatus and ER. FEBS Lett 508:201–209

Viotti C (2016) ER to Golgi-dependent protein secretion: the conventional pathway. Methods Mol Biol 1459:3–29

Wang T, Hong W (2002) Interorganellar regulation of lysosome positioning by the Golgi apparatus through Rab34 interaction with Rab-interacting lysosomal protein. Mol Biol Cell 13:4317–4332

Wang D, Lou J, Ouyang C, Chen W, Liu Y, Liu X, Cao X, Wang J, Lu L (2010) Ras-related protein Rab10 facilitates TLR4 signaling by promoting replenishment of TLR4 onto the plasma membrane. Proc Natl Acad Sci USA 107:13806–13811

Wang S, Ma Z, Xu X, Wang Z, Sun L, Zhou Y, Lin X, Hong W, Wang T (2014) A role of Rab29 in the integrity of the trans-Golgi network and retrograde trafficking of mannose-6-phosphate receptor. PLoS One 9:e96242

Wang P, Liu H, Wang Y, Liu O, Zhang J, Gleason A, Yang Z, Wang H, Shi A, Grant BD (2016) RAB-10 promotes EHBP-1 bridging of filamentous actin and tubular recycling endosomes. PLoS Genet 12:e1006093

Weide T, Bayer M, Koster M, Siebrasse JP, Peters R, Barnekow A (2001) The Golgi matrix protein GM130: a specific interacting partner of the small GTPase rab1b. EMBO Rep 2:336–341

White J, Johannes L, Mallard F, Girod A, Grill S, Reinsch S, Keller P, Tzschaschel B, Echard A, Goud B, Stelzer EH (1999) Rab6 coordinates a novel Golgi to ER retrograde transport pathway in live cells. J Cell Biol 147:743–760

Wilson BS, Nuoffer C, Meinkoth JL, McCaffery M, Feramisco JR, Balch WE, Farquhar MG (1994) A Rab1 mutant affecting guanine nucleotide exchange promotes disassembly of the Golgi apparatus. J Cell Biol 125:557–571

Witkos TM, Lowe M (2017) Recognition and tethering of transport vesicles at the Golgi apparatus. Curr Opin Cell Biol 47:16–23

Wong M, Munro S (2014) Membrane trafficking. The specificity of vesicle traffic to the Golgi is encoded in the golgin coiled-coil proteins. Science 346:1256898

Wu YW, Tan KT, Waldmann H, Goody RS, Alexandrov K (2007) Interaction analysis of prenylated Rab GTPase with Rab escort protein and GDP dissociation inhibitor explains the need for both regulators. Proc Natl Acad Sci USA 104:12294–12299

Wu YW, Goody RS, Abagyan R, Alexandrov K (2009) Structure of the disordered C terminus of Rab7 GTPase induced by binding to the Rab geranylgeranyl transferase catalytic complex reveals the mechanism of Rab prenylation. J Biol Chem 284:13185–13192

Wu YW, Oesterlin LK, Tan KT, Waldmann H, Alexandrov K, Goody RS (2010) Membrane targeting mechanism of Rab GTPases elucidated by semisynthetic protein probes. Nat Chem Biol 6:534–540

Yadav S, Linstedt AD (2011) Golgi positioning. Cold Spring Harb Perspect Biol 3. https://doi.org/10.1101/cshperspect.a005322

Yang M, Chen T, Han C, Li N, Wan T, Cao X (2004) Rab7b, a novel lysosome-associated small GTPase, is involved in monocytic differentiation of human acute promyelocytic leukemia cells. Biochem Biophys Res Commun 318:792–799

Young J, Stauber T, del Nery E, Vernos I, Pepperkok R, Nilsson T (2005) Regulation of microtubule-dependent recycling at the trans-Golgi network by Rab6A and Rab6A'. Mol Biol Cell 16:162–177

Young J, Menetrey J, Goud B (2010) RAB6C is a retrogene that encodes a centrosomal protein involved in cell cycle progression. J Mol Biol 397:69–88

Zhen Y, Stenmark H (2015) Cellular functions of Rab GTPases at a glance. J Cell Sci 128:3171–3176

Zheng JY, Koda T, Fujiwara T, Kishi M, Ikehara Y, Kakinuma M (1998) A novel Rab GTPase, Rab33B, is ubiquitously expressed and localized to the medial Golgi cisternae. J Cell Sci 111 (Pt 8):1061–1069

Part II
Golgi and Centriole Positioning, Interactions and Dynamics

Chapter 7
Positioning of the Centrosome and Golgi Complex

Amos Orlofsky

Abstract For over a century, the centrosome has been an organelle more easily tracked than understood, and the study of its peregrinations within the cell remains a chief underpinning of its functional investigation. Increasing attention and new approaches have been brought to bear on mechanisms that control centrosome localization in the context of cleavage plane determination, ciliogenesis, directional migration, and immunological synapse formation, among other cellular and developmental processes. The Golgi complex, often linked with the centrosome, presents a contrasting case of a pleiomorphic organelle for which functional studies advanced somewhat more rapidly than positional tracking. However, Golgi orientation and distribution has emerged as an area of considerable interest with respect to polarized cellular function. This chapter will review our current understanding of the mechanism and significance of the positioning of these organelles.

List of Abbreviations

ADAP	Adhesion- and degranulation-promoting adapter protein
Akt	v-akt murine thymoma viral oncogene homolog 1
Arf1	ADP-ribosylation factor 1
Arp2/3	Actin-related protein 2/3
APC	Antigen-presenting cell
BARS	BFA-dependent ADP-ribosylation substrate
BICD2	Bicaudal D2
CD3	Cluster of differentiation 3
Cdc42	Cell division control protein 42 homolog
Cdk1	Cyclin-dependent kinase 1
CENPF	Centromere protein F

A. Orlofsky (✉)
Department of Biological Sciences and Geology, Queensborough Community College, City University of NY, Bayside, NY, USA

Department of Pathology, Albert Einstein College of Medicine, Bronx, NY, USA
e-mail: amos.orlofsky@einstein.yu.edu

© Springer Nature Switzerland AG 2019
M. Kloc (ed.), *The Golgi Apparatus and Centriole*, Results and Problems in Cell Differentiation 67, https://doi.org/10.1007/978-3-030-23173-6_7

CEP	Centrosomal protein
CLIP-170	Cytoplasmic linker protein 170 alpha-2
C-Nap1	Centrosomal Nek2-associated protein 1
DAG	Diacylglycerol
Dlg1	Discs large-1
DOCK	Dedicator of cytokinesis
EGF	Epidermal growth factor
ELMO	Engulfment and cell motility
EMS	Endomesodermal
ERM	Ezrin-radixin-moesin
Gαi	G protein alpha i
GAS2L1	Growth arrest specific 2-like 1
GEF	Guanine nucleotide exchange factor
GKAP	Guanylate kinase-associated protein
GM130	Golgi membrane protein 130 kD
GMAP	Golgi microtubule-associated protein
GOLPH3	Golgi phosphoprotein 3
GRASP55	Golgi reassembly-stacking protein of 55 kDa
GRASP65	Golgi reassembly-stacking protein of 65 kDa
HS1	Hematopoietic lineage cell-specific protein 1
IQGAP	IQ motif-containing GTPase activating protein
IS	Immunological synapse
KASH	Klarsicht, ANC-1, Syne homology
KIF25	kinesin family member 25
LIMK	LIM domain kinase
LIS1	Lissencephaly-1
Mst2	Mammalian Ste20-like 2
MYO18A	Myosin XVIIIA
Ncd	Nonclaret disjunctional
Nde1	Nuclear distribution E homolog 1
NE	Nuclear envelope
Nek2A	NIMA-related kinase 2A
NNE	Notochord/neural/endoderm
Num1	Nuclear migration protein 1
NuMA	Nuclear mitotic apparatus
PAR	Partitioning defective
PCP	Planar cell polarity
PI3 kinase	Phosphatidylinositol-3 kinase
PITPNA	Phosphatidylinositol transfer protein alpha
PITPNB	Phosphatidylinositol transfer protein beta
PKC	Protein kinase C
PKD	Protein kinase D
Plk1	Polo-like kinase 1
PP1	Protein phosphatase 1

PPP2CA	Protein phosphatase 2A, catalytic subunit alpha
Rac	Ras-related C3 botulinum toxin substrate
RanGTP	GTP-bound form of *ras*-related nuclear protein
Rap1	Ras-associated protein 1
Rho	*Ras* homologous
ROCK	Rho-associated protein kinase
RPE1	Retinal pigmented epithelium-1
Slk	Ste20-like kinase
STK25	Serine/threonine protein kinase 25
STRAD	STE20-related kinase adapter protein
SUN	Sad1 and UNC-84
TBCCD1	Tubulin binding cofactor C domain-containing protein
Wnt	Wingless/integrated
Zyg-12	Zygote defective-12

7.1 Introduction

A signal feature of eukaryotic cells is the regulated geometry of their intracellular organization, which in turn supports their spatially organized interactions with neighboring cells and substrates and helps guide the orientation of cell division. Of the many subcellular structures that take part in this ballet, the centrosome is unique in the variety of positions it can adopt and the corresponding variety of structures with which it is functionally associated, including microtubule asters, mitotic spindles, cilia and flagella, immunological synapses, developing neurites, and the lamellipodia and pseudopodia of migrating cells. Centrosome positioning, then, may be a key control point for many cell behaviors and developmental processes that depend on the processing of spatial information.

While the centrosome has, historically, held pride of place in theories of cell organization, a growing body of studies indicates that the Golgi complex, often closely associated with the centrosome, also has important organizing function, acting either independently of or in coordination with centrosomes. Golgi and centrosome positioning may therefore both individually and interactively serve as focal points for geometrically coordinated cellular functions. Here we review the current understanding of the control of the localization of these organelles.

7.2 Centrosome Positioning: An Overview

Centrosomes are found in most animals and many protists, as well as lower plants and certain fungi, and consist of a pair of centrioles (mother and daughter) surrounded by pericentriolar material, except in dividing cells, in which mother and daughter centrioles separate to form distinct centrosomes. Centrosomes are commonly observed in one of three positions: (1) as basal bodies at the plasma membrane, where they serve to assemble and anchor cilia/flagella; (2) adjacent to nuclei (paranuclear), with an orientation that may be modified by polarizing stimuli; (3) in the spindle poles of dividing cells. The phylogenetic distribution of these structures suggests that the location in basal bodies is primordial and was most likely present in the last eukaryotic common ancestor (Ross and Normark 2015; Woodland and Fry 2008). However, in some taxa, including higher fungi, nematodes and arthropods, centrosomes are present but cilia/flagella are rare or absent. In general, the different observed centrosome positions are due to repositioning events and not to the formation of new centrosomes. It was previously thought that basal body centrosomes in multiciliated cells could arise de novo (Klos Dehring et al. 2013), in which case their initial positioning might be specified independently of preexisting centrosomes. However, more recent live-imaging studies demonstrated that basal body centrosomes in multiciliated cells arise from procentriolar extensions from parental daughter centrioles (Al Jord et al. 2014). Consequently, with rare exceptions [as in the case of the mouse early embryo (Courtois et al. 2012)], all centrosomes arise from other centrosomes and have initial positions determined by the location of the parent structure. Centrosome localization, then, is governed by repositioning.

Several of the more common scenarios of centrosome repositioning, which will be discussed in detail later, are summarized in Fig. 7.1. In nonpolarized interphase cells, the centrosome maintains a paranuclear position. In prophase, the centrosome undergoes disjunction to form separate mother and daughter centrosomes, which migrate to opposite ends of the nucleus, where they form spindle poles, which are subject to adjustments in positioning that specify the plane of cell division. Centrosomes resume their paranuclear position in telophase. Especially (but not exclusively) in postmitotic cells, the paranuclear centrosome may dock at the plasma membrane, where the mother centriole will form the basal body of a primary cilium, a structure that has emerged as a key regulator of a wide variety of cellular and developmental processes (Walz 2017). Polarizing events can modulate centrosome position in a variety of ways. A migratory stimulus leads to centrosome orientation in the axis of migration, either in front of or behind the nucleus. Epithelial polarization repositions the centrosome to the apical surface, and its location at that surface may be further specified by planar cell polarity signaling. Certain epithelia develop multiple, motile cilia that are structurally and functionally distinct from the primary cilium. In lymphocytes polarized by recognition of antigen, the centrosome migrates to the surface at which recognition has taken place (immunological synapse), while in developing neurons, centrosomes may migrate to regions of axon formation.

Even this brief summary may suffice to convey the impression of the centrosome as a remarkably multipurpose structure. Yet in many of these scenarios, our understanding of the function of the organelle is still very incomplete. That understanding will undoubtedly develop within the framework provided by studies of the regulation of centrosome position, which will now be discussed further.

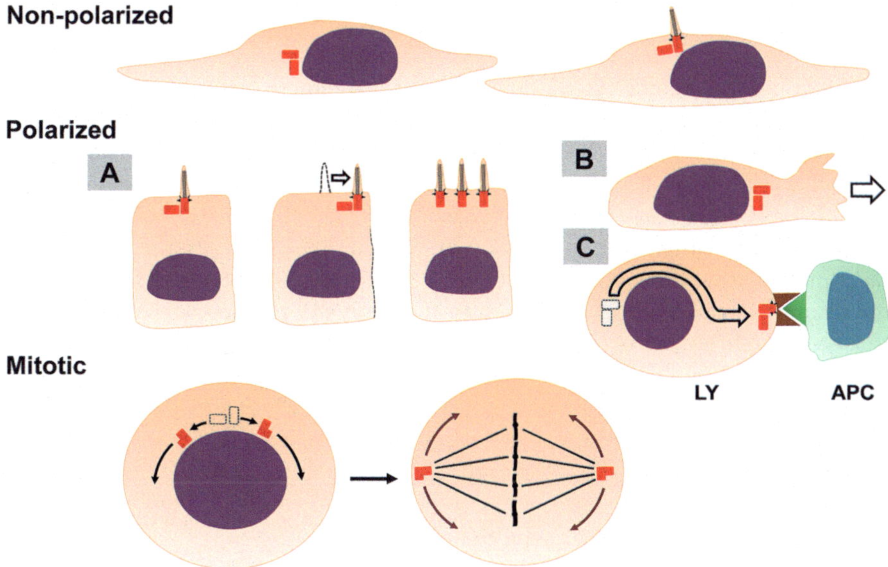

Fig. 7.1 Regulation of centrosome positioning. Non-polarized cells have paranuclear centrosomes. In the non-polarized cell on the right the centrosome is nucleating a primary cilium. Polarized cells: Panel (**a**) shows three epithelial cells with apicobasal polarity. In the middle cell, planar cell polarity signals have shifted the position of the cilium. The cell on the right is multiciliated. (**b**) migrating cell. (**c**) Lymphocyte (LY) and antigen-presenting cell (APC) form an immunological synapse, to which the centrosome migrates (arrow). In mitotic cells, centrosomes disjoin and migrate in prophase (left) and form spindle poles that can reorient at metaphase (right)

7.3   Positioning of the Interphase Centrosome in Nonpolarized Cells

A large number of studies of nonpolarized cells in interphase have shown that the paranuclear centrosome typically occupies a position close to the cell centroid, i.e., the geometric center of the cell. This central position, from which the name of the organelle derives, is rapidly achieved in cells, such as fertilized eggs, in which centrosomes are initially off-center (see below), indicating an active centering force. As will be discussed later, centration, once achieved, can be maintained even when cells are responding to polarizing migratory stimuli, implying the existence of restraining forces. This restraint is shown even more clearly by the difference in motility between mother and daughter centrioles in enucleated cytoplasts: the daughter wanders, while the mother, which is the dominant centriole with respect to nucleation of microtubules (MTs), is much less mobile (Piel et al. 2000). The daughter, in turn, is restrained by linkage to the mother, as shown by the correlation of their movements. Centrosomes in these cells, then, are subject to centering forces at all times. Two obvious potential sources for such centering forces are the structures with which the paranuclear centrosome is most clearly associated: MTs and nuclei. These will be considered in turn.

7.3.1 Microtubules and Centration

Since the centrosome nucleates an aster of MTs that radiate to the cell periphery, it is natural to suppose that the microtubules might exert forces to maintain the positioning of the organelle at the cell centroid. Such forces may consist of active pushing or pulling, as well as passive tethering. Early studies in a kidney epithelial cell line probed these ideas by a local application of an MT-depolymerizing agent, nocodazole, to disrupt MTs in a region of the cell. When actomyosin contractility was simultaneously inhibited to permit specific visualization of MT forces, the centrosome was observed to move away from the site of nocodazole application, consistent with an active localization of the centrosome by MT pulling forces (Burakov et al. 2003). Subsequently, evidence for a pulling mechanism was obtained in the one-cell embryo of *Caenorhabditis elegans*, in which the male pronucleus, flanked by a pair of paranuclear centrosomes, migrates to the center of the egg (Kimura and Onami 2005). The authors compared computational models of centration by either MT pushing or pulling forces and identified a qualitative feature of the pulling model that was well matched by in vivo data.

A logical candidate for mediation of an MT pulling force is the minus-end-directed motor protein, cytoplasmic dynein 1 (henceforth dynein). Since dynein moves cargo centrally along MTs, if the motor is anchored (that is, its "cargo" is immobile), then it would be expected to pull MTs toward the periphery. In *C. elegans* one-cell embryos, the centrosome-centering force was shown to be dynein dependent (Kimura and Onami 2005), although actomyosin forces may also contribute (Goulding et al. 2007). Other studies have also shown that dynein inhibition leads to aberrant centrosome localization in interphase (Burakov et al. 2003; Wu et al. 2011). These findings, as well as simulation studies (Zhu et al. 2010), lend support to a model of positioning based on dynein pulling forces. However, they do not rule out contributions from forces generated by myosin or from MT pushing against cytoplasmic structures, which may work in concert with pulling forces (Zhu et al. 2010). Indeed, the observation that, in enucleated cytoplasts, MT depolymerization does not disturb the position of the stable mother centriole indicates a role for additional forces (Piel et al. 2000). The authors further observed that daughter centriole motions were differently affected by actomyosin or MT disruption, suggesting different roles for these cytoskeletal elements: actomyosin appears to regulate the range of daughter movement, while MTs mediate sudden changes in direction. Myosin motors, however, are not essential to maintaining centration (Burakov et al. 2003). Overall, the evidence points to an important role for dynein in the active maintenance of centration, while the roles of other force generators are less defined.

A mechanism that actively maintains the centrosome at the cell centroid (or at any specified position in the cytoplasm) must be able to generate asymmetric restoring forces: that is, displacement from the center must result in a net centering force. How might dynein pulling forces be configured to provide this asymmetry? The answer depends greatly on where the dynein is. Since, as discussed later, dynein localizes to the cell cortex in mitosis to exert force on the spindle, one possibility is that cortical dynein is also the force generator in interphase. The difficulty, however, is that a displaced (off-center) aster would be expected to make more contact with proximal than with distal cortical sites, and cortical dynein should in this case be a decentering force. One proposal to address this issue, supported by cell-free studies of asters of polymerized tubulin in chambers coated with dynein, is that slipping of MT ends as they abut the cortex tends to move the cortical pulling sites away from the proximal end, leading to an anisotropic (distally biased) MT distribution that supports centration by dynein pulling forces (Laan et al. 2012). A second hypothesis is that the cortical sites that anchor dynein are limited, such that no matter where the aster might migrate within the cell, MTs will remain in sufficient excess to ensure continual occupancy of all such sites, and consequently a constant net centering force will be maintained (Grill and Hyman 2005). However, there is little direct evidence for pulling by cortical dynein in nonpolarized cells in interphase. Live imaging of dynein traffic during interphase in HeLa cells does not provide clear evidence of cortical dynein accumulation (Kobayashi and Murayama 2009).

A third mechanism, which currently has the most experimental support in studies of interphase centrosomes, is that MT pulling forces are dependent on MT length, so that centrosome departure from the centroid will generate a force imbalance to restore centration. Early evidence for length-dependent pulling came from Hamaguchi and Hiramoto in a study of sand dollar eggs. By using UV irradiation to locally restore MT growth in eggs treated with an MT-depolymerizing agent, the authors were able to generate a sperm aster that centered itself within the irradiated region, and that did so by following the longest astral MT rays, consistent with length-dependent MT pulling (Hamaguchi and Hiramoto 1986). The growing astral rays appear to exert force prior to reaching the cell cortex, suggesting that the length-dependent forces are not cortically localized. Indeed, in contrast to the cortical dynein-based models, a mechanism based on length-dependent MT pulling forces requires force generators distributed along the length of MTs. One possibility is that cytoplasmically anchored dynein is distributed along the MT length. However, there is currently no evidence for such uniformly distributed cytoplasmic anchors. A second possibility was introduced by further studies of pronuclear centrosome centration in *C. elegans* one-cell embryos (Kimura and Kimura 2011). The authors first noted that a cortically based pulling force model was unlikely in this system since MTs oriented in the direction of centrosomal movement were not long enough to reach the facing cortical regions. Furthermore, centration was shown to depend on a dynein subunit responsible for the movement of organelles such as lysosomes and yolk granules along MTs but that is not required to generate dynein pulling force. Variations in speed of centration were shown to correlate with variations in speed of organellar movement, and interference with organellar motility suppressed centrosome centration (Kimura and Kimura 2011). The basis of the dynein-dependent MT pulling force, then, is most likely not anchored dynein but rather the reactive drag forces generated continuously along the MTs by organellar movement (Fig. 7.2). This model has recently received further support from studies of sperm aster centration in sea urchin eggs (Tanimoto et al. 2016). As with *C. elegans* one-cell embryos,

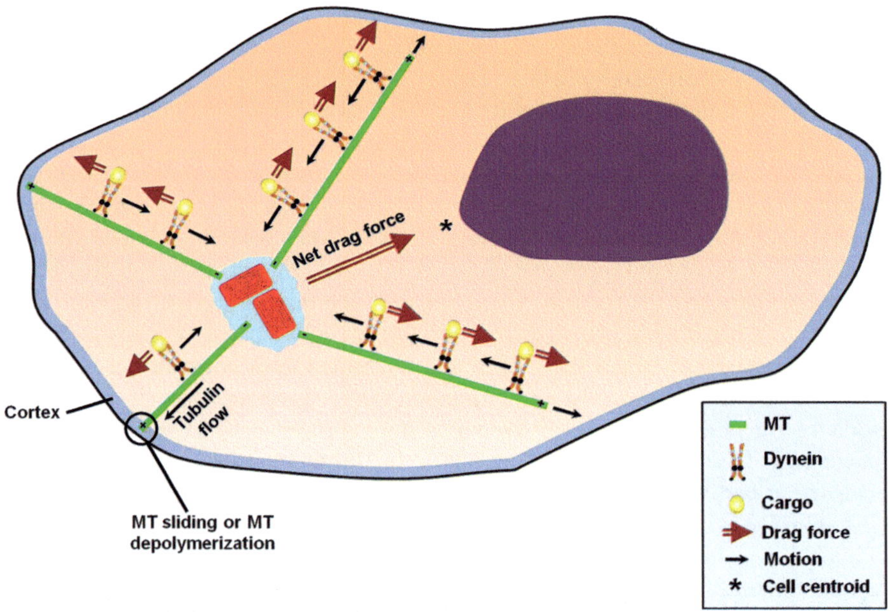

Fig. 7.2 Length-dependent force model for centrosome centration. The centrosome, with centrioles (red) embedded in MT-nucleating pericentriolar material (blue), is depicted as being moved to the cell centroid (asterisk) by a net force derived from the sum of drag forces generated by the minus end-directed movement of dynein motors bearing cytoplasmic cargoes. MTs closer to the centroid (on the right) are longer and hence bear more dynein and generate greater force. These MTs have not yet reached the cortex and are therefore moving (as well as polymerizing) toward cortex. The shorter MTs on the left have reached the cortex, and their pulling action depends on tubulin flow resulting from either sliding or depolymerization at the cortex

centration took place long before MT growth could reach the opposing cortex, arguing against pulling based on cortical anchors. Aster movement was dynein dependent, and laser ablation of one astral side induced movement toward the nonablated side, consistent with a pulling mechanism. However, a cautionary note emerges from another recent study in this system: by using microinjected dynein-binding magnetic particles to create a cortically localized pulling "cap" in the egg, and thereby measure pulling forces, it was shown that the astral centering forces that center the male pronucleus are three times stronger during the centration process than afterward (Salle et al. 2018). It is possible, then, that the "drag force" mechanism represents a specialized adaptation to meet the demand for large and rapid nuclear movement in fertilized eggs. In this case, the extension of the drag force model to interphase centration in other systems may require further investigation.

Regardless of the site from which dynein pulls, the centrosome cannot move unless the MT plus end either moves or is continually removed. Both of these mechanisms have been described in the context of cortical dynein forces exerted in mitosis, where they are referred to, respectively, as sliding models and capture-shrinkage or end-on models (Guild et al. 2017; Omer 2018). Experimental evidence for either model in interphase cells is limited. However, in the study, described earlier, in which centrosome movement was observed after local nocodazole application, the addition of taxol, which prevents MT depolymerization, abolished centrosome motion (Burakov et al. 2003), suggesting that MT shortening may be necessary for the centering process.

While the majority of studies of interphase centrosome positioning have been interpreted in terms of a pulling force model, studies in some systems point to a role for MT pushing forces. Studies in the slime mold *Dictyostelium* suggest that pushing by plus-end-directed kinesin motors contributes to centrosome movement (Brito et al. 2005). Pushing force may also be generated by the growth of MTs that either are anchored at or abut against peripheral structures. In vitro studies of polymerized-tubulin asters demonstrate that substantial pushing forces can be generated in this fashion (Holy et al. 1997). In *Drosophila* oocytes, nuclear centration appears to be driven by MT pushing (Zhao et al. 2012). This system has unusual features, however, as the centrosomes are positioned between the nucleus and the proximal cellular end from which displacement occurs, possibly precluding a pulling force-based mechanism. More evidence for pushing comes from fission yeast: when the yeast nucleus (and associated spindle pole body, the yeast equivalent of a centrosome) are experimentally displaced from the cell center, nuclear return occurs when a growing MT reaches the near end of the cell and then continues to grow against it, indicating that this MT is pushing the nucleus and the centrosome back toward the center (Tolic-Norrelykke et al. 2005).

Further studies in the yeast system, however, suggest that pushing and pulling forces may be tailored to specific aspects of cell organization. The MTs in the yeast are aligned with the long axis of the cell, except in meiotic prophase, when they adopt an astral geometry that may be more comparable to the MT array observed in other systems. Interestingly, in meiotic yeast cells with astral MTs, experiments using laser cutting of MTs indicated that dynein-dependent pulling rather than pushing forces controlled spindle pole body/nuclear movement (Vogel et al. 2009). When nondividing mitotic cells, with axially aligned MTs, were genetically manipulated to adopt an astral geometry similar to meiotic cells, the nucleus and the centrosome were displaced, suggesting that pushing forces may be inadequate for centrosome centration in the presence of astral MTs (Tolic-Norrelykke et al. 2005). Overall, then, dynein-mediated MT pulling forces appear to have widespread importance for regulating interphase centrosome position, while contributions from other forces, such as MT pushing, may be tailored to particular contexts of mechanical constraint. With respect to cellular organizing forces, architecture is likely to dictate strategy.

7.3.2 Nucleus–Centrosome Interaction

The close apposition of the nucleus and the centrosome in nonpolarized cells (and in some polarized cells) suggests that the nucleus may anchor and restrain the centrosome and also that nuclear motion might enforce centrosome repositioning. Two questions then arise. First, what is the evidence of mechanisms linking the organelles? Second, if they are mechanically linked, who is the cart and who is the horse?

Early evidence for the physical attachment of the nucleus and the centrosome came from cell fractionation studies. Homogenates of liver or spleen [tissues that are mostly unciliated (D'Angelo and Franco 2011)] often contain centrosomes attached to nuclei, and these attachments survive nonionic detergent and nuclease treatment (Bornens 1977; Nadezhdina et al. 1979). Additional evidence derives from the observation of organelle motions. Nuclei and centrosomes frequently comigrate [reviewed in Burakov and Nadezhdina (2013)], and when centrosomes reposition, they often do so while maintaining apposition to the nuclear envelope (NE), as for example when moving toward their eventual spindle pole positions during prophase. Furthermore, when centrosome rotary motion is observed in cells of the slime mold *Dictyostelium*, the attached nucleus rotates with the centrosome, as detected by nucleolar movement (Koonce et al. 1999). Fibrous structures that link the centrosome and the nucleus have been described in *Dictyostelium* (Omura and Fukui 1985) and also in the green alga *Chlamydomonas* (Wright et al. 1989), but similar structures in metazoa have not been described. Pericentriolar material, however, may contain components that can associate with the outer nuclear membrane, as suggested by the observation of skeletal muscle cells, in which centrosomes are independent of the nucleus, which instead becomes coated with a matrix of pericentriolar proteins that serve as the MT-nucleating center (Srsen et al. 2009; Tassin et al. 1985).

More recent studies have revealed a variety of molecular links between nuclei and centrosomes. Many of these potential mechanisms involve members of the SUN protein family, localized to the inner nuclear membrane, interacting with various members of the KASH family in the outer nuclear membrane. Cytoplasmic KASH domains are capable of a variety of interactions with actin, tubulin, dynein, and other cytoskeletal components, and the depletion of some KASH proteins, such as various members of the nesprin family in mammalian cells or ZYG-12 in nematodes, leads to the separation of centrosomes from nuclei (Burakov and Nadezhdina 2013; Obino et al. 2016). Depletion of emerin, an NE protein that binds tubulin, generates a similar phenotype in mammalian cells (Salpingidou et al. 2007), as does nuclear lamin knockdown in cells grown on circular micropatterns (Hale et al. 2011). It has been proposed that nesprin-4 linkage of the nucleus and the centrosome may be used as a mechanism to regulate centrosome positioning (Roux et al. 2009). Actin filaments have also recently been implicated in centrosome binding to the nucleus (Obino et al. 2016), and centrosomes can nucleate actin (Farina et al. 2016). It remains unclear, however, whether there is any direct molecular bridge between the two organelles. The centrosome is typically separated from the NE by a distance of >1 µm, and while this space might conceivably be bridged by, for example, dimerization of putative linkers such as ZYG-12, the association with the NE may instead be dependent on dynamic connections via MTs or actin cytoskeleton (Burakov and Nadezhdina 2013; Tikhonenko et al. 2013). The latter view is consistent with experiments in which centrosomes become separated from nuclei upon the depletion of either dynein (Robinson et al. 1999) or NE proteins that bind to dynein, including ZYG-12 (Malone et al. 2003) in nematodes and nuclear-pore-associated Bicaudal D2 (BICD2) in human cells (Splinter et al. 2010). If linkage is indeed based on dynamic cytoskeletal connection, these links might serve as restraining tethers or even to convey displacing forces, but they might alternatively serve as tracks for circumnuclear centrosome navigation or as scaffolds for signaling complexes.

If mechanisms of linkage have yet to be fully elucidated, there is even less information with respect to the transmission of restraining or displacing forces between the two organelles. Centrosome centration is maintained in enucleated cells (Bornens 2012; Piel et al. 2000), indicating that the nucleus is dispensable for centering. In migrating cells, nuclear movement to the rear does not perturb the location of the centrosome at the cell centroid (Gomes et al. 2005), and the motions of the centrosome in migrating cells are not coordinated with those of the nucleus (Yvon et al. 2002). However, it is noteworthy that when dynein is prevented from pulling on the centrosome, the rearward nuclear movement now does seem to pull back the centrosome (Gomes et al. 2005). This suggests that the nucleus can exert sufficient force to displace the centrosome and that therefore under normal conditions (with active dynein) it at least provides a restraining force. There is, overall, little evidence for displacing force conveyed by the nucleus. Indeed, observations of deformation in nuclei rotating with centrosomes in *Dictyostelium* were interpreted to suggest that the nucleus is a passive partner, following centrosome motion rather than initiating it (Koonce et al. 1999). Imaging of centrosome migration to an immunological synapse also shows the deforming nucleus as an apparently passive rather than active participant [movie S2 in Kuhn and Poenie (2002)]. Furthermore, even when apposed to the nucleus, centrosomes must be capable of motion unlinked to nuclear motion, else it would be impossible for two replicated centrosomes, after disjunction, to move in opposite directions along the NE. Overall, then, current evidence is more consistent with a view of the nucleus as, at most, a dynamic anchor rather than an active positioner of the centrosome.

7.4 Centrosome Disjunction and Positioning During Mitosis

The spindle is the central organizing feature of dividing cells. In many respects, the spindle can be viewed as a well-conserved machine performing a stereotyped function. In other respects, however, the spindle is far from stereotyped: in its orientation/positioning and also in its centriole asymmetry, the spindle can be loaded with information that is highly context dependent and critical both for symmetrical cell division and for a host of asymmetrical developmental processes. The repositioning of the centrosome during mitosis, then, tracks the role of this organelle not only in the generation of spindle poles but also as a mediator of developmental and tissue-organizing information.

7.4.1 Spindle Formation 1: Prior to Nuclear Envelope Breakdown

At the onset of mitosis, a first step in the preparation for spindle formation is the disjunction of mother and daughter centrosomes. Mother and daughter centrioles remain closely linked until late G2 phase. This linkage is controlled by the phosphorylation state of several linker proteins, including rootletin and C-Nap1 (Wang et al. 2014). The centrosomal kinase Nek2A phosphorylates the linker proteins, but during most of interphase Nek2A is associated with the phosphatase PP1, which maintains the linker proteins in a dephosphorylated state and prevents their disassembly. Nek2A-PP1 association is controlled by the kinase Mst2. At the G2/M transition, the linker phosphorylation state is shifted in response to signals that drive the cell into mitosis. Specifically, the accumulation of cyclin B during G2 phase is a critical event leading to the activation of the cyclin-dependent kinase Cdk1, which localizes to the centrosome. At the same time, the scaffold protein CEP192 is recruited to the centrosome and in turn recruits the kinase Aurora A, as well its effector kinase, Plk1 (the terms used are for vertebrate homologs of components generally conserved in animals and fungi). Activated Plk1 phosphorylates Mst2, inhibiting association of Nek2A with PP1 and thereby leading to linker protein phosphorylation and linker disassembly (Mardin et al. 2011). In many cancer cell lines, signaling through the EGF receptor can activate Nek2A independently of Plk1 (Mardin et al. 2013).

While the Nek2A mechanism seems sufficient to disjoin the centrioles, it can be supplemented by a motor-driven mechanism mediated by Eg5, a member of the kinesin-5 family of plus-end-directed MT motors (Mardin et al. 2010). In contrast to the pulling forces generated by anchored, minus-end-directed dynein, an anchored plus-end-directed motor bound to a centrosome-nucleated MT would be expected to push the centrosome. Eg5 does not require anchoring, however, because it forms a tetramer that is capable of simultaneously binding two antiparallel MTs and moving along both of them, effectively causing the two MTs to slide over each other in opposite directions (Kapitein et al. 2005). Therefore, the current view is that Eg5 acts in between the two separating centrosomes to simultaneously push them in opposite directions, forcing them apart (De Simone et al. 2016). The Eg5 mechanism is capable of enforcing centrosome disjunction in the absence of Nek2A activity (Mardin et al. 2010). This result implies that Eg5 must itself be regulated to prevent premature separation, and this is apparently accomplished by making the recruitment of Eg5 to the centrosome dependent on Plk1 activation (Mardin et al. 2011; Smith et al. 2011). In further support of an MT-based mechanism of disjunction, the centrosomal protein GAS2L1 was shown to enhance MT binding to the centrosome and also to mediate disjunction (Au et al. 2017).

Recently, a motor has been identified in human cells that may oppose Eg5 and act to maintain centrosome tethering (Decarreau et al. 2017). Kif25 is in the kinesin-14 family, whose members, unlike most kinesins, are minus end directed. Kif25 localizes to centrosomes and, like Eg5, has a tetrameric structure, suggesting that it may act via a similar antiparallel sliding mechanism, but in the opposing direction. Kif25 depletion results in premature centrosome disjunction as early as S phase (Decarreau et al. 2017), suggesting that a dynamic force balance is already generated at the centrosome at the time of its duplication.

Once unlinked at the G2/M boundary, mother and daughter centrioles, each forming a centrosome with pericentriolar material and each nucleating an MT aster, begin to migrate in opposite directions over the NE during prophase. A live-imaging study in newt lung cells showed that this migration involves fluctuating motions: the movements of the two centrosomes were not correlated, and included both forward and backward motions, but the frequency and average speed of the forward movements were greater (Waters et al. 1993). In all systems studied, including *Drosophila* and *C. elegans* embryos, as well as mammalian cells, dynein is implicated as a force generator for prophase migration (Ferenz et al. 2010; Gonczy et al. 1999; Raaijmakers et al. 2012; Robinson et al. 1999; Sharp et al. 2000; Tanenbaum et al. 2008; Vaisberg et al. 1993). In mammalian cells, however, a major role is played by Eg5, so that the role of dynein only became evident, in a cell type-dependent fashion, when Eg5 was partly inhibited or when cells were selected for Eg5 independence (Raaijmakers et al. 2012; Tanenbaum et al. 2008). As in the case of centrosome disjunction, the action of Eg5 during prophase is most likely by a sliding microtubule mechanism that pushes both centrosomes simultaneously (Fig. 7.3). Elegant support for this idea comes from the observation that centrosome movement in prophase ceases to be Eg5 dependent (while remaining dynein dependent) if only one centrosome is present in the cell (Raaijmakers et al. 2012). On the other hand, Eg5 is dispensable for mitosis in *C. elegans* and for the prophase of *Drosophila* embryos (Ferenz et al. 2010), while dynein is essential for prophase centrosome separation in both of these systems (Gonczy et al. 1999; Robinson et al. 1999).

The universal importance of dynein pulling forces for driving centrosome separation raises the question of where these forces are generated. Since the movement takes place along the NE, and dynein can localize to the envelope (Splinter et al. 2010), envelope-bound dynein is a logical candidate. BICD2 is a motor adaptor that localizes to the NE during G2 phase and binds both dynein and kinesin-1, while CENPF is an NE protein that binds the dynein adapter, Nde1 (Bolhy et al. 2011; Splinter et al. 2010). In mammalian cells, the depletion of either BICD2 or CENPF inhibited prophase centrosome separation, implicating NE-associated dynein as a critical motor (Raaijmakers et al. 2012). How dynein in this location can create asymmetric force to move centrosomes is unclear. One possibility is the generation of asymmetric length-dependent force: astral MTs may grow longer in the "forward" direction (away from the other centrosome) because, in the backward direction, encounters with MTs from the opposing aster may initiate catastrophes and limit MT growth (Raaijmakers et al. 2012).

The action of NE-localized dynein is expected to maintain centrosome–nucleus association during migration, but this function of the motor seems to be opposed by kinesin-1, a plus-end-directed motor. When dynein was depleted, kinesin-1 mediated extensive separation of the nucleus from the centrosomes, moving all these structures toward the cell periphery in a manner that depended on its association with

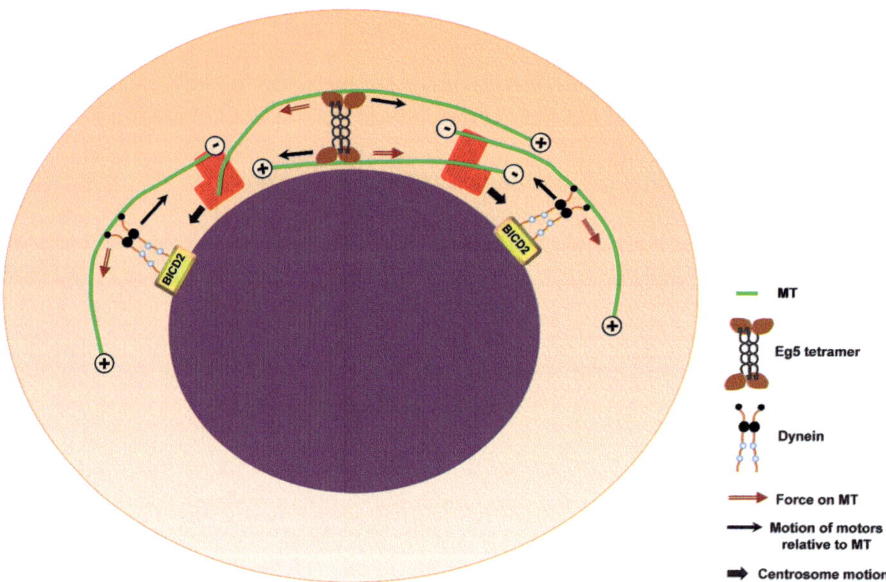

Fig. 7.3 Model of prophase centrosome migration in mammalian cells. Plus-end-directed Eg5 tetramers perform sliding action to exert simultaneous outward pushing force on both centrosomes. A second centrosome-separating force is generated by the minus end-directed walking of dynein motors anchored to nuclear envelope-bound BICD2. Each centriole bears a lateral protuberance representing the procentriole acquired in S phase

the NE via BICD2 (Splinter et al. 2010). On the other hand, kinesin-1 depletion in cells with inhibited Eg5 (in which case centrosome separation is dynein dependent) promoted centrosome separation (Raaijmakers et al. 2012). Balanced interaction among multiple motors, then, appears to regulate centrosome positioning during mammalian prophase.

In *Drosophila* embryos, on the other hand, several studies implicate the cortex as a key site of force generation for migrating centrosomes (Buttrick et al. 2008; Cao et al. 2010; Cytrynbaum et al. 2005). Ncd, a plus-end-directed member of the kinesin-14 family, provides an opposing force (Sharp et al. 2000). As with NE-localized dynein, the question of how asymmetric force is generated by cortical dynein remains unsettled, and the mechanism may depend on features of embryo geometry and cortical organization that are specific to this system and not widely applicable (Cytrynbaum et al. 2005; Tanenbaum and Medema 2010). A dynein-independent mechanism may also exist in this system since some migration persists in the absence of dynein (Robinson et al. 1999). This mechanism may be linked to the extension of cortical actin, potentially driving the motion of astral MTs that track cortical actin flow (Cao et al. 2010). A concept of coupling between cortical actin and force-mediating MT has also been developed in studies of centrosome migration around the male pronucleus of the *C. elegans* one-cell embryo. In this system, migration was impaired upon the removal of either NE-localized dynein (by the depletion of

ZYG-12) or cortical dynein (by the depletion of Gα proteins required for cortical localization), and both cortical dynein and centrosome migration were shown to spatially correlate with cortical actomyosin flow (De Simone et al. 2016), consistent with actin-MT coupling.

7.4.2 Spindle Formation 2: After Nuclear Envelope Breakdown

At the conclusion of prophase, NE breakdown takes place, permitting the assembly of a stable bipolar spindle. Centrosome migration to opposite poles is often complete prior to envelope breakdown, but in some cells this is not the case and migration is completed in the absence of the envelope (Silkworth et al. 2012). In mammalian cells, this late migration is actomyosin mediated and may involve the movement of MTs bound to cortical regions that flow in response to myosin activity (Rosenblatt et al. 2004). Following the placement of centrosomes at opposing poles, studies in multiple systems, including yeast, flies, and vertebrates, have identified motors that regulate the intercentrosomal distance as spindle assembly is completed in prometaphase (Sharp et al. 2000; Tanenbaum and Medema 2010; van Heesbeen et al. 2014; Yount et al. 2015; Yukawa et al. 2018). Pushing and pulling forces, which respectively promote outward and inward movement of centrosomes, are generated, respectively, by Eg5 and kinesin-14, each of which can bind to antiparallel microtubules to create sliding action. A second outward force is generated by the plus-end-directed kinesin-12, which pushes by a sliding action on parallel MTs (Drechsler and McAinsh 2016), while, at least in vertebrate systems, a second inward force is mediated by dynein, which exerts a pulling force from within the spindle (van Heesbeen et al. 2014). The contribution of cortical dynein to prometaphase spindle assembly is not clear in vertebrate cells, although it plays a role in *Drosophila* embryos (Sharp et al. 2000) and also in fungi (Fink et al. 2006). During metaphase, on the other hand, cortical dynein assumes a crucial role in the rotations that adjust spindle orientation (Gallini et al. 2016; Grill and Hyman 2005; Kiyomitsu and Cheeseman 2012; Kotak et al. 2012), as discussed below.

While forces exerted within the spindle regulate centrosome separation, and hence spindle length, centrosomes are governed by forces outside the spindle to stabilize and adjust the spindle with respect to both rotational orientation and also longitudinal position. A recent detailed study of spindle dynamics using live imaging of frog embryonic epithelium shows a clear temporal separation between a first phase (comprising prometaphase and early metaphase) in which a steady rotation repositions the spindle, followed by a second phase, in late metaphase, in which orientation is maintained with little net rotation (Larson and Bement 2017; Roszko et al. 2006). In the second phase, rotary oscillations take place, steadily dampening as anaphase approaches, suggesting an important role for stabilizing forces. A recent investigation of spindle maintenance in *C. elegans* one-cell embryos in metaphase revealed remarkable accuracy, robustness, and stability of positioning, with respect to both centering and also axial orientation (Pecreaux et al. 2016), consistent with spindle centering observed in other systems (Howard and Garzon-Coral 2017). Injection of the one-cell embryos with superparamagnetic beads allowed the direct measurement of spindle-position-maintaining forces using magnetic tweezers (Garzon-Coral et al. 2016). The authors interpreted their results as consistent with a model in which each centrosome is stabilized by damped springs that oppose transverse displacement with a force sufficient to dampen thermal forces but weak enough to permit adjustments.

What are these stabilizing forces? Potential candidates include the astral MT-mediated forces that were discussed earlier in the context of interphase centrosome centration. These include (1) MT pushing forces, driven by MT polymerization at cortical sites; (2) MT length-dependent drag forces mediated by dynein-driven cargo transport; (3) MT pulling forces dependent on cortically localized dynein. Evidence has been obtained to support all three of these mechanisms as stabilizers of spindle position. The authors of the magnetic tweezer study in *C. elegans* embryos proposed an MT pushing mechanism because of its consistency with the spring-like properties they observed, as well as the agreement of the drag forces they measured with values predicted by the model (Garzon-Coral et al. 2016; Howard and Garzon-Coral 2017). This mechanism, however, requires that MTs extend to the cortex, and therefore it has been argued that spindles forming in larger cells, such as eggs, must instead be stabilized by MT length-dependent drag forces (Minc et al. 2011; Wuhr et al. 2010). This is especially the case for metaphase spindles, which in *Xenopus* and zebrafish zygotes have asters whose radius is no more than 10% of the zygote radius (Mitchison et al. 2012).

The third mechanism, MT pulling via cortically localized dynein, is supported by numerous studies demonstrating that dynein is localized to cortical anchoring sites during metaphase and anaphase and that cortical dynein is vital for correct spindle orientation [reviewed in Kotak and Gonczy (2013)]. The anchoring structure for dynein in the cortex is a ternary complex comprised (using vertebrate terms) of Gαi, LGN, and NuMA. This complex only forms during mitosis since during interphase NuMA is stored in the nucleus. Two recent studies, one in human cells and one in the *C. elegans* embryo, found that optogenetic targeting of NuMA to the plasma membrane was sufficient to recruit dynein and locally attract the spindle, confirming the potency of the NuMA-dynein mechanism as a means of controlling spindle orientation (Fielmich et al. 2018; Okumura et al. 2018). NuMA also has MT-binding capability, which may facilitate dynein action but may also provide an alternative means of exerting pulling force via MT shortening (Grishchuk et al. 2005). Inhibition of MT depolymerization with taxol only slightly reduced the spindle-repositioning effect of optogenetically localized NuMA, indicating that while dynein-independent MT shortening may be an important contributor to spindle control (Seldin et al. 2016), it is likely not the major mechanism (Okumura et al. 2018).

While the evidence for the role of cortical dynein as a stabilizing force is compelling, there remains the question, as discussed earlier for interphase cells, of how an MT pulling force that is not length dependent can generate the asymmetric force distribution needed to restore the orientation of a deviated spindle. An answer was provided by a study of the regulation of dynein distribution in the mitotic cortex of human cells (Kiyomitsu and Cheeseman 2012). The authors found that NuMA, and hence dynein, is symmetrically polarized to the regions of lateral cortex that surround the spindle poles, via a mechanism based on chromosome-localized RanGTP that excludes LGN/NuMA complexes from the cortical regions near the spindle midzone. This localization of pulling force provides a means to stabilize the spindle against rotational deviation, but there remained the question of how off-centering motions along the longitudinal axis were corrected. The authors addressed this issue with the demonstration that centrosomal proximity to the cortex (within 2 µm) leads to the local depletion of dynein due to the action of centrosomal Plk1, which phosphorylates NuMA, as well as dynactin, a dynein cofactor (Kiyomitsu and Cheeseman 2012). Consequently, displacement of the spindle toward one pole leads to reduced dynein activity at that pole and a net restoring force to recenter the spindle (Fig. 7.4). More recent findings have extended this work to show that anaphase spindle elongation is driven by further NuMA accumulation, resulting from a shifting balance in NuMA phosphorylation state that depends on the kinase CDK1 and the phosphatase PPP2CA (Kiyomitsu and Cheeseman 2013; Kotak et al. 2013; Kotak and Gonczy 2014).

Is dynein, then, the only pulling motor that stabilizes spindle poles, and are MTs the only instrument for the conduction of force (pushing or pulling) to the centrosome? The recent discovery that centrosomes are surrounded by actin clouds in vivo and can nucleate actin filaments in vitro, using Arp2/3 and the centrosome-localized nucleation-promoting factor WASH (Farina et al. 2016), encourages the exploration of actomyosin as a force generator potentially acting directly on centrosomes. Indeed, a study of spindle positioning in frog gastrula epithelial cells revealed that proper positioning in the plane of the epithelium required both a basal pull generated by MTs and an apical pull that was dependent on actin and myosin-2 (Woolner and Papalopulu 2012). However, it is not clear whether the apical pull on the centrosome reflects the direct action of myosin-2 as a force generator since this myosin was shown to be required for the apically directed cortical flow of actin filaments in these cells. Therefore, the role of myosin-2 may be to apically position actin filaments that then influence the spindle by an as yet unknown, MT-independent mechanism. The authors note in this regard the evidence for subcortical actin structures that appear to exert a pulling force on spindles (Fink et al. 2011).

A final consideration with respect to spindle stabilization is that corrective positioning is not only a matter of regulation of forces. In the late anaphase of HeLa cells, off-centering of the spindle can be corrected by asymmetrical addition of plasma membrane at the polar membrane closest to the spindle, by means of the chromosome-generated RanGTP gradient, which locally depletes anillin to promote local membrane blebbing (Kiyomitsu and Cheeseman 2013). Essentially, instead of pulling the spindle back, the cell extends its boundary asymmetrically in order to restore spindle centrality. It is ultimately architecture, and not forces per se, that matter.

In late anaphase (anaphase B), the spindle poles move apart. Eg5 continues at this stage to represent the principal outward force, so that centrosome separation is most

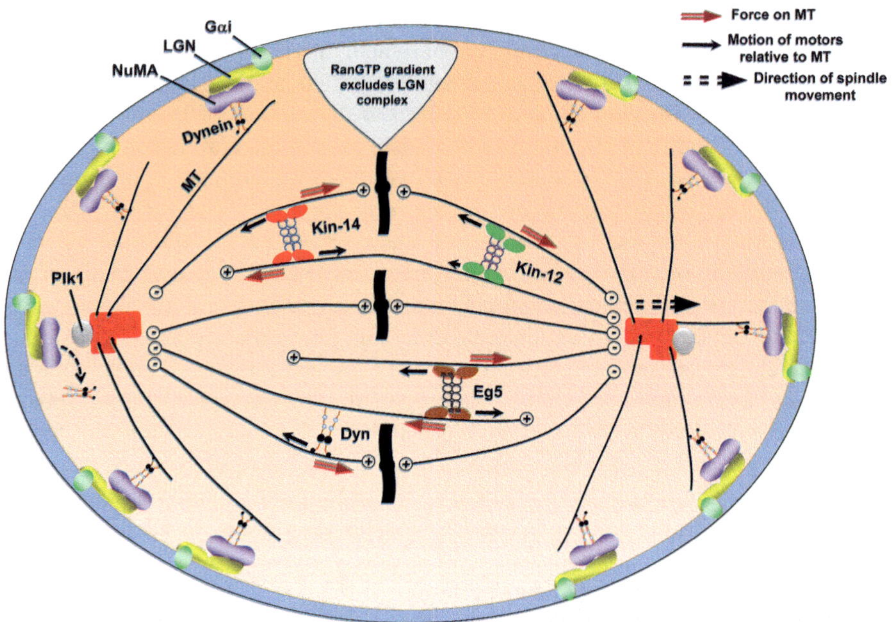

Fig. 7.4 Mechanisms that govern centrosome position during spindle assembly and stabilization. Centrosome separation (spindle length) is governed by the balance between separating motors (kinesin-12 and Eg5) and contracting motors (kinesin-14 and dynein). Note that kinesin-12 generates its force by moving more rapidly on the upper of the two parallel MT to which it is bound. Centrosome positions are also stabilized by the pulling force of dynein bound to cortical ternary complexes. These complexes are excluded from the midzone regions by a gradient of RanGTP. The contribution of NuMA-MT interaction is not depicted. Longitudinal (left-right) off-centering of the spindle is in the process of being corrected by the depletion of dynein pulling force at the left pole due to the proximity of centrosome-bound Plk1. The consequent correcting motion is shown as a dashed double arrow on the right

likely initiated by the limitation of some opposing process. Studies in *Drosophila* embryos suggest that the opposing process is kinesin-13-mediated minus-end MT depolymerization at the spindle poles. Cyclin B degradation at the onset of anaphase leads to the expression of the kinesin-13 antagonist, Patronin, and the consequent stabilization of minus-ends generates outward MT growth (Brust-Mascher et al. 2015; Wang et al. 2013). The findings, however, are also consistent with a model based on plus-end MT stabilization (Brust-Mascher et al. 2015).

7.4.3 The Mitotic Centrosome as Information Reader: Spindle Orientation

Spindles must not only form and be stabilized; they must also be correctly oriented and positioned in order to specify the appropriate plane of cell division. The control of division plane in symmetric divisions determines the geometry of growth in a tissue, while in asymmetrical divisions the division plane is critical for a variety of developmental processes, which include axis formation, fate determination, morphogenesis, and tissue regeneration. In all of these situations, then, the spindle, through its centrosomes, must receive external information in order to be repositioned either by rotation or by translational (longitudinal) motion.

The migration of centrosomes in prophase results in the fixing of initial points for spindle pole formation. Conceivably, then, in a symmetrical cell lacking anisotropic signals to reorient the spindle, the centrosome migration takes place in a stereotyped symmetrical manner, and the resulting spindle, undergoing no adjustment, will adopt a predictable default orientation. Since each centrosome migrates 90° in this case, each division plane is predicted to be rotated 90° relative to the preceding mitosis (Strome 1993). This default scenario, sometimes referred to as Sachs's rule, is observed to occur in many settings, such as early cleavage stages in ascidians (Negishi and Nishida 2017) and later embryonic divisions in *Spiralia* (Brun-Usan et al. 2017). This default "rule" is not in fact adequate to specify a spindle orientation in three dimensions: it cannot determine, for example, whether a cell in a monolayer divides parallel or perpendicular to the monolayer. Nevertheless, the rule is useful since it raises the question of how deviations from this default take place. Such deviations must occur, for instance, whenever a cell undergoes successive divisions along the same axis, as happens, for example, when an epithelial or neural stem cell divides repeatedly along an axis perpendicular to the epithelial plane. Deviations from default can potentially arise at three stages in centrosome positioning: (1) in interphase, by repositioning of the interphase centrosome; (2) in prophase, by asymmetric migration of the disjoined centrosomes; (3) later in mitosis, by rotation of the spindle. Cell division within an epithelial monolayer, as modeled with cultured MDCK cells, offers an example of all three deviations (Reinsch and Karsenti 1994). Since the spindles of the dividing cells lie in the epithelial plane, the centrosome at the conclusion of anaphase is lateral to the nucleus, where the default scenario requires it to remain until disjunction. However, epithelial cells have apicobasal polarity and form primary cilia at the apical surface during interphase. Consequently, during telophase, the centrosome migrates apically and remains there during interphase. A second deviation from default occurs in the ensuing prophase migration: instead of a stereotyped 90° arc, the authors found that the migratory paths of the two centrosomes were in fact highly variable and often asymmetrical, resulting in a variety of positions in prometaphase. Since the final metaphase spindle was robustly parallel to the epithelial plane (within 10°), it is likely that spindle rotation (a third deviation) also took place. More recent studies in chick neuroepithelium demonstrated similarly that, following apical centrosome disjunction, highly variable centrosome separations were corrected by spindle z-plane rotation early in metaphase to generate a planar orientation, while at the same time x-y rotations (within the plane) appeared to be random (Peyre et al. 2011). Similar observations with respect to centrosome migration during mitosis have been made in other epithelial cell types (Rattner and Berns 1976; Zeligs and Wollman 1979), and a similar rotational correction has been observed in algae (Bisgrove and Kropf 1998).

Examples of asymmetric centrosome migration can also be found in systems of asymmetric cell division. In the mitoses of *Drosophila* larval neuroblasts, the spindle is oriented along the apicobasal axis, and an asymmetric division generates an apical neuroblast and a basal neuroglial progenitor cell. The neuroblast then repeats this division in the same orientation, in violation of the default process. In the prophase of this mitosis, the disjoined centrosomes take very different paths: the daughter centrosome makes only short-range movements and remains in its apical position, while the mother centrosome begins a series of rapid, irregular motions culminating in a basal localization (Rebollo et al. 2007), thereby creating the axis that determines the spindle orientation. The asymmetric division of *Drosophila* male germline stem cells offers a similar example, although in this case the centrosome disjunction occurs in interphase rather than prophase (Yamashita et al. 2007). A third example occurs in the embryos of the ascidian *Phallusia mammillata*. In the dorsal midline of the blastula of this species, NNE cells have a spindle aligned with the animal–vegetal axis, producing an NN cell on the animal side and an E cell on the vegetal (Fig. 7.5). The E cell, with an interphase centrosome facing the vegetal pole, then divides with a spindle again on the animal–vegetal axis, representing a deviation from default. As in the case of *Drosophila* neuroblasts and germline stem cells, one of the disjoined E cell centrosomes remains at its original vegetal location during prophase, while the second centrosome moves 180° to the animal side, providing for the necessary spindle orientation (Fig. 7.5) (Negishi and Yasuo 2015).

In contrast to these systems in which atypical centrosome migration plays an important role, a number of deviations from default spindle positioning can be primarily attributed to spindle rotation in metaphase. One example is the division of the NN cell in the ascidian embryo (Fig. 7.5) (Negishi and Yasuo 2015). A second example takes place in the stem cell niche of *Drosophila* neuroepithelium. Embryonic neuroblasts (distinct from the larval neuroblasts discussed above) initially form a spindle parallel to the epithelial plane (similar to their neighboring epidermoblasts). This spindle then rotates 90° so that it lies perpendicular to the epithelial plane, leading to an asymmetric division between an apical daughter, which retains stem cell character, and a basal neural progenitor cell (Kaltschmidt et al. 2000). In its subsequent divisions, however, the apical daughter directly generates perpendicular spindles without the need for rotation (Rebollo et al. 2009). A third example is found in the keratinocytes of developing mouse epidermis. Individual basal keratinocytes are able to choose between a spindle orientation parallel to the basement membrane and a perpendicular orientation leading to the stratification of the epithelium (Poulson and Lechler 2010). The authors showed that interphase centrosomes did not migrate in this system and also that spindle orientation was random at prometaphase. They inferred that final spindle orientations must be determined by rotation in metaphase, and they supported this interpretation by demonstrating metaphase rotations in cultured keratinocytes. A similar use of metaphase rotation to correct random prometaphase spindle alignment occurs during the asymmetric division of the *Drosophila* sensory organ precursor cell (Bellaiche et al. 2001). Finally, the first three rounds of cell division in *C. elegans* embryos afford three instances of 90° rotation, although in each case the rotating object is not the spindle but rather a combined nucleus-centrosome

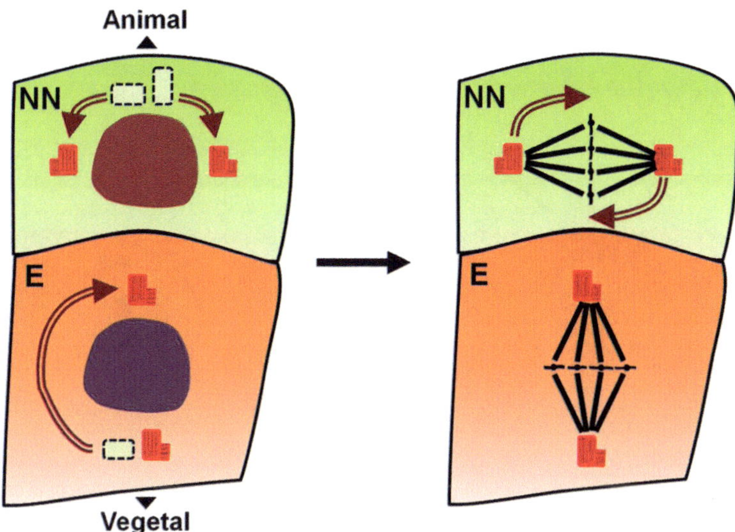

Fig. 7.5 Deviations from default spindle positioning in an ascidian embryo. The dorsal midline of the blastula of *Phallusia mammillata* contains pairs of bilaterally symmetric NN and E cells. The right NN and right E cells are depicted in dorsal view. The NN cell deviates from default positioning by rotation of its spindle, while the E cell deviation is the result of asymmetric centrosome migration in prophase. In both cells, the final spindle orientation is identical to that of their mother NNE cell (i.e. parallel to the animal-vegetal axis)

complex, just prior to NE breakdown. These rotations occur in the zygotic division, which creates an AB cell and a P1 cell, the division of P1, creating P2 and EMS, and the division of EMS (Fig. 7.6) (Galli and van den Heuvel 2008). The nucleus and centrosomes usually rotate as a unit, but in some cases the motions of the two centrosomes are not coordinated (Hyman and White 1987). Notably, for both the P1 and EMS rotations, the choice between clockwise and counterclockwise rotation was random (Hyman and White 1987), as was also the case for the ascidian NN cell mentioned above. However, in the ascidian case, this choice of rotational direction was correlated between the two bilaterally symmetric NN cells (Negishi and Yasuo 2015), suggesting a role for intercellular or tissue-level signaling in the control of spindle rotation.

These dramatic spindle rotations might conceivably be governed by special mechanisms required for specific developmental events. The evidence suggests that, at least with respect to the mechanism of force generation, this is not the case: the NuMA-dynein cortical pulling mechanism used to stabilize metaphase spindles in most cells is also used in cases in which rotational reorientation is required. However, in the asymmetric cells in which these reorienting spindles are observed, the mechanism has an added feature: the force generators are asymmetrically localized to cell cortex. The asymmetric localization of LGN and NuMA has been implicated in the spindle rotations observed in *Drosophila* embryonic neuroblasts (Yoshiura et al. 2012), mouse epidermis (Poulson and Lechler 2010;

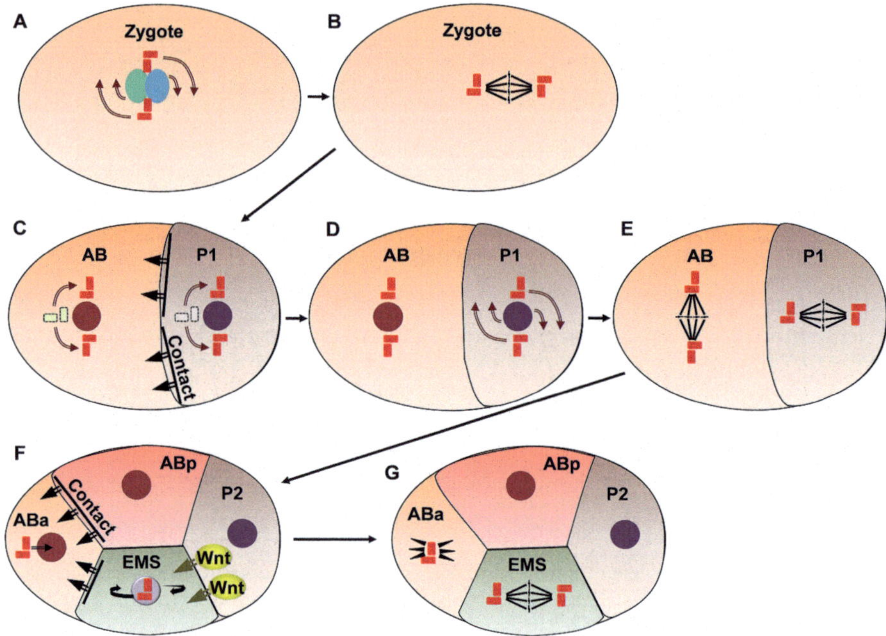

Fig. 7.6 Spindle positioning in the *C. elegans* embryo. The embryo is seen in side view, with anterior to the left. In the zygote (**a**) and P1 cell (**d**), paired arrows indicate the rotation of the nucleus-centrosome complex. In the zygote, this complex includes the two pronuclei. (**b**) After rotation, the complex moves posteriorly to create an off-center spindle. (**c–f**) The AB cell generates ABa and ABp, while P1 generates P2 and EMS. Arrows in AB (**c**), P1 (**c**), and ABa (**f**) indicate the path of centrosome migration. (**c**, **f**) Signaling via cell-cell contact is indicated by black bars linked to short arrows that indicate the direction of instruction. (**c**) contact with P1 controls spindle direction in AB. (**f**) contact with both the larger ABp and the smaller EMS controls the orientation of the ABa spindle, which points into the page along the left-right axis (**g**). Wnt signals generated by P2 (**f**) lead to rotation of the nucleus-centrosome complex in EMS (black arrows), generating a spindle along the anterior-posterior axis (**g**)

Williams et al. 2014), and the *C. elegans* one-cell embryo (Park and Rose 2008). Asymmetric NuMA (but not LGN) localization is also implicated in the *Drosophila* sensory organ precursor (Segalen et al. 2010). NuMA is also required, but without asymmetric localization, for rotation in the *C. elegans* EMS cell (Heppert et al. 2018; Liro and Rose 2016). In ascidian NN cells, an asymmetric cytoplasmic distribution of dynein is implicated in spindle rotation, but whether or how this dynein is anchored has yet to be determined (Negishi and Yasuo 2015). Finally, translational and rotational spindle positioning may be governed by similar pathways: in budding yeast, a translational movement of the spindle into the mother bud neck is mediated by cortical dynein anchored by Num1 (Omer et al. 2018), which functions analogously to NuMA, although it is not a NuMA structural homlog (Greenberg et al. 2018).

These studies imply that anisotropic information, supplied by either intrinsic or extrinsic signals, flows to spindle pole centrosomes via the asymmetric localization of force generators. What links the information-bearing signals to motor localization? Two conserved pathways of polarity-regulating proteins have been broadly implicated in cell polarization that governs spindle orientation in metazoans: the planar cell polarity (PCP) pathway, which is commonly activated by Wnt ligands, and the PAR (partitioning-defective) pathway (Devenport 2016; Lu and Johnston 2013; McCaffrey and Macara 2012; Smith et al. 2017). Components of these pathways (polarity proteins) can support cell polarization by self-organizing into distinct, mutually excluding cortical domains. PCP signaling, for example, is required for NuMA localization and spindle orientation in the *Drosophila* sensory organ precursor, as well as zebrafish gastrula (Segalen et al. 2010). PAR signaling is required for spindle orientation in *C. elegans* one-cell embryos (Gotta et al. 2003), and differences in PAR signaling account for the distinct spindle orientations in *Drosophila* embryonic neuroblasts compared to their epithelial neighbor cells (McCaffrey and Macara 2012). Cortical PAR complexes can asymmetrically localize the ternary complex through their ability to bind LGN via the adapter Inscuteable (Mapelli and Gonzalez 2012). Cooperation between the PCP and PAR pathways is necessary for spindle control in both *Drosophila* sensory organ precursor (Besson et al. 2015) and developing mouse epidermis (Williams et al. 2014). However, while the control of motor localization and activity is necessary for correct spindle orientation, it may not be sufficient. It is possible that independent mechanisms predetermine the final spindle axis and thereby control the extent of rotation, as has been suggested for the *C. elegans* embryo (Hyman and White 1987). Recent optogenetic studies in the *C. elegans* embryo addressed the question of NuMA sufficiency by targeting NuMA to specific cortical regions (Fielmich et al. 2018). In the zygote, mislocalized NuMA initially reoriented the spindle, but in late mitosis the correct orientation was restored, consistent with the notion of an independent axis-determining constraint. However, in two-cell embryos, this restoration did not occur. This system may provide unique opportunities to examine the interaction between distinct polarity-determining mechanisms.

While dynein-independent forces may contribute to spindle repositioning, evidence for them is currently limited. One well-studied example is provided by the *C. elegans* one-cell embryo, in which the spindle undergoes a translational shift from the cell center toward the posterior end, generating asymmetric daughter cells of unequal size (Fig. 7.6b, c). This spindle movement is driven in part by force generated by cytoplasmic flow, in addition to a contribution from MT-dependent dynein pulling force (Saturno et al. 2017). Specifically, initial asymmetry in the cell (created by sperm entry at the posterior end) leads to asymmetric distribution of cortical actomyosin-based contractility, producing anteriorly directed cortical flow that in turn results in a posteriorly directed flow of interior cytoplasm (Hird and White 1993), which helps to push the spindle posteriorly. Similar off-centering of the spindle is likely to be required in other situations of unequal cleavage, as commonly occurs in many protostomes, by mechanisms yet to be determined. Recent further studies in the *C. elegans* embryo provide evidence that myosin-driven cortical flow can control spindle orientation in a manner that is independent of dynein and MT-dependent forces (Sugioka and Bowerman 2018). In this study, Sugioka and Bowerman showed that, in two-cell embryos, the spindle orientation in AB cells is unaffected by either the depletion of LGN (and hence cortical dynein) or the depolymerization of MTs but is dependent on actin and myosin. Furthermore, treatments (discussed below) that induced accurate spindle orientation in isolated AB cells also triggered asymmetric myosin flow similar to that observed in vivo, suggesting that the spindle was oriented by flow.

Some progress has been made, then, with respect to how information is "read" into mitotic centrosomes. But what is the nature of this information? Anisotropic cues are potentially available from cell geometry or internal organization (intrinsic cues), as well as from a host of external sources, including adhesion patterns, intercellular junctions, signals from neighboring cells, and tissue-scale mechanical forces. In zygotes, spindle orientation can be governed by both intrinsic information, such as the animal–vegetal axis in eggs (Sardet et al. 2007), and extrinsic cues, such as the sperm entry site in *C. elegans* (Galli and van den Heuvel 2008). In the *C. elegans* system, some of the cues and signaling pathways that control spindle orientation have begun to be elucidated. In particular, in the four-cell embryo, the P2 cell signals to the neighboring EMS cell via the paracrine factor Wnt (Fig. 7.6f) and, in a separate, parallel pathway, the receptor Mes-1 in order to drive a 90° spindle rotation (Liro and Rose 2016). Members of the Wnt family are initiating signals for PCP polarity signaling, so similar cues from neighboring cells may play a role in many systems in which PCP signaling has been shown to control spindle orientation.

Can asymmetric information be conveyed other than by intercellular signal transduction? Adhesions and intercellular junctions are a potential source of such information, but they are experimentally difficult to assess since, for example, the disruption of physical contact may have nonspecific effects. In their study of *C. elegans* embryos, Sugioka and Bowerman developed an interesting alternative approach: the use of adherent beads as substitutes for neighboring cells (Sugioka and Bowerman 2018). When isolated AB cells were paired with a bead, their spindle oriented parallel to the bead, exactly as if the bead were a P1 cell. They next examined the daughter of AB, ABa, which is flanked by two cells of unequal size and normally divides along a left-right axis that is parallel to its neighbors but orthogonal to the spindle of its mother AB cell. Remarkably, correct spindle orientation was recovered in an isolated ABa cell only when it was paired with unequal beads resembling its normal neighbors. While these data do not yet reveal the nature of the signals conveyed to ABa, they are compelling evidence that cell–cell adhesive contact per se, in the absence of any specific extrinsic signal, can supply sufficient information to guide spindle orientation. Interestingly, the authors proceeded to employ the same method to examine the EMS cell, which as we have seen is normally instructed by P2 to rotate its spindle. When EMS was paired with a bead plus cells other than P2, its spindle was instructed by the bead in a manner similar to AB and ABa: that is, it oriented parallel to the bead. But when EMS was paired with a bead plus P2, it rotated its spindle to point toward P2, exactly as in situ, and this orientation was Wnt-dependent. Thus, EMS responded both to contact and to Wnt, but when the two signals conflicted, the Wnt signal was dominant. This study, then, provides the first evidence of a possible hierarchy of extrinsic cues, in which one cue may override another.

Cues from adjacent cells were also explored for their potential to control the asymmetric migration of prophase centrosomes in the E cells of the ascidian embryo (Negishi and Yasuo 2015). Ablation of neighbor or near-neighbor cells generated a variety of effects on centrosome migration. The interpretation of the authors was that asymmetric migration was likely governed by an intrinsic signal related to E-cell shape, which in turn was affected by the ablation of nearby cells, rather than by signaling derived from cell–cell contact.

Cells often need to adjust spindle orientation based on information that is anisotropic but not necessarily asymmetric. A cell dividing in a monolayer faces this situation, for example, in determining the direction for its spindle within the plane of the monolayer. It would seem both mechanically and energetically favorable for the cell to divide across its longest axis, and a century ago this was proposed as a rule by Hertwig (1884). But if the cell is sensing its shape as a cue, what is the source of information? One possibility is that the cell collects global shape data via its MT asters; another is that it identifies landmarks, such as junctions or adhesions. To experimentally distinguish these mechanisms is challenging since alteration of adhesion, for example, is likely to alter shape. By using micropatterned substrates, however, it is possible to maintain cell shape while varying adhesion pattern, and the results clearly identify adhesions, rather than global shape, as a cue for spindle orientation (Thery et al. 2005). A more recent study has identified tricellular junctions, which accumulate NuMA, as a cue (Bosveld et al. 2016, 2018). The problem arises that when the cell enters mitosis, it rounds up, profoundly altering its shape, adhesions, and junctions. However, there is compelling evidence that the original information is not quite lost: retraction fibers persist, generating cortical cues that can guide spindle orientation (Fink et al. 2011; Thery et al. 2005). These cues likely involve actin filaments, which may indirectly recruit NuMA and dynein via Slk-activated ERM proteins (Machicoane et al. 2014). Nevertheless, while these cortical cues might suffice to enforce a planar spindle orientation, it seems that they are not the only input to the spindle. A recent study of the columnar epithelium in *Drosophila* gastrulae revealed that when the rounding up of the mitotic columnar cells was inhibited by blocking myosin activity, the spindle oriented perpendicular to the plane, as if obeying a Hertwig-rule geometric instruction (dividing so as to bisect the long axis) rather than the cortical cues, even though the cortical localization of LGN (Pins in *Drosophila*) remained unaffected by myosin depletion (Chanet et al. 2017). This study represents, then, another example of the potentially hierarchical nature of orientation cues, as in the case of the override by Wnt signaling in the *C. elegans* EMS cell. It seems plausible that in stratified epithelia such as mammalian epidermis, in which spindles perpendicular to the plane are necessary for tissue renewal, a similar override may be mediated by PCP signaling (Poulson and Lechler 2010; Williams et al. 2014). The notion of cell geometric cues operating in the absence of cortical cues is also consistent with studies of spindle orientation in large eggs (Minc et al. 2011). From this point of view, the centrosome at the spindle pole represents an integration point for multiple streams of information flow. How these streams of input are sorted, organized, and prioritized will be a fascinating area for further investigation.

7.4.4 The Mitotic Centrosome as Information Reader: Centrosome Asymmetry

While the symmetry of the spindle around its equator is apparent, it is never absolute: mother and daughter centrosomes have distinct histories and features (Piel et al. 2000), and this asymmetry can potentially be exploited as a means for an asymmetrically dividing cell to collect and transmit information from asymmetric cues. It is of interest, for example, to determine in cases of asymmetric stem cell division whether one daughter cell preferentially acquires a mother or daughter centrosome. Remarkably, in five out of the six such systems that have been examined, asymmetric centrosomal inheritance indeed takes place. In mitoses of *Drosophila* neuroblasts (Januschke et al. 2011, 2013) and female germline stem cells (Salzmann et al. 2014), the daughter cell that retains stem character also retains the daughter centrosome. On the other hand, in mitoses of *Drosophila* male germline stem cells (Yamashita et al. 2007), as well as the radial glial progenitor cells of developing mammalian neocortex (Wang et al. 2009) and also mammalian embryonic stem cells (Habib et al. 2013), the daughter cell with stem character retains the mother centrosome. Interestingly, retention of the mother centrosome in embryonic stem cells is triggered by asymmetric Wnt signaling (Habib et al. 2013). In addition, in budding yeast, the mother cell retains the mother spindle pole body (Lengefeld and Barral 2018). The lone counterexample among the stem cell systems is the granule neuron progenitor of developing mouse cerebellum (Chatterjee 2018).

Neither the mechanism nor the functional significance of asymmetry in centrosomal inheritance is as yet well understood. The molecular nature of the asymmetry is incompletely characterized and, as a consequence, tools for its manipulation are still largely lacking. The mother centrosome, as discussed later, is responsible for the formation of primary cilia, and cells inheriting the mother centrosome can form such cilia more quickly, perhaps leading to differential signal transduction in daughter cells (Anderson and Stearns 2009; Paridaen et al. 2013). Another possibility is that the centrosomes carry distinct cargo to the two daughter cells. Proteins destined for proteasomal breakdown have been observed to specifically associate with one of the two centrosomes in the cell (Fuentealba et al. 2008), as have selected mollusc embryonic mRNAs (Lambert and Nagy 2002), although it was not determined if these were mother or daughter centrosomes. When asymmetric centrosome inheritance in the neocortex was disrupted by the depletion of ninein, a MT-nucleating protein found preferentially in mother centrosomes, radial glial progenitor cells were lost from the ventricular zone (Wang et al. 2009), suggesting that important functional distinctions exist between older and younger centrosomes. The impact of centrosomal inheritance on daughter cell fates is a promising area for future investigation.

7.5 Centrosome Positioning and Cell Polarity

While the mitotic centrosome serves as a focal point for the control of mitotic polarity, the interphase centrosome responds to and transmits information guiding polarized cellular activity. The interphase centrosome is uniquely suited to this function as it is distinctly localized and persistently unitary: although it contains two centrioles that can be separated by several μm in G1 phase, this intercentriolar distance is fairly stable (within approximately 1 μm) and, perhaps more importantly, the mother centriole, which performs the bulk of interphase MT nucleation, maintains a stable position in the absence of external stimuli (Piel et al. 2000).

The polarized functions associated with centrosome positioning include directional responses to migratory stimuli, apicobasal polarity in epithelia, morphogenetic polarity in development, planar polarity in ciliary positioning, and polarized cell–cell interaction in the case of the immunological synapse. Interacting with several of these processes is a polarized cellular feature that is present in most eukaryotic cells: the primary cilium, whose formation is organized by the mother centrosome that serves as its basal body. While the functions of primary cilia are still in an early stage of characterization, this structure may represent the most direct and unambiguous link between the centrosome and cellular polarity. Consideration of the control of ciliation with respect to centrosome positioning will serve as a useful introduction to a further discussion of polarized functions.

7.5.1 Centrosome Migration During Cilia Assembly and Disassembly

Formation of a primary cilium has classically been viewed as a hallmark of postmitotic cells in vertebrate systems. However, proliferating cells can form cilia, although they must be transiently removed during mitosis (Plotnikova et al. 2009; Wheatley et al. 1996). Cilia are also not uniformly detected in postmitotic tissues (D'Angelo and Franco 2011) or on cultured cells upon cell cycle exit (Vorobjev and Chentsov Yu 1982; Wheatley et al. 1996), and cell culture settings can affect cilium formation in ways that are cell type dependent and not understood (Strugnell et al. 1996). Thus, both ciliation and deciliation are subject to complex regulation by mechanisms that are only beginning to be investigated.

Ciliation involves the transition of the mother centriole into a basal body docked at the plasma membrane and therefore requires centrosome positioning at the membrane. While specific signals that trigger this centrosome movement have not been elucidated, some evidence has been obtained with respect to the forces involved. In *C. elegans*, a primary cilium forms in the dendrites of sensory neurons, requiring centrosome migration from cell body to dendrite. A recent study showed this migration to be partly dynein dependent, suggesting a role for MT pulling forces (Li et al. 2017). On the other hand, MT pushing forces have been implicated in studies of centrosome migration during ciliation induced by serum withdrawal in micropatterned retinal epithelial (RPE1) cells. The authors demonstrated that centrosome migration in the apical direction was associated with MT stabilization and the formation of dense MT bundles oriented between the centrosome and the basal pole of the cell, suggesting that MTs were pushing off the basal cortex to move the centrosome apically (Pitaval et al. 2017). Consistently, migration was enhanced by treatments that stabilized MTs or promoted MT growth. They also showed that the depletion of a specific centrosomal protein, the distal appendage protein CEP164, prevented both MT stabilization and migration. Since CEP164 has also been linked to ciliogenesis effector functions and to the regulation of epithelial-mesenchymal transition (Slaats et al. 2014), this finding may provide an avenue for the investigation of centrosomal functions that link ciliogenesis with other cellular pathways. In addition, the authors found that migration was dependent on kinesin-2, which hints at a second possible pushing mechanism. Kinesin-2 is plus end directed and so, if anchored, could potentially exert pushing force toward the centrosome. Kinesin-1, also plus end directed, interacts with nesprin-4, which is bound to the NE by a SUN–KASH interaction. This interaction has been shown to generate a large centrosome displacement from the nucleus in nesprin-4-overexpressing cells (Roux et al. 2009). Several studies have implicated nesprins in cilia formation (Dawe et al. 2009; Pitaval et al. 2017). Finally, the migration to the plasma membrane may also be regulated at its final stage: the docking of centrioles to the membrane [see Wang and Dynlacht (2018) for a recent review of the docking mechanism]. An interesting series of recent studies suggests that centrosome migration during ciliogenesis depends on docking signals mediated through meckelin, a receptor that may interact with PCP signaling and actin modulation (Adams et al. 2012; Dawe et al. 2007b, 2009).

While the ciliating centrosome must always reach the plasma membrane, there are cell types, including fibroblasts and also retinal epithelial cells, in which it does not reach the cell surface. In these cells, vesicles are recruited to the centrosome within the first hour of serum withdrawal and fuse to form a ciliary vesicle docked to the mother centriole, which initiates the growth of axonemal microtubules into this vesicle while the assembly is still cytoplasmic (Kobayashi et al. 2014; Lu et al. 2015; Schmidt et al. 2012; Westlake et al. 2011). When the elongated vesicle finally fuses with the plasma membrane, the result is a deeply invaginated pocket, with the centriole/basal body at its base. In contrast to this "intracellular pathway," in some epithelial cell types the centrosome first migrates to the cell surface, then docks at the plasma membrane and initiates cilium formation (extracellular pathway) (Wang and Dynlacht 2018). Regulation of ciliation, then, is not only a matter of "when" but also "how." Interestingly, the depletion of centriole proteins needed for the formation of subdistal appendages, a structure unique to the mother centriole, abolishes the intracellular pathway. The mutant mother centriole separates from the daughter (in contrast to normal ciliogenesis, in which the two centrioles retain their association), moves away from the Golgi complex (which may normally provide vesicles needed for ciliary vesicle formation), and docks at the plasma membrane to generate a cell surface cilium (Mazo et al. 2016). These studies provide important new tools for the investigation of centrosome dynamics.

The basal bodies generated by the intracellular pathway typically remain close to the nucleus, and it remains to be established to what extent, in these cells, centrosome association with the nucleus is altered. In ciliated fibroblasts, or other cells using this pathway, retention of the basal body in a paranuclear position might facilitate migratory responses that depend on centrosome-nucleated MTs, as well as transitions in and out of mitosis, by eliminating the need for "shuttling" to and from the cell surface. Even in ciliated proliferating epithelial cells (Alieva and Vorobjev 2004; Reinsch and Karsenti 1994), as well as in the micropatterned retinal epithelial cells treated by serum withdrawal (Pitaval et al. 2017), the final apical position of the centrosome is also not very distant from the nucleus, and most if not all of the centrosome migration may consist of movement along the NE. Certainly in ciliated epithelial and neuronal tissues in situ, apically localized basal bodies can be considerably separated from the nucleus (Alieva and Vorobjev 2004; Carvajal-Gonzalez et al. 2016a; Spear and Erickson 2012), but this positioning likely reflects a complex program of apicobasal polarization (Rodriguez-Boulan and Macara 2014) that has not been fully recapitulated in the cell culture systems that have been used to study ciliogenesis. The extension of these studies to systems that more fully capture epithelial maturation will likely yield important insight into mechanisms that govern centrosome positioning.

The reverse movement of centrosomes during deciliation is not well studied. Deciliation can be triggered by mitogenic stimulation, for example, via the PDGF receptor, which leads to the activation of the Aurora A kinase (Nielsen et al. 2015). As discussed earlier with respect to mitotic entry, Aurora A localizes to centrosomes in G2, coincident with a wave of ciliary resorption (Plotnikova et al. 2009). Aurora A is both necessary and sufficient for deciliation, and it is also active early in G1 phase after mitogenic stimulation, at the time of another wave of resorption (Plotnikova et al. 2009). Intriguingly, while the cilium is completely disassembled prior to mitotic entry, the same is not always true for the ciliary vesicle. An endocytosed remnant vesicle remains attached to the mother centriole throughout mitosis, enhancing the previously discussed advantage possessed by the mother centriole with respect to the rapidity of reciliation in the daughter cell to which that centriole passes (Paridaen et al. 2013). This persistence of docking through mitosis is not universal: it was observed in two mammalian cell lines, as well as developing neocortex, but in the latter case persistence waned as development progressed (Paridaen et al. 2013).

7.5.2 *Centrosome Positioning During Directional Migration*

Directional migration is the polarized function that has been most intensively studied with respect to linkage to centrosome positioning. The systems in which this linkage has been examined include slime molds, fish keratocytes, and mammalian neurons, astrocytes, fibroblasts, leukocytes, and epithelial cells, among others. While many years of effort have revealed mechanisms through which centrosome position and migratory polarization may be connected, a number of fundamental questions remain unsettled.

One question is whether there is such a thing as a nonpolarized cell, with respect to migratory direction. Experimentally, polarization is assessed in response to a stimulus. But are unstimulated cells devoid of any substructure (such as an oriented centrosome) that biases the direction of migration? Early studies in the slime mold *Dictyostelium*, which moves by an amoeboid process of pseudopod extension, suggested that the centrosome did not bias the cell: in a randomly migrating cell, the production of a new pseudopod, in a new direction, was not preceded by an orientation of the centrosome in the new direction. In fact, after new pseudopod initiation, an average of 12 s elapsed before centrosome orientation to the new direction began (Ueda et al. 1997). This centrosome orientation, however, was necessary for the pseudopod to persist. When the cells were induced to turn using a chemotactic stimulus, centrosome orientation lagged similarly. The results were consistent with random (nonpolarized) choices of direction, followed by centrosome-mediated stabilization. More recent studies with *Dictyostelium*, however, showed that the "random" migration is not a random walk because turns in one direction tend to be followed by turns in the other, resulting in a degree of persistence (Li et al. 2008), as later shown also for amoeboid migration in T cells (Liu et al. 2015). Amoeboid migration in newt eosinophils also shows directional persistence, with a leading centrosome, and persistence was lost upon centrosome ablation (Koonce et al. 1984). The authors of the *Dictyostelium* study suggested that the cell has an "intrinsic vector" that biases its next change of direction and hypothesized that this vector may be based on centrosome position (Li et al. 2008). Compelling evidence for an intrinsic vector was obtained in studies of neutrophil-like HL-60 cells, when "undirected" migration was induced by uniform treatment with a chemokinetic factor (Xu et al. 2007). The authors first determined the nucleus–centrosome axis prior to treatment. The subsequent-induced migration displayed a clear directional bias with respect to this axis. Remarkably, this bias was not along the axis, but rather to its left, implying that migratory direction was influenced by a chiral substructure of which centrosome position represented a component (Fig. 7.7a, b). Unlike most mammalian cells, neutrophils share with *Dictyostelium* an amoeboid style of locomotion. Might a chiral centrosome-based entity direct alternating turn polarities during amoeboid migration, perhaps by rotations or inversions of the chiral structure? Do similar "intrinsic vectors" or chiral substructures exist in other cells, including those that undergo mesenchymal locomotion, involving lamellipod extension? In fact, multiple cell lines show evidence of chirality during polarization on micropatterns and in 3D cultures, suggesting that chirality may be a universal cell property (Fig. 7.7c, d) (Chen et al. 2012; Chin et al. 2018; Wan et al. 2013). Further investigation in this area promises to be quite interesting.

A second basic question is: to what position do centrosomes move when a cell polarizes in response to a directional migratory stimulus? Many early studies of centrosome behavior, employing wound (scratch) migration assays in a variety of

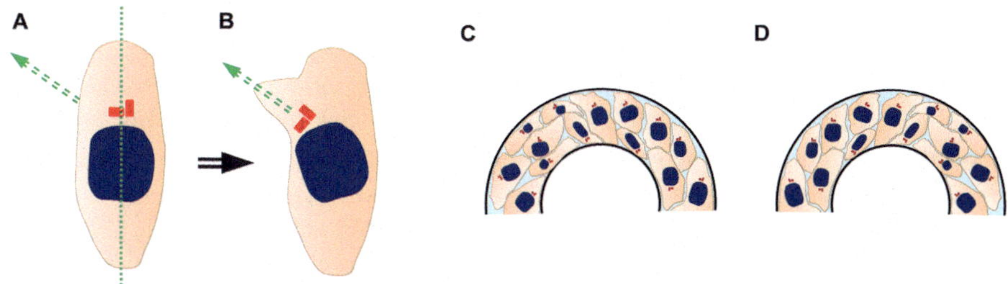

Fig. 7.7 Centrosome positioning linked to chirality, Part 1. (**a**, **b**) Centrosome position defines chirality in polarizing HL-60 neutrophil-like cells (Xu et al. 2007). The axis formed by the nucleus and centrosome prior to polarization is shown as a dotted line in (**a**). An "intrinsic vector" of polarization (green arrow) is inferred from the subsequent axis of polarization (**b**) formed upon treatment with fMLP. (**c**, **d**) Chiral patterning in mammalian cells grown on ring-shaped micropatterns (half the micropattern is shown) (Wan et al. 2011). Different cell types display either a counterclockwise (**c**) or clockwise (**d**) alignment. Centrosomes are shown in red

mammalian cells, observed that centrosomes became positioned on the side of the nucleus nearest the leading edge of the migrating cell (Etienne-Manneville and Hall 2001; Euteneuer and Schliwa 1992; Gotlieb et al. 1981; Gundersen and Bulinski 1988; Kupfer et al. 1982). Initially, these observations suggested a movement of the centrosome to a position "ahead of the nucleus," from which it could support migration. However, more careful examination in epithelial cells showed that throughout the wound response, the centrosome remained at the cell centroid, or sometimes slightly behind it, either in the presence or absence of nuclei (Euteneuer and Schliwa 1992). The authors suggested that the shift in the nucleus–centrosome relationship was unimportant and that instead the centrosome should be seen as simply tracking the cell centroid, which gradually moved in the direction of the extending lamellipod at the leading edge. Similar findings were later obtained with fibroblasts (Gomes et al. 2005). The relevance of being "ahead" of the nucleus was further downgraded by the discovery that in these wound assays, the centrosome could be found either ahead of or behind the nucleus, depending on the cell line, the substrate, or the expression of cadherins (Dupin et al. 2009; Schutze et al. 1991; Yvon et al. 2002). Subsequently, the centrosome (or the Golgi complex, which is generally adjacent to the centrosome) was found to be in the rear of migrating cells observed in situ in the lateral line primordium of zebrafish and also in cells on one-dimensional micropatterned tracks (Pouthas et al. 2008). In these systems, the Golgi complex/ centrosome appears to be considerably behind the cell centroid. The authors observed that as they widened the micropatterned track on which cells migrated, thus approaching the two-dimensional conditions used in the earlier wound studies, the Golgi position gradually tended to shift from rear to front, confirming the importance of cellular environment with respect to the organization of cellular polarity. Migratory behavior on one-dimensional tracks showed similar properties to migration in three-dimensional matrices, in which rear-positioned centrosomes were also observed (Doyle et al. 2009). However, front-facing centrosomes are not only found in wound assays: they are also observed, for example, in migrating neurons in developing neocortex (Dantas et al. 2016) and in a majority of migrating epithelial cells in the peripheral cornea (Silverman et al. 2017), among others [reviewed in Luxton and Gundersen (2011)]. Thus, centrosome positioning is highly context-dependent during migration, as has also been observed in *Dictyostelium* (Sameshima et al. 1988). It should also be emphasized that in most of these studies, the reported organelle localizations are statistical, with considerable variation among cells. In a recent in vivo study in medaka fish, centrosomes in migrating leukocytes were observed to continually change their position from front to rear, in a manner that correlated with localized actin accumulation (Crespo et al. 2014). Similar switching between front and rear, but with a pronounced rearward preference, was also recently observed with respect to the position of the Golgi complex in breast epithelial cells migrating on one-dimensional micropatterns (Natividad et al. 2018). The authors also noted that a substantial fraction of these cells did not switch but appeared instead to be committed to either a front or rear Golgi position. Cells, then, may be able to choose not only between alternative centrosome/Golgi positions but also between "committed" and "plastic" positioning. Interestingly, when cells on one-dimensional tracks were amputated at each end, so that the centrosome was closer to one cut end, 78% of cells migrated toward the cut end that was distal to the centrosome, suggesting that the cell "interprets" the centrosome position as the rear end (Zhang and Wang 2017). This finding comports with unpublished studies that argue for rear-end migratory steering based on observations of asymmetries in myosin distribution and actin flow in cells that persistently turn as they migrate (Allena et al. 2018). It is possible that a default mechanism positions the centrosome rearward to participate in directional control but that final positioning depends on the interaction of this mechanism with additional regulatory networks, polarizing stimuli, and/or mechanical constraints in ways that have yet to be elucidated.

A third question is how the positioning of the centrosome is controlled during migration. The forces acting on the centrosome are most likely MT pulling forces, mediated by dynein, rather than actomyosin forces [although these have been implicated in the case of glial-guided neuronal migration (Solecki et al. 2009)]. In wound models using multiple cell types, centrosome reorientation toward the wound is abolished by MT-depolymerizing drugs, even at low concentrations that preserve the MT cytoskeleton, while treatments that depolymerize actin, or that either inhibit or activate the upstream actomyosin regulators Rac and Rho kinase, had no effect on centrosome position, although they efficiently blocked lamellipodia formation and cell migration (Etienne-Manneville and Hall 2001; Gotlieb et al. 1983; Liao et al. 1995; Palazzo et al. 2001). During polarization, a prominent network of stabilized MTs with modified tubulin forms between the centrosome and the leading edge. However, this stabilized network is not required for centrosome reorientation (Palazzo et al. 2001). Inhibition of dynein by antibody injection or by dynamitin overexpression leads to mislocalization of the centrosome in wound models, consistent with a role for dynein pulling forces in centrosome orientation (Etienne-Manneville and Hall 2001; Palazzo et al. 2001). Subsequent live-imaging studies suggested that dynein is needed in order to free the centrosome from nuclear restraint: in dynein-inhibited cells, the application of a polarizing stimulus causes the centrosome to move backward from the cell centroid, accompanying the nucleus as an apparently passive passenger, in contrast to its normal positioning ahead of the nucleus (Gomes et al. 2005). The rearward movement of the nucleus, which is characteristic of wound models with front-facing centrosomes, was not affected by dynein inhibition, indicating that centrosome reorientation is independent of nuclear positioning, consistent with the observed dependence of nuclear, but not centrosome movement on actomyosin (Gomes et al. 2005). As discussed earlier, however, dynein inhibition also leads to centrosome off-centering in nonmigrating cells (Burakov et al. 2003; Wu et al. 2011), and therefore the effects of dynein inhibition in polarizing cells could reflect effects on a baseline positioning mechanism, such as cytoplasmic astral pulling forces. It remained uncertain from these studies, then, whether polarization generates a new source of dynein pulling force.

Support for a polarized dynein mechanism came from studies of cortical localization of dynein in migrating cells. In front-facing wound models, polarization is accompanied by the accumulation of dynein, and its cofactors dynactin and LIS1, at the leading edge (Dujardin et al. 2003; Manneville et al. 2010). A component of this accumulation consists of puncta of dynein that are independent of actin polymerization, consistent with a role in centrosome positioning. These results suggested a novel pathway of cortical dynein localization in polarizing cells. Subsequent studies forged connections between leading-edge dynein and polarity pathway signals that are critical for both centrosome orientation and cell migration. Previous work in wound models had established that the Rho family GTPase, Cdc42, acts as a key upstream signal in these systems (Etienne-Manneville and Hall 2001; Palazzo et al. 2001), linking detection of spatial cues, perhaps provided by integrins (Etienne-Manneville and Hall 2001), with downstream polarity pathway signals that ultimately govern centrosome positioning and migration. Upon wounding, Cdc42 becomes activated, highly expressed and partly localized to the leading edge (Etienne-Manneville and Hall 2001; Palazzo et al. 2001), where in astrocytes it forms a complex with the polarity proteins PAR-6 and PKCζ, each of which is essential for centrosome orientation (Etienne-Manneville and Hall 2001). This complex in turn recruits the polarity protein Dlg1 to microtubules at the leading edge, which in turn recruits and binds to the scaffold GKAP. GKAP binds dynein, which also localizes to leading edge microtubules (Manneville et al. 2010). Both Dlg1-GKAP and GKAP-dynein interactions are essential to centrosome polarization in the astrocyte wound model (Manneville et al. 2010). Besides recruiting Dlg1, the PAR-6–PKCζ complex also recruits the MT-binding protein APC, which interacts with Dlg1 to promote centrosome orientation, perhaps by the enhancement of dynein localization (Etienne-Manneville et al. 2005; Manneville et al. 2010) or by the promotion of MT cortical anchoring or growth (Reilein and Nelson 2005). In addition to the PAR pathway, signaling through PCP pathway components via Wnt ligands has also been implicated in centrosome and Golgi repositioning in a fibroblast wound model (Schlessinger et al. 2007).

These studies in wounded monolayers provide strong evidence that the positioning of interphase centrosomes is regulated through the polarity complexes that have been established as general mediators of cell polarization throughout the animal kingdom (Iden and Collard 2008; McCaffrey and Macara 2012). In wounded fibroblast cultures, another frequent component of polarity complexes, PAR-3, was also shown to govern centrosome reorientation via an interaction with cortical dynein (Schmoranzer et al. 2009). Interestingly, in this case, the localization of the complexes was at cell–cell boundaries rather than at the leading edge, suggesting that centrosome position in polarizing cells may require orchestration of pulling forces from multiple directions. In *Drosophila*, PAR-3 can recruit the centrosome to the cell cortex if normal PCP signaling is disturbed (Jiang et al. 2015). In the migrating leukocytes in medaka fish, centrosome orientation depended on the interaction of PKCζ with both PAR-3 and PAR-6 (Crespo et al. 2014). Disruption of these interactions led to the preferential localization of the centrosome toward the front. Collectively, these studies point to the possibility of unraveling links between specific polarity pathways and positional choices that shape the internal organization of the migrating cell.

Finally, one may ask: what is the functional significance of centrosome positioning during cell migration? While locomotion per se requires only the actomyosin network, many studies have shown a requirement of the MT network for directionality of migration (Euteneuer and Schliwa 1986; Malech et al. 1977; Yoo et al. 2012). However, there are notable counterexamples (Euteneuer and Schliwa 1984, 1986), and even when directionality is shown to depend on MTs and dynein, these results do not provide sufficient evidence for a role of centrosome positioning. A core difficulty is that experimental perturbation of centrosome movement inevitably involves disturbance of the MT network. The observation that migrating cells can change Golgi/centrosome positioning without changing direction (Crespo et al. 2014; Natividad et al. 2018) demonstrates that organelle positioning per se cannot be the only determinant of migratory polarity. It may be that centrosome positioning in migrating cells has functions unrelated to steering. One possible function is to control the orientation of primary cilia. Primary cilia orient in the direction of migration in a number of cell types (Christensen et al. 2008; Mirvis et al. 2018). Multiple receptors and polarity pathway components localize to the cilium and can potentially mediate the sensing of chemoattractants, extracellular matrix components, or other stimuli related to directional migration (Christensen et al. 2008; Mirvis et al. 2018). Nevertheless, directed migration can take place in unciliated cells (Boehlke et al. 2015). A second possibility is that centrosome positioning changes in order to coordinate regional actomyosin activity, consistent with the observations in medaka fish leukocytes (Crespo et al. 2014). Finally, perhaps centrosome position in some systems does not govern direction but rather reinforces an axis of migration, for example by nucleating axial MTs to enhance vesicular flow from rear to front. In that case, axial centering of the centrosome may be more critical than an anterior-posterior position. Overall, with respect to centrosome positioning in migrating cells, we have learned much about "how" and much less about "why."

7.5.3 Centrosome Positioning During Morphogenesis

Developmental processes at the tissue or organ level are frequently asymmetric, implying a potential role for cellular polarization. Centrosome repositioning has been shown to be associated with both epithelial and neuronal morphogenesis in studies conducted both in vivo and in vitro. In several of these systems, the evidence points to an important morphogenetic role for centrosome localization.

7.5.3.1 Epithelial Morphogenesis 1: Apicobasal Polarity

The maturation of epithelial tissue requires the establishment of apicobasal polarity (Rodriguez-Boulan and Macara 2014). This process entails centrosome repositioning to the apical surface, commonly resulting in primary cilium formation, as discussed earlier. However, the apical translocation of centrosomes takes place in unciliated epithelia as well, as, for example, in the gut and wing epithelia of *Drosophila* (Carvajal-Gonzalez et al. 2016a), the intestinal epithelium of *C. elegans* (Feldman and Priess 2012), and also in mammalian epithelial cells under certain culture conditions (Boehlke et al. 2015; Chin et al. 2018). In mouse duodenal columnar epithelium, the apical centrosome is on average 1.5 μm from the surface, and cilia are not in evidence (Taverna et al. 2016). These apical, nonciliated centrosomes are not associated with the majority of the cell's MTs (Meads and Schroer 1995). Indeed, in *C. elegans* intestine, the apically migrating centrosome seems to play an active role in handing off its MT-nucleating function to newly forming noncentrosomal MT-nucleation centers (Feldman and Priess 2012). Apical translocation of centrosomes can be observed in cultured epithelial cells as they progress to confluency, or when they are confined as single cells on micropatterns, and in each case the centrosome movement is MT dependent (Buendia et al. 1990; Pitaval et al. 2017). These monolayer cultures, however, do not capture certain morphogenetic events such as lumen formation. When epithelial cells are confined to a circular micropattern, a lumen forms in the developing cyst, accompanied by apical centrosome translocation (Rodriguez-Fraticelli et al. 2012). Both processes are dependent on PKCζ, part of the polarity pathway that regulates centrosome positioning in migrating cells. Apical translocation was also prevented by the activation of basolateral actomyosin when confinement was reduced by widening the micropattern, suggesting that both cytoskeletal and polarity pathways control centrosome positioning. These findings are consistent with earlier work using 3D epithelial cultures, in which lumen formation was shown to require signaling via polarity pathway members previously found to control centrosome positioning both in migrating cells and during primary ciliogenesis, including Cdc42, PAR-3, and PAR-6 (Bryant et al. 2010). Similarly, studies in *C. elegans* intestinal epithelium strongly implicate PAR-3, and to a lesser degree PAR-6, in apical centrosome migration (Feldman and Priess 2012). Basally located centrosomes in these cells first migrated to lateral positions with respect to the nucleus and then moved to the lateral membranes near PAR-3 foci before proceeding apically. This final apical movement was PAR-3 dependent. Apical translocation in the nematode intestine was not actin mediated, although it was dependent on MTs, dynein, and unc-83, a NE KASH protein (Feldman and Priess 2012). Collectively, these studies are consistent with the notion of a polarity-pathway-mediated MT-dependent mechanism for the positioning of centrosomes that may have common features in multiple settings that involve cell polarization.

7.5.3.2 Epithelial Morphogenesis 2: Planar Cell Polarity and the Positioning of Cilia

While epithelial polarization establishes an apicobasal axis perpendicular to the epithelial plane, positioning along an orthogonal axis, parallel to the plane, is controlled by the PCP pathway. PCP signaling generates asymmetric cortical domains within the plane, and these domains can then orchestrate asymmetric planar positioning of cellular structures, including centrosomes. Most studies of PCP effects on centrosomes have focused on the positioning and rotational orientation of ciliary basal bodies. For example, in the hair cells of the developing mouse cochlea, the primary cilium, known as the kinocilium, is initially centrally located at the apical cell surface. Over a period of several days, the cilium, together with the microvilli-like stereocilia that provide mechanosensation, moves to the lateral edge of the cell surface. This off-centering is dependent on core PCP proteins, which are asymmetrically localized to the medial and lateral domains of the cell [reviewed in Jones and Chen (2008)]. Subsequent studies have revealed a widespread involvement of PCP signaling in the localization of primary cilia, including cells of the lens, neuroepithelium, and node (Borovina et al. 2010; Hashimoto et al. 2010; Sugiyama et al. 2010). Notably, similar PCP signaling leads to off-centering of the apical centrosome of unciliated epithelial cells in the *Drosophila* wing, suggesting that the link between the PCP pathway and the positioning of epithelial centrosomes is widely conserved (Carvajal-Gonzalez et al. 2016b).

PCP signals also govern the localization of motile cilia in postmitotic multiciliated cells. Ciliogenesis in multiciliated cells is a fundamentally different process from primary ciliogenesis since the basal bodies derive primarily from deuterosomes generated from the daughter centrosome (Al Jord et al. 2014). The migration of these basal bodies to the plasma membrane for docking is governed by actomyosin forces (Dawe et al. 2007a). Docking at the membrane depends upon the apical enrichment of cortical actin, mediated by RhoA, PCP signaling, and the interaction between Rac1, the Rac-GEF ELMO-DOCK1, and ezrin, among other factors (Epting et al. 2015; Ioannou et al. 2013; Pan et al. 2007; Park et al. 2008). After the docking step, PCP signaling remains necessary for the alignment of basal bodies with respect to their rotational orientation around the axis of the cilium. The basal body has a single, asymmetric basal foot that projects laterally and determines the direction of beating for the cilium. Rotational orientation around the ciliary axis specifies the direction of the foot, and the alignment of this orientation across the epithelium is critical for the organization of fluid flow and mucociliary clearance (Spassky and Meunier 2017). Studies in amphibian larval epidermis and in mammalian ependyma show that coordinated orientation is regulated by the coupling of PCP signaling with instruction from the hydrodynamic forces generated by fluid flow (Guirao et al. 2010; Mitchell et al. 2007) and is also dependent on both actin and MT cytoskeleton (Clare et al. 2014; Werner et al. 2011). In addition to rotational orientation, ependymal cells also display translational polarity of their cilia: the cilia are initially clustered near the center of the apical plane and subsequently migrate as a patch toward the anterior edge (Mirzadeh et al. 2010). This movement is reminiscent of the PCP-directed migration of primary cilia, and remarkably it appears to be directed by the primary cilium present in the radial glial cell that is the embryonic precursor of the ependymal cell. This primary cilium responds to PCP signals, and perhaps also fluid flow, by moving off-center to the same region that will later be occupied by the multiciliated patch, and the presence of the primary cilium is a prerequisite for the later patch movement (Boutin et al. 2014; Ohata and Alvarez-Buylla 2016; Ohata et al. 2015). Another intriguing feature of this system is that depletion of different PCP proteins yields two classes of polarity defects: either an alteration in patch shape and rotational orientations or a failure to coordinate migration at a tissue level (Boutin et al. 2014). Further work in this area is likely to provide important insights into the interactions of PCP, cytoskeleton, and symmetry breaking in basal body positioning.

While motile cilia are primarily found in multiciliated cells, in some instances primary cilia possess motility and can generate flow. A particularly interesting example of such a motile primary cilium is found in the node, the mammalian term for a transient structure located at the posterior notochord in gastrula-stage vertebrate embryos. Ventral node cells express primary cilia that, as a consequence of PCP signaling and mechanical strain, localize asymmetrically to the posterior edge of the cell (Antic et al. 2010; Borovina et al. 2010; Chien et al. 2018; Hashimoto et al. 2010; Minegishi et al. 2017). Possibly as a result of this posterior shift on a rounded cell surface, the repositioned cilia are tilted downward in a posterior direction. Unlike other motile cilia, which move by beating through a plane, these primary cilia have clockwise, rotary movement. The combination of posterior tilt and clockwise rotation on a ventral surface results in the generation of net leftward flow since strokes to the left are further from the surface, and therefore opposed by less drag, than strokes to the right (Fig. 7.8a, b) (Nonaka et al. 2005). There is considerable evidence that the left-right developmental axis of vertebrates depends on this asymmetric flow, which may asymmetrically trigger mechanosensory cilia lateral to the node (Okada et al. 2005; Schweickert et al. 2007; Tisler et al. 2017; Yuan et al. 2015; Zinski et al. 2018). In this instance, we can see the simultaneous harvesting of the multiple capacities of the interphase centrosome to generate asymmetry: it is single, it can polarize to one surface, it can respond to planar polarity within that surface, and, as a basal body, it can generate both ciliary tilt and ciliary motion in a biased (clockwise) direction.

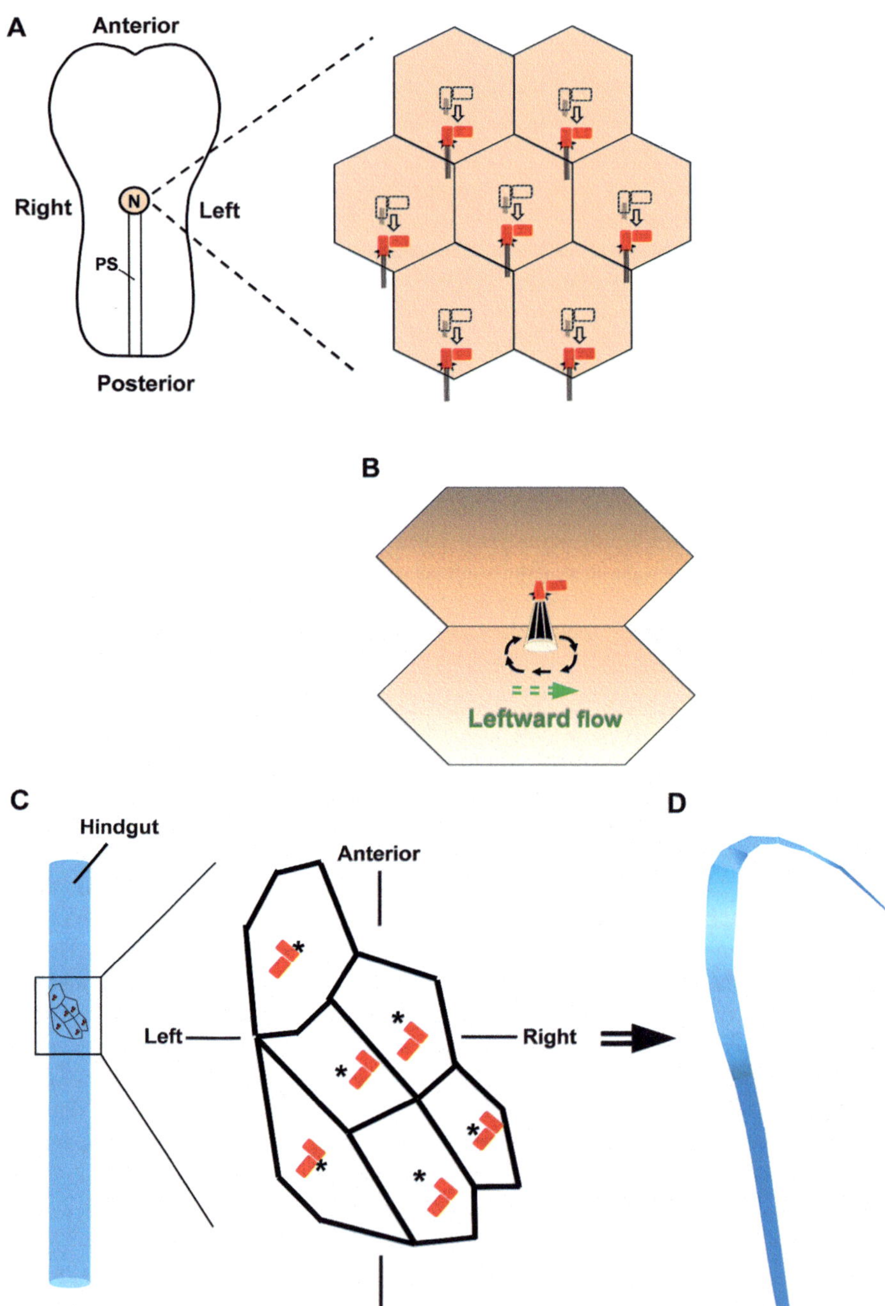

Fig. 7.8 Centrosome positioning linked to chirality, Part 2. (**a**, **b**) Centrosome positioning determines left-right chirality in development. (**a**) Ventral schematic view of the mouse gastrula. N, Node. PS, primitive streak. The expanded view on the right shows epithelial cells in which basal bodies have shifted posteriorly (open arrows), with posterior tilt. (**b**) Perspective view showing that clockwise rotary motion of the cilium drives net leftward flow, because the leftward stroke is further

7.5.3.3 Epithelial Morphogenesis 3: Polarization in the Context of Cell–Cell Contact

Stable contact among adjoining cells, generally characterized by cadherin-based adherens junctions, is a defining feature of epithelial tissues but one that did not initially receive great attention in studies of cell polarization. This deficiency has been remedied, and work on the interaction between contact and polarity promises to yield important insight into tissue organization, organogenesis, and pathogenesis (as in the case of metastatic tumor progression).

When an epithelial cell positions its centrosome, what influence does it receive from its junctions with its epithelial neighbors? Multiple studies have shown a correlation between junctional composition and centrosome localization. When a pair of mammalian epithelial cells was plated in agarose microwells that enforced contact without permitting migration, the centrosome positions of the two cells were initially uncorrelated but within 6 h became oriented to face away from each other (Desai et al. 2009). Both the formation and the maintenance of this oriented state were dependent on E-cadherin function, suggesting that the centrosomes might be instructed by the intercellular junction. Similar dependence of centrosome orientation on cadherins has been observed in other cell types (Dupin et al. 2009). Instructions from junctions might involve polarity pathway signals: as discussed earlier, centrosome orientation in migrating fibroblasts is controlled by PAR-3 localized to cell–cell contact regions, where it colocalizes with MTs and dynein (Schmoranzer et al. 2009). Consistent results were recently obtained using pairs of breast epithelial cells plated on micropatterns that permitted contact but not locomotion. When epithelial-mesenchymal transition (EMT) was modeled in this system by treatment with TGFβ, centrosomes switched from being polarized toward the junction [similar to the lumen-forming system discussed earlier (Rodriguez-Fraticelli et al. 2012)] to the reverse (away from the junction), and this switch in polarity was accompanied by changes in the composition of the junction: E-cadherin, PAR-3, and α-catenin were reduced, and Par-3 levels controlled centrosome repositioning (Burute et al. 2017). Since MTs can stabilize adherens junctions (Meng et al. 2008), it was notable that TGFβ treatment also abrogated the accumulation of stabilized MTs at the junction, strengthening the correlation between weakening of the junction and reorientation of the centrosome. Consistently, when TGFβ-treated cells were freed from the micropattern restriction and permitted to move, the doublets separated, but this was prevented by treatments that stabilized MTs.

Fig. 7.8 (continued) from the cell surface. (**c, d**) Chiral cellular architecture associated with chiral morphogenesis in *Drosophila* embryonic hindgut (Taniguchi et al. 2011). The hindgut, shown in dorsal view, is initially symmetrical (**c**) and then makes a left-handed rotation (**d**). Prior to rotation, hindgut epithelial cells (**c**, expansion of boxed region) display a chiral morphology associated with off-centering of the centrosome, which adopts a mean position to the posterior right of the cell centroid (asterisks)

The correlation of junctional weakening with centrosome release and repositioning has recently been observed in detailed studies of the delamination of differentiating neurons in the neuroepithelium of chick spinal cord (Das and Storey 2014; Kasioulis et al. 2017). The centrosome of the undifferentiated neuroepithelial cell is a basal body nucleating a primary cilium at the apical, ventricular surface. Adjacent neuroepithelial cells are linked via subapical N-cadherin, whose expression depends on contact by centrosome-nucleated MTs. The onset of neuronal differentiation involves an MT-dependent, EMT-like transition (Singh and Solecki 2015), in which N-cadherin is downregulated and the apical tip of the cell is withdrawn from the ventricular surface. Remarkably, the centrosome does not leave its apical post until a very late stage, when the remaining apical process of the cell, through which the centrosome must travel, is barely wide enough to permit its passage, suggesting an important role for the centrosome as an orchestrator of junctional dissolution and apical release (Kasioulis et al. 2017).

During in vivo morphogenesis, epithelial cells have junctions with multiple neighbors, and therefore in order to polarize correctly, the cell likely needs to integrate information from all these junctions with additional signals that may be derived from chemotactic gradients, mechanical forces from neighboring cell movements, the stiffness of the matrix (Burute et al. 2017), and the presence of a contact-free edge (as in the wound response). Whether centrosome positioning plays a role in this integration is largely unexplored, although studies of the migration of epithelial cell collectives point to a vital role for cadherin-based junctions. The migrating collective contains a group of leader cells that polarizes, in a cooperative cadherin-dependent manner, and then instructs the follower cells by mechanisms that are also mediated by cadherin-based junctions (Mayor and Etienne-Manneville 2016; Venhuizen and Zegers 2017). Studies of this communication have focused on the actin cytoskeleton (Malinova and Huveneers 2018). The role of the centrosome and microtubule network has received less attention, perhaps because follower cells do not always display centrosome orientation with respect to the collective migratory direction (Reffay et al. 2011; Theveneau et al. 2010). For example, cells at any border of the collective tend to orient their centrosomes toward the free space (Reffay et al. 2011). However, polarization of the Golgi complex in follower cells has been observed in collectively migrating keratinocytes (Lang et al. 2018). It is also noteworthy that cells in the collective do not have uniform migratory direction: regions within the collective typically contain groups of cells with distinct migratory orientation, correlated to some extent with the avoidance of shear stress (Tambe et al. 2011; Zaritsky et al. 2015). Whether polarization in these "subcollectives" involves centrosome positioning has not been studied. Centrosome positioning in collectives is highly dynamic, in both leader and follower cells (Reffay et al. 2011), which may indicate responsiveness to multiple cues. It is possible that smaller collectives with more defined cues may be helpful for the analysis of centrosome positioning. For example, when small numbers of epithelial cells are permitted to form migrating chains on micropatterns, their centrosomes are all front facing, except for the trailing cell, which has an opposite orientation (Li and Wang 2018). Interestingly, the persistence of the chain through turns depends on PCP signaling, consistent with a centrosome/MT/junction-mediated mechanism.

Studies of morphogenesis in vivo may offer the best opportunities for assessing the integration of spatial cues. For example, epithelial organogenesis often results in a chiral structure, involving asymmetry in all three developmental axes. A potential role for the centrosome in chiral organogenesis is suggested by studies of the epithelium of the *Drosophila* hindgut, which during development rotates to the left (so that the former dorsal side faces left) (Taniguchi et al. 2011). This rotation requires left-right symmetry breaking. The authors found that, prior to the rotation, hindgut epithelial cells displayed left-right bias with respect to centrosome position, cell geometry, and E-cadherin intensity at cell boundaries (Fig. 7.8c, d). Disruption of either E-cadherin or actin signaling abolished both cell geometry bias and hindgut laterality. A mutation in the unconventional myosin MyoID reversed the cell geometry bias, the E-cadherin intensity bias, and the hindgut laterality. Remarkably, in mosaic hindgut whose cells were either wild type or mutant for MyoID, the left-right bias with respect to E-cadherin intensity at cell boundaries was wild type if both cells at that boundary were wild type, reversed if both cells were mutant, and unbiased if the cells were mixed. Similar findings were obtained with respect to the bias in cell geometry (Hatori et al. 2014). Chirality of tissues, then, appears to be determined at the level of cells. Similarly, a study of centrosome position in vascular endothelium observed a chiral bias in vivo that could be reversed by PKCα activity in vitro. When endothelial cells of mixed chirality were cocultured, they retained their conflicting biases and could not form normal junctions with each other, again demonstrating chiral determination at the cellular level (Fan et al. 2018).

The *Drosophila* hindgut study strongly suggests that cytoskeletal chirality can drive chiral organogenesis, but whether the initial determination of left-right polarity occurs in the actin or MT networks remains unsettled. Actin networks can evolve into chiral patterns even when MTs are removed by nocodazole treatment (Tee et al. 2015), and the chiral alignments observed in cells grown on ring-shaped micropatterns, or as spheroids in 3D culture, were reversed by the disturbance of the actin cytoskeleton (Chin et al. 2018; Wan et al. 2013). In contrast, nocodazole had no effect on the chirality of micropatterned cells, although a slightly higher dose appeared to abolish (not reverse) chirality in the spheroid system. Overall, these results suggest actin as the driver of chirality. On the other hand, the "intrinsic vector" of HL-60 cells discussed earlier, the first demonstration of chirality in cultured cells, can be abolished by MT depolymerization, or by the inhibition of either dynein or PKCζ (Xu et al. 2007). Moreover, it has been reported that this chirality in HL-60 polarity is eliminated by the expression of a mutant tubulin comparable to tubulin mutations that disrupt left-right chirality in plants, nematodes, and frogs (Lobikin et al. 2012). These studies in HL-60 cells clearly support an MT-based chiral mechanism. The elucidation of the chirality-determining entities within cells, and their interaction with actin, MTs, and centrosomes, has emerged as a key challenge for cell biologists.

7.5.3.4 Neuronal Morphogenesis

Polarization is a major theme of neurogenesis. In developing cortex, newborn neurons must polarize correctly in order to migrate between the apical ventricular zone and the cortical plate and also to form axons with proper orientation. The role of centrosome positioning in these events, particularly in axon formation, has been studied by several investigators. Initially, it was observed that in cultured hippocampal neurons, the centrosome is positioned at the lamellipodium that initiates axonogenesis. Axon development was shown to be associated with stable MT formation and to be dependent on MTs and on an intact centrosome (de Anda et al. 2005). Notably, when cytokinesis blockade was used to generate neurons with two centrosomes, the cells formed two axon-like extensions, strongly implying a centrosome role in axon formation. Subsequent studies in embryonic neocortex showed that centrosomes in newborn multipolar neurons migrate from their initial basal position to an apical position just prior to the apical outgrowth of the neurite that becomes the axon (de Anda et al. 2010). Axon formation was again dependent on intact centrosome and MTs, and was associated with stable MT formation. However, the extension of the neurite did not require the continued presence of the centrosome at its base: shortly after neurite initiation, cells reverted from multipolar to bipolar morphology while beginning migration to the cortical plate, and the centrosome at this time moved basally to the leading process. The findings are consistent with a role for centrosome positioning either in the selection of sites for axon outgrowth or alternatively in supporting early stages of growth in neurites that have been selected to become axons. A recent study in the same system supports the latter model (Sakakibara et al. 2014). The authors observed that multipolar neurons, prior to axon extension, formed many processes that varied in the persistence of their growth. The centrosome migrated toward whichever process was currently most persistent, sometimes switching several times, as different processes assumed the "dominant" role. Eventually one of these dominant processes developed into an axon (Fig. 7.9). From this point of view, the translocation of the centrosome from axon to leading process in the bipolar stage is simply explained by the dominance of the leading process, which is thicker than the axon. More generally, this model is consistent with studies of polarization in migrating cells, which show, in systems as diverse as *Dictyostelium* and mammalian wound models, that protrusion comes first and centrosome repositioning second (Ueda et al. 1997; Wong and Gotlieb 1988). Finally, studies of zebrafish sensory neuron development lend further support to a role for centrosome positioning in axonogenesis (Andersen and Halloran 2012). These neurons generate two types of axon: central and peripheral. The outgrowth of the peripheral axon correlated with centrosome migration toward its base and was strongly inhibited by laser ablation of the centrosome. Remarkably, the authors found that a LIM homeodomain transcription factor that was necessary for peripheral axon initiation was also necessary for the centrosome migration toward the axon, as well as for normal centrosome motility. Collectively, these studies highlight the unique opportunities that these neuronal systems are

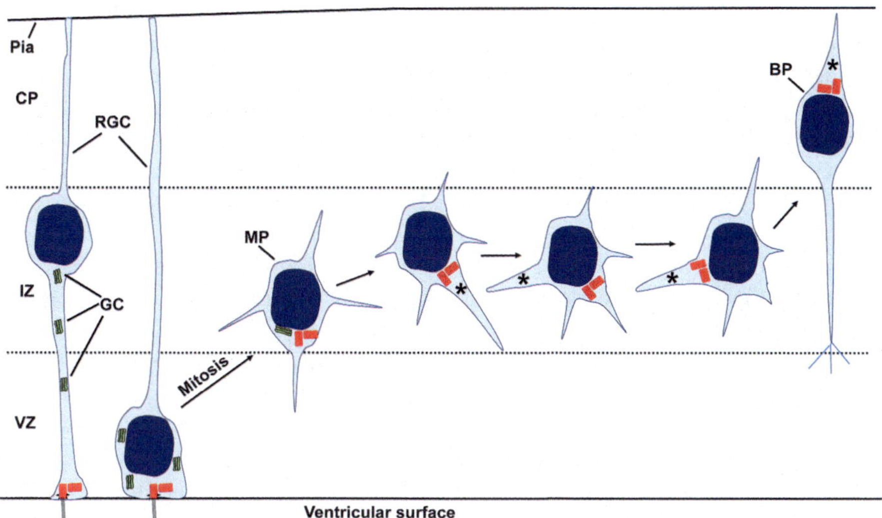

Fig. 7.9 Centrosome and Golgi positioning during neurogenesis in developing neocortex. The radial glial cell (RGC) on the left extends from the apical ventricular surface through the ventricular zone (VZ), intermediate zone (IZ) and cortical plate (CP) to the basal pial surface. Multiple stacks of Golgi complex (GC) are localized apical to the nucleus and away from the centrosome/basal body (Taverna et al. 2016). The GC remains apical to the nucleus as the nucleus migrates to the apical end to undergo mitosis. Asymmetric mitosis of the RGC generates a differentiating progenitor cell that delaminates and develops into a multipolar neuron (MP) in the IZ. Upon delamination, Golgi complex and centrosome adopt a perinuclear localization. The study of Sakakibara et al. (2014) indicates that the MP forms a series of short-lived "dominant" processes (asterisks), and that the centrosome repositions to the base of each newly-formed dominant process. Finally, one such process develops into an axon and the MP develops into a bipolar neuron (BP) that migrates into the CP, with the centrosome positioned at the base of the leading process

likely to provide for defining and investigating specific centrosome roles in morphogenesis.

7.5.4 *Centrosome Positioning in Polarized Immune Cell Interactions*

The hallmark of adaptive immunity is specificity. For a lymphocyte encountering antigen, this specificity is manifested, first, in the selectivity of its recognition and, second, in the specific targeting of its response to the antigen-presenting cell (APC) rather than neighboring cells. A notable feature of the lymphocyte–APC interaction, enhancing both the specificity and efficiency of lymphocyte response, is the immunological synapse (IS), a temporary (minutes to hours), micron-sized zone of close contact between the apposed plasma membranes of the two cells, within which a highly organized set of supramolecular clusters, including antigen receptor complexes and costimulatory and adhesion receptors, acts to facilitate both signaling between the cells and targeted delivery of secretory products. The nature of these products depends on the cells involved. Cytotoxic T cells (CTL) and natural killer cells form synapses that deliver cytotoxic granule contents to target cells, while T-helper cells deliver cytokines through synapses with APC, and B cells acquiring antigen from APC secrete lysosomal contents to aid in antigen extraction (Bustos-Moran et al. 2016; de la Roche et al. 2016; Dustin and Choudhuri 2016; Huse 2017; Tolar 2017).

The formation of the IS is a polarizing event for the lymphocyte and is accompanied by the translocation of several organelles, including the centrosome and Golgi complex, to a position near the synapse. It is, in fact, typically a repolarizing event since the lymphocyte, prior to synapse formation, migrates over the surface of the cognate cell in search of antigen. Lymphocytes migrate with polarized, rear-facing centrosomes, so when they encounter antigen and initiate synapse formation, the repolarizing centrosome is obliged to migrate from the cell tail (uropod) around the nucleus and then to the cell surface, where it remains for the duration of the synapse. What is accomplished by this migration? At the synapse, centrosome-nucleated MTs, many of which are stabilized (Serrador et al. 2004), extend either parallel to the cell surface, providing potential tracks for the organization of receptor clusters, or interiorly, where the Golgi complex is located, potentially facilitating the delivery of secretory products. In addition, the centrosome, which is separated from the plasma membrane by only a 20-nm-wide space that is substantially depleted of cortical actin, may play a more direct role in secretion. In CTL, cytolytic granules cluster around the centrosome as it polarizes and they arrive at the IS with the centrosome, which is localized in the center of the synapse (Ritter et al. 2015; Stinchcombe et al. 2006). Consistently, CTL killing efficiency is reduced by the inhibition of PKCζ (Davenport et al. 2018), which is required to stabilize centrosome polarization (Yuseff et al. 2011), but is not affected by the inhibition of plus-end-directed movement on MTs, suggesting that direct centrosome-mediated delivery of granules is both necessary and sufficient for CTL function (Stinchcombe et al. 2006).

The motors and the signaling pathways that drive centrosome migration to the synapse have been the subject of numerous studies [reviewed in Martin-Cofreces and Sanchez-Madrid (2018)] (Fig. 7.10). MT integrity and dynein activity are critically important for migration (Martin-Cofreces et al. 2008; Nath et al. 2016; Wang et al. 2017; Yi et al. 2013). Dynein localizes to the IS by binding to the scaffolding protein ADAP that is linked to both T-cell receptor and integrin signaling (Combs et al. 2006). Dynein at the IS is complexed with Nde1, which was found to be essential for centrosome migration (Nath et al. 2016). MTs extend from the IS to the migrating centrosome, and these MTs straighten and shorten as the centrosome moves linearly toward the IS. Collectively, these studies suggest that cortically anchored dynein moves the centrosome by pulling forces that involve a "capture-shrinkage" MT-mediated mechanism, although an alternative proposal is that the dynein-pulled MTs, instead of shortening, slide along and past the IS surface toward the back of the cell (Kuhn and Poenie 2002; Nath et al. 2016). The strength of the dynein pulling force was made evident in a study employing an optical trap approach

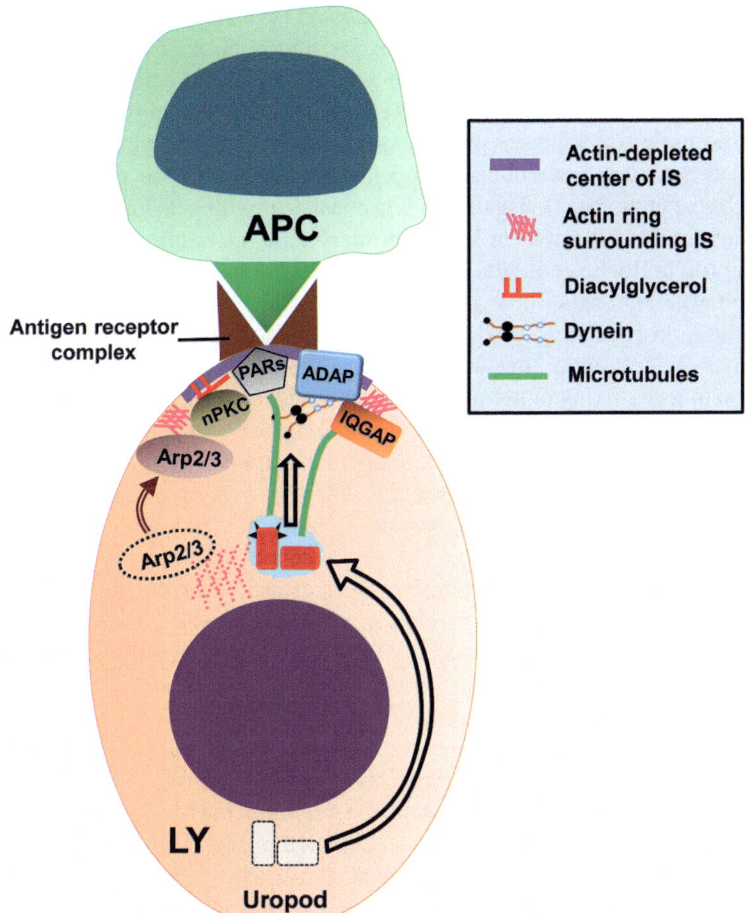

Fig. 7.10 Potential mechanisms of centrosome positioning at the immunological synapse. Following antigen receptor activation, centrosome repositioning (open arrows) includes a migration from the uropod to the opposite side of the nucleus, followed by movement to and docking at the plasma membrane of the synapse. This repositioning requires the presence, among others, of ADAP, IQGAP, PAR pathway proteins, and novel PKCs (nPKCs), as well as the relocation of Arp2/3 from nucleus to synapse. Each of these represents a potentially distinct mechanism for centrosome movement. The scaffold ADAP provides a cortical anchoring site for dynein. The expanding actin ring at the synapse may pull MT via mediators such as IQGAP. PAR signals can potentially recruit dynein. Synaptic diacylglycerol recruits nPKCs and dynein. Arp2/3 relocation may release the centrosome from tethering by perinuclear actin. *APC* antigen-presenting cell. *LY*, lymphocyte

to ensure that the centrosome's initial position in the uropod was precisely opposite to that of APC contact. In several cases, this led to a "frustrated" centrosome that was unable to find its way around the nucleus, so that the pulling forces instead forced the IS to deeply invaginate into the T cell (Yi et al. 2013). This study also demonstrated that centrosome migration is biphasic, with an initial, rapid phase that brings the centrosome within 2 μm of the surface, followed by a slower docking phase that is especially sensitive to dynein inhibition. Specific control of this docking step is indicated by its dependence on Lck, a kinase recruited to the activated antigen receptor complex (Tsun et al. 2011). Remarkably, the docking of the centrosome greatly resembles docking for primary ciliogenesis, with the mother centriole oriented to the surface and subdistal appendages contacting the plasma membrane (Stinchcombe et al. 2015). This finding is consistent with previous studies showing functional similarities between IS and primary cilia, including localized hedgehog signaling (de la Roche et al. 2013) and the use of the intraflagellar transport system (Finetti et al. 2009, 2014).

IS formation is accompanied by a profound remodeling of the local actin cytoskeleton, including actin clearance in the central region of the IS, as well as the formation of a dense ring of peripheral actin that surrounds this central area (Hammer et al. 2018). IQGAP, part of a complex linking actin and MT, localizes to the peripheral ring, leading to the suggestion that actomyosin forces might pull on synaptic MTs as actin moves peripherally, thereby moving the centrosome to the synapse (Stinchcombe et al. 2006). In support of this model, both IQGAP and CLIP-170, which binds both MTs and IQGAP, are necessary for centrosome migration to the synapse, as are the actin-regulating proteins Rap1 and cofilin (Lim et al. 2018; Wang et al. 2017). Inhibition of myosin II reduces centrosome migration to the IS, especially when combined with nocodazole, but myosin II localizes to the uropod, the opposite side of the cell from the IS (Liu et al. 2013). Altogether, these studies indicate a role for actin in centrosome migration but do not establish that actomyosin forces act on the organelle. More recently, some insight into this issue has been provided by studies that give the actin–centrosome relationship a new twist. Obino and coworkers demonstrated that centrosome migration in B cells is associated with a reduction in Arp2/3-mediated actin nucleation at the centrosome and a concomitant increase in Arp2/3 at the synapse, where it is recruited by HS1, a member of the cortactin family, and presumably contributes to IS actin remodeling (Obino et al. 2016). Remarkably, migration was entirely independent of these events at the synapse and strictly dependent on the loss of Arp2/3 nucleation activity at the centrosome, which was sufficient to induce centrosome–nucleus separation in the absence of antigen. The authors propose that migration requires the release of the centrosome from actin tethers that restrain it at the nucleus and that this release entails an antigen-induced shift of actin-nucleating activity from the centrosome to the cell surface. This model accounts for previous findings of migration dependence on actin regulation, without invoking actomyosin force. It will be important to determine whether a similar Arp2/3-based mechanism contributes to the regulation of centrosome repositioning in other settings. It should also be borne in mind that the degree of centrosome separation from the nucleus is variable in many studies of IS formation, and therefore, even if Arp2/3 removal from the centrosome increases its mobility, this may not necessarily imply a complete unlinking of the centrosome and the nucleus. In this regard, Kuhn and coworkers generated time-lapse sequences of centrosomes that, following their arrival at the plasma membrane, oscillated laterally between two APC contact sites (Kuhn and Poenie 2002). These images convey a strong impression of linked nucleus-centrosome movement. Conceivably, since actomyosin has been linked to nuclear movement in migrating cells, it may play a role in this context.

Signaling pathways that govern centrosome migration through these cytoskeletal mechanisms have also begun to be elucidated, beginning with the antigen receptor complex, several of whose components have been shown to be required for the polarization of the centrosome (Kuhne et al. 2003). Of the many effectors of this complex, phospholipase C-γ has been implicated in polarization, as has its product, diacylglycerol (DAG), which is also required for dynein recruitment to the region of antigen receptor activation in the plasma membrane (Quann et al. 2009). Remarkably, dynein is recruited within 10 s of DAG production, followed a few seconds later by centrosome migration. Subsequent studies from the same group revealed that DAG exerted its effects via the activation of the novel PKCs: PKC-ε, PKC-η, and PKC-θ, all of which are recruited to the site of activation (Quann et al. 2011). PKC-ε and PKC-η act to recruit PKC-θ, and consistently, PKC-θ and either of PKC-ε or PKC-η are necessary for centrosome migration.

In addition to the DAG pathway, dynein localization may also be driven by polarity pathway signaling similar to what was described earlier with respect to migrating cells. Specifically, PAR-1b, PAR-3, and PKCζ are each required for centrosome migration to the synapse (Bertrand et al. 2010; Lin et al. 2009; Reversat et al. 2015; Yuseff et al. 2011). PAR-3 has in addition been shown to be necessary for dynein localization to the synapse, and specifically for the prolonged residence time, in the central area of the synapse, of dynein puncta to which PAR-3 is colocalized (Reversat et al. 2015). PAR-3 and PKCζ have also been linked to synapse functionality (Bertrand et al. 2010; Reversat et al. 2015; Yuseff et al. 2011). What lies upstream of these polarity proteins? As discussed earlier, Cdc42 frequently regulates this pathway, and it has indeed been linked to centrosome migration in the same systems in which PAR-3 and PKCζ were implicated (Reversat et al. 2015). Cdc42 is activated during IS formation by signaling from β1-integrin and from the CD3 component of the antigen receptor complex, mediated by PI3 kinase and Akt (Carlin et al. 2011; Makrogianneli et al. 2009). However, two studies failed to find a Cdc42 requirement for centrosome migration (Chemin et al. 2012; Gomez et al. 2007). It is possible that the involvement of the polarity signals in migration is variable. A second possibility is that the primary role of Cdc42 is the promotion of actin remodeling (Chemin et al. 2012) and that another upstream signal is responsible for polarity pathway activation. This signal may in fact come from the DAG pathway. PKC-θ binds to and phosphorylates PKCζ, and PKCζ mediates the PKC-θ activation of NF-κB in T cells (Gruber et al. 2008). There may be no setting that places as much demand on the centrosome to travel far, and quickly, as its migration to the synapse, and it will therefore be no surprise if coordination among several mechanisms is required to accomplish it.

7.6 Positioning of the Golgi Complex

The vertebrate Golgi complex consists of a series of cisternal stacks that are laterally connected by tubules to form a unified ribbon structure. In most vertebrate cells, the complex is found in close proximity to the centrosome, except during mitosis, when it is dispersed into vesicles. It is natural to suppose that this alignment reflects a need to maximize the general efficiency of vesicular trafficking: for a structure that packages cargo for transit, the region in which "roads" originate would seem an appropriate location. However, it is not clear whether this view is correct. The maintenance of the ribbon structure is MT dependent, and treatment with nocodazole disassembles the ribbon into individual cisternal stacks (mini-stacks), which are dispersed through the cytoplasm with little or no polarization to the centrosomal region (note that this dispersal into mini-stacks is not to be confused with the mitotic dispersal into individual vesicles). Notably, this dispersal does not compromise the global delivery of ER-derived cargo to the cell surface but instead generates a specific defect in the polarized delivery of apical markers to the apical surface of epithelial cells (Eilers et al. 1989; Rindler et al. 1987; Rogalski et al. 1984). These early findings suggested that polarity, rather than general efficiency, is the key feature of the unified ribbon, but the use of nocodazole was problematic due to its likely broad impact on polarized function. More recent studies have supported the initial findings using genetic approaches that disassemble the ribbon into dispersed mini-stacks without affecting the MT cytoskeleton. Several of these studies target golgins, a nonhomologous family of coiled-coil proteins that localize to the Golgi matrix and attach to Golgi membranes (Witkos and Lowe 2015). The golgins golgin-84, GM130, GMAP-210, and golgin-160 have each been shown to be required for maintaining the Golgi ribbon: in each case, depletion of the golgin results in dispersed mini-stacks but not in reduced global trafficking efficiency (for golgin-84, global trafficking was reduced absolutely, but not relative to levels of Golgi membrane in the cell) (Diao et al. 2003; Puthenveedu et al. 2006; Yadav et al. 2009). In wound models, however, the depletion of either GMAP-210 or golgin-160 abolished polarized vesicle delivery to the leading edge and more generally prevented cell polarization: depleted cells failed to orient the centrosome and Golgi complex toward the leading edge and did not migrate to the wound (Yadav et al. 2009). Further support for an important role of Golgi positioning in cell polarity derives from the finding that nearly half of the MT in RPE1 cells are nucleated by the Golgi complex (Efimov et al. 2007). In contrast to the astral distribution of centrosome-nucleated MT, Golgi-nucleated MTs are primarily directed toward the Golgi-facing cell perimeter, corresponding to the leading edge of a migrating cell (with front-facing centrosomes). In addition, specific depletion of the Golgi-nucleated MTs generated dispersed mini-stacks, again resulting in a specific defect in polarized trafficking to the leading edge, as well as an associated failure of directional migration (Miller et al. 2009).

Collectively, these studies establish that the principal functional consequence of Golgi localization to the centrosomal region is to support polarized cellular function through asymmetrical targeting of vesicular traffic. It remains, however, to be determined whether this is equally the case in the many instances in which the centrosome and Golgi complex are oriented toward the rear of a migrating cell. If the mechanism of migration is based on rearward plasma membrane flow, balanced by rear-to-front vesicular traffic, which is likely to be the case at least for amoeboid migration (O'Neill et al. 2018; Tanaka et al. 2017), then the principal function of Golgi positioning may be to support vesicular membrane flow by providing stabilized Golgi-nucleated MTs in appropriate orientation. In this case, support for endocytic flow at the rear may be as valuable as support for exocytosis at the front. Increased exocytosis has been demonstrated at the front of migrating cells (Schmoranzer et al. 2003).

Localization of the Golgi complex to the centrosomal region is far from universal. A system of dispersed stacks is found in plants, invertebrates, and many fungi (Wei and Seemann 2010). Indeed, dispersed stacks can also be found in vertebrates. They occur, for example, in the parietal cells of gastric epithelium (Gunn et al. 2011) and in uroepithelial cells (Kreft et al. 2010), in each case perhaps reflecting a need for distributed Golgi activity created by the anatomical features of those cells. They may be even more common in early development: in gastrulating zebrafish embryos, dispersed Golgi stacks are found in most cells until segmentation stages (Sepich and Solnica-Krezel 2016). They are also found throughout the cell cycle in embryonic fibroblasts deficient in BARS, the protein that mediates the fission of Golgi tubules necessary for ribbon disassembly prior to mitosis. The absence of a ribbon allows these mutant fibroblasts to pass the G2 checkpoint for Golgi disassembly (Colanzi et al. 2007). Collectively, these findings indicate that transitions between dispersed and compact Golgi localization take place in vertebrate cells. The dispersed mini-stack configuration shows some resemblance to the dispersed mini-stacks that occur as an intermediate form during the reassembly of the Golgi ribbon that begins in late mitosis (Miller et al. 2009). It is conceivable, then, that the dispersed stacks that are found, for example, in gastrulae represent an arrested stage in postmitotic reassembly. Cell division occurs very rapidly in these embryos, and since cell migrations at this stage are stereotyped and may be robustly cued, the energy saved by not having to continually recenter the stacks may be more valuable than a polarized Golgi ribbon.

Some details of the mechanism that converts dispersed stacks into a ribbon have emerged. In late mitosis, isolated vesicles fuse to form cisternae and assemble into mini-stacks by a process that is MT independent (Tang et al. 2008). These stacks then collect into clusters, still dispersed, whose formation is mediated by the dynein-dependent movement of the stacks toward the minus end of Golgi-nucleated MTs. Finally, the clusters move toward the centrosome via centrosome-nucleated MTs (Hurtado et al. 2011; Miller et al. 2009). Dynein is recruited to Golgi membranes by golgin-160, and this binding was shown to be necessary for the repositioning of Golgi membranes to the centrosome (Yadav et al. 2012). The authors further showed that golgin-160 localization to Golgi membranes is in turn dependent on the small GTP-binding protein Arf1. Golgin-160 binds to Arf1 in its GTP-bound activated state, which is dependent on the guanine nucleotide exchange factor GBF1.

Consistently, the depletion of GBF1 caused the dissociation of golgin-160 and the dispersal of the Golgi ribbon (Yadav et al. 2012), although these effects could also be due to other Arf1 functions (Jackson 2018), such as the regulation of GMAP-210, which plays a role in maintaining Golgi localization to the centrosomal region (Rios et al. 2004). Finally, upon recruitment to the centrosome, mini-stacks assemble into the ribbon structure, the maintenance of which is dependent on the Golgi membrane proteins GRASP65 and GRASP55 (Feinstein and Linstedt 2008; Puthenveedu et al. 2006), although perhaps not in all cells (Hurtado et al. 2011). The maintenance of the ribbon near the centrosome during interphase is an active process, as evidenced by the dispersal observed upon the disabling of dynein (Vaisberg et al. 1996; Xu et al. 2002). Actin depolymerization, in contrast, has little effect (Ho et al. 1989; Xu et al. 2002). The disassembly of the ribbon, which is essential for mitotic entry, reverses the assembly sequence (Valente and Colanzi 2015). First, the deactivation of Arf1 and the phosphorylation of GRASP55 and GRASP65 in G2 phase promote the fragmentation of the ribbon into mini-stacks, a process mediated by BARS. The isolated stacks then go through unstacking and vesiculation steps in prophase.

The observation of dispersed Golgi stacks distant from centrosomes raises the question of whether cells may be able to modulate Golgi positioning independent of centrosome location. Given the need of neurons to precisely regulate highly polarized structures over large distances, it is perhaps not surprising that these cells provide the lone examples to date of such independent control. One such example is found in the radial glial cells of developing neocortex. These cells extend a thin apical process to the ventricular surface, at which the centrosome is localized, as well as a thin basal process in the opposite direction that contacts the pial surface. The nucleus cycles between the basal and apical ends. A recent study of Golgi localization in these cells found that the complex consisted of dispersed mini-stacks that were separate from the centrosome and whose positioning adjusted in accord with nuclear migration (Taverna et al. 2016). The stacks were always apical (occasionally lateral) to the nucleus and were distributed throughout the apical region, but always at least 8 μm (average of 20 μm) from the centrosome (Fig. 7.9). As the nucleus moved apically, the stacks shortened, adjusting their position to remain apical to the nucleus. The authors demonstrated the functional significance of this localization by showing that Golgi-derived N-linked glycans were found only on the apical and not the basal process. However, when the cell divides asymmetrically, it gives rise to a basal progenitor cell that delaminates from the apical surface and loses apicobasal polarity, and in this daughter cell the Golgi complex and centrosome were reunited next to the nucleus (Fig. 7.9). The separation of the Golgi complex from the centrosome may be a general feature of columnar epithelia since it was also observed by the authors in duodenal mucosa.

How is Golgi positioning controlled in radial glial cells? The apical distribution may be controlled by centrosome-nucleated MTs since inhibition of dynein moved the Golgi stacks basally, while kinesin inhibition did the reverse (Taverna et al. 2016). A more recent study provides further mechanistic insight (Xie et al. 2018). In the absence of two proteins (PITPNA and PITPNB) that redundantly stimulate the production of phosphatidylinositol-4-phosphate (PI4P), the apically distributed

Golgi stacks relocalize to a perinuclear position as the cell loses its radial alignment. This shift in Golgi position was not due to a loss of apical contact with the ventricle. Instead, both Golgi apical positioning and the maintenance of radial glial polarity could be attributed to a requirement for PI4P-dependent Golgi recruitment of the PI4P-binding protein GOLPH3 and to the presence of the unconventional myosin MYO18A, which has previously been shown to complex with GOLPH3 to promote Golgi orientation toward the leading edge in migrating cells (Xing et al. 2016). The results suggest a contribution of actomyosin forces to apical Golgi distribution in these cells. Further studies in this system, then, may provide unique insight into the coordination of multiple mechanisms for Golgi positioning.

A second example of Golgi positional regulation comes from studies of neuronal formation of dendritic arbors. Cultured hippocampal neurons first form dendrites symmetrically, in several directions, and then asymmetrically select one to be the major (principal) dendrite, which will extend, widen, and form several levels of branches. A study of Golgi localization during this process concluded that the perinuclear Golgi complex orients to face the direction in which the principal dendrite will later develop (Horton et al. 2005) (Fig.7.11a, b). Notably, when the overexpression of GRASP65 was used to cause Golgi vesiculation and dispersal, the polarization of dendritic development was inhibited. More recently, the depletion of the Golgi-localized kinase STK25, or its associated adaptor, STRAD, was shown to abolish Golgi polarization in newborn adult hippocampal neurons, resulting in the formation of abnormal neurites, each of which displayed mislocalized Golgi membranes at its base (Rao et al. 2018). Collectively, these findings indicate an instructive role for Golgi orientation in neurons and also suggest a distinction between the morphogenetic roles of centrosome and Golgi complex: as discussed earlier, centrosome orientation is associated with axonogenesis (de Anda et al. 2005), whereas the Golgi complex polarizes to the primary dendrite but not toward the axon (Horton et al. 2005). Golgi polarization in support of arbor development does not end there, however. When the trafficking of ER-derived cargo was studied by live imaging, cargo was observed to arrest and accumulate at branchpoints within the principal dendrite (Horton and Ehlers 2003). The branchpoints were shown to contain small stacks of cisternae termed Golgi outposts. These outposts were rare in other dendrites, and many neurons lacked them altogether (Horton and Ehlers 2003; Horton et al. 2005). Similar outposts have been demonstrated in *Drosophila* sensory neurons, where they have been implicated in MT nucleation and the maintenance of dendritic branch dynamics (Ori-McKenney et al. 2012; Ye et al. 2007). The orientation of the outpost may bias the direction of MT nucleation to regulate the placement of dendritic branches and thereby shape arbor morphogenesis (Yalgin et al. 2015). Golgi outposts have also been shown to selectively process cargo that is diverted from the perinuclear Golgi complex (Jeyifous et al. 2009).

Recent studies in hippocampal cultures uncovered a mechanism for Golgi outpost formation (Quassollo et al. 2015). The perinuclear Golgi ribbon generates tubules, which, within seconds, enter the principal dendrite and undergo fission, followed by the distal migration of the severed Golgi membrane, bearing markers of both *cis-* and *trans*-Golgi, into the dendrite. Tubule fission was shown to be essential for outpost

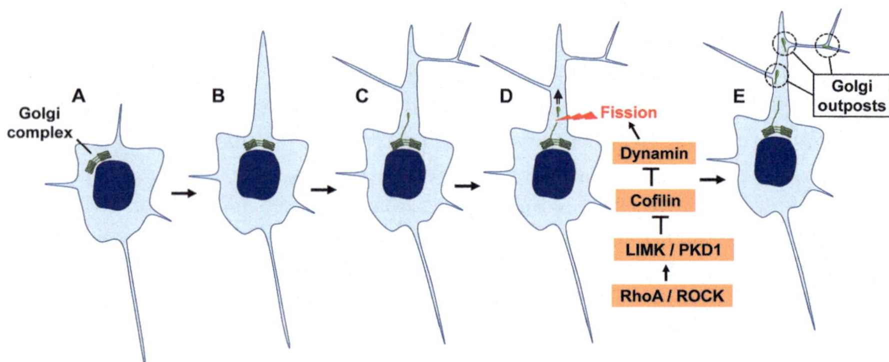

Fig. 7.11 Golgi polarization and outpost formation in hippocampal neurons. (**a, b**) The Golgi complex in cultured hippocampal neurons becomes polarized to the site of formation of the principal dendrite (Horton et al. 2005). (**c–e**) Golgi outpost formation (Quassollo et al. 2015). A tubular extension of the Golgi complex enters the principal dendrite (**c**) and is then cleaved by a mechanism dependent on dynamin and on the inhibition of cofilin activity via RhoA/ROCK/LIMK/ PKD1 signaling (**d**). The distal fragment of the tubule then migrates to a dendritic branchpoint, forming a Golgi outpost (**e**)

formation. Fission was dependent on signaling from RhoA through ROCK, LIMK1, PKD1, cofilin, and dynamin, similar to the pathway responsible for other Golgi fission events (Salvarezza et al. 2009) and implying an actin-based fission mechanism (Fig. 7.11c–f). Orientation of the tubular membrane to the dendrite was independent of this pathway, however, and may instead be guided by dynein, which has been shown to be necessary for correct dendritic localization of Golgi outposts in *Drosophila* neurons (Kelliher et al. 2018; Lin et al. 2015; Zheng et al. 2008).

The absence of Golgi outposts in many neurons raises the question of whether the placement of distal Golgi elements may take other forms. Recently, a newly designed Golgi tracker was used in hippocampal cultures to demonstrate that most dendrites, in most neurons, contain structures smaller than Golgi outposts, termed Golgi satellites by the authors, that accumulate cargo, possess glycosylation machinery, and engage in transport activities (Mikhaylova et al. 2016). The relationship of these structures to Golgi complexes is still very unclear, and much interesting work lies ahead. As with the centrosome, unraveling of the mechanisms by which Golgi complexes, or their derivatives, detect and respond to spatial cues promises to illuminate many fundamental issues in cellular organization.

7.7 Perspective: Positioning as Clue to Centrosome Function

The nature of the centrosome makes it well suited to studies of position but ill-suited to studies of function. The last two decades have yielded fascinating results with respect to the "where" and "how" of centrosome positioning, but the very fascination speaks to the limitation of these approaches: they excite interest partly because they remain the major evidence for hypotheses with respect to "why." Outside of its primordial function as the basal body for cilia, it is difficult to identify roles for the centrosome that are supported by clear mechanistic evidence. Laser ablation studies provide support for a centrosome requirement in certain functions related to cell polarity (Andersen and Halloran 2012; Cowan and Hyman 2004; Koonce et al. 1984; Wakida et al. 2010), but centrosomes seem to be dispensable for most developmental processes in *Drosophila* (Gogendeau and Basto 2010), as well as spindle formation in vertebrate cells (Khodjakov et al. 2000). A general understanding of centrosome function, then, is still elusive and may be greatly informed by positioning studies.

Outside of ciliation, the most widespread observation linking centrosome position to potential function is its nucleation of MT asters, and for several decades, the discussion of centrosome function has focused on its role as an "organizer" of MTs. From this notion, it is not difficult to develop a more general hypothesis of the centrosome as an organizer of cell architecture. Support for a broad organizing function comes from studies in *Chlamydomonas*, a green alga with highly stereotyped organellar geometry. These organelles, including the nucleus, contractile vacuole, and eyespot, all maintain chirally asymmetric positions that can be structurally traced, via asymmetrically deployed cytoskeletal elements, to the chiral asymmetry of the two flagella and their two asymmetric basal bodies (Holmes and Dutcher 1989). Furthermore, experiments with mutant strains provide evidence that the centriole controls nuclear position and spindle orientation and also that mother centrioles control the position of their daughters (Feldman et al. 2007; Feldman and Marshall 2009; Marshall 2012). Notably, one of these mutations affects a protein whose homolog, TBCCD1, is localized to centrosomes both in mammalian cells and in the protist *Trypanosoma brucei*. Depletion of TBCCD1 in mammalian cells caused the displacement of the centrosome from the nucleus, as well as Golgi dispersal, and inhibited directional migration, without affecting centrosomal MT-nucleation function (Goncalves et al. 2010). In trypanosomes, TBCCD1 depletion disrupted cell organization, including loss of centriole-kinetoplast linkage (Andre et al. 2013).

While these findings provide support for a view of the algal centrosome as a cell organizer, this view may not be an adequate description of centrosome function in metazoa. Metazoan cells often have an organization that is regulated, rather than stereotyped, and even when it is stereotyped, as in the embryogenesis of many taxa, it often must vary within a cell lineage. This regulation requires collection and organization of vectorial information, and this organized information must be represented in structures that, either individually or collectively, contain the precision, robustness, and degrees of freedom necessary to adequately convey these processed vectorial cues to effectors that govern cell shape, organization, polarized function, and plane of division. Cells may have several such "information processors," which may be either distinct or overlapping, condensed or dispersed, but centrosomes, or their centriole components, present themselves as attractive candidates.

A number of the positioning studies reviewed here shed light on the potential for centrosomes to fulfill this role. Among these, for example, are the studies of astral centration. If regulatory cues are delivered to centrosomes, they presumably must be overlaid on a baseline of preexisting cytoskeletal forces acting on the organelle. The centered interphase aster, likely governed by the cytoplasmic length-scaled forces supported by multiple studies, may represent such a baseline, and strongly enforced baseline positioning by such forces might be an impediment to a cued response. It is interesting in this regard that Salle and coworkers observed a steep decline in centering forces after centration had been achieved in sea urchin eggs (Salle et al. 2018). The authors noted that that this decline might be due to a depletion of cytoplasmic cargo, but it conceivably may represent a regulatory event. In either case, the finding suggests the possibility that most cells have astral baseline forces that are low compared to the systems (chiefly eggs) in which these forces have been largely studied, and this reduction of baseline forces may provide more sensitivity to cues. The question arises as to whether there are in fact significant astral centering forces in most cells. One study observed centrosomes to be displaced from the cell centroid in confluent fibroblast monolayers and also found that centrosomes could be moved away from the centroid by adhesion to micropatterns that enforced a polarized cell shape (Hale et al. 2011), suggesting that, even in the absence of overt, extrinsic polarizing signals, geometric or other cues may be generally dominant.

Positioning studies have provided clues to the nature of the information that may be flowing to the centrosome, as well as to the forces mediating this flow. Our understanding of both the content and the forces is more advanced for mitotic spindle rotation than for the interphase centrosome, but the available evidence suggests there may be considerable commonality. For example, in multiple systems, the information delivered to the spindle poles includes instruction from neighboring cells in the form of junctional polarity pathway signals. The interphase centrosome also seems to receive such content, as seen in the effect of PAR-3, localized to lateral cortex, on polarizing centrosomes both in polarizing epithelial cells and in polarizing fibroblasts at the edge of a wound, and also as seen in the role of the polarity pathway in the formation of the immunological synapse. A second example is provided by the role of cell–cell contact. The role of contact in spindle orientation was demonstrated by Sugioka and Bowerman using bead–cell interactions in *C. elegans* embryos, while, with respect to interphase centrosome positioning, the important role of cadherins has been demonstrated in models of epithelial polarization, as well as collective cell migration. In spite of this progress, however, the elucidation of cues, and the likely cross-talk between them, is at an early stage.

Dynein appears to be the major force generator, acting via centrosome-nucleated MTs to control both spindle rotation and interphase centrosome positioning, although other forces, including kinesins and MT pushing forces, have also been implicated in both mitosis and interphase. Cortical anchors for dynein have been described for force communication to spindle poles, while in interphase different anchoring mechanisms appear to be responsible for cortical dynein localization during directional migration and during immunological synapse formation. Elaboration of the distinctions and similarities among these various mechanisms, and how they interact with MT dynamics to modulate force, represents an overarching challenge for the field. For migrating cells, the challenge is amplified by the mystery of why centrosomes move to different locations in different cells or at different times in the same cell. However, this puzzle is also likely to be an opportunity, if studies of positional switching by centrosomes during migration can be leveraged to uncover underlying mechanisms.

The view of the centrosome as an information processor depends greatly on determining how much information it is able to simultaneously receive and report: that is, in the degrees of freedom of its manipulation by forces (or other signals). Positioning studies are potentially critical for addressing this issue, but there is the question of, first, what aspect of positioning to measure and, second, what experimental system provides an appropriate test. Two lines of investigation discussed earlier have provided intriguing hints. First, measuring centrosome position in vivo in a bilaterally symmetric animal allows the use of developmental axes, coupled with the cell centroid, as a coordinate system. Taniguchi and coworkers exploited this idea to demonstrate a chiral specification of centrosome position in developing *Drosophila* hindgut, an organ that subsequently undergoes a chiral (left-right asymmetric) rotation (Taniguchi et al. 2011). While the authors showed that asymmetry in E-cadherin intensity correlated with centrosome position and implicated E-cadherin in chiral organogenesis, the mechanism by which asymmetric information is detected and propagated in this tissue, and specifically the role of the centrosome in this mechanism, remains to be determined. The second line of investigation, examining cell polarization in vitro, simply uses the nuclear centroid and the centrosome position to generate an axis, which is then compared either to the direction in which the cell subsequently moves or, in micropatterned cultures, to the axis of the micropattern (a radial axis or longitudinal axis in a circular or linear micropattern, respectively). These studies have consistently demonstrated chirality related to centrosome positioning. Collectively, these approaches raise the crucial and unanswered question as to the nature of the chiral center(s) in cells. Intriguingly, the direction of left-right asymmetry in micropatterned culture is cell specific (Wan et al. 2013), which should be a useful starting point for genetic and other approaches to investigate this issue.

The notion of the centrosome as a potential chiral center, or a component thereof, inevitably leads to the consideration of the features of the centrosome, and its components, that might participate in the transmission of chiral information (Marshall 2012; Regolini 2013). While a discussion of centriolar structure, as well as emerging details of the pericentriolar material, is beyond the scope of this review, there is one aspect of centrosome structure that may be directly related to the transmission of force to the organelle: the rotational orientation of the centrosome, considered as a potential means to convey symmetry-breaking information. Outside the special case of multiciliated cells, which are uncommon, is there evidence that a centrosome can possess an adjustable rotational specification? Since the two centrioles are linked, a rotation of one could lead to the revolution of the other, and therefore the spatial relationship of mother and daughter provides one index of rotational orientation. Studies of the cochlea kinocilium, as it migrates over the apical surface in response to PCP signals, show that in these primary cilia, the mother and daughter centrioles adopt a fixed, uniform orientation within the apical plane in response to PCP signals, implying a PCP-specified rotational orientation about the axis orthogonal to the plane (Jones et al. 2008). The mother–daughter relationship was very stable over several days of slow migration, and this stability was shown to the result of confining forces, which might potentially include forces that regulate orientation (Lepelletier et al. 2013). In addition to rotation within the cell surface plane, there is in principle the possibility of rotation about other axes. For example, the pointing of primary cilia in the direction of migration might represent the pointing of the mother centriole, which would require rotational specification about two axes. Alternatively, since these cilia are often near the front edge of the nucleus, the downward slope of the cell surface might contribute to their tilt, as has been proposed for the posterior tilt of nodal cilia. Nevertheless, an EM study of randomly migrating fibroblasts observed that mother centrioles, some nucleating a cilium, were predominantly oriented parallel to the substrate, while the daughter centriole was predominantly perpendicular to the substrate (Albrecht-Buehler and Bushnell 1979). Centrioles aligned in this way would necessarily possess specified orientation with respect to three axes of rotation. Further support for a regulated rotation of centrioles comes from time-lapse studies of cultured epithelial cells, which demonstrated that primary cilia move spontaneously, changing their angle with respect to the cell surface (Battle et al. 2015). This motion is correlated with the motion of the underlying basal body and indicates a hinge point below the basal body. The motion is energy and myosin dependent, indicating that centrioles can be actively reoriented. We currently lack convenient methods to examine centriole orientation. The recent advent of ultrastructure expansion microscopy may provide some new avenues to assess centrosome rotation as a chiral mechanism (Gambarotto et al. 2019).

Finally, centrosome positioning may provide insight into the effector mechanisms of this organelle. An early example is the study of Raff and Glover, who observed that the migration of centrosomes to the posterior pole of *Drosophila* embryos, in the absence of nuclei, was sufficient to elicit the organization of pole bud cells (Raff and Glover 1989). More recent examples also focus on the importance of centrosome proximity. As discussed earlier, spindle pole approach to the polar cortex in metaphase initiates the depletion of cortical dynein to restore spindle position (Kiyomitsu and Cheeseman 2012). In the *C. elegans* embryo, centrosome contact with the cortex initiates embryo polarization by inducing local actomyosin relaxation, via an unknown mechanism (Saturno et al. 2017). Why proximity is important in all these cases is unclear, but one possibility is that the asymmetry of centrosomal structure facilitates localized, asymmetric deployment of critical factors (such as Plk1 in the spindle pole) to initiate these responses. The localization of Cdk1 activity on the centrosome may permit the organelle to initiate chemical waves that control cell cycle progression and potentially other functions in frog embryos (Ishihara et al. 2014). Asymmetries that enhance the centrosome's ability to "read" may, then, also enhance its ability to "write." We appear to have only begun to tap the novel mechanisms of this remarkable organelle.

References

Adams M, Simms RJ, Abdelhamed Z, Dawe HR, Szymanska K, Logan CV, Wheway G, Pitt E, Gull K, Knowles MA, Blair E, Cross SH, Sayer JA, Johnson CA (2012) A meckelin-filamin A interaction mediates ciliogenesis. Hum Mol Genet 21:1272–1286

Al Jord A, Lemaitre AI, Delgehyr N, Faucourt M, Spassky N, Meunier A (2014) Centriole amplification by mother and daughter centrioles differs in multiciliated cells. Nature 516:104–107

Albrecht-Buehler G, Bushnell A (1979) The orientation of centrioles in migrating 3T3 cells. Exp Cell Res 120:111–118

Alieva IB, Vorobjev IA (2004) Vertebrate primary cilia: a sensory part of centrosomal complex in tissue cells, but a "sleeping beauty" in cultured cells? Cell Biol Int 28:139–150

Allena GM, Leeb KC, Barnharta EL, Tsuchidaa MA, Wilson CA, Gutierrez E, Groisman A, Mogilnerd A, Theriot JA (2018) Cell mechanics at the rear act to steer the direction of cell migration. BioRxiv. https://doi.org/10.1101/443408

Andersen EF, Halloran MC (2012) Centrosome movements in vivo correlate with specific neurite formation downstream of LIM homeodomain transcription factor activity. Development 139:3590–3599

Anderson CT, Stearns T (2009) Centriole age underlies asynchronous primary cilium growth in mammalian cells. Curr Biol 19:1498–1502

Andre J, Harrison S, Towers K, Qi X, Vaughan S, McKean PG, Ginger ML (2013) The tubulin cofactor C family member TBCCD1 orchestrates cytoskeletal filament formation. J Cell Sci 126:5350–5356

Antic D, Stubbs JL, Suyama K, Kintner C, Scott MP, Axelrod JD (2010) Planar cell polarity enables posterior localization of nodal cilia and left-right axis determination during mouse and Xenopus embryogenesis. PLoS One 5:e8999

Au FK, Jia Y, Jiang K, Grigoriev I, Hau BK, Shen Y, Du S, Akhmanova A, Qi RZ (2017) GAS2L1 is a centriole-associated protein required for centrosome dynamics and disjunction. Dev Cell 40:81–94

Battle C, Ott CM, Burnette DT, Lippincott-Schwartz J, Schmidt CF (2015) Intracellular and extracellular forces drive primary cilia movement. Proc Natl Acad Sci USA 112:1410–1415

Bellaiche Y, Gho M, Kaltschmidt JA, Brand AH, Schweisguth F (2001) Frizzled regulates localization of cell-fate determinants and mitotic spindle rotation during asymmetric cell division. Nat Cell Biol 3:50–57

Bertrand F, Esquerre M, Petit AE, Rodrigues M, Duchez S, Delon J, Valitutti S (2010) Activation of the ancestral polarity regulator protein kinase C zeta at the immunological synapse drives polarization of Th cell secretory machinery toward APCs. J Immunol 185:2887–2894

Besson C, Bernard F, Corson F, Rouault H, Reynaud E, Keder A, Mazouni K, Schweisguth F (2015) Planar cell polarity breaks the symmetry of PAR protein distribution prior to mitosis in Drosophila sensory organ precursor cells. Curr Biol 25:1104–1110

Bisgrove SR, Kropf DL (1998) Alignment of centrosomal and growth axes is a late event during polarization of Pelvetia compressa zygotes. Dev Biol 194:246–256

Boehlke C, Janusch H, Hamann C, Powelske C, Mergen M, Herbst H, Kotsis F, Nitschke R, Kuehn EW (2015) A cilia independent role of Ift88/Polaris during cell migration. PLoS One 10: e0140378

Bolhy S, Bouhlel I, Dultz E, Nayak T, Zuccolo M, Gatti X, Vallee R, Ellenberg J, Doye V (2011) A Nup133-dependent NPC-anchored network tethers centrosomes to the nuclear envelope in prophase. J Cell Biol 192:855–871

Bornens M (1977) Is the centriole bound to the nuclear membrane? Nature 270:80–82

Bornens M (2012) The centrosome in cells and organisms. Science 335:422–426

Borovina A, Superina S, Voskas D, Ciruna B (2010) Vangl2 directs the posterior tilting and asymmetric localization of motile primary cilia. Nat Cell Biol 12:407–412

Bosveld F, Markova O, Guirao B, Martin C, Wang Z, Pierre A, Balakireva M, Gaugue I, Ainslie A, Christophorou N, Lubensky DK, Minc N, Bellaiche Y (2016) Epithelial tricellular junctions act as interphase cell shape sensors to orient mitosis. Nature 530:495–498

Bosveld F, Wang Z, Bellaiche Y (2018) Tricellular junctions: a hot corner of epithelial biology. Curr Opin Cell Biol 54:80–88

Boutin C, Labedan P, Dimidschstein J, Richard F, Cremer H, Andre P, Yang Y, Montcouquiol M, Goffinet AM, Tissir F (2014) A dual role for planar cell polarity genes in ciliated cells. Proc Natl Acad Sci USA 111:E3129–E3138

Brito DA, Strauss J, Magidson V, Tikhonenko I, Khodjakov A, Koonce MP (2005) Pushing forces drive the comet-like motility of microtubule arrays in Dictyostelium. Mol Biol Cell 16:3334–3340

Brun-Usan M, Marin-Riera M, Grande C, Truchado-Garcia M, Salazar-Ciudad I (2017) A set of simple cell processes is sufficient to model spiral cleavage. Development 144:54–62

Brust-Mascher I, Civelekoglu-Scholey G, Scholey JM (2015) Mechanism for anaphase B: evaluation of "slide-and-cluster" versus "slide-and-flux-or-elongate" models. Biophys J 108:2007–2018

Bryant DM, Datta A, Rodriguez-Fraticelli AE, Peranen J, Martin-Belmonte F, Mostov KE (2010) A molecular network for de novo generation of the apical surface and lumen. Nat Cell Biol 12:1035–1045

Buendia B, Bre MH, Griffiths G, Karsenti E (1990) Cytoskeletal control of centrioles movement during the establishment of polarity in Madin-Darby canine kidney cells. J Cell Biol 110:1123–1135

Burakov AV, Nadezhdina ES (2013) Association of nucleus and centrosome: magnet or velcro? Cell Biol Int 37:95–104

Burakov A, Nadezhdina E, Slepchenko B, Rodionov V (2003) Centrosome positioning in interphase cells. J Cell Biol 162:963–969

Burute M, Prioux M, Blin G, Truchet S, Letort G, Tseng Q, Bessy T, Lowell S, Young J, Filhol O, Thery M (2017) Polarity reversal by centrosome repositioning primes cell scattering during epithelial-to-mesenchymal transition. Dev Cell 40:168–184

Bustos-Moran E, Blas-Rus N, Martin-Cofreces NB, Sanchez-Madrid F (2016) Orchestrating lymphocyte polarity in cognate immune cell-cell interactions. Int Rev Cell Mol Biol 327:195–261

Buttrick GJ, Beaumont LM, Leitch J, Yau C, Hughes JR, Wakefield JG (2008) Akt regulates centrosome migration and spindle orientation in the early *Drosophila melanogaster* embryo. J Cell Biol 180:537–548

Cao J, Crest J, Fasulo B, Sullivan W (2010) Cortical actin dynamics facilitate early-stage centrosome separation. Curr Biol 20:770–776

Carlin LM, Evans R, Milewicz H, Fernandes L, Matthews DR, Perani M, Levitt J, Keppler MD, Monypenny J, Coolen T, Barber PR, Vojnovic B, Suhling K, Fraternali F, Ameer-Beg S, Parker PJ, Thomas NS, Ng T (2011) A targeted siRNA screen identifies regulators of Cdc42 activity at the natural killer cell immunological synapse. Sci Signal 4:ra81

Carvajal-Gonzalez JM, Mulero-Navarro S, Mlodzik M (2016a) Centriole positioning in epithelial cells and its intimate relationship with planar cell polarity. Bioessays 38:1234–1245

Carvajal-Gonzalez JM, Roman AC, Mlodzik M (2016b) Positioning of centrioles is a conserved readout of Frizzled planar cell polarity signalling. Nat Commun 7:11135

Chanet S, Sharan R, Khan Z, Martin AC (2017) Myosin 2-induced mitotic rounding enables columnar epithelial cells to interpret cortical spindle positioning cues. Curr Biol 27:3350–3358

Chatterjee A, Chinnappa K, Ramanan N, Mani S (2018) Centrosome inheritance does not regulate cell fate in granule neuron progenitors of the developing cerebellum. Cerebellum 17:685–691

Chemin K, Bohineust A, Dogniaux S, Tourret M, Guegan S, Miro F, Hivroz C (2012) Cytokine secretion by CD4+ T cells at the immunological synapse requires Cdc42-dependent local actin remodeling but not microtubule organizing center polarity. J Immunol 189:2159–2168

Chen TH, Hsu JJ, Zhao X, Guo C, Wong MN, Huang Y, Li Z, Garfinkel A, Ho CM, Tintut Y, Demer LL (2012) Left-right symmetry breaking in tissue morphogenesis via cytoskeletal mechanics. Circ Res 110:551–559

Chien YH, Srinivasan S, Keller R, Kintner C (2018) Mechanical strain determines cilia length, motility, and planar position in the left-right organizer. Dev Cell 45:316–330

Chin AS, Worley KE, Ray P, Kaur G, Fan J, Wan LQ (2018) Epithelial cell chirality revealed by three-dimensional spontaneous rotation. Proc Natl Acad Sci USA 115:12188–12193

Christensen ST, Pedersen SF, Satir P, Veland IR, Schneider L (2008) The primary cilium coordinates signaling pathways in cell cycle control and migration during development and tissue repair. Curr Top Dev Biol 85:261–301

Clare DK, Magescas J, Piolot T, Dumoux M, Vesque C, Pichard E, Dang T, Duvauchelle B, Poirier F, Delacour D (2014) Basal foot MTOC organizes pillar MTs required for coordination of beating cilia. Nat Commun 5:4888

Colanzi A, Hidalgo Carcedo C, Persico A, Cericola C, Turacchio G, Bonazzi M, Luini A, Corda D (2007) The Golgi mitotic checkpoint is controlled by BARS-dependent fission of the Golgi ribbon into separate stacks in G2. EMBO J 26:2465–2476

Combs J, Kim SJ, Tan S, Ligon LA, Holzbaur EL, Kuhn J, Poenie M (2006) Recruitment of dynein to the Jurkat immunological synapse. Proc Natl Acad Sci USA 103:14883–14888

Courtois A, Schuh M, Ellenberg J, Hiiragi T (2012) The transition from meiotic to mitotic spindle assembly is gradual during early mammalian development. J Cell Biol 198:357–370

Cowan CR, Hyman AA (2004) Centrosomes direct cell polarity independently of microtubule assembly in C. elegans embryos. Nature 431:92–96

Crespo CL, Vernieri C, Keller PJ, Garre M, Bender JR, Wittbrodt J, Pardi R (2014) The PAR complex controls the spatiotemporal dynamics of F-actin and the MTOC in directionally migrating leukocytes. J Cell Sci 127:4381–4395

Cytrynbaum EN, Sommi P, Brust-Mascher I, Scholey JM, Mogilner A (2005) Early spindle assembly in Drosophila embryos: role of a force balance involving cytoskeletal dynamics and nuclear mechanics. Mol Biol Cell 16:4967–4981

D'Angelo A, Franco B (2011) The primary cilium in different tissues-lessons from patients and animal models. Pediatr Nephrol 26:655–662

Dantas TJ, Carabalona A, Hu DJ, Vallee RB (2016) Emerging roles for motor proteins in progenitor cell behavior and neuronal migration during brain development. Cytoskeleton (Hoboken) 73:566–576

Das RM, Storey KG (2014) Apical abscission alters cell polarity and dismantles the primary cilium during neurogenesis. Science 343:200–204

Davenport AJ, Cross RS, Watson KA, Liao Y, Shi W, Prince HM, Beavis PA, Trapani JA, Kershaw MH, Ritchie DS, Darcy PK, Neeson PJ, Jenkins MR (2018) Chimeric antigen receptor T cells form nonclassical and potent immune synapses driving rapid cytotoxicity. Proc Natl Acad Sci USA 115:E2068–E2076

Dawe HR, Farr H, Gull K (2007a) Centriole/basal body morphogenesis and migration during ciliogenesis in animal cells. J Cell Sci 120:7–15

Dawe HR, Smith UM, Cullinane AR, Gerrelli D, Cox P, Badano JL, Blair-Reid S, Sriram N, Katsanis N, Attie-Bitach T, Afford SC, Copp AJ, Kelly DA, Gull K, Johnson CA (2007b) The

Meckel-Gruber Syndrome proteins MKS1 and meckelin interact and are required for primary cilium formation. Hum Mol Genet 16:173–186

Dawe HR, Adams M, Wheway G, Szymanska K, Logan CV, Noegel AA, Gull K, Johnson CA (2009) Nesprin-2 interacts with meckelin and mediates ciliogenesis via remodelling of the actin cytoskeleton. J Cell Sci 122:2716–2726

de Anda FC, Pollarolo G, Da Silva JS, Camoletto PG, Feiguin F, Dotti CG (2005) Centrosome localization determines neuronal polarity. Nature 436:704–708

de Anda FC, Meletis K, Ge X, Rei D, Tsai LH (2010) Centrosome motility is essential for initial axon formation in the neocortex. J Neurosci 30:10391–10406

de la Roche M, Ritter AT, Angus KL, Dinsmore C, Earnshaw CH, Reiter JF, Griffiths GM (2013) Hedgehog signaling controls T cell killing at the immunological synapse. Science 342:1247–1250

de la Roche M, Asano Y, Griffiths GM (2016) Origins of the cytolytic synapse. Nat Rev Immunol 16:421–432

De Simone A, Nedelec F, Gonczy P (2016) Dynein transmits polarized actomyosin cortical flows to promote centrosome separation. Cell Rep 14:2250–2262

Decarreau J, Wagenbach M, Lynch E, Halpern AR, Vaughan JC, Kollman J, Wordeman L (2017) The tetrameric kinesin Kif25 suppresses pre-mitotic centrosome separation to establish proper spindle orientation. Nat Cell Biol 19:384–390

Desai RA, Gao L, Raghavan S, Liu WF, Chen CS (2009) Cell polarity triggered by cell-cell adhesion via E-cadherin. J Cell Sci 122:905–911

Devenport D (2016) Tissue morphodynamics: translating planar polarity cues into polarized cell behaviors. Semin Cell Dev Biol 55:99–110

Diao A, Rahman D, Pappin DJ, Lucocq J, Lowe M (2003) The coiled-coil membrane protein golgin-84 is a novel rab effector required for Golgi ribbon formation. J Cell Biol 160:201–212

Doyle AD, Wang FW, Matsumoto K, Yamada KM (2009) One-dimensional topography underlies three-dimensional fibrillar cell migration. J Cell Biol 184:481–490

Drechsler H, McAinsh AD (2016) Kinesin-12 motors cooperate to suppress microtubule catastrophes and drive the formation of parallel microtubule bundles. Proc Natl Acad Sci USA 113: E1635–E1644

Dujardin DL, Barnhart LE, Stehman SA, Gomes ER, Gundersen GG, Vallee RB (2003) A role for cytoplasmic dynein and LIS1 in directed cell movement. J Cell Biol 163:1205–1211

Dupin I, Camand E, Etienne-Manneville S (2009) Classical cadherins control nucleus and centrosome position and cell polarity. J Cell Biol 185:779–786

Dustin ML, Choudhuri K (2016) Signaling and polarized communication across the T cell immunological synapse. Annu Rev Cell Dev Biol 32:303–325

Efimov A, Kharitonov A, Efimova N, Loncarek J, Miller PM, Andreyeva N, Gleeson P, Galjart N, Maia AR, McLeod IX, Yates JR 3rd, Maiato H, Khodjakov A, Akhmanova A, Kaverina I (2007) Asymmetric CLASP-dependent nucleation of noncentrosomal microtubules at the trans-Golgi network. Dev Cell 12:917–930

Eilers U, Klumperman J, Hauri HP (1989) Nocodazole, a microtubule-active drug, interferes with apical protein delivery in cultured intestinal epithelial cells (Caco-2). J Cell Biol 108:13–22

Epting D, Slanchev K, Boehlke C, Hoff S, Loges NT, Yasunaga T, Indorf L, Nestel S, Lienkamp SS, Omran H, Kuehn EW, Ronneberger O, Walz G, Kramer-Zucker A (2015) The Rac1 regulator ELMO controls basal body migration and docking in multiciliated cells through interaction with Ezrin. Development 142:174–184

Etienne-Manneville S, Hall A (2001) Integrin-mediated activation of Cdc42 controls cell polarity in migrating astrocytes through PKCzeta. Cell 106:489–498

Etienne-Manneville S, Manneville JB, Nicholls S, Ferenczi MA, Hall A (2005) Cdc42 and Par6-PKCzeta regulate the spatially localized association of Dlg1 and APC to control cell polarization. J Cell Biol 170:895–901

Euteneuer U, Schliwa M (1984) Persistent, directional motility of cells and cytoplasmic fragments in the absence of microtubules. Nature 310:58–61

Euteneuer U, Schliwa M (1986) The function of microtubules in directional cell movement. Ann N Y Acad Sci 466:867–886

Euteneuer U, Schliwa M (1992) Mechanism of centrosome positioning during the wound response in BSC-1 cells. J Cell Biol 116:1157–1166

Fan J, Ray P, Lu Y, Kaur G, Schwarz JJ, Wan LQ (2018) Cell chirality regulates intercellular junctions and endothelial permeability. Sci Adv 4:eaat2111

Farina F, Gaillard J, Guerin C, Coute Y, Sillibourne J, Blanchoin L, Thery M (2016) The centrosome is an actin-organizing centre. Nat Cell Biol 18:65–75

Feinstein TN, Linstedt AD (2008) GRASP55 regulates Golgi ribbon formation. Mol Biol Cell 19:2696–2707

Feldman JL, Marshall WF (2009) ASQ2 encodes a TBCC-like protein required for mother-daughter centriole linkage and mitotic spindle orientation. Curr Biol 19:1238–1243

Feldman JL, Priess JR (2012) A role for the centrosome and PAR-3 in the hand-off of MTOC function during epithelial polarization. Curr Biol 22:575–582

Feldman JL, Geimer S, Marshall WF (2007) The mother centriole plays an instructive role in defining cell geometry. PLoS Biol 5:e149

Ferenz NP, Gable A, Wadsworth P (2010) Mitotic functions of kinesin-5. Semin Cell Dev Biol 21:255–259

Fielmich LE, Schmidt R, Dickinson DJ, Goldstein B, Akhmanova A, van den Heuvel S (2018) Optogenetic dissection of mitotic spindle positioning in vivo. elife 7:e38198

Finetti F, Paccani SR, Riparbelli MG, Giacomello E, Perinetti G, Pazour GJ, Rosenbaum JL, Baldari CT (2009) Intraflagellar transport is required for polarized recycling of the TCR/CD3 complex to the immune synapse. Nat Cell Biol 11:1332–1339

Finetti F, Patrussi L, Masi G, Onnis A, Galgano D, Lucherini OM, Pazour GJ, Baldari CT (2014) Specific recycling receptors are targeted to the immune synapse by the intraflagellar transport system. J Cell Sci 127:1924–1937

Fink G, Schuchardt I, Colombelli J, Stelzer E, Steinberg G (2006) Dynein-mediated pulling forces drive rapid mitotic spindle elongation in *Ustilago maydis*. EMBO J 25:4897–4908

Fink J, Carpi N, Betz T, Betard A, Chebah M, Azioune A, Bornens M, Sykes C, Fetler L, Cuvelier D, Piel M (2011) External forces control mitotic spindle positioning. Nat Cell Biol 13:771–778

Fuentealba LC, Eivers E, Geissert D, Taelman V, De Robertis EM (2008) Asymmetric mitosis: unequal segregation of proteins destined for degradation. Proc Natl Acad Sci USA 105:7732–7737

Galli M, van den Heuvel S (2008) Determination of the cleavage plane in early C. elegans embryos. Annu Rev Genet 42:389–411

Gallini S, Carminati M, De Mattia F, Pirovano L, Martini E, Oldani A, Asteriti IA, Guarguaglini G, Mapelli M (2016) NuMA phosphorylation by Aurora-A orchestrates spindle orientation. Curr Biol 26:458–469

Gambarotto D, Zwettler FU, Le Guennec M, Schmidt-Cernohorska M, Fortun D, Borgers S, Heine J, Schloetel JG, Reuss M, Unser M, Boyden ES, Sauer M, Hamel V, Guichard P (2019) Imaging cellular ultrastructures using expansion microscopy (U-ExM). Nat Methods 16:71–74

Garzon-Coral C, Fantana HA, Howard J (2016) A force-generating machinery maintains the spindle at the cell center during mitosis. Science 352:1124–1127

Gogendeau D, Basto R (2010) Centrioles in flies: the exception to the rule? Semin Cell Dev Biol 21:163–173

Gomes ER, Jani S, Gundersen GG (2005) Nuclear movement regulated by Cdc42, MRCK, myosin, and actin flow establishes MTOC polarization in migrating cells. Cell 121:451–463

Gomez TS, Kumar K, Medeiros RB, Shimizu Y, Leibson PJ, Billadeau DD (2007) Formins regulate the actin-related protein 2/3 complex-independent polarization of the centrosome to the immunological synapse. Immunity 26:177–190

Goncalves J, Nolasco S, Nascimento R, Lopez Fanarraga M, Zabala JC, Soares H (2010) TBCCD1, a new centrosomal protein, is required for centrosome and Golgi apparatus positioning. EMBO Rep 11:194–200

Gonczy P, Pichler S, Kirkham M, Hyman AA (1999) Cytoplasmic dynein is required for distinct aspects of MTOC positioning, including centrosome separation, in the one cell stage Caenorhabditis elegans embryo. J Cell Biol 147:135–150

Gotlieb AI, May LM, Subrahmanyan L, Kalnins VI (1981) Distribution of microtubule organizing centers in migrating sheets of endothelial cells. J Cell Biol 91:589–594

Gotlieb AI, Subrahmanyan L, Kalnins VI (1983) Microtubule-organizing centers and cell migration: effect of inhibition of migration and microtubule disruption in endothelial cells. J Cell Biol 96:1266–1272

Gotta M, Dong Y, Peterson YK, Lanier SM, Ahringer J (2003) Asymmetrically distributed C. elegans homologs of AGS3/PINS control spindle position in the early embryo. Curr Biol 13:1029–1037

Goulding MB, Canman JC, Senning EN, Marcus AH, Bowerman B (2007) Control of nuclear centration in the C. elegans zygote by receptor-independent Galpha signaling and myosin II. J Cell Biol 178:1177–1191

Greenberg SR, Tan W, Lee WL (2018) Num1 versus NuMA: insights from two functionally homologous proteins. Biophys Rev 10:1631–1636

Grill SW, Hyman AA (2005) Spindle positioning by cortical pulling forces. Dev Cell 8:461–465

Grishchuk EL, Molodtsov MI, Ataullakhanov FI, McIntosh JR (2005) Force production by disassembling microtubules. Nature 438:384–388

Gruber T, Fresser F, Jenny M, Uberall F, Leitges M, Baier G (2008) PKCtheta cooperates with atypical PKCzeta and PKCiota in NF-kappaB transactivation of T lymphocytes. Mol Immunol 45:117–126

Guild J, Ginzberg MB, Hueschen CL, Mitchison TJ, Dumont S (2017) Increased lateral microtubule contact at the cell cortex is sufficient to drive mammalian spindle elongation. Mol Biol Cell 28:1975–1983

Guirao B, Meunier A, Mortaud S, Aguilar A, Corsi JM, Strehl L, Hirota Y, Desoeuvre A, Boutin C, Han YG, Mirzadeh Z, Cremer H, Montcouquiol M, Sawamoto K, Spassky N (2010) Coupling between hydrodynamic forces and planar cell polarity orients mammalian motile cilia. Nat Cell Biol 12:341–350

Gundersen GG, Bulinski JC (1988) Selective stabilization of microtubules oriented toward the direction of cell migration. Proc Natl Acad Sci USA 85:5946–5950

Gunn PA, Gliddon BL, Londrigan SL, Lew AM, van Driel IR, Gleeson PA (2011) The Golgi apparatus in the endomembrane-rich gastric parietal cells exist as functional stable mini-stacks dispersed throughout the cytoplasm. Biol Cell 103:559–572

Habib SJ, Chen BC, Tsai FC, Anastassiadis K, Meyer T, Betzig E, Nusse R (2013) A localized Wnt signal orients asymmetric stem cell division in vitro. Science 339:1445–1448

Hale CM, Chen WC, Khatau SB, Daniels BR, Lee JS, Wirtz D (2011) SMRT analysis of MTOC and nuclear positioning reveals the role of EB1 and LIC1 in single-cell polarization. J Cell Sci 124:4267–4285

Hamaguchi M, Hiramoto Y (1986) Analysis of the role of astral rays in pronuclear migration in sand dollar eggs by the colcemid-UV method. Dev Growth Differ 28:143–156

Hammer JA, Wang J, Saeed M, Pedrosa A (2018) Origin, organization, dynamics, and function of actin and actomyosin networks at the T cell immunological synapse. Annu Rev Immunol 37:201–224. https://doi.org/10.1146/annurev-immunol-042718-041341

Hashimoto M, Shinohara K, Wang J, Ikeuchi S, Yoshiba S, Meno C, Nonaka S, Takada S, Hatta K, Wynshaw-Boris A, Hamada H (2010) Planar polarization of node cells determines the rotational axis of node cilia. Nat Cell Biol 12:170–176

Hatori R, Ando T, Sasamura T, Nakazawa N, Nakamura M, Taniguchi K, Hozumi S, Kikuta J, Ishii M, Matsuno K (2014) Left-right asymmetry is formed in individual cells by intrinsic cell chirality. Mech Dev 133:146–162

Heppert JK, Pani AM, Roberts AM, Dickinson DJ, Goldstein B (2018) A CRISPR tagging-based screen reveals localized players in Wnt-directed asymmetric cell division. Genetics 208:1147–1164

Hertwig O (1884) Das problem der befruchtung und der isotropie des eies. eine theorie der vererbung. Jena Z Med Naturwiss 18:276–318

Hird SN, White JG (1993) Cortical and cytoplasmic flow polarity in early embryonic cells of Caenorhabditis elegans. J Cell Biol 121:1343–1355

Ho WC, Allan VJ, van Meer G, Berger EG, Kreis TE (1989) Reclustering of scattered Golgi elements occurs along microtubules. Eur J Cell Biol 48:250–263

Holmes JA, Dutcher SK (1989) Cellular asymmetry in *Chlamydomonas reinhardtii*. J Cell Sci 94:273–285

Holy TE, Dogterom M, Yurke B, Leibler S (1997) Assembly and positioning of microtubule asters in microfabricated chambers. Proc Natl Acad Sci USA 94:6228–6231

Horton AC, Ehlers MD (2003) Dual modes of endoplasmic reticulum-to-Golgi transport in dendrites revealed by live-cell imaging. J Neurosci 23:6188–6199

Horton AC, Racz B, Monson EE, Lin AL, Weinberg RJ, Ehlers MD (2005) Polarized secretory trafficking directs cargo for asymmetric dendrite growth and morphogenesis. Neuron 48:757–771

Howard J, Garzon-Coral C (2017) Physical limits on the precision of mitotic spindle positioning by microtubule pushing forces: mechanics of mitotic spindle positioning. Bioessays. https://doi.org/10.1002/bies.201700122

Hurtado L, Caballero C, Gavilan MP, Cardenas J, Bornens M, Rios RM (2011) Disconnecting the Golgi ribbon from the centrosome prevents directional cell migration and ciliogenesis. J Cell Biol 193:917–933

Huse M (2017) Mechanical forces in the immune system. Nat Rev Immunol 17:679–690

Hyman AA, White JG (1987) Determination of cell division axes in the early embryogenesis of Caenorhabditis elegans. J Cell Biol 105:2123–2135

Iden S, Collard JG (2008) Crosstalk between small GTPases and polarity proteins in cell polarization. Nat Rev Mol Cell Biol 9:846–859

Ioannou A, Santama N, Skourides PA (2013) Xenopus laevis nucleotide binding protein 1 (xNubp1) is important for convergent extension movements and controls ciliogenesis via regulation of the actin cytoskeleton. Dev Biol 380:243–258

Ishihara K, Nguyen PA, Wuhr M, Groen AC, Field CM, Mitchison TJ (2014) Organization of early frog embryos by chemical waves emanating from centrosomes. Philos Trans R Soc Lond Ser B Biol Sci 369(1650):20130454. https://doi.org/10.1098/rstb.2013.0454

Jackson CL (2018) Activators and effectors of the small G protein Arf1 in regulation of Golgi dynamics during the cell division cycle. Front Cell Dev Biol 6:29

Januschke J, Llamazares S, Reina J, Gonzalez C (2011) Drosophila neuroblasts retain the daughter centrosome. Nat Commun 2:243

Januschke J, Reina J, Llamazares S, Bertran T, Rossi F, Roig J, Gonzalez C (2013) Centrobin controls mother-daughter centriole asymmetry in Drosophila neuroblasts. Nat Cell Biol 15:241–248

Jeyifous O, Waites CL, Specht CG, Fujisawa S, Schubert M, Lin EI, Marshall J, Aoki C, de Silva T, Montgomery JM, Garner CC, Green WN (2009) SAP97 and CASK mediate sorting of NMDA receptors through a previously unknown secretory pathway. Nat Neurosci 12:1011–1019

Jiang T, McKinley RF, McGill MA, Angers S, Harris TJ (2015) A Par-1-Par-3-centrosome cell polarity pathway and its tuning for isotropic cell adhesion. Curr Biol 25:2701–2708

Jones C, Chen P (2008) Primary cilia in planar cell polarity regulation of the inner ear. Curr Top Dev Biol 85:197–224

Jones C, Roper VC, Foucher I, Qian D, Banizs B, Petit C, Yoder BK, Chen P (2008) Ciliary proteins link basal body polarization to planar cell polarity regulation. Nat Genet 40:69–77

Kaltschmidt JA, Davidson CM, Brown NH, Brand AH (2000) Rotation and asymmetry of the mitotic spindle direct asymmetric cell division in the developing central nervous system. Nat Cell Biol 2:7–12

Kapitein LC, Peterman EJ, Kwok BH, Kim JH, Kapoor TM, Schmidt CF (2005) The bipolar mitotic kinesin Eg5 moves on both microtubules that it crosslinks. Nature 435:114–118

Kasioulis I, Das RM, Storey KG (2017) Inter-dependent apical microtubule and actin dynamics orchestrate centrosome retention and neuronal delamination. eLife 6(6):e26215. https://doi.org/10.7554/eLife.26215

Kelliher MT, Yue Y, Ng A, Kamiyama D, Huang B, Verhey KJ, Wildonger J (2018) Autoinhibition of kinesin-1 is essential to the dendrite-specific localization of Golgi outposts. J Cell Biol 217:2531–2547

Khodjakov A, Cole RW, Oakley BR, Rieder CL (2000) Centrosome-independent mitotic spindle formation in vertebrates. Curr Biol 10:59–67

Kimura K, Kimura A (2011) Intracellular organelles mediate cytoplasmic pulling force for centrosome centration in the *Caenorhabditis elegans* early embryo. Proc Natl Acad Sci USA 108:137–142

Kimura A, Onami S (2005) Computer simulations and image processing reveal length-dependent pulling force as the primary mechanism for *C. elegans* male pronuclear migration. Dev Cell 8:765–775

Kiyomitsu T, Cheeseman IM (2012) Chromosome- and spindle-pole-derived signals generate an intrinsic code for spindle position and orientation. Nat Cell Biol 14:311–317

Kiyomitsu T, Cheeseman IM (2013) Cortical dynein and asymmetric membrane elongation coordinately position the spindle in anaphase. Cell 154:391–402

Klos Dehring DA, Vladar EK, Werner ME, Mitchell JW, Hwang P, Mitchell BJ (2013) Deuterosome-mediated centriole biogenesis. Dev Cell 27:103–112

Kobayashi T, Murayama T (2009) Cell cycle-dependent microtubule-based dynamic transport of cytoplasmic dynein in mammalian cells. PLoS One 4:e7827

Kobayashi T, Kim S, Lin YC, Inoue T, Dynlacht BD (2014) The CP110-interacting proteins Talpid3 and Cep290 play overlapping and distinct roles in cilia assembly. J Cell Biol 204:215–229

Koonce MP, Cloney RA, Berns MW (1984) Laser irradiation of centrosomes in newt eosinophils: evidence of centriole role in motility. J Cell Biol 98:1999–2010

Koonce MP, Kohler J, Neujahr R, Schwartz JM, Tikhonenko I, Gerisch G (1999) Dynein motor regulation stabilizes interphase microtubule arrays and determines centrosome position. EMBO J 18:6786–6792

Kotak S, Gonczy P (2013) Mechanisms of spindle positioning: cortical force generators in the limelight. Curr Opin Cell Biol 25:741–748

Kotak S, Gonczy P (2014) NuMA phosphorylation dictates dynein-dependent spindle positioning. Cell Cycle 13:177–178

Kotak S, Busso C, Gonczy P (2012) Cortical dynein is critical for proper spindle positioning in human cells. J Cell Biol 199:97–110

Kotak S, Busso C, Gonczy P (2013) NuMA phosphorylation by CDK1 couples mitotic progression with cortical dynein function. EMBO J 32:2517–2529

Kreft ME, Di Giandomenico D, Beznoussenko GV, Resnik N, Mironov AA, Jezernik K (2010) Golgi apparatus fragmentation as a mechanism responsible for uniform delivery of uroplakins to the apical plasma membrane of uroepithelial cells. Biol Cell 102:593–607

Kuhn JR, Poenie M (2002) Dynamic polarization of the microtubule cytoskeleton during CTL-mediated killing. Immunity 16:111–121

Kuhne MR, Lin J, Yablonski D, Mollenauer MN, Ehrlich LI, Huppa J, Davis MM, Weiss A (2003) Linker for activation of T cells, zeta-associated protein-70, and Src homology 2 domain-containing leukocyte protein-76 are required for TCR-induced microtubule-organizing center polarization. J Immunol 171:860–866

Kupfer A, Louvard D, Singer SJ (1982) Polarization of the Golgi apparatus and the microtubule-organizing center in cultured fibroblasts at the edge of an experimental wound. Proc Natl Acad Sci USA 79:2603–2607

Laan L, Pavin N, Husson J, Romet-Lemonne G, van Duijn M, Lopez MP, Vale RD, Julicher F, Reck-Peterson SL, Dogterom M (2012) Cortical dynein controls microtubule dynamics to generate pulling forces that position microtubule asters. Cell 148:502–514

Lambert JD, Nagy LM (2002) Asymmetric inheritance of centrosomally localized mRNAs during embryonic cleavages. Nature 420:682–686

Lang E, Polec A, Lang A, Valk M, Blicher P, Rowe AD, Tonseth KA, Jackson CJ, Utheim TP, Janssen LMC, Eriksson J, Boe SO (2018) Coordinated collective migration and asymmetric cell division in confluent human keratinocytes without wounding. Nat Commun 9:3665

Larson ME, Bement WM (2017) Automated mitotic spindle tracking suggests a link between spindle dynamics, spindle orientation, and anaphase onset in epithelial cells. Mol Biol Cell 28:746–759

Lengefeld J, Barral Y (2018) Asymmetric segregation of aged spindle pole bodies during cell division: mechanisms and relevance beyond budding yeast. Bioessays 40:e1800038

Lepelletier L, de Monvel JB, Buisson J, Desdouets C, Petit C (2013) Auditory hair cell centrioles undergo confined Brownian motion throughout the developmental migration of the kinocilium. Biophys J 105:48–58

Li D, Wang YL (2018) Coordination of cell migration mediated by site-dependent cell-cell contact. Proc Natl Acad Sci USA 115:10678–10683

Li L, Norrelykke SF, Cox EC (2008) Persistent cell motion in the absence of external signals: a search strategy for eukaryotic cells. PLoS One 3:e2093

Li W, Yi P, Zhu Z, Zhang X, Ou G (2017) Centriole translocation and degeneration during ciliogenesis in Caenorhabditis elegans neurons. EMBO J 36:2553–2566

Liao G, Nagasaki T, Gundersen GG (1995) Low concentrations of nocodazole interfere with fibroblast locomotion without significantly affecting microtubule level: implications for the role of dynamic microtubules in cell locomotion. J Cell Sci 108:3473–3483

Lim WM, Ito Y, Sakata-Sogawa K, Tokunaga M (2018) CLIP-170 is essential for MTOC repositioning during T cell activation by regulating dynein localisation on the cell surface. Sci Rep 8:17447

Lin J, Hou KK, Piwnica-Worms H, Shaw AS (2009) The polarity protein Par1b/EMK/MARK2 regulates T cell receptor-induced microtubule-organizing center polarization. J Immunol 183:1215–1221

Lin CH, Li H, Lee YN, Cheng YJ, Wu RM, Chien CT (2015) Lrrk regulates the dynamic profile of dendritic Golgi outposts through the golgin Lava lamp. J Cell Biol 210:471–483

Liro MJ, Rose LS (2016) Mitotic spindle positioning in the ems cell of Caenorhabditis elegans requires LET-99 and LIN-5/NuMA. Genetics 204:1177–1189

Liu X, Kapoor TM, Chen JK, Huse M (2013) Diacylglycerol promotes centrosome polarization in T cells via reciprocal localization of dynein and myosin II. Proc Natl Acad Sci USA 110:11976–11981

Liu X, Welf ES, Haugh JM (2015) Linking morphodynamics and directional persistence of T lymphocyte migration. J R Soc Interface 12(106):20141412. https://doi.org/10.1098/rsif.2014.1412

Lobikin M, Wang G, Xu J, Hsieh YW, Chuang CF, Lemire JM, Levin M (2012) Early, nonciliary role for microtubule proteins in left-right patterning is conserved across kingdoms. Proc Natl Acad Sci USA 109:12586–12591

Lu MS, Johnston CA (2013) Molecular pathways regulating mitotic spindle orientation in animal cells. Development 140:1843–1856

Lu Q, Insinna C, Ott C, Stauffer J, Pintado PA, Rahajeng J, Baxa U, Walia V, Cuenca A, Hwang YS, Daar IO, Lopes S, Lippincott-Schwartz J, Jackson PK, Caplan S, Westlake CJ (2015) Early steps in primary cilium assembly require EHD1/EHD3-dependent ciliary vesicle formation. Nat Cell Biol 17:228–240

Luxton GW, Gundersen GG (2011) Orientation and function of the nuclear-centrosomal axis during cell migration. Curr Opin Cell Biol 23:579–588

Machicoane M, de Frutos CA, Fink J, Rocancourt M, Lombardi Y, Garel S, Piel M, Echard A (2014) SLK-dependent activation of ERMs controls LGN-NuMA localization and spindle orientation. J Cell Biol 205:791–799

Makrogianneli K, Carlin LM, Keppler MD, Matthews DR, Ofo E, Coolen A, Ameer-Beg SM, Barber PR, Vojnovic B, Ng T (2009) Integrating receptor signal inputs that influence small Rho GTPase activation dynamics at the immunological synapse. Mol Cell Biol 29:2997–3006

Malech HL, Root RK, Gallin JI (1977) Structural analysis of human neutrophil migration. Centriole, microtubule, and microfilament orientation and function during chemotaxis. J Cell Biol 75:666–693

Malinova TS, Huveneers S (2018) Sensing of cytoskeletal forces by asymmetric adherens junctions. Trends Cell Biol 28:328–341

Malone CJ, Misner L, Le Bot N, Tsai MC, Campbell JM, Ahringer J, White JG (2003) The C. elegans hook protein, ZYG-12, mediates the essential attachment between the centrosome and nucleus. Cell 115:825–836

Manneville JB, Jehanno M, Etienne-Manneville S (2010) Dlg1 binds GKAP to control dynein association with microtubules, centrosome positioning, and cell polarity. J Cell Biol 191:585–598

Mapelli M, Gonzalez C (2012) On the inscrutable role of Inscuteable: structural basis and functional implications for the competitive binding of NuMA and Inscuteable to LGN. Open Biol 2:120102

Mardin BR, Lange C, Baxter JE, Hardy T, Scholz SR, Fry AM, Schiebel E (2010) Components of the Hippo pathway cooperate with Nek2 kinase to regulate centrosome disjunction. Nat Cell Biol 12:1166–1176

Mardin BR, Agircan FG, Lange C, Schiebel E (2011) Plk1 controls the Nek2A-PP1gamma antagonism in centrosome disjunction. Curr Biol 21:1145–1151

Mardin BR, Isokane M, Cosenza MR, Kramer A, Ellenberg J, Fry AM, Schiebel E (2013) EGF-induced centrosome separation promotes mitotic progression and cell survival. Dev Cell 25:229–240

Marshall WF (2012) Centriole asymmetry determines algal cell geometry. Curr Opin Plant Biol 15:632–637

Martin-Cofreces NB, Sanchez-Madrid F (2018) Sailing to and docking at the immune synapse: role of tubulin dynamics and molecular motors. Front Immunol 9:1174

Martin-Cofreces NB, Robles-Valero J, Cabrero JR, Mittelbrunn M, Gordon-Alonso M, Sung CH, Alarcon B, Vazquez J, Sanchez-Madrid F (2008) MTOC translocation modulates IS formation and controls sustained T cell signaling. J Cell Biol 182:951–962

Mayor R, Etienne-Manneville S (2016) The front and rear of collective cell migration. Nat Rev Mol Cell Biol 17:97–109

Mazo G, Soplop N, Wang WJ, Uryu K, Tsou MF (2016) Spatial control of primary ciliogenesis by subdistal appendages alters sensation-associated properties of cilia. Dev Cell 39:424–437

McCaffrey LM, Macara IG (2012) Signaling pathways in cell polarity. Cold Spring Harb Perspect Biol 4:a009654

Meads T, Schroer TA (1995) Polarity and nucleation of microtubules in polarized epithelial cells. Cell Motil Cytoskeleton 32:273–288

Meng W, Mushika Y, Ichii T, Takeichi M (2008) Anchorage of microtubule minus ends to adherens junctions regulates epithelial cell-cell contacts. Cell 135:948–959

Mikhaylova M, Bera S, Kobler O, Frischknecht R, Kreutz MR (2016) A dendritic Golgi satellite between ERGIC and retromer. Cell Rep 14:189–199

Miller PM, Folkmann AW, Maia AR, Efimova N, Efimov A, Kaverina I (2009) Golgi-derived CLASP-dependent microtubules control Golgi organization and polarized trafficking in motile cells. Nat Cell Biol 11:1069–1080

Minc N, Burgess D, Chang F (2011) Influence of cell geometry on division-plane positioning. Cell 144:414–426

Minegishi K, Hashimoto M, Ajima R, Takaoka K, Shinohara K, Ikawa Y, Nishimura H, McMahon AP, Willert K, Okada Y, Sasaki H, Shi D, Fujimori T, Ohtsuka T, Igarashi Y, Yamaguchi TP, Shimono A, Shiratori H, Hamada H (2017) A Wnt5 activity asymmetry and intercellular signaling via PCP proteins polarize node cells for left-right symmetry breaking. Dev Cell 40:439–452 e434

Mirvis M, Stearns T, James Nelson W (2018) Cilium structure, assembly, and disassembly regulated by the cytoskeleton. Biochem J 475:2329–2353

Mirzadeh Z, Han YG, Soriano-Navarro M, Garcia-Verdugo JM, Alvarez-Buylla A (2010) Cilia organize ependymal planar polarity. J Neurosci 30:2600–2610

Mitchell B, Jacobs R, Li J, Chien S, Kintner C (2007) A positive feedback mechanism governs the polarity and motion of motile cilia. Nature 447:97–101

Mitchison T, Wuhr M, Nguyen P, Ishihara K, Groen A, Field CM (2012) Growth, interaction, and positioning of microtubule asters in extremely large vertebrate embryo cells. Cytoskeleton 69:738–750

Nadezhdina ES, Fais D, Chentsov YS (1979) On the association of centrioles with the interphase nucleus. Eur J Cell Biol 19:109–115

Nath S, Christian L, Tan SY, Ki S, Ehrlich LI, Poenie M (2016) Dynein separately partners with NDE1 and dynactin to orchestrate T cell focused secretion. J Immunol 197:2090–2101

Natividad RJ, Lalli ML, Muthuswamy SK, Asthagiri AR (2018) Golgi stabilization, not its front-rear bias, is associated with EMT-enhanced fibrillar migration. Biophys J 115:2067–2077

Negishi T, Nishida H (2017) Asymmetric and unequal cell divisions in ascidian embryos. Results Probl Cell Differ 61:261–284

Negishi T, Yasuo H (2015) Distinct modes of mitotic spindle orientation align cells in the dorsal midline of ascidian embryos. Dev Biol 408:66–78

Nielsen BS, Malinda RR, Schmid FM, Pedersen SF, Christensen ST, Pedersen LB (2015) PDGFRbeta and oncogenic mutant PDGFRalpha D842V promote disassembly of primary cilia through a PLCgamma- and AURKA-dependent mechanism. J Cell Sci 128:3543–3549

Nonaka S, Yoshiba S, Watanabe D, Ikeuchi S, Goto T, Marshall WF, Hamada H (2005) De novo formation of left-right asymmetry by posterior tilt of nodal cilia. PLoS Biol 3:e268

O'Neill PR, Castillo-Badillo JA, Meshik X, Kalyanaraman V, Melgarejo K, Gautam N (2018) Membrane flow drives an adhesion-independent amoeboid cell migration mode. Dev Cell 46:9–22 e24

Obino D, Farina F, Malbec O, Saez PJ, Maurin M, Gaillard J, Dingli F, Loew D, Gautreau A, Yuseff MI, Blanchoin L, Thery M, Lennon-Dumenil AM (2016) Actin nucleation at the centrosome controls lymphocyte polarity. Nat Commun 7:10969

Ohata S, Alvarez-Buylla A (2016) Planar organization of multiciliated ependymal (E1) cells in the brain ventricular epithelium. Trends Neurosci 39:543–551

Ohata S, Herranz-Perez V, Nakatani J, Boletta A, Garcia-Verdugo JM, Alvarez-Buylla A (2015) Mechanosensory genes Pkd1 and Pkd2 contribute to the planar polarization of brain ventricular epithelium. J Neurosci 35:11153–11168

Okada Y, Takeda S, Tanaka Y, Belmonte JI, Hirokawa N (2005) Mechanism of nodal flow: a conserved symmetry breaking event in left-right axis determination. Cell 121:633–644

Okumura M, Natsume T, Kanemaki MT, Kiyomitsu T (2018) Dynein-Dynactin-NuMA clusters generate cortical spindle-pulling forces as a multi-arm ensemble. eLife 7:e36559

Omer S, Greenberg SR, Lee WL (2018) Cortical dynein pulling mechanism is regulated by differentially targeted attachment molecule Num1. eLife 7:e36745

Omura F, Fukui Y (1985) Dictyostelium MTOC: structure and linkage to the nucleus. Protoplasma 127:212–221

Ori-McKenney KM, Jan LY, Jan YN (2012) Golgi outposts shape dendrite morphology by functioning as sites of acentrosomal microtubule nucleation in neurons. Neuron 76:921–930

Palazzo AF, Joseph HL, Chen YJ, Dujardin DL, Alberts AS, Pfister KK, Vallee RB, Gundersen GG (2001) Cdc42, dynein, and dynactin regulate MTOC reorientation independent of Rho-regulated microtubule stabilization. Curr Biol 11:1536–1541

Pan J, You Y, Huang T, Brody SL (2007) RhoA-mediated apical actin enrichment is required for ciliogenesis and promoted by Foxj1. J Cell Sci 120:1868–1876

Paridaen JT, Wilsch-Brauninger M, Huttner WB (2013) Asymmetric inheritance of centrosome-associated primary cilium membrane directs ciliogenesis after cell division. Cell 155:333–344

Park DH, Rose LS (2008) Dynamic localization of LIN-5 and GPR-1/2 to cortical force generation domains during spindle positioning. Dev Biol 315:42–54

Park TJ, Mitchell BJ, Abitua PB, Kintner C, Wallingford JB (2008) Dishevelled controls apical docking and planar polarization of basal bodies in ciliated epithelial cells. Nat Genet 40:871–879

Pecreaux J, Redemann S, Alayan Z, Mercat B, Pastezeur S, Garzon-Coral C, Hyman AA, Howard J (2016) The mitotic spindle in the one-cell *C. elegans* embryo is positioned with high precision and stability. Biophys J 111:1773–1784

Peyre E, Jaouen F, Saadaoui M, Haren L, Merdes A, Durbec P, Morin X (2011) A lateral belt of cortical LGN and NuMA guides mitotic spindle movements and planar division in neuroepithelial cells. J Cell Biol 193:141–154

Piel M, Meyer P, Khodjakov A, Rieder CL, Bornens M (2000) The respective contributions of the mother and daughter centrioles to centrosome activity and behavior in vertebrate cells. J Cell Biol 149:317–330

Pitaval A, Senger F, Letort G, Gidrol X, Guyon L, Sillibourne J, Thery M (2017) Microtubule stabilization drives 3D centrosome migration to initiate primary ciliogenesis. J Cell Biol 216:3713–3728

Plotnikova OV, Pugacheva EN, Golemis EA (2009) Primary cilia and the cell cycle. Methods Cell Biol 94:137–160

Poulson ND, Lechler T (2010) Robust control of mitotic spindle orientation in the developing epidermis. J Cell Biol 191:915–922

Pouthas F, Girard P, Lecaudey V, Ly TB, Gilmour D, Boulin C, Pepperkok R, Reynaud EG (2008) In migrating cells, the Golgi complex and the position of the centrosome depend on geometrical constraints of the substratum. J Cell Sci 121:2406–2414

Puthenveedu MA, Bachert C, Puri S, Lanni F, Linstedt AD (2006) GM130 and GRASP65-dependent lateral cisternal fusion allows uniform Golgi-enzyme distribution. Nat Cell Biol 8:238–248

Quann EJ, Merino E, Furuta T, Huse M (2009) Localized diacylglycerol drives the polarization of the microtubule-organizing center in T cells. Nat Immunol 10:627–635

Quann EJ, Liu X, Altan-Bonnet G, Huse M (2011) A cascade of protein kinase C isozymes promotes cytoskeletal polarization in T cells. Nat Immunol 12:647–654

Quassollo G, Wojnacki J, Salas DA, Gastaldi L, Marzolo MP, Conde C, Bisbal M, Couve A, Caceres A (2015) A RhoA signaling pathway regulates dendritic Golgi outpost formation. Curr Biol 25:971–982

Raaijmakers JA, van Heesbeen RG, Meaders JL, Geers EF, Fernandez-Garcia B, Medema RH, Tanenbaum ME (2012) Nuclear envelope-associated dynein drives prophase centrosome separation and enables Eg5-independent bipolar spindle formation. EMBO J 31:4179–4190

Raff JW, Glover DM (1989) Centrosomes, and not nuclei, initiate pole cell formation in Drosophila embryos. Cell 57:611–619

Rao S, Kirschen GW, Szczurkowska J, Di Antonio A, Wang J, Ge S, Shelly M (2018) Repositioning of somatic Golgi apparatus is essential for the dendritic establishment of adult-born hippocampal neurons. J Neurosci 38:631–647

Rattner JB, Berns MW (1976) Centriole behavior in early mitosis of rat kangaroo cells (PTK2). Chromosoma 54:387–395

Rebollo E, Sampaio P, Januschke J, Llamazares S, Varmark H, Gonzalez C (2007) Functionally unequal centrosomes drive spindle orientation in asymmetrically dividing Drosophila neural stem cells. Dev Cell 12:467–474

Rebollo E, Roldan M, Gonzalez C (2009) Spindle alignment is achieved without rotation after the first cell cycle in Drosophila embryonic neuroblasts. Development 136:3393–3397

Reffay M, Petitjean L, Coscoy S, Grasland-Mongrain E, Amblard F, Buguin A, Silberzan P (2011) Orientation and polarity in collectively migrating cell structures: statics and dynamics. Biophys J 100:2566–2575

Regolini MF (2013) Centrosome: is it a geometric, noise resistant, 3D interface that translates morphogenetic signals into precise locations in the cell? Ital J Anat Embryol 118:19–66

Reilein A, Nelson WJ (2005) APC is a component of an organizing template for cortical microtubule networks. Nat Cell Biol 7:463–473

Reinsch S, Karsenti E (1994) Orientation of spindle axis and distribution of plasma membrane proteins during cell division in polarized MDCKII cells. J Cell Biol 126:1509–1526

Reversat A, Yuseff MI, Lankar D, Malbec O, Obino D, Maurin M, Penmatcha NV, Amoroso A, Sengmanivong L, Gundersen GG, Mellman I, Darchen F, Desnos C, Pierobon P, Lennon-Dumenil AM (2015) Polarity protein Par3 controls B-cell receptor dynamics and antigen extraction at the immune synapse. Mol Biol Cell 26:1273–1285

Rindler MJ, Ivanov IE, Sabatini DD (1987) Microtubule-acting drugs lead to the nonpolarized delivery of the influenza hemagglutinin to the cell surface of polarized Madin-Darby canine kidney cells. J Cell Biol 104:231–241

Rios RM, Sanchis A, Tassin AM, Fedriani C, Bornens M (2004) GMAP-210 recruits gamma-tubulin complexes to cis-Golgi membranes and is required for Golgi ribbon formation. Cell 118:323–335

Ritter AT, Asano Y, Stinchcombe JC, Dieckmann NM, Chen BC, Gawden-Bone C, van Engelenburg S, Legant W, Gao L, Davidson MW, Betzig E, Lippincott-Schwartz J, Griffiths GM (2015) Actin depletion initiates events leading to granule secretion at the immunological synapse. Immunity 42:864–876

Robinson JT, Wojcik EJ, Sanders MA, McGrail M, Hays TS (1999) Cytoplasmic dynein is required for the nuclear attachment and migration of centrosomes during mitosis in Drosophila. J Cell Biol 146:597–608

Rodriguez-Boulan E, Macara IG (2014) Organization and execution of the epithelial polarity programme. Nat Rev Mol Cell Biol 15:225–242

Rodriguez-Fraticelli AE, Auzan M, Alonso MA, Bornens M, Martin-Belmonte F (2012) Cell confinement controls centrosome positioning and lumen initiation during epithelial morphogenesis. J Cell Biol 198:1011–1023

Rogalski AA, Bergmann JE, Singer SJ (1984) Effect of microtubule assembly status on the intracellular processing and surface expression of an integral protein of the plasma membrane. J Cell Biol 99:1101–1109

Rosenblatt J, Cramer LP, Baum B, McGee KM (2004) Myosin II-dependent cortical movement is required for centrosome separation and positioning during mitotic spindle assembly. Cell 117:361–372

Ross L, Normark BB (2015) Evolutionary problems in centrosome and centriole biology. J Evol Biol 28:995–1004

Roszko I, Afonso C, Henrique D, Mathis L (2006) Key role played by RhoA in the balance between planar and apico-basal cell divisions in the chick neuroepithelium. Dev Biol 298:212–224

Roux KJ, Crisp ML, Liu Q, Kim D, Kozlov S, Stewart CL, Burke B (2009) Nesprin 4 is an outer nuclear membrane protein that can induce kinesin-mediated cell polarization. Proc Natl Acad Sci USA 106:2194–2199

Sakakibara A, Sato T, Ando R, Noguchi N, Masaoka M, Miyata T (2014) Dynamics of centrosome translocation and microtubule organization in neocortical neurons during distinct modes of polarization. Cereb Cortex 24:1301–1310

Salle J, Xie J, Ershov D, Lacassin M, Dmitrieff S, Minc N (2018) Asymmetric division through a reduction of microtubule centering forces. J Cell Biol 218(3):771. https://doi.org/10.1083/jcb.201807102

Salpingidou G, Smertenko A, Hausmanowa-Petrucewicz I, Hussey PJ, Hutchison CJ (2007) A novel role for the nuclear membrane protein emerin in association of the centrosome to the outer nuclear membrane. J Cell Biol 178:897–904

Salvarezza SB, Deborde S, Schreiner R, Campagne F, Kessels MM, Qualmann B, Caceres A, Kreitzer G, Rodriguez-Boulan E (2009) LIM kinase 1 and cofilin regulate actin filament population required for dynamin-dependent apical carrier fission from the trans-Golgi network. Mol Biol Cell 20:438–451

Salzmann V, Chen C, Chiang CY, Tiyaboonchai A, Mayer M, Yamashita YM (2014) Centrosome-dependent asymmetric inheritance of the midbody ring in Drosophila germline stem cell division. Mol Biol Cell 25:267–275

Sameshima M, Imai Y, Hashimoto Y (1988) The position of the microtubule-organizing center relative to the nucleus is independent of the direction of cell migration in Dictyostelium discoideum. Cell Motil Cytoskeleton 9:111–116

Sardet C, Paix A, Prodon F, Dru P, Chenevert J (2007) From oocyte to 16-cell stage: cytoplasmic and cortical reorganizations that pattern the ascidian embryo. Dev Dyn 236:1716–1731

Saturno DM, Castanzo DT, Williams M, Parikh DA, Jaeger EC, Lyczak R (2017) Sustained centrosome-cortical contact ensures robust polarization of the one-cell C. elegans embryo. Dev Biol 422:135–145

Schlessinger K, McManus EJ, Hall A (2007) Cdc42 and noncanonical Wnt signal transduction pathways cooperate to promote cell polarity. J Cell Biol 178:355–361

Schmidt KN, Kuhns S, Neuner A, Hub B, Zentgraf H, Pereira G (2012) Cep164 mediates vesicular docking to the mother centriole during early steps of ciliogenesis. J Cell Biol 199:1083–1101

Schmoranzer J, Kreitzer G, Simon SM (2003) Migrating fibroblasts perform polarized, microtubule-dependent exocytosis towards the leading edge. J Cell Sci 116:4513–4519

Schmoranzer J, Fawcett JP, Segura M, Tan S, Vallee RB, Pawson T, Gundersen GG (2009) Par3 and dynein associate to regulate local microtubule dynamics and centrosome orientation during migration. Curr Biol 19:1065–1074

Schutze K, Maniotis A, Schliwa M (1991) The position of the microtubule-organizing center in directionally migrating fibroblasts depends on the nature of the substratum. Proc Natl Acad Sci USA 88:8367–8371

Schweickert A, Weber T, Beyer T, Vick P, Bogusch S, Feistel K, Blum M (2007) Cilia-driven leftward flow determines laterality in Xenopus. Curr Biol 17:60–66

Segalen M, Johnston CA, Martin CA, Dumortier JG, Prehoda KE, David NB, Doe CQ, Bellaiche Y (2010) The Fz-Dsh planar cell polarity pathway induces oriented cell division via Mud/NuMA in Drosophila and zebrafish. Dev Cell 19:740–752

Seldin L, Muroyama A, Lechler T (2016) NuMA-microtubule interactions are critical for spindle orientation and the morphogenesis of diverse epidermal structures. elife 5:e12504

Sepich DS, Solnica-Krezel L (2016) Intracellular Golgi Complex organization reveals tissue specific polarity during zebrafish embryogenesis. Dev Dyn 245:678–691

Serrador JM, Cabrero JR, Sancho D, Mittelbrunn M, Urzainqui A, Sanchez-Madrid F (2004) HDAC6 deacetylase activity links the tubulin cytoskeleton with immune synapse organization. Immunity 20:417–428

Sharp DJ, Brown HM, Kwon M, Rogers GC, Holland G, Scholey JM (2000) Functional coordination of three mitotic motors in Drosophila embryos. Mol Biol Cell 11:241–253

Silkworth WT, Nardi IK, Paul R, Mogilner A, Cimini D (2012) Timing of centrosome separation is important for accurate chromosome segregation. Mol Biol Cell 23:401–411

Silverman E, Zhao J, Merriam JC, Nagasaki T (2017) Intracellular position of centrioles and the direction of homeostatic epithelial cell movements in the mouse cornea. J Histochem Cytochem 65:83–91

Singh S, Solecki DJ (2015) Polarity transitions during neurogenesis and germinal zone exit in the developing central nervous system. Front Cell Neurosci 9:62

Slaats GG, Ghosh AK, Falke LL, Le Corre S, Shaltiel IA, van de Hoek G, Klasson TD, Stokman MF, Logister I, Verhaar MC, Goldschmeding R, Nguyen TQ, Drummond IA, Hildebrandt F, Giles RH (2014) Nephronophthisis-associated CEP164 regulates cell cycle progression, apoptosis and epithelial-to-mesenchymal transition. PLoS Genet 10:e1004594

Smith E, Hegarat N, Vesely C, Roseboom I, Larch C, Streicher H, Straatman K, Flynn H, Skehel M, Hirota T, Kuriyama R, Hochegger H (2011) Differential control of Eg5-dependent centrosome separation by Plk1 and Cdk1. EMBO J 30:2233–2245

Smith P, Azzam M, Hinck L (2017) Extracellular regulation of the mitotic spindle and fate determinants driving asymmetric cell division. Results Probl Cell Differ 61:351–373

Solecki DJ, Trivedi N, Govek EE, Kerekes RA, Gleason SS, Hatten ME (2009) Myosin II motors and F-actin dynamics drive the coordinated movement of the centrosome and soma during CNS glial-guided neuronal migration. Neuron 63:63–80

Spassky N, Meunier A (2017) The development and functions of multiciliated epithelia. Nat Rev Mol Cell Biol 18:423–436

Spear PC, Erickson CA (2012) Apical movement during interkinetic nuclear migration is a two-step process. Dev Biol 370:33–41

Splinter D, Tanenbaum ME, Lindqvist A, Jaarsma D, Flotho A, Yu KL, Grigoriev I, Engelsma D, Haasdijk ED, Keijzer N, Demmers J, Fornerod M, Melchior F, Hoogenraad CC, Medema RH, Akhmanova A (2010) Bicaudal D2, dynein, and kinesin-1 associate with nuclear pore complexes and regulate centrosome and nuclear positioning during mitotic entry. PLoS Biol 8: e1000350

Srsen V, Fant X, Heald R, Rabouille C, Merdes A (2009) Centrosome proteins form an insoluble perinuclear matrix during muscle cell differentiation. BMC Cell Biol 10:28

Stinchcombe JC, Majorovits E, Bossi G, Fuller S, Griffiths GM (2006) Centrosome polarization delivers secretory granules to the immunological synapse. Nature 443:462–465

Stinchcombe JC, Randzavola LO, Angus KL, Mantell JM, Verkade P, Griffiths GM (2015) Mother centriole distal appendages mediate centrosome docking at the immunological synapse and reveal mechanistic parallels with ciliogenesis. Curr Biol 25:3239–3244

Strome S (1993) Determination of cleavage planes. Cell 72:3–6

Strugnell GE, Wang AM, Wheatley DN (1996) Primary cilium expression in cells from normal and aberrant human skin. J Submicrosc Cytol Pathol 28:215–225

Sugioka K, Bowerman B (2018) Combinatorial contact cues specify cell division orientation by directing cortical myosin flows. Dev Cell 46:257–270 e255

Sugiyama Y, Stump RJ, Nguyen A, Wen L, Chen Y, Wang Y, Murdoch JN, Lovicu FJ, McAvoy JW (2010) Secreted frizzled-related protein disrupts PCP in eye lens fiber cells that have polarised primary cilia. Dev Biol 338:193–201

Tambe DT, Hardin CC, Angelini TE, Rajendran K, Park CY, Serra-Picamal X, Zhou EH, Zaman MH, Butler JP, Weitz DA, Fredberg JJ, Trepat X (2011) Collective cell guidance by cooperative intercellular forces. Nat Mater 10:469–475

Tanaka M, Kikuchi T, Uno H, Okita K, Kitanishi-Yumura T, Yumura S (2017) Turnover and flow of the cell membrane for cell migration. Sci Rep 7:12970

Tanenbaum ME, Medema RH (2010) Mechanisms of centrosome separation and bipolar spindle assembly. Dev Cell 19:797–806

Tanenbaum ME, Macurek L, Galjart N, Medema RH (2008) Dynein, Lis1 and CLIP-170 counteract Eg5-dependent centrosome separation during bipolar spindle assembly. EMBO J 27:3235–3245

Tang D, Mar K, Warren G, Wang Y (2008) Molecular mechanism of mitotic Golgi disassembly and reassembly revealed by a defined reconstitution assay. J Biol Chem 283:6085–6094

Taniguchi K, Maeda R, Ando T, Okumura T, Nakazawa N, Hatori R, Nakamura M, Hozumi S, Fujiwara H, Matsuno K (2011) Chirality in planar cell shape contributes to left-right asymmetric epithelial morphogenesis. Science 333:339–341

Tanimoto H, Kimura A, Minc N (2016) Shape-motion relationships of centering microtubule asters. J Cell Biol 212:777–787

Tassin AM, Maro B, Bornens M (1985) Fate of microtubule-organizing centers during myogenesis in vitro. J Cell Biol 100:35–46

Taverna E, Mora-Bermudez F, Strzyz PJ, Florio M, Icha J, Haffner C, Norden C, Wilsch-Brauninger M, Huttner WB (2016) Non-canonical features of the Golgi apparatus in bipolar epithelial neural stem cells. Sci Rep 6:21206

Tee YH, Shemesh T, Thiagarajan V, Hariadi RF, Anderson KL, Page C, Volkmann N, Hanein D, Sivaramakrishnan S, Kozlov MM, Bershadsky AD (2015) Cellular chirality arising from the self-organization of the actin cytoskeleton. Nat Cell Biol 17:445–457

Thery M, Racine V, Pepin A, Piel M, Chen Y, Sibarita JB, Bornens M (2005) The extracellular matrix guides the orientation of the cell division axis. Nat Cell Biol 7:947–953

Theveneau E, Marchant L, Kuriyama S, Gull M, Moepps B, Parsons M, Mayor R (2010) Collective chemotaxis requires contact-dependent cell polarity. Dev Cell 19:39–53

Tikhonenko I, Magidson V, Graf R, Khodjakov A, Koonce MP (2013) A kinesin-mediated mechanism that couples centrosomes to nuclei. Cell Mol Life Sci 70:1285–1296

Tisler M, Thumberger T, Schneider I, Schweickert A, Blum M (2017) Leftward flow determines laterality in conjoined twins. Curr Biol 27:543–548

Tolar P (2017) Cytoskeletal control of B cell responses to antigens. Nat Rev Immunol 17:621–634

Tolic-Norrelykke IM, Sacconi L, Stringari C, Raabe I, Pavone FS (2005) Nuclear and division-plane positioning revealed by optical micromanipulation. Curr Biol 15:1212–1216

Tsun A, Qureshi I, Stinchcombe JC, Jenkins MR, de la Roche M, Kleczkowska J, Zamoyska R, Griffiths GM (2011) Centrosome docking at the immunological synapse is controlled by Lck signaling. J Cell Biol 192:663–674

Ueda M, Graf R, MacWilliams HK, Schliwa M, Euteneuer U (1997) Centrosome positioning and directionality of cell movements. Proc Natl Acad Sci USA 94:9674–9678

Vaisberg EA, Koonce MP, McIntosh JR (1993) Cytoplasmic dynein plays a role in mammalian mitotic spindle formation. J Cell Biol 123:849–858

Vaisberg EA, Grissom PM, McIntosh JR (1996) Mammalian cells express three distinct dynein heavy chains that are localized to different cytoplasmic organelles. J Cell Biol 133:831–842

Valente C, Colanzi A (2015) Mechanisms and regulation of the mitotic inheritance of the Golgi complex. Front Cell Dev Biol 3:79

van Heesbeen RG, Tanenbaum ME, Medema RH (2014) Balanced activity of three mitotic motors is required for bipolar spindle assembly and chromosome segregation. Cell Rep 8:948–956

Venhuizen JH, Zegers MM (2017) Making heads or tails of it: cell-cell adhesion in cellular and supracellular polarity in collective migration. Cold Spring Harb Perspect Biol 9:a027854

Vogel SK, Pavin N, Maghelli N, Julicher F, Tolic-Norrelykke IM (2009) Self-organization of dynein motors generates meiotic nuclear oscillations. PLoS Biol 7:e1000087

Vorobjev IA, Chentsov Yu S (1982) Centrioles in the cell cycle. I. Epithelial cells. J Cell Biol 93:938–949

Wakida NM, Botvinick EL, Lin J, Berns MW (2010) An intact centrosome is required for the maintenance of polarization during directional cell migration. PLoS One 5:e15462

Walz G (2017) Role of primary cilia in non-dividing and post-mitotic cells. Cell Tissue Res 369:11–25

Wan LQ, Ronaldson K, Guirguis M, Vunjak-Novakovic G (2013) Micropatterning of cells reveals chiral morphogenesis. Stem Cell Res Ther 4:24

Wang L, Dynlacht BD (2018) The regulation of cilium assembly and disassembly in development and disease. Development 145:dev151407. https://doi.org/10.1242/dev.151407

Wang X, Tsai JW, Imai JH, Lian WN, Vallee RB, Shi SH (2009) Asymmetric centrosome inheritance maintains neural progenitors in the neocortex. Nature 461:947–955

Wang H, Brust-Mascher I, Civelekoglu-Scholey G, Scholey JM (2013) Patronin mediates a switch from kinesin-13-dependent poleward flux to anaphase B spindle elongation. J Cell Biol 203:35–46

Wang G, Jiang Q, Zhang C (2014) The role of mitotic kinases in coupling the centrosome cycle with the assembly of the mitotic spindle. J Cell Sci 127:4111–4122

Wang JC, Lee JY, Christian S, Dang-Lawson M, Pritchard C, Freeman SA, Gold MR (2017) The Rap1-cofilin-1 pathway coordinates actin reorganization and MTOC polarization at the B cell immune synapse. J Cell Sci 130:1094–1109

Waters JC, Cole RW, Rieder CL (1993) The force-producing mechanism for centrosome separation during spindle formation in vertebrates is intrinsic to each aster. J Cell Biol 122:361–372

Wei JH, Seemann J (2010) Unraveling the Golgi ribbon. Traffic 11:1391–1400

Werner ME, Hwang P, Huisman F, Taborek P, Yu CC, Mitchell BJ (2011) Actin and microtubules drive differential aspects of planar cell polarity in multiciliated cells. J Cell Biol 195:19–26

Westlake CJ, Baye LM, Nachury MV, Wright KJ, Ervin KE, Phu L, Chalouni C, Beck JS, Kirkpatrick DS, Slusarski DC, Sheffield VC, Scheller RH, Jackson PK (2011) Primary cilia membrane assembly is initiated by Rab11 and transport protein particle II (TRAPPII) complex-dependent trafficking of Rabin8 to the centrosome. Proc Natl Acad Sci USA 108:2759–2764

Wheatley DN, Wang AM, Strugnell GE (1996) Expression of primary cilia in mammalian cells. Cell Biol Int 20:73–81

Williams SE, Ratliff LA, Postiglione MP, Knoblich JA, Fuchs E (2014) Par3-mInsc and Galphai3 cooperate to promote oriented epidermal cell divisions through LGN. Nat Cell Biol 16:758–769

Witkos TM, Lowe M (2015) The golgin family of coiled-coil tethering proteins. Front Cell Dev Biol 3:86

Wong MK, Gotlieb AI (1988) The reorganization of microfilaments, centrosomes, and microtubules during in vitro small wound reendothelialization. J Cell Biol 107:1777–1783

Woodland HR, Fry AM (2008) Pix proteins and the evolution of centrioles. PLoS One 3:e3778

Woolner S, Papalopulu N (2012) Spindle position in symmetric cell divisions during epiboly is controlled by opposing and dynamic apicobasal forces. Dev Cell 22:775–787

Wright RL, Adler SA, Spanier JG, Jarvik JW (1989) Nucleus-basal body connector in Chlamydomonas: evidence for a role in basal body segregation and against essential roles in mitosis or in determining cell polarity. Cell Motil Cytoskeleton 14:516–526

Wu J, Misra G, Russell RJ, Ladd AJ, Lele TP, Dickinson RB (2011) Effects of dynein on microtubule mechanics and centrosome positioning. Mol Biol Cell 22:4834–4841

Wuhr M, Tan ES, Parker SK, Detrich HW 3rd, Mitchison TJ (2010) A model for cleavage plane determination in early amphibian and fish embryos. Curr Biol 20:2040–2045

Xie Z, Hur SK, Zhao L, Abrams CS, Bankaitis VA (2018) A Golgi lipid signaling pathway controls apical Golgi distribution and cell polarity during neurogenesis. Dev Cell 44:725–740 e724

Xing M, Peterman MC, Davis RL, Oegema K, Shiau AK, Field SJ (2016) GOLPH3 drives cell migration by promoting Golgi reorientation and directional trafficking to the leading edge. Mol Biol Cell 27:3828–3840

Xu Y, Takeda S, Nakata T, Noda Y, Tanaka Y, Hirokawa N (2002) Role of KIFC3 motor protein in Golgi positioning and integration. J Cell Biol 158:293–303

Xu J, Van Keymeulen A, Wakida NM, Carlton P, Berns MW, Bourne HR (2007) Polarity reveals intrinsic cell chirality. Proc Natl Acad Sci USA 104:9296–9300

Yadav S, Puri S, Linstedt AD (2009) A primary role for Golgi positioning in directed secretion, cell polarity, and wound healing. Mol Biol Cell 20:1728–1736

Yadav S, Puthenveedu MA, Linstedt AD (2012) Golgin160 recruits the dynein motor to position the Golgi apparatus. Dev Cell 23:153–165

Yalgin C, Ebrahimi S, Delandre C, Yoong LF, Akimoto S, Tran H, Amikura R, Spokony R, Torben-Nielsen B, White KP, Moore AW (2015) Centrosomin represses dendrite branching by orienting microtubule nucleation. Nat Neurosci 18:1437–1445

Yamashita YM, Mahowald AP, Perlin JR, Fuller MT (2007) Asymmetric inheritance of mother versus daughter centrosome in stem cell division. Science 315:518–521

Ye B, Zhang Y, Song W, Younger SH, Jan LY, Jan YN (2007) Growing dendrites and axons differ in their reliance on the secretory pathway. Cell 130:717–729

Yi J, Wu X, Chung AH, Chen JK, Kapoor TM, Hammer JA (2013) Centrosome repositioning in T cells is biphasic and driven by microtubule end-on capture-shrinkage. J Cell Biol 202:779–792

Yoo SK, Lam PY, Eichelberg MR, Zasadil L, Bement WM, Huttenlocher A (2012) The role of microtubules in neutrophil polarity and migration in live zebrafish. J Cell Sci 125:5702–5710

Yoshiura S, Ohta N, Matsuzaki F (2012) Tre1 GPCR signaling orients stem cell divisions in the Drosophila central nervous system. Dev Cell 22:79–91

Yount AL, Zong H, Walczak CE (2015) Regulatory mechanisms that control mitotic kinesins. Exp Cell Res 334:70–77

Yuan S, Zhao L, Brueckner M, Sun Z (2015) Intraciliary calcium oscillations initiate vertebrate left-right asymmetry. Curr Biol 25:556–567

Yukawa M, Yamada Y, Yamauchi T, Toda T (2018) Two spatially distinct kinesin-14 proteins, Pkl1 and Klp2, generate collaborative inward forces against kinesin-5 Cut7 in *S. pombe*. J Cell Sci 131(1):jcs210740. https://doi.org/10.1242/jcs.210740

Yuseff MI, Reversat A, Lankar D, Diaz J, Fanget I, Pierobon P, Randrian V, Larochette N, Vascotto F, Desdouets C, Jauffred B, Bellaiche Y, Gasman S, Darchen F, Desnos C, Lennon-Dumenil AM (2011) Polarized secretion of lysosomes at the B cell synapse couples antigen extraction to processing and presentation. Immunity 35:361–374

Yvon AM, Walker JW, Danowski B, Fagerstrom C, Khodjakov A, Wadsworth P (2002) Centrosome reorientation in wound-edge cells is cell type specific. Mol Biol Cell 13:1871–1880

Zaritsky A, Welf ES, Tseng YY, Angeles Rabadan M, Serra-Picamal X, Trepat X, Danuser G (2015) Seeds of locally aligned motion and stress coordinate a collective cell migration. Biophys J 109:2492–2500

Zeligs JD, Wollman SH (1979) Mitosis in rat thyroid epithelial cells in vivo. II. Centrioles and pericentriolar material. J Ultrastruct Res 66:97–108

Zhang J, Wang YL (2017) Centrosome defines the rear of cells during mesenchymal migration. Mol Biol Cell 28:3240–3251

Zhao T, Graham OS, Raposo A, St Johnston D (2012) Growing microtubules push the oocyte nucleus to polarize the Drosophila dorsal-ventral axis. Science 336:999–1003

Zheng Y, Wildonger J, Ye B, Zhang Y, Kita A, Younger SH, Zimmerman S, Jan LY, Jan YN (2008) Dynein is required for polarized dendritic transport and uniform microtubule orientation in axons. Nat Cell Biol 10:1172–1180

Zhu J, Burakov A, Rodionov V, Mogilner A (2010) Finding the cell center by a balance of dynein and myosin pulling and microtubule pushing: a computational study. Mol Biol Cell 21:4418–4427

Zinski J, Tajer B, Mullins MC (2018) TGF-beta family signaling in early vertebrate development. Cold Spring Harb Perspect Biol 10(6):a033274. https://doi.org/10.1101/cshperspect.a033274

Chapter 8
Centriole Positioning: Not Just a Little Dot in the Cell

Angel-Carlos Roman, Sergio Garrido-Jimenez, Selene Diaz-Chamorro, Francisco Centeno, and Jose Maria Carvajal-Gonzalez

Abstract Organelle positioning as many other morphological parameters in a cell is not random. Centriole positioning as centrosomes or ciliary basal bodies is not an exception to this rule in cell biology. Indeed, centriole positioning is a tightly regulated process that occurs during development, and it is critical for many organs to function properly, not just during development but also in the adulthood. In this book chapter, we overview our knowledge on centriole positioning in different and highly specialized animal cells like photoreceptor or ependymal cells. We will also discuss recent advances in the discovery of molecular pathways involved in this process, mostly related to the cytoskeleton and the cell polarity pathways. And finally, we present quantitative methods that have been used to assess centriole positioning in different cell types although mostly in epithelial cells.

8.1 About Centriole, Centrosome, and Basal Body

In animals, the size, structure, protein composition, function, number and positioning of centrioles, centrosomes, and ciliary basal bodies have an impact on many aspects of development and physiology. Indeed, abnormalities in these centriole, centrosome, and basal body features are implicated in many diseases including, diabetes, obesity, microcephaly, and cancer (Bettencourt-Dias et al. 2011; Braun and Hildebrandt 2017; Lopes et al. 2018; Vaisse et al. 2017; Volta and Gerdes 2017). More specifically related to ciliary defects, abnormal ciliary functions will lead to ciliopathies, such as primary ciliary dyskinesia (PCD), Meckel syndrome (MKS), nephronophthisis (NPHP), and Joubert syndrome (JBTS) (Reiter and Leroux 2017). In this book chapter, we will review our knowledge on centrioles, centrosomes, and basal bodies positioning in animal cells and discuss existing methods to analyze and compare centriole positioning.

A.-C. Roman · S. Garrido-Jimenez · S. Diaz-Chamorro · F. Centeno ·
J. M. Carvajal-Gonzalez (✉)
Facultad de Ciencias, Departamento de Bioquímica, Biología Molecular y Genética,
Universidad de Extremadura, Badajoz, Spain
e-mail: jmcarvaj@unex.es

© Springer Nature Switzerland AG 2019
M. Kloc (ed.), *The Golgi Apparatus and Centriole*, Results and Problems in Cell
Differentiation 67, https://doi.org/10.1007/978-3-030-23173-6_8

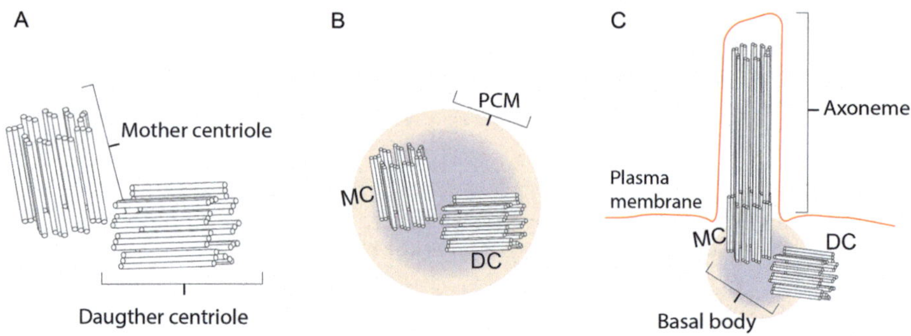

Fig. 8.1 Centriole, centrosome and basal body. Mother and daughter centrioles (**a**) as part of the centrosome in (**b**) and basal body of a cilium in (**c**). *PCM* pericentriolar matrix, *MC* mother centriole, *DC* daughter centriole

The centriole is a microtubule-based structure conserved through evolution that is fundamental to form centrosomes and cilia (Marshall 2001). In a centriole, microtubules are arranged in a cylindrical ninefold symmetric configuration, a structure commonly known as a barrel-shaped, which is maintained in many eukaryotes (Fig. 8.1). The size is below the resolution of regular microscopes being 100–250 nm in diameter and 100–400 nm in length. Apart from this general shape and composition, centrioles also maintain a conserved number, two per cell, in cells from different species. Based on their "age" and composition, one centriole is considered as a mother centriole (the oldest inherited during mitosis) and the other one as a daughter centriole.

The combination of two centrioles, mother and daughter, surrounded by a pericentriolar matrix (PCM) conforms to the canonical centrosome (Fig. 8.1). This PCM has recently been described as multiple radial layers of proteins around the pair of centriole (Mennella et al. 2014). In dividing cells, centrosomes are the organelles at the poles of the spindle ensuring the proper allocation of chromosomes among cells. During interphase, centrosomes act as the major microtubule organizing center (MTOC), nucleating cytoplasmic microtubules (MTs) from the surrounding pericentriolar material (PCM) (Bornens 2012; Werner et al. 2017).

The mother centriole sits at the heart of the basal body, which is a highly organized structure essential for the formation of cilia (Fig. 8.1). Cilia are microtubule-based organelles that project from the surface of cells and serve both motile (secondary cilia) and sensory functions (primary cilia) (Wang and Dynlacht 2018). During cilia formation, basal bodies dock to a cellular membrane through their distal appendages (also known as transition fibers). Then the mother centriole's microtubule-growing end extends and forms the cilium skeleton (the axoneme). Primary cilia share similar structural characteristics with motile cilia, both consisting of nine microtubule doublets forming the ciliary axoneme (9 + 0). Motile cilia, however, also contain a central microtubule doublet (9 + 2), radial spokes, and a nexin ring, allowing for movement and thus functions, such as propulsion or clearing of mucus and debris, not experienced in primary cilia.

An exception to the role of a pair of centrioles per cell appears in very specialized cells known as multiciliated cells, where centriole duplication is promoted by a transcriptional program led by transcription factors like Forkhead Box J1 (Foxj1). In these multiciliated cells, the number of centrioles will increase up to 200–300 per cell.

During development and homeostasis, centrioles as centrosomes or basal bodies play critical functions for cell division in the mitotic spindle, for cell motility (centrosome), in signaling (cilia-related) and in cilia directional movement (cilia-related). To better convey the variety of centriolar functions, four examples of centrioles and their cellular positioning in epithelial cells are:

1. The role in oriented cell division perfectly exemplifies the importance of centrioles/centrosomes during cells division. In this cellular process, the mitotic spindle is aligned within the plane of the epithelium through an interaction between astral microtubules growing from the centrosome and capture sites that are located on the cortical surface of the cell. Oriented cell division has an important function in morphogenesis of epithelial tissues determining its axial elongation and ultimately determining organ size and shape.
2. To highlight the role of centrioles in cell migration, one could look at the developmental process, namely, the convergent extension. This developmental process is necessary to increase the length and narrowing of a field of cells, for example, during gastrulation. Centrosomes will be polarized toward the migrating direction, so that the entire cell polarity axis is shifted in the same direction. As a result, a cell migrates and intercalates within the plane of the epithelium. This process will determine the entire body size.
3. The proper signaling of the Sonic hedgehog (Shh) pathway is based on partitioning of the Shh receptor and downstream signaling component inside or outside the ciliary axoneme. In the presence of Shh, Smoothened (Smo) is translocated inside the ciliary axoneme liberating the transcription factors downstream. Proper Shh signaling is key for embryo development.
4. A more specific function for centrioles related to cilia is found in tissues with very specific functions. In the airway epithelium, multiciliated cells projecting their cilia toward the lumen of the airway will produce a directional mucus flow out of the respiratory system, which is important to remove dust or undesired toxic particles. To properly achieve this movement at the single cell level, centrioles as basal bodies should be arranged so that the orientation of all the ciliary basal body points toward the same side of the cell. In the airway epithelium, centrioles' orientation is coordinated to create the directional fluid flow.

As mentioned above, centrioles/centrosome plays a key role in the control of cell motility and shape, oriented cell divisions, and cilia positioning. For all these processes, a proper centriole/centrosome positioning or changes in positioning is necessary (Elric and Etienne-Manneville 2014).

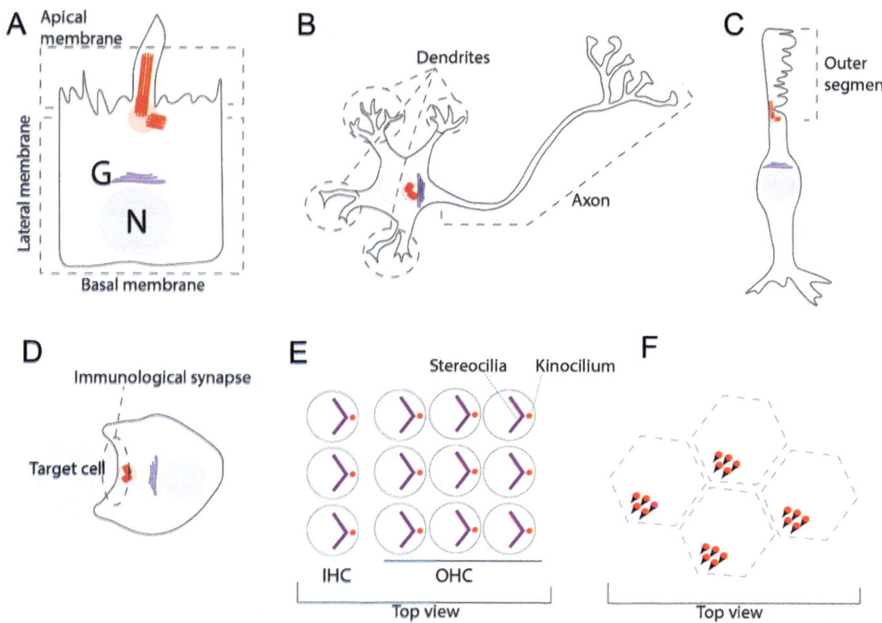

Fig. 8.2 Centriole as centrosome or ciliary basal body positioning in different cell types. A distinct centriole positioning is depicted in epithelial cells with apical, lateral, and basal membrane (**a**), a neuron with dendrites and the axon (**b**), a photoreceptor with the outer segment (**c**), an immune cell with a target cells (**d**), a group of sensory cells in inner ear (**e**), and a group of ependymal cells (**f**). *IHC* inner hair cells, *OHC* outer hair cells. It must be described. *N* nucleus, *G* Golgi, but it must be explained

8.2　Centrosome and Basal Body Positioning in Highly Specialized Cell Types

8.2.1　*Centriole Positioning in Neurons*

In a classic description of centriole/centrosome positioning, these are located close to the nucleus, near the center of the cell, a distribution common for mesenchymal cells or neurons (Fig. 8.2). A neuron is a highly polarized cell that has a unique thin and elongated membrane projection named axon and one or more projections named dendrites. In brain morphogenesis, neuronal proliferation, migration, and polarization are key processes dependent on centrosome and microtubules (Baas and Lin 2011; Conde and Caceres 2009; Kuijpers and Hoogenraad 2011). In proliferating neurons, the initial centrosome is duplicated and the two newly formed centrosomes participate in the assembly and organization of the mitotic spindle (Glotzer 2009; Lim et al. 2009). As mentioned above, mutations on several genes coding for centrosome proteins cause microcephaly. Mutations in Centromere Protein J (CENPJ), Centrosomal Protein 152 (CEP152) and Centrosomal Protein 63 (CEP63) all cause primary microcephaly-associated proteins (Bond et al. 2002; Guernsey et al. 2010; Sir et al. 2011). CENPJ regulates centriole duplication and elongation, and CEP63 and CEP152 colocalize at the proximal end of the maternal centriole and are required for centriole duplication, suggesting that the centrosomes are essential in the proliferative phase regulating brain size. Recently, other two microcephaly inducer-proteins have been identified, WD Repeat Domain 62 (WDR62) and Abnormal Spindle Microtubule Assembly (ASPM) (Jayaraman et al. 2016). Wdr62 and Aspm localize to the proximal end of the mother centriole and are required together with CEP63 to localize CENPJ at the apical region of the centriole.

Migration allows neurons to achieve their final position in the brain, and it is a complex process where migrating neurons undergo important morphological changes and strong shifts in centrosome positioning. In most of the migrating neurons, centrosome is located ahead of the nucleus and coupled by a cage-like structure of perinuclear microtubules. Moreover, these centrosome perinuclear microtubules are extended outward in the migratory leading process (Kuijpers and Hoogenraad 2011). The first migration process is called nuclear translocation. Lissencephaly-1 (Lis1) and dynein, a microtubule minus-end-directed motor, are responsible for coupling nuclear translocation and centrosome movement (Tsai et al. 2007). Microtubules are oriented with their minus end toward the centrosome, and dynein is concentrated at an expansion in the leading process and near the nucleus. Therefore, dynein may pull the centrosome forward toward the expansion and the nucleus toward the centrosome (Sakakibara et al. 2013; Tsai et al. 2007). However, the nuclear translocation independent of centrosome positioning also has been shown (Distel et al. 2010; Umeshima et al. 2007). Therefore, it was proposed that a concentrated actomyosin system in front of a translocating nucleus pulls the nucleus. It is also plausible that cooperative force generation by dynein and actomyosin is used in nuclear translocation (Solecki et al. 2009).

Neurons polarize and develop an axon and several dendrites during and after neuronal migration. Neuronal microtubules appear predominantly bundled, presenting both parallel and antiparallel configurations. Whereas in axons most of the microtubules are oriented with their plus ends away from the soma, microtubules in dendrites show mixed polarity, with a large fraction of microtubule plus ends oriented toward the soma. Before polarization, the centrosome is usually localized to the site of axon outgrowth (Zmuda and Rivas 1998), and several reports identified the centrosome location as a robust predictor of axonal fate (de Anda et al. 2005). Interestingly, most of the microtubules in mature neurons are not connected to the centrosome, and removal of the centrosome affected neither axon growth nor neuronal microtubule organization and morphogenesis (Basto et al. 2006; Nguyen et al. 2011; Stiess et al. 2010). A current point of view is that noncentrosomal microtubules are responsible for dendritic and axon formation. So, at early stages of neuronal development in nonpolarized cells, newly formed axons already contain microtubules of opposite polarity, suggesting that the establishment of uniform plus-end-out microtubules occurs during axon formation and depending on external signals (Yau et al. 2016). An essential component of all microtubules is the protein γ-tubulin, which, together with γ-tubulin complex proteins, assembles into large γ-tubulin ring complexes that function as microtubule nucleator (Kuijpers and Hoogenraad 2011; Teixido-Travesa et al. 2012). Moreover, recently it was demonstrated that augmin (a multisubunit protein complex for lateral nucleation of microtubules) and γ-tubulin ring complexes are crucial for microtubule organization in postmitotic neurons, generating the highly bundled neuronal microtubule network and ensuring uniform plus end-out microtubule polarity in axons. While γ-tubulin ring complex is essential for the density and organization of the microtubule network, augmin complex regulates the polarity of axonal microtubules (Sanchez-Huertas et al. 2016).

8.2.2 Centriole Positioning in Immune Cells

This central and close to the nucleus positioning of the centriole varies sometimes due to external stimuli in some cells, such as immune cells (Fig. 8.2). This centriolar movement inside the cell is key for the immune system function. The immune system is composed of diverse types of cells whose principal role is the maintenance of homeostasis and integrity of all the tissues. Immune cells are essential in the protection of organisms against infectious diseases and cancer, participating, also, in key processes such as the development, reproduction or wounds healing (Sattler 2017; Stinchcombe Jane and Griffiths Gillian 2014). The regulation of immune response depends mainly on close interactions between immune cells and those known as antigen presenting cells (APC). Those interactions not only involve receptor-target recognition but also change in protein organization (polarization of receptors) and cell morphology. This rearrangement of cell components forms an area of specialization denominated as the immunological synapse formed by membrane domains known as supramolecular activation clusters (SMAC). For instance, CD4 T cells have a ring of integrin proteins that surround a central region with T cell receptors and signaling proteins and a distal ring with actin and actin-associated proteins (Angus and Griffiths 2013; Grakoui et al. 1999; Monks et al. 1998; Stinchcombe Jane and Griffiths Gillian 2014).

The cell–cell interactions trigger the migration of centrosome from the periphery of the nucleus toward the immunological synapse and its binding to the plasma membrane (Vertii and Doxsey 2016; Roig-Martinez et al. 2019), which is a process initially observed in cytotoxic T cells (Stinchcombe et al. 2006) and recently in CD4 and NK cells (Stinchcombe et al. 2011; Ueda et al. 2011) (Fig. 8.2). Centrosome polarization toward the immunological synapse reorganizes the intracellular actin and microtubule cytoskeleton, which is essential for the transmission of signals and the traffic of secretion vesicles toward the area of contact/interaction (Stinchcombe et al. 2006, 2011). Furthermore, this polarization controls other processes that occur during the immunological synapses, such as the processing of newly synthesized proteins and the endocytic vesicles recycling (Ambuj et al. 2014; Angus and Griffiths 2013; Stinchcombe Jane and Griffiths Gillian 2014).

8.2.3 Centriole Positioning in Specialized Epithelial Cells

Centrioles occupy other specific locations that differ between cell types and cellular contexts. For instance, in mammalian epithelial cells, centrioles as a part of the basal body are located near the apical surface (Spassky and Meunier 2017) (Fig. 8.2). However, this localization can also evolve due to the specific functions acquired by epithelial cells in a tissue. Clear examples for such a centriole positioning are found in hair cells in the inner ear or ependymal and choroidal ciliated cells in the brain ventricles.

8.2.3.1 Centriole Positioning in Hair Cell of the Cochlea

The mammalian auditory organ is composed of sensory cells that detect sound-evoked vibrations. This perception is mediated by the hair bundle, located at their apical surface, a mechanosensitive structure formed of large and stiff microvilli known as stereocilia. At the hair bundle, vertex is located in the kinocilium, a specialized primary cilium. The stereocilia are arranged in rows of increasing height with the kinocilium located adjacent to and central with respect to the tallest row (Nayak et al. 2007). The stereocilia are filled from top to bottom with numerous, highly cross-linked F-actin filaments. A proportion of these F-actin filaments pass through the base of each stereocilium that extends into a dense network of cytoskeletal proteins that underlies each hair bundle. The kinocilium is based in a basal body, the mother centriole, located under the cell surface (Jones and Chen 2008; Satir et al. 2010). A daughter centriole is connected to the basal body by intercentriolar linkers (Bornens 2012). Finally, the kinocilium is attached to the vertex of the forming hair bundle by fibrous links that connect it to the tallest row of stereocilia (Michel et al. 2005; Webb et al. 2011).

During development, numerous short microvilli appear on the apical surface surrounding a single centrally located kinocilium. The mother and daughter centrioles undergo confined random movements on short timescales at the apical surface of hair cells, and the centriole movements are consistent with the Brownian motion constrained by a radial force (Lepelletier et al. 2013). This allows the kinocilium to migrate toward the periphery of the cell surface (Fig. 8.2). Microvilli adjacent to the kinocilium thicken and elongate, eventually forming a V-shaped bundle of stereocilia, with the kinocilium at its vertex (Fig. 8.2). However, little is known about the mechanisms that coordinate this special planar cell polarity. It has shown that the ablation of *Lis1* (a well-established microtubule regulator) disrupts centrosome anchoring and the normal V-shape of hair bundles, accompanied by defects in the pericentriolar matrix and microtubule organization (Sipe et al. 2013). Recently, it has been demonstrated the interdependency and interaction between the integrin $\alpha 8$ (Itga8) and the protocadherin-15a (Pcdh15a) present in ciliary complex. The absence of Itga8 or Pcdh15a affects kinocilia elongation and/or maintenance (Goodman and Zallocchi 2017).

8.2.3.2 Centriole Positioning in Ependymal and Choroidal Ciliated Cells

The choroid plexus is a tissue that protrudes into brain ventricles. Its epithelium consists of choroid plexus epithelial cells that produce the cerebrospinal fluid (Damkier et al. 2013). These cells also secrete ligands important for brain physiology, and regulate the selective protein crossing from vascular tissue to cerebrospinal fluid (Reboldi et al. 2009; Redzic et al. 2005). However, the main function of the ependymal cells, lining the ventricles of the brain and central canal of the spinal cord, is to move cerebrospinal fluid. Because they derive from the dorsal neuroepithelium and form a continuous monolayer with the ependyma, ependymal epithelium is sometimes described as choroidal or modified ependyma. However, choroid plexus epithelial cells and ependymal cells are distinct in many aspects. Thus, ependymal cells contain hundreds of motile (secondary) cilia (9 + 2) that beat in a coordinate manner to allow the circulation of the cerebrospinal fluid (Narita and Takeda 2015). In contrast, choroid plexus epithelial cells have one or two dozens of nonmotile (9 + 0) cilia (Narita et al. 2010) which exhibit transient motility only around the perinatal period, with low and small beating frequency and amplitude and random orientation (Narita et al. 2012). Also, there are differences in the mechanism of ciliary formation of cilia between choroid plexus epithelial and ependymal cells. Thus, in a knockout mouse for Cadherin EGF LAG seven-pass G-type receptor 2 (*Celsr2*), an impairment of ciliogenesis is observed in ependymal but not in choroid plexus epithelial cells (Tissir et al. 2010). Both ependymal and choroid plexus epithelial cells share origin. The precursor cells, termed radial glia, already establish orientation of cilia prior to the generation of choroid plexus epithelial and ependymal cells (Mirzadeh et al. 2010). In both cell types, multiciliogenesis is associated with the induction of transcription factors, Foxj1 and regulatory factor × 3 (*Rf×3*). The formation of multiple cilia in choroid plexus epithelial cells occurs shortly after the cells differentiate from the neuroepithelium during organogenesis (about embryonic day 11 in mice) (Narita and Takeda 2015). However, ependymal cells undergo multiciliogenesis after birth to establish hundreds of motile cilia in 2 weeks. Finally, a combination of planar polarity signaling and hydrodynamic forces (by the emerging fluid flow) regulates docking and rotational orientation of the ependymal motile cilia (Guirao et al. 2010), and important planar polarity components such as CELSR1, CELSR2 and CELSR3 and *VANGL2* (Vang-like protein 2) are essential for ciliary orientation and function (Tissir et al. 2010).

8.2.4 Centriole Positioning in Photoreceptors

Although photoreceptors are neuroepithelial cells which generate electrical responses when stimulated by light, their centriole positioning resembles one of the classical epithelial cells (Fig. 8.2). They contain a distinctive photosensory organelle derived from a primary nonmotile cilium, named the outer segment (OS). The OS is connected to the cell soma via a thin, eccentrically positioned bridge, called the connecting cilium, which is connected to the basal body. This basal body is composed of nine doublets of microtubules (9 + 0) derived from the mother centriole, and it remains anchored to the periciliary membrane by transition fibers (Pazour and Bloodgood 2008; Rosenbaum and Witman 2002; Wei et al. 2013). The outer segment consists of stacks of membranous discs which are organized around a microtubule-based axoneme and which contain proteins required for phototransduction, such as the photopigment opsin. The connecting cilium, which is partly equivalent to the transition zone in other cilia types, joins the outer segment with the inner segment (Khanna 2015).

The major stages of outer segment morphogenesis are similar across all species, and its initial stages closely parallel to the morphogenesis of primary cilia in other cell types were described in great detail (Sedmak and Wolfrum 2011). Outer segment morphogenesis begins with the maturation of the basal body as it migrates toward the distal end of the inner segment. This basal body consists of the mother and daughter centrioles. In photoreceptors, the first step in cilium formation is the attachment of an intracellular ciliary vesicle to the distal end of the mother centriole, and transition fibers, projected from the mother centriole, likely mediate this vesicle's attachment. Axonemal extension occurs next and causes the ciliary vesicle to invaginate and form the ciliary sheath. The basal body ciliary vesicle structure then docks to the plasma membrane, and the outer membrane of the sheath fuses with the plasma membrane. It is likely that upon this fusion, the ciliary sheath becomes the membrane region often referred to as the periciliary membrane. The next steps are oriented to the building of photoreceptor discs. Primary cilium and outer segment morphogenesis and architecture became a subject of renewed attention because humans suffer from a variety of inherited ciliopathies and syndromic diseases that often include problems with vision (Fliegauf et al. 2007). For example, ablation of *Macf1* (Microtubule actin crosslinking factor 1) in the developing retina abolishes ciliogenesis, and basal bodies fail to dock to ciliary vesicles or to migrate apically. Moreover, deletion of *Macf1* in adult photoreceptors causes reversal of basal body docking and loss of outer segments, reflecting a continuous requirement for Macf1 function (May-Simera et al. 2016).

8.3 Molecular Pathways Underlying Centrosome Positioning During Interphase

In recent years, molecular mechanisms that control centrosome positioning have been elucidated. Pathways including cytoskeletal and polarity proteins have been demonstrated to be important for understanding this process (Carvajal-Gonzalez et al. 2016a; Tang and Marshall 2012). In the next paragraphs, we will describe some of these mechanisms.

8.3.1   The Cytoskeleton as a Key Player in Centriole, Centrosome, and Cilia Positioning

Several studies indicate that centriole positioning depends on both tubulin and actin dynamics as well as motor proteins like dynein or myosin. They play a role on the generation of forces that pull centrosome to the correct position in many cell types during interphase (Buendia et al. 1990; Euteneuer and Schliwa 1985; Zhu et al. 2010). According to this, in some cellular processes, disruption of actin and microtubules cytoskeleton using cytochalasin D or nocodazole and cold causes the arrest of centrioles' movement (Garrido-Jimenez et al. 2018; Piel et al. 2000). These experiments have revealed that actin polymerization is required to maintain the centrioles in restricted areas in epithelial cells (Boisvieux-Ulrich et al. 1990; Buendia et al. 1990; Garrido-Jimenez et al. 2018).

One of the main structures in which the centrosome interacts with elements of the cytoskeleton is the basal body of the ciliated cells. This structure (more specifically the basal foot) acts as an anchoring point of the network of microtubules that allows cytoskeleton to exercise mechanical forces caused by its depolymerization (Hagiwara et al. 2000). Recent studies in this field propose that some basal body proteins play a key role on cilia basal bodies migration to the cell surface, a process which is dependent on the actin cytoskeleton (Dawe et al. 2007a; Lemullois et al. 1988). For instance, in ciliated cells, centrioles dock in the apical surface to become basal bodies. In addition, basal body proteins MKS3 (Meckel Syndrome Type 3) have been recently proposed as a regulating element of the position of the basal body (Abdelhamed et al. 2015; Dawe et al. 2007b).

8.3.2   Polarity Pathways and Centriole, Centrosome, and Cilia Positioning

During organism development, cell coordination is essential for proper morphogenesis and patterning. Thus, directional information drives cells to form oriented cellular structures like cilia, kinocilia and outer segments of photoreceptors, among others. During development, two types of polarity are stablished, planar cell polarity (PCP) and apical-basolateral polarity. (Campanale et al. 2017; Gray et al. 2011; Wallingford 2012).

8.3.2.1   Apical-Basolateral Polarity

In epithelial cells, apical-basolateral (Ap-Bl) polarity refers to the generation of two different membrane domains: one is the apical membrane, in contact with the environment, and the other the basal membrane, anchored to extracellular matrix (ECM). Apical-basal polarity is a result of the differential distribution of protein complexes, phospholipids and components of the cytoskeleton (Goldstein and Macara 2007; Rodriguez-Boulan and Macara 2014). This polarized distribution is essential to regulate cellular functions such as directional transport of ions in epithelial cells or the transmission of the electrical impulse in neurons. In vertebrates, apical and basolateral membranes are separated by tight and/or adherent junctions, maintaining this cellular asymmetry (Bryant and Mostov 2008; Mostov et al. 2003).

Several molecular pathways that regulate the Ap-Bl polarity establishment have been clarified in the last decades, and they reflect the presence of, at least, three interacting protein complexes. First, the main regulator of the Ap-Bl polarity is the Par complex, consisting of Partitioning defective protein 3 (Par3), Partitioning defective protein 6 (Par6), and atypical protein kinase C (aPCK) (Goldstein and Macara 2007; St Johnston and Sanson 2011). This complex provides polarity information during the first cell cycle, and is key for proper asymmetrical distribution of cell-fate regulators. A second complex for polarity is formed by Crumbs, Stardust and Pals1-associated tight junction (PATJ), and its function is more restricted to epithelial cells, defining apical membrane. Finally, the last complex consists of Scribble, Discs large (Dlg), and Lethal giant larvae (Lgl), located in the basal membrane. The interactions of these three protein complexes act through negative regulation, defining the apical-basolateral surfaces of epithelial cells (Tanentzapf and Tepass 2003).

The Ap-Bl polarity has been related to centrosome positioning through Partitioning defective 6 homolog gamma (Par6γ) function in centrosomal protein recruitment. In this sense, loss of Par6γ impairs the centrosomal recruitment of proteins (Dormoy et al. 2013). In addition, it has been shown that apical moving of centrosome during epithelial polarization is a process dependent on Par3. Thus, embryos lacking Par3 fail to localize their centrosomes apically, affecting cilia positioning (Feldman and Priess 2012). Par complex also regulates centrosome anchoring at the cell cortex through its interaction with NuMA (nuclear mitotic apparatus), Pins (partner of inscuteable), and G proteins, which have a clear implication in asymmetric cell division (Galli et al. 2011). This pathway has been recently related with primary cilium movement to the apical surface of the cell. Deletion of G proteins and Pins impairs the migration of the kinocilium in cochlear cells and disrupts hair orientation and shape (Ezan et al. 2013).

8.3.2.2   Planar Cell Polarity (PCP)

Over the last years, growing evidences related centriole positioning with a second cell polarity mechanism orthogonal to Ap-Bl axis, named planar cell polarity (PCP). This mechanism governs the coordinated polarization within the plane of a cell sheet (Goodrich and Strutt 2011; Guirao et al. 2010; Ybot-Gonzalez et al. 2007). PCP signaling is a well-conserved mechanism based on the asymmetric distribution of protein complexes, and drives morphogenetic changes across the tissue through mutual exclusion of these proteins at opposite sides of the cell. In Drosophila, PCP is controlled by genes including *frizzled* (Fz), *van gogh* (Vang, Vangl in vertebrates), and *flamingo* (Fmi, Celsr in vertebrates); the cytoplasmic components *dishevelled* (Dsh, Dvl in vertebrates), *diego* (Dgo), and *prickle* (Pk), and PCP effectors including, among others, *inturned* (In), *fuzzy* (Fy), and *fritz* (Frtz, Wdpcp in vertebrates), implicated in early stages of centriole positioning during ciliogenesis (Adler 2012; Park et al. 2006; Wallingford 2012; Wang et al. 2017).

The Frizzled PCP signaling pathway described above is established through the asymmetric segregation of the protein complex comprised of Fz-Dsh-Fmi, located on the distal side of the cell, and the protein complex composed of Vang-Pk-Fmi, enriched on the proximal region. Differential distribution of proximal and distal complexes generates a pattern that propagates throughout the tissue and it is involved in many cellular processes such as gastrulation, neural tube closure and patterning (Curtin et al. 2003; Kibar et al. 2001; Wallingford et al. 2000). The amplification of asymmetry along the tissue is driven by repulsive interactions between the Vang-Pk-Fmi complex and the F-Dsh-Fmi complex inside the cell (Jenny et al. 2005), whereas cell–cell communication is performed through the direct interaction of transmembrane proteins Vang and Fz.

In vertebrate ciliated cells, PCP have been widely related with centrosome and basal body positioning, revealing its key role in generation of directional fluid flow in a variety of ciliated cells (Mitchell et al. 2007, 2009; Wallingford 2010). For instance, disruption of Dvl function results in a randomization of planar polarity in multiciliated cell, impairing the cilia beating and directional fluid flow across the epithelium (Park et al. 2008). In brain, Celsr 2/3, Vangl 2 and Dvl are implicated in polarization of ciliated ependymal cells, and disruption of these proteins causes hydrocephalus (Guirao et al. 2010; Hirota et al. 2010; Tissir et al. 2010). Moreover, recent studies relate Vangl2 to the primary cilia localization to the posterior apical membrane of neuroepithelial cells (Borovina et al. 2010). The essential role of PCP in centrioles/basal bodies positioning and ciliogenesis is evident in the fact that a wide variety of syndromes has been related to alterations such as heterotaxy or *situs inversus* of PCP proteins (Pennekamp et al. 2015). PCP have been also related to the eccentric position of primary cilia and with defects in the orientation of kinocilium in the organ of Corti of inner ear (Axelrod 2008; Jones and Chen 2008). In addition, mice with knockout of Fltp (Flattop, a gene transcriptionally activated during PCP acquisition) show basal bodies and ciliogenesis defects in multiciliated lung cells (Gegg et al. 2014).

Many advances in the knowledge of Fz-PCP were made in *Drosophila* wings. In this model system, it has been recently described that the positioning of centrosomes is polarized toward the distal side of each cell under the control of PCP pathway, demonstrating that PCP also controls the position of centrioles in nonciliated cells. Gain- and loss-of function experiments of PCP core components have revealed alterations in the positioning of the centrioles when the distribution of PCP proteins was impaired (Carvajal-Gonzalez et al. 2016a, b). In addition, unlike described in vertebrates, lack of known PCP effectors *frtz*, *drock* (Rho-kinase), and *mwh* (multiple wing hair) does not alter centriole positioning in *Drosophila* (Garrido-Jimenez et al. 2018).

8.4 Methods to Measure Centrioles Polarity

The analyses of phenotypes associated with centrioles, centrosomes, basal bodies, or MTOCs have been focused on the quantification of their number (1), size/shape (2), and/or position (3). The most relevant factor that is related to cell polarity is its relative position within the cell, here referred as centriole positioning, but we will also briefly describe the techniques that are used for the quantification of their number, size and shape.

Differences in the number of centrioles within cells have been described by Theodor Boveri in 1887, including its potential association with aneuploidy and tumoral progression (Holland and Cleveland 2009). Then, in the last few years, several studies showed the role of centrosome amplification in different types of cancer (Levine et al. 2017). Very often, clinical studies use manual quantification in a set of immunohistochemical images with markers for centriole components like pericentrin or gamma-tubulin, (Denu et al. 2016) while others use software-assisted quantification (Levine et al. 2017). Finally, there is also the possibility to use indirect measures such as Western blot (Iemura et al. 2007) to quantify the relative quantity levels of centrosome components in different samples. In summary, the analysis of centrosome number is mainly based on imaging studies, and novel automatic or semiautomatic technologies will be able to screen phenotypes involving not only centrosome number but also its associated shape (Marteil et al. 2018).

Nevertheless, it is fair to consider that the relative position of the centriole/centrosome/basal body/MTOC within a cell is the most important parameter to define defects in its polarization. It is important to define the different aspects that this type of complex quantification contains. First, this centriole position is relative to the reference points or organelles, so we require the quantification of at least another cellular structure that we assume that is fixed (a reference position). Potential possibilities for these references include the nucleus, the centroid (geometric center of the cell), or the plasma membrane, among others. The choice of the reference in the cell will depend on our previous knowledge of the cell type and biology, and it will probably affect the final quantification, even when there are no studies that measure this effect. Then, we can quantify the distance between the centriole and the reference (distance-based measurements) as a proxy for centriole polarization. The most used technique that quantifies this distance is called Average Basal Body Positioning (ABP), and it uses two references, the most anterior point and the most posterior point in the cell. These are defined as -1.0 and $+1.0$, respectively, and the x-position of the centriole within this Anterior–Posterior axis is considered as a measure of its polarization. The average for a set of cells generates a single score that quantifies centriole polarization (Minegishi et al. 2017). As an alternative, the Polarity Index (Burute et al. 2017) estimates the relative distance (in the x-axis) between the nucleus and the centrosome normalized by the size of the nucleus by using a new coordinate system that links the nuclei of neighboring cells as the x-axis. However, these distance-based measurements do not fully describe the relative position of the centriole, as multiple locations can have the same distance (Euclidean or single axis) to the references (Fig. 8.3).

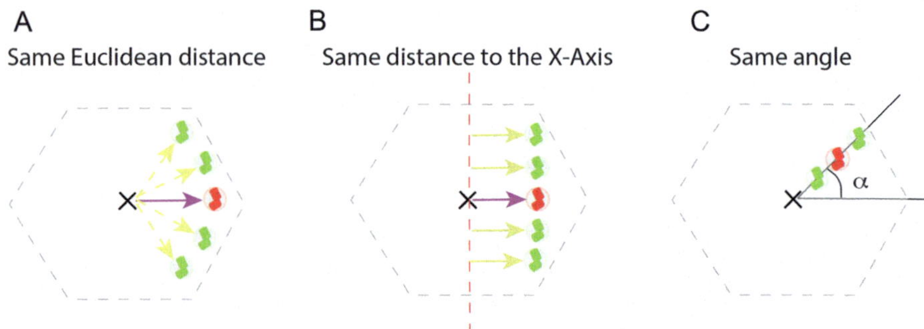

Fig. 8.3 Centriole positioning quantitative methods related referred to the center of the cell (centroid). The distance methods (**a** and **b**) can quantify different measures like Euclidean distance (**a**) or a single-axis distance (**b**). Nevertheless, in the figure it is shown how multiple centrosomes (in green) will become equal to the test sample (in red) in terms of these distances even with different relative positions within the cell. This also occurs with angle-based measures (**c**), when multiple centrosomes (in green) share the same angle (α) with the test sample (in green)

Instead, these multiple locations will form different angles with respect to the reference point. While distance-based measurements only require one reference, angle-based measurements require at least two references in order to obtain an angular score. A primary example of this technique is the Quantile (Q) method (Taniguchi et al. 2011). In this case, the references are the Anterior–Posterior and the Left–Right axes of the cell as they cross by its centroid. This defines four potential angular positions of the centrosome within a cell (Left Anterior, Left Posterior, Right Anterior and Right Posterior). This simple classification turns into a measure of centriole polarization by the calculation of the percentage of cells located in these four types. There are alternative angle-based methods as the direct angular quantification either between the nucleus-centrosome vector and the central cavity axis in epiblast cells (Burute et al. 2017) or the cell centroid–centriole vector and the proximal–distal axis in *Drosophila* pupal wing cells (Carvajal-Gonzalez et al. 2016b). In parallel to distance-based measures, the same angle-based score might reflect different centriole positions (Fig. 8.3).

Thus, neither distance-based nor angle-based measures by themselves are able to fully explain differences in centriole polarization. Novel techniques that address both position and orientation at the same time are required to obtain clearer centriole phenotypes associated to polarity defects. In this way, the Representative Polarized Centriole Distribution (RPCD) constitutes an effort to combine position and orientation in a single score (Garrido-Jimenez et al. 2018). The idea underlying this technique is to generate a statistical model to represent a specific centriole polarized system having as reference the cell centroid and a theoretical normalized shape of the cell. In the case of *Drosophila* wings, the normalized cell was modeled as a regular hexagon (Garrido-Jimenez et al. 2018) but this could be changed to different polygons in other systems. In every case, the size of single cells is normalized using the real area and the predicted area of the polygon. Finally, statistical analyses will return if a specific set of sample cells is significantly different from the RPCD or not. This technique is powerful enough to detect phenotypes that were not detected in distance-based or angle-based methods (Garrido-Jimenez et al. 2018) but it also presents several inconveniences, as we are required to generate a library (a RPCD) of the specific biological model we want to use in order to test it. Nevertheless, these RPCDs can be iteratively improved as we increase the number of used cells, and the binary classification (RPCD or not RPCD) can be potentially expanded to, for example, different stages during the development.

In summary, there is a repertoire of imaging techniques to assess the centriole/centrosome/MTOC polarity in a cell sample. We need to remark that these techniques are used only for translational polarity. In the case of rotational polarity, the standard analysis quantifies the relative angles between the basal feet and the basal bodies in each multiciliated cell. Specifically, the use of scores like the Index of Alignment [I_a, (Herawati et al. 2016)] can define up to four stages during multiciliated tracheal cell differentiation (Floret, Scatter, Partial Alignment and Alignment). In the next few years, we should obtain additional techniques that implement geometrical as well as simulation and modeling tools in order to further dissect centriole polarity phenotypes in multiple biological models.

Acknowledgments This work was supported by BFU2014-54699-P and BFU2017-85547-P grants from the Ministry of Economy to J.M. C-G. and GR15164 and GR18116 from Junta de Extremadura to F.C. Á.-C.R. S.G-J. was a recipient of a Fellowship from the Universidad de Extremadura, and S.D-Ch was a recipient of a Fellowship from Junta de Extremadura. All Spanish funding is cosponsored by the European Union FEDER program.

J.M. C-G was a recipient of an "Atraccion y Retencion de talento" contract from the GOBEX (Extremadura government) and was recipient of a Ramón y Cajal contract (RYC-2015-17867).

Competing Financial Interests The authors declare no competing financial interests.

References

Abdelhamed ZA, Natarajan S, Wheway G, Inglehearn CF, Toomes C, Johnson CA, Jagger DJ (2015) The Meckel-Gruber syndrome protein TMEM67 controls basal body positioning and epithelial branching morphogenesis in mice via the non-canonical Wnt pathway. Dis Model Mech 8:527–541

Adler PN (2012) The frizzled/stan pathway and planar cell polarity in the Drosophila wing. Curr Top Dev Biol 101:1–31

Ambuj K, Vidya R, Rao S, Rituraj P (2014) Role of centrosome in regulating immune response. Curr Drug Targets 15:558–563

Angus KL, Griffiths GM (2013) Cell polarisation and the immunological synapse. Curr Opin Cell Biol 25:85–91

Axelrod JD (2008) Basal bodies, kinocilia and planar cell polarity. Nat Genet 40:10

Baas PW, Lin S (2011) Hooks and comets: the story of microtubule polarity orientation in the neuron. Dev Neurobiol 71:403–418

Basto R, Lau J, Vinogradova T, Gardiol A, Woods CG, Khodjakov A, Raff JW (2006) Flies without centrioles. Cell 125:1375–1386

Bettencourt-Dias M, Hildebrandt F, Pellman D, Woods G, Godinho SA (2011) Centrosomes and cilia in human disease. Trends Genet 27:307–315

Boisvieux-Ulrich E, Lainé M-C, Sandoz D (1990) Cytochalasin D inhibits basal body migration and ciliary elongation in quail oviduct epithelium. Cell Tissue Res 259:443–454

Bond J, Roberts E, Mochida GH, Hampshire DJ, Scott S, Askham JM, Springell K, Mahadevan M, Crow YJ, Markham AF, Walsh CA, Woods CG (2002) ASPM is a major determinant of cerebral cortical size. Nat Genet 32:316–320

Bornens M (2012) The centrosome in cells and organisms. Science 335:422–426

Borovina A, Superina S, Voskas D, Ciruna B (2010) Vangl2 directs the posterior tilting and asymmetric localization of motile primary cilia. Nat Cell Biol 12:407

Braun DA, Hildebrandt F (2017) Ciliopathies. Cold Spring Harb Perspect Biol 9:a028191

Bryant DM, Mostov KE (2008) From cells to organs: building polarized tissue. Nat Rev Mol Cell Biol 9:887–901

Buendia B, Bré MH, Griffiths G, Karsenti E (1990) Cytoskeletal control of centrioles movement during the establishment of polarity in Madin-Darby canine kidney cells. J Cell Biol 110:1123–1135

Burute M, Prioux M, Blin G, Truchet S, Letort G, Tseng Q, Bessy T, Lowell S, Young J, Filhol O, Thery M (2017) Polarity reversal by centrosome repositioning primes cell scattering during epithelial-to-mesenchymal transition. Dev Cell 40:168–184

Campanale JP, Sun TY, Montell DJ (2017) Development and dynamics of cell polarity at a glance. J Cell Sci 130:1201–1207

Carvajal-Gonzalez JM, Mulero-Navarro S, Mlodzik M (2016a) Centriole positioning in epithelial cells and its intimate relationship with planar cell polarity. Bioessays 38:1234–1245

Carvajal-Gonzalez JM, Roman A-C, Mlodzik M (2016b) Positioning of centrioles is a conserved readout of Frizzled planar cell polarity signalling. Nat Commun 7:11135

Conde C, Caceres A (2009) Microtubule assembly, organization and dynamics in axons and dendrites. Nat Rev Neurosci 10:319–332

Curtin JA, Quint E, Tsipouri V, Arkell RM, Cattanach B, Copp AJ, Henderson DJ, Spurr N, Stanier P, Fisher EM, Nolan PM, Steel KP, Brown SDM, Gray IC, Murdoch JN (2003) Mutation of Celsr1 disrupts planar polarity of inner ear hair cells and causes severe neural tube defects in the mouse. Curr Biol 13:1129–1133

Damkier HH, Brown PD, Praetorius J (2013) Cerebrospinal fluid secretion by the choroid plexus. Physiol Rev 93:1847–1892

Dawe HR, Farr H, Gull K (2007a) Centriole/basal body morphogenesis and migration during ciliogenesis in animal cells. J Cell Sci 120:7

Dawe HR, Smith UM, Cullinane AR, Gerrelli D, Cox P, Badano JL, Blair-Reid S, Sriram N, Katsanis N, Attie-Bitach T, Afford SC, Copp AJ, Kelly DA, Gull K, Johnson CA (2007b) The Meckel–Gruber Syndrome proteins MKS1 and meckelin interact and are required for primary cilium formation. Hum Mol Genet 16:173–186

de Anda FC, Pollarolo G, Da Silva JS, Camoletto PG, Feiguin F, Dotti CG (2005) Centrosome localization determines neuronal polarity. Nature 436:704–708

Denu RA, Zasadil LM, Kanugh C, Laffin J, Weaver BA, Burkard ME (2016) Centrosome amplification induces high grade features and is prognostic of worse outcomes in breast cancer. BMC Cancer 16:47

Distel M, Hocking JC, Volkmann K, Koster RW (2010) The centrosome neither persistently leads migration nor determines the site of axonogenesis in migrating neurons in vivo. J Cell Biol 191:875–890

Dormoy V, Tormanen K, Sütterlin C (2013) Par6γ is at the mother centriole and controls centrosomal protein composition through a Par6α-dependent pathway. J Cell Sci 126:860–870

Elric J, Etienne-Manneville S (2014) Centrosome positioning in polarized cells: common themes and variations. Exp Cell Res 328:240–248

Euteneuer U, Schliwa M (1985) Evidence for an involvement of actin in the positioning and motility of centrosomes. J Cell Biol 101:96–103

Ezan J, Lasvaux L, Gezer A, Novakovic A, May-Simera H, Belotti E, Lhoumeau A-C, Birnbaumer L, Beer-Hammer S, Borg J-P, Le Bivic A, Nürnberg B, Sans N, Montcouquiol M (2013) Primary cilium migration depends on G-protein signalling control of subapical cytoskeleton. Nat Cell Biol 15(9):1107–1115

Feldman JL, Priess JR (2012) A role for the centrosome and PAR-3 in the hand-off of MTOC function during epithelial polarization. Curr Biol 22:575–582

Fliegauf M, Benzing T, Omran H (2007) When cilia go bad: cilia defects and ciliopathies. Nat Rev Mol Cell Biol 8:880–893

Galli M, Muñoz J, Portegijs V, Boxem M, Grill SW, Heck AJR, van den Heuvel S (2011) aPKC phosphorylates NuMA-related LIN-5 to position the mitotic spindle during asymmetric division. Nat Cell Biol 13:1132

Garrido-Jimenez S, Roman A-C, Alvarez-Barrientos A, Carvajal-Gonzalez JM (2018) Centriole planar polarity assessment in Drosophila wings. Development 145:dev169326

Gegg M, Böttcher A, Burtscher I, Hasenoeder S, Van Campenhout C, Aichler M, Walch A, Grant SGN, Lickert H (2014) Flattop regulates basal body docking and positioning in mono- and multiciliated cells. eLife 3:e03842

Glotzer M (2009) The 3Ms of central spindle assembly: microtubules, motors and MAPs. Nat Rev Mol Cell Biol 10:9–20

Goldstein B, Macara IG (2007) The PAR proteins: fundamental players in animal cell polarization. Dev Cell 13:609–622

Goodman L, Zallocchi M (2017) Integrin alpha8 and Pcdh15 act as a complex to regulate cilia biogenesis in sensory cells. J Cell Sci 130:3698–3712

Goodrich LV, Strutt D (2011) Principles of planar polarity in animal development. Development 138:1877

Grakoui A, Bromley SK, Sumen C, Davis MM, Shaw AS, Allen PM, Dustin ML (1999) The immunological synapse: a molecular machine controlling T cell activation. Science 285:221–227

Gray RS, Roszko I, Solnica-Krezel L (2011) Planar cell polarity: coordinating morphogenetic cell behaviors with embryonic polarity. Dev Cell 21:120–133

Guernsey DL, Jiang H, Hussin J, Arnold M, Bouyakdan K, Perry S, Babineau-Sturk T, Beis J, Dumas N, Evans SC, Ferguson M, Matsuoka M, Macgillivray C, Nightingale M, Patry L, Rideout AL, Thomas A, Orr A, Hoffmann I, Michaud JL, Awadalla P, Meek DC, Ludman M, Samuels ME (2010) Mutations in centrosomal protein CEP152 in primary microcephaly families linked to MCPH4. Am J Hum Genet 87:40–51

Guirao B, Meunier A, Mortaud S, Aguilar A, Corsi J-M, Strehl L, Hirota Y, Desoeuvre A, Boutin C, Han Y-G, Mirzadeh Z, Cremer H, Montcouquiol M, Sawamoto K, Spassky N (2010) Coupling between hydrodynamic forces and planar cell polarity orients mammalian motile cilia. Nat Cell Biol 12:341

Hagiwara H, Kano A, Aoki T, Ohwada N, Takata K (2000) Localization of γ–tubulin to the basal foot associated with the basal body extending a cilium. Histochem J 32:669–671

Herawati E, Taniguchi D, Kanoh H, Tateishi K, Ishihara S, Tsukita S (2016) Multiciliated cell basal bodies align in stereotypical patterns coordinated by the apical cytoskeleton. J Cell Biol 214:571–586

Hirota Y, Meunier A, Huang S, Shimozawa T, Yamada O, Kida YS, Inoue M, Ito T, Kato H, Sakaguchi M, Sunabori T, Nakaya M-A, Nonaka S, Ogura T, Higuchi H, Okano H, Spassky N, Sawamoto K (2010) Planar polarity of multiciliated ependymal cells involves the anterior migration of basal bodies regulated by non-muscle myosin II. Development 137:3037

Holland AJ, Cleveland DW (2009) Boveri revisited: chromosomal instability, aneuploidy and tumorigenesis. Nat Rev Mol Cell Biol 10:478–487

Iemura K, Kamemura K, Miwa M (2007) Assessment of the centrosome amplification by quantification of gamma-tubulin in Western blotting. Anal Biochem 371:256–258

Jayaraman D, Kodani A, Gonzalez DM, Mancias JD, Mochida GH, Vagnoni C, Johnson J, Krogan N, Harper JW, Reiter JF, Yu TW, Bae BI, Walsh CA (2016) Microcephaly proteins

Wdr62 and Aspm define a mother centriole complex regulating centriole biogenesis, apical complex, and cell fate. Neuron 92:813–828

Jenny A, Reynolds-Kenneally J, Das G, Burnett M, Mlodzik M (2005) Diego and Prickle regulate Frizzled planar cell polarity signalling by competing for Dishevelled binding. Nat Cell Biol 7:691

Jones C, Chen P (2008) Primary cilia in planar cell polarity regulation of the inner ear. Curr Top Dev Biol 85:197–224

Khanna H (2015) Photoreceptor sensory cilium: traversing the ciliary gate. Cells 4:674–686

Kibar Z, Vogan KJ, Groulx N, Justice MJ, Underhill DA, Gros P (2001) Ltap, a mammalian homolog of Drosophila Strabismus/Van Gogh, is altered in the mouse neural tube mutant Looptail. Nat Genet 28:251

Kuijpers M, Hoogenraad CC (2011) Centrosomes, microtubules and neuronal development. Mol Cell Neurosci 48:349–358

Lemullois M, Boisvieux-Ulrich E, Laine M-C, Chailley B, Sandoz D (1988) Development and functions of the cytoskeleton during ciliogenesis in metazoa. Biol Cell 63:195–208

Lepelletier L, de Monvel JB, Buisson J, Desdouets C, Petit C (2013) Auditory hair cell centrioles undergo confined Brownian motion throughout the developmental migration of the kinocilium. Biophys J 105:48–58

Levine MS, Bakker B, Boeckx B, Moyett J, Lu J, Vitre B, Spierings DC, Lansdorp PM, Cleveland DW, Lambrechts D, Foijer F, Holland AJ (2017) Centrosome amplification is sufficient to promote spontaneous tumorigenesis in mammals. Dev Cell 40(313–322):e315

Lim HH, Zhang T, Surana U (2009) Regulation of centrosome separation in yeast and vertebrates: common threads. Trends Cell Biol 19:325–333

Lopes CAM, Mesquita M, Cunha AI, Cardoso J, Carapeta S, Laranjeira C, Pinto AE, Pereira-Leal JB, Dias-Pereira A, Bettencourt-Dias M, Chaves P (2018) Centrosome amplification arises before neoplasia and increases upon p53 loss in tumorigenesis. J Cell Biol 217:2353–2363

Marshall WF (2001) Centrioles take center stage. Curr Biol 11:R487–R496

Marteil G, Guerrero A, Vieira AF, de Almeida BP, Machado P, Mendonca S, Mesquita M, Villarreal B, Fonseca I, Francia ME, Dores K, Martins NP, Jana SC, Tranfield EM, Barbosa-Morais NL, Paredes J, Pellman D, Godinho SA, Bettencourt-Dias M (2018) Over-elongation of centrioles in cancer promotes centriole amplification and chromosome missegregation. Nat Commun 9:1258

May-Simera HL, Gumerson JD, Gao C, Campos M, Cologna SM, Beyer T, Boldt K, Kaya KD, Patel N, Kretschmer F, Kelley MW, Petralia RS, Davey MG, Li T (2016) Loss of MACF1 abolishes ciliogenesis and disrupts apicobasal polarity establishment in the retina. Cell Rep 17:1399–1413

Mennella V, Agard DA, Huang B, Pelletier L (2014) Amorphous no more: subdiffraction view of the pericentriolar material architecture. Trends Cell Biol 24:188–197

Michel V, Goodyear RJ, Weil D, Marcotti W, Perfettini I, Wolfrum U, Kros CJ, Richardson GP, Petit C (2005) Cadherin 23 is a component of the transient lateral links in the developing hair bundles of cochlear sensory cells. Dev Biol 280:281–294

Minegishi K, Hashimoto M, Ajima R, Takaoka K, Shinohara K, Ikawa Y, Nishimura H, McMahon AP, Willert K, Okada Y, Sasaki H, Shi D, Fujimori T, Ohtsuka T, Igarashi Y, Yamaguchi TP, Shimono A, Shiratori H, Hamada H (2017) A Wnt5 activity asymmetry and intercellular signaling via PCP proteins polarize node cells for left-right symmetry breaking. Dev Cell 40 (439–452):e434

Mirzadeh Z, Han YG, Soriano-Navarro M, Garcia-Verdugo JM, Alvarez-Buylla A (2010) Cilia organize ependymal planar polarity. J Neurosci 30:2600–2610

Mitchell B, Jacobs R, Li J, Chien S, Kintner C (2007) A positive feedback mechanism governs the polarity and motion of motile cilia. Nature 447:97

Mitchell B, Stubbs JL, Huisman F, Taborek P, Yu C, Kintner C (2009) The PCP pathway instructs the planar orientation of ciliated cells in the Xenopus larval skin. Curr Biol 19:924–929

Monks CRF, Freiberg BA, Kupfer H, Sciaky N, Kupfer A (1998) Three-dimensional segregation of supramolecular activation clusters in T cells. Nature 395:82

Mostov K, Su T, ter Beest M (2003) Polarized epithelial membrane traffic: conservation and plasticity. Nat Cell Biol 5:287

Narita K, Takeda S (2015) Cilia in the choroid plexus: their roles in hydrocephalus and beyond. Front Cell Neurosci 9:39

Narita K, Kawate T, Kakinuma N, Takeda S (2010) Multiple primary cilia modulate the fluid transcytosis in choroid plexus epithelium. Traffic 11:287–301

Narita K, Kozuka-Hata H, Nonami Y, Ao-Kondo H, Suzuki T, Nakamura H, Yamakawa K, Oyama M, Inoue T, Takeda S (2012) Proteomic analysis of multiple primary cilia reveals a novel mode of ciliary development in mammals. Biol Open 1:815–825

Nayak GD, Ratnayaka HS, Goodyear RJ, Richardson GP (2007) Development of the hair bundle and mechanotransduction. Int J Dev Biol 51:597–608

Nguyen MM, Stone MC, Rolls MM (2011) Microtubules are organized independently of the centrosome in Drosophila neurons. Neural Dev 6:38

Park TJ, Haigo SL, Wallingford JB (2006) Ciliogenesis defects in embryos lacking inturned or fuzzy function are associated with failure of planar cell polarity and Hedgehog signaling. Nat Genet 38:303

Park TJ, Mitchell BJ, Abitua PB, Kintner C, Wallingford JB (2008) Dishevelled controls apical docking and planar polarization of basal bodies in ciliated epithelial cells. Nat Genet 40:871–879

Pazour GJ, Bloodgood RA (2008) Targeting proteins to the ciliary membrane. Curr Top Dev Biol 85:115–149

Pennekamp P, Menchen T, Dworniczak B, Hamada H (2015) Situs inversus and ciliary abnormalities: 20 years later, what is the connection? Cilia 4:1–1

Piel M, Meyer P, Khodjakov A, Rieder CL, Bornens M (2000) The respective contributions of the mother and daughter centrioles to centrosome activity and behavior in vertebrate cells. J Cell Biol 149:317–330

Reboldi A, Coisne C, Baumjohann D, Benvenuto F, Bottinelli D, Lira S, Uccelli A, Lanzavecchia A, Engelhardt B, Sallusto F (2009) C-C chemokine receptor 6-regulated entry of TH-17 cells into the CNS through the choroid plexus is required for the initiation of EAE. Nat Immunol 10:514–523

Redzic ZB, Preston JE, Duncan JA, Chodobski A, Szmydynger-Chodobska J (2005) The choroid plexus-cerebrospinal fluid system: from development to aging. Curr Top Dev Biol 71:1–52

Reiter JF, Leroux MR (2017) Genes and molecular pathways underpinning ciliopathies. Nat Rev Mol Cell Biol 18:533–547

Rodriguez-Boulan E, Macara IG (2014) Organization and execution of the epithelial polarity programme. Nat Rev Mol Cell Biol 15:225–242

Roig-Martinez M, Saavedra-Lopez E, Casanova PV, Cribaro GP, Barcia C (2019) The MTOC/Golgi complex at the T cell immunological synapse. In: Kloc M (ed) The Golgi apparatus and centriole: functions, interactions and role in disease. Springer, Heidelberg

Rosenbaum JL, Witman GB (2002) Intraflagellar transport. Nat Rev Mol Cell Biol 3:813–825

Sakakibara A, Ando R, Sapir T, Tanaka T (2013) Microtubule dynamics in neuronal morphogenesis. Open Biol 3:130061

Sanchez-Huertas C, Freixo F, Viais R, Lacasa C, Soriano E, Luders J (2016) Non-centrosomal nucleation mediated by augmin organizes microtubules in post-mitotic neurons and controls axonal microtubule polarity. Nat Commun 7:12187

Satir P, Pedersen LB, Christensen ST (2010) The primary cilium at a glance. J Cell Sci 123:499–503

Sattler S (2017) The role of the immune system beyond the fight against infection. In: Sattler S, Kennedy-Lydon T (eds) The immunology of cardiovascular homeostasis and pathology. Springer International, Cham, pp 3–14

Sedmak T, Wolfrum U (2011) Intraflagellar transport proteins in ciliogenesis of photoreceptor cells. Biol Cell 103:449–466

Sipe CW, Liu L, Lee J, Grimsley-Myers C, Lu X (2013) Lis1 mediates planar polarity of auditory hair cells through regulation of microtubule organization. Development 140:1785–1795

Sir JH, Barr AR, Nicholas AK, Carvalho OP, Khurshid M, Sossick A, Reichelt S, D'Santos C, Woods CG, Gergely F (2011) A primary microcephaly protein complex forms a ring around parental centrioles. Nat Genet 43:1147–1153

Solecki DJ, Trivedi N, Govek EE, Kerekes RA, Gleason SS, Hatten ME (2009) Myosin II motors and F-actin dynamics drive the coordinated movement of the centrosome and soma during CNS glial-guided neuronal migration. Neuron 63:63–80

Spassky N, Meunier A (2017) The development and functions of multiciliated epithelia. Nat Rev Mol Cell Biol 18:423

St Johnston D, Sanson B (2011) Epithelial polarity and morphogenesis. Curr Opin Cell Biol 23:540–546

Stiess M, Maghelli N, Kapitein LC, Gomis-Ruth S, Wilsch-Brauninger M, Hoogenraad CC, Tolic-Norrelykke IM, Bradke F (2010) Axon extension occurs independently of centrosomal microtubule nucleation. Science 327:704–707

Stinchcombe Jane C, Griffiths Gillian M (2014) Communication, the centrosome and the immunological synapse. Philos Trans R Soc Lond B Biol Sci 369:20130463

Stinchcombe JC, Majorovits E, Bossi G, Fuller S, Griffiths GM (2006) Centrosome polarization delivers secretory granules to the immunological synapse. Nature 443:462

Stinchcombe JC, Salio M, Cerundolo V, Pende D, Arico M, Griffiths GM (2011) Centriole polarisation to the immunological synapse directs secretion from cytolytic cells of both the innate and adaptive immune systems. BMC Biol 9:45

Tanentzapf G, Tepass U (2003) Interactions between the crumbs, lethal giant larvae and bazooka pathways in epithelial polarization. Nat Cell Biol 5:46–52

Tang N, Marshall WF (2012) Centrosome positioning in vertebrate development. J Cell Sci 125:4951–4961

Taniguchi K, Maeda R, Ando T, Okumura T, Nakazawa N, Hatori R, Nakamura M, Hozumi S, Fujiwara H, Matsuno K (2011) Chirality in planar cell shape contributes to left-right asymmetric epithelial morphogenesis. Science 333:339–341

Teixido-Travesa N, Roig J, Luders J (2012) The where, when and how of microtubule nucleation - one ring to rule them all. J Cell Sci 125:4445–4456

Tissir F, Qu Y, Montcouquiol M, Zhou L, Komatsu K, Shi D, Fujimori T, Labeau J, Tyteca D, Courtoy P, Poumay Y, Uemura T, Goffinet AM (2010) Lack of cadherins Celsr2 and Celsr3 impairs ependymal ciliogenesis, leading to fatal hydrocephalus. Nat Neurosci 13:700–707

Tsai JW, Bremner KH, Vallee RB (2007) Dual subcellular roles for LIS1 and dynein in radial neuronal migration in live brain tissue. Nat Neurosci 10:970–979

Ueda H, Morphew MK, McIntosh JR, Davis MM (2011) CD4+ T-cell synapses involve multiple distinct stages. Proc Natl Acad Sci USA 108:17099

Umeshima H, Hirano T, Kengaku M (2007) Microtubule-based nuclear movement occurs independently of centrosome positioning in migrating neurons. Proc Natl Acad Sci USA 104:16182–16187

Vaisse C, Reiter JF, Berbari NF (2017) Cilia and obesity. Cold Spring Harb Perspect Biol 9: a028217

Vertii A, Doxsey S (2016) The centrosome: a phoenix organelle of the immune response. Single Cell Biol 5:1

Volta F, Gerdes JM (2017) The role of primary cilia in obesity and diabetes. Ann N Y Acad Sci 1391:71–84

Wallingford JB (2010) Planar cell polarity signaling, cilia and polarized ciliary beating. Curr Opin Cell Biol 22:597–604

Wallingford JB (2012) Planar cell polarity and the developmental control of cell behavior in vertebrate embryos. Annu Rev Cell Dev Biol 28:627–653

Wallingford JB, Rowning BA, Vogeli KM, Rothbächer U, Fraser SE, Harland RM (2000) Dishevelled controls cell polarity during Xenopus gastrulation. Nature 405:81

Wang L, Dynlacht BD (2018) The regulation of cilium assembly and disassembly in development and disease. Development 145:dev151407

Wang Y, Naturale VF, Adler PN (2017) Planar cell polarity effector fritz interacts with dishevelled and has multiple functions in regulating PCP. G3 (Bethesda) 7:1323–1337

Webb SW, Grillet N, Andrade LR, Xiong W, Swarthout L, Della Santina CC, Kachar B, Muller U (2011) Regulation of PCDH15 function in mechanosensory hair cells by alternative splicing of the cytoplasmic domain. Development 138:1607–1617

Wei Q, Xu Q, Zhang Y, Li Y, Zhang Q, Hu Z, Harris PC, Torres VE, Ling K, Hu J (2013) Transition fibre protein FBF1 is required for the ciliary entry of assembled intraflagellar transport complexes. Nat Commun 4:2750

Werner S, Pimenta-Marques A, Bettencourt-Dias M (2017) Maintaining centrosomes and cilia. J Cell Sci 130:3789–3800

Yau KW, Schatzle P, Tortosa E, Pages S, Holtmaat A, Kapitein LC, Hoogenraad CC (2016) Dendrites in vitro and in vivo contain microtubules of opposite polarity and axon formation correlates with uniform plus-end-out microtubule orientation. J Neurosci 36:1071–1085

Ybot-Gonzalez P, Savery D, Gerrelli D, Signore M, Mitchell CE, Faux CH, Greene NDE, Copp AJ (2007) Convergent extension, planar-cell-polarity signalling and initiation of mouse neural tube closure. Development 134:789–799

Zhu J, Burakov A, Rodionov V, Mogilner A (2010) Finding the cell center by a balance of dynein and myosin pulling and microtubule pushing: a computational study. Mol Biol Cell 21:4418–4427

Zmuda JF, Rivas RJ (1998) The Golgi apparatus and the centrosome are localized to the sites of newly emerging axons in cerebellar granule neurons in vitro. Cell Motil Cytoskeleton 41:18–38

Chapter 9
The MTOC/Golgi Complex at the T-Cell Immunological Synapse

Meritxell Roig-Martinez, Elena Saavedra-Lopez, Paola V. Casanova, George P. Cribaro, and Carlos Barcia

Abstract T cells effectively explore the tissue in search for antigens. When activated, they dedicate a big amount of energy and resources to arrange a complex structure called immunological synapse (IS), containing a particular distribution of molecules defined as supramolecular activation clusters (SMACs), and become polarized toward the target cell in a manner that channels the information specifically. This arrangement is symmetrical and requires the polarization of the MTOC and the Golgi to be operational, especially for the proper delivery of lytic granules and the recycling of molecules three dimensionally segregated at the clustered interface. Alternatively, after the productive encounter, T cells need to rearrange again to newly navigate through the tissue, changing back to a motile state called immunological kinapse (IK). In this IK state, the MTOC and the Golgi apparatus are repositioned and recruited at the back of the T cell to facilitate motility, while the established symmetry of the elements of the SMACs is broken and distributed in a different pattern. Both states, IS and IK, are interchangeable and are mainly orchestrated by the MTOC/Golgi complex, being critical for an effective immune response.

9.1 Introduction

T cells patrol the tissues searching for foreign antigens or neoantigens (Lu and Robbins 2016; Ariotti et al. 2012; Yarchoan et al. 2017). They have to constantly move through different environments and tissues and travel by intricate surroundings that require high adaptability (Weninger et al. 2014). This feature of great cellular motility involves the constant reorganization of the cytoskeleton involving microtubules and actin fibers (Dustin 2007). This active cytoskeletal arrangement is controlled by dynamics of the microtubule-organizing center (MTOC), which is positioned differently in each condition (Dustin 2010).

M. Roig-Martinez · E. Saavedra-Lopez · P. V. Casanova · G. P. Cribaro · C. Barcia (✉)
Institut de Neurociències, Neuroimmunity Research Group, Department of Biochemistry and Molecular Biology, School of Medicine, Lab M2-107, Universitat Autònoma de Barcelona, Barcelona, Spain
e-mail: carlos.barcia@uab.es

© Springer Nature Switzerland AG 2019
M. Kloc (ed.), *The Golgi Apparatus and Centriole*, Results and Problems in Cell Differentiation 67, https://doi.org/10.1007/978-3-030-23173-6_9

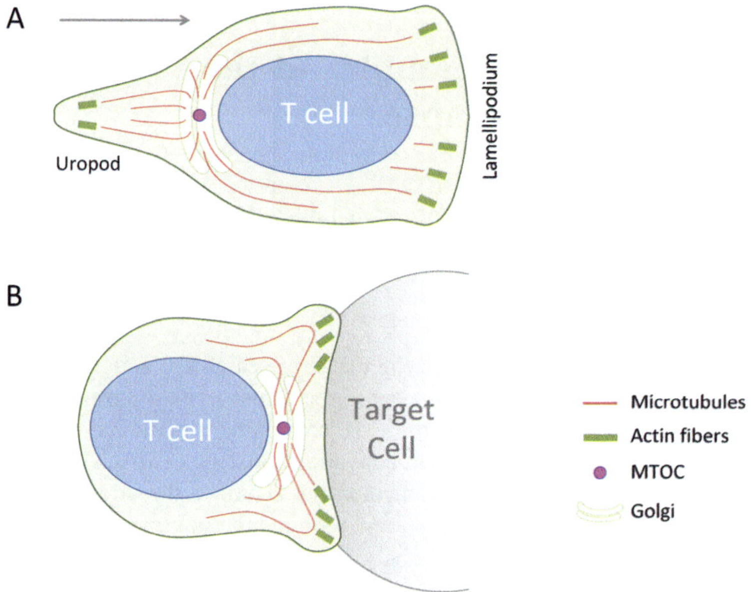

Fig. 9.1 T cells kinapse vs synapse. T cells alternate between kinapse (**a**) and synapse state (**b**). With low antigen sensing, T cells maintain a motile state (**a**) (arrow indicates the direction), with the configuration of a leading lamella and lamellipodium, as well as a trailing uropod. In this state, the MTOC and Golgi are kept behind the cell nucleus. After T cells get activated, during IS formation with a target cell (**b**), the MTOC and Golgi are positioned toward the interface with the target cell. Original figure by authors

Particularly, T cells alternate between two states, the immunological synapse (IS) and the immunological kinapse (IK) (Fig. 9.1), according to the level of antigen sensing (Moreau et al. 2012, 2015). On the one hand, IS is formed as a productive interaction with target cells when the level of antigen is sufficient to trigger T-cell activation. In this state, T cells undergo a dramatic arrangement with a high degree of polarization. On the other hand, IK is a motile state, arranged when the sensing of antigen is low.

9.2 Formation of Immunological Synapses

An IS is formed when T cells engage with antigen-presenting cells (APC) (Monks et al. 1998). The interaction between the two cells generates a synaptic cleft that is considered crucial for the specificity of the immune response (Mitxitorena et al. 2015; Dustin 2005). This space is created with the characteristic arrangement of different molecules configuring a secluded chamber where communication between both cells occurs (Grakoui et al. 1999). The initiation of IS formation takes place when T cells encounter an antigen on the target cell (Huppa and Davis 2003). This encounter

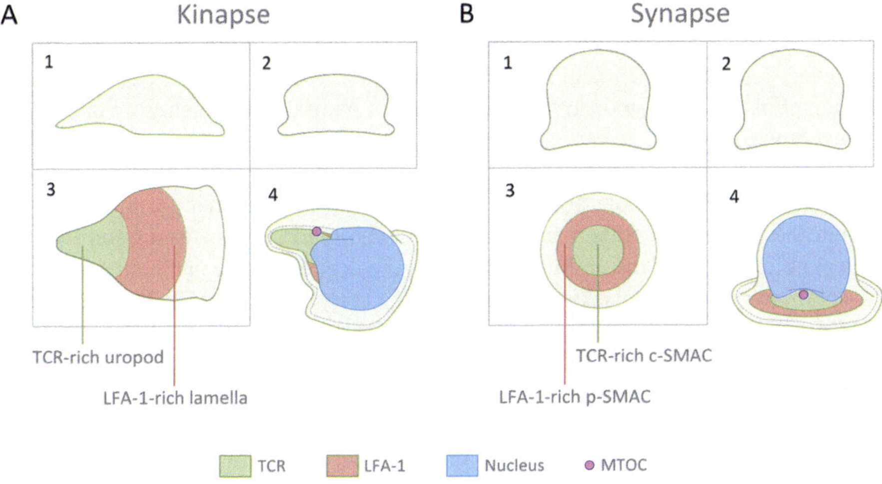

Fig. 9.2 Structure of immunological synapse and kinapse. A multiview projection of a T cell in kinapse state (**a**) and synapse state (**b**). Frontal and lateral views can be observed in 1 and 2, while 3 shows the interacting interface. A 3D interpretation is depicted in 4. In kinapse (**a**), the interface shows a particular distribution, displaying TCR molecules relegated to the uropod while LFA-1 is distributed in the lamella. In contrast, at the synaptic interface (**b**), TCR is concentrated at the c-SMAC, surrounded by a LFA-1 ring forming the p-SMAC. In the 3D depiction, the distribution of the interface can be seen accordingly, with the position of the MTOC in relation to the cell nucleus. Original figure by authors

triggers a response that activates T cells through the phosphorylation of tyrosine kinases such as zeta-chain-associated protein kinase 70 (Zap-70) or lymphocyte-specific protein tyrosine kinase (Lck) (Lee et al. 2002; Blanchard et al. 2002; Holdorf et al. 2002). This activation initiates a cascade of signaling that entails the transformation of the whole cell involving a cytoskeletal arrangement and a massive polarization (Kuhn and Poenie 2002). The organization of the cytoskeleton is accompanied by the physical distribution of molecules, such as T-cell receptor (TCR) that becomes clustered at the center of the IS interface forming the so-called central supramolecular activation cluster (c-SMAC) (Monks et al. 1998). Simultaneously, adhesion molecules such as lymphocyte function-associated antigen 1 (LFA-1) are clustered as a ring at the external border of the interface forming the peripheral SMAC (p-SMAC) (Kaizuka et al. 2007) (Fig. 9.2). Outer to that ring, other molecules such as cluster of differentiation 45 (CD45) accumulate forming the distal SMAC (d-SMAC) (Yu et al. 2013; Johnson et al. 2000). This configuration generates an isolated area externally sealed by the adhesion ring of LFA-1, bound to intercellular adhesion molecule 1 (ICAM-1) at the target cell, where specific molecules, such as cytotoxic granules, can be delivered specifically without altering bystander cells (Mitxitorena et al. 2015; Huse et al. 2006; Dustin et al. 2010).

Although the function of the IS is not completely understood, the fact that configures a specific polarization toward the target cell has been considered essential for a proper immune response. Since the first description (Monks et al. 1998), the visualization of the IS containing SMAC has been mostly based on in vitro experiments, where it can be analyzed and examined in high detail (Cai et al. 2017). However, despite the technical difficulties, signatures of IS can also be seen in tissue (Barcia et al. 2006), demonstrating that this is a natural phenomenon with physiological consequences.

For IS to happen, T cells suffer this arrangement involving the morphological change of the cytoskeleton, positioning the microtubule-organizing center (MTOC) toward the target cell (Billadeau et al. 2007). This position creates a complete and radical polarization of the entire cell where some organelles are gathered toward the synaptic interface (Franciszkiewicz et al. 2013) (Figs. 9.1 and 9.2). The role of the MTOC is fundamental to modify the position of the Golgi apparatus and fundamentally the nucleus location (Vicente-Manzanares and Sanchez-Madrid 2004; Kloc et al. 2014).

Among these organelles, mitochondria are also dragged toward the IS by the cytoskeleton rearrangement and it is thought that functions as the absorbent of calcium at this level and intervenes in the synaptic dynamics (Maccari et al. 2016). In fact, calcium channels such as ORAI and PMC are specifically localized at the IS to control calcium trafficking (Quintana et al. 2011).

Several molecules have been described to be implicated in the positioning of the MTOC at the IS. The lipid second messenger diacylglycerol (DAG) plays a fundamental role in this process being centered at the IS and involving the reorientation of MTOC (Chauveau et al. 2014; Huse et al. 2013). Additionally, nuclear distribution E homolog 1 (NDE-1) and dynactin also take part on this polarization, particularly, NDE-1 in the translocation of the MTOC whereas dynactin allows the translocation of lytic granules (Nath et al. 2016). In the latter, the MTOC is so close to the membrane that may contact the plasma membrane (Stinchcombe et al. 2006).

Thus, the MTOC drags the Golgi apparatus along with the microtubules creating a hemispherical dome. The T cell adopts this shape by alpha-tubulin ribs (Kuhn and Poenie 2002), which are driven through a Lck-dependent mechanism by the gamma-tubulin cluster at the MTOC (Tsun et al. 2011). This structure facilitates a balance of the compression-tension able to hold the space at the interface area and channel the information, fundamentally carried by granules.

Importantly, since the MTOC sets the position of the Golgi apparatus, its location becomes critical for the orientation of endocytosis and secretory domains. This situation of the Golgi at the IS is especially relevant for the recycling of numerous molecules, such as TCR-CD3 complexes, of which the polarized secretion at the SMAC is very active (Carpier et al. 2018). Interestingly, this polarization of the MTOC-Golgi complex occurs prior to the release of cytolytic granules, which indicates the importance of the IS cellular arrangement for an effective immune response (Stinchcombe et al. 2001). Particularly, the granules are transported through microtubules and not by F-actin, which remains excluded from the c-SMAC. Then, granules are transported minus-end through long microtubules toward the IS and transported plus-end through the short microtubules (Stinchcombe et al. 2006; Kloc et al. 2014).

9.3 The Formation of Immunological Kinapse

IK, in contrast with IS, is formed when the antigen recognition is low, so T cells increase their motility forming an asymmetrical interacting interface (Dustin 2007; Moreau et al. 2012). In this case, the MTOC is located toward the back of the cell, the uropod, while the front, the lamellipodium, navigates through the environment being highly sensitive to antigen (Negulescu et al. 1996; Kloc et al. 2014).

T cells interchange between the two states, IS and IK, in a symmetry-forming and symmetry-breaking cycle, which is balanced according to the antigen recognition (Dustin 2008), being the MTOC position a key factor organizing the architecture of the cell and the position of the organelles (Dustin 2007).

In the kinaptic state, the interface is asymmetric, segregating the molecules of the d-SMAC to the leading edge or lamellipodium and adhesion molecules (analogously to the p-SMAC) to the lamellar focal point, while the TCR molecules (the elements conforming the c-SMAC) are relegated to the uropod (Evans et al. 2009; Dustin 2008; Fritzsche and Dustin 2018). This arrangement is regulated by protein kinase C-Θ (PKCΘ), playing a critical role in the T-cell motility and localization of the MTOC toward the uropod (Cannon et al. 2013). This is essential for the interchange between these two states that T cells undergo (Dustin 2008). From the IK state, the IS formation is triggered when a high load of antigen is recognized, reducing the motility and arranging a synaptic interface, which is symmetrically driven by the centriole position (Stinchcombe et al. 2006).

Motility of cells depends on different signaling pathways involving the arrangement of the cytoskeleton and stress fibers. The activation of small GTPases such as cell division control protein 42 (Cdc42) and Ras homologous (Rho) protein is critical for the motility and its direction. Cdc42 is considered the center of polarity and intervenes in the position of the MTOC (Etienne-Manneville 2004). A classical example was defined in fibroblasts, by using in vitro experiments. After an artificial physical damage is done, fibroblasts tend to close that gap by forming a cellular scar, a phenomenon called wound healing. During this process, fibroblasts orient to the wound and move toward this gap (Vicente-Manzanares et al. 2009). In this process, the MTOC is located toward the leading edge, between the lamella and the nucleus (Fig. 9.3). This position is assumed to happen in most of the mammal motile cells but it does not seem very well defined in some leukocytes, including myeloid cells (Crespo et al. 2014) and brain macrophage-like microglia (Lively and Schlichter 2013).

Although in myeloid cells the MTOC position in the front seems to lead the direction, the perinuclear position of the MTOC interchanges and oscillates between the front and the back, probably as a result of the changes in the rotation of the nucleus when cells crawl (Crespo et al. 2014). In lymphocytes, however, it appears to be settled that the MTOC position is maintained in the back of the cell, the uropod when migrating, a position that seems to be adequate for the motility and the incursion into narrow spaces (Weninger et al. 2014). One useful example, with its obvious limitations, of the advantage of this distribution for the alternation between IS and IK, is considering the alpha-tubulin cytoskeleton ribs and its dynamics as an

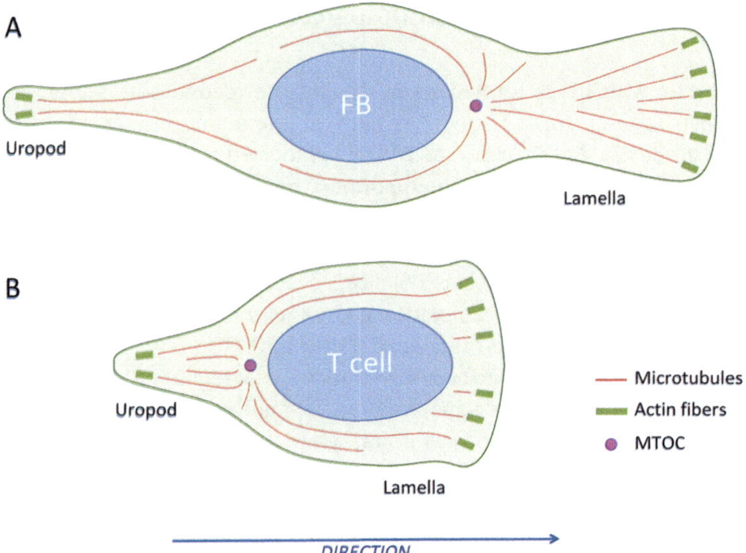

Fig. 9.3 Distinct pattern of MTOC position in kinaptic lymphocytes. This illustration is depicting the position of the MTOC in motile fibroblasts (**a**) and lymphocytes (**b**). The MTOC is positioned toward the leading edge in many cell types, such as fibroblasts (**a**), and appears displaced in front of the nucleus, oriented to the area where the leading lamella is arranged. Microtubules appear distributed radially from the MTOC to the actin-rich lamella and the trailing uropod at the back of the cell. However in motile T cells (**b**), the MTOC presents a different position, being behind the nucleus driving the motility of T cells in kinapsis mode. Original figure by authors

umbrella. The open umbrella would resemble the IS position whereas the motility and IK situation would be the closed umbrella (Ratner et al. 1997). In this way, organelles are contained at this space facilitating the motility.

Importantly, these dynamics are physiologically relevant and can be seen in the tissue, and an imbalance between the two states, IS and IK, could be apparent in pathological conditions (Diaz et al. 2018) reflecting the difficulty of T cells to detect antigens in particular areas.

Acknowledgements This work was supported by grants from the Spanish Ministry of Economy and Competitiveness, and the European Regional Development Fund (*Fondo Europeo de Desarrollo Regional*, FEDER) (Reference Grants: RYC-2010-06729, SAF2013-45178-P, and SAF2015-64123-P), Generalitat de Catalunya (Reference Grant: 2014 SGR-984) and by the *Asociación Española Contra el Cancer* (AECC).

References

Ariotti S, Beltman JB, Chodaczek G, Hoekstra ME, Van Beek AE, Gomez-Eerland R, Ritsma L, Van Rheenen J, Maree AF, Zal T, De Boer RJ, Haanen JB, Schumacher TN (2012) Tissue-resident memory CD8+ T cells continuously patrol skin epithelia To quickly recognize local antigen. Proc Natl Acad Sci U S A 109:19739–19744

Barcia C, Thomas CE, Curtin JF, King GD, Wawrowsky K, Candolfi M, Xiong WD, Liu C, Kroeger K, Boyer O, Kupiec-Weglinski J, Klatzmann D, Castro MG, Lowenstein PR (2006) In vivo mature immunological synapses forming SMACs mediate clearance of virally infected astrocytes from the brain. J Exp Med 203:2095–2107

Billadeau DD, Nolz JC, Gomez TS (2007) Regulation of T-cell activation by the cytoskeleton. Nat Rev Immunol 7:131–143

Blanchard N, Di Bartolo V, Hivroz C (2002) In the immune synapse, ZAP-70 controls T cell polarization and recruitment of signaling proteins but not formation of the synaptic pattern. Immunity 17:389–399

Cai E, Marchuk K, Beemiller P, Beppler C, Rubashkin MG, Weaver VM, Gerard A, Liu TL, Chen BC, Betzig E, Bartumeus F, Krummel MF (2017) Visualizing dynamic microvillar search and stabilization during ligand detection by T cells. Science 356:eaal3118

Cannon JL, Asperti-Boursin F, Letendre KA, Brown IK, Korzekwa KE, Blaine KM, Oruganti SR, Sperling AI, Moses ME (2013) PKCtheta regulates T cell motility via ezrin-radixin-moesin localization to the uropod. PLoS One 8:e78940

Carpier JM, Zucchetti AE, Bataille L, Dogniaux S, Shafaq-Zadah M, Bardin S, Lucchino M, Maurin M, Joannas LD, Magalhaes JG, Johannes L, Galli T, Goud B, Hivroz C (2018) Rab6-dependent retrograde traffic of LAT controls immune synapse formation and T cell activation. JExp Med 215:1245–1265

Chauveau A, Le Floc'h A, Bantilan NS, Koretzky GA, Huse M (2014) Diacylglycerol kinase alpha establishes T cell polarity by shaping diacylglycerol accumulation at the immunological synapse. Sci Signal 7:ra82

Crespo CL, Vernieri C, Keller PJ, Garre M, Bender JR, Wittbrodt J, Pardi R (2014) The PAR complex controls the spatiotemporal dynamics of F-actin and the MTOC in directionally migrating leukocytes. J Cell Sci 127:4381–4395

Diaz LR, Saavedra-Lopez E, Romarate L, Mitxitorena I, Casanova PV, Cribaro GP, Gallego JM, Perez-Valles A, Forteza-Vila J, Alfaro-Cervello C, Garcia-Verdugo JM, Barcia C Sr, Barcia C Jr (2018) Imbalance of immunological synapse-kinapse states reflects tumor escape to immunity in glioblastoma. JCI Insight 3:120757

Dustin ML (2005) A dynamic view of the immunological synapse. Semin Immunol 17:400–410

Dustin ML (2007) Cell adhesion molecules and actin cytoskeleton at immune synapses and kinapses. Curr Opin Cell Biol 19:529–533

Dustin ML (2008) T-cell activation through immunological synapses and kinapses. Immunol Rev 221:77–89

Dustin ML (2010) The immunological synapse. In: Bradshaw RA, Dennis EA (eds) Handbook of cell signaling. Academic, London

Dustin ML, Chakraborty AK, Shaw AS (2010) Understanding the structure and function of the immunological synapse. Cold Spring Harb Perspect Biol 2:a002311

Etienne-Manneville S (2004) Cdc42–the centre of polarity. J Cell Sci 117:1291–1300

Evans R, Patzak I, Svensson L, De Filippo K, Jones K, Mcdowall A, Hogg N (2009) Integrins in immunity. J Cell Sci 122:215–225

Franciszkiewicz K, Le Floc'h A, Boutet M, Vergnon I, Schmitt A, Mami-Chouaib F (2013) CD103 or LFA-1 engagement at the immune synapse between cytotoxic T cells and tumor cells promotes maturation and regulates T-cell effector functions. Cancer Res 73:617–628

Fritzsche M, Dustin ML (2018) Organization of immunological synapses and kinapses. In: Putterman C, Cowburn D, Almo S (eds) *Structural biology in immunology*. Academic, London

Grakoui A, Bromley SK, Sumen C, Davis MM, Shaw AS, Allen PM, Dustin ML (1999) The immunological synapse: a molecular machine controlling T cell activation. Science 285:221–227

Holdorf AD, Lee KH, Burack WR, Allen PM, Shaw AS (2002) Regulation of Lck activity by CD4 and CD28 in the immunological synapse. Nat Immunol 3:259–264

Huppa JB, Davis MM (2003) T-cell-antigen recognition and the immunological synapse. Nat Rev Immunol 3:973–983

Huse M, Lillemeier BF, Kuhns MS, Chen DS, Davis MM (2006) T cells use two directionally distinct pathways for cytokine secretion. Nat Immunol 7:247–255

Huse M, Le Floc'h A, Liu X (2013) From lipid second messengers to molecular motors: microtubule-organizing center reorientation in T cells. Immunol Rev 256:95–106

Johnson KG, Bromley SK, Dustin ML, Thomas ML (2000) A supramolecular basis for CD45 tyrosine phosphatase regulation in sustained T cell activation. Proc Natl Acad Sci U S A 97:10138–10143

Kaizuka Y, Douglass AD, Varma R, Dustin ML, Vale RD (2007) Mechanisms for segregating T cell receptor and adhesion molecules during immunological synapse formation in Jurkat T cells. Proc Natl Acad Sci U S A 104:20296–20301

Kloc M, Kubiak JZ, Li XC, Ghobrial RM (2014) The newly found functions of MTOC in immunological response. J Leukoc Biol 95:417–430

Kuhn JR, Poenie M (2002) Dynamic polarization of the microtubule cytoskeleton during CTL-mediated killing. Immunity 16:111–121

Lee KH, Holdorf AD, Dustin ML, Chan AC, Allen PM, Shaw AS (2002) T cell receptor signaling precedes immunological synapse formation. Science 295:1539–1542

Lively S, Schlichter LC (2013) The microglial activation state regulates migration and roles of matrix-dissolving enzymes for invasion. J Neuroinflammation 10:75

Lu YC, Robbins PF (2016) Cancer immunotherapy targeting neoantigens. Semin Immunol 28:22–27

Maccari I, Zhao R, Peglow M, Schwarz K, Hornak I, Pasche M, Quintana A, Hoth M, Qu B, Rieger H (2016) Cytoskeleton rotation relocates mitochondria to the immunological synapse and increases calcium signals. Cell Calcium 60:309–321

Mitxitorena I, Saavedra E, Barcia C (2015) Kupfer-type immunological synapses in vivo: raison D'etre of SMAC. Immunol Cell Biol 93:51–56

Monks CR, Freiberg BA, Kupfer H, Sciaky N, Kupfer A (1998) Three-dimensional segregation of supramolecular activation clusters in T cells. Nature 395:82–86

Moreau HD, Lemaitre F, Terriac E, Azar G, Piel M, Lennon-Dumenil AM, Bousso P (2012) Dynamic in situ cytometry uncovers T cell receptor signaling during immunological synapses and kinapses in vivo. Immunity 37:351–363

Moreau HD, Lemaitre F, Garrod KR, Garcia Z, Lennon-Dumenil AM, Bousso P (2015) Signal strength regulates antigen-mediated T-cell deceleration by distinct mechanisms to promote local exploration or arrest. Proc Natl Acad Sci U S A 112:12151–12156

Nath S, Christian L, Tan SY, Ki S, Ehrlich LI, Poenie M (2016) Dynein separately partners with NDE1 and dynactin To orchestrate T cell focused secretion. J Immunol 197:2090–2101

Negulescu PA, Krasieva TB, Khan A, Kerschbaum HH, Cahalan MD (1996) Polarity of T cell shape, motility, and sensitivity to antigen. Immunity 4:421–430

Quintana A, Pasche M, Junker C, Al-Ansary D, Rieger H, Kummerow C, Nunez L, Villalobos C, Meraner P, Becherer U, Rettig J, Niemeyer BA, Hoth M (2011) Calcium microdomains at the immunological synapse: how ORAI channels, mitochondria and calcium pumps generate local calcium signals for efficient T-cell activation. EMBO J 30:3895–3912

Ratner S, Sherrod WS, Lichlyter D (1997) Microtubule retraction into the uropod and its role in T cell polarization and motility. J Immunol 159:1063–1067

Stinchcombe JC, Bossi G, Booth S, Griffiths GM (2001) The immunological synapse of CTL contains a secretory domain and membrane bridges. Immunity 15:751–761

Stinchcombe JC, Majorovits E, Bossi G, Fuller S, Griffiths GM (2006) Centrosome polarization delivers secretory granules to the immunological synapse. Nature 443:462–465

Tsun A, Qureshi I, Stinchcombe JC, Jenkins MR, De La Roche M, Kleczkowska J, Zamoyska R, Griffiths GM (2011) Centrosome docking at the immunological synapse is controlled by Lck signaling. J Cell Biol 192:663–674

Vicente-Manzanares M, Sanchez-Madrid F (2004) Role of the cytoskeleton during leukocyte responses. Nat Rev Immunol 4:110–122

Vicente-Manzanares M, Ma X, Adelstein RS, Horwitz AR (2009) Non-muscle myosin II takes centre stage in cell adhesion and migration. Nat Rev Mol Cell Biol 10:778–790

Weninger W, Biro M, Jain R (2014) Leukocyte migration in the interstitial space of non-lymphoid organs. Nat Rev Immunol 14:232–246

Yarchoan M, Johnson BA 3rd, Lutz ER, Laheru DA, Jaffee EM (2017) Targeting neoantigens to augment antitumour immunity. Nat Rev Cancer 17:209–222

Yu Y, Smoligovets AA, Groves JT (2013) Modulation of T cell signaling by the actin cytoskeleton. J Cell Sci 126:1049–1058

Chapter 10
Semi-Intact Cell System for Reconstituting and Analyzing Cellular Golgi Dynamics

Fumi Kano and Masayuki Murata

Abstract Morphology of Golgi apparatus changes frequently and diversely depending on various cellular conditions and these changes correlate with the balance between membrane inflow and outflow at the Golgi via vesicular transports. In a previous study, we introduced a semi-intact cell system suitable for the reconstitution of morphological changes that organelles undergo as well as the vesicular transport between them. Semi-intact cells are cells that have undergone plasma membrane permeabilization by the cholesterol-dependent pore-forming cytolysin, streptolysin O (SLO). Permeabilization enables the introduction of various molecules into the cells, as well as the substitution of the original cytosol with an exogenously made cytosol prepared from cells in various stages of cell cycle, differentiation, and disease progression. Coupled with a green fluorescent protein (GFP)-visualization technique, this cell-based system enables the analysis of the molecular mechanisms underlying biological processes that are highly dependent on the integrity of the intracellular architecture. In this chapter, we present a variety of reconstitution assays concerning biological reactions pertaining to the Golgi apparatus.

F. Kano (✉)
Cell Biology Center, Institute of Innovative Research, Tokyo Institute of Technology, Yokohama, Kanagawa, Japan

Laboratory of Frontier Image Analysis, Graduate School of Arts and Science, The University of Tokyo, Tokyo, Japan
e-mail: kano.f.aa@m.titech.ac.jp

M. Murata (✉)
Cell Biology Center, Institute of Innovative Research, Tokyo Institute of Technology, Yokohama, Kanagawa, Japan

Department of Life Sciences, Graduate School of Arts and Sciences, The University of Tokyo, Tokyo, Japan

Laboratory of Frontier Image Analysis, Graduate School of Arts and Science, The University of Tokyo, Tokyo, Japan
e-mail: mmurata@bio.c.u-tokyo.ac.jp

© Springer Nature Switzerland AG 2019
M. Kloc (ed.), *The Golgi Apparatus and Centriole*, Results and Problems in Cell Differentiation 67, https://doi.org/10.1007/978-3-030-23173-6_10

10.1 Introduction

In mammalian cells, the Golgi apparatus forms a cisternal, ribbon-like structure. This structure can be found near the nucleus during interphase, but it is drastically changed during mitosis. The Golgi apparatus is fragmented and dispersed throughout the cytoplasm during mitosis that enables equal partitioning to two daughter cells (Shorter and Warren 2002; Colanzi et al. 2003; Altan-Bonnet et al. 2004). In addition to cell cycle-dependent morphological changes, the morphology of the Golgi apparatus changes frequently and diversely depending on various cellular conditions, including DNA damage (Farber-Katz et al. 2014), apoptosis (Chiu et al. 2002; Mancini et al. 2000), neurodegenerative diseases (Ayala and Colanzi 2017; Rendón et al. 2013), cancer (Petrosyan 2015), and cellular response to pathogens (Jesenberger et al. 2000; Reiling et al. 2013; Hansen et al. 2017). While the Golgi apparatus maintains its unique structure during interphase, it connects with the endoplasmic reticulum (ER) and various late-endosomal compartments through a dynamic membranous flow involving vesicular transport. This morphological homeostasis of the Golgi apparatus appears to be a characteristic state of the physiological equilibrium maintained by the balance between inflow and outflow of the membranes and components by vesicular transport. In other words, changes in Golgi morphology might be the result of many biological reactions occurring in a "concerted" fashion within the membrane-trafficking network around the Golgi apparatus. In addition, not only the structures of the organelles and the cytoskeleton, but also their proper relative spatial position is crucial for membrane trafficking and cellular organelle dynamics. Accordingly, in order to investigate the mechanisms of morphological changes and vesicular transport involved in the Golgi apparatus in mammalian cells, the optimal reconstitution assays must be identified, in which the concerted process may be dissected into several elementary reactions morphologically and biochemically, with the spatiotemporal information of the biological reactions also included. In this review, we describe the reconstitution assays relevant to cell cycle-dependent morphological changes observed in the Golgi and ER. In addition, we discuss the vesicular transport between the two organelles, Golgi disassembly induced by biological toxins, and the Golgi targeting of Rab6. Semi-intact cell systems coupled with green fluorescent protein (GFP)-visualization techniques could be a promising approach for addressing the above-mentioned obstacles. In this review, we focus on semi-intact cell systems including in-cell reconstitution systems for biochemical examination of factors involved in organelle morphology and function.

10.2 Semi-Intact Cell Systems

Semi-intact cells are cells with permeabilized plasma membranes and intact organelles and cytoskeletons. Cells are permeabilized using one of the cholesterol-dependent cytolysins, streptolysin O (SLO) (Fig. 10.1) (Alouf 1980). SLO binds

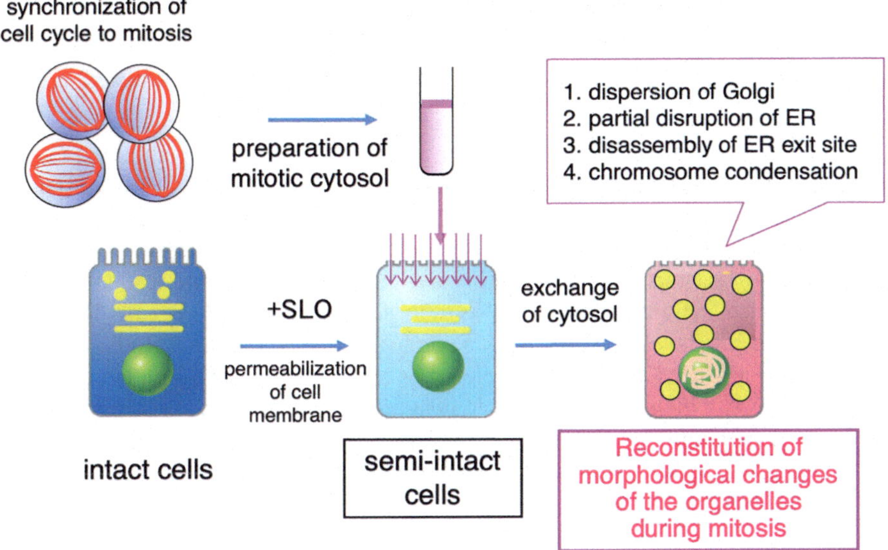

Fig. 10.1 Scheme of reconstitution of the morphological changes of organelles during mitosis using the semi-intact cell system

the plasma membrane in a cholesterol-dependent manner, and oligomerizes to form pores in the plasma membrane. The SLO pores are approximately 30 nm in diameter (Bhakdi et al. 1985; Sekiya et al. 1993), which allows both the leakage of cytosol from permeabilized cells as well as exogenous introduction of various molecules, including proteins, nucleotides, and membrane-impermeable small and mid-sized chemical compounds, into cells. Having structural integrity, SLO-mediated semi-intact cells can be used as cell-type "test tubes" for investigating the function of exogenously introduced molecules under intracellular conditions. Furthermore, the cytosol can be substituted with cytosol prepared from cells in various stages of cell cycle, differentiation, and disease progression (Fig. 10.1), which permits elucidation of the context-dependent behavior of exogenously introduced molecules. Owing to its advantages, the semi-intact cell system is suitable for the reconstitution and investigation of cell cycle-dependent membrane dynamics processes, which specifically require integrity of cytoskeletons and spatial relationship between organelles.

10.3   Reconstitution of Cell Cycle-Dependent Morphological Changes of the Golgi and the ER Network as well as Vesicular Transport Between the Two Organelles

The Golgi apparatus and ER maintain their unique structures during interphase while vesicular membrane transports allow continuous influx and efflux of substances between the two organelles. In mitosis, the Golgi and ER undergo drastic simultaneous changes. As these organelles are connected via vesicular transport, it is difficult to separate specific biological processes for examining the roles of specific molecules in cells. Furthermore, analysis of these processes is further complicated by asynchronous cell cycle progression. In semi-intact cells, the intracellular environment can be stabilized at fixed cell cycle status by incubating cells with interphase or mitotic cytosol, which allows morphological and biochemical dissection of complicated processes involved in organelle dynamics. Therefore, by using a semi-intact cell system, we reconstituted several cell cycle-dependent morphological changes of the Golgi apparatus and ER.

10.3.1   Reconstitution of Cell Cycle-Dependent Golgi Disassembly in Semi-Intact Cells

In order to reconstitute the mitotic Golgi disassembly process, we used a MDCK cell line, MDCK-GT, that constitutively expresses a GFP-tagged Golgi resident protein, namely, mouse galactosyltransferase (GT) fused with GFP (GT-GFP). Semi-intact MDCK-GT cells were prepared by first incubating cells with SLO and subsequently with *Xenopus* mitotic extracts (Kano et al. 2000b). Confocal microscopic observation revealed that the Golgi apparatus was first fragmented into large vesicles that were associated with microtubules under the apical plasma membrane. This process was followed by further dispersion of the Golgi membrane as small vesicles into the cytoplasm (Fig. 10.2). These processes were induced by *Xenopus* mitotic extracts, but not by interphase extracts. Based on confocal microscopic observation, we classified the disassembly process into three stages. Stage I (intact) was characterized by a perinuclear, intact Golgi that resides near the nucleus; stage II (punctate) was characterized by fragmented membranous structures associated with apical microtubules; and stage III (dispersed) was characterized by completely dispersed Golgi membranes (Fig. 10.2).

Next, we performed a Golgi disassembly assay in the presence of kinase-specific inhibitors or specific kinase immuno-depleted cytosol, and identified the kinase that was required for each of the two elementary processes. Mitogen-activated protein kinase kinase (MEK) was responsible for stage I to II, and cdc2 kinase was responsible for stage II to III (Fig. 10.2) (Kano et al. 2000b). Interestingly, MEK was identified to be required for mitotic Golgi disassembly prior to the action of cdc2, a master kinase in mitosis. Several reports support this progressive model of

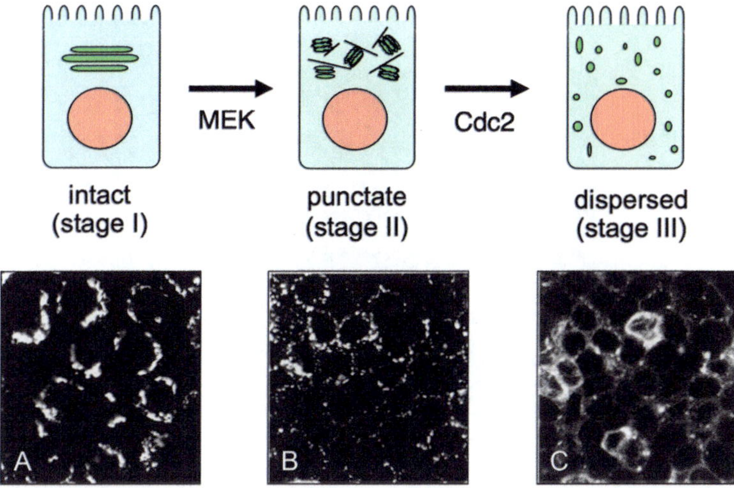

Fig. 10.2 Morphological and biochemical dissection of mitotic Golgi disassembly. The intact Golgi apparatus (stage I) was first fragmented into large vesicles that are associated with the microtubules underneath the apical membrane (stage II), in a MEK-dependent manner, and further dispersed throughout the cytosol (stage III) in a cdc2-dependent manner. Lower panel shows the confocal microscopy images of cells in each stage

mitotic Golgi disruption and the possible role of MEK in disconnecting the Golgi cisternae before disruption. For example, MEK1 or extracellular-activated protein kinase 2 (Erk2) phosphorylates Golgi reassembly stacking protein 55 (GRASP55), a matrix protein that links the Golgi cisternae (Short et al. 2005), thereby inhibiting the stacking of Golgi cisternae, which ultimately results in their complete disconnection (Feinstein and Linstedt 2007; Jesch et al. 2001).

10.3.2   *Reconstitution of Partial ER Disruption During Mitosis and Reformation of the ER Network During Interphase in Semi-Intact Cells*

We examined the cell cycle-dependent morphological changes of ER using a semi-intact cell system. The morphology of the ER, a polygonal network at the cytoplasm that connects to cisternae near the nucleus, was visualized in CHO-HSP cells, Chinese hamster ovary (CHO) cells constitutively expressing GFP-conjugated heat shock protein 47 (GFP-HSP47). HSP47 is an ER chaperone involved in the maturation and transport of collagen (Ishida and Nagata 2011). Fluorescent microscopy revealed that the network structure of ER was maintained during mitosis, which is consistent with findings from other studies (Puhka et al. 2012; Voeltz et al. 2002). In addition, we found partial disruption of the ER network in mitotic cells. Next, we reconstituted the partial disruption of the ER network in semi-intact CHO-HSP cells

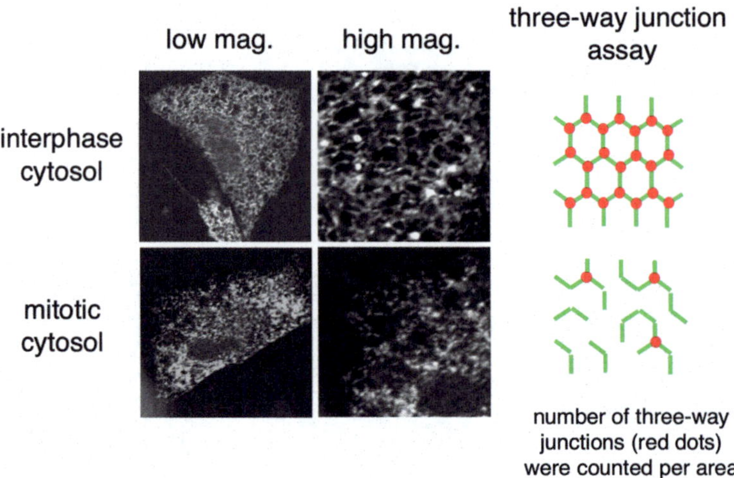

Fig. 10.3 Partial disruption of the ER network by mitotic cytosol and its morphometric assay. Semi-intact CHO-HSP cells were permeabilized with SLO and incubated with interphase or mitotic cytosol. Cells were observed using confocal microscopy. Morphological changes of the ER network were quantified by counting the number of three-way junctions (red dots) of ER network per area

in the presence of cytosol derived from mitotic HeLa cells. We preincubated CHO-HSP cells with nocodazole to eliminate the effect of microtubules on ER morphology. Subsequently, we permeabilized the CHO-HSP cells to generate semi-intact CHO-HSP cells and incubated them with cytosol that was prepared from mitotic HeLa cells. Interestingly, we observed a cdc2 kinase-dependent disruption of the polygonal ER network in the presence of mitotic cytosol (Fig. 10.3). Depolymerization of the microtubules by nocodazole was essential for the partial disruption of mitotic ER networks, suggesting that microtubule depolymerization was required for ER network disruption at the onset of mitosis.

Additional microscopic observation of ER dynamics revealed that mitotic cytosol-induced disruption of the ER network was not a consequence of repressed tubulation and/or bifurcation, but a consequence of inhibited ER tubule fusion. Therefore, we hypothesized that the fusion event could be inactivated in a cdc2 kinase-dependent manner. One of the cdc2 substrates with a role in membrane fusion is p47, a cofactor of p97, also known as valosin-containing protein (VCP). The p97/p47 complex mediates the fusion events involved in Golgi reassembly and ER network formation (Kondo et al. 1997; Uchiyama et al. 2002), and phosphorylation of p47 by cdc2 at Ser140 was reported to be essential for Golgi disassembly during mitosis (Uchiyama et al. 2003). Therefore, we examined whether a recombinant protein of p47NP, a mutant p47 with a Ser to Ala substitution at position 140, would inhibit partial ER disruption in an ER disruption assay and found that the addition of p47NP in combination with mitotic cytosol resulted in ER network maintenance, suggesting a crucial role for cdc2-mediated p47 phosphorylation in partial disruption of the ER network during mitosis (Kano et al. 2005b). Interestingly, we found that

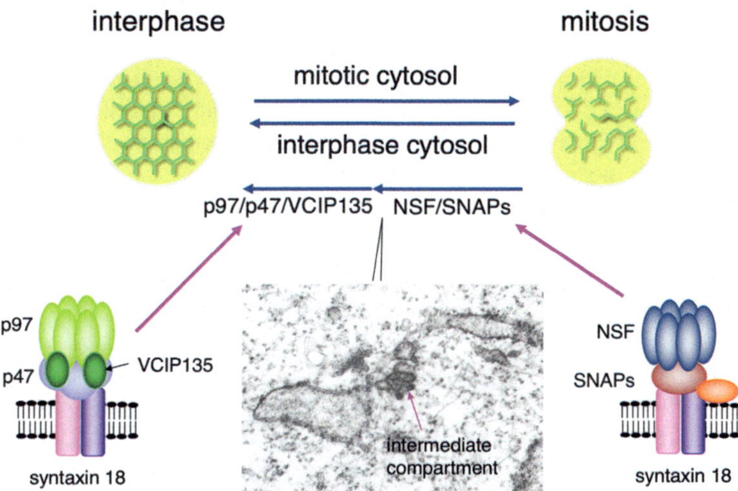

Fig. 10.4 Model for the cell cycle-dependent morphological changes in the ER network. The ER network in semi-intact cells was partially disrupted by mitotic cytosol. Subsequent incubation with interphase cytosol induced reformation of the ER network structure. The reformation was dissected into two processes: (1) Connection of disrupted ER tubules into intermediate compartments by the NSF and SNAP complex; and (2) A p97/p47/VCIP135 induced complete fusion of the ER tubules to create the network structure of ER

washing out the mitotic cytosol and further incubating the cells with interphase cytosol resulted in the reformation of the polygonal ER network that was disrupted by the mitotic cytosol (Kano et al. 2005a). We hypothesized that the fusion between ER tubules would be reactivated in the presence of interphase cytosol; however, the p97/p47 complex-mediated fusion event did not sufficiently induce full reformation of the ER network.

In order to investigate the biochemical requirements for the reformation step, we focused on the role of the N-ethylmaleimide-sensitive factor (NSF) and p97 complexes, which are well-known cellular fusion machineries (Malhotra et al. 1988; Rabouille et al. 1995; Meyer 2005). To this end, we prepared N-ethylmaleimide (NEM)-treated interphase cytosol [NEM(I)] because NEM inactivates ATPases including NSF and p97. Next, we screened for cytosolic proteins, which restored the three-way junction of the ER network in the presence of NEM(I). Using the reformation assay and subsequent morphometric analyses based on three-way junction assays (Fig. 10.3, right), we found that this process only required NSF and its cofactor α- and γ-synaptosomal nerve-associated protein (SNAP) (referred to as NSFs) and p97/p47 (referred to as p97s) + p97/p47 complex-interacting protein p135 (VCIP135) as exogenous factors (Fig. 10.4). It was unexpected that factors required and sufficient for polygonal ER network formation were only NSFs and p97s + VCIP135, because it had previously been reported that, in addition to the vesicle fusion process, the tubulation/bifurcation process was required for ER network formation (Dreier and Rapoport 2000). Furthermore, we showed that the two fusion processes mediated by NSFs and p97s + VCIP135 functioned sequentially during ER network reformation. The first membrane fusion process was mediated by NSFs to create the membranous intermediates that connect the disrupted ER tubules, which were visualized using electron microscopy (Fig. 10.4). The subsequent process responsible for complete fusion of the ER tubules and formation of the ER network was shown to be mediated by p97s + VCIP135 (Fig. 10.4). In addition, using the reformation assay, we demonstrated that syntaxin 18 is a common t-soluble NSF attachment protein receptor (SNARE) involved in both fusion reactions that occur during ER network reformation.

10.3.3 Reconstitution of Disassembly of ER Exit Sites During Mitosis and Anterograde and Retrograde Vesicular Transports Between the ER and Golgi in Semi-Intact Cells

ER exit sites (ERES) are domain structures generated on the ER network and known to be the budding sites of transport vesicles destined to the Golgi apparatus (Jensen and Schekman 2011). We found that GFP-fused Yip1A (also known as Yip1 domain family member 5 [YIPf5]), a mammalian homolog of yeast YPT-interacting protein (Yip) 1, accumulated at the ERES during interphase, and diffused throughout the ER network at the onset of mitosis in CHO cells (Kano et al. 2004). Thus, we visualized ERES in CHO-YIP cells constitutively expressing Yip1A-GFP as an ERES marker, and reconstituted the mitotic disassembly of the ERES using semi-intact CHO-YIP cells (Kano et al. 2004) (Fig. 10.5a). We found that addition of mitotic cytosol into semi-intact cells induced ERES disassembly, which was inhibited by the nonphosphorylatable form of p47, p47NP. This indicated that cdc2-dependent phosphorylation of p47 might be crucial for ERES disassembly. In all, at the onset of mitosis, cdc2-dependent phosphorylation of p47 is likely to be, at least partly, a key regulator of Golgi fragmentation, partial ER network disruption, and ERES disassembly.

As the Golgi apparatus and ER are connected through anterograde- and retrograde-vesicular transport during interphase, their perturbation under patholog-ical conditions or during mitosis would be expected to not only affect the quantity and quality of the protein/lipid components that flow in and out of the organelles, but also the overall organelle morphology (Miles et al. 2001; Zaal et al. 1999). Thus, quantitative analysis of vesicular transport between the Golgi and ER under inter-phase or mitotic conditions would elucidate the regulatory mechanisms underlying mitosis-induced morphological changes in organelles. In order to further analyze this issue, we first established CHO-GT cells; although most of the GT-GFP appears to localize to the Golgi during interphase, GT-GFP is recycled between the ER and Golgi through vesicular transport. Next, we reconstituted anterograde- or retrograde-vesicular transport between the Golgi and ER of semi-intact CHO-GT cells, in the

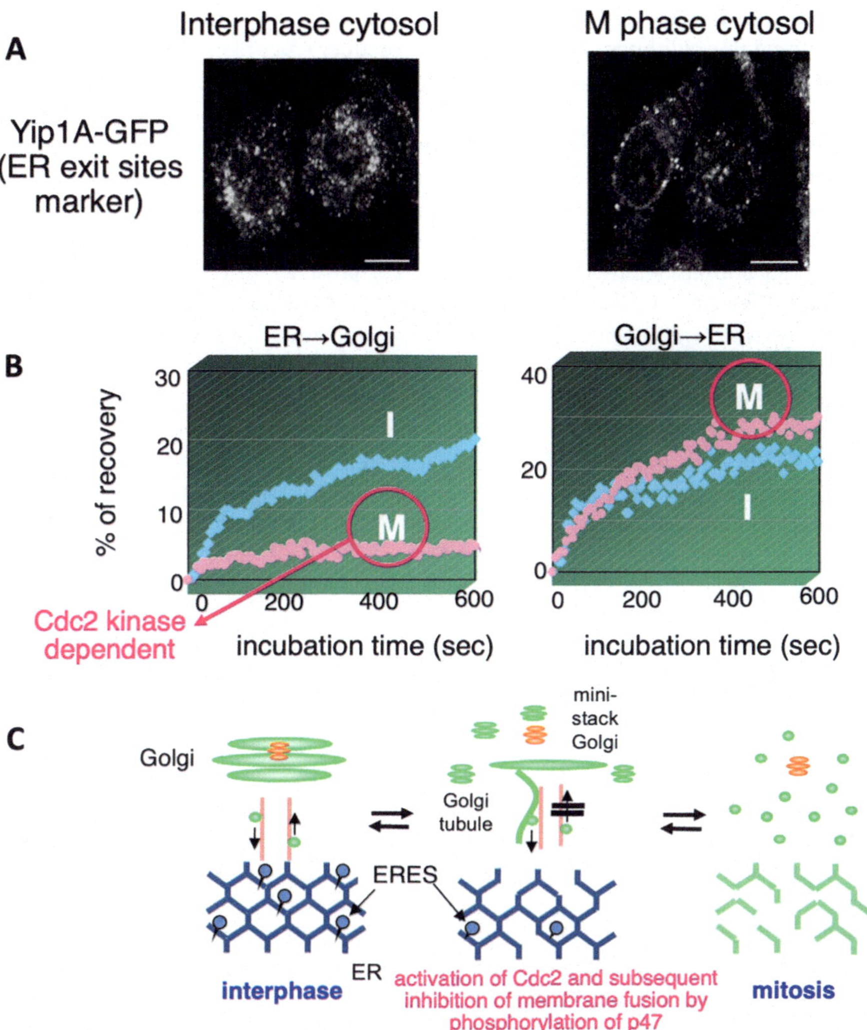

Fig. 10.5 Disassembly of ER exit sites by mitotic cytosol, and ER-to-Golgi and Golgi-to-ER transport in semi-intact cells with interphase or mitotic cytosols. (**a**) Mitotic cytosol induced the disassembly of ER exit site marker Yip1A-GFP throughout the ER network from the dotted domain structures, but interphase cytosol did not. (**b**) Using GT-GFP, which recycles between the ER and Golgi as a marker, the extent of anterograde and retrograde transport between the ER and Golgi was quantified using the fluorescence recovery after photobleaching (FRAP) method. In the presence of mitotic cytosol (M) in semi-intact cells, ER-to-Golgi anterograde transport was significantly inhibited in a cdc2-dependent manner. (**c**) The model for the coupling of cell cycle-dependent morphological changes of the ER and Golgi, with vesicular transport in mammalian cells

presence of interphase or mitotic cytosol (Murata and Kano 2012). We conducted quantitative analysis of the vesicular transport of GT-GFP in semi-intact cells using the fluorescence recovery after photobleaching (FRAP) method. In brief, we bleached GT-GFP fluorescence at a specific region of the Golgi in semi-intact cells by repetitive laser illumination under the microscope. After bleaching, we quantified the extent of fluorescence recovery resulting from anterograde transport of GT-GFP from the ER to the Golgi. In order to measure the extent of retrograde GT-GFP transport from the Golgi to the ER, we bleached the fluorescence in the ER network and quantified the extent of fluorescence recovery at the bleached region. Analysis of fluorescence recovery kinetics at the Golgi and ER showed inhibition of anterograde transport but not of retrograde transport by mitotic cytosol (Fig. 10.5b). In addition, we showed normal anterograde transport in the presence of cdc2-depleted mitotic cytosol, suggesting a cdc2-dependent repression of anterograde transport in the mitotic cytosol. These findings are consistent with the above finding that ERES, the budding region of anterograde transport vesicles from the ER, was disrupted in the presence of mitotic cytosol in a cdc2 kinase-dependent manner. Interestingly, the retrograde transport assay further revealed that the mitotic cytosol likely enhanced the retrograde transport of GT-GFP compared to the interphase cytosol, suggesting that the translocation of certain Golgi components to the ER might be facilitated at the onset of mitosis.

10.3.4 Schematic Model of Mitosis-Induced Morphological Changes in the Golgi and ER Coupled with Vesicular Transport

Golgi architecture is coupled with vesicular transport between the ER and Golgi. For example, inhibition of anterograde transport from the ER through microinjection or overexpression of the dominant negative type of Sar1, GTP-restricted Sar1, causes loss of the Golgi structure near the nucleus and subsequently results in cytoplasmic dispersion (Miles et al. 2001). Morphological changes in the Golgi and vesicular transport process share common protein factors owing to similarities in the elementary steps of these processes, including membrane curvature, elongation, and fusion. Therefore, reconstitution of these phenomena would be helpful for understanding the function of each factor unique to each phenomenon. Based on our findings and those of others (Bisel et al. 2008; Colanzi et al. 2003; Sütterlin and Colanzi 2010), we proposed a model for the coupling of cell cycle-dependent morphological changes in the ER and Golgi, with vesicular transport in mammalian cells (Fig. 10.5c). In this model, just before or at the very beginning of mitosis, MEK activation is required for the separation of Golgi cisternae into large vesicles, which might correspond to the splitting and bipolar movement of centrosomes to the opposite sides of the nucleus to create the mitotic spindle (Rabouille and Kondylis 2007). Indeed, Golgi fragmentation is often observed near split centrosomes. During the earlier stages of mitosis, cdc2 kinase is activated and phosphorylates a variety of proteins in the cells, including p47. The phosphorylation of p47 induces Golgi disassembly and simultaneous partial disruption of the ER network, as well as ERES disassembly. ERES disassembly inhibits ER-to-Golgi anterograde transport, but does not affect Golgi-to-ER retrograde transport. In consequence, the efflux of some Golgi components surpasses the influx of other cytosolic components, thus enhancing Golgi disassembly during the early phase of mitosis. Disruption of the ER network and vesiculation of Golgi membranes enables the dispersion of the components of these organelles throughout the cell, resulting in their equal distribution into two daughter cells.

10.4 Dissection of BFA-Induced Disassembly of the Golgi Using Semi-Intact Cells

Another concerted morphological change in the Golgi is the brefeldin A (BFA)-induced absorption of the Golgi into the ER. The fungal metabolite BFA inhibits several members of ADP ribosylation factors (Arfs) and guanine nucleotide exchange factors (Casanova 2007), and inhibits membrane export from the ER in vivo (Lippincott-Schwartz et al. 1989). BFA-induced dynamic morphological changes of the Golgi have been extensively investigated in living cells using GT-GFP as a Golgi marker (Sciaky et al. 1997). Light microscopic observation of BFA-treated cells revealed the formation of many Golgi tubules within ~10 min, which persisted for 5–10 min before rapid fusion with the ER. BFA-induced Golgi disassembly is characterized by the formation of prominent, long tubules from the Golgi apparatus. These BFA-induced Golgi tubules are proposed to be transient membrane structures that merely accentuate the normally observed constitutive Golgi-to-ER retrograde transport. Several studies (Lippincott-Schwartz et al. 1990; Lippincott-Schwartz et al. 1991; Wood et al. 1991) have shown that Golgi disassembly can be dissected into two elementary reactions: Golgi tubule formation followed by fusion of the Golgi tubules with the ER membrane (Fig. 10.6a). This fusion leads to the quick relocation of mainly *cis*- or medial-Golgi components to the ER. BFA-dependent Golgi disassembly involves calmodulin (Figueiredo and Brown 1995), cytoplasmic phospholipase A_2 (Figueiredo et al. 1998), lysophosphatidic acid-specific acyltransferase (Schmidt and Brown 2009), kinesin as the microtubule-associated motor protein (Klausner et al. 1992), ATP as energy source, and the fusogenic protein NSF (Sciaky et al. 1997; Fukunaga et al. 1998). However, in BFA-treated cells, the formation of Golgi tubules and the fusion between Golgi tubules and the ER membrane occur in a concerted fashion at different time points, which makes the morphological dissection of these processes and the investigation of the biochemical requirements of each elementary reaction challenging.

In order to independently investigate the factors that are required for Golgi tubulation or fusion, we reconstituted BFA-induced Golgi disassembly in semi-intact CHO-GT cells. At first, we reconstituted the Golgi disassembly in the presence of cytosol derived from L5178Y mouse lymphoma cells and ATP-regenerating systems and then dissected the whole process biochemically and morphologically into two independent steps in semi-intact CHO-GT cells (Kano et al. 2000a): Golgi

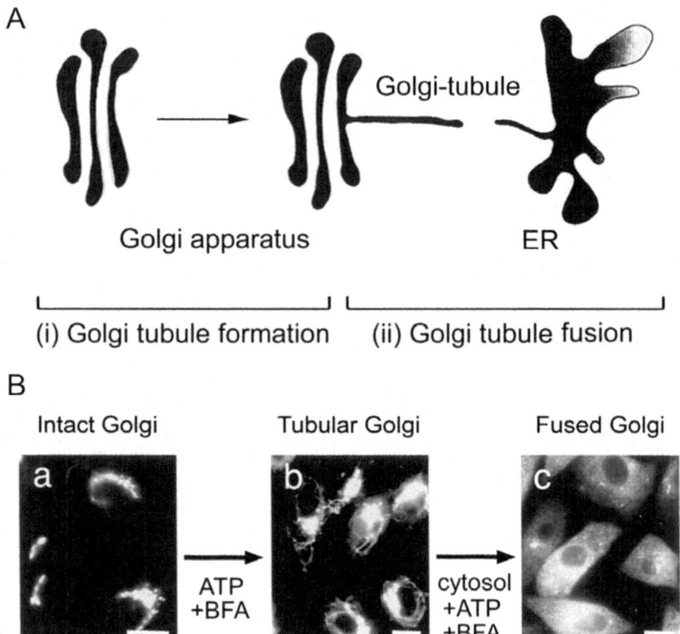

Fig. 10.6 Reconstitution of BFA-induced Golgi disassembly in semi-intact cells. (**a**) BFA induced the formation of long tubules from the Golgi apparatus and the subsequent absorption of the Golgi apparatus to the ER. These elementary processes were independently reconstituted in semi-intact cells using the Golgi tubule formation assay and Golgi tubule fusion assay. (**b**) Formation of the Golgi tubules was induced in the presence of ATP and BFA, and the fusion and absorption of the Golgi tubules to the ER was induced in the presence of cytosol of L5178Y cells, ATP, and BFA, indicating that the latter process required cytosolic factors

tubule formation and subsequent fusion of Golgi tubules with the ER membrane. For this analysis, we classified Golgi morphology as normal perinuclear ribbon-like "intact Golgi," >5 μm long "tubular Golgi," and "fused Golgi" with the Golgi components absorbed to the ER (Fig. 10.6b). In consequence, the morphological changes to the Golgi upon BFA treatment were expressed as the percentage of cells having intact, tubular, or fused Golgi. Based on this morphometric analysis, we developed assays for Golgi disassembly and for the two dissected elementary processes of Golgi tubule formation and Golgi tubule fusion to the ER.

NEM treated cytosol, referred to as NEM-cytosol, induced normal Golgi tubule formation, but did not induce Golgi fusion, owing to the potential inactivation of NSF, a protein that mediates intracellular membrane fusion. Golgi tubule fusion assay using NEM-cytosol containing various recombinant proteins (e.g., active or inactive form of NSF) confirmed the essential role of NSF in the fusion process. In addition, Golgi tubule formation assay showed facilitated motility and bifurcation of Golgi tubules in the presence of NEM-cytosol, without any effect on tubule number and length.

As cytoskeletons and organelles remain almost intact, the semi-intact cell system is suitable for the investigation of the effect of microtubules on each elementary process during BFA-induced Golgi disassembly. First, Golgi disassembly assays were performed using nocodazole-treated semi-intact CHO-GT cells with partially disrupted microtubules. Interestingly, BFA-induced Golgi disassembly occurred even in microtubule-disrupted semi-intact cells; however, the rate of Golgi tubule formation was slower compared to that of nocodazole-untreated cells. In contrast, Golgi tubule fusion occurred at a similar rate in nocodazole-treated and -untreated cells. These findings suggested that microtubules affect Golgi tubule formation, but is not required for fusion to the ER membrane in BFA-induced Golgi disassembly. Therefore, we were able to easily investigate the role of specific intracellular proteins, cytoskeleton, or the interaction of both in the biological reactions reconstituted in semi-intact cells.

10.5   Reconstitution of Golgi Targeting of Rab6 Using Semi-Intact Cells

The remaining integrity and spatial organization of the organelles and cytoskeleton in semi-intact cells make them suitable systems for reconstituting the targeting of proteins to specific intracellular sites. Proteins localize to specific organelles where they have specific functions, and the underlying mechanisms of protein localization are diverse. For example, while proteins that localize to the nucleus, mitochondria, or ER possess a "signal sequence," which is an amino acid tag required for localization, other proteins do not have this tag sequence. The monomeric GTPase Rab family proteins, an example of the latter, localize to various organelles and are involved in vesicular transport and communication among organelles. The large Rab family comprises 60 different proteins found in humans, and they all share a common GTP binding site and lipid modification site; however, different intracellular localization sites have been described for each Rab protein (Stenmark 2009; Wennerberg et al. 2005). For example, Rab1 localizes to the ER and Golgi apparatus, whereas Rab2 localizes to the *cis*-Golgi apparatus, and Rab5 localizes to the endosome, each involved in the regulation of vesicular transport processes in their respective organelles. No amino acid tag sequence has been reported for the localization of these proteins. In order to address this issue, we reconstituted organelle targeting using the semi-intact cell system (Matsuto et al. 2015) (Fig. 10.7). For this, we selected Rab6A, which localizes to the Golgi apparatus of animal cells (Beranger et al. 1994; Martinez et al. 1997).

First, we performed bacterial transformation and purification of Rab6 protein labeled with glutathione S-transferase (GST-Rab6). When purified GST-Rab6 protein was added to semi-intact CHO cells, the protein did not localize to the Golgi apparatus. However, when GST-Rab6 was added to these cells in the presence of cytosol of mouse lymphoma-derived L5178Y cells containing an ATP-regenerating

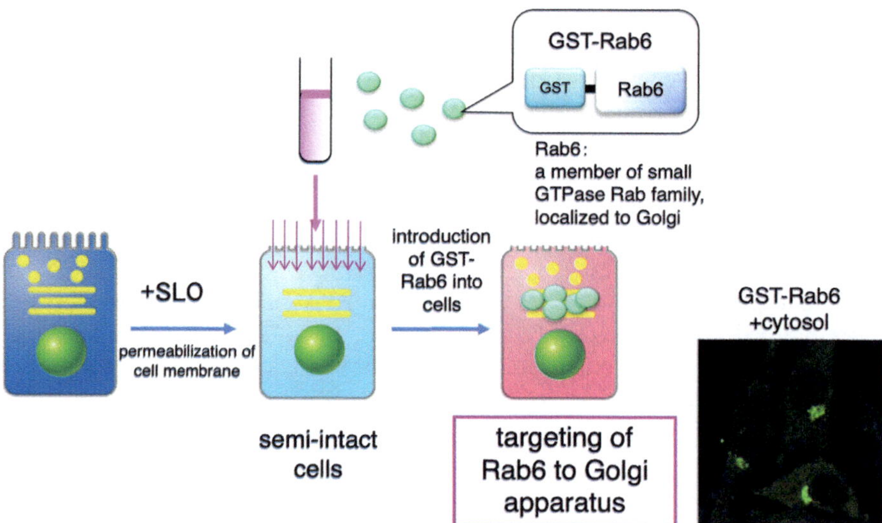

Fig. 10.7 Reconstitution of the Golgi targeting of Rab6 in semi-intact cells. Cells were permeabilized with SLO, and the GST-tagged recombinant Rab6 (GST-Rab6) was introduced into the semi-intact cells. Intracellular localization of GST-Rab6 was examined by performing immunofluorescent analysis using anti-GST antibody. Accumulation of GST-Rab6 was observed only in the presence of cytosol (GST-Rab6A + cytosol)

system, immunofluorescence analysis using the anti-GST antibody demonstrated GST-Rab6 localization to the Golgi apparatus (Fig. 10.7, GST-Rab6 + cytosol). This finding suggested the existence of a factor in the L5178Y cytosol that is essential for GST-Rab6 localization to the Golgi apparatus. Furthermore, a protein group that binds to GST-Rab6 was recovered from the L5178Y cytosol via co-immunoprecipitation and identified as bicaudal D homolog 2 (BICD2) via mass spectrometry. BICD2 is one of the adapter proteins connecting dynein, a microtubule motor protein, to transport vesicles (Matanis et al. 2002; Hoogenraad et al. 2001). Moreover, GST-Rab6 localization to the Golgi apparatus was inhibited in the BICD2-depleted cytosol by immune depletion and in the anti-BICD2 antibody-containing cytosol. Therefore, the involvement of BICD2 in GST-Rab6 localization to the Golgi was confirmed. Interestingly, microtubule integrity remained uninfluenced by BICD2-dependent GST-Rab6 localization to the Golgi, indicating that BICD2 might be involved in the localization to Golgi via a mechanism that differed from that of the adapter of the microtubule motor protein.

The screening method applied for the identification of a factor involved in Rab6 protein localization to the Golgi apparatus described here could be easily applied for screening a regulating factor of subcellular localization for other Rab family proteins. Moreover, this method has potential applications in the research on organelle localization for other proteins (Kano et al. 2011). Our findings showed that exploiting the features of the semi-intact cell assay system, otherwise known as the "cell-type test tube," is a good approach for investigating the mechanisms for a broad range of protein localization and targeting.

10.6 Conclusion

As mentioned above, the semi-intact cell system is a unique and powerful cellular platform for the investigation of molecular mechanisms underlying the morphological and functional regulation of organelles. Recently, we have developed a disease model of cells using the cell-resealing technique, which is a modified version of the semi-intact cell system. After introduction of pathological cytosol obtained from the tissues of disease model animals, into semi-intact cells, the repair of the injured plasma membrane via endocytosis, exocytosis, or bleb formation, is triggered by the addition of $CaCl_2$ (Andrews et al. 2014; Cooper and McNeil 2015). For example, we established the diabetic hepatocyte model by adding the liver cytosol of db/db diabetic model mouse into semi-intact cells (Kano et al. 2012, 2017). Interestingly, diabetic model cells mimicked the state of "insulin resistance" at the cellular level, with decreased transcriptional repression of gluconeogenic genes and aberrant glucose production (Kano et al. 2017). It would be interesting to observe the functional and morphological changes of the Golgi apparatus along with other organelles in such disease model cells, since it would highlight the undiscovered involvement of organelles in diseases and might shed light on the organelle as a novel disease marker. In fact, we found that phosphatidylinositol-3-phosphate, a lipid enriched in early endosomes, is depleted in diabetic model cells, which affects endocytosis and signal transduction (Kano et al. 2012). In order to compare normal and disease model cells, we believe that the quantification of organelle morphology and function is required. This comparison is becoming easier owing to continuous improvements in image recognition technologies (Wortzel et al. 2017; Sugawara et al. 2012). In combination with novel technologies, the semi-intact cell system should serve as a powerful tool for investigating the various aspects of organelle morphology and function.

References

Alouf JE (1980) Streptococcal toxins (streptolysin O, streptolysin S, eryth-rogenic toxins). Pharmacol Ther 11(3):661–717

Altan-Bonnet N, Sougrat R, Lippincott-Schwartz J (2004) Molecular basis for Golgi maintenance and biogenesis. Curr Opin Cell Biol 16(4):364–372

Andrews NW, Almeida PE, Corrotte M (2014) Damage control: cellular mechanisms of plasma membrane repair. Trends Cell Biol 24:734–742

Ayala I, Colanzi A (2017) Alterations of Golgi organization in Alzheimer's disease: a cause or a consequence? Tissue Cell 49(2. Pt A):133–140

Beranger F, Paterson H, Powers S, de Gunzburg J, Hancock JF (1994) The effector domain of Rab6, plus a highly hydrophobic C terminus, is required for Golgi apparatus localization. Mol Biol Cell 14:744–758

Bhakdi S, Tranum-Jensen J, Sziegoleit A (1985) Mechanism of membrane damage by streptolysin-O. Infect Immun 47(1):52–60

Bisel B, Wang Y, Wei JH, Xiang Y, Tang D, Miron-Mendoza M, Yoshimura S, Nakamura N, Seemann J (2008) ERK regulates Golgi and centrosome orientation towards the leading edge through GRASP65. J Cell Biol 182(5):837–843

Casanova JE (2007) Regulation of Arf activation: the Sec7 family of guanine nucleotide exchange factors. Traffic 8(11):1476–1485

Chiu R, Novikov L, Mukherjee S, Shields D (2002) A caspase cleavage fragment of p115 induces fragmentation of the Golgi apparatus and apoptosis. J Cell Biol 159(4):637–648

Colanzi A, Suetterlin C, Malhotra V (2003) Cell-cycle-specific Golgi fragmentation: how and why? Curr Opin Cell Biol 15(4):462–467

Cooper ST, McNeil PL (2015) Membrane repair: mechanisms and pathophysiology. Physiol Rev 95:1205–1240

Dreier L, Rapoport TA (2000) In vitro formation of the endoplasmic reticulum occurs independently of microtubules by a controlled fusion reaction. J Cell Biol 148(5):883–898

Farber-Katz SE, Dippold HC, Buschman MD, Peterman MC, Xing M, Noakes CJ, Tat J, Ng MM, Rahajeng J, Cowan DM, Fuchs GJ, Zhou H, Field SJ (2014) DNA damage triggers Golgi dispersal via DNA-PK and GOLPH3. Cell 156(3):413–427

Feinstein TN, Linstedt AD (2007) Mitogen-activated protein kinase kinase 1-dependent Golgi unlinking occurs in G2 phase and promotes the G2/M cell cycle transition. Mol Biol Cell 18 (2):594–604

Figueiredo PD, Brown WJ (1995) A role for calmodulin in organelle membrane tubulation. Mol Biol Cell 6:871–887

Figueiredo PD, Drecktrah D, Katzenellenbogen JA, Strang M, Brown WJ (1998) Evidence that phospholipase A2 activity is required for the Golgi complex and *trans* Golgi network membrane tubulation. Proc Natl Acad Sci U S A 95:8642–8647

Fukunaga T, Furuno A, Hatsuzawa K, Tani K, Ymamoto A, Tagaya M (1998) NSF is required for the brefeldin A-promoted disassembly of the Golgi apparatus. FEBS Lett 18:237–240

Hansen MD, Johnsen IB, Stiberg KA, Sherstova T, Wakita T, Richard GM, Kandasamy RK, Meurs EF, Anthonsen MW (2017) Hepatitis C virus triggers Golgi fragmentation and autophagy through the immunity-related GTPase M. Proc Natl Acad Sci U S A 114(17):E3462–E3471

Hoogenraad CC, Akhmanova A, Howell SA, Dortland BR, De Zeeuw CI, Willemsen R, Visser P, Grosveld F, Galjart N (2001) Mammalian Golgi-associated Bicaudal-D2 functions in the dynein-dynactin pathway by interacting with these complexes. EMBO J 20(15):4041–4054

Ishida Y, Nagata K (2011) Hsp47 as a collagen-specific molecular chaperone. Methods Enzymol 499:167–182

Jensen D, Schekman R (2011) COPII-mediated vesicle formation at a glance. J Cell Sci 124 (Pt 1):1–4

Jesch SA, Lewis TS, Ahn NG, Linstedt AD (2001) Mitotic phosphorylation of Golgi reassembly stacking protein 55 by mitogen-activated protein kinase ERK2. Mol Biol Cell 12(6):1811–1817

Jesenberger V, Procyk KJ, Yuan J, Reipert S, Baccarini M (2000) Salmonella-induced caspase-2 activation in macrophages: a novel mechanism in pathogen-mediated apoptosis. J Exp Med 192 (7):1035–1046

Kano F, Sako Y, Tagaya M, Yanagida T, Murata M (2000a) Reconstitution of brefeldin A-induced Golgi-tubulation and fusion with the ER in semi-intact CHO cells. Mol Biol Cell 11:3073–3087

Kano F, Takenaka K, Yamamoto A, Nagayama K, Nishida E, Murata M (2000b) MEK and Cdc2 kinase are sequentially required for Golgi disassembly in MDCK cells by the mitotic Xenopus extracts. J Cell Biol 149(2):357–368

Kano F, Tanaka AR, Yamauchi S, Kondo H, Murata M (2004) Cdc2 kinase-dependent disassembly of endoplasmic reticulum (ER) exit sites inhibits ER-to-Golgi vesicular transport during mitosis. Mol Biol Cell 15(9):4289–4298

Kano F, Kondo H, Yamamoto A, Kaneko Y, Uchiyama K, Hosokawa N, Nagata K, Murata M (2005a) NSF/SNAPs and p97/p47/VCIP135 are sequentially required for cell cycle-dependent reformation of the ER network. Genes Cells 10(10):989–999

Kano F, Kondo H, Yamamoto A, Tanaka AR, Hosokawa N, Nagata K, Murata M (2005b) The maintenance of the endoplasmic reticulum network is regulated by p47, a cofactor of p97, through phosphorylation by cdc2 kinase. Genes Cells 10(4):333–344

Kano F, Arai T, Matsuto M, Hayashi H, Sato M, Murata M (2011) Hydrogen peroxide depletes phosphatidylinositol-3-phosphate from endosomes in a p38 MAPK-dependent manner and perturbs endocytosis. Biochim Biophys Acta 1813(5):784–801

Kano F, Nakatsu D, Noguchi Y, Yamamoto A, Murata M (2012) A resealed-cell system for analyzing pathogenic intracellular events: perturbation of endocytic pathways under diabetic conditions. PLoS One 7(8):e44127

Kano F, Noguchi Y, Murata M (2017) Establishment and phenotyping of disease model cells created by cell-resealing technique. Sci Rep 7:15167

Klausner RD, Donaldson JG, Lippincott-Schwartz J (1992) Brefeldin A: insights into the control of membrane traffic and organelle structure. J Cell Biol 116:1071–1080

Kondo H, Rabouille C, Newman R, Levine TP, Pappin D, Freemont P, Warren G (1997) p47 is a cofactor for p97-mediated membrane fusion. Nature 388(6637):75–78

Lippincott-Schwartz J, Yuan LC, Bonifacino JS, Klausner RD (1989) Rapid redistribution of Golgi proteins into the ER in cells treated with brefeldin A: evidence for membrane cycling from Golgi to ER. Cell 56:801–813

Lippincott-Schwartz J, Donaldson JG, Schweizer A, Berger EG, Hauri HP, Yuan LC, Klausner RD (1990) Microtubule-dependent retrograde transport of proteins into the ER in the presence of brefeldin A suggests an ER recycling pathway. Cell 60(5):821–836

Lippincott-Schwartz J, Yuan L, Tipper C, Amherdt M, Orci L, Klausner RD (1991) Brefeldin A's effects on endosomes, lysosomes, and the TGN suggest a general mechanism for regulating organelle structure and membrane traffic. Cell 67(3):601–616

Malhotra V, Orci L, Glick BS, Block MR, Rothman JE (1988) Role of an N-ethylmaleimide-sensitive transport component in promoting fusion of transport vesicles with cisternae of the Golgi stack. Cell 54(2):221–227

Mancini M, Machamer CE, Roy S, Nicholson DW, Thornberry NA, Casciola-Rosen LA, Rosen A (2000) Caspase-2 is localized at the Golgi complex and cleaves golgin-160 during apoptosis. J Cell Biol 149(3):603–612

Martinez O, Antony C, Pehau-Arnaudet G, Berger EG, Salamero J, Goud B (1997) GTP-bound forms of rab6 induce the redistribution of Golgi proteins into the endoplasmic reticulum. Proc Natl Acad Sci U S A 94:1828–1833

Matanis T, Akhmanova A, Wulf P, Del Nery E, Weide T, Stepanova T, Galjart N, Grosveld F, Goud B, De Zeeuw CI, Barnekow A, Hoogenraad CC (2002) Bicaudal-D regulates COPI-independent Golgi-ER transport by recruiting the dynein-dynactin motor complex. Nat Cell Biol 4(12):986–992

Matsuto M, Kano F, Murata M (2015) Reconstitution of the targeting of Rab6A to the Golgi apparatus in semi-intact HeLa cells: a role of BICD2 in stabilizing Rab6A on Golgi membranes and a concerted role of Rab6A/BICD2 interactions in Golgi-to-ER retrograde transport. Biochim Biophys Acta 1853(10):2592–2609

Meyer HH (2005) Golgi reassembly after mitosis: the AAA family meets the ubiquitin family. Biochim Biophys Acta 1744(3):481–492

Miles S, McManus H, Forsten KE, Storrie B (2001) Evidence that the entire Golgi apparatus cycles in interphase HeLa cells: sensitivity of Golgi matrix proteins to an ER exit block. J Cell Biol 155 (4):543–555

Murata M, Kano F (2012) Semi-intact cell system: application to the nalysis of membrane trafficking beween the endoplasmic reticulum and the Golgi apparatus and of cell cycle-dependent changes in the morphology of these organelles. In: Weigert R (ed) Crosstalk and integration of membrane trafficking pathways, IntechOpen. https://doi.org/10.5772/31440

Petrosyan A (2015) Onco-Golgi: is fragmentation a gate to cancer progression? Biochem Mol Biol J 1(1):16

Puhka M, Joensuu M, Vihinen H, Belevich I, Jokitalo E (2012) Progressive sheet-to-tubule transformation is a general mechanism for endoplasmic reticulum partitioning in dividing mammalian cells. Mol Biol Cell 23(13):2424–2432

Rabouille C, Kondylis V (2007) Golgi ribbon unlinking: an organelle-based G2/M checkpoint. Cell Cycle 6(22):2723–2729

Rabouille C, Levine TP, Peters JM, Warren G (1995) An NSF-like ATPase, p97, and NSF mediate cisternal regrowth from mitotic Golgi fragments. Cell 82(6):905–914

Reiling JH, Olive AJ, Sanyal S, Carette JE, Brummelkamp TR, Ploegh HL, Starnbach MN, Sabatini DM (2013) A CREB3-ARF4 signalling pathway mediates the response to Golgi stress and susceptibility to pathogens. Nat Cell Biol 15(12):1473–1485

Rendón WO, Martínez-Alonso E, Tomás M, Martínez-Martínez N, Martínez-Menárguez JA (2013) Golgi fragmentation is Rab and SNARE dependent in cellular models of Parkinson's disease. Histochem Cell Biol 139(5):671–684

Schmidt JA, Brown WJ (2009) Lysophosphatidic acid acyltransferase 3 regulates Golgi complex structure and function. J Cell Biol 186(2):211–218

Sciaky N, Presley J, Smith C, Zaal KJM, Cole N, Moreira JE, Terasaki M, Siggia E, Lippincott-Schwartz J (1997) Golgi tubule traffic and the effects of brefeldin A visualized in living cells. J Cell Biol 139:1137–1155

Sekiya K, Satoh R, Danbara H, Futaesaku Y (1993) A ring-shaped structure with a crown formed by streptolysin O on the erythrocyte membrane. J Bacteriol 175(18):5953–5961

Short B, Haas A, Barr FA (2005) Golgins and GTPases, giving identity and structure to the Golgi apparatus. Biochim Biophys Acta 1744(3):383–395

Shorter J, Warren G (2002) Golgi architecture and inheritance. Annu Rev Cell Dev Biol 18:379–420

Stenmark H (2009) Rab GTPases as coordinators of vesicle traffic. Nat Rev Mol Cell Biol 10:513–525

Sugawara T, Nakatsu D, Kii H, Maiya N, Adachi A, Yamamoto A, Kano F, Murata M (2012) PKCδ and ε regulate the morphological integrity of the ER-Golgi intermediate compartment (ERGIC) but not the anterograde and retrograde transports via the Golgi apparatus. Biochim Biophys Acta 1823(4):861–875

Sütterlin C, Colanzi A (2010) The Golgi and the centrosome: building a functional partnership. J Cell Biol 188(5):621–628

Uchiyama K, Jokitalo E, Kano F, Murata M, Zhang X, Canas B, Newman R, Rabouille C, Pappin D, Freemont P, Kondo H (2002) VCIP135, a novel essential factor for p97/p47-mediated membrane fusion, is required for Golgi and ER assembly in vivo. J Cell Biol 159 (5):855–866

Uchiyama K, Jokitalo E, Lindman M, Jackman M, Kano F, Murata M, Zhang X, Kondo H (2003) The localization and phosphorylation of p47 are important for Golgi disassembly-assembly during the cell cycle. J Cell Biol 161(6):1067–1079

Voeltz GK, Rolls MM, Rapoport TA (2002) Structural organization of the endoplasmic reticulum. EMBO Rep 3(10):944–950

Wennerberg K, Rossman KL, Der CJ (2005) The Ras superfamily at a glance. J Cell Sci 118:843–846

Wood SA, Park JE, Brown WJ (1991) Brefeldin A causes a microtubule-mediated fusion of the trans-Golgi network and early endosomes. Cell 67(3):591–600

Wortzel I, Koifman G, Rotter V, Seger R, Porat Z (2017) High throughput analysis of Golgi structure by imaging flow cytometry. Sci Rep 7:788

Zaal KJ, Smith CL, Polishchuk RS, Altan N, Cole NB, Ellenberg J, Hirschberg K, Presley JF, Roberts TH, Siggia E, Phair RD, Lippincott-Schwartz J (1999) Golgi membranes are absorbed into and reemerge from the ER during mitosis. Cell 99(6):589–601

**Part III
Role of Centriole and Golgi
in the Organization of Cell, Embryo
and Organ Geometry**

Chapter 11
The Centrosome as a Geometry Organizer

Marco Regolini

Abstract 'Does the geometric design of centrioles imply their function? Several principles of construction of a microscopically small device for locating the directions of signal sources in microscopic dimensions: it appears that the simplest and smallest device that is compatible with the scrambling influence of thermal fluctuations, as are demonstrated by Brownian motion, is a pair of cylinders oriented at right angles to each other. Centrioles locate the direction of hypothetical signals inside cells' (Albrecht-Buehler G, Cell Motil, 1:237–245; 1981).

Despite a century of devoted efforts (articles on the centrosome always begin like this) its role remains vague and nebulous: does the centrosome suffer from bad press? Likely it does, it has an unfair image problem. It is dispensable in mitosis, but a fly zygote, artificially deprived of centrosomes, cannot start its development; its sophisticated architecture (200 protein types, highly conserved during evolution) constitutes an enigmatic puzzle; centrosome reduction in gametogenesis is a challenging brainteaser; its duplication cycle (only one centrosome per cell) is more complicated than chromosomes. Its striking geometric design (two ninefold symmetric orthogonal centrioles) shows an interesting correspondence with the requirements of a cellular compass: a reference system organizer based on a pair of orthogonal goniometers; through its two orthogonal centrioles, the centrosome may play the role of a cell geometry organizer: it can establish a finely tuned geometry, inherited and shared by all cells. Indeed, a geometrical and informational primary role for the centrosome has been ascertained in *Caenorhabditis elegans* zygote: the sperm centrosome locates its polarity factors. The centrosome, through its aster of microtubules, possesses all the characteristics necessary to operate as a biophysical geometric compass: it could recognize cargoes equipped with topogenic sequences and drive them precisely to where they are addressed (as hypothesized by Albrecht-Buehler nearly 40 years ago). Recently, this geometric role of the centrosome has been rediscovered by two important findings; in the Kupffer's vesicle (the laterality organ of zebrafish), chiral cilia orientation and rotational movement have been described: primary cilia, in left and right halves of the Kupffer's vesicle, are symmetrically oriented relative to the midline and rotate in reverse direction. In mice

M. Regolini (✉)
AudioLogic, Department of Bioengineering and Mathematical Modeling, Milano, Italy

© Springer Nature Switzerland AG 2019
M. Kloc (ed.), *The Golgi Apparatus and Centriole*, Results and Problems in Cell
Differentiation 67, https://doi.org/10.1007/978-3-030-23173-6_11

node (laterality organ) left and right perinodal cells can distinguish flow directionality through their primary cilia: primary cilium, ninefold symmetric, is strictly
connected to the centrosome that is located immediately under it (basal body).
Kupffer's vesicle histology and mirror behaviour of mice perinodal cells suggest
primary cilia are enantiomeric geometric organelles. What is the meaning of the
geometric design of centrioles and centrosomes? Does it imply their function?

11.1 Introduction

The most important characteristic of *centrioles* is their extraordinary chemical
stability and longevity.

Basal Bodies Basal bodies are centrioles at the base of eukaryotic cilia and flagella:
through their distal appendages, basal bodies anchor cilia and flagella to cytoskeletal
microtubules (MTs), besides organizing the assemblage and function of the (nine
fold) axoneme of cilia/flagella.

Centrosomes Centrosomes are membrane-free organelles of metazoan cells,
roughly spherical (rather polyhedral); they are made up of a pair of orthogonal
centrioles, embedded and immersed in an orderly structured matrix of proteins, the
Peri Centriolar Material (PCM): from the surface of this grid several molecular
platforms, named γ-Tubulin Ring Complexes (γ-TuRCs), nucleate an aster of
MTs: these radiate from the centrosome PCM lattice to the cell cortex; this aster is
built around the centrosome during mitosis, and comprises a lot of MTs. Some of
these MTs are nucleated directly by centrosome γ-TuRCs, while others start from the
wall of centrosomal MTs (Sánchez-Huertas and Lüders 2015) through augmin-
dependent microtubule organizing centres (MTOCs). Centrosomes, as just said,
comprise two polarized centrioles, orthogonal during S, G2 and M phases; one,
axial, named 'Mother centriole' (MC) is equipped with two sets of appendages,
distal and subdistal; the second, named 'Daughter centriole' (DC) is located at the
proximal end of the MC, shows distal outgrowths or ribs, but is appendage free.

Centrosome Duplication Cycle Centrosome and primary cilium, like the nucleus,
are the only subcellular organelles present in single copy: during S phase, the two
centrioles disengage, the distal one, the 'Daughter', matures acquiring nine distal
and nine subdistal appendages, thus becoming itself a new mature 'Mother' centriole; from each 'Mother' centriole a new 'Daughter' centriole arises, orthogonally,
using its Mother as a platform, not as a template. The PCM proteins are recruited
around this new centrosome; thus, it is formed, as just seen, by the (former)
ex-Daughter centriole, now transformed in a (new) actual Mother centriole, and its
own orthogonally joined Daughter. This is no doubt, a very complicated, singular
and unique process.

Cilia Motile cilia and flagella show 9 + 2 geometry (nine peripheral MT doublets,
plus two inner MTs). In contrast, the primary cilia (9 + 0 symmetry, no inner MTs)

are non-motile, except in a few cells of the innermost part of the mouse node (the mouse laterality organ). Cells have only one primary cilium, as only one centrosome: cilium and PCM are resorbed and degraded before mitosis. The primary cilia act as the external antennae (chemical signalling pathways and mechanosensitive functions). Besides MC, also the DC plays a fundamental role in the process of cilium assembly (Loukil et al. 2017).

Bilateral Symmetry Bilateral Symmetry, in terms of evolution, is a successful and very ancient trait shared by almost all Metazoa (Bilateria). It is ideal for the balance and mechanical stability of locomotive systems (walking, running, swimming, flying and sidewinding) and for the differential analysis of the perceived stimuli by sensorineural apparatuses (stereoscopy and stereo acoustics: detection of visual and acoustic stimuli source location).

Symmetry Breaking Mirror symmetry, however, is not a good choice for internal organs not in charge of motion or perception. It is early broken to solve architectural anatomical problems (dramatic in narrow snake bodies) and functional challenges (hydrodynamics of double circulation and gut length). Thus, during development, the bilateral symmetry is, in a first step, early imposed and later, very soon cancelled in visceral organs.

Node Node is the laterality organ in mice (responsible for establishing left–right asymmetry), as Hensen's node in birds and Kupffer's vesicle in zebrafish. It consists of about 250 cells formed, during gastrulation, at the midline, near the posterior end of the notochord. Each cell of the node possesses a primary cilium.

11.2 Are Centrioles and Centrosome Enantiomeric (Then Geometric) Structures?

Bilateria, seen from the outside, are mirror symmetric, but their visceral organs are mostly unpaired and not bilaterally symmetric. In Vertebrates, heart and spleen are usually located in the left side (*situs viscerum solitus*), but liver, stomach and pancreas are right sided, while the gut and lungs are not symmetric and not sagittal. Unexpectedly, the complete inversion of viscera (*situs viscerum totalis*) does not create any abnormality whereas the partial, incomplete reverse disposition of a limited number of internal organs, known as *situs viscerum inversus*, represents a severe pathology.

How is asymmetry of visceral organs established? Nonaka and co-workers (2002) suggested that in mice, a ciliary flow produced by the node may create a morphogen gradient. Node cells possess on their apical side one primary cilium: primary cilia of the innermost cells (pit cells) are motile whereas peripheral cells (called crown cells) have immotile cilia; pit cells produce a fluid flow sensed by crown cells. Two hypotheses about this mechanism have been proposed: (1) perinodal cells can be bent and mechanically stretched, opening polycystin-2 (PKD2) cation channels;

(2) perinodal cells can sense molecular gradients (although never identified). Both mechanism trigger the expression of laterality genes only in the left side. In the mouse node, motile primary cilia of inner cells rotate clockwise (viewed from above their apical side), and so in zebrafish Kupffer's vesicle (Ferreira et al. 2018; Okabe et al. 2008): thus, a fluid flow directed towards the left side of the node is produced; here monociliated (left) perinodal crown cells likely work as mechanosensitive devices. The PKD2 cation channels, localized on their cilia, are stretch activated when cilia are bent by the nodal flow (Yoshiba and Hamada 2014): thus, Nodal expression (laterality gene cascade) is induced in their cytoplasm and expressed only in the left lateral plate mesoderm, realizing the usual asymmetry of visceral organs (*situs solitus*). To ascertain the role of the nodal flow, Nonaka et al. (2002) and Yoshiba et al. (2012) cultured in wide flow chambers several mouse embryos, fixed in parallel rows. These embryos have been correctly and coordinately oriented so that crown cells of each node could be exposed and subjected together to artificial laminar flows, alternately rightward and leftward. The right perinodal cells, which also have ciliary-positioned PKD2 cation channels (Yoshiba et al. 2012), responded to artificially induced rightward flows: Nodal expression was induced in their cytoplasm and Nodal pathway expression activated in the right lateral plate meso-derm just like left cells responded to leftward flows. By the morphogen hypothesis, the hypothesized chemosensors should be mirror positioned on left and right crown cell cilia. Thus, we must conclude that perinodal cells can distinguish flow direc-tionality, showing mirror differential sensibility to oppositely directed flows: left crown cells are excited by leftward flows, right cells by rightward. Moreover, both left and right perinodal cells, excited by proper flows, trigger asymmetric expression of the Nodal cascade, but inversely: right crown cells activate the Nodal pathway in the right lateral plate mesoderm, producing *situs inversus*, the mirror image of *situs solitus*: internal organs are mirror shaped and reversed from their usual positions; one genomic pathway, two bilaterally symmetric realizations. Perinodal cell mirror behaviour suggests that primary cilia are enantiomeric geometric organelles.

In the zebrafish laterality organ (Kupffer's vesicle), Ferreira et al. (2018) mea-sured motile and immotile cilia orientation relative to the midline. In the left and right side of the early (3 somites stage) vesicle, primary cilia orientation is markedly mirror symmetric: $+14°$ in the left side, $-21°$ in the right for motile cilia. The immotile cilia of left and right hemisphere showed a similar divergent orientation of ~40°, overall more dextral. Later in development (8–14 somites stage) cilia orientation rotates ~20° towards the right, showing a dextral orientation over the whole vesicle, yet maintaining the same angular difference between the left and right side. Cilia of laterality organs in mice and fish appear, morphologically and phys-iologically, mirror symmetric.

What is the meaning of this geometric design of centrioles and centrosomes? Do they perform any geometric task inside the cell? Histological bilateral symmetry of actual tissues and organs (like the zebrafish Kupffer's vesicle) together with the mirror functioning of mouse perinodal cells strongly suggests that chirality of centrioles/centrosomes is central in left–right patterning; organs can be 'actually' implemented as left handed (sinistral) or right handed (dextral) depending on the side from which each organ derives: mother centrioles appear as chiral tools, bilaterally symmetric, responsible for driving (mirror symmetrically and autonomously) growth and form of organs in Bilateria. Then, one possibility (as crazy as logical, but not counterintuitive) is that, in left and right halves of Bilateria, left ('levo', '+') and right ('dextro', '−') enantiomeric mother centrioles (and then basal bodies, primary cilia, centrosomes) do exist. This may be the reason why organs developed from left-sided buds are precise mirror images of the same organs grown from right-sided primordia. After all, centrioles show the highest micro- macro-scale correlation: they organize pericentriolar material, asters and sister asters, and, in mitosis, by the peculiar centrosome duplication cycle, the cytoskeleton of daughter cells is patterned upon that of their mother.

11.3 'On Growth and Form': 3D Geometry of Organs and Organisms

D'Arcy Wentworth Thompson, Scottish biologist, mathematician and pioneer of mathematical biology (more cited than read) worked on the concept of allometry (the effect of scale and shear on size and shape of organisms); he wrote that an organism is so complex a thing, and growth so complex a phenomenon, that for growth to be so uniform and constant in all the parts as to keep the whole shape unchanged would indeed be an unlikely and an unusual circumstance. He observed that rates vary, proportions change, and the whole configuration alters accordingly with the effects of scale on the shape of animals and plants.

The famous sheep Dolly was produced by a nucleus transfer from a mammary gland cell of a Finn Dorset sheep into the cytoplasm of an enucleated oocyte of a Scottish Blackface: Dolly showed the morphological features of the Finn Dorset sheep, the DNA donor. How can DNA organize anisotropic growth to realize precise shape of organs and organisms? Production and achievement of the typical and reproducible species-specific 3D shape in Metazoa cannot be obtained by messy growth of cells: on the contrary it is a geometric and highly anisotropic, reproducible process. It is not random, but attentively planned: memorized and coded in the genome, it is replicated from the zygote to the adult organism: and always in the same way, billion and billion times in each species. Management of directions, precise and noise-resistant traffic of intra and extracellular signals are crucial: cell migration, adhesion, apical constriction and accurate orientation of extracellular matrix fibres are mechanisms scrupulously planned; they assure highly geometrical and functional results. The same global orientation relative to the sagittal plane is maintained, common and unchanged, in organs far away (giraffe and elephant). Collagen fibrils in the cornea, for example, are meticulously oriented to guarantee the capability of transmitting light. The transparency and clearness of the cornea really derives from the geometric design of the extracellular matrix, a grid composed of orthogonal sheets of parallel fibrils. Metazoa are organisms made up of billions of cells and dozens of complex organs. To organize such a great numbers of cells in forecast dispositions, they establish, already in the zygote, the axes of their intrinsic geometry; a quasi-Cartesian reference system locates polarity complexes in the cell cortex. Eventually other local 'private' systems of 3D coordinates are established, as in limbs and appendages. in addition, almost all animals are bilaterally symmetric (metazoans are, in the great majority, bilaterians: a clear sagittal plane divides the whole organism in two mirror symmetric, left and right, halves). Coordinating between them various reference systems to build an orderly assembled multicellular organism (where 'multi' means 10^{14}) at first glance may appear a tremendous challenge, but, as we will see, it is not *'mission impossible'*. In the zygote of *Caenorhabditis elegans* the sperm centrosome locates polarity factors (Zonies et al. 2010). Do metazoan cells use the centrosome as a geometrical and informational organizer of the 3D architecture of their cytoskeleton? Is this the mechanism that manages the common and shared geometry of far cells and organs? Is this the mechanism that orients and accurately tunes growth and development of extremely complicated organs (inner ears with cochlea and semi-circular canals)? Gradients of chemicals or morphogens, isotropically diffusing, are incapable of self-orienting towards preferred favourite directions. Moreover, they are intrinsically unstable: thermal fluctuations prevent them from organizing in great detail 3D reproducible arrangements of cells with the observed precision of single-cell width (Abouchar and co-workers 2014). Diffusing chemicals are randomly moved and propelled by thermal energy: the well-known chemical gradients of Planar Cell Polarity (PCP) signals (Decapentaplegic, Hedgehog, Engrailed, Wingless) act in two-dimensional (2D) fields (planarly). These fields have small size (less than 1 mm^2, a few hundreds of cell diameters) and are rapidly degraded. At the same time, the size of chemical gradients is incompatible with the small dimensions of single cells and cannot be realized within the cytoplasm. Furthermore, PCP cellular receptors such as Frizzled, Flamingo, Van Gogh, Prickle, Dishevelled, or Diego are too much widespread on the cell cortex to finely tune, by themselves, the precise location of landmarks: division planes, directional movements (gastrulation and neurulation) and, above all, accurate arrangements of extracellular fibres (cornea) need a more detailed cell map. The fruit fly *Drosophila melanogaster* has three pairs of imaginal leg discs, whose morphology is different from each other (Schubiger et al. 2012); then, diverse are the adult legs (above all for the angles between the different parts: coxa, trochanter, femur, tibia and tarsus). Given the similar spatial diffusion of morphogens (although with some difference in diverse leg discs), how can chemicals, working on 2D discs, exactly control the three-dimensional eversion of legs with different articular 3D angles? What mechanism shapes the reproducible well oriented and anisotropic directionality of morphogen fields in insect imaginal discs? In plants, devoid of centrosomes, anatomy of flat organs is more 2D (planar) than 3D: they realize leaves, petals and very simple structures (phloem and xylem vascular tubular bundles). Also Planarians (flatworms), lacking centrosomes, can build only rudimentary 2D organs. The plant anatomy is not comparable to Metazoa; it is remarkable that, without centrosomes, besides three-dimensionality, also bilateral symmetry is missing: it is enough to check the different curvature near the petiole of the borders of (apparently symmetric) leaves, how veins start (in an alternate fashion) from the central vein in leaves, sepals and petals, or the asymmetry of planarian eyes. We cannot compare the design of veins (tracheae) in *Drosophila* left and right wings (realized with the precision of one cell level: see ahead Abouchar et al. 2014) with the gross layout of veins in petals of zygomorphic flowers like orchids (which, at first glance, could appear mirror symmetric). Moreover, planarians, without a centrosome, are not able to perform gastrulation (Azimzadeh 2014), the most important process in Metazoa development.

11.4   Cell and Tissue Local Geometries

The ratio between cell diameters and organism sizes is about 10^6. Making very similar shapes, and precise angles (bilaterally symmetric) distant 10^4–10^5 cell diameters (legs, arms, kidneys, ears, eyes) cannot be sustained by morphogen gradients. Abouchar et al. (2014) described impressively the precision of development in the fruit fly *D. melanogaster*: 'Developmental processes in multicellular organisms occur in fluctuating environments and are prone to noise, yet they produce complex patterns with astonishing reproducibility. We have measured the left-right and inter-individual precision of bilaterally symmetric fly wings across the natural range of genetic and environmental conditions and found that wing vein patterns are specified with identical spatial precision and are reproducible to within a single-cell width. The early fly embryo operates at a similar degree of reproducibility, suggesting the overall spatial precision of morphogenesis in *Drosophila* performs at the single-cell level. Could development be operating at the physical limit of what a biological system can achieve?' Similar considerations may be done for the cytoplasm inside the cell, where the average size of proteins is 10^4 times smaller than the cell diameter. Free directional movements of proteins are impeded by large membranous organelle (Golgi and endoplasmic reticulum). In addition, as said, cell dimensions are too small for gradients to be established inside. Bicoid, Nanos, Hunchback and Caudal form clear gradients in fertilized *D. melanogaster* eggs, whose length (0.5 mm) is no doubt unusual, 10^2 times a normal cell diameter. How can different cargoes be driven to their forecast destination so fast and precisely? (Gáspár and Ephrussi 2017). Such quick precision requires mechanisms different than diffusion it seems necessary, as suggested by Albrecht-Buehler, an informational cytoskeleton (i.e. made up of molecularly labelled fibres, distinguishable each other through different receptors); mitochondrial TIM TOM complexes and Snare (SNAp receptor) proteins follow similar labelling processes. The centrosome aster wires geometrically the cell cortex (something like an underground railway network): γ-Turc heterogeneity, recently shown in flies (Tovey et al. 2018), may be the basis capable of differentiating astral MTs, recognizing cargoes equipped with molecular labels; different γ-Turcs can match their own 'labelled' macromolecules, DNA coded for precise cortical locations.

As already said, dorsal–ventral and anterior–posterior orthogonal axes (a quasi-Cartesian reference system) are imagined to be, round about arbitrarily, immediately positioned in the zygote (in agreement with the entry point of the sperm or randomly oriented). These axes are always perpendicular to themselves. As a matter of fact, first blastomere mitoses occur following programmed and species-invariable pattern of orthogonal division planes. After all, orthogonality cannot arise by chance, it must be sustained by any molecular structure. In addition, a third axis, medial–lateral, is created, on top of that, and it too is orthogonal to both other axes. A credible mechanism capable of building right angles in Metazoa has never been unveiled. In Prokaryotes orthogonality of Z-rings (in respect with the main cell axis) is sustained by FtsZ proteins, homologues to Eukaryote tubulins. The centrosomes too are made up of two perpendicular centrioles composed of tubulin blades. During mitosis, as seen, a new centriole (Daughter) arises orthogonally from a pre-existing mature centriole (Mother). Likely centrosomes possess the unique orthogonal machinery in Metazoa.

During development, step by step other reference (orthogonal) systems are created and differently oriented: at the beginning of zygote development the general reference system for the whole body is established. Later other reference systems are generated for arms, legs, fingers, nails, each one possessing its own orientation, strictly forecast, programmed and imposed: the agreement with the upstream (previously established) reference systems is attentively controlled. In flies, each appendage (legs, wings, drumsticks/halteres, antennae) grows with its own axes (its 'private' system of coordinates). Yet these 'private' reference systems are correctly oriented in respect with the sagittal plane of the whole body: the proximal–distal axis has a precise (mirror symmetric) 3D orientation with the body planes (check, e.g. the bilateral symmetric 3D tilting of insect's legs—coxa, trochanter and femur—respectively to the sagittal plane). As known, in flies, legs derive from groups of only 20–30 cells (initial imaginal discs). At pupariation, discs are already composed of 50,000 cells disposed in concentric rings corresponding to the several leg segments (the outermost ring corresponding to the more proximal segment, the coxa, the innermost to the more distal part, the tarsus). During metamorphosis appendages evert through their stalk showing an astonishing ripple effect of precise angles and tilting, from flat 2D discs to 3D legs. The *Drosophila* mutants defective for DSas-4, a key centriole protein, develop up to the adult stage. After using up maternal provisions of DSas-4 are almost completely lacking in centrosomes (Basto et al. 2006). The adult mutant fly shows an individual with monstrous deformities (see on the Internet the comparative images of wild type vs. mutant in Basto's free article 'Fly without Centrioles'): the shape, tilt and anomalous curvature of the wings certainly impede flight and the abnormal angle between coxae and body cannot allow walking movements. In conclusion: what is the link between morphogenesis (literally 'shape generation') and centrosome?

Formation of limb's ectodermal appendages (hairs, feathers, nails, teeth) can better explicate this concept. Cells are capable of orienting themselves in accord with: (1) 'cardinal' points of the whole organism (the organism's general coordinates: head–tail or superior–inferior, front–rear or anterior–posterior, sagittal plane or medial–lateral); (2) 'local' reference points, characteristic of each limb (proximal–distal); (3) 'private' own reference points of the appendage itself: each finger, tooth, tusk and hair (in animal furs, human eyebrows, moustaches and beards) shows a characteristic orientation of its own local axes. Yet all these reference systems are each other coordinated. For example, the diverse direction of the medial- and lateral-most hairs in the same human eyebrow or the correct orientation of each tooth along the dental arch. In the founder cells of each initial bud, the orientation of the reference system changes in respect with the body's planes, following genetic species-specific programmed patterns. Birds' legs and wings have their own proper axes whose tilts are coordinated relative to the general body's planes. The flight feathers of the wings (remiges) and tail (rectrices) have a precisely fixed orientation relative either to the general body planes and axes and to local limb axes. This orientation is critical in controlling flight. Nonetheless each feather is shaped around its own local (private) three axes: proximal–distal, from the large intradermic portion of the rachis to its subtle apex; dorsal–ventral, from the dorsal outer convex to the ventral inner concave face; anterior–posterior, between the two lateral sharp edges. All its components—rachis, barbs, barbules, barbicels—are built in respect with their 'private' local points of reference, the feather's own axes and the body axes. Note that each feather is not itself mirror symmetric: it does not have an own plane of symmetry (it cannot be divided into two mirror corresponding halves). Nonetheless it is bilaterally symmetric in respect with the feather which is its contralateral counterpart.

This is an interesting issue: within the whole organism, its limbs and appendages, we can see different (general and local) reference systems but only a unique plan of symmetry, the sagittal plane of the entire body: local planes of mirror symmetry do not exist (our fingers are not bilaterally symmetric).

11.5   A Cellular Reference System Organizer: An Overview

How do Metazoa organize their 3D geometry?

Many growing processes or developmental programs can hardly be modelled without an intrinsic cellular compass. To coordinate 10^{14} cells it is necessary a biological mechanism capable of recognizing and memorizing the general 3D coordinates of the whole body besides its local points of reference. An example of this assertion is the 'polonaise' movement during the formation of the primitive streak in chicken embryos when cells from the posterior zone move towards the midline, here they turn in the direction of the centre of the epiblast, and are replaced by cells hailing from the lateral posterior marginal zone. During this process two large, mirror-symmetric (clockwise/counterclockwise) rotational fluxes of flowing cells appear: two opposite circular parades of all the cells of the epiblast occur, side by side along the primitive streak, just resembling a 'polonaise' dance. How many different (orderly disposed above all) morphogen fields must there exist for gener-ating such complex directional symmetric migrations of cells continuously changing direction? Moreover: how and by what upstream process are these morphogen fields positioned? Not only: while this single layer of cells moves and concentrates towards the posterior zone of the monolayered epiblast, the single sheet streak is subject to a multi-layered transformation (from 2D to 3D) becoming a clearly visible thick structure.

'How do PAR proteins become asymmetrically localized in response to the varied cues that polarize cells? How does a conserved set of molecular interactions provide the basis for cells with distinct structural and functional properties? The organization of polarity mechanisms into positive- and negative-feedback loops raises the chicken-and-egg question of how asymmetry is initially established. In some cases, such as in the *C. elegans* zygote and blastomeres, symmetry-breaking cues provided by the sperm-derived centrosome and Rho GTPase regulators are reasonably well understood' (Nance and Zallen 2011). PAR proteins and PCP signalling complexes are landmarks unable to self-locate in forecast locations: at the same time cell cortex does not have the capability of self-compartmentalizing for attracting specific landmarks in forecast locations. The reproducibility is indeed the fundamental property of living beings: how can polarity complexes be always correctly positioned in the same locations in billions of cells? They need any upstream noise resistant mechanism able to carry out patterned, DNA coded and reproducible morphological geometric programmes.

Planarians (flatworms) lack centrosomes (Azimzadeh et al. 2012) but possess centrioles: 'The embryogenesis of freshwater planarians is equally intriguing: cleavage of the fertilized egg was described as 'anarchic' by early developmental biologists. No overt gastrulation or epiboly has been described in these embryos' (Alvarado 2004). Without the centrosome, a real, actual three-dimensionality is difficult to be reached.

The role of the centrosome and its aster in *C. elegans* zygote is similar to a cellular reference system organizer. In the zygote it generates the body general axes, mapping and wiring the cell cortex of the zygote; through the typical centrosome duplication cycle. During mitosis each new MC is firstly oriented as the older MC; then a common polarity is transmitted (and shared) to zygote offspring (blastomeres).

Inside the cell, a theoretical organizer of cell geometry must be really connected to the cortex for mapping distinct compartments. Moreover, each wiring way to the cortex must be distinguished from the others and identified by univocal signals. To have an idea of a reference system organizer inside the cell, we can think to a router, or rather a switch. This well-known (pre-wireless) electronic device is physically and really connected (wired) to several computers. It recognizes and matches coded address signals (input) driving them to their physical addressed destination (output): so, coded packets or frames of data are received, processed and forwarded to the forecast computer.

Mapping, wiring and matching are then the 'core businesses' of this kind of interface. Like a router, the centrosome can realize a univocal, one to one, noise resistant correspondence. The DNA-coded signals (molecular topogenic 'labels') intended for a defined cortical compartment may match with microtubular pathway receptors (target) specifying the same compartment. What are the genetic geometric codes? They consist of 'topogenic' DNA sequences, added to cargoes to correctly drive and address them. The topogenic sequences of TIM/TOM mitochondrial complexes operate this way, and so do SNARE and SNAP proteins. The mechanism of matching and recognition is the same for antigens and antibodies, signals and receptors, codons and anticodons, enzymes and their substrates. As already said, γ-Turc heterogeneity has been ascertained in flies. Thus an idea arises: this could be the proper, real, authentic primary task of the centrosome, driving cargoes that have been targeted for a given compartment and equipped with the corresponding label for that compartment (Regolini et al. 2018). The shape (tertiary structure) of the topogenic sequence (label) permits it to match only and uniquely with the corresponding cytoskeletal receptor (γ-Turc) of the forecast MT pathway (one to one correspondence between cargo labels and cytoskeletal receptors). Centrosome involvement in mitosis is a secondary task, that other organelles can perform. But, as every hypothesis, this idea must be proven.

Let's go on studying an ideal spherical reference system organizer.

Building a Spherical Reference System Organizer with Two Orthogonally Arranged Protractors

A spherical reference system is clearly represented by a globe: it consists of an axis, 'z' (the earth's axis) and an equatorial plane orthogonal to the 'z' axis; a second axis ('x') may be traced in any position on the equatorial plane, but, once positioned, it is unmovable and fixed. The measures of the angle 'φ' (longitude) are taken starting from the 'x' axis in a counter-clockwise direction; a radial axis 'r', arising from the intersection 'O' (origin) between the equatorial plane and the 'z' axis, reaches every point 'P' of the space; 'P' coordinates then are: 'φ', the longitude, 'θ', the latitude and 'r' length, the distance from the origin 'O'.

Protractors/goniometers are more suitable and adequate instruments to use in geometric drawings, while in our model we are not interested in measuring angle amplitude: other similar tools must be taken into consideration. For example a wall clock, a compass or a wind rose, more similar to a cross section of a centriole; after being orientated (relative to geographical coordinates, to a Cartesian system, to a side of a squared sheet or even randomly) these instruments maintain their fixed oriented position. They divide the plane (2D) around themselves into several sectors (nine in our model); each sector is permanently identified by and through the corresponding tool's figure (hours or cardinal points). Centrioles cross sections clearly resemble protractors, clocks or compasses: they have nine triplets or blades of MTs, arranged with the characteristic ninefold symmetry, equally separated and distanced from each other. Several works on Protists (see ahead) have shown that the triplets of centrioles are not equivalent, but diverse and orderly sequenced exactly like clock's figures are. From here on we will take into consideration only nine-marked goniometers.

For understanding a spherical reference system, as said, the globe is a good example: one protractor lies on the equatorial plane, the other is on a meridian plane, orthogonal to the first. The first protractor, 'horizontal', is orthogonal to the 'z' axis. Its knobs (or marks, nine in our case) subdivide the space around it into several meridian sectors or 3D wedges (longitude). The other protractor manages the latitude, the 'θ' coordinate, is vertical and perpendicular to the first. It is polarized and oriented like a wall clock: its top–bottom axis (with the figure '12' at the top) is the vertical 'z' axis, the 'globe pinwheel'. In the classic spherical reference system, the 'θ' coordinate goes from $0°$ to $180°$ (in a globe from $0°$ to $+/- 90°$, i.e. $0°–90°$ North/ $0°–90°$ South): the second protractor is then composed of two halves, two mirror hemi-goniometers facing each other (in fact a globe is usually equipped with only one half-goniometer). In our 'nine-marked goniometer' one half shows on its border four marks ($+40°$; $+80°$; $+120°$; $+160°$) ordered in a clockwise fashion starting from the top. The second half shows the same four marks, but counter-clockwise ordered from the top: $-40°$; $-80°$; $-120°$; $-160°$; both goniometers share the poles, i.e. $0°$ and $180°$. These eight marks are symmetrically positioned: they divide the space into five parallel 'horizontal' segments or discs: two polar caps and three segments. It is not necessary the centres of the two goniometers, equatorial and meridian, coincide, as it occurs in a globe, because of the cylindrical shape of centrioles.

Let's now combine in 3D the spatial subdivision obtained by these two goniometers. Like on the globe, the surface of a sphere is subdivided in a grid of many irregular quadrilaterals; each one defined by the intersections of two consecutive meridians and parallels. In our 'nine-marked' spherical reference system, each one of these nine meridian 'vertical' wedges is sectioned into five 'horizontal' parts. Thus, the space is subdivided into 45 spherical sectors or pyramids with the apex at the centre. Each base faces and subtends a vertex solid angle of $4\pi/45$ steradians. In a cell with a diameter of 10 μm (radius: 5 μm; surface: approximately 314 μm^2) the base of a pyramid ($4\,\pi r^2/45$), corresponds to about 7 μm^2 (a circle with a radius of 1.5 μm, or a square with a side of 2.6 μm). This ordered subdivision of the cell cortex into 45 compartments gives an idea of the order of magnitude and precision of the cell's finely tuned polarity and high noise-resistance, much better than chemical gradients. An ultimate cortical subdivision of each compartment is obtained through other markers like SNARE proteins. It is worth noting that astral MTs radiate from the centrosome surface with a centrifugal direction like that of the local radius. Their nucleating units (γ-TuRCs) are then necessarily perpendicular to the local radius; their declivity must be the sum of two inclinations, longitudinal and latitudinal. Within the centrosome, centrioles are not 2D flat discs as goniometers, but barrel-shaped 3D cylinders, and this is a great advantage. One centriole is 'vertical', contains the 'z' axis and can easily create nine 3D wedges in the PCM much better than a 2D disc. Mennella (2014) showed that pericentrin/PLP (a PCM protein) 'forms elongated fibers in the PCM and support the notion that centriole symmetry is not confined to its perimeter but acts as an organizing principle that extends into the PM. The coiled-coil [centrosomal] protein Cep152/Asl has similar orientation and distribution to pericentrin/PLP'. Similarly, the other 'horizontal' centriole can organize the PCM in five parallel horizontal segments. This is likely the meaning of two orthogonal centrioles, their geometric capability of erecting scaffoldings within the cell (MT aster) organizing 3D architecture in Metazoa. Through such a mechanism, development can really operate at the physical limit reachable by a biological system (Abouchar et al. 2014).

11.6   Centriole Informational Architecture

Are centrioles real goniometers?

Non-equivalence of triplets, that is, an individual structural and molecular diverse shape capable of distinguishing each from the other, is the most important issue to discuss. Only if the nine triplets are non-equivalent, different as are the numbers of a goniometer, the figures of a clock or the marks of a compass, can a centriole work as an 'informational' or 'communicative' device.

Protists possess a centriole/basal body whose distal appendages are connected to the cytoskeleton: centrioles organize the cytoskeleton and anomalies in centriolar components cause anomalies in the cytoskeleton (Feldman 2007). Centrioles operate also as basal bodies at the base of the flagellum: they organize the complex trafficking in its axoneme, composed of 9 MT doublets, not triplets. The cross section of cilia shows the typical 9 + 2 schematic pattern: 9 MT doublets circumferentially arranged with the usual ninefold symmetry besides two extra MT doublets at the centre.

Does rotational polarity (ordered non-equivalence of centriolar MT blades) exist in Protist? Many researchers have ascertained that centriolar distal appendages are biochemically and morphologically distinct: these findings have been recently confirmed by Tassin et al. (2016) in *Paramecium*: 'The basal body circumferential polarity is marked by the asymmetrical organization of its associated appendages'. In Ciliates, centrioles show structural and functional polarities transmitted and conferred upon that of the cell (Beisson and Jerka-Dziadosz (1999), Geimer and Melkonian (2004). Several studies on Protists have demonstrated the nine triplets of their centrioles/basal bodies are not only different (non-equivalent), but also ordered in a given pattern of sequence (not randomly arranged). The circumferential polarity in the arrangement of basal-body triplets is accordant with the disposition of the cytoskeleton. Protists contain only centrioles/basal bodies, each at the base of a cilium or flagellum and do not have centrosomes with two orthogonal centrioles. The centriole orthogonality, however, appears during the assembly of a new centriole, which arises and grows perpendicularly to a pre-existing one. Centriole formation not de novo but from a pre-existing mature centriole used as a platform is an ancient and conserved mechanism. This unique mechanism may correctly orientate the ordered sequence of triplets in the new arising centriole corresponding to those of the pre-existing one. The cilia and flagella of the same organism (*Chlamydomonas* has two flagella, *Paramecium* 4000) must vibrate coordinately, then centrioles/basal bodies must share the same orientation: so the new centriole is patterned upon its mother's oriented architecture to correctly according to its arrangement in the cytoskeleton and with the other centriole/s. Deficiency of centrin (a centriolar protein) in Paramecium affects the geometry of basal-body duplication and causes their mislocalization within the cell (Ruiz et al. 2006). This scheme of replication is conserved also in high Metazoa: the cytoskeleton of a new arising cell is patterned upon that of its mother to transmit a shared polarity. The biflagellate unicellular green alga *Chlamydomonas reinhardtii* has an ordered location of its organelles, cilia, oral apparatus, nucleus, chloroplasts, pyrenoid, eyespot, excretory vacuoles. Its 'anatomy' depends on the disposition of distinct cytoskeletal fibres (four cruciform rootlets, two thick ones, made up of four MTs, and two thin ones, made up of two MTs). The cell appears clearly polarized: an apical–basal axis from the cilia to the pyrenoid and a dorsal–ventral axis, orthogonal to the first, from the nucleus to the oral apparatus. Also a transverse axis of asymmetry (orthogonal to the other two axes) is established because of the asymmetric position of the eyespot. In Protists, the cytoskeleton is not self-assembled or randomly located in the cytoplasm: cyto-skeletal fibres disposition is organized by centrioles.

Structural anomalies in centrioles cause disorders in the cytoskeleton (cruciform fibres lose their normal composition and orthogonal disposition). *Chlamydomonas* has two apical flagella, whose movements are coordinated during planar 2D strokes, and during conical–helical 3D rotations. The axoneme of its motile flagella is composed of the canonical 9 + 2 MT doublets. These blades are not equivalent. Electron microscopy has allowed each one of the nine doublets to be distinguished: each one shows its own morphology that characterizes itself, being different from the shape of the others. Moreover, each blade has its own fixed location relative to the others. Electron microscopy (Hoops and Witman 1983) has highlighted circumfer-ential asymmetry of the axoneme, which corresponds with an even more marked circumferential asymmetry of the basal body/centriole: it is possible to distinguish each one of its nine orderly sequenced triplets. In the unicellular green alga *Chlamydomonas* its two centrioles/basal bodies are rotated of 180° between them-selves to coordinate and emphasize flagellar propulsion. We can observe the con-nection between each triplet and the fibres of the cytoskeleton: the striated fibres of the 'distal connector, tie and fasten triplets 9-1-2 of both basal bodies. The thick cruciform fibres are attached to triplets 3 and 4, the thin ones are linked to triplet 8. The shape of any structure, in addition, is necessarily the consequence of its molecular composition. In *C. reinhardtii* the polypeptide VFL1 coded by the gene vfl1 binds only to triplet 1 (Silflow 2001) confirming the molecular nature of circumferential asymmetry.

Feldman (2007) studied the unicellular green alga *Chlamydomonas* to screen for mutants that could not swim towards light. They found 'a set of mutants in which the centrioles and flagella are displaced from their normal location within the cell... suggesting centrioles play a role in positioning other structures within the cell, such as the nucleus'; they also observed that 'in these cells, which contain two centrioles differing in age, the older centriole plays a role in positioning the newer centriole'. The authors conclude that 'cells may have a way to propagate spatial patterns from one generation to the next'. This is the essential, vital, indispensable capability of centrioles of transmitting shared and common geometry through generation of cells in multicellular Metazoa.

Inside the basal body of C. *reinhardtii* Geimer and Melkonian (2004) described a ring-like structure as an early marker of radial polarity. They called this filament 'an "acorn-like" asymmetric structure, positioned in the inner distal part of the basal-body, adhering in highly inter-individual reproducible and invariable manner to triplets 2-1-9-8-7'. Using electron microscopy, these authors showed another structure, shaped like the uppercase letter 'V' in contact with triplets 9, 5 and 4. They proposed that the cartwheel (Sas-6 protein: Kitagawa 2011) is the biochemical structural basis of the ninefold radial symmetry of the triplets and that the acorn-like belt, together with the 'V' structure, might play an equally important role. These structures impose constant and reproducible molecular differences (rotational polarity) on the microtubular triplets. They lead to the ordered circumferential assembly of basal body-associated fibres and hence to the general planned polarity of the whole cell.

The process through which centrioles are built occurs in two separate steps. Firstly, the structural base (the cartwheel) for the molecular architecture of ninefold symmetry is established. Eventually triplets are assembled: the order of appearance of the nine 'A' MTs ('A' is the name of the innermost microtubule of the triplet, the first of the three MTs of the triplet to be built) in the nascent procentriole occurs in a random manner, not orderly sequenced (Guichard 2010). Only in a second step rotational polarity is imposed (the acorn-like ring of Geimer and Melkonian 2004). It is worth underlining that the newly formed procentriole is immediately oriented and connected to the cytoskeleton through its triplets. Thus, the rotational polarity of the arising centriole is patterned upon that of the preexisting one. The Protist *Paramecium tetraurelia* (Beisson and Jerka-Dziadosz 1999) 'has a very high number of basal-bodies and cilia, about 4000, accompanied by their ability to organize themselves together (Feldman 2007): each one connects to the others in about 70 regular rows, always with the same orientation (in order to move with coordination and synchronism) just using their triplet's mark. Each new basal body arises at right angle to an old one, then straightens up and rotates to acquire a precise and coordinated orientation with the complex cytoskeleton. The triplet 9 links to the "postciliary ribs", triplet 4 is attached to the "transverse ribbon", and the triplets 5-6-7 are connected to the "kinetodesmal fibres"'.

Cilia (often hundreds in cells of ciliated epithelia) are formed by their own centriole. These centrioles/basal bodies are assembled de novo; while the centrosome and its centrioles are not duplicated de novo. This de novo formation happens only in ciliated epithelia and parthenogenetic oocytes. During mitosis each daughter cell inherits only one centrosome whose duplication cycle is too complicated to be without reason. Firstly, Mother and Daughter centrioles disengage; Daughter centriole matures to become itself a Mother centriole, acquiring distal and subdistal appendages: now the dividing cell possesses two MC and no DC. immediately later a new centriole (a new Daughter) is assembled orthogonally to each MC (used as a platform that helps the assembly of the new DC. Thus, the high complex process of centriole duplication is part of a mechanism capable of transmitting information. The cytoskeleton of a new arising cell is patterned and oriented upon that of its mother. This is the mechanism that can sustain the shared orientation of the centrosomes in neighbouring cells. Through this mechanism (step by step, from the zygote to the adult organism) shared coordinated general and local points of reference (general and local coordinates) can be transmitted. Shared and coordinated orientation of cells is indispensable to build reproducible multicellular structures, made of cells that move and change position without losing their orientation: from the movements of 'polonaise', gastrulation and neurulation to arrangements like two symmetric set of three semi-circular canals correctly oriented in three orthogonal planes.

'In Mammals the circumferential, morphological, structural and molecular asymmetry of centrioles can only be inferred from ciliated epithelia. Although the circumferential anisotropy of centrioles is difficult to ascertain within the centrosome, its existence can be inferred from the properties they express during ciliogenesis, be it the formation of a primary cilium or of bona fide 9 + 2 cilia in ciliated epithelia, some of which at least derive directly from the centrioles. As in Ciliates and flagellates, these basal-bodies nucleate appendages of various molecular compositions (basal foot, striated rootlets, alarm sheets etc., which anchor the basal-body to the membrane and to the cytoskeleton) and these nucleations arise at specific sites of the basal-body cylinder; in particular the basal foot is located on triplets 5 and 6 corresponding to the side of the effective stroke of the cilium. What is remarkable is that basal feet develop before the basal-bodies reach their membrane site and before they acquire their functional orientation' (Beisson and Jerka-Dziadosz 1999).

Centriole enantiomerism, shown in mammals (mice) and fishes, together with γ-Turc heterogeneity, shown in flies, are impressively consistent with the hypothesis that centriolar triplets are not-equivalent in all Metazoa.

11.7 Modelling the Centrosome

In Ciliates, as previously said, triplets are clearly different (not equivalent) because of their distinct ultrastructure and diverse connections to the cytoskeleton. In each species centrioles/basal bodies show standard, sequential and ordered set-up of triplets, each distinguishable from the other. What biological process can sort and put nine different structures in single file, always following the same sequence in constant invariable order? Unidimensionality is the key of life: primary structure of DNA, RNA and proteins, made up of different units, is capable of realizing very complex 3D structures. A patterned sequential order of nine different structures can be realized by a linear polymer (a polynucleotide or a polypeptide). It can be easily memorized and coded in the genome and eventually arranged in a ring or any circular structure capable of decorating the nine radial spokes of the centriole. The 'acorn-like' ring was identified by Geimer and Melkonian (2004) in the ultrastructure of centriole/basal body of *Chlamydomonas* and RNAs were found in centrioles of *Spisula solidissima* (an Atlantic surf clam) by Alliegro and co-workers (2006). Taken together, these two findings allow to hypothesize that the ordered sequence of marks is sustained by an informational ring complex, be it an RNA, a ribonucleoprotein or a polypeptide. This hypothesis raises a consequent problem. A ring is 2D while the centrosome is spherical, rather polyhedral, indeed 3D. This implies that a building mechanism exists, which, starting from the orthogonal arrangement of MC and DC within the centrosome, is able to organize the local 3D declivity of the molecular platforms (γ-TuRCs) from which astral MTs arise radially. The γ-TuRCs declivity is the cause of the radial orientation of aster MTs, The radial orientation of astral MTs requires that the γ-TuRCs, which they start from, be positioned on precisely tilted platforms on the centrosome PCM. The fact that Sas-6 dimers possess the remarkable capability of building angles of 140° (and the supplemental of 40°) is not a small thing (see also Mennella 2014).

11.8   Bilateral Symmetry

General Considerations

Bilateral symmetry is an evolutionarily very successful trait shared by almost all Metazoa: a fundamental, basic property of their locomotive system and sensorineural apparatus that drives locomotion movements. The bilateral symmetry is the simplest and the most efficient way for controlling the direction of motion and localizing perceived signals (balance of differential stimulation of two bilateral receptors). Eventually, during development, in some internal organs not in charge of motion, symmetry is broken: the process of symmetry breaking is quite complex because difficult architectural anatomical problems must be solved such as the heart and great vessels. To show another example, in some Echinoderms (starfish), bilaterally symmetric larvae acquire a strange, singular pseudo-radial symmetry through a unique developmental process. Many species of this phylum have adopted an unusual circular pentamerism, very different from the straight metamerism of segments in flies or somites in vertebrates: starfish amplify enormously left-right asymmetry, during development the left side of the body grows much more than the right one (Morris 2007).

Mirror symmetry of two objects (in 2D as in 3D) is much more simple than one can imagine, because it involves the marking of only one coordinate. If two objects are mirror image with respect to a plane (let's say the plane 'yz') each point P (+x; +y; +z) of the first object, is symmetric to a point P' with coordinates −x; +y; +z: only their 'x' coordinates are opposite, i.e. the marking of the 'x' coordinate changes from '+' to '−'; in a spherical reference system, only the marking of the 'φ' coordinate changes.

A role in mirror symmetry for centrioles and centrosomes has often come into play and they have been accredited many times with this task. Several authors summoned them suspecting they are the actors of a special chiral mechanism for consistently specifying the difference between left and right sides. 'What sub-cellular component is responsible for the crucial orientation event that defines leftward?' (Brown and Wolpert 1990). Vandenberg and Levin (2009) suggested 'the coordination of the three axes is performed by a cytoskeletal organizing center such as the centriole or basal-body: a sharp midline separation is evident already after the first cell cleavage in *Xenopus laevis* and left and right blastomeres inherit immediately differential chiral information, then transmitted to the progeny'. In the *Xenopus* zygote Danilchik et al. (2006) filmed in vivo 'a circumferential asymmetry consisting of rotational equatorial cortical movements (pharmacologically induced) oriented in a single fixed direction in the zygote and in parthenogenetically activated oocytes'. After the first mitosis, the two blastomeres show similar equatorial cortical movements but with opposite direction (clockwise/counter-clockwise). In the same way the intriguing clockwise/counter-clockwise 'polonaise' movements of hundreds of cells are evident in chick, mouse and human epiblast: their behaviour resembles very closely that of persons divided in two groups, one equipped with normal compasses, the other with opposite polarized compasses. Both groups follow the same driving instructions based on cardinal points, but cover two symmetric pathways, depending on the type of compass they have. Xu et al. (2007) and colleagues proposed 'an intrinsically chiral structure "perhaps the centrosome" serving as a template for managing polarity in the absence of spatial cues: such a template could help to determine left–right asymmetry and mirror planar polarity in development'.

However, a credible biologic mechanism capable of explaining and realizing bilateral symmetry has never been proposed. Errors in symmetry breaking cause 'heterotaxia' confirming the left or right determination of somatic cells. In the rare cases of 'right' cardiac isomerism, two 'right' atria develop while in 'left' cardiac isomerism two 'left' atria are formed. Surprisingly, in both cases, the two atria are mirror images of each other (bilateral symmetry) (Hildreth et al. 2009).

In amphioxus, manipulating the Nodal cascade, Li et al. (2017) obtained '2-left' or '2-right' phenotypes. The '2-left' mutants duplicate the wild type left-sided organs and lose the right-sided ones. Instead of their typical mouth only on the left side, the '2-left' mutants have two, mirror symmetric, 'paired bilateral mouths' as authors observed. Amphioxus larvae are bilaterally symmetric and show 'bilaterally programmed differences [asymmetries] between their left and right sides': which authors? authors demonstrated 'a system in which asymmetric expression of the signal gene *Nodal* is controlled by positive and negative feedback loops with its own inhibitors'. When this system is disrupted, embryos develop mirror-image symmetry with two 'left' sides or two 'right' sides; in larvae the mouth is positioned on the left half of the body. Manipulating *Nodal* gene expression, 'embryos develop a consistent 'two-left-side' phenotype with a mouth opening on each side'; mouths, however, are mirror images of each other. A mechanism must exist that builds left-handed structures in the left half and right-handed in the right side. The same gene cascade, invariably, expressed in the left side, produces only and always left-handed organs and so in the right half.

The inverse, opposite rotational polarity of the MC may be the key for bilateral symmetry in Metazoa. This assertion appears attractive and tempting. As seen, mice perinodal cells prove to be adept at distinguishing flow directionality; in Kupffer's vesicle, primary cilia in left and right half, are mirror oriented relative to the midline; Kupffer's vesicle histology and perinodal cell mirror behaviour suggest primary cilia are enantiomeric geometric organelles. Likely the easiest possibility of producing animals formed by two mirror symmetric halves consists in the inverse polarization of the Mother centriole (chirality of the MC). Through an opposite rotational sequence, a circumferentially chiral organelle is assembled. During organogenesis, centrioles transmit their intrinsic enantiomorphism to each symmetrical organ, carrying on symmetrically each developmental program (cell division, migration, adhesion).

The first two blastomeres are likely already patterned one left and one right (and then their progeny will be left or right). This is true for normal embryos, while in experimental surgical separation of the first blastomeres the subsequent early development depends on the contact between cells in that the neighbouring cell can be absent or present, although killed. Roux, in the early 1900s, in his famous experiments, destroyed without removing one cell of a 2-cell frog embryo, and only one-half embryo, left or right, developed. In contrast, a complete splitting, like in twins, or surgically operated embryos, results in the development of two complete organisms, although sometimes smaller than normal.

Many experiments on regenerating limbs have shown that left and right cells have a circumferential asymmetric polarity. The adjacent cells 'sense' neighbouring signals. If developing left and right tissues normally not adjacent are juxtaposed, duplications arise. Bryant et al. (1981) polar coordinate model forecasts the orientation of duplicated limbs. The graft of a 'left' limb bud (or a left regeneration blastema) on the contralateral right stump, causes three areas of re-growth that produce three limbs; a central 'left' limb, composed of the transplanted 'left' cells, maintains its circumferential value and the characteristic 'left-handed' tilt and orientation of the three axes. In addition two abnormal external right limbs grow, composed of stump right cells: these two supernumerary limbs conserve the Anterior–Posterior and Dorsal–Ventral axes proper of the right limb. However, their Proximal–Distal axes are rotated by +/− 90° relatively to the sagittal plane. It seems that cells try to normalize circumferential cell–cell contacts inducing the arising of correctly polarized cells interpolated between graft and stump cells.

Drosophila is manifestly mirror-symmetric: however, in embryos, first divisions are syncytial; this could appear in disagreement with the previous hypotheses about left-right patterning already in the first divisions. However, all syncytial mitotic divisions are absolutely non-random or anarchic: they are strictly spatially controlled by the centrosomes that remain joined to their own nuclei. As a consequence, if centrosomes are artificially damaged, development immediately stops. The first four divisions generate 16 nuclei that remain radially equidistant from the centre to form a sphere. Subsequently, during cycles 4–6, nuclei distance themselves along the Anterior–Posterior axis. Their spherical disposition becomes elliptic and marginal nuclei are symmetrically equidistant from the cortex. During stages 7–10 a symmetric migration towards the cortex occurs. The movements of nuclei in relation to the embryo cortex are mediated by forces acting on the centrosomes rather than on the nuclei themselves. Beisson and Jerka-Dziadosz (1999) observed that the asters are presumably the main target of such forces. The ordered separation of nuclei, the controlled asymmetry of their arrangement in anterior–posterior direction, and their division into two halves (right and left) have been highlighted in a study based on the diverse beginning and speed of mitosis 14 (Foe 1989), The domains have been defined whose cells, with common developmental fate, start to divide at the same time: 25 left and 25 right domains with different form and extent are evident. All the domains, as described by the author, also occur in pairs. 'Whether paired or not, every domain is bilaterally symmetric. A clear midline is evident both dorsally and ventrally' (Figs. 1A, 1B, 1C, 3A, 3B, 3C in Foe 1989: http://dev.biologists.org/content/107/1/1.long).

Surprisingly this compartmentalization of the embryo cortex reflects and corresponds with that which the centrosome arranges on the cell cortex surface.

Strachan and Read (2010) observed that '*Drosophila* Ultrabithorax mutants instead of drumsticks/halteres develop an additional pair of wings that are built with the same correct bilateral symmetry; a mutation in the *apterous* gene causes a phenotype lacking wings: by inserting in the mutant fly embryo the human ortholog gene LHX2, the normal wild-type phenotype is rescued, with two bilaterally symmetric wings and two halters'. The words 'bilaterally symmetric' are extremely important: they confirm that when a genetic developmental program is started, it is always carried out mirror symmetrically. It is realized left- or right-handed, depending on the side where it is implemented and performed. In mutants, be they wild-types or rescued individuals, appendages are built always symmetric. Are genetic instructions symmetrically translated by left and right MC? Indeed, right and left wings reproduce, mirror-like, the same shape, edge curvature, compartments and arrangement of veins (tracheae) and their anastomoses with the astonishing precision of one cell level: as already said, Abouchar et al. (2014) have shown that 'developmental processes in multicellular organisms occur in fluctuating environments (variable temperature and thermal fluctuations) and are prone to noise, yet they produce complex patterns with astonishing reproducibility'. The authors measured 'the left/right precision of bilaterally symmetric fly wings across the natural range of genetic and environmental conditions (14 °C–30 °C)' and found that 'wing vein patterns were specified with identical spatial precision and were reproducible to within a single-cell width in right and left wings; the spatial precision of morphogenesis in *Drosophila* performs at the single-cell level'.

11.9 Centrioles and Gametogenesis

Rodents are often taken in consideration to contradict the precedent hypotheses: do they constitute an exception? Their first blastomeres undoubtedly lack centrioles and centrosomes, but fundamental (rudimental) parts of them are conferred by the spermatozoon (Goto 2010; Jin 2012; Coelho 2013), so that they are not completely devoid of 'sub-procentriolar-structures'. Azimzadeh (2014) exploring the evolutionary history of centrosomes, proposes that likely during evolution distinct separate tasks of centrosomal components (separation of chromosomes in mitosis, construction of the cytoskeleton, assemblage of cilia) have been united in one organelle, the centrosome itself (convergent evolution). A similar process (evo-devo) of addition of roles occurs in zygotes after germ cells have arranged for centrioles and centrosomes to be reduced. The subsets of centrosome and centriole components, called 'immature' or 'rudimentary' procentriole-like structures or 'atypical centrioles', are supplied by sperms to the zygote, and eventually, they are assembled in mature centrosomes (in the zygote itself or in blastomeres after 3–4 mitoses).

Indeed, centrosome behaviour in mitosis, as seen, is unique and quite complicated, but in meiosis it is a real mystery. As is known, each gamete provides the zygote with its chromatids (pronuclei): immediately after fertilization, they fuse, and, before the first division of the zygote, are duplicated during the usual mitotic S phase. So one would expect each gamete to supply its Mother centriole: during the S phase of the first division, it could assemble its own Daughter to prepare one of the two centrosomes of the zygote necessary to enter the first M phase. However, this is not the case, this process does not occur; only one gamete (usually the male, apart parthenogenesis) provides two centrioles: one is a quasi-mature centriole, the other one is more rudimental, unrefined and rough ('work in progress', frequently named 'procentriole').

Is this strange process correlated with left/right patterning?

The reason for this process is not clear: prevention of unwanted or unchecked parthenogenesis (Washitani-Nemoto 1994; Manandhar 2000) is realized by extrusion of centrosomes and centrioles during oogenesis. Then, why are centrioles and centrosomes reduced also during male spermiogenesis. As seen, Kupffer's vesicle histology and mirror behaviour of mice perinodal cells suggest that primary cilia are enantiomeric geometric organelles: what is the meaning of centriole enantiomerism? Does it imply a role in left/right patterning? Thinking of the centrosome as the 3D geometry organizer, this process could have an interesting interpretation. Before the first zygotic division, a unique, fundamental, one and for all process must occur, two different MCs, one left and one right, are assembled and (this process no longer can take place in descendant cells) from the left cells will originate only left cells equipped with the left MCs and the same will happen in the right cells (first blastomeres splitting, twins, and surgical manipulations, already discussed, are not considered here). Thus the process of centrosome and centriole reduction may be a mechanism to 'undress' centrioles. The 'naked' Mother centrioles can be tailored as left or right, avoiding the confusion of left or right centriole supply. Thus, sin left and right halves of Bilateria, left ('levo', '+') and right ('dextro', '−') enantiomeric mother centrioles (and then basal bodies, primary cilia, centrosomes) are orderly imposed.

Apparently, this process may occur in the very first zygotic division or immediately after as in some mammals and rodents where centrosomes appear at the 32/64 cells stage. Yet, in 2- and 4-cell mouse embryos, already after the initial zygote division inter-blastomere gene expression differences have been described (Biase et al. 2014). Mouse unfertilized oocytes, although clearly polarized, after fertilization appear to lose their polarity. However, recently, Ajduk and Zernicka-Goetz (2016), reviewing polarity and division orientation in Metazoa cleavage, observed that in mouse 'is likely that some, unknown yet, components, localized in a polarized way along the animal-vegetal axis, still affect cell divisions and developmental potential of the blastomeres'.

Avidor-Reiss (2015) suggested that mice spermatozoa may contribute a degenerated centriole that, after maturation to a complete centriole, functions in the zygote to assemble Daughter centrioles. They observe: 'There is a general belief that mice spermatozoa do not contribute actual centrioles to the zygote, but several recent studies suggest a paternal centriolar structure is contributed to the zygote. Injection of mouse spermatozoa into a cat ovum results in the formation of asters, at the base of the sperm nucleus, immediately after fertilization, suggesting the mouse sperm does provide an MT organizing center. Such fertilization of cat ovum by mice sperm results in the formation of bipolar spindles and successful zygote division. This study shows that the mice ovum is different from that of cat or other mammalian ova in that it forms astral MTs, not in the presence of an evident centriolar structure but through any paternal centriolar material (set of centriolar components) contributed to the zygote by the sperm' (see also Jin 2012). Another study on the protein Speriolin, a sperm centriole protein, also suggests that a centriolar structure may be present in mouse spermatozoa (Goto 2010). Spindle formation in the mouse zygote requires Plk4 (Coelho 2013): 'since Plk4 is the master regulator of centriole formation, this observation confirms that a centriolar structure is present in mice sperm. Furthermore, failure to eliminate maternal centrioles has been reported to result in multipolar mitotic spindles in *C. elegans* zygotes; therefore, maternal centriole loss or inactivation is essential for normal embryo development'.

11.10　Conclusions

Here I have tried to propose credible hypotheses, biochemically well-founded and compatible with evolutionary mechanisms, to explain how the centrosome can drive 10^{14} cells, putting them together. Hypotheses that show how the centrosome can play its geometric role for realizing reproducible and species-specific anisotropic directions of growth, reachable only by geometric instructions, DNA coded and shared by all cells. Through the centrosome, cells are able to orient themselves coordinately in different organs, decode and translate genomic geometric instructions in real forecast 3D locations. Now these hypotheses must be experimentally tested, undoubtedly a difficult task, given that in a cell there is only one centrosome.

References

Abouchar L, Petkova MD, Steinhardt CR, Gregor T (2014) Fly wing vein patterns have spatial reproducibility of a single cell. J R Soc Interface 11:20140443

Ajduk A, Zernicka-Goetz M (2016) Polarity and cell division orientation in the cleavage embryo: from worm to human. Mol Hum Reprod 22(10):691–703

Alliegro MC, Alliegro MA, Palazzo RE (2006) Centrosome-associated RNA in surfclam oocytes. Proc Natl Acad Sci U S A 103:9034–9038

Alvarado AS (2004) Planarians. Curr Biol 14:R737–R738

Avidor-Reiss T (2015) Atypical centrioles during sexual reproduction. Front Cell Dev Biol 3:21

Azimzadeh J (2014) Exploring the evolutionary history of centrosomes. Philos Trans R Soc B 369:20130453

Azimzadeh J, Wong ML, Downhour DM, Sánchez Alvarado A, Marshall WF (2012) Centrosome loss in the evolution of planarians. Science 335(6067):461–463

Basto R, Lau J, Vinogradova T, Gardiol A, Woods CG, Khodjakov A, Raff JW (2006) Flies without centrioles. Cell 125(7):1375–1386

Beisson J, Jerka-Dziadosz M (1999) Polarities of the centriolar structure: morphogenetic consequences. Biol Cell 91:367–378

Biase FH, Cao X, Zhong S (2014) Cell fate inclination within 2-cell and 4-cell mouse embryos revealed by single-cell RNA sequencing, vol 24. Cold Spring Harbor Laboratory Press, Cold Spring Harbor, pp 1787–1796

Brown NA, Wolpert L (1990) The development of handedness in left/right asymmetry. Development 109:1–9

Bryant SV, French V, Bryant PJ (1981) Distal regeneration and symmetry. Science 212:993–1002

Coelho PA (2013) Spindle formation in the mouse embryo requires Plk4 in the absence of centrioles. DevCell 27:586–597

Danilchik MV, Brown EE, Riegert K (2006) Intrinsic chiral properties of the Xenopus egg cortex: an early indicator of left-right asymmetry? Development 15:4517–4526

Feldman JL (2007) The mother centriole plays an instructive role in defining cell geometry. PLoS Biol (5):e149

Ferreira RR, Vilfan A, Pakula G, Supatto W, Vermot J (2018) Chiral cilia orientation in the left-right organizer. bioRxiv. https://doi.org/10.1101/252502

Foe VE (1989) Mitotic domains reveal early commitment of cells in Drosophila embryos. Development 107:1–22

Gáspár I, Ephrussi A (2017) RNA localization feeds translation. Science 357(6357):1235–1236

Geimer S, Melkonian M (2004) The ultrastructure of the *Chlamydomonas reinhardtii* basal apparatus: identification of an early marker of radial asymmetry inherent in the basal body. J Cell Sci 117:2663–2674

Goto M (2010) Speriolin is a novel human and mouse sperm centrosome protein. HumReprod 25:1884–1894

Guichard P (2010) Procentriole assembly revealed by cryo-electron tomography. EMBO J 29:1565–1572

Hildreth V, Webb S, Chaudhry B (2009) Left cardiac isomerism in the Sonic hedgehog null mouse. J Anat 214:894–904

Hoops H, Witman G (1983) Outer doublet heterogeneity reveals structural polarity related to beat direction in Chlamydomonas flagella. J Cell Biol 97:902–908

Jin YX (2012) Cat fertilization by mouse sperm injection. Zygote 20:371–378

Kitagawa D (2011) Structural basis of the 9-fold symmetry of centrioles. Cell 144:364–375

Li G, Liu X, Xing C, Zhang H, Shimeld SM, Wanga Y (2017) Cerberus–Nodal–Lefty–Pitx signalling cascade controls left–right asymmetry in amphioxus. Proc Natl Acad Sci U S A 114:3684–3689

Loukil A, Tormanen K, Sütterlin C (2017) The daughter centriole controls ciliogenesis by regulating Neurl-4 localization at the centrosome. J Cell Biol 216(5):1287–1300

Manandhar G (2000) Centrosome reduction during mammalian spermiogenesis. Curr Top Dev Biol 49:343–363

Mennella V (2014) Amorphous no more: subdiffraction view of the Pericentriolar Material architecture. Trends Cell Biol 24:188–197

Morris VB (2007) Origins of radial symmetry identified in an echinoderm during adult development and the inferred axes of ancestral bilateral symmetry. Proc Biol Sci 27:1511–1516

Nance J, Zallen JA (2011) Elaborating polarity: PAR proteins and the cytoskeleton. Development 138(5):799–809

Nonaka S, Shiratori H, Saijoh Y, Hamada H (2002) Determination of left-right patterning of the mouse embryo by artificial nodal flow. Nature 418:96–99

Okabe N, Xu B, Burdine RD (2008) Fluid dynamics in zebrafish Kupffer's vesicle. Dev Dyn 237:3602–3612

Regolini M, Barvitenko N, Lawen A, Muhammad A, Pantaleo A, Saldanha C, Skverchinskaya E, Tuszynski JA (2018) Integration of intracellular signaling: biological analogues of wires, processors and memories organized by a centrosome 3D reference system. Biosystems. https://doi.org/10.1016/j.biosystems.2018.08.007

Ruiz F, Garreau N, Klotz C, Koll F, Beisson J (2006) Centrin deficiency in Paramecium affects the geometry of basal-body duplication. Curr Biol 2097:106

Sánchez-Huertas S, Lüders J (2015) The augmin connection in the geometry of microtubule network. Curr Biol 25(7):R294–R299

Schubiger G, Schubiger M, Sustar A (2012) The three leg imaginal discs of Drosophila: "Vive ladifference". Dev Biol 369:76–90

Silflow C (2001) The Vfl1 protein in Chlamydomonas localizes in a rotationally asymmetric pattern at the distal ends of the basal bodies. J Cell Biol 153:63–74

Strachan T, Read A (2010) Human molecular genetics. Garland Science, New York, Chaps 9.3, 10 and p 334

Tassin AM, Lemullois M, Aubusson-Fleury A (2016) *Paramecium tetraurelia* basal body structure. Cilia 5:6

Tovey CA, Tubman CE, Conduit PC (2018) γ-Turc heterogeneity revealed by analysis of Mozart1. Curr Biol 28:2314–2323

Vandenberg LN, Levin M (2009) Perspectives and open problems in the early phases of left-right patterning. Semin Cell Dev Biol 20:456–463

Washitani-Nemoto S (1994) Artificial parthenogenesis in starfish eggs: behavior of nuclei and chromosomes resulting in tetraploidy of parthenogenotes produced by the suppression of polar body extrusion. Dev Biol 163:293–301

Xu J, Van Keymeulen A, Wakida NM, Carlto P, Berns MW, Bourne HR (2007) Polarity reveals intrinsic cell chirality. Proc Natl Acad Sci U S A 104:9296–9300

Yoshiba S, Hamada H (2014) Roles of cilia, fluid flow, and Ca2+ signalling in breaking of left–right symmetry. Trends Genet 30:10–17

Yoshiba S, Shiratori HY, Kuo IY, Aiko Kawasumi A, Shinohara K, Nonaka S, Asai Y, Sasaki G, Belo JA, Sasaki H, Nakai J, Dworniczak B, Ehrlich BE, Pennekamp P, Hamada H (2012) Cilia at the node of mouse embryos sense fluid flow for left-right determination via Pkd2. Science 338:226–231

Zonies S, Motegi F, Hao Y, Seydoux G (2010) Symmetry breaking and polarization of the C. elegans zygote by the polarity protein PAR-2. Development 137(10):1669–1677

Chapter 12
Coordination of Embryogenesis by the Centrosome in *Drosophila melanogaster*

Caitlyn Blake-Hedges and Timothy L. Megraw

Abstract The first 3 h of *Drosophila melanogaster* embryo development are exemplified by rapid nuclear divisions within a large syncytium, transforming the zygote to the cellular blastoderm after 13 successive cleavage divisions. As the syncytial embryo develops, it relies on centrosomes and cytoskeletal dynamics to transport nuclei, maintain uniform nuclear distribution throughout cleavage cycles, ensure generation of germ cells, and coordinate cellularization. For the sake of this review, we classify six early embryo stages that rely on processes coordinated by the centrosome and its regulation of the cytoskeleton. The first stage features migration of one of the female pronuclei toward the male pronucleus following maturation of the first embryonic centrosomes. Two subsequent stages distribute the nuclei first axially and then radially in the embryo. The remaining three stages involve centrosome-actin dynamics that control cortical plasma membrane morphogenesis. In this review, we highlight the dynamics of the centrosome and its role in controlling the six stages that culminate in the cellularization of the blastoderm embryo.

12.1 The Development of the Syncytial Embryo: Six Key Steps

Drosophila early embryo development occurs in a large syncytium in 13 rapid and synchronous nuclear cleavage cycles with 10–13 min separating each mitosis. These divisions occur over approximately 2 h, culminating in roughly 6000 nuclei that cellularize in interphase of cycle 14 to form the cellular blastoderm (Foe et al. 1993; Foe and Alberts 1983). During these early cleavage divisions, the centrosome coordinates cytoskeletal dynamics that are essential for proper development.

The centrosome is the major microtubule organizing center (MTOC) in most animal cells and is composed of two centrioles surrounded by the pericentriolar material (PCM) where microtubule assembly occurs. This coordination of

C. Blake-Hedges (✉) · T. L. Megraw
Department of Biomedical Sciences, College of Medicine, Florida State University,
Tallahassee, FL, USA
e-mail: cb16j@my.fsu.edu; timothy.megraw@med.fsu.edu

© Springer Nature Switzerland AG 2019
M. Kloc (ed.), *The Golgi Apparatus and Centriole*, Results and Problems in Cell
Differentiation 67, https://doi.org/10.1007/978-3-030-23173-6_12

microtubule production results in a polar microtubule array with the minus ends of the microtubules anchored at the centrosomes and plus ends that can rapidly grow and shrink. In the syncytial embryo, the centrosome is the only known MTOC. The centrosome has only recently been identified as an actin filament organizing center (Farina et al. 2016) but whether this is the case in the early embryo remains to be determined.

Here we describe the six key cell biological and developmental stages that rely on the centrosome and cytoskeletal dynamics during early embryo development. The first stage involves the maturation of the two centrioles contributed by the sperm, migration of one female pronucleus toward the male pronucleus, and the first zygotic division (Fig. 12.1a). The second stage consists of nuclear migrations that distribute the nuclei axially (Fig. 12.1b). The third stage is a perpendicular nuclear migration toward the cortex that generates the syncytial blastoderm (Fig 12.1c). The fourth, fifth, and sixth stages involve cortical membrane reorganization around each nucleus to generate cells (Fig. 12.1d–e). Each stage utilizes the centrosome in very different modes to organize the nuclei, assist in mitotic divisions, and/or form the first embryonic cells.

The first stage occurs during the initiation of embryogenesis, triggered by sperm entry through the anterior micropyle during fertilization in the uterus. Two paternally supplied centrioles mature and replicate utilizing maternally supplied PCM and centriolar proteins to form the first two embryonic centrosomes (Blachon et al. 2014). These centrosomes nucleate microtubules, termed the sperm aster, that assist in the migration of one female pronucleus toward the male pronucleus (Fig. 12.1a) (Callaini and Riparbelli 1996; Riparbelli et al. 2000). The first zygotic division is orchestrated by the newly formed centrosome pair, and four subsequent cleavage cycles precede the remaining centrosome-dependent stages.

During the second stage, axial nuclear migration, the early nuclei distribute evenly along the anterior-posterior (A-P) axis during cleavage cycles 4–7 (Fig. 12.1b). Localized actomyosin cortical contractions produce cytoplasmic streaming that assists in this nuclear migration (Royou et al. 2002; von Dassow and Schubiger 1994; Wheatley et al. 1995).

The third stage, cortical nuclear migration, positions the majority of the nuclei evenly along the cortex during cleavage cycles 7–9 (Fig 12.1c). Asymmetric microtubules nucleate preferentially toward the interior of the embryo to facilitate in this nuclear migration (Baker et al. 1993). A subset of nuclei, known as the yolk nuclei, remain in the interior of the embryo, complete error-prone replications that result in polyploid nuclei, and eventually lose their centrosomes (Fig. 12.1d) (Foe et al. 1993; Foe and Alberts 1983). Little is known of the molecular regulators of these nuclear migrations, but their function in positioning the nuclei is necessary for subsequent developmental stages (Niki and Okada 1981; Niki 1984; Okada 1982; Hatanaka and Okada 1991).

In the fourth stage, the nuclei that arrive in the posterior pole plasm during cortical migration are the first to cellularize, doing so during cleavage cycles 9–10. These nuclei cellularize before the remainder of the embryo to form the pole cells (primordial germ cells), the future gametic cells of the adult fly (Fig. 12.1d) (Foe and Alberts

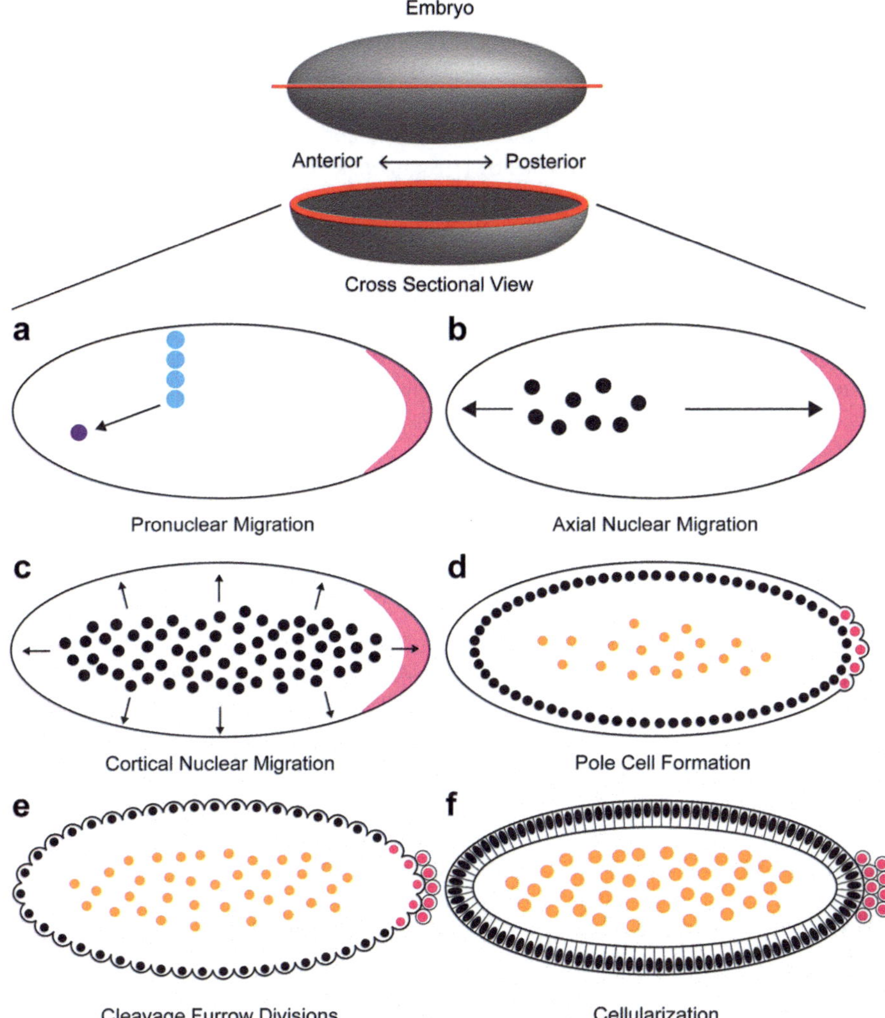

Fig. 12.1 The six stages of the *Drosophila* syncytial embryo that rely on centrosome-cytoskeletal dynamics. Each stage is viewed as a cross section through the anterior-posterior axis (red ring). (**a**) One female pronucleus (blue) migrates toward the male pronucleus (purple) to form the first zygotic nucleus. The pole plasm (pink), which is localized to the posterior of the oocyte during oogenesis, is present at the posterior pole of the embryo. (**b**) During cleavage cycles 4–7, the nuclei migrate along the anterior-posterior axis. (**c**) During cleavage cycles 7–9, a majority of the nuclei migrate to the cortex. (**d**) During cleavage cycles 9–10, a subset of nuclei at the posterior pole cellularize to form the pole cells (pink circles). The yolk nuclei (orange) remain in the interior of the embryo. (**e**) The final four cleavage divisions (10–13) occur at the cortex where membrane invaginations surround each dividing nucleus. (**f**) After the 13th cleavage cycle, the cortical nuclei form distinct cellular membranes during interphase of cycle 14

1983). The pole plasm contains germ cell-specific proteins and mRNAs that are localized to the posterior of the oocyte in a microtubule-dependent manner during oogenesis (Lantz et al. 1999; Mahowald 2001). The pole plasm, which is necessary and sufficient to drive pole cell cellularization, is contained in polar granules that transport to the nuclei dependent on centrosomes and microtubules (Illmensee and Mahowald 1974; Lerit and Gavis 2011; Shamanski and Orr-Weaver 1991). The centrosomes coordinate reorganization of the plasma membrane to surround each nucleus as it divides, until the membrane is pinched off to form separate cells (Fig. 12.1d) (Raff and Glover 1989).

During the fifth stage, the remaining cortical nuclei complete four final divisions that utilize centrosome-dependent actin-microtubule dynamics to reorganize the cortical plasma membrane (Fig. 12.1e). The membrane dynamics resemble the organization of the posterior membrane during the fourth stage, but the membrane does not seal or close to form new cells. These membrane arrangements are termed pseudo-cleavage furrows, or Rappaport furrows (Raff and Glover 1988; Ede and Counce 1956; Turner and Mahowald 1976; Foe and Alberts 1983). These final divisions are important for increasing nuclear numbers and priming the embryo for cellularization.

The final stage succeeds the 13th division and occurs in the 70-min-long interphase of cleavage cycle 14 (Foe and Alberts 1983). The cortical nuclei are surrounded by long membrane invaginations rich in actin and cytokinetic components that cleave at the base to form cells (Fullilove and Jacobson 1971; Warn and Robert-Nicoud 1990; Young et al. 1991) (Fig. 12.1f). The centrosomes and microtubules assist in the membrane invaginations and eventual cell formation. This last step transitions the syncytial embryo to the cellular blastoderm (Zalokar and Erk 1976; Foe and Alberts 1983).

12.2 The Structure of the Embryonic Centrosome and Regulation of Microtubule Assembly

The embryonic centrosome is organized into a pair of centrioles surrounded by the PCM from which microtubules are nucleated and regulated. The embryonic centrioles have a canonical structure similar to differentiated tissue and mammalian centrioles with slight variation in length and the number of radial microtubules. *Drosophila* centrioles do not contain distal and subdistal appendages, structural features found on vertebrate mother centrioles (Callaini et al. 1997). The structure of the syncytial embryo centrioles remains constant throughout all of embryogenesis and into the larval stages, indicating that the structure is not unique to the specificities of the syncytial embryo (Callaini et al. 1997; González et al. 1998).

The embryonic centrioles are ~0.2 µm wide and long, composed of nine doublet microtubules that are all equal in length (Fig. 12.2) (Debec et al. 1999; Moritz et al. 1995; Lattao et al. 2017). The centrioles contain a "cartwheel" structure with a

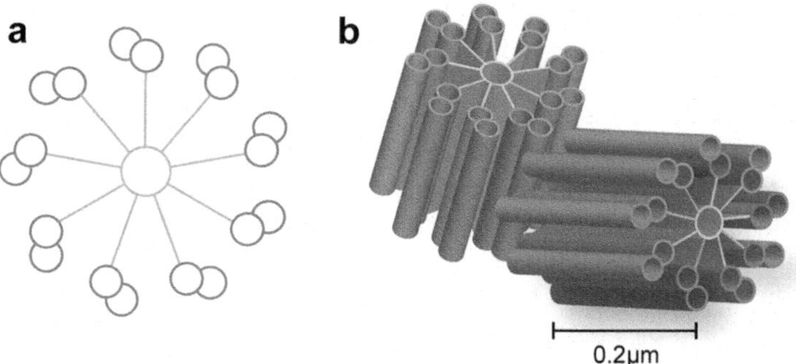

Fig. 12.2 The centrioles of the syncytial *Drosophila* embryo. (**a**) Top view of a centriole containing nine peripheral microtubule doublets (gray) connected through radial spokes to the central hub (brown). (**b**) The centrosome contains two pairs of centrioles that are each as long as they are wide, about 0.2 µm. The centrioles are orthogonal to one another, and the daughter centriole is located at the proximal end of the mother

central hub linked to each peripheral doublet through radial spokes along the entire length (Fig. 12.2) (Debec et al. 1999; Callaini et al. 1997). Differentiated tissues, such as wing epidermal cells (Tucker et al. 1986) and ommatidia sensory bristles (Mogensen et al. 1993), contain microtubule triplets absent of the cartwheel structure, while midgut epithelial cells and rhabdomeric cells contain microtubule doublets with the cartwheel structure (Gottardo et al. 2015). Therefore, over development, some specified cells have different centriolar microtubule compositions. The reasons for these differences and any differential functions they might impart are not known.

The embryo is primed to construct centrioles thanks to the maternal supply of centriolar components, but centriole formation is blocked without the sperm centrioles. This block in activation of centriole biogenesis can be bypassed in unfertilized embryos through overexpression of centriole assembly/replication proteins such as Spindle assembly abnormal 6 (Sas-6) (Peel et al. 2007; Rodrigues-Martins et al. 2007a), Anastral spindle 2 (Ana2), Asterless (Asl) (Stevens et al. 2010), or Polo like kinase 4 (Plk4) (Peel et al. 2007; Rodrigues-Martins et al. 2007b), which drives de novo formation of centrioles.

While embryos are permissive for de novo centriole assembly, ovaries are not. Overexpression of either Spindle assembly abnormal 4 (Sas-4), Sas-6, or Plk4 during oogenesis still allows for the destruction of centrioles in the oocyte, resulting in unfertilized embryos initially absent of centrioles (Peel et al. 2007). De novo centriole formation is never seen in wildtype unfertilized embryos, indicating inhibitory mechanisms that limit centriole formation until fertilization. Dynein plays a negative regulatory role in centriole formation, as a dominant negative form of Dynein Heavy Chain 64C (Dhc64C) in unfertilized embryos causes de novo centriole formation (Belecz et al. 2001).

Most de novo centrioles maintain the typical embryonic architecture, such as during Plk4 overexpression (Rodrigues-Martins et al. 2008), but in some cases the centrioles have abnormal structures, such as during Sas-6 overexpression or in dominant negative Dynein heavy chain *Dhc64C* embryos. Tube-like structures rather than bona fide centrioles are produced, suggesting a precursory role in centriole biogenesis for Sas-6 and Dynein (Rodrigues-Martins et al. 2007a; Belecz et al. 2001). Although most of these embryos contain a large number of de novo centrioles, it is unclear whether the centrioles can replicate on their own. Rodrigues-Martins et al. showed that de novo centrioles from Plk4 overexpression can form procentrioles (Rodrigues-Martins et al. 2007b), but Peel et al. concluded no replication of de novo centrioles through live imaging of fluorescently tagged and overexpressed Plk4, Sas-4, or Sas-6 embryos (Peel et al. 2007). However, de novo centrioles can recruit PCM and nucleate microtubules (Peel et al. 2007; Rodrigues-Martins et al. 2007a, b, 2008; Stevens et al. 2010; Belecz et al. 2001).

One of the principle components of the PCM, γ-Tubulin, is the main microtubule nucleator at the centrosome. γ-Tubulin is expressed as two isoforms. γTUB37C, a maternal isoform, is expressed only in the ovaries and embryos. γTUB23C is the ubiquitous isoform, but the two isoforms are functionally redundant (Wilson et al. 1997). γ-Tubulin assembles into at least two different complexes that are important for microtubule nucleation and anchoring of the microtubule minus ends to MTOCs. The γ-Tubulin small complex (γ-TuSC) and the γ-Tubulin ring complex (γ-TuRC) are composed of γ-Tubulin complex proteins (GCPs) that contain grip domains which associate with γ-Tubulin and with other GCPs (Gunawardane et al. 2000; Oakley 2000; Farache et al. 2018; Lin et al. 2015; Kollman et al. 2015; Oakley et al. 2015).

γ-Tubulin is essential for syncytial embryo development (Tavosanis et al. 1997). *γTUB37C* mutants cannot nucleate astral microtubules and PCM recruitment is disrupted although spindles can still form (Wilson and Borisy 1998; Llamazares et al. 1999). γ-Tubulin complexes are recruited to the centrosome by the PCM component Centrosomin (Cnn) by the Centrosomin Motif 1 (CM1) domain of Cnn (Zhang and Megraw 2007; Chen et al. 2017).

In somatic tissues, Cnn is required for γ-Tubulin accumulation at the centrosome as well as astral microtubule production and PCM recruitment (Megraw et al. 1999; 2001; Mahoney et al. 2006). In the embryonic cleavage divisions of *cnn* maternal-effect mutants, PCM components such as γ-Tubulin are severely depleted at the centrosomes, but the centrioles can support a reduced amount of microtubule assembly, evident by small astral microtubules (Zhang and Megraw 2007). However, once the centrioles are lost, particularly in the later cleavage cycles, the centrioles cannot be properly maintained at the spindle poles, resulting in no detectable microtubule asters (Lucas and Raff 2007). Other PCM components are absent from the centrosome or are transiently recruited in *cnn* mutants, consistent with its primary role in recruiting PCM components (Zhang and Megraw 2007; Lucas and Raff 2007; Megraw et al. 1999; Vaizel-Ohayon and Schejter 1999).

The microtubules nucleated from γ-Tubulin are stabilized by Transforming acidic coiled-coil protein (Tacc) and Minispindles (Msps), which form a complex that is required for microtubule assembly and regulates astral microtubule length. Tacc-Msps localize at the centrosomes, and Tacc appears to recruit Msps. *tacc* mutants display a reduction in Msps localization at the centrosomes and a reduction in astral microtubules. Overexpression of Tacc causes a greater density of astral microtubules and more Msps recruitment to the centrosome than wildtype (Gergely et al. 2000a; Lee et al. 2001).

Tacc-Msps localization at the centrosomes is dependent on Aurora A (AurA) and the CM1 domain of Cnn (Gergely et al. 2000b; Barros et al. 2005; Cullen and Ohkura 2001; Lee et al. 2001; Giet et al. 2002; Zhang and Megraw 2007). *aurA* mutants display less localization of Tacc-Msps and shorter astral microtubules (Giet et al. 2002), while null *cnn* or CM1 domain mutants ($cnn^{\Delta 1}$) still partially recruit Tacc-Msps (Zhang and Megraw 2007). The CM1 of Cnn domain also recruits γ-Tubulin, and recent work has revealed that Msps orthologs (Stu2 and Alp14) directly binds to γ-Tubulin complex proteins to assist in microtubule nucleation through their tumor overexpressed gene (TOG) domains (Gunzelmann et al. 2018; Flor-Parra et al. 2018; Nithianantham et al. 2018). Therefore, reduced Tacc-Msps in *cnn* mutants may be due to reduced γ-Tubulin localization at the centrosomes.

For broader coverage of the centrosome and MTOCs, see Centrosomal and Non-centrosomal Microtubule-Organizing Centers (MTOCs) in *Drosophila melanogaster* (Tillery et al. 2018).

12.3   Fertilization and the First Zygotic Division

Embryogenesis of the zygote begins with syngamy of the haploid female and male pronuclei at fertilization. Sperm entry, pronuclear migration, and the first zygotic division all occur within 15 min and rely on the complementary contributions of the paternal centrioles and maternally supplied PCM components (Foe et al. 1993). The sperm enters the egg in the uterus and female meiosis, arrested in metaphase I, is activated by passage through the oviduct (Von Stetina and Orr-Weaver 2011). Meiosis produces four pronuclei that are arranged in a row perpendicular to the cortex (Fig. 12.3a–c). The sperm supplies two centrioles that immediately recruit maternal PCM components and assemble astral microtubules that stretch toward the cortex, termed the sperm aster (Fig. 12.3b–c) (Callaini and Riparbelli 1996; Riparbelli et al. 2000). The female pronucleus farthest from the cortex (and closest to the sperm aster) migrates toward the male pronucleus along the sperm aster until the two pronuclei are in apposition (Fig. 12.3c–d). The first zygotic division proceeds utilizing the newly matured centrioles and their templated daughter centrioles that form the first embryonic centrosomes (Loppin et al. 2015).

Two centrioles are supplied by the sperm: a larger "giant" centriole (GC), derived from the basal body, and the smaller unconventional centriole, referred to as the PCL (proximal centriole-like) (Fig. 12.3a) (Blachon et al. 2009; Blachon et al. 2014). The PCL is unconventional in that it lacks centriolar microtubules (Khire et al. 2016). During centriole maturation, maternal PCM components including Cnn, Asl, γ-Tubulin, Spindle defective 2 (Spd2), Pericentrin-like protein (Plp), and Centrosomal protein 190kD (CP190) localize at the GC and PCL upon sperm

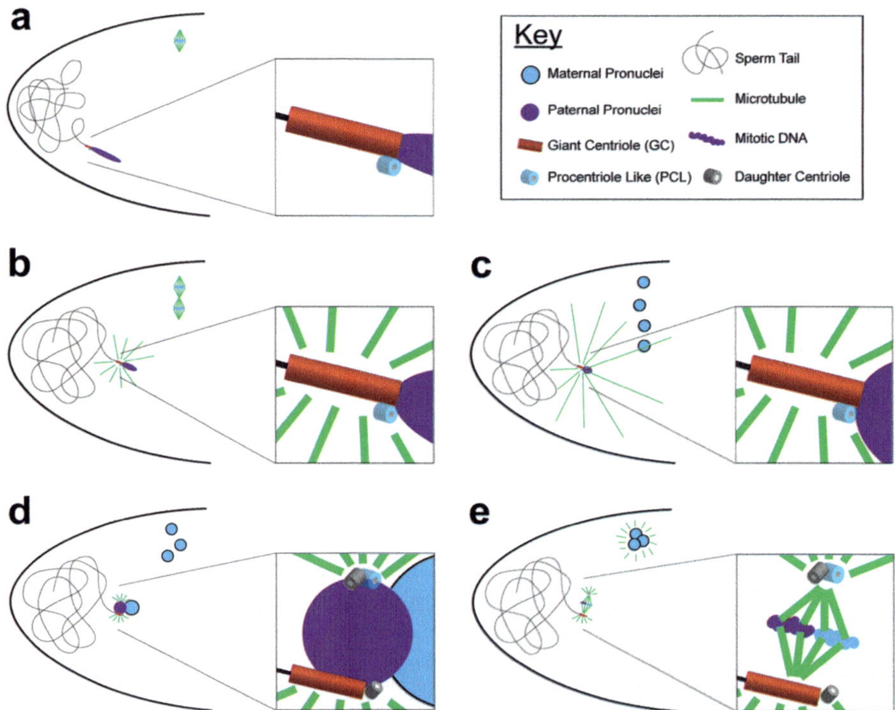

Fig. 12.3 Fertilization in the syncytial embryo utilizes paternally supplied centrioles for pronuclear migration and the first zygotic division. (**a**) The sperm enters the egg from the anterior micropyle, while the female chromosomes reactivate meiosis I. The sperm supplies two centrioles, the larger GC and smaller PCL. (**b**) The GC and PCL nucleate microtubules, termed the sperm aster, while the female chromosomes complete meiosis II. (**c**) The microtubules of the sperm aster reach the female pronucleus furthest from the cortex to facilitate in its migration. (**d**) The female pronucleus migrates toward the male pronucleus and the sperm aster diminishes. The GC and PCL separate to opposite poles of the male pronucleus, divide, and nucleate astral microtubules to prepare for the division. (**e**) The centrioles replicate to form two functioning centrosomes that aid in the gonomeric division of the female and male pronuclei. The remaining female pronuclei condense into polar bodies and utilize cytoskeletal elements to keep them separate and inactive. Figure based on Loppin et al. (2015)

entry (Blachon et al. 2014; Callaini et al. 1999; Khire et al. 2016). PCM recruitment to sperm centrioles is reliant on Spd, as *spd2* mutants recruit Asl but fail to recruit Cnn and have impaired sperm aster formation and pronuclear migration (Dix and Raff 2007). This requirement of Spd2 for sperm centriole maturation differs from somatic cells where loss of Spd2 only partially impedes PCM recruitment, allowing for significant centrosome activity to remain (Dix and Raff 2007; Giansanti et al. 2008).

Following PCM recruitment/maturation of the sperm centrioles and the completion of female meiosis, the sperm aster facilitates in the migration of one female pronucleus to the male pronucleus (Fig. 12.3c–d) (Loppin et al. 2015). The three remaining female pronuclei become polar bodies, which do not divide and remain throughout the syncytial embryo stages (Fig. 12.3c–d) (Dävring and Sunner 1973).

MTOC function and sperm aster assembly are essential for pronucleus migration. Loss of Asl or Spd2 disrupts sperm aster formation and pronuclear migration fails (Blachon et al. 2014; Varmark et al. 2007; Dix and Raff 2007). Loss of the PCM component, Asp, or the spindle microtubule regulators, Polo or Wispy (Wisp), produces a weaker phenotype where the sperm aster forms but does not fully extend toward the cortex, also preventing pronuclear migration (Riparbelli et al. 2002; Riparbelli et al. 2000; Brent et al. 2000). Loss of Tacc, which regulates microtubule stability, results in pronuclear migration failure. *tacc* mutants that survive to later stages of embryogenesis display diminished aster and spindle microtubules, suggesting failure of pronuclear migration may be due to a diminished sperm aster (Gergely et al. 2000b).

The Linker of Nucleoskeleton and Cytoskeleton (LINC) complex helps maintain centrosome-nucleus connections (Hieda 2017) and facilitates female pronuclear migration in zebrafish and *Caenorhabditis elegans* embryos (Lindeman and Pelegri 2012; Malone et al. 2003). However, in *Drosophila*, Klarischt (Klar) and Muscle-Specific Protein 300 kDa (Msp300), both LINC complex components, are not necessary for pronuclear migration (Technau and Roth 2008). In *C. elegans,* the LINC complex cooperates with microtubule motor proteins to assist in pronuclear migration (Meyerzon et al. 2009), and while the LINC complex is not necessary in *Drosophila*, motors are needed. What tethers nuclei to the cytoskeleton during pronuclear migration and other nuclear movements in early *Drosophila* embryos remains unknown.

The kinesins Non-claret disjunctional (Ncd), Subito (Sub), and Kinesin-like protein at 3A (Klp3A) all play roles in pronuclear migration. The female pronucleus migrates in a minus-end-directed manner, and Ncd, a microtubule minus-end-directed motor, assists in this migration in conjunction with an isoform of α-Tubulin, αTub67C (Komma and Endow 1997). αTub67C is a maternal-specific isoform of α-Tubulin with the special property of conveying faster microtubule assembly (Venkei et al. 2006). Sub is a kinesin involved in antiparallel microtubule bundling and *sub* mutants display similar phenotypes to *polo, wispy,* and *α-tub67C* mutants, where the mitotic spindles do not form properly and the embryos arrest early without any zygotic divisions (Giunta et al. 2002; Cesario et al. 2006). Therefore, Sub may play a role in attaching the female pronucleus to the sperm aster through microtubule interactions. Loss of Klp3A prevents pronuclear migration, but Klp3A is a plus-end-directed motor, implicating an indirect role in pronuclear migration. Klp3A recruits Polo, suggesting that it may regulate the formation of the sperm aster, which is necessary for pronuclear migration (Glover 2005; Williams et al. 1997).

During migration the pronuclei swell until the nuclei are in apposition to one another, resulting in a slightly larger female pronucleus (Fig. 12.3d) (Callaini and Riparbelli 1996). The paternal centrioles separate to opposite poles once the pronuclei are apposed and template daughter centrioles to form two functioning centrosomes that aid in the first zygotic division (Fig. 12.3d–e) (Blachon et al. 2014).

This division is gonomeric because the female and male chromosomes remain separated on the metaphase plate until telophase when they join to form two diploid zygotic nuclei (Fig. 12.3e) (Callaini and Riparbelli 1996).

In order to properly replicate and recruit PCM proteins, the PCL requires unique components such as Proteome of centrioles 1 (Poc1). In *poc1* mutant testis, the PCL does not assemble and sperm contain only a GC. In *poc1* paternal effect mutant embryos, PCM proteins are recruited to the GC only (because no PCL is delivered with the sperm), resulting in monopolar spindles that contain only the GC and its replicated daughter centriole (Khire et al. 2016).

The ultrastructure of each centriole goes through multiple changes, and centriolar components such as Anastral spindle 1 (Ana1), Ana2, Asl, Sas-4, and Sas-6 are stripped away during spermatogenesis (Khire et al. 2016; Blachon et al. 2014). These components are maternally supplied in the embryo, and at least Asl is recruited to the sperm centrioles, while Sas-4 and Sas-6 remain absent from the GC and PCL. However, when the centrioles replicate, Sas-4 and Sas-6 are present at the newly formed daughter centrioles (Blachon et al. 2014).

For more on *Drosophila* fertilization, see a recent review (Loppin et al. 2015).

12.4 The Syncytial Embryo Employs an Adapted Cell Cycle

Due to the accelerated pace of nuclear divisions, the syncytial embryo involves a modified cell cycle that does not utilize gap phases but only S and M phases until cellularization (Glover et al. 1989). Because of the fast transitions from mitosis to interphase, there is also a severe reduction in transcription until the maternal to zygotic transition (MZT) during the tenth cleavage division (Lamb and Laird 1976; McKnight et al. 1977; Zalokar 1976; Edgar and Schubiger 1986). Instead, the embryo relies on the activities of maternally supplied proteins and mRNAs to execute the syncytial nuclear divisions (O'Farrell 2015; Lasko 2012).

The major cell cycle regulators, Cyclin-dependent kinases (Cdks) and Cyclins, are important in managing the timing of mitosis in the syncytial blastoderm. Cdks and Cyclins cooperate to regulate the timing of protein activation during various stages of the cell cycle. Typically, Cdk and Cyclin levels are regulated through temporal expression, but the lack of transcription in the early embryo results in a modified mechanism to support cleavage cycle regulation. In the embryo, Cdks diffuse through the embryo in waves, causing subsequent waves of mitosis (Deneke et al. 2016), whereas cyclins are locally degraded at the centrosomes to prevent global destruction that would halt further mitotic cycles (Huang and Raff 1999; Raff et al. 2002).

The syncytial nuclei divide in a synchronous wave regulated by Cdk1 that propagates along the A-P axis. Cdk1 forms complexes with Cyclin A and B to regulate entry into mitosis. Because diffusion would take too long for mitotic activation in the embryo, Cdk1 propagates throughout the embryo as a wave that signals the nuclei to enter mitosis. Particularly in the later cleavage cycle stages,

Cdk1 waves spread throughout the embryo during S phase to trigger mitosis, resulting in a subsequent wave of mitosis of the nuclei (Deneke et al. 2016).

Both Cyclin A and B are involved in regulating the syncytial embryo cell cycle. Cyclin A is localized at the nucleus and regulates the duration of the entire cell cycle, as decreased Cyclin A results in a longer cell cycle, while the mitotic index remains the same (Edgar et al. 1994; Stiffler et al. 1999). Cyclin B plays a more complex role, regulating not only specific mitotic stages but also microtubule length and nuclear velocity in the migration stages. Cyclin B localizes to the spindle microtubules during metaphase and astral microtubules in later mitotic stages (Huang and Raff 1999; Stiffler et al. 1999). Decreasing Cyclin B levels results in longer astral and spindle microtubules, as well as centrosome detachment from their respective nuclei. Increasing Cyclin B levels causes shorter astral and spindle microtubules, resulting in nuclear spacing defects (Stiffler et al. 1999).

Cyclin B is destroyed during metaphase and this destruction is localized to the spindle microtubules, starting at the centrosomes (Huang and Raff 1999) and catalyzed by Cdc20 (Fizzy (Fzy))-dependent Anaphase-promoting complex (APC) activation (Raff et al. 2002; Sigrist et al. 1995). Fzy is localized to the centrosomes, spindle microtubules, and kinetochores during the start of mitosis and begins to disappear during metaphase. The localization of Fzy to the centrosome is microtubule-dependent as colcemid treatment (a microtubule depolymerizer) causes Fzy to localize strictly to the kinetochores. It is hypothesized that the Fzy-APC complexes are activated at the centrosomes and spread to the kinetochores due to the localization of Fzy (Raff et al. 2002) and that centrosome and spindle attachment is necessary for Cyclin B destruction (Wakefield et al. 2000).

When centrosomes detach from the spindle and Cyclin B is not destroyed, spindles arrest in mitosis, which is also seen in *sas-4* mutants (spindle arrest, absent of centrioles), supporting a role for the centrosomes and Cyclin B destruction (Stevens et al. 2007; Wakefield et al. 2000). Pan gu (png) forms a complex with Giant nuclei (Gnu) and Plutonium (Plu) to regulate Cyclin B levels, specifically in the early embryo. It is hypothesized that this complex works to stabilize Cyclin B, as Cyclin B levels are decreased in either *png*, *glu*, or *plu* mutants (Fenger et al. 2000). These mutants also display DNA replication without division resulting in polyploid nuclei, as well as centrosome detachment from the spindles. (Freeman et al. 1986; Freeman and Glover 1987; Elfring et al. 1997; Shamanski and Orr-Weaver 1991). This phenotype is also seen in embryos lacking Cyclin B, supporting the idea of localized Cyclin B destruction at the centrosome (Stiffler et al. 1999). The detached centrosomes continue to replicate, uncoupled from the DNA replication cycle, indicating a mechanism by which the centrosome and DNA replication cycles can be uncoupled (Freeman et al. 1986; Freeman and Glover 1987; Elfring et al. 1997; Shamanski and Orr-Weaver 1991).

In aphidicolin-injected embryos (DNA replication inhibitor) that have a prolonged S phase, centrosomes separate from the nuclei, and over-replication of centrosomes occurs (Raff and Glover 1988; Debec et al. 1996). Also in *dhc64C* mutants, centrosomes separate from the nuclei and continue to replicate, leading to excessive centrosome replication (Belecz et al. 2001). Similarly, in *mcph1* mutants where S phase is prolonged causing an increase in the DNA replication cycle length, the length of the centrosome replication cycle stays the same, resulting in excessive centriole replication (Brunk et al. 2007).

Centriole replication throughout the syncytial embryo mitoses is regulated by the same components involved in initial centriole formation during fertilization. The centriole assembly proteins, Sas-4, Sas-6, and Plk4, are required for the formation of the daughter centriole during centrosome replication. In *sas*-4, *sas*-6, or *plk4* maternal mutants, centrioles are not formed, and embryos arrest early with very few divisions that have abnormally shaped anastral spindles. Additionally, PCM components such as Cnn are not recruited to the spindle poles (Rodrigues-Martins et al. 2008; Stevens et al. 2007). Overexpression of either Sas-6 or Plk4 causes excessive centrosome replication, due to their role in centriole biogenesis (Peel et al. 2007; Rodrigues-Martins et al. 2007b). Excessive centrosome replication due to Sas-6 overexpression is exacerbated by the loss of Centriole Coiled Coil Protein 110 kDa (CP110). *cp110* mutants cause excessive centrosome replication when either Asl or Ana2 was overexpressed, which do not display abnormal centrosome replication on their own (Franz et al. 2013). Therefore, CP110 negatively regulates centrosome duplication through Sas-6, Asl, and Ana2.

12.5   Centrosome-Nucleus Association

As in most cell types and organisms, centrosomes are closely linked to the nuclei of the syncytial embryo through microtubule interactions, motor proteins, and microtubule-associated proteins (MAPs). Close centrosome-nuclear localization allows for rapid assembly of the mitotic spindle during the quick transitions of the cleavage cycle, as well as aiding in force mechanisms for nuclear migration and positioning. Due to the syncytial nature of the embryo, this association is important to prevent centrosomes from drifting away from their respective nuclei, which is less of a concern in the containment of a cell. Free centrosomes can disrupt nuclear divisions as well as prevent proper nuclear positioning; therefore, the syncytial embryo requires unique mechanisms to keep the centrosome-nuclear association intact.

Nuclear envelope breakdown in the embryo deviates from the canonical cell cycle, as the centrosomes remain extremely close to the nuclei. During prophase, the nuclei become indented near the centrosomes; during prometaphase, portions of the nuclear envelope breakdown at these indents, theoretically due to the astral microtubules piercing the nucleus. Remnants of the nuclear envelope continue to surround the spindle until telophase, when it fully breaks down and reforms at the two newly separated nuclei (Stafstrom and Staehelin 1984; Paddy et al. 1996; Rothwell and Sullivan 2000).

The LINC complex is an obvious candidate for nuclear attachment to the centrosome; however, loss of the LINC complex components Klar, Msp300, or Klaroid (Koi) does not display any obvious centrosome-nuclear attachment defects

(Archambault and Pinson 2010; Technau and Roth 2008). Rather, the microtubule motor protein dynein and PCM proteins are necessary for this attachment. *dhc64c* mutants display detached centrosomes from interphase/prophase nuclei as well as spindle poles, and centrosomes often fail to separate properly during prophase (Robinson et al. 1999). Loss of Polo also causes interphase/prophase centrosome detachment, particularly during the cortical migration stage. This may be due to the recapturing of detached centrosomes during spindle formation when they are still close by, but during the migration stage, the nuclei move too far for the centrosomes to be recaptured, resulting in monopolar spindles. Centrosome detachment is not highly penetrant in *polo* mutants, but this phenotype is exacerbated by the overexpression of Microtubule-associated protein 205 (Map 205) or Greatwall (Gwl) (Archambault et al. 2007; Archambault et al. 2008). Map 205 sequesters Polo to microtubules during interphase, and Gwl antagonizes Polo via inhibition of the regulatory subunit of Protein phosphatase 2A, Twins (Wang et al. 2011).

Mutations in the gene for another MAP, Mars (HURP homolog), shows centrosome detachment from prophase nuclei when depleted, but more often centrosomes detach from the mitotic spindle (Zhang et al. 2009). Centrosome detachment from the spindle is also seen in *asp* mutants, evident by monopolar spindles (González et al. 1990). A syncytial embryo specific MAP, Toucan (Toc), localizes to the nuclear envelope and centrosomes during interphase and the spindle microtubules during mitosis. Mutant *toc* embryos display detached centrosomes from spindles and defective spindle formation. Astral microtubules remain intact and these embryos typically arrest early on in a metaphase state, indicating a specialized role for Toc in regulating syncytial mitotic spindles (Debec et al. 2001; Mirouse et al. 2005).

Microcephalin (MCPH1), which localizes to the centrosomes and spindle, is necessary for centrosome-spindle pole attachment as mutant *mcph1* embryos display detached centrosomes and monopolar spindles. These mutants arrest early on in a metaphase state but also have a delayed S phase, which results in uncoupling of the centrosome and cell cycles. Desynchronization of these cycles can result in over-replication of centrosomes, which may be the cause of the detached centrosomes (Brunk et al. 2007).

For more on centrosome attachment in the syncytial embryo, see Free Centrosomes: Where Do They Come From? (Archambault and Pinson 2010).

12.6 Axial Nuclear Migration Distributes Nuclei along the A-P Axis

The first 3–4 cleavage cycles are skewed toward the anterior end of the embryo where syngamy occurs (Karr 1991). Before the nuclei migrate toward the cortex and form the blastoderm, they organize into a uniform distribution along the A-P axis through axial nuclear migration (also known as axial expansion) (Fig. 12.1b) (Hatanaka and Okada 1991). Axial nuclear migration occurs during cleavage cycles

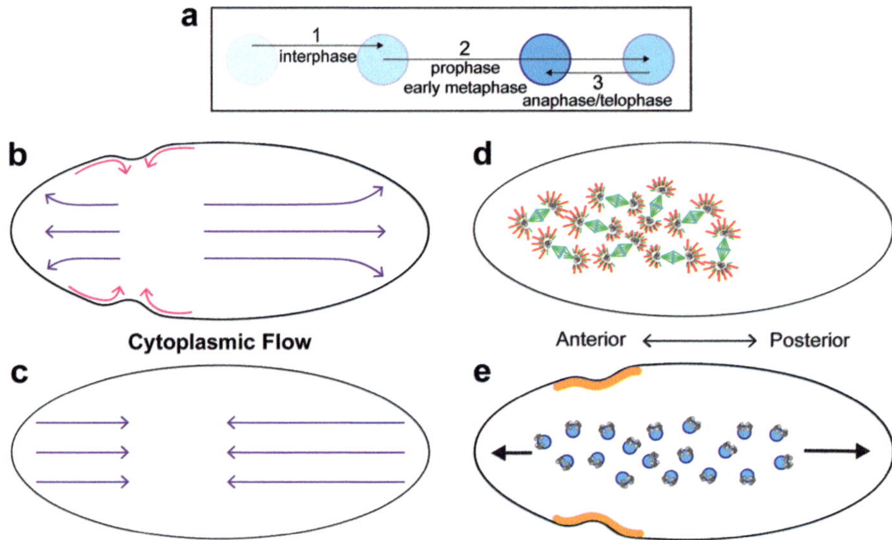

Fig. 12.4 During axial nuclear migration the nuclei migrate along the A-P axis due to localized contractions at the anterior cortex. (**a**) The nuclei (blue) slowly begin to migrate along the A-P axis during interphase. During prophase/early metaphase, the nuclei migrate faster along the A-P axis. The nuclei slightly retract along the A-P axis during anaphase/telophase. (**b**) Cross-sectional view of an embryo along the A-P axis during interphase of axial nuclear migration. The deep cytoplasm (purple arrows) moves toward the poles, while the peripheral cytoplasm (pink arrows) converges at the constriction point, which is slightly anterior at the cortex. (**c**) Cross-sectional view of an embryo along the A-P axis during anaphase/telophase of axial nuclear migration. The deep cytoplasm retracts and moves inwards away from the poles. (**d**) Cross-sectional view of an embryo along the A-P axis during metaphase of cleavage cycle 4. The centrosomes nucleate actin asters (red) that are more intense at the nuclei toward the poles and less intense at the inward nuclei. (**e**) Cross-sectional view of an embryo along the A-P axis during interphase of cleavage cycle 5. Actin and Myosin II (orange) localize to the constriction point, while the nuclei make an overall migration toward the anterior and posterior poles

4–7 and relies on cytoskeletal dynamics that are regulated through the cell cycle and the centrosome.

Actin and non-muscle myosin II (MyoII) localize at the anterior cortex in a cell cycle-dependent manner, where they control physical contractions of the embryo that assist in the axial nuclear migration (Fig. 12.4d) (Royou et al. 2002; von Dassow and Schubiger 1994; Wheatley et al. 1995). It is hypothesized that this contraction causes cytoplasmic streaming that forces the nuclei to migrate along the A-P axis, which is further supported by the cytoplasmic movements during these stages (Fig. 12.4b) (von Dassow and Schubiger 1994). A second hypothesis is that the actin network causes the cytoplasm to become stiffer in the middle of the embryo and looser toward the poles. Interacting microtubules from neighboring centrosomes repel the nuclei away from one another, forcing the nuclei to expand laterally along the A-P axis during these contractions. (Foe et al. 1993; von Dassow and Schubiger

1994). There is support for both of these hypotheses, and both may contribute to the nuclear movements during axial nuclear migration.

Three distinct phases of nuclear movements transpire during each cleavage cycle 4–7 at different stages of the cleavage cycle (Fig. 12.4a). First, the nuclei slowly begin to migrate toward the poles along the A-P axis at the end of interphase. In the second phase, this movement rapidly increases during prophase and early metaphase. Finally, nuclear movement slows down and they regress slightly along the A-P axis, away from the poles, during anaphase and telophase (Baker et al. 1993; von Dassow and Schubiger 1994).

The specificity of cell cycle stages during these nuclear movements suggests a role for cell cycle regulation in axial nuclear migration. In support of this, increasing overall Cyclin B levels in the embryo decreases the velocity of nuclear movements, and decreasing overall Cyclin B levels in the embryo increases the velocity of nuclear movements (Stiffler et al. 1999). Therefore, Cyclin B regulates the nuclear velocity during these three phases of nuclear migration.

The cytoplasm contained inside the entire embryo displays movements that mimic nuclear migration. As the nuclei move toward the pole, the cytoplasm deep in the middle of the embryo moves outward toward the poles as well (Fig. 12.4b). The cytoplasm at the periphery of the cortex flows toward the middle of the embryo, and the two distinct waves, from the posterior and anterior, converge slightly anterior at the cortex, at what is termed the constriction point. Because of the opposing flows of the deep cytoplasm and peripheral cytoplasm, this movement is referred to as fountain streaming. During the final phase of each nuclear movement when the nuclei slightly retract, the cytoplasm also retracts toward the middle of the embryo (Fig. 12.4c) (von Dassow and Schubiger 1994).

Axial nuclear migration relies on actin filaments and not microtubules as colchicine treatment (a microtubule depolymerizer) does not affect A-P nuclear distribution, but cytochalasin (prevents actin polymerization) does inhibit it (Zalokar and Erk 1976; Hatanaka and Okada 1991). Filamentous actin (F-actin) appears to nucleate from the centrosomes starting at metaphase, growing through telophase, and dispersing in interphase. These actin asters are greatest at the outward nuclei, closest to the poles, and weakest at the inward nuclei, furthest from the poles (Fig. 12.4d). Because the outward nuclei move more than the inward nuclei, it is suggested that these actin asters facilitate in axial nuclear migration (von Dassow and Schubiger 1994).

Loss of Grandchildless N26 (Gs(1)N26), Grandchildless N41 (Gs(1)N41), or Paralog (Par), all of which have not been mapped to a physical locus, cause actin to appear as a uniform layer over the cortex with rough aggregates and defective axial nuclear migration (Hatanaka and Okada 1991). This indicates they play a role in regulating actin distribution during axial nuclear migration.

Actin, together with MyoII which is also necessary for axial nuclear migration, shows a distinct localization progressing through cleavage cycles 4–7 (Kiehart et al. 1990; Wheatley et al. 1995; Royou et al. 2002). MyoII localizes at the cortex, slightly anterior, during interphase of the axial nuclear cleavage divisions (Fig. 12.4e). MyoII localizes to the anterior constriction site starting at interphase

4 and increases in intensity during the following interphase cycles until interphase 7, where it starts to disperse along the cortex. By interphase 8, when the nuclei are evenly distributed along the A-P axis, MyoII appears almost entirely disperse at the cortex. Actin also cycles to the cortex in a similar manner (Royou et al. 2002).

The cell cycle-dependent localization of MyoII is regulated by Cdk1 and Cyclin B. Localized degradation of Cyclin B during late anaphase inactivates Cdk1 (Su et al. 1998), and increasing Cyclin B levels prevents MyoII cortical localization, while inhibiting Cdk1 results in abnormal localization of cortical MyoII (Royou et al. 2002). However, cortical MyoII localization is not reliant on actin, as cytochalasin or latrunculin (prevents actin polymerization) injection does not disrupt MyoII localization (Chodagam et al. 2005). Additionally, in mutants for the regulatory light chain of MyoII, Spaghetti squash (Sqh), cortical actin localization is not disrupted, indicating it is independent of MyoII (Royou et al. 2002).

Antimyosin antibody injection or *sqh* mutants display defective axial nuclear migration (Kiehart et al. 1990; Wheatley et al. 1995), specifically, the primary activating phosphorylation site of Sqh, Serine 21 (Karess et al. 1991), is necessary for this migration. Phosphorylation site mutants that either mimic phosphorylation or prevent phosphorylation of Serine 21 both hinder axial nuclear expansion, indicating proper regulation of Sqh phosphorylation is necessary for axial nuclear migration (Jordan and Karess 1997). This site is phosphorylated by Rho kinase (Rok), and inhibition of Rok with Y-27632 also hinders axial nuclear expansion as well as MyoII distribution (Royou et al. 2002).

The centrosome, apparently acting through the centrosomal protein CP190, also plays a role in regulating axial nuclear migration. *cp190* mutants display defective axial nuclear migration and MyoII localization, but actin organization remains intact. The constitutively active Sqh phosphomimetic mutant can partially rescue this phenotype, implicating the centrosome in MyoII regulation. Although originally identified as a centrosomal protein, CP190 is best known as a chromatin insulator (Kellogg et al. 1989; Pai et al. 2004). CP190 localization is regulated during the cell cycle as it localizes to the centrosomes during mitosis and is nuclear during interphase (Chodagam et al. 2005). These localization dynamics may regulate the localization of MyoII during axial nuclear migration. Two unmapped genes regulate axial nuclear migration and affect centrosome localization: *shackleton* (*shkl*) and *out of sync* (*oosy*) mutants display defective axial nuclear migration, and centrosome loss from the spindles. *oosy* mutants also display asynchronous cleavage divisions (Yohn et al. 2003). Therefore, the timing of axial nuclear migration relies on cell cycle regulators that may involve the centrosome in an unknown way.

Overall, axial nuclear migration is a poorly understood process that is regulated by Rho1-dependent actin-MyoII dynamics and has an unclear connection to centrosomes. It is critical for the timely delivery of nuclei to the germ plasm during the cycle 10 window of pole cell formation. It also spreads the nuclei along the A-P axis to establish an even migration to the cortex.

12.7 Cortical Nuclear Migration Positions the Nuclei at the Cortex

During cleavage cycles 7–9, the majority of nuclei migrate from the interior of the embryo to the cortex (Fig. 12.1c) (Foe and Alberts 1983). A subset of nuclei, the yolk nuclei, fall back into the interior of the embryo between cleavage cycles 8 and 9 (Foe et al. 1993). Little is known about the function of these nuclei, but they are involved in yolk digestion (Bownes 1982) as well as the future development of the midgut (Walker et al. 2000).

The yolk nuclei asynchronously divide twice, then complete two rounds of DNA replication without divisions to become polyploid (Zalokar and Erk 1976; Foe and Alberts 1983). During these divisions, the centrosomes display defects in mitotic spindle organization that ultimately leads to centrosome loss at the yolk nuclei. During the first asynchronous division, a majority of the nuclei display defective centrosome separation, resulting in "V"-shaped spindles that resemble monopolar spindles. These aberrant spindles result in defective DNA segregation, yet the centrosomes continue to replicate, and a second round of abnormal divisions occurs. The replicated centrosomes also do not separate, as mother and daughter centrosomes remain close to one another (Callaini and Dallai 1991; Riparbelli and Callaini 2003). By the second yolk nuclei division, most of the centrosomes have detached from the spindles. The centrosomes appear normal, as CP190, Asp, Pavarotti, and γ-Tubulin all remain localized there. However, Cyclin B localization is disrupted as it does not associate with the aberrant spindles but weakly localizes to the centrosomes during anaphase and telophase. It does not localize to the entire spindle pole but rather the inner core of the centrosomes (Callaini and Dallai 1991; Riparbelli and Callaini 2003).

In contrast with axial nuclear migration, cortical nuclear migration relies on microtubules as colchicine treatment inhibits cortical nuclear migration, and does not rely on actin, as cytochalasin treatment does not inhibit cortical nuclear migration (Zalokar and Erk 1976). Centrosomes appear to carry nuclei during this migration, as centrosomes dissociated from their respective nuclei continue to migrate to the cortex (Raff and Glover 1989).

It was initially hypothesized that the centrosomes nucleate microtubules that connect to the cortex to pull the nuclei by forces produced by the microtubules (Wolf 1980). However, analysis of the microtubule arrays during cortical nuclear migration has disputed this theory. The majority of microtubules project inwards, rather than to the cortex, and interact with the microtubules of the yolk nuclei (Fig. 12.5). Shorter microtubules project toward the cortex, but do not reach the membrane (Fig. 12.5). Microtubules of neighboring nuclei, as well as nuclei at the opposite side of the embryo, also interact with one another (Fig. 12.5) (Baker et al. 1993).

The density of astral microtubules is far greater in the cortical nuclear migration stages compared to those during the axial nuclear migration stages (Baker et al. 1993). In *tacc* mutants, the astral microtubules are reduced or absent, and the nuclei

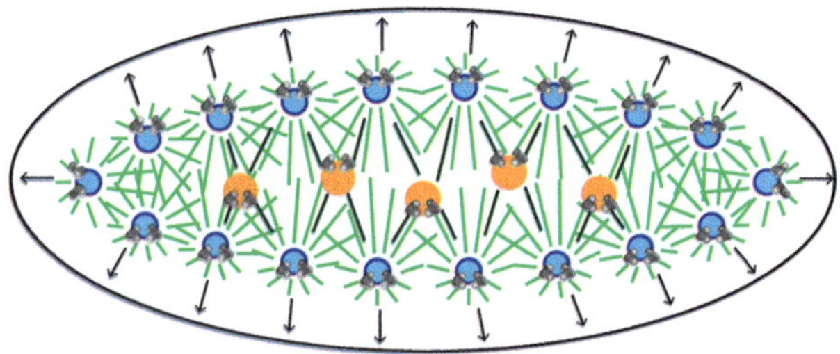

Fig. 12.5 Cortical nuclear migration relies on microtubules for proper migration. Cross-sectional view of a stage 7 embryo where the nuclei (blue) are migrating toward the cortex. The centrosomes (gray) emanate small astral microtubules (green) that stretch toward the cortex and long microtubules that stretch toward the yolk nuclei (orange). The yolk nuclei emanate microtubules (dark green) that interact with the migrating nuclei

do not migrate to the cortex (Gergely et al. 2000b). The current hypothesis is that antiparallel microtubule interactions provide a pushing force that directs the centrosomes, with their nuclei as passengers, toward the cortex (Fig. 12.5) (Baker et al. 1993; Raff and Glover 1989).

Very little is known about the molecular regulators of cortical nuclear migration, but it is coordinated with the cleavage cycle. The nuclear velocity is greatest during telophase when astral microtubules are most abundant, further supporting their necessity during this migration (Foe and Alberts 1983; Baker et al. 1993). As mentioned earlier, higher levels of cyclin B reduce the speed of nuclear migration, but this could be due to the decrease in microtubule length caused by the increased levels (Stiffler et al. 1999). In *oosy* mutants, the cleavage divisions are not synchronous and cortical migration is defective, suggesting that this synchrony is also important for proper migration (Yohn et al. 2003).

Overall, while little is understood about how cortical nuclear migration is regulated, it appears to be driven by astral microtubules assembled at centrosomes that grow toward the center of the embryo and which push the nuclei outward toward the cortex as they grow.

12.8 Pole Cells Cellularize Before the Other Nuclei

During cortical nuclear migration, a subset of nuclei reaches the posterior cortex and pole plasm during cleavage cycle 9. These nuclei will cellularize at cycle 10, becoming pole cells, while the majority of nuclei at the cortex continue to divide (Fig. 12.1d) (Foe and Alberts 1983). Pole cells will develop into the germ cells of the fly and contain germ cell-specific components. The germ plasm is localized to the posterior end of the oocyte during oogenesis and is contained in polar granules in the early embryo that are anchored by the actin cytoskeleton (Fig. 12.6a) (Lantz et al. 1999; Mahowald 2001).

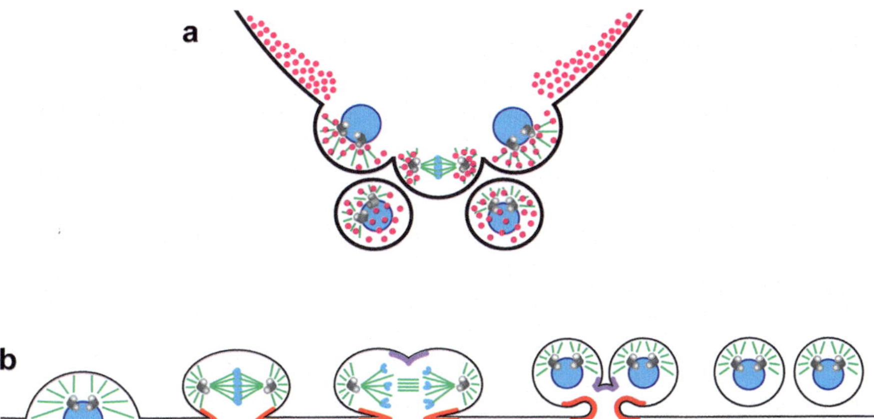

Fig. 12.6 Formation for the pole cells at the posterior cortex relies on membrane invaginations to cleave the cells as the nuclei divide. (**a**) The pole plasm is contained in polar granules (pink) that transport along the astral microtubules (green) toward the nuclei (blue) located at the cortex. (**b**) During prophase, the nuclei reach the posterior cortex and the centrosomes (gray) impinge on the plasma membrane to form the pole bud. (**c**) During metaphase, the BF (red) constricts beneath the furrow. (**d**) During anaphase, the BF remains and the AF (purple) forms above the chromosomes. (**e**) The two newly formed nuclei are cleaved from the plasma membrane at the bud furrow and anaphase furrow. (**f**) Two pole cells are formed after cleavage and remain localized at the posterior cortex

The pole plasm is sufficient to initiate pole cell formation, as transplantation of the pole plasm to the anterior cortex stimulates abnormal pole cell formation at the anterior pole (Illmensee and Mahowald 1974). Oskar (Osk), a germ cell-specific protein required to localize other germ cell-specific components, is also sufficient to cause abnormal pole cell formation when transplanted to the anterior pole (Ephrussi and Lehmann 1992). Germ-cell less (Gcl) is necessary for pole cell formation, but is not sufficient for the formation of pole cells when transplanted to the anterior pole (Jongens et al. 1992). However, when transplanted together with Anillin (encoded by *scraps*), which regulates actomyosin contractile rings and is also necessary for pole cell formation, they can stimulate abnormal pole cell formation at the anterior pole (Cinalli and Lehmann 2013; Field et al. 2005).

Gcl also plays a role in limiting pole cell formation to the posterior pole. Gcl promotes actomyosin organization downstream of the Rho1 pathway to constrict the plasma membrane and form the pole cells at the posterior pole (Cinalli and Lehmann 2013). The Arf-GEF Stepkke (Step) inhibits the Rho1-actomyosin pathway, and during pole cell formation stages, it is equally distributed around the cortical membrane of the embryo, preventing any cellular formation. Loss of Step activity results in abnormal pole cell formation at the anterior cortex due to loss of Rho1 inhibition. However, the posterior pole is distinct from the remainder of the embryonic membrane due to the presence of Gcl, which locally inhibits Step activity to allow for Rho1-mediated actomyosin membrane constriction. Loss of Step in a *gcl* mutant background allows for proper pole cell formation (Lee et al. 2015).

Another determinant of pole cell formation is proper axial nuclear migration. During interphase of cleavage cycle 9, the first nuclei reach the posterior pole in the proper time window for pole cell formation to begin. If axial nuclear migration is defective and the nuclei do not reach the posterior pole until a time point after cleavage cycle 10, pole cell formation is inhibited (Niki and Okada 1981; Niki 1984; Okada 1982; Hatanaka and Okada 1991). This is likely due to the degradation of germ plasm components before the nuclei can reach the posterior pole. When pole cell formation fails in *gs(1)N26* or *gs(1)N441* mutants, which display defective axial nuclear migration, the localized pole plasm components either degrade or delocalize from the posterior pole (Iida and Kobayashi 2000). In *shkl* or *oosy* mutant embryos, which also display defective axial expansion, pole cell numbers are lower than wildtype, further supporting the necessity of axial nuclear migration and timely arrival of nuclei at the posterior pole plasm for pole cell formation (Yohn et al. 2003).

When the nuclei reach the posterior membrane during cleavage cycle 9, they enter prophase and plasma membrane protrusions, called pole buds, form above each nuclei (Fig. 12.6b) (Foe et al. 2000; Warn et al. 1985; Cinalli and Lehmann 2013). As the nuclei begin to divide, the pole buds protrude farther from the membrane to surround the metaphase spindles (Fig. 12.6b). Membrane furrows, termed the bud furrows (BFs), form at the edge of the pole buds and constrict beneath the chromosomes at the basal membrane (Fig. 12.6b). Once the nuclei progress into anaphase, a second furrow, the anaphase furrow (AF), forms above the spindle opposite to the BF. Similar to the cytokinetic furrow, it forms in between the dividing nuclei (Fig. 12.6b). Once mitosis is complete, the AF constricts to separate the two nuclei into separate cells, while the BF constricts, liberating the nascent pole cells from the embryo (Fig. 12.6b) (Cinalli and Lehmann 2013).

Known cytokinetic factors Anillin, MyoII, Peanut (Pnut), and Rho guanine nucleotide exchange factor 2 (RhoGEF2) all localize to the actin-rich BFs in preparation for cleavage of the plasma membrane (Padash Barmchi et al. 2005; Warn et al. 1985; Field and Alberts 1995; Young et al. 1991). Anillin and MyoII also localize to the actin-rich AFs, but further analysis is required to determine other components of the AF. Rho1 functions upstream of cytokinetic components, and inhibiting Rho1 or Rok prevents pole bud formation, specifically diminishing Anillin localization at the BFs (Cinalli and Lehmann 2013). Anillin is necessary for cleavage of the pole cells, as *scraps* mutants display BFs that retract and never form pole cells (Field et al. 2005). *diaphanous (dia)* (a Formin downstream of Rho1) or *rhogef2* mutants display defective BF cleavage due to the disruption of the actomyosin contractile ring, as actin and MyoII are absent at the BFs (Afshar et al. 2000; Padash Barmchi et al. 2005).

The rate of BF constriction is regulated by Gcl, as overexpressing Gcl causes over-constriction of the BFs, resulting in the displacement of somatic nuclei and increased pole cell numbers (Cinalli and Lehmann 2013; Jongens et al. 1994). *gcl* mutants display under-constriction of the BFs, resulting in decreased pole cell numbers (Robertson et al. 1999; Cinalli and Lehmann 2013). In *gcl* mutants, the AF constricts to separate the cells but the BF never constricts, even though Anillin is present at both, preventing the formation of the pole cells. Therefore, Gcl is necessary for BF cleavage, but not AF cleavage (Cinalli and Lehmann 2013).

Both actin and microtubules are necessary for proper pole cell formation as injection of either Colcemid or cytochalasin inhibits pole cell formation (Raff and Glover 1989; Cinalli and Lehmann 2013). Colcemid treatment prevents AF cleavage but not BF cleavage and the nuclei arrest in metaphase, resulting in large pole cell-like cells with inappropriate DNA content (Cinalli and Lehmann 2013). Cytochalasin injection after pole cell formation causes the cells to collapse, indicating actin is necessary for pole cell stabilization (Raff and Glover 1989).

Centrosomes alone are sufficient to produce pole cells. Active centrosomes dissociated from nuclei during aphidicolin injection still migrate to the posterior pole and produce pole buds and reorganize actin. Pole cells form that are indistinguishable from normal pole cells except that they lack nuclei (Raff and Glover 1989). The centrosomal protein CP110 regulates pole cell formation. Neuralized E3 ubiquitin protein ligase 4 (Neurl4) localizes to the centrosome and downregulates CP110 levels in mammalian cells (Al-Hakim et al. 2012; Li et al. 2012). In *neurl4* mutant embryos, CP110 levels are elevated at centrosomes compared to wildtype as well as enriched at foci distinct from centrosomes. *neurl4* mutants display reduced pole cell numbers with abnormal morphologies due to CP110 overexpression, as the phenotype was partially rescued in *neurl4* mutants with the addition of one mutant copy of *cp110*, resulting in less abnormal pole cells (Jones and Macdonald 2015). How CP110 impacts the centrosome and affects pole cell formation is unclear.

Centrosome separation during mitosis regulates the formation and size of the BFs. *gcl* mutants display disrupted centrosome separation specifically in the pole cells and not somatic cells, which leads to shallow BFs (Lerit et al. 2017). These smaller BFs often retract, preventing their cellularization which leads to lower numbers of pole cells (Cinalli and Lehmann 2013). The percentage of pole buds with defective centrosome separation in *gcl* mutants correlates with the number of *gcl* mutant embryos that lack pole cells (Lerit et al. 2017; Robertson et al. 1999). Some nuclei overcome this BF defect and form pole cells, but they contain an abnormal number of centrosomes and multipolar spindles (Lerit et al. 2017). The CM1 domain of Cnn and also the Kinesin-5 motor protein (Klp61F), a plus-end-directed motor, are necessary for centrosome separation (Heck et al. 1993; Zhang and Megraw 2007), and *cnn*$^{\Delta 1}$ mutants or RNAi-mediated knockdown of Klp61F results in shallow, abnormally shaped BFs and reduced pole cell numbers (Lerit et al. 2017).

Defective centrosome separation also disrupts astral microtubule organization, which assists in the transportation of the polar granules that contain germ cell-specific mRNAs and proteins toward the cortical nuclei (Fig. 12.6a) (Lerit and Gavis 2011; Lerit et al. 2017). Polar granules migrate along the astral microtubules toward the centrosomes during interphase and remain localized around the centrosomes during mitosis until they segregate into separate cells during membrane cleavage (Fig. 12.6a) (Lerit and Gavis 2011). In *gcl* or *Klp61f* mutants where centrosomes do not properly separate and microtubules are not properly organized, polar granules still migrate from the posterior cortex, but their distribution around the nuclei is aberrant resulting in a reduced number of granules in pole cells (Lerit et al. 2017).

Microtubules are necessary for polar granule transport, as colcemid treatment prevents their movement, but actin is not as cytochalasin or latrunculin does not disrupt their transport (Lerit and Gavis 2011). In *png* mutants, active centrosomes travel to the posterior end without their respective nuclei (Shamanski and Orr-Weaver 1991), and polar granules are still trafficked along their astral microtubules, indicating centrosomes and astral microtubules are sufficient for not only pole cell formation but also transport of the polar granules. Cnn, Tacc, and AurA all regulate astral microtubule length and stability (Gergely et al. 2000b; Giet et al. 2002; Megraw et al. 1999), and loss of any of these three proteins results in impaired polar granules transport. Additionally, fewer pole cells are formed, and those that do form contain either a reduced level of polar granules or none at all (Lerit and Gavis 2011).

Polar granules rely on Dynein for trafficking along the astral microtubules in a minus-end-directed manner toward the centrosomes. In *dhc64c* mutants, astral microtubules remain intact, but the directed movement of polar granules is inhibited, resulting in a reduction of pole plasm in pole cells and reduced pole cell numbers (Lerit and Gavis 2011). Overexpression of Dynactin 2, p150 subunit (DCTN2-p150) inhibits dynein function (Burkhardt et al. 1997), resulting in cessation of polar granule motility. Conversely, the plus-end-directed motor kinesin does not function in trafficking polar granules, as *kinesin heavy chain* (*khc*) mutants display proper polar granule movement (Lerit and Gavis 2011). Therefore, the transport of polar granules relies on minus-end motors but not plus-end.

12.9   The Cortical Cleavage Cycles

After cortical nuclear migration, four final cleavage divisions (cycles 10–13) occur at the cortex before cellularization in interphase of cycle 14. These divisions form actin-rich furrows that resemble pole cell BFs, but the nuclei do not cleave off to form separate cells (Fig. 12.7a) (Raff and Glover 1988; Ede and Counce 1956; Turner and Mahowald 1976; Foe and Alberts 1983). Instead, these actin furrows are called pseudo-cleavage furrows (also known as cortical cleavage or Rappaport furrows). They are analogous to the furrows that form in-between adjacent nuclei from overlapping astral microtubules discovered by Rappaport in sand dollar embryos (Rappaport 1961). The centrosome regulates the formation of these furrows.

When nuclei reach the cortex during interphase of cleavage cycle 10, the centrosomes organize the plasma membrane to form a cortical bud, similar to the pole bud but not as protruding (Fig. 12.7a) (Raff and Glover 1988; Ede and Counce 1956; Turner and Mahowald 1976; Foe and Alberts 1983). Filamentous actin is rearranged from an evenly distributed layer along the cortical membrane to highly localized pockets directly over the centrosomes, termed actin caps (Fig. 12.7b) (Karr and Alberts 1986; Kellogg et al. 1988; Warn et al. 1984). Actin caps are necessary for even nuclear distribution along the cortex as disrupted cap formation due to

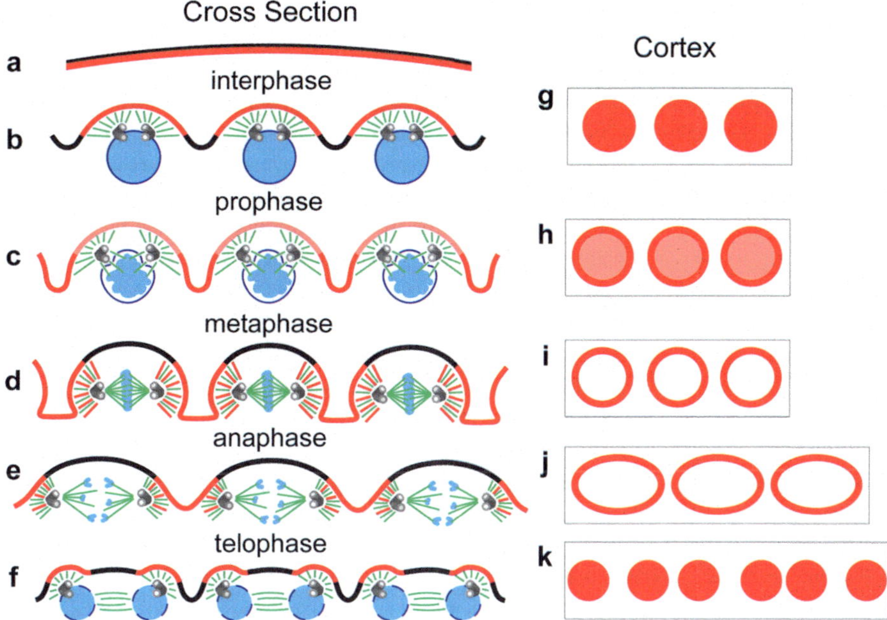

Fig. 12.7 The cortical cleavage cycles (**a**) During the cortical cleavage cycles, actin (red) is rearranged from an even distributed layer at the cortex, to pseudo-cleavage furrows that surround the dividing nuclei (blue). The centrosomes (gray) nucleate microtubules (thinner green lines) that assist in this actin redistribution. Astral actin filaments (thicker red lines) also surround the centrosomes. (**b**) At the cortex, actin is rearranged to form a cap above the nuclei (dark red) during interphase. As the nuclei divide, the actin cap expands (light red) into to surround the nuclei as they divide. During telophase, the separated chromosomes begin to form new actin caps

cytochalasin treatment or mutants that affect actin organization leads to abnormal clustering of nuclei (Stevenson et al. 2001; Zalokar and Erk 1976; Callaini et al. 1992). As the nuclei enter mitosis and the centrosomes separate, the actin caps expand and distribute into furrows of invaginated membrane (Fig. 12.7a–b). As mitosis proceeds, the furrows surround the spindles and then begin to recede during telophase to form actin caps over the centrosomes of the newly divided nuclei as they proceed into the next cleavage cycle (Fig. 12.7a–b) (Karr and Alberts 1986; Kellogg et al. 1988; Warn et al. 1984).

Actin cap expansion is necessary for furrow formation and relies on a number of actin nucleators to organize actin at the furrows. The Arp2/3 complex aids in actin nucleation to create branched actin filaments through the regulation of SCAR/WAVE proteins (Pollitt and Insall 2009). Actin-related protein 3 (Arp3), a component of the Arp2/3 complex, localizes to the furrows and between actin caps dependent on the centrosomal protein Scrambled (Sced) (Stevenson et al. 2002). Scrambled localizes to the same locations as Arp3, as well as the centrosomes, independent of microtubules. In *sced* mutants, the actin caps do not expand preventing furrow formation, which leads to spindle fusion and chromosome segregation errors (Stevenson et al. 2001). Loss of Actin-related protein 2/3 complex, subunit 1 (Arpc1), another component of the Arp2/3 complex, displays the same phenotype as *sced* mutants, suggesting Sced may recruit Arp2/3 to actin cap margins to aid in actin polymerization at the furrows (Stevenson et al. 2002; Zallen et al. 2002). It is unclear how Sced may recruit Arp2/3, but further studies have revealed that SCAR also localizes to the furrows and is required for furrow assembly (Zallen et al. 2002). Therefore, Sced may interact with SCAR to recruit Arp2/3 to the furrows, implicating a role for the centrosome in actin polymerization at the furrows.

Rho1 inhibition also prevents actin cap expansion and is an upstream regulator of the formin Dia that nucleates actin filaments (Cao et al. 2010; Watanabe et al. 1997). Dia localizes between the actin caps and furrows, specifically the tips of the advancing furrows (Afshar et al. 2000), and *dia* mutants do not form furrows due to defective actin cap expansion (Cao et al. 2010; Afshar et al. 2000; Webb et al. 2009). Dia is also required for the localization of cytokinetic components such as MyoII, Anillin, Pnut, and Adenomatous polyposis coli 2 (Apc2) to the invaginated furrows, and between the actin caps, as they weakly localize between the caps, but are absent from furrows (Afshar et al. 2000; Webb et al. 2009).

As the cleavage furrows form, new membrane and actin is supplied through recycling endosome (RE)-derived vesicles that are localized at the centrosome and transported along astral microtubules to the growing furrow (Swanson and Poodry 1981; Mermall et al. 1994; Mermall and Miller 1995; Rothwell et al. 1999; Riggs et al. 2003). The recruitment of these vesicles relies on Nuclear fallout (Nuf), which is an adaptor protein that links Rab11 to microtubule-based motors (dynein and Kinesin-1) for trafficking on microtubules (Riggs et al. 2003). Loss of either Nuf or Rab11 disrupts vesicle-based membrane recruitment and transport of furrow components, leading to disrupted actin furrows (Riggs et al. 2003; Rothwell et al. 1999). Nuf/Rab11 complexes localize Discontinuous actin hexagon (Dah), which is required for furrow formation, to the furrows, as *nuf* or *rab11* mutants display abnormal localization of Dah (Riggs et al. 2003; Rothwell et al. 1999; Zhang et al. 2000). Both *nuf* and *rab11* also display disruption in RhoGEF2 localization at the furrow, and injection of active RhoA (the mammalian ortholog of Rho1) in *nuf* mutants rescues the furrow phenotype. However, in *nuf* mutants, Rho1 and Dia localization is normal, suggesting their localization does not rely on RE vesicle transport, but RhoGEF2 localization does (Cao et al. 2008).

Endocytosis at the membrane occurs from interphase until metaphase, where it is inhibited to distribute membrane from RE-derived vesicles to the furrows as they are growing. Endocytosis once again occurs during telophase as the furrows are regressing (Sokac and Wieschaus 2008a; Rikhy et al. 2015). Dynamin, encoded by *shibire* (*shi*), and Clathrin are involved in endocytic vesicle formation and localize to the cleavage furrows and between actin caps at times of invagination, while localizing to the spindles during metaphase. In *shi* mutants, furrow formation is disrupted, and vesicles cannot sever from the plasma membrane, while Dia, Anillin, Pnut, involved in actin remodeling, are no longer localized to the furrows (Rikhy et al. 2015).

The cleavage furrows contain similar components to cytokinetic furrows (Miller and Kiehart 1995), including the centralspindlin complex, and yet cytokinesis does not occur (Crest et al. 2012; Minestrini et al. 2003). This is due to the absence of the cytokinetic regulator of Rho1, RhoGEF Pebble (Pbl), at the furrows and central spindle in the syncytial embryo. Instead, RhoGEF2 activates Rho1 locally at the cleavage furrows. Introducing ectopic active RhoA into the embryo during the cortical cleavage stages induces cytokinetic furrow formation over the central spindle, indicating that the machinery is in place but that spatial Rho1 is deterministic of the site of furrow formation (Crest et al. 2012). Rho1 and another major actin regulator Cdc42 play an antagonistic role in furrow formation. Constitutively active Cdc42 or dominant negative Rho1 both disrupt actin furrows and MyoII localization, but the microtubules remain intact (Crawford et al. 1998).

Centrosomes are sufficient for actin rearrangement and furrow ingression. In aphidicolin-injected embryos where only the centrosomes reach the cortex, cortical buds are formed, and actin is rearranged in a cell cycle-dependent manner (Raff and Glover 1989; Yasuda et al. 1991). Microtubules are required for actin reorganization in a specific time window. Colchicine injection during anaphase disrupts actin rearrangement to the furrows, but injection during interphase or telophase does not disrupt actin rearrangement. During telophase, robust astral microtubules of neighboring centrosomes overlap, possibly defining the furrow position for the subsequent cleavage cycle (Riggs et al. 2007). The polar bodies lack centrosomes but form microtubule projections similar to the astral microtubules of the centrosome and when the polar bodies lie close to the cortex, they rearrange the cortical actin (Foe et al. 2000).

There is also an accumulation of actin at the centrosomes that form aster-like filaments as the nuclei divide, appearing to emanate from the centrosome similar to astral microtubules (Fig. 12.7a). These actin asters appear to be reliant on proper centrosome activity and microtubules, as *cnn* mutants that disrupt the PCM and astral microtubules, or colchicine treatment also disrupts actin aster formation (Riparbelli et al. 2007). However, the role of actin asters in cortical cleavage divisions remains unclear.

The furrows provide a physical barrier between the dividing nuclei that prevents centrosomes from interacting with the spindles of neighboring nuclei (Kotadia et al. 2010; Sullivan et al. 1993b). When furrows are disrupted, mitotic spindles of neighboring nuclei aberrantly interact, which causes centrosomes to dissociate from their respective nuclei. The nuclei recede into the embryo leaving a patch devoid of a nucleus at the cortex (Sullivan et al. 1993a; Sullivan et al. 1993b). This process, which occurs in multiple mutant backgrounds when cleavage furrows are disrupted, is known as nuclear fallout. *cnn*, *arpc1*, *scar*, and *dia* mutants all display fused spindles and nuclear fallout due to the absence of furrow formation (Stevenson et al. 2002; Zallen et al. 2002; Afshar et al. 2000; Megraw et al. 1999).

The mechanism of nuclear fallout is a protective checkpoint that prevents damaged nuclei from multiplying and integrating into the future embryo by removing them from the cortex (O'Farrell et al. 2004). Due to the absence of gap phases and decrease in transcription, the syncytial embryo has an altered response to cell cycle checkpoint mechanisms. For example, DNA replication arrest due to aphidicolin injection does not block nuclei from entering mitosis (Raff and Glover 1988; Foe and Alberts 1983). This is especially important during the cortical cleavage cycles, due to the increase in cell cycle length and the MZT.

During cleavage cycles 10–13, the cell cycle increases in length from roughly 8 to 21 min (Foe and Alberts 1983). The increase in transcription that starts during cleavage cycle 10 requires a longer cell cycle, especially in S phase (Lamb and Laird 1976; McKnight et al. 1977; Zalokar 1976; Edgar and Schubiger 1986; Shermoen et al. 2010). The increase in length during these cycles is reliant on *grapes* (*grp*, Chk1 homolog) and meiotic 41 (*mei-41*, ATR homolog) as *grp* or *mei-41* mutants the lengthening of the cleavage cycles fails (Sibon et al. 1997; Sibon et al. 1999). As a result, *grp* or *mei-41* mutants also arrest during the cleavage division cycles with damaged nuclei and nuclear fallout (Fogarty et al. 1994; Fogarty et al. 1997). Centrosomes lose their function indicated by the loss of γ-Tubulin and γTuRC components from the centrosomes and the inability to separate chromosomes during mitosis (Sibon et al. 2000). Centrosome inactivation and nuclear fallout also occur during aphidicolin injection as well as a treatment with a variety of DNA-damaging agents (Raff and Glover 1988; Sibon et al. 2000; Takada et al. 2003).

This pathway of centrosome inactivation is regulated by *loki* (Lok, also known *mnk*, a Chk2 homolog), which localizes to centrosomes and spindles. During DNA damage, this localization increases, and Lok also accumulates at the nuclei (Takada et al. 2003). Lok causes mRNA nuclear retention after DNA damage, including mRNAs that encode for centrosomal proteins, such as γ-Tubulin ring protein 91 (Grip91); this mRNA nuclear retention causes nuclear fallout (Iampietro et al. 2014). In *lok* mutants, when DNA is damaged through either Bleomycin (induces DNA damage) injection or in a *grp* mutant background, centrosomes do not inactivate (Takada et al. 2003).

Multiple PCM proteins also play an important role in regulating cleavage furrow formation, and their dysfunction can result in fused spindles, aneuploid nuclei, and nuclear fallout. *cnn* mutants are maternal-effect lethal, arresting in the cleavage furrow stages and displaying a failure in furrow ingression that leads to fused metaphase spindles and colliding nuclei in telophase and inevitable nuclear fallout (Megraw et al. 1999; Vaizel-Ohayon and Schejter 1999). In *cnn*$^{\Delta 1}$ mutants where astral microtubules can still form, some furrows still ingress (Zhang and Megraw 2007), but in CM2 domain *cnn* mutants (*cnn*^{b4}), astral microtubules are present and furrow ingression was severely impaired. The CM2 domain of Cnn interacts with Centrocortin (Cen), and *cen* mutants display aberrant actin organization at the furrows as well as spindle fusions, indicating it interacts with *cnn* to regulate actin organization at the cleavage furrow (Kao and Megraw 2009).

sponge (*spg*, a Rho GEF family protein) mutants have defective actin caps and furrows leading to aberrant spindle interactions (Postner et al. 1992; Riparbelli et al. 2007). Loss of Eb1, which binds the plus ends of microtubules, does not result in defective actin localization at the furrows. Instead, furrows are partially invaginated, resulting in severe spindle defects and loss of nuclei due to nuclear fallout (Rogers et al. 2002; Webb et al. 2009). Nuf, which supplies membrane and other components to the invaginating furrows, localizes to the centrosome during prophase dependent on microtubules (Rothwell et al. 1998; Riggs et al. 2003). Because REs display a pericentriolar accumulation, this localization of Nuf is important for vesicle transport. *nuf* mutants display incomplete actin furrows, which results in spindle fusions (Rothwell et al. 1998).

12.10 Centrosome Separation During the Cortical Cleavage Cycles

As nuclei enter mitosis, the centrosomes begin to separate at prophase to form bipolar spindles. Centrosome separation is especially important in the cortical cleavage cycles due to the membrane invagination and spreading of actin that is orchestrated in conjunction with centrosome movement. It was hypothesized that centrosome separation guides actin cap expansion and furrow invagination as multiple centrosomal components are necessary for proper actin reorganization. Free centrosomes are sufficient to reorganize actin at the membrane (Raff and Glover 1989; Yasuda et al. 1991), and the cycling of actin structures parallels centrosomes' movements (Karr and Alberts 1986). However, in colchicine-treated embryos, centrosome separation fails, but actin caps still expand (Stevenson et al. 2001; Cao et al. 2010). Therefore, it appears that actin spreading may guide the centrosomes, as either latrunculin, cytochalasin, or jasplakinolide (actin filament stabilizer) injection both inhibit centrosome separation (Cao et al. 2010; Stevenson et al. 2001). Jasplakinolide also shrinks the actin caps, indicating actin turnover is required for centrosome separation (Cao et al. 2010).

It is hypothesized that the astral microtubules of centrosomes interact with actin to allow for proper centrosome separation. In support of this, Apc2, which stabilizes microtubules with Eb1 through Dia (Wen et al. 2004) and interacts with actin at actin-microtubule interaction sites (McCartney and Peifer 2000), plays a role in centrosome separation. Apc2 localizes to the actin caps and furrows where astral microtubules are interacting with the actin furrows (McCartney et al. 2001; Webb et al. 2009). *apc2* mutants display centrosome separation defects as well as defective furrow formation (Webb et al. 2009; Buttrick et al. 2008). Apc2 localization is reliant on Akt1 and Shaggy (Sgg, a Zw3 homolog), which is upstream of Akt1 (Shaw et al. 1997). *Akt1* mutants have normal actin and microtubule organization, yet they display centrosome separation defects and nuclear fallout due to disruption in Apc2 localization. Introduction of one copy of a *sgg* mutant allele into *akt1* mutants rescues this phenotype. Therefore, Apc2 regulates centrosome separation through microtubule-actin interactions, dependent on Akt1 and Sgg (Buttrick et al. 2008).

αTub67C is enriched at the interpolar microtubules that embrace the nuclear envelope, and *αTub67C* mutants display shorter microtubules. αTub67C mutants arrest in mitosis due to defective centrosome separation. These interpolar microtubules must rapidly assemble, and αTub67C appears to be necessary for this rapid growth as cells depleted of αTub67C will eventually grow long microtubules from other isoforms of α-Tubulin. Therefore, αTub67C is hypothesized to quickly nucleate these interpolar microtubules to push the centrosomes apart during centrosome separation (Venkei et al. 2006).

Centrosome separation also relies on multiple microtubule-based motor proteins to provide the force that moves them apart. Dynein is required for proper centrosome separation, as *dhc64c* mutant embryos display centrosomes that do not fully separate (Robinson et al. 1999). Klp61f is required to keep the centrosomes separated during metaphase but is not required for their initial separation. *klp61f* mutants display normal centrosome separation in prometaphase, but the centrosomes slide back together before the metaphase spindles can form. However, in *ncd klp61f* double mutant embryos, the centrosomes separate and remain separated until telophase indicating an opposing force on the minus- and plus-end-directed motors, respectively, that keeps the metaphase centrosomes separated. However, in these double mutant backgrounds, the distance between daughter nuclei is abnormally short, indicating a role for these motors in internuclear spacing as well as centrosome separation (Sharp et al. 1999).

During centrosome separation, Cnn fibers, hypothesized to be intercentrosomal microtubules, connect the centrosomes and persist into late anaphase, while $cnn^{\Delta 1}$ mutants display defects in centrosome separation (Zhang and Megraw 2007). In *sced* or *sponge* mutants, where furrows do not form, centrosome separation is also delayed (Stevenson et al. 2001; Postner et al. 1992). *arpc1* or *dia* mutants as well as Rho1 inhibition by C3 exotransferase, which display abnormal furrow formation, all result in defective centrosome separation (Cao et al. 2010). Loss of Daughterless-like (Dal), a protein of unknown function, also results in defective centrosome separation resulting in spindle fusion and nuclear fallout, but only during the cortical cleavage cycles (Sullivan et al. 1990).

In summary, centrosome separation in the early embryo is necessary for proper cleavage furrow formation during the cortical cleavage cycles. Centrosome separation is reliant on actin, microtubules, microtubule motors, and a variety of proteins involved in cleavage furrow formation. The mechanisms that force the centrosomes apart during mitosis still remain unclear, but the variety of factors involved suggest it is a complicated process that requires further investigation.

12.11 Cellularization Transitions the Syncytial Embryo to the Cellular Blastoderm

After 13 successive cleavage divisions, the cortical nuclei remain in interphase for roughly 70 min as cellular membranes surround each nucleus to form the multicellular embryo. The process of cellularization begins identically to the cortical cleavage cycles: the membrane protrudes forming a cortical bud and actin cap above each nucleus and its associated pair of centrosomes (Fig. 12.8a) (Foe and Alberts 1983). The actin cap is localized to microvillar projections that densely decorate the cortical buds at the onset of cellularization (Fig. 12.8a) (Turner and Mahowald 1979; Fullilove and Jacobson 1971). Furrows begin to form, and the hairpin-shaped tip of each furrow, the furrow canal (FC), is enriched in components necessary for cellularization (Fig. 12.8b) (Fullilove and Jacobson 1971; Warn and Robert-Nicoud 1990; Young et al. 1991). Immediately after furrow ingression, basal cell junctions assemble below the FC composed of E-cadherin, α-catenin, and β-catenin in preparation for cell formation (Hunter and Wieschaus 2000; Müller and Wieschaus 1996).

As the furrow ingresses, the nuclei elongate and extend into the embryo, resulting in oblong nuclei (Lecuit and Wieschaus 2000; Knoblich 2000). This change in nuclear shape is due to a basally extending microtubule basket formed by the astral microtubules (Fig. 12.8b–e) (Callaini and Anselmi 1988; Kellogg et al. 1991). Nocodazole treatment prevents nuclear shape change, supporting a role for these microtubule baskets in nuclear elongation during furrow ingression (Brandt et al. 2006). These microtubule baskets also rely on actin, as cytochalasin treatment disrupts microtubule basket organization (Edgar et al. 1987).

Unlike the cortical cleavage cycles, cellularization requires the nuclei to be present as aphidicolin-injected embryos do not cellularize (Raff and Glover 1989). This is most likely due to the requirement of zygotic transcription that starts during the cortical cleavage cycles and decreases at the end of cellularization (Lamb and Laird 1976; McKnight et al. 1977; Zalokar 1976; Edgar and Schubiger 1986) as α-amanitin injection (a transcription inhibitor) blocks cellularization (Edgar et al. 1986). Most gene products required for cellularization are maternally supplied but some necessary proteins are zygotically transcribed, indicating they have a specific role in this process (Mazumdar and Mazumdar 2002).

The furrows slowly progress until they reach the basal end of the nuclei, where progression then rapidly increases with furrows ultimately reaching a depth of approximately 35 μm (Lecuit and Wieschaus 2000; Foe and Alberts 1983). An actomyosin contractile ring forms at the FC as it invaginates and begins to contract once the furrows have passed the nuclei. This results in a change from the hexagonal array of actin that surrounds the nuclei to a ring shape localization in preparation for cell closure (Fig. 12.8f–g) (Theurkauf 1994). The contractile rings then pinch off the membranes at the end of cellularization to form single cells (Fig. 12.8e).

Because furrow invagination is so extensive and demanding of membrane resources, the membrane present at the cortex does not provide a sufficient supply needed for cellularization, and new sources of membrane are required (Figard et al. 2013). Two different deposits of membrane are supplied to the furrows during different time points of cellularization.

The first deposit is derived from the microvillar actin projections on the surface of the cortex above each nucleus (Fig. 12.8a–d). These microvillar projections assemble before cellularization, and are depleted by its completion (Fabrowski et al. 2013; Fullilove and Jacobson 1971). It is estimated that the microvilli contain about half of the membrane necessary for furrow formation (Figard et al. 2013). New membrane is

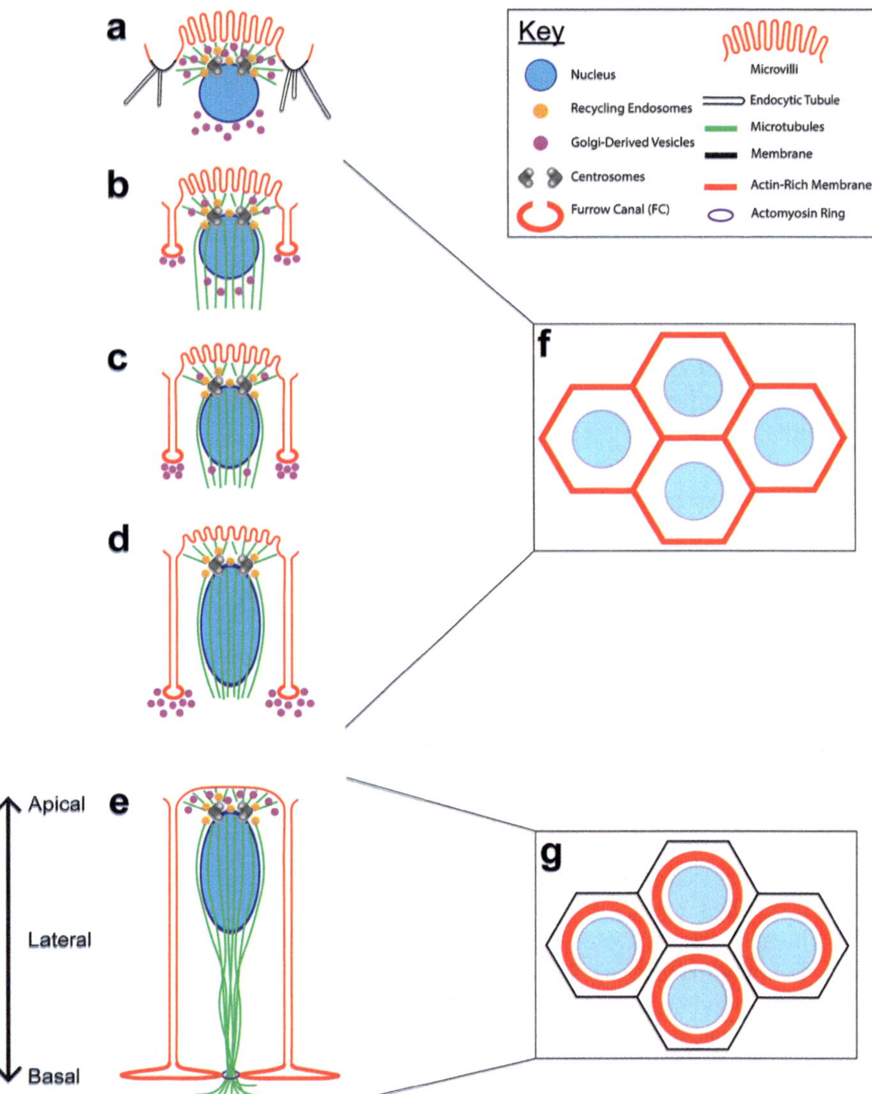

Fig. 12.8 Cellularization of the nuclei requires a deep invagination of the membrane which eventually cleaves to form separate cells. (**a**) During prophase, the centrosomes impinge on the plasma membrane to form a cortical bud over the nucleus. The REs remains apical to the nucleus near the centrosomes and Golgi-derived vesicles localize at the basal end of the nucleus. Long microvillar projections rich in actin form above each nucleus. (**b**) The furrow canal forms, slowly invaginating away from the cortex. The microtubules begin to form around the nucleus as it elongates. Golgi-derived vesicles localize to the furrow canal. The microvilli slowly regress into the furrows. (**c**) The furrow continues to grow, and the microtubules surround the nucleus as it continues to elongate. More Golgi-derived vesicles begin to localize at the furrow canal and the microvilli continue to recede as the membrane invaginates. (**d**) The nucleus is fully elongated, and the microtubules surround it in a basket shape. The Golgi-derived vesicles are all localized at the furrow canal. The microvilli are almost depleted as the furrow canal reaches the basal end of the nucleus. (**e**) The furrow canal spreads to form a ring at the bottom of the nucleus that pinches off the

added at the apical end of the furrows and old membrane is pushed basally into the furrows, rather than the addition of membrane to the FC, supporting the role for microvillar membrane addition (Lecuit and Wieschaus 2000). Microvillar depletion also mimics the kinetics of furrow ingression, starting slow as the furrows reach the bottom of the nuclei and speeding up toward the end of cellularization. Therefore, the microvilli are unfolded and pulled directly into the invaginating furrows (Figard et al. 2013).

Abl tyrosine kinase (Abl) is involved in the regulation of microvillar density and length as *abl* mutants display longer microvilli than wildtype at the onset of cellularization which do not diminish over time as in wildtype. Actin also abnormally accumulates at the apical end of the furrows due to enrichment of the actin nucleators, Arp3 and Dia, resulting in excessive F-action nucleation and abnormal furrow formation. These defects are due to abnormal localization of Enabled (Ena) at the apical cortex and abnormal actin accumulation, which is normally disperse during cellularization, as *ena abl* mutants have less cortical actin accumulation. *abl* mutants also display disrupted microtubule baskets, a further indication that actin regulation plays a role in their formation (Grevengoed et al. 2003).

The second source of membrane is supplied by Golgi-derived vesicles and REs. Centrosomes assist in trafficking these vesicles which contain the proteins and membrane needed for furrow invagination along the microtubules (Mazumdar and Mazumdar 2002). Centrosomes are necessary for furrow invagination and loss of astral microtubules disrupts furrow invagination (Zalokar and Erk 1976; Foe and Alberts 1983). During early cleavage cycles, the Golgi is localized in disperse puncta at the cortex of the embryo in close proximity to the ER (Ripoche et al. 1994; Stanley et al. 1997). Once the nuclei reach the cortex and transcription increases, separate ER and Golgi structures segregate to single nuclei, with each nucleus containing a single ER/Golgi system. This segregation is dependent on microtubules, as nocodazole treatment disrupts ER/Golgi compartmentalization, indicating a role for the centrosome (Frescas et al. 2006). Golgi-derived vesicles either fuse to the apical membrane and endocytose to be sorted by the RE for transport, or traffic directly to the membrane (LaLonde et al. 2006; Lee and Harris 2014). Golgi-derived vesicle and RE transport rely on separate trafficking pathways to transport different components to the invaginating furrows.

Golgi-derived vesicles supply membrane and components that are required for the rapid extension of the furrows. Brefeldin A, which inhibits Golgi vesicle transport, inhibits furrow progression in the final, fast stage of cellularization (Sisson et al. 2000; Frescas et al. 2006). Transport of these vesicles is dependent on microtubules, as colcemid or colchicine treatment during the slow cellularization

---

Fig. 12.8 (continued) membrane to form a cell. The Golgi-derived vesicles move away from the furrow canal to the apical side of the nucleus. (**f**) Cortical view of cellularization during **b–d**. Actin surrounds the nuclei forming a hexagonal pattern at neighboring junctions. (**g**) Cortical view of cellularization during e. The actin surrounding each nucleus begins to constrict into rings

stage stalls vesicle movements at the basal end of the nuclei, preventing furrow invagination (Lecuit and Wieschaus 2000; Sisson et al. 2000; Foe and Alberts 1983; Zalokar and Erk 1976). Golgi vesicle transport is not dependent on actin as cytochalasin injection does not disrupt Golgi vesicle movement during cellularization (Sisson et al. 2000).

The dynein-dynactin complex in association with Cytoplasmic linker protein 190 (CLIP190), which links vesicles to microtubules (Lantz and Miller 1998), assists in transporting Golgi-derived vesicles along the microtubules. *dhc64C* mutants do not form furrows and Golgi-derived vesicle movement is blocked, while the microtubules remain intact (Papoulas et al. 2005). Dhc64C, DCTN2-p150, and CLIP190 all associate at the Golgi and specifically bind the Golgi-associated protein Lava Lamp (Lva), which acts as an adaptor for Dynein-Dynactin vesicular trafficking (Sisson et al. 2000; Papoulas et al. 2005). In support of this, *lva* mutants that cannot bind Dhc64C, DCTN2-p150, or CLIP190 display impaired furrow progression and inhibition of Golgi vesicle movement. Lva is necessary for CLIP190-dependent microtubule-vesicle attachment, as *lva* mutants that cannot bind dynein inhibit CLIP190 localization to the Golgi and FC, resulting in impaired furrow formation (Papoulas et al. 2005).

The second class of vesicles, REs, are necessary for all stages of cellularization. REs localize intermediately near the centrosomes as the vesicles are being sorted to transport membrane and protein to the furrows (van Ijzendoorn 2006). Rab5 and Dynamin are necessary for the initial endocytosis and budding of REs from the apical membrane, as overexpressing a dominant negative variant of Rab5 or *shi* mutant backgrounds display impaired furrow ingression (Pelissier et al. 2003). An intermediate stage of the RE pathway displays tubular membrane projections from the FC, but only at the onset of cellularization (Fig. 12.8a) (Sokac and Wieschaus 2008a). In *shi* mutants, these endocytic tubules are longer than normal, indicating stalled trafficking of REs (Sokac and Wieschaus 2008a; Su et al. 2013; Sherlekar and Rikhy 2016).

The regulation of the REs follows similar dynamics to RE trafficking during the cortical cleavage cycles. Rab11, Dynamin, and Nuf are required for vesicle trafficking and dominant negative Rab11, *shi*, and *nuf* mutants all display impaired furrow invagination, while *shi* mutants display Rab11 vesicles halted at the centrosome (Pelissier et al. 2003; Rothwell et al. 1998). Rab11 and Nuf also localize RhoGEF2 to the furrows to drive furrow invagination through the same mechanisms outlined in the cortical cleavage cycles section. (Riggs et al. 2003; Cao et al. 2008).

Rab11, Nuf, and Dynamin regulate the localization of Slow as molasses (Slam), which is both maternally supplied and zygotically transcribed, to the FCs (Acharya et al. 2014). Slam is necessary for Rho1, RhoGEF2, and MyoII localization to the furrows and *slam* mutants display disrupted furrow invagination (Wenzl et al. 2010; Acharya et al. 2014). Centrosomes are also necessary for Slam localization to the furrows as nuclei-less centrosomes still localize Slam, and ablating centrosomes disrupts Slam localization (Acharya et al. 2014). Therefore, Slam utilizes centrosome-dependent RE vesicle transport for localization at the FCs.

Slam plays a redundant role with Nullo, another zygotically transcribed gene product that is necessary for furrow invagination and stabilization, in Dia localization (Sokac and Wieschaus 2008a, b; Hunter and Wieschaus 2000). *slam nullo* double mutants, but not single mutants of either, display disruption of Dia localization at the FCs (Acharya et al. 2014). Dia localization is also dependent on RhoGEF2, both of which localize to the FCs before invagination, suggesting they aid in FC formation. Mutants for *rhogef2* and *dia* display areas absent of furrows and misshapen FCs. When either *rhogef2* or *dia* mutants are combined with *nullo* mutants, the furrow defects are stronger, suggesting Nullo acts in a pathway separate from RhoGEF2 and Dia (Großhans et al. 2005). Therefore, Nullo may coordinate furrow invagination through F-actin regulation by Slam, RhoGEF2, and Dia.

Rho1, Rok, and RhoGEF2 are all required for the final stage of cellularization. RhoGEF2 localizes Rho1 to the furrows for localized Rok activation (Padash Barmchi et al. 2005). *rhogef2* and *rok* mutants display similar defects in cellularization, defective furrow invagination (Dawes-Hoang et al. 2005), and dominant negative *rho1* mutants do not cellularize (Crawford et al. 1998). The Rho1 pathway regulates MyoII activation, which is necessary for contractile ring constriction (Xue and Sokac 2016). Anillin is known to link Rho1, actin, and MyoII during cytokinesis (Piekny and Glotzer 2008) and is necessary for MyoII localization as well as Pnut localization to the FCs. In *anillin* mutants, FC morphology is abnormal, contractile rings fail to form, and cellularization does not occur, indicating it may link Rho1, actin, and MyoII during cellularization (Field et al. 2005; Thomas and Wieschaus 2004).

Bottleneck (Bnk), a zygotically transcribed gene product, regulates the constriction of the contractile ring toward the end of cellularization. Bnk localizes to the FCs during the slow phases of cellularization and disappears during the fast stage when constriction begins. In *bnk* mutants, the hexagonal actin rings constrict before the furrows are past the nuclei resulting in bottle-shaped nuclei (Schejter and Wieschaus 1993; Theurkauf 1994). Therefore, bnk is a negative regulator of actomyosin constriction during cellularization.

For more on *Drosophila* cellularization, see How one becomes many: blastoderm cellularization in *Drosophila melanogaster* (Mazumdar and Mazumdar 2002).

12.12   Summary

The centrosome plays an important role in regulating the dynamics of early development in the *Drosophila* syncytial embryo at multiple stages. The centrosome regulates the cytoskeletal elements, microtubules and actin, in varying contexts to allow for each step of embryogenesis to properly occur. The first three centrosome-dependent stages of embryogenesis are nuclear migrations that utilize cytoskeletal components. The remaining three stages utilize centrosome-dependent actin regulation to form furrows that separate the nuclei as they divide and eventually form cellular membranes. Defective centrosomes can disrupt the cell cycle, mitotic cleavage divisions, and nuclear positioning, resulting in morphogenic defects and embryonic lethality. The syncytial embryonic cell cycle is modified to address the rapid nuclear divisions, resulting in changes to the regulation of the centrosome. The central roles of the centrosome in the early embryo contrast with the ability of zygotic development to be accomplished successfully without functional centrosomes. All of the processes controlled by the centrosome in the early embryo are not well understood, so much remains for investigators to discover.

References

Acharya S, Laupsien P, Wenzl C, Yan S, Großhans J (2014) Function and dynamics of *slam* in furrow formation in early Drosophila embryo. Dev Biol 386(2):371–384

Afshar K, Stuart B, Wasserman SA (2000) Functional analysis of the *Drosophila* diaphanous FH protein in early embryonic development. Development 127(9):1887–1897

Al-Hakim AK, Bashkurov M, Gingras AC, Durocher D, Pelletier L (2012) Interaction proteomics identify NEURL4 and the HECT E3 ligase HERC2 as novel modulators of centrosome architecture. Mol Cell Proteomics 11(6):M111.014233

Archambault V, Pinson X (2010) Free centrosomes: where do they all come from? Fly 4 (2):172–177

Archambault V, Zhao X, White-Cooper H, Carpenter AT, Glover DM (2007) Mutations in *Drosophila Greatwall/Scant* reveal its roles in mitosis and meiosis and interdependence with polo kinase. PLoS Genet 3(11):e200

Archambault V, D'Avino PP, Deery MJ, Lilley KS, Glover DM (2008) Sequestration of polo kinase to microtubules by phosphopriming-independent binding to map 205 is relieved by phosphorylation at a CDK site in mitosis. Genes Dev 22(19):2707–2720

Baker J, Theurkauf WE, Schubiger G (1993) Dynamic changes in microtubule configuration correlate with nuclear migration in the preblastoderm *Drosophila* embryo. J Cell Biol 122 (1):113–121

Barros TP, Kinoshita K, Hyman AA, Raff JW (2005) Aurora a activates D-TACC-Msps complexes exclusively at centrosomes to stabilize centrosomal microtubules. J Cell Biol 170(7):1039–1046

Belecz I, Gonzalez C, Puro J, Szabad J (2001) Dominant-negative mutant dynein allows spontaneous centrosome assembly, uncouples chromosome and centrosome cycles. Curr Biol 11 (2):136–140

Blachon S, Cai X, Roberts KA, Yang K, Polyanovsky A, Church A, Avidor-Reiss T (2009) A proximal centriole-like structure is present in *Drosophila* spermatids and can serve as a model to study centriole duplication. Genetics 182(1):133–144

Blachon S, Khire A, Avidor-Reiss T (2014) The origin of the second centriole in the zygote of Drosophila melanogaster. Genetics 197(1):199–205

Bownes M (1982) Embryogenesis. A handbook of *Drosophila*. Elsevier, Amsterdam

Brandt A, Papagiannouli F, Wagner N, Wilsch-Bräuninger M, Braun M, Furlong EE, Loserth S, Wenzl C, Pilot F, Vogt N, Lecuit T, Krohne G, Grosshans J (2006) Developmental control of nuclear size and shape by Kugelkern and Kurzkern. Curr Biol 16(6):543–552

Brent AE, MacQueen A, Hazelrigg T (2000) The *Drosophila wispy* gene is required for RNA localization and other microtubule-based events of meiosis and early embryogenesis. Genetics 154(4):1649–1662

Brunk K, Vernay B, Griffith E, Reynolds NL, Strutt D, Ingham PW, Jackson AP (2007) Microcephalin coordinates mitosis in the syncytial *Drosophila* embryo. J Cell Sci 120. (Pt 20:3578–3588

Burkhardt JK, Echeverri CJ, Nilsson T, Vallee RB (1997) Overexpression of the dynamitin (p50) subunit of the dynactin complex disrupts dynein-dependent maintenance of membrane organelle distribution. J Cell Biol 139(2):469–484

Buttrick GJ, Beaumont LM, Leitch J, Yau C, Hughes JR, Wakefield JG (2008) Akt regulates centrosome migration and spindle orientation in the early *Drosophila melanogaster* embryo. J Cell Biol 180(3):537–548

Callaini G, Anselmi F (1988) Centrosome splitting during nuclear elongation in the *Drosophila* embryo. Exp Cell Res 178(2):415–425

Callaini G, Dallai R (1991) Abnormal behavior of the yolk centrosomes during early embryogenesis of *Drosophila melanogaster*. Exp Cell Res 192(1):16–21

Callaini G, Riparbelli MG (1996) Fertilization in *Drosophila melanogaster*: centrosome inheritance and organization of the first mitotic spindle. Dev Biol 176(2):199–208

Callaini G, Dallai R, Riparbelli MG (1992) Cytochalasin induces spindle fusion in the syncytial blastoderm of the early *Drosophila* embryo. Biol Cell 74(3):249–254

Callaini G, Whitfield WG, Riparbelli MG (1997) Centriole and centrosome dynamics during the embryonic cell cycles that follow the formation of the cellular blastoderm in *Drosophila*. Exp Cell Res 234(1):183–190

Callaini G, Riparbelli MG, Dallai R (1999) Centrosome inheritance in insects: fertilization and parthenogenesis. Biol Cell 91(4–5):355–366

Cao J, Albertson R, Riggs B, Field CM, Sullivan W (2008) Nuf, a Rab11 effector, maintains cytokinetic furrow integrity by promoting local actin polymerization. J Cell Biol 182 (2):301–313

Cao J, Crest J, Fasulo B, Sullivan W (2010) Cortical actin dynamics facilitate early-stage centrosome separation. Curr Biol 20(8):770–776

Cesario JM, Jang JK, Redding B, Shah N, Rahman T, McKim KS (2006) Kinesin 6 family member Subito participates in mitotic spindle assembly and interacts with mitotic regulators. J Cell Sci 119(Pt 22):4770–4780

Chen JV, Buchwalter RA, Kao LR, Megraw TL (2017) A splice variant of centrosomin converts mitochondria to microtubule-organizing centers. Curr Biol 27(13):1928–1940.e1926

Chodagam S, Royou A, Whitfield W, Karess R, Raff JW (2005) The centrosomal protein CP190 regulates myosin function during early *Drosophila* development. Curr Biol 15(14):1308–1313

Cinalli RM, Lehmann R (2013) A spindle-independent cleavage pathway controls germ cell formation in *Drosophila*. Nat Cell Biol 15(7):839–845

Crawford JM, Harden N, Leung T, Lim L, Kiehart DP (1998) Cellularization in *Drosophila melanogaster* is disrupted by the inhibition of rho activity and the activation of Cdc42 function. Dev Biol 204(1):151–164

Crest J, Concha-Moore K, Sullivan W (2012) RhoGEF and positioning of rappaport-like furrows in the early *Drosophila* embryo. Curr Biol 22(21):2037–2041

Cullen CF, Ohkura H (2001) Msps protein is localized to acentrosomal poles to ensure bipolarity of *Drosophila* meiotic spindles. Nat Cell Biol 3(7):637–642

Dävring L, Sunner M (1973) Female meiosis and embryonic mitosis in *Drosophila melanogaster*. I. Meiosis and fertilization. Hereditas 73(1):51–64

Dawes-Hoang RE, Parmar KM, Christiansen AE, Phelps CB, Brand AH, Wieschaus EF (2005) *folded gastrulation*, cell shape change and the control of myosin localization. Development 132 (18):4165–4178

Debec A, Kalpin RF, Daily DR, McCallum PD, Rothwell WF, Sullivan W (1996) Live analysis of free centrosomes in normal and aphidicolin-treated *Drosophila* embryos. J Cell Biol 134 (1):103–115

Debec A, Marcaillou C, Bobinnec Y, Borot C (1999) The centrosome cycle in syncytial *Drosophila* embryos analyzed by energy filtering transmission electron microscopy. Biol Cell 91 (4–5):379–391

Debec A, Grammont M, Berson G, Dastugue B, Sullivan W, Couderc JL (2001) Toucan protein is essential for the assembly of syncytial mitotic spindles in *Drosophila melanogaster*. Genesis 31 (4):167–175

Deneke VE, Melbinger A, Vergassola M, Di Talia S (2016) Waves of Cdk1 activity in S phase synchronize the cell cycle in *Drosophila* embryos. Dev Cell 38(4):399–412

Dix CI, Raff JW (2007) *Drosophila* Spd-2 recruits PCM to the sperm centriole, but is dispensable for centriole duplication. Curr Biol 17(20):1759–1764

Ede DA, Counce SJ (1956) A cinematographic study of the embryology of *Drosophila melanogaster*. Roux Arch Dev Biol 148(4):402–415

Edgar BA, Schubiger G (1986) Parameters controlling transcriptional activation during early *Drosophila* development. Cell 44(6):871–877

Edgar BA, Kiehle CP, Schubiger G (1986) Cell cycle control by the nucleo-cytoplasmic ratio in early *Drosophila* development. Cell 44(2):365–372

Edgar BA, Odell GM, Schubiger G (1987) Cytoarchitecture and the patterning of fushi tarazu expression in the *Drosophila* blastoderm. Genes Dev 1(10):1226–1237

Edgar BA, Sprenger F, Duronio RJ, Leopold P, O'Farrell PH (1994) Distinct molecular mechanism regulate cell cycle timing at successive stages of *Drosophila* embryogenesis. Genes Dev 8 (4):440–452

Elfring LK, Axton JM, Fenger DD, Page AW, Carminati JL, Orr-Weaver TL (1997) *Drosophila* PLUTONIUM protein is a specialized cell cycle regulator required at the onset of embryogenesis. Mol Biol Cell 8(4):583–593

Ephrussi A, Lehmann R (1992) Induction of germ cell formation by oskar. Nature 358 (6385):387–392

Fabrowski P, Necakov AS, Mumbauer S, Loeser E, Reversi A, Streichan S, Briggs JA, De Renzis S (2013) Tubular endocytosis drives remodelling of the apical surface during epithelial morphogenesis in *Drosophila*. Nat Commun 4:2244

Farache D, Emorine L, Haren L, Merdes A (2018) Assembly and regulation of γ-tubulin complexes. Open Biol 8(3)

Farina F, Gaillard J, Guérin C, Couté Y, Sillibourne J, Blanchoin L, Théry M (2016) The centrosome is an actin-organizing centre. Nat Cell Biol 18(1):65–75

Fenger DD, Carminati JL, Burney-Sigman DL, Kashevsky H, Dines JL, Elfring LK, Orr-Weaver TL (2000) PAN GU: a protein kinase that inhibits S phase and promotes mitosis in early *Drosophila* development. Development 127(22):4763–4774

Field CM, Alberts BM (1995) Anillin, a contractile ring protein that cycles from the nucleus to the cell cortex. J Cell Biol 131(1):165–178

Field CM, Coughlin M, Doberstein S, Marty T, Sullivan W (2005) Characterization of anillin mutants reveals essential roles in septin localization and plasma membrane integrity. Development 132(12):2849–2860

Figard L, Xu H, Garcia HG, Golding I, Sokac AM (2013) The plasma membrane flattens out to fuel cell-surface growth during *Drosophila* cellularization. Dev Cell 27(6):648–655

Flor-Parra I, Iglesias-Romero AB, Chang F (2018) The XMAP215 ortholog Alp14 promotes microtubule nucleation in fission yeast. Curr Biol 28(11):1681–1691.e1684

Foe VE, Alberts BM (1983) Studies of nuclear and cytoplasmic behaviour during the five mitotic cycles that precede gastrulation in *Drosophila* embryogenesis. J Cell Sci 61:31–70

Foe VE, Odell GM, Edgar BA (1993) Mitosis and morphogenesis in the *Drosophila* embryo: point and counterpoint. In: Bate M, Martinez Arias A (eds) The Development of *Drosophila melanogaster*, vol 1. Cold Spring Harbor Press, Cold Spring Harbor, NY

Foe VE, Field CM, Odell GM (2000) Microtubules and mitotic cycle phase modulate spatiotemporal distributions of F-actin and myosin II in *Drosophila* syncytial blastoderm embryos. Development 127(9):1767–1787

Fogarty P, Kalpin RF, Sullivan W (1994) The *Drosophila* maternal-effect mutation *grapes* causes a metaphase arrest at nuclear cycle 13. Development 120(8):2131–2142

Fogarty P, Campbell SD, Abu-Shumays R, Phalle BS, Yu KR, Uy GL, Goldberg ML, Sullivan W (1997) The *Drosophila grapes* gene is related to checkpoint gene *chk1/rad27* and is required for late syncytial division fidelity. Curr Biol 7(6):418–426

Franz A, Roque H, Saurya S, Dobbelaere J, Raff JW (2013) CP110 exhibits novel regulatory activities during centriole assembly in *Drosophila*. J Cell Biol 203(5):785–799

Freeman M, Glover DM (1987) The *gnu* mutation of *Drosophila* causes inappropriate DNA synthesis in unfertilized and fertilized eggs. Genes Dev 1(9):924–930

Freeman M, Nüsslein-Volhard C, Glover DM (1986) The dissociation of nuclear and centrosomal division in *gnu*, a mutation causing giant nuclei in *Drosophila*. Cell 46(3):457–468

Frescas D, Mavrakis M, Lorenz H, Delotto R, Lippincott-Schwartz J (2006) The secretory membrane system in the *Drosophila* syncytial blastoderm embryo exists as functionally compartmentalized units around individual nuclei. J Cell Biol 173(2):219–230

Fullilove SL, Jacobson AG (1971) Nuclear elongation and cytokinesis in Drosophila montana. Dev Biol 26(4):560–577

Gergely F, Karlsson C, Still I, Cowell J, Kilmartin J, Raff JW (2000a) The TACC domain identifies a family of centrosomal proteins that can interact with microtubules. Proc Natl Acad Sci U S A 97(26):14352–14357

Gergely F, Kidd D, Jeffers K, Wakefield JG, Raff JW (2000b) D-TACC: a novel centrosomal protein required for normal spindle function in the early *Drosophila* embryo. EMBO J 19 (2):241–252

Giansanti MG, Bucciarelli E, Bonaccorsi S, Gatti M (2008) *Drosophila* SPD-2 is an essential centriole component required for PCM recruitment and astral-microtubule nucleation. Curr Biol 18(4):303–309

Giet R, McLean D, Descamps S, Lee MJ, Raff JW, Prigent C, Glover DM (2002) *Drosophila* Aurora a kinase is required to localize D-TACC to centrosomes and to regulate astral microtubules. J Cell Biol 156(3):437–451

Giunta KL, Jang JK, Manheim EA, Subramanian G, McKim KS (2002) *subito* encodes a kinesin-like protein required for meiotic spindle pole formation in *Drosophila melanogaster*. Genetics 160(4):1489–1501

Glover DM (2005) Polo kinase and progression through M phase in *Drosophila*: a perspective from the spindle poles. Oncogene 24(2):230–237

Glover DM, Alphey L, Axton JM, Cheshire A, Dalby B, Freeman M, Girdham C, Gonzalez C, Karess RE, Leibowitz MH (1989) Mitosis in *Drosophila* development. J Cell Sci Suppl 12:277–291

González C, Saunders RD, Casal J, Molina I, Carmena M, Ripoll P, Glover DM (1990) Mutations at the asp locus of *Drosophila* lead to multiple free centrosomes in syncytial embryos, but restrict centrosome duplication in larval neuroblasts. J Cell Sci 96(Pt 4):605–616

González C, Tavosanis G, Mollinari C (1998) Centrosomes and microtubule organisation during *Drosophila* development. J Cell Sci 111(Pt 18):2697–2706

Gottardo M, Callaini G, Riparbelli MG (2015) The *Drosophila* centriole—conversion of doublets into triplets within the stem cell niche. J Cell Sci 128(14):2437–2442

Grevengoed EE, Fox DT, Gates J, Peifer M (2003) Balancing different types of actin polymerization at distinct sites: roles for Abelson kinase and enabled. J Cell Biol 163(6):1267–1279

Großhans J, Wenzl C, Herz HM, Bartoszewski S, Schnorrer F, Vogt N, Schwarz H, Müller HA (2005) RhoGEF2 and the formin Dia control the formation of the furrow canal by directed actin assembly during *Drosophila* cellularisation. Development 132(5):1009–1020

Gunawardane RN, Lizarraga SB, Wiese C, Wilde A, Zheng Y (2000) γ-tubulin complexes and their role in microtubule nucleation. Curr Top Dev Biol 49:55–73

Gunzelmann J, Rüthnick D, Lin TC, Zhang W, Neuner A, Jäkle U, Schiebel E (2018) The microtubule polymerase Stu2 promotes oligomerization of the γ-TuSC for cytoplasmic microtubule nucleation. elife 7

Hatanaka K, Okada M (1991) Retarded nuclear migration in *Drosophila* embryos with aberrant F-actin reorganization caused by maternal mutations and by cytochalasin treatment. Development 111(4):909–920

Heck MM, Pereira A, Pesavento P, Yannoni Y, Spradling AC, Goldstein LS (1993) The kinesin-like protein KLP61F is essential for mitosis in *Drosophila*. J Cell Biol 123(3):665–679

Hieda M (2017) Implications for diverse functions of the LINC complexes based on the structure. Cell 6(1)

Huang J, Raff JW (1999) The disappearance of cyclin B at the end of mitosis is regulated spatially in *Drosophila* cells. EMBO J 18(8):2184–2195

Hunter C, Wieschaus E (2000) Regulated expression of nullo is required for the formation of distinct apical and basal adherens junctions in the *Drosophila* blastoderm. J Cell Biol 150 (2):391–401

Iampietro C, Bergalet J, Wang X, Cody NA, Chin A, Lefebvre FA, Douziech M, Krause HM, Lécuyer E (2014) Developmentally regulated elimination of damaged nuclei involves a Chk2-dependent mechanism of mRNA nuclear retention. Dev Cell 29(4):468–481

Iida T, Kobayashi S (2000) Delocalization of polar plasm components caused by *grandchildless* mutations, *gs(1)N26* and *gs(1)N441*, in *Drosophila melanogaster*. Develop Growth Differ 42 (1):53–60

Illmensee K, Mahowald AP (1974) Transplantation of posterior polar plasm in *Drosophila*. Induction of germ cells at the anterior pole of the egg. Proc Natl Acad Sci U S A 71 (4):1016–1020

Jones J, Macdonald PM (2015) Neurl4 contributes to germ cell formation and integrity in *Drosophila*. Biol Open 4(8):937–946

Jongens TA, Hay B, Jan LY, Jan YN (1992) The *germ cell-less* gene product: a posteriorly localized component necessary for germ cell development in *Drosophila*. Cell 70(4):569–584

Jongens TA, Ackerman LD, Swedlow JR, Jan LY, Jan YN (1994) *Germ cell-less* encodes a cell type-specific nuclear pore-associated protein and functions early in the germ-cell specification pathway of *Drosophila*. Genes Dev 8(18):2123–2136

Jordan P, Karess R (1997) Myosin light chain-activating phosphorylation sites are required for oogenesis in *Drosophila*. J Cell Biol 139(7):1805–1819

Kao LR, Megraw TL (2009) Centrocortin cooperates with centrosomin to organize *Drosophila* embryonic cleavage furrows. Curr Biol 19(11):937–942

Karess RE, Chang XJ, Edwards KA, Kulkarni S, Aguilera I, Kiehart DP (1991) The regulatory light chain of nonmuscle myosin is encoded by *spaghetti-squash*, a gene required for cytokinesis in *Drosophila*. Cell 65(7):1177–1189

Karr TL (1991) Intracellular sperm/egg interactions in *Drosophila*: a three-dimensional structural analysis of a paternal product in the developing egg. Mech Dev 34(2–3):101–111

Karr TL, Alberts BM (1986) Organization of the cytoskeleton in early *Drosophila* embryos. J Cell Biol 102(4):1494–1509

Kellogg DR, Mitchison TJ, Alberts BM (1988) Behaviour of microtubules and actin filaments in living *Drosophila* embryos. Development 103(4):675–686

Kellogg DR, Field CM, Alberts BM (1989) Identification of microtubule-associated proteins in the centrosome, spindle, and kinetochore of the early *Drosophila* embryo. J Cell Biol 109(6 Pt 1):2977–2991

Kellogg DR, Sullivan W, Theurkauf W, Oegema K, Raff JW, Alberts BM (1991) Studies on the centrosome and cytoplasmic organization in the early *Drosophila* embryo. Cold Spring Harb Symp Quant Biol 56:649–662

Khire A, Jo KH, Kong D, Akhshi T, Blachon S, Cekic AR, Hynek S, Ha A, Loncarek J, Mennella V, Avidor-Reiss T (2016) Centriole remodeling during spermiogenesis in *Drosophila*. Curr Biol 26(23):3183–3189

Kiehart DP, Ketchum A, Young P, Lutz D, Alfenito MR, Chang XJ, Awobuluyi M, Pesacreta TC, Inoué S, Stewart CT (1990) Contractile proteins in *Drosophila* development. Ann N Y Acad Sci 582:233–251

Knoblich JA (2000) Epithelial polarity: the ins and outs of the fly epidermis. Curr Biol 10(21): R791–R794

Kollman JM, Greenberg CH, Li S, Moritz M, Zelter A, Fong KK, Fernandez JJ, Sali A, Kilmartin J, Davis TN, Agard DA (2015) Ring closure activates yeast γTuRC for species-specific microtubule nucleation. Nat Struct Mol Biol 22(2):132–137

Komma DJ, Endow SA (1997) Enhancement of the ncdD microtubule motor mutant by mutants of *αTub67C*. J Cell Sci 110(Pt 2):229–237

Kotadia S, Crest J, Tram U, Riggs B, Sullivan W (2010) Blastoderm formation and cellularisation in *Drosophila melanogaster*. eLS

LaLonde M, Janssens H, Yun S, Crosby J, Redina O, Olive V, Altshuller YM, Choi SY, Du G, Gergen JP, Frohman MA (2006) A role for phospholipase D in *Drosophila* embryonic cellularization. BMC Dev Biol 6:60

Lamb MM, Laird CD (1976) Increase in nuclear poly(A)-containing RNA at syncytial blastoderm in *Drosophila melanogaster* embryos. Dev Biol 52(1):31–42

Lantz VA, Miller KG (1998) A class VI unconventional myosin is associated with a homologue of a microtubule-binding protein, cytoplasmic linker protein-170, in neurons and at the posterior pole of *Drosophila* embryos. J Cell Biol 140(4):897–910

Lantz VA, Clemens SE, Miller KG (1999) The actin cytoskeleton is required for maintenance of posterior pole plasm components in the *Drosophila* embryo. Mech Dev 85(1–2):111–122

Lasko P (2012) mRNA localization and translational control in *Drosophila* oogenesis. Cold Spring Harb Perspect Biol 4(10)

Lattao R, Kovács L, Glover DM (2017) The centrioles, centrosomes, basal bodies, and cilia of *Drosophila melanogaster*. Genetics 206(1):33–53

Lecuit T, Wieschaus E (2000) Polarized insertion of new membrane from a cytoplasmic reservoir during cleavage of the *Drosophila* embryo. J Cell Biol 150(4):849–860

Lee DM, Harris TJ (2014) Coordinating the cytoskeleton and endocytosis for regulated plasma membrane growth in the early *Drosophila* embryo. BioArchitecture 4(2):68–74

Lee MJ, Gergely F, Jeffers K, Peak-Chew SY, Raff JW (2001) Msps/XMAP215 interacts with the centrosomal protein D-TACC to regulate microtubule behaviour. Nat Cell Biol 3(7):643–649

Lee DM, Wilk R, Hu J, Krause HM, Harris TJ (2015) Germ cell segregation from the *Drosophila* Soma is controlled by an inhibitory threshold set by the Arf-GEF Steppke. Genetics 200 (3):863–872

Lerit DA, Gavis ER (2011) Transport of germ plasm on astral microtubules directs germ cell development in *Drosophila*. Curr Biol 21(6):439–448

Lerit DA, Shebelut CW, Lawlor KJ, Rusan NM, Gavis ER, Schedl P, Deshpande G (2017) Germ cell-less promotes centrosome segregation to induce germ cell formation. Cell Rep 18 (4):831–839

Li J, Kim S, Kobayashi T, Liang FX, Korzeniewski N, Duensing S, Dynlacht BD (2012) Neurl4, a novel daughter centriole protein, prevents formation of ectopic microtubule organizing centres. EMBO Rep 13(6):547–553

Lin TC, Neuner A, Schiebel E (2015) Targeting of γ-tubulin complexes to microtubule organizing centers: conservation and divergence. Trends Cell Biol 25(5):296–307

Lindeman RE, Pelegri F (2012) Localized products of *futile cycle/lrmp* promote centrosome-nucleus attachment in the zebrafish zygote. Curr Biol 22(10):843–851

Llamazares S, Tavosanis G, Gonzalez C (1999) Cytological characterisation of the mutant phenotypes produced during early embryogenesis by null and loss-of-function alleles of the γTub37C gene in *Drosophila*. J Cell Sci 112(Pt 5):659–667

Loppin B, Dubruille R, Horard B (2015) The intimate genetics of *Drosophila* fertilization. Open Biol 5(8)

Lucas EP, Raff JW (2007) Maintaining the proper connection between the centrioles and the pericentriolar matrix requires *Drosophila* centrosomin. J Cell Biol 178(5):725–732

Mahoney NM, Goshima G, Douglass AD, Vale RD (2006) Making microtubules and mitotic spindles in cells without functional centrosomes. Curr Biol 16(6):564–569

Mahowald AP (2001) Assembly of the *Drosophila* germ plasm. Int Rev Cytol 203:187–213

Malone CJ, Misner L, Le Bot N, Tsai MC, Campbell JM, Ahringer J, White JG (2003) The *C. elegans* hook protein, ZYG-12, mediates the essential attachment between the centrosome and nucleus. Cell 115(7):825–836

Mazumdar A, Mazumdar M (2002) How one becomes many: blastoderm cellularization in *Drosophila melanogaster*. BioEssays 24(11):1012–1022

McCartney BM, Peifer M (2000) Teaching tumour suppressors new tricks. Nat Cell Biol 2(4):E58–E60

McCartney BM, McEwen DG, Grevengoed E, Maddox P, Bejsovec A, Peifer M (2001) *Drosophila* APC2 and armadillo participate in tethering mitotic spindles to cortical actin. Nat Cell Biol 3 (10):933–938

McKnight SL, Bustin M, Miller OL (1977) Electron microscope analysis of chromosome metabolism in *Drosophila melanogaster embryo*. Cold Spring Harb Symp Quant Biol 41:741–754

Megraw TL, Li K, Kao LR, Kaufman TC (1999) The centrosomin protein is required for centrosome assembly and function during cleavage in Drosophila. Development 126(13):2829–2839

Megraw TL, Kao LR, Kaufman TC (2001) Zygotic development without functional mitotic centrosomes. Curr Biol 11(2):116–120

Mermall V, Miller KG (1995) The 95F unconventional myosin is required for proper organization of the *Drosophila* syncytial blastoderm. J Cell Biol 129(6):1575–1588

Mermall V, McNally JG, Miller KG (1994) Transport of cytoplasmic particles catalysed by an unconventional myosin in living *Drosophila* embryos. Nature 369(6481):560–562

Meyerzon M, Gao Z, Liu J, Wu JC, Malone CJ, Starr DA (2009) Centrosome attachment to the *C. elegans* male pronucleus is dependent on the surface area of the nuclear envelope. Dev Biol 327(2):433–446

Miller KG, Kiehart DP (1995) Fly division. J Cell Biol 131(1):1–5

Minestrini G, Harley AS, Glover DM (2003) Localization of Pavarotti-KLP in living *Drosophila* embryos suggests roles in reorganizing the cortical cytoskeleton during the mitotic cycle. Mol Biol Cell 14(10):4028–4038

Mirouse V, Dastugue B, Couderc JL (2005) The *Drosophila* toucan protein is a new mitotic microtubule-associated protein required for spindle microtubule stability. Genes Cells 10 (1):37–46

Mogensen MM, Tucker JB, Baggaley TB (1993) Multiple plasma membrane-associated MTOC systems in the acentrosomal cone cells of *Drosophila* ommatidia. Eur J Cell Biol 60(1):67–75

Moritz M, Braunfeld MB, Fung JC, Sedat JW, Alberts BM, Agard DA (1995) Three-dimensional structural characterization of centrosomes from early *Drosophila* embryos. J Cell Biol 130 (5):1149–1159

Müller HA, Wieschaus E (1996) *armadillo, bazooka,* and *stardust* are critical for early stages in formation of the zonula adherens and maintenance of the polarized blastoderm epithelium in *Drosophila*. J Cell Biol 134(1):149–163

Niki Y (1984) Developmental analysis of the *grandchildless (gs(1)N26)* mutation in *Drosophila melanogaster*: abnormal cleavage patterns and defects in pole cell formation. Dev Biol 103 (1):182–189

Niki Y, Okada M (1981) Isolation and characterization of *grandchildless-like* mutants in *Drosophila melanogaster*. Roux Arch Dev Biol 190(1):1–10

Nithianantham S, Cook BD, Beans M, Guo F, Chang F, Al-Bassam J (2018) Structural basis of tubulin recruitment and assembly by microtubule polymerases with tumor overexpressed gene (TOG) domain arrays. elife 7

O'Farrell PH (2015) Growing an embryo from a single cell: a hurdle in animal life. Cold Spring Harb Perspect Biol 7(11)

O'Farrell PH, Stumpff J, Su TT (2004) Embryonic cleavage cycles: how is a mouse like a fly? Curr Biol 14(1):R35–R45

Oakley BR (2000) γ-Tubulin. Curr Top Dev Biol 49:27–54

Oakley BR, Paolillo V, Zheng Y (2015) γ-Tubulin complexes in microtubule nucleation and beyond. Mol Biol Cell 26(17):2957–2962

Okada M (1982) Loss of the ability to form pole cells in *Drosophila* embryos with artificially delayed nuclear arrival at the posterior pole. Prog Clin Biol Res 85(Pt A):363–372

Padash Barmchi M, Rogers S, Häcker U (2005) DRhoGEF2 regulates actin organization and contractility in the *Drosophila* blastoderm embryo. J Cell Biol 168(4):575–585

Paddy MR, Saumweber H, Agard DA, Sedat JW (1996) Time-resolved, in vivo studies of mitotic spindle formation and nuclear lamina breakdown in *Drosophila* early embryos. J Cell Sci 109 (Pt 3):591–607

Pai CY, Lei EP, Ghosh D, Corces VG (2004) The centrosomal protein CP190 is a component of the *gypsy* chromatin insulator. Mol Cell 16(5):737–748

Papoulas O, Hays TS, Sisson JC (2005) The golgin lava lamp mediates dynein-based Golgi movements during *Drosophila* cellularization. Nat Cell Biol 7(6):612–618

Peel N, Stevens NR, Basto R, Raff JW (2007) Overexpressing centriole-replication proteins in vivo induces centriole overduplication and de novo formation. Curr Biol 17(10):834–843

Pelissier A, Chauvin JP, Lecuit T (2003) Trafficking through Rab11 endosomes is required for cellularization during *Drosophila* embryogenesis. Curr Biol 13(21):1848–1857

Piekny AJ, Glotzer M (2008) Anillin is a scaffold protein that links RhoA, actin, and myosin during cytokinesis. Curr Biol 18(1):30–36

Pollitt AY, Insall RH (2009) WASP and SCAR/WAVE proteins: the drivers of actin assembly. J Cell Sci 122(Pt 15):2575–2578

Postner MA, Miller KG, Wieschaus EF (1992) Maternal effect mutations of the *sponge* locus affect actin cytoskeletal rearrangements in *Drosophila melanogaster* embryos. J Cell Biol 119 (5):1205–1218

Raff JW, Glover DM (1988) Nuclear and cytoplasmic mitotic cycles continue in *Drosophila* embryos in which DNA synthesis is inhibited with aphidicolin. J Cell Biol 107(6 Pt 1):2009–2019

Raff JW, Glover DM (1989) Centrosomes, and not nuclei, initiate pole cell formation in *Drosophila* embryos. Cell 57(4):611–619

Raff JW, Jeffers K, Huang JY (2002) The roles of Fzy/Cdc20 and Fzr/Cdh1 in regulating the destruction of cyclin B in space and time. J Cell Biol 157(7):1139–1149

Rappaport R (1961) Experiments concerning the cleavage stimulus in sand dollar eggs. J Exp Zool 148:81–89

Riggs B, Rothwell W, Mische S, Hickson GR, Matheson J, Hays TS, Gould GW, Sullivan W (2003) Actin cytoskeleton remodeling during early *Drosophila* furrow formation requires recycling endosomal components nuclear-fallout and Rab11. J Cell Biol 163(1):143–154

Riggs B, Fasulo B, Royou A, Mische S, Cao J, Hays TS, Sullivan W (2007) The concentration of Nuf, a Rab11 effector, at the microtubule-organizing center is cell cycle regulated, dynein-dependent, and coincides with furrow formation. Mol Biol Cell 18(9):3313–3322

Rikhy R, Mavrakis M, Lippincott-Schwartz J (2015) Dynamin regulates metaphase furrow formation and plasma membrane compartmentalization in the syncytial *Drosophila* embryo. Biol Open 4(3):301–311

Riparbelli MG, Callaini G (2003) Assembly of yolk spindles in the early *Drosophila* embryo. Mech Dev 120(4):441–454

Riparbelli MG, Callaini G, Glover DM (2000) Failure of pronuclear migration and repeated divisions of polar body nuclei associated with MTOC defects in *polo* eggs of *Drosophila*. J Cell Sci 113(Pt 18):3341–3350

Riparbelli MG, Callaini G, Glover DM, Avides MC (2002) A requirement for the abnormal spindle protein to organise microtubules of the central spindle for cytokinesis in *Drosophila*. J Cell Sci 115(Pt 5):913–922

Riparbelli MG, Callaini G, Schejter ED (2007) Microtubule-dependent organization of subcortical microfilaments in the early *Drosophila* embryo. Dev Dyn 236(3):662–670

Ripoche J, Link B, Yucel JK, Tokuyasu K, Malhotra V (1994) Location of Golgi membranes with reference to dividing nuclei in syncytial *Drosophila* embryos. Proc Natl Acad Sci U S A 91 (5):1878–1882

Robertson SE, Dockendorff TC, Leatherman JL, Faulkner DL, Jongens TA (1999) *germ cell-less* is required only during the establishment of the germ cell lineage of *Drosophila* and has activities which are dependent and independent of its localization to the nuclear envelope. Dev Biol 215 (2):288–297

Robinson JT, Wojcik EJ, Sanders MA, McGrail M, Hays TS (1999) Cytoplasmic dynein is required for the nuclear attachment and migration of centrosomes during mitosis in *Drosophila*. J Cell Biol 146(3):597–608

Rodrigues-Martins A, Bettencourt-Dias M, Riparbelli M, Ferreira C, Ferreira I, Callaini G, Glover DM (2007a) DSAS-6 organizes a tube-like centriole precursor, and its absence suggests modularity in centriole assembly. Curr Biol 17(17):1465–1472

Rodrigues-Martins A, Riparbelli M, Callaini G, Glover DM, Bettencourt-Dias M (2007b) Revisiting the role of the mother centriole in centriole biogenesis. Science 316 (5827):1046–1050

Rodrigues-Martins A, Riparbelli M, Callaini G, Glover DM, Bettencourt-Dias M (2008) From centriole biogenesis to cellular function: centrioles are essential for cell division at critical developmental stages. Cell Cycle 7(1):11–16

Rogers SL, Rogers GC, Sharp DJ, Vale RD (2002) *Drosophila* EB1 is important for proper assembly, dynamics, and positioning of the mitotic spindle. J Cell Biol 158(5):873–884

Rothwell WF, Fogarty P, Field CM, Sullivan W (1998) Nuclear-fallout, a drosophila protein that cycles from the cytoplasm to the centrosomes, regulates cortical microfilament organization. Development 125(7):1295–1303

Rothwell WF, Sullivan W (2000) The centrosome in early *Drosophila* embryogenesis. Curr Top Dev Biol 49:409–447

Rothwell WF, Zhang CX, Zelano C, Hsieh TS, Sullivan W (1999) The *Drosophila* centrosomal protein Nuf is required for recruiting Dah, a membrane associated protein, to furrows in the early embryo. J Cell Sci 112(Pt 17):2885–2893

Royou A, Sullivan W, Karess R (2002) Cortical recruitment of nonmuscle myosin II in early syncytial *Drosophila* embryos: its role in nuclear axial expansion and its regulation by Cdc2 activity. J Cell Biol 158(1):127–137

Schejter ED, Wieschaus E (1993) *bottleneck* acts as a regulator of the microfilament network governing cellularization of the *Drosophila* embryo. Cell 75(2):373–385

Shamanski FL, Orr-Weaver TL (1991) The *Drosophila plutonium* and *pan gu* genes regulate entry into S phase at fertilization. Cell 66(6):1289–1300

Sharp DJ, Yu KR, Sisson JC, Sullivan W, Scholey JM (1999) Antagonistic microtubule-sliding motors position mitotic centrosomes in *Drosophila* early embryos. Nat Cell Biol 1(1):51–54

Shaw M, Cohen P, Alessi DR (1997) Further evidence that the inhibition of glycogen synthase kinase-3beta by IGF-1 is mediated by PDK1/PKB-induced phosphorylation of Ser-9 and not by dephosphorylation of Tyr-216. FEBS Lett 416(3):307–311

Sherlekar A, Rikhy R (2016) Syndapin promotes pseudocleavage furrow formation by actin organization in the syncytial *Drosophila* embryo. Mol Biol Cell 27(13):2064–2079

Shermoen AW, McCleland ML, O'Farrell PH (2010) Developmental control of late replication and S phase length. Curr Biol 20(23):2067–2077

Sibon OC, Stevenson VA, Theurkauf WE (1997) DNA-replication checkpoint control at the *Drosophila* midblastula transition. Nature 388(6637):93–97

Sibon OC, Laurençon A, Hawley R, Theurkauf WE (1999) The *Drosophila* ATM homologue Mei-41 has an essential checkpoint function at the midblastula transition. Curr Biol 9 (6):302–312

Sibon OC, Kelkar A, Lemstra W, Theurkauf WE (2000) DNA-replication/DNA-damage-dependent centrosome inactivation in *Drosophila* embryos. Nat Cell Biol 2(2):90–95

Sigrist S, Jacobs H, Stratmann R, Lehner CF (1995) Exit from mitosis is regulated by *Drosophila fizzy* and the sequential destruction of cyclins A, B and B3. EMBO J 14(19):4827–4838

Sisson JC, Field C, Ventura R, Royou A, Sullivan W (2000) Lava lamp, a novel peripheral golgi protein, is required for *Drosophila melanogaster* cellularization. J Cell Biol 151(4):905–918

Sokac AM, Wieschaus E (2008a) Local actin-dependent endocytosis is zygotically controlled to initiate *Drosophila* cellularization. Dev Cell 14(5):775–786

Sokac AM, Wieschaus E (2008b) Zygotically controlled F-actin establishes cortical compartments to stabilize furrows during *Drosophila* cellularization. J Cell Sci 121(11):1815–1824

Stafstrom JP, Staehelin LA (1984) Dynamics of the nuclear envelope and of nuclear pore complexes during mitosis in the *Drosophila* embryo. Eur J Cell Biol 34(1):179–189

Stanley H, Botas J, Malhotra V (1997) The mechanism of Golgi segregation during mitosis is cell type-specific. Proc Natl Acad Sci U S A 94(26):14467–14470

Stevens NR, Raposo AA, Basto R, St Johnston D, Raff JW (2007) From stem cell to embryo without centrioles. Curr Biol 17(17):1498–1503

Stevens NR, Dobbelaere J, Brunk K, Franz A, Raff JW (2010) *Drosophila* Ana2 is a conserved centriole duplication factor. J Cell Biol 188(3):313–323

Stevenson VA, Kramer J, Kuhn J, Theurkauf WE (2001) Centrosomes and the scrambled protein coordinate microtubule-independent actin reorganization. Nat Cell Biol 3(1):68–75

Stevenson V, Hudson A, Cooley L, Theurkauf WE (2002) Arp2/3-dependent pseudocleavage [correction of psuedocleavage] furrow assembly in syncytial *Drosophila* embryos. Curr Biol 12(9):705–711

Stiffler LA, Ji JY, Trautmann S, Trusty C, Schubiger G (1999) Cyclin A and B functions in the early *Drosophila* embryo. Development 126(23):5505–5513

Su TT, Sprenger F, DiGregorio PJ, Campbell SD, O'Farrell PH (1998) Exit from mitosis in *Drosophila* syncytial embryos requires proteolysis and cyclin degradation, and is associated with localized dephosphorylation. Genes Dev 12(10):1495–1503

Su J, Chow B, Boulianne GL, Wilde A (2013) The BAR domain of amphiphysin is required for cleavage furrow tip-tubule formation during cellularization in *Drosophila* embryos. Mol Biol Cell 24(9):1444–1453

Sullivan W, Minden JS, Alberts BM (1990) *daughterless-abo-like*, a *Drosophila* maternal-effect mutation that exhibits abnormal centrosome separation during the late blastoderm divisions. Development 110(2):311–323

Sullivan W, Daily DR, Fogarty P, Yook KJ, Pimpinelli S (1993a) Delays in anaphase initiation occur in individual nuclei of the syncytial *Drosophila* embryo. Mol Biol Cell 4(9):885–896

Sullivan W, Fogarty P, Theurkauf W (1993b) Mutations affecting the cytoskeletal organization of syncytial *Drosophila* embryos. Development 118(4):1245–1254

Swanson MM, Poodry CA (1981) The *shibire(ts)* mutant of *Drosophila*: a probe for the study of embryonic development. Dev Biol 84(2):465–470

Takada S, Kelkar A, Theurkauf WE (2003) *Drosophila* checkpoint kinase 2 couples centrosome function and spindle assembly to genomic integrity. Cell 113(1):87–99

Tavosanis G, Llamazares S, Goulielmos G, Gonzalez C (1997) Essential role for γ-tubulin in the acentriolar female meiotic spindle of *Drosophila*. EMBO J 16(8):1809–1819

Technau M, Roth S (2008) The *Drosophila* KASH domain proteins Msp-300 and Klarsicht and the SUN domain protein Klaroid have no essential function during oogenesis. Fly (Austin) 2(2):82–91

Theurkauf WE (1994) Actin cytoskeleton. Through the bottleneck. Curr Biol 4(1):76–78

Thomas JH, Wieschaus E (2004) *src64* and *tec29* are required for microfilament contraction during *Drosophila* cellularization. Development 131(4):863–871

Tillery MML, Blake-Hedges C, Zheng Y, Buchwalter RA, Megraw TL (2018) Centrosomal and non-centrosomal microtubule-organizing centers (MTOCs) in *Drosophila melanogaster*. Cell 7(9)

Tucker JB, Milner MJ, Currie DA, Muir JW, Forrest DA, Spencer MJ (1986) Centrosomal microtubule-organizing centers and a switch in the control of protofilament number for cell surfacing-associated microtubules during *Drosophila* wing morphogenesis. Eur J Cell Biol 41:279–289

Turner FR, Mahowald AP (1976) Scanning electron microscopy of *Drosophila* embryogenesis. 1. The structure of the egg envelopes and the formation of the cellular blastoderm. Dev Biol 50 (1):95–108

Turner FR, Mahowald AP (1979) Scanning electron microscopy of *Drosophila* embryogenesis. III. Formation of the head and caudal segments. Dev Biol 68(1):96–109

Vaizel-Ohayon D, Schejter ED (1999) Mutations in centrosomin reveal requirements for centrosomal function during early *Drosophila* embryogenesis. Curr Biol 9(16):889–898

van Ijzendoorn SC (2006) Recycling endosomes. J Cell Sci 119(Pt 9):1679–1681

Varmark H, Llamazares S, Rebollo E, Lange B, Reina J, Schwarz H, Gonzalez C (2007) Asterless is a centriolar protein required for centrosome function and embryo development in *Drosophila*. Curr Biol 17(20):1735–1745

Venkei Z, Gáspár I, Tóth G, Szabad J (2006) alpha4-tubulin is involved in rapid formation of long microtubules to push apart the daughter centrosomes during early *Drosophila* embryogenesis. J Cell Sci 119(Pt 15):3238–3248

von Dassow G, Schubiger G (1994) How an actin network might cause fountain streaming and nuclear migration in the syncytial *Drosophila* embryo. J Cell Biol 127(6 Pt 1):1637–1653

Von Stetina JR, Orr-Weaver TL (2011) Developmental control of oocyte maturation and egg activation in metazoan models. Cold Spring Harb Perspect Biol 3(10):a005553

Wakefield JG, Huang JY, Raff JW (2000) Centrosomes have a role in regulating the destruction of cyclin B in early *Drosophila* embryos. Curr Biol 10(21):1367–1370

Walker JJ, Lee KK, Desai RN, Erickson JW (2000) The *Drosophila melanogaster* sex determination gene *sisA* is required in yolk nuclei for midgut formation. Genetics 155(1):191–202

Wang P, Pinson X, Archambault V (2011) PP2A-twins is antagonized by greatwall and collaborates with polo for cell cycle progression and centrosome attachment to nuclei in *Drosophila* embryos. PLoS Genet 7(8):e1002227

Warn RM, Robert-Nicoud M (1990) F-actin organization during the cellularization of the *Drosophila* embryo as revealed with a confocal laser scanning microscope. J Cell Sci 96 (Pt 1):35–42

Warn RM, Magrath R, Webb S (1984) Distribution of F-actin during cleavage of the *Drosophila* syncytial blastoderm. J Cell Biol 98(1):156–162

Warn RM, Smith L, Warn A (1985) Three distinct distributions of F-actin occur during the divisions of polar surface caps to produce pole cells in *Drosophila* embryos. J Cell Biol 100 (4):1010–1015

Watanabe N, Madaule P, Reid T, Ishizaki T, Watanabe G, Kakizuka A, Saito Y, Nakao K, Jockusch BM, Narumiya S (1997) p140mDia, a mammalian homolog of *Drosophila* diaphanous, is a target protein for Rho small GTPase and is a ligand for profilin. EMBO J 16(11):3044–3056

Webb RL, Zhou MN, McCartney BM (2009) A novel role for an APC2-diaphanous complex in regulating actin organization in *Drosophila*. Development 136(8):1283–1293

Wen Y, Eng CH, Schmoranzer J, Cabrera-Poch N, Morris EJ, Chen M, Wallar BJ, Alberts AS, Gundersen GG (2004) EB1 and APC bind to mDia to stabilize microtubules downstream of Rho and promote cell migration. Nat Cell Biol 6(9):820–830

Wenzl C, Yan S, Laupsien P, Grosshans J (2010) Localization of RhoGEF2 during *Drosophila* cellularization is developmentally controlled by Slam. Mech Dev 127(7–8):371–384

Wheatley S, Kulkarni S, Karess R (1995) Drosophila nonmuscle myosin II is required for rapid cytoplasmic transport during oogenesis and for axial nuclear migration in early embryos. Development 121(6):1937–1946

Williams BC, Dernburg AF, Puro J, Nokkala S, Goldberg ML (1997) The *Drosophila* kinesin-like protein KLP3A is required for proper behavior of male and female pronuclei at fertilization. Development 124(12):2365–2376

Wilson PG, Borisy GG (1998) Maternally expressed γTub37CD in *Drosophila* is differentially required for female meiosis and embryonic mitosis. Dev Biol 199(2):273–290

Wilson PG, Zheng Y, Oakley CE, Oakley BR, Borisy GG, Fuller MT (1997) Differential expression of two gamma-tubulin isoforms during gametogenesis and development in Drosophila. Dev Biol 184(2):207–221

Wolf R (1980) Migration and division of cleavage nuclei in the gall midge, *Wachtliella persicariae*: II. Origin and ultrastructure of the migration cytaster. Roux Arch Dev Biol 188(1):65–73

Xue Z, Sokac AM (2016) Back-to-back mechanisms drive actomyosin ring closure during *Drosophila* embryo cleavage. J Cell Biol 215(3):335–344

Yasuda GK, Baker J, Schubiger G (1991) Independent roles of centrosomes and DNA in organizing the *Drosophila* cytoskeleton. Development 111(2):379–391

Yohn CB, Pusateri L, Barbosa V, Lehmann R (2003) *l(3)malignant* brain tumor and three novel genes are required for *Drosophila* germ-cell formation. Genetics 165(4):1889–1900

Young PE, Pesacreta TC, Kiehart DP (1991) Dynamic changes in the distribution of cytoplasmic myosin during *Drosophila* embryogenesis. Development 111(1):1–14

Zallen JA, Cohen Y, Hudson AM, Cooley L, Wieschaus E, Schejter ED (2002) SCAR is a primary regulator of Arp2/3-dependent morphological events in *Drosophila*. J Cell Biol 156(4):689–701

Zalokar M (1976) Autoradiographic study of protein and RNA formation during early development of *Drosophila* eggs. Dev Biol 49(2):425–437

Zalokar M, Erk I (1976) Division and migration of nuclei during early embryogenesis of *Drosophila melanogaster*. J Microsc Biol Cell 25:97–106

Zhang J, Megraw TL (2007) Proper recruitment of γ-tubulin and D-TACC/Msps to embryonic *Drosophila* centrosomes requires Centrosomin Motif 1. Mol Biol Cell 18(10):4037–4049

Zhang CX, Rothwell WF, Sullivan W, Hsieh TS (2000) Discontinuous actin hexagon, a protein essential for cortical furrow formation in *Drosophila*, is membrane associated and hyperphosphorylated. Mol Biol Cell 11(3):1011–1022

Zhang G, Breuer M, Förster A, Egger-Adam D, Wodarz A (2009) Mars, a *Drosophila* protein related to vertebrate HURP, is required for the attachment of centrosomes to the mitotic spindle during syncytial nuclear divisions. J Cell Sci 122(Pt 4):535–545

Chapter 13
Centrosomes in Branching Morphogenesis

Sofia J. Araújo

Abstract The centrosome, a major microtubule organizer, has important functions in regulating the cytoskeleton as well as the position of cellular structures and orientation of cells within tissues. The centrosome serves as the main cytoskeleton-organizing centre in the cell and is the classical site of microtubule nucleation and anchoring. For these reasons, centrosomes play a very important role in morphogenesis, not just in the early stages of cell divisions but also in the later stages of organogenesis. Many organs such as lung, kidney and blood vessels develop from epithelial tubes that branch into complex networks. Cells in the nervous system also form highly branched structures in order to build complex neuronal networks. During branching morphogenesis, cells have to rearrange within tissues though multicellular branching or through subcellular branching, also known as single-cell branching. For highly branched structures to be formed during embryonic development, the cytoskeleton needs to be extensively remodelled. The centrosome has been shown to play an important role during these events.

13.1 Centrosomes in Branching Morphogenesis

The general role of centrosomes during embryonic development has mainly been studied in regard to cell division. However, it is becoming clear that subcellular events that rely on centrosome positioning, movement and activity are key to many developmental processes (Tang and Marshall 2012). From cell migration to force generation and both multicellular and subcellular branching, the role of the centrosome at the subcellular level can influence many morphogenetic processes. Branching morphogenesis has been extensively studied in many model organisms and distinct organs and despite our knowledge on many of the molecular players involved in branching, the centrosome's role in branching morphogenesis has only been reported clearly in a few of these systems.

S. J. Araújo (✉)
Department of Genetics, Microbiology and Statistics, School of Biology, University of Barcelona, Barcelona, Spain
e-mail: sofiajaraujo@ub.edu

© Springer Nature Switzerland AG 2019

M. Kloc (ed.), *The Golgi Apparatus and Centriole*, Results and Problems in Cell Differentiation 67, https://doi.org/10.1007/978-3-030-23173-6_13

13.2 The Active Role of the Centrosome in Tubulogenesis

13.2.1 The Tracheal System

Tracheal system development has been broadly studied in the fruit fly *Drosophila melanogaster* and is an important model for the analysis of branching morphogenesis the developmental process that gives rise to many vertebrate organs including the lung, vascular system, kidney and pancreas (Affolter and Caussinus 2008; Manning and Krasnow 1993; Ghabrial et al. 2003; Hayashi and Kondo 2018).

During *Drosophila* embryogenesis, the tracheal primordial cells appear as 10 pairs of tracheal placodes, of about 80 cells each, that invaginate into the body cavity while maintaining epithelial integrity, and undergo stereotyped branching and fusion processes to form a continuous network of tubular epithelium providing all tissues with the necessary amount of oxygen (Fig. 13.1a) (reviewed in Hayashi and Kondo 2018). Interestingly, embryonic tracheal development proceeds without the need for cell division, which makes it an excellent system to study branching morphogenesis in post-mitotic cells.

A key process in branching tubulogenesis is the invagination of the epithelial placodes, which converts flat cellular sheets into three-dimensional structures and positions some of the cells of these sheets below the surface on which they were originally positioned. Tracheal invagination starts by the apical constriction in a small group of cells at the centre of the placode. These cells begin the internalization events which are then followed by distinct rearrangements of the adjacent cells in the dorsal and ventral part of this placode. This process is regulated by the activity of the Trachealess (Trh) transcription factor and epidermal growth factor receptor (EGFR) signalling (Nishimura et al. 2007; Ogura et al. 2018). In invaginating placodes at stage 11 of *Drosophila* embryogenesis, the nucleus is basally located, the apical domain faces the future lumen, and centrosomes localize within a subapical domain (Brodu et al. 2010). Here, an apical array of short microtubules (MTs), organized into a meshwork, forms a cap-like structure at the apical domain whereas long MT fibres are distributed along the basolateral cell domain. As microtubule organizing centres (MTOCs), these centrosomes colocalize with gamma-tubulin at this stage. However, this colocalization changes as invagination progresses and by embryonic stage 13, MTOCs relocate, from the centrosome to the apical membrane, by a two-step process controlled by the transcription factor Trh. The first step involves a Spastin-mediated microtubule release from the centrosome, followed by Pio-pio (Pio) facilitated anchoring of microtubules to the apical plasma membrane (Brodu et al. 2010). This dynamic centrosomal, a two-step mechanism, controls MT-network reorganization during in vivo development and these changes are essential for tracheal branching morphogenesis (Brodu et al. 2010). However, despite these studies of the centrosomal involvement during invagination and pre-liminary lumen formation stages, centrosome localization and dynamics have not yet been studied during cell migration stages that occur after placode invagination. During these stages, the main chemoattractant responsible for cell migration is the

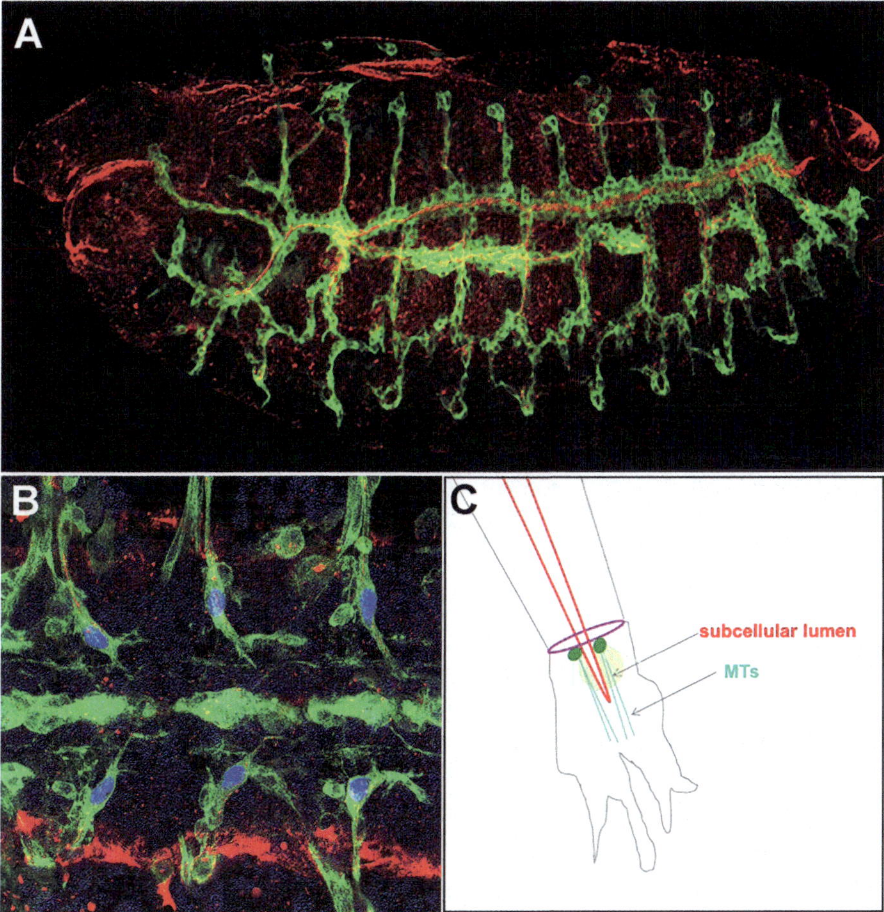

Fig. 13.1 *Drosophila* tracheal system. (**a**) The tracheal system of an embryonic stage 16 *Drosophila* embryo, with all tracheal cells labelled in green (UASGFP expression under the control of btlGAL4) and tracheal lumen in red by wheat germ aglutinin (WGA). (**b**) Detail of the ganglionic branch terminal cells (TCs) marked with an antibody against *Drosophila* Serum Response Factor (DSRF) (blue), tracheal cells in green and lumen in red as in (**a**). (**c**) Schematic representation of the involvement of centrosomes (green dots) in subcellular lumen (red) formation: two centrosomes localize apically to the junction of the TC with the stalk cell. There, they are active MTOCs and provide the cytoskeletal structure necessary to start forming the ingrowing subcellular lumen. The MTs that emanate from this centrosome-pair grow towards the basolateral membrane (tip) of the TC, forming two tracks through which membrane can be delivered and the new lumen built

FGF homologue Branchless (Bnl) (Sutherland et al. 1996). Bnl activates the FGF receptor (FGFR) Breathless (Btl) on tracheal tip cells, which leads to the concerted cell migration towards the Bnl source (Klämbt et al. 1992; Lee et al. 1996). These tip cells actively migrate, extending filopodia towards the chemoattractant until they reach their targets.

Centrosome position has been shown to be important for many cell migratory events (Barker et al. 2016). In mammalian cell in vitro systems, the involvement of the centrosome in cell movement has been analysed and in some cases it has been shown that centrosome amplification is able to increase the activity of the Rho GTPase Rac1 in the cell promoting migratory invasive events (Godinho et al. 2014). In other cases, excess centrosomes perturb the cytoskeleton resulting in reduced migration (Kushner et al. 2016). Thus, it will be interesting to know if centrosomes are also involved in the tracheal migratory steps that lead to both multicellular and single-cell branching events.

In *Drosophila*, after most of the tracheal migration is accomplished, specific cells at strategic positions within the tracheal network differentiate into different cell types (Fig. 13.1b). Of these, two very special cell types in the *Drosophila* tracheal system are able to form lumina within the cytoplasm of one cell. These are the tip or terminal cells (TCs) that form a luminal space inside their cytoplasm as they elongate (Fig. 13.1b); and the fusion cells (FCs) that mediate the fusion between anastomosing branches of this complex system in order to assure network continuity (Sigurbjörnsdóttir et al. 2014). These two cell types form a type of lumen called subcellular due to its 'intracellular' characteristics, which arises by de novo growth of an apical membrane towards the inside of the cell (Sigurbjörnsdóttir et al. 2014). Tracheal TC lumen formation depends on cytoplasmic extension, asymmetric actin accumulation and vesicle trafficking guided by the cytoskeleton (Gervais and Casanova 2010; Schottenfeld-Roames and Ghabrial 2012; Schottenfeld-Roames et al. 2014). The beginning of TC subcellular lumen formation relies on centrosomes that localize to the apical junction between the terminal and the stalk cell. From this centrosome pair, microtubules are organized in the direction of cell elongation and form two tracks that are thought to facilitate membrane delivery and concomitant lumen formation (Fig. 13.1c) (Ricolo et al. 2016). Subsequent lumen extension requires microtubule polymerization and asymmetric actin accumulation at the basolateral tip of the terminal cell (Gervais and Casanova 2010). An increase in centrosome number is able to induce the growth of additional lumina inside the terminal cell and centrosome loss impairs terminal tubulogenesis (Ricolo et al. 2016). So, at embryonic stages, the centrosome plays an active role in TC subcellular lumen formation, the first branching step of the TC. In fusion cell anastomosis and subcellular lumen formation, the participation of the centrosome has not yet been analysed.

At larval stages, tracheal TCs ramify extensively to form many new terminal branches, long cytoplasmic extensions that grow towards oxygen-starved cells and then form a cytoplasmic, membrane-bound lumen, creating tiny tubes inside one cell in an intricate single-cell branching process. The localization of cytoskeletal markers in larval TCs suggests that the overall principles of growth that start at embryonic stages are maintained at larval stages (Sigurbjörnsdóttir et al. 2014). Acetylated MTs line the gas-filled tube and extend beyond it or branch off it into filopodia (Jayanandanan et al. 2014). The gamma-tubulin remains restricted to the apical

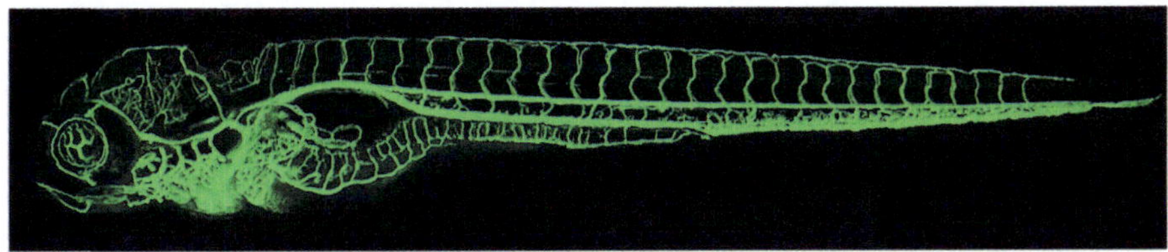

Fig. 13.2 Vertebrate vasculature. The vascular system of a 4-day-old zebrafish embryo vascular system expressing EGFP under the control of the endothelial specific promoter of the *flk1* gene (TG: flk1:EGFP)

membrane after lumen formation, suggesting that MT orientation remains the same as during lumen elongation (Gervais and Casanova 2010). And End-binding protein 1 (EB1), a marker for polymerizing MT tips, shows growth of MTs throughout branches (Schottenfeld-Roames and Ghabrial 2012). However, the debate is still open in regard to both the positioning and possible participation of centrosomes at this stage.

13.2.2 Vertebrate Vasculature

Like the fruit fly tracheal system, the vertebrate vasculature is a highly ramified branched organ (Fig. 13.2) (Kotini et al. 2018). The vasculature ensures the proper transport and distribution of relevant molecules to every organ. The initial vessels are generated during embryonic development by a process called 'vasculogenesis' and these serve as the substrate for the formation of most of the fine, branched vascular networks that are formed later. Like the tracheal system, growth of new vessels requires directional migration, cell rearrangements and cell shape changes, and it is mostly stereotyped in the early developing embryo. Unlike the tracheal system, vascular system branching requires endothelial cell (EC) division. After vasculogenesis, new branches arise through a process called 'angiogenesis'. Angiogenesis is responsible for the addition of new vessels to the already formed vascular system in response to the physiological needs of growing tissues and organs. There are two modes of angiogenesis called 'sprouting angiogenesis' (SA) and 'intussusceptive angiogenesis' (IA) (Makanya et al. 2009). SA involves the formation of tip cells that form numerous filopodial extensions and explore the environment to react to several positive and negative guidance cues; IA implicates the insertion of new branching points within the tubular branches of the network or the splitting of existing branches into finer ones. Both processes require considerable changes in cell shape and fast cytoskeletal reorganization (Ochoa-Espinosa and Affolter 2012).

There are many differences and as many similarities between cells of the vertebrate vascular system and the invertebrate tracheal system. Animal blood circulation does not contain tree-like blunt-ended branches such as those present in insect tracheal systems, because branching capillaries need to anastomose with each other in order to ensure full blood circulation. However, parallels can be drawn regarding branching mechanisms due to equal needs for extensive cytoskeletal remodelling during branching events.

As in tracheal cells, endothelial branching requires cell movement. In many cases, centrosomes orient and sustain migratory polarity via nucleation of MTs and centrosome reorientation relative to the nucleus is believed to be required for proper migration (Tang and Marshall 2012). In cell culture, ECs establish a centrosome-forward orientation, with the centrosome in front of the nucleus relative to migration direction (Luxton and Gundersen 2011).

Furthermore, in ECs, centrosomes reorient within cells as they begin sprouting angiogenesis during tube formation and centrosome number regulation is necessary for appropriate blood vessel sprouting (Gierke and Wittmann 2012; Kushner et al. 2014). However, non-centrosomal MTs have also been reported during sprouting angiogenesis (Martin et al. 2018).

Another vascular branching mechanism where centrosomes have been reported to play an active role is in oriented cell divisions (Tang and Marshall 2012). As such, orientation of spindles along the long axis of developing tubes is seen in blood vessel development (Zeng et al. 2007) and EC cell divisions occur in coordination with tubular architecture avoiding disruptions in the tubular network (Aydogan et al. 2015). Hence, centrosome positioning during cell division is important for vascular architecture.

Notably, as in the tracheal system, the centrosome is important for EC lumen formation as shown by the localization of centrosomes adjacent to the vacuoles needed to build the lumen (Davis et al. 2007) and by the centrosomal localization at junctional membranes and lumen initiation between adjacent cells (Rodríguez-Fraticelli et al. 2012). However, it is as yet unclear what happens to the EC lumen in conditions where supernumerary centrosomes are present in the EC cytoplasm as may happen during cancer progression.

13.3　Centrosomes in Axonal Growth Specification

The extraordinary morphological transformations neurons undergo as they migrate, extend axons and dendrites and establish synaptic connections, imply a precisely regulated process of structural reorganization and dynamic remodelling of the cytoskeleton (Fig. 13.3a). Mutations in genes encoding centrosomal proteins cause severe neurodevelopmental disorders (NDDs) (Bonini et al. 2017) leading to several diseases, such as lissencephaly, microcephaly and schizophrenia and centrosomal proteins have been shown to be involved in neurodegeneration (Diaz-Corrales et al. 2005; Madero-Perez et al. 2018). For this reason, unravelling the importance of centrosomal participation in axonal growth and branching is of foremost importance.

Many studies on the neuronal centrosome have focused on neuronal migration, a developmental phase many neurons undertake prior to axonal and dendritic development (Kuijpers and Hoogenraad 2011). In migrating neurons, most microtubules

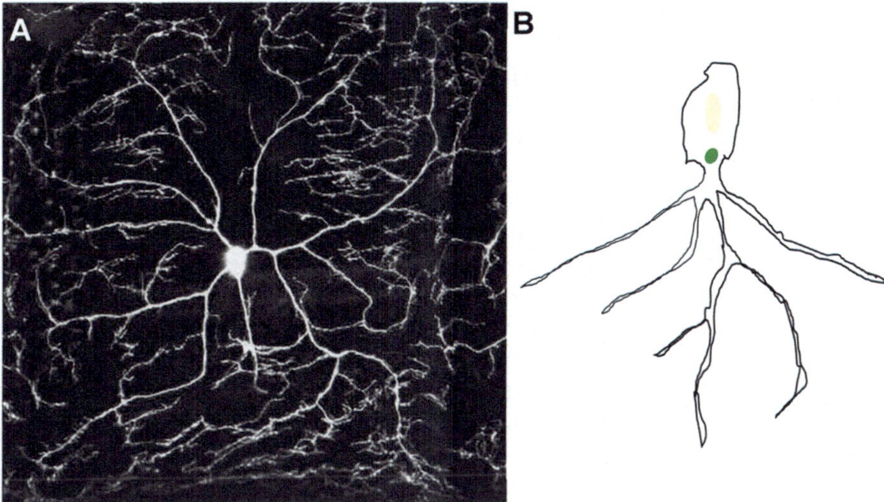

Fig. 13.3 Neuronal branching. (**a**) *Drosophila* third instar larval c4da neuron with long, branched dendrites; in some of these neurons, branching has not been shown to depend on centrosomes. (**b**) Schematic representation of a zebrafish Rohon-Beard (RB) sensory neuron, showing centrosome localization during axonal extension (adapted from Andersen and Halloran 2012)

are attached to the centrosome, an interaction necessary for pulling along the cell body and nucleus as the neuron moves to its final destination (Higginbotham and Gleeson 2007). But what happens to this centrosome during branching, once the neuron finishes migration and starts growing an axon and dendrites?

Neurons are highly polarized cells that generally extend a single thin, long axon, which transmits signals, and multiple shorter dendrites, which are specialized in receiving signals. Both axon and dendrites are rich in microtubules (Baas et al. 2016). The microtubule arrays within these processes are essential for providing architectural support, for enabling axons and dendrites to take on different shapes and branching patterns, and allowing for organelle distribution within the whole differentiated cell (Conde and Cáceres 2009). They are indispensable for the complex neuronal architecture. However, most of the highly organized microtubules in axons and dendrites are not attached to the centrosome or any recognizable organizing structure. Instead, the microtubules are free at both ends, and take on various lengths within the axon and dendrites (Baas et al. 2016). In both vertebrate and invertebrate neurons, axons have a uniform arrangement of microtubules with plus ends distal to the cell body (plus-end-out), and dendrites have equal numbers of plus- and minus-end-out microtubules (Conde and Cáceres 2009; Kapitein and Hoogenraad 2015).

Initial research on centrosomes and their role as MTOCs in neurons resulted in the proposal that the neuronal centrosome acts as the initiator of microtubules for both the axon and dendrites (Ahmad and Baas 1995; Ahmad et al. 1998). After this pioneer role of the centrosome, a consequence of earlier developmental stages, MTs are released and transported into the axon and dendrites by molecular motor proteins. Several observations have, since then, associated the centrosome with the primary site of axonal extension.

Cultured cerebellar granule neurons, after they stop migrating, grow a single axon, followed by a second axon to attain a bipolar morphology. Afterwards, they extend several short dendrites from the cell body. In these neurons, the centrosome is first positioned near where the initial axon develops and then moves to where the secondary axon develops, suggesting that the position of the centrosome is related to the development of each of the two axonal extensions (Zmuda and Rivas 1998).

It was reported that in cultured hippocampal neurons, the axon consistently arose from the first immature neurite forming after the final mitotic division of the neuroblast, and that the Golgi and endosomes (which generally cluster together with the centrosome) clustered in the location where the first neurite formed (De Anda et al. 2005). These observations are consistent with a previous report on cerebellar granule neurons described by Zmuda and Rivas (1998). Interestingly, these authors also found that light inactivation of the centrosome prevented normal polarization of *Drosophila* neurons (De Anda et al. 2005). Moreover, they observed that a small number of hippocampal neurons had two centrosomes and such neurons consistently formed two axons. It is interesting that most neurons normally have a single centrosome and according to the 'one-neurite-only' growth signal, the centrosome may be the originator of this signal (Craig and Banker 1994; De Anda et al. 2005). Thus, it has been proposed that the singularity of the axon and the singularity of the centrosome are somehow related, with cells with one centrosome extending one axon and cells with two centrosomes extending two. Nonetheless, in neurons such as cerebellar granule neurons, one centrosome is also capable of inducing the sprouting of two axons by changing its position (Zmuda and Rivas 1998). Centrosomal movement was also shown to be important for neurite formation in vivo. Using live imaging of zebrafish Rohon-Beard (RB) sensory neurons, a spatiotemporal relationship between centrosome position and the formation of RB axons was established (Fig. 13.3b) (Andersen and Halloran 2012).

Conflicting evidence correlates the role of the centrosome as an active MTOC with inhibition of neurite outgrowth. Centlein, a protein required for centrosome cohesion, is also a microtubule-associated protein (MAP) exerting its function by stabilizing microtubules. It was found that overexpression of centlein inhibited neurite outgrowth, associating centlein and the centrosome as negative regulators of neurite formation (Jing et al. 2016). Furthermore, many non-centrosomal MT networks have been associated with axonal extension (Stiess et al. 2010; Conde and Cáceres 2009; Sanchez-Huertas et al. 2016; Nguyen et al. 2011).

Taken together, these many observations indicate that there is no 'one-size-fits-all' scenario for the location of the neuronal centrosome or its function as the main MTOC during axonal specification. Even so, it would certainly appear that the centrosome is an important structure in the neuron, for axonal formation, at least during the early stages of development.

13.4 Centrosomes and Dendritic Arborization

Many of the most fundamental differences between axons and dendrites directly or indirectly result from distinct patterns of microtubule orientation in each type of process. In the axon, nearly all of the microtubules are oriented with their plus ends distal to the cell body, whereas in the dendrite, the microtubules have a mixed pattern of orientation (Delandre et al. 2016). In addition, unlike the case with the axon, dendrites are almost always multiple in numbers (Fig. 13.3a), and it would be hard to visualize a centrosome that could be so mobile in the cell body as to move from dendrite to dendrite and then back to the axon to serve each neurite one at a time. Interestingly, it was reported several years ago what appears to be streams of microtubules flowing from the centrosome into developing dendrites of cultured hippocampal neurons, with a location roughly centralized among the dendrites (Sharp et al. 1995). So, one could envisage that the centrosome may act as the main MTOC and then MT fragmentation could lead to the varied pattern of MT polarity within dendrites. Microtubule fragmentation occurs during dendritic specification and growth as was shown from the genetic disruption of MT severing enzymes such as spastin or katanin, all necessary for correct dendritic patterning (Yu et al. 2008). Nonetheless, localization of gamma-tubulin to axons and dendrites, as well as the close tie between microtubule polarity and gamma-tubulin activity, argues that many microtubules in mature neurons are likely to be generated locally at the dendrite rather than transported from the cell body (Nguyen et al. 2014).

The centrosome importance in dendritic arborization has also been shown as a nucleator of other factors/complexes necessary for dendrite formation. One of them is the cell division cycle 20-anaphase promoting complex/cyclosome (Cdc20-APC/ C) complex, whose subcellular location at the centrosome was shown to be critical for its ability to drive dendrite development (Kim et al. 2009).

Other structures in addition to the centrosome can function as MTOCs. Membranous organelles such as the nuclear envelope, endosomes, mitochondria and Golgi can also act as nucleation sites as have been shown for many cell types (Petry and Vale 2015). In addition, the Golgi apparatus tends to cluster at the centrosome, because membranous elements that comprise the Golgi are transported by cytoplasmic dynein toward minus ends of microtubules (Corthesy-Theulaz et al. 1992). So, many of the molecular components used by centrosomes to organize MT nucleation can also be found at the Golgi.

In *Drosophila* sensory neurons, MT nucleation was observed at Golgi fragments situated within dendrites and called Golgi outposts that also share composition with centrosomes (Ori-Mckenney et al. 2012). In *Drosophila*, Centrosomin (Cnn) and Pericentrin-like protein (Plp) link MT nucleation to dendritic Golgi outposts (Yalgin et al. 2015; Ori-Mckenney et al. 2012), and both can also be found at the Golgi (Sanders and Kaverina 2015). However, removing Golgi outposts from dendrites did not prevent gamma-tubulin MT nucleation at these dendrites arguing that gamma-tubulin in dendrites is associated with some other internal structure (Nguyen et al. 2014). Furthermore, MTs can nucleate from other MTs, an event mediated by the

Augmin complex and this has been shown to control neuronal microtubule polarity (Sanchez-Huertas and Luders 2015; Sanchez-Huertas et al. 2016).

Again, as it seems, a 'one-model-fits-all' appears not to exist regarding MT nucleation in dendrites. Dendritic arbours are very diverse and this diversity may be controlled by distinct ways of MT polymerization and branching.

13.5 Other Branching Organs

Centrosomes are also important for the development of branching structures in other organs. In kidney tubules, similarly to the vascular system, cells primarily divide along the proximal–distal (longitudinal) axis of the epithelium, leading to lengthening of the tubule, while maintaining a constant diameter. In mouse and rat models for polycystic kidney disease, a ciliopathy, the orientation of tubule epithelial cell divisions is randomized, leading to increased tubular diameter and subsequent cysts showing that the control of the orientation of cell division is crucial (Fischer et al. 2006). In mouse knockout models and in human renal tissue from polycystic kidney disease patients in vivo, supernumerary centrosomes were found in cells (Battini et al. 2008). The presence of supernumerary centrosomes was detected in normal tubular cells in these mutant conditions, suggesting that centrosome amplification is an early event that precedes cyst formation (Battini et al. 2008). In addition, centrosome amplification was sufficient to induce rapid cystogenesis both during development and after ischemic renal injury (Dionne et al. 2018). Thus, despite centrosome amplification not having been shown as the direct cause of these renal pathologies, it is likely that changes in centrosome number affect the orientation of tubular cell divisions and induce defects in cilia, leading to the appearance of cysts.

Longitudinally oriented cell divisions also occur in the developing lung, and using a mathematical model, it was recently demonstrated that a change in airway shape can be explained entirely on the basis of the distribution of spindle angles, without requiring oriented changes in other cellular processes, such as proliferation or cell shape (Tang et al. 2011). There are, of course, other cases, such as in mouse lung development, where centrosome dynamics directly affects branching by affecting cell proliferation (Schnatwinkel and Niswander 2012).

13.6 Conclusions

It has been shown in various types of cells that centrosomal localization is important for branching morphogenesis. The location of the centrosome may be functionally important not just as a MTOC but also as a provider of factors necessary for cytoskeletal rearrangements. Furthermore, the centrosome may be responsible for the localization of Golgi outposts or the transport of Golgi-derived vesicles. Another possibility is that the centrosome is important to gather together various proteins that form the pericentriolar material (PCM). For example, the PCM is rich in kinases (Hames et al. 2005), and hence the centrosome could act as a processing centre to phosphorylate functionally important proteins. Alternatively, the PCM might act as a sink for various proteins that would otherwise be widely distributed in the cytoplasm or as a hub for different protein machinery such as the proteasome (Avidor-Reiss and Gopalakrishnan 2013; Puram et al. 2013).

Thus, in branching morphogenesis, the centrosome is important developmentally, and may also be required in more mature differentiated cells in which the centrosome appears to have become vestigial. Perhaps under certain circumstances, the centrosome is reactivated to enable the cell to meet a particular challenge, such as breaking its symmetry or changing its branching pattern in response to disease or during regeneration.

References

Affolter M, Caussinus E (2008) Tracheal branching morphogenesis in Drosophila: new insights into cell behaviour and organ architecture. Development 135:2055–2064

Ahmad FJ, Baas PW (1995) Microtubules released from the neuronal centrosome are transported into the axon. J Cell Sci 108 (Pt 8):2761–2769

Ahmad FJ, Echeverri CJ, Vallee RB, Baas PW (1998) Cytoplasmic dynein and dynactin are required for the transport of microtubules into the axon. J Cell Biol 140:391–401

Andersen EF, Halloran MC (2012) Centrosome movements in vivo correlate with specific neurite formation downstream of LIM homeodomain transcription factor activity. Development 139:3590–3599

Avidor-Reiss T, Gopalakrishnan J (2013) Building a centriole. Curr Opin Cell Biol 25:72–77

Aydogan V, Lenard A, Denes AS, Sauteur L, Belting H-G, Affolter M (2015) Endothelial cell division in angiogenic sprouts of differing cellular architecture. Biology Open 4:1259–1269

Baas PW, Rao AN, Matamoros AJ, Leo L (2016) Stability properties of neuronal microtubules. Cytoskeleton (Hoboken) 73:442–460

Barker AR, Mcintosh KV, Dawe HR (2016) Centrosome positioning in non-dividing cells. Protoplasma 253:1007–1021

Battini L, Macip S, Fedorova E, Dikman S, Somlo S, Montagna C, Gusella GL (2008) Loss of polycystin-1 causes centrosome amplification and genomic instability. Hum Mol Genet 17:2819–2833

Bonini SA, Mastinu A, Ferrari-Toninelli G, Memo M (2017) Potential role of microtubule stabilizing agents in neurodevelopmental disorders. Int J Mol Sci 18(8). https://doi.org/10.3390/ijms18081627

Brodu V, Baffet AD, Le Droguen P-M, Casanova J, Guichet A (2010) A developmentally regulated two-step process generates a noncentrosomal microtubule network in Drosophila tracheal cells. Dev Cell 18:790–801

Conde C, Cáceres A (2009) Microtubule assembly, organization and dynamics in axons and dendrites. Nat Rev Neurosci 10:319–332

Corthesy-Theulaz I, Pauloin A, Pfeffer SR (1992) Cytoplasmic dynein participates in the centrosomal localization of the Golgi complex. J Cell Biol 118:1333–1345

Craig AM, Banker G (1994) Neuronal polarity. Annu Rev Neurosci 17:267–310

Davis GE, Koh W, Stratman AN (2007) Mechanisms controlling human endothelial lumen formation and tube assembly in three-dimensional extracellular matrices. Birth Defects Res C Embryo Today 81:270–285

De Anda FC, Pollarolo G, Da Silva JS, Camoletto PG, Feiguin F, Dotti CG (2005) Centrosome localization determines neuronal polarity. Nature 436:704–708

Delandre C, Amikura R, Moore AW (2016) Microtubule nucleation and organization in dendrites. Cell Cycle 15:1685–1692

Diaz-Corrales FJ, Asanuma M, Miyazaki I, Miyoshi K, Ogawa N (2005) Rotenone induces aggregation of gamma-tubulin protein and subsequent disorganization of the centrosome: relevance to formation of inclusion bodies and neurodegeneration. Neuroscience 133:117–135

Dionne LK, Shim K, Hoshi M, Cheng T, Wang J, Marthiens V, Knoten A, Basto R, Jain S, Mahjoub MR (2018) Centrosome amplification disrupts renal development and causes cystogenesis. J Cell Biol 217:2485–2501

Fischer E, Legue E, Doyen A, Nato F, Nicolas J-F, Torres V, Yaniv M, Pontoglio M (2006) Defective planar cell polarity in polycystic kidney disease. Nat Genet 38:21–23

Gervais L, Casanova J (2010) In vivo coupling of cell elongation and lumen formation in a single cell. Curr Biol 20:359–366

Ghabrial A, Luschnig S, Metzstein M, Krasnow M (2003) Branching morphogenesis of the Drosophila tracheal system. Annu Rev Cell Dev Biol 19:623–647

Gierke S, Wittmann T (2012) EB1-recruited microtubule +TIP complexes coordinate protrusion dynamics during 3D epithelial remodeling. Curr Biol 22:753–762

Godinho SA, Picone R, Burute M, Dagher R, Su Y, Leung CT, Polyak K, Brugge JS, Théry M, Pellman D (2014) Oncogene-like induction of cellular invasion from centrosome amplification. Nature 510(7503):167–171

Hames RS, Crookes RE, Straatman KR, Merdes A, Hayes MJ, Faragher AJ, Fry AM (2005) Dynamic recruitment of Nek2 kinase to the centrosome involves microtubules, PCM-1, and localized proteasomal degradation. Mol Biol Cell 16:1711–1724

Hayashi S, Kondo T (2018) Development and function of the Drosophila tracheal system. Genetics 209:367–380

Higginbotham HR, Gleeson JG (2007) The centrosome in neuronal development. Trends Neurosci 30:276–283

Jayanandanan N, Mathew R, Leptin M (2014) Guidance of subcellular tubulogenesis by actin under the control of a synaptotagmin-like protein and Moesin. Nat Commun 5:3036

Jing Z, Yin H, Wang P, Gao J, Yuan L (2016) Centlein, a novel microtubule-associated protein stabilizing microtubules and involved in neurite formation. Biochem Biophys Res Commun 472:360–365

Kapitein LC, Hoogenraad CC (2015) Building the neuronal microtubule cytoskeleton. Neuron 87:492–506

Kim A, Puram S, Bilimoria P, Ikeuchi Y, Keough S, Wong M, Rowitch D, Bonni A (2009) A centrosomal Cdc20-Apc pathway controls dendrite morphogenesis in postmitotic neurons. Cell 136:322–336

Klämbt C, Glazer L, Shilo BZ (1992) Breathless, a Drosophila FGF receptor homolog, is essential for migration of tracheal and specific midline glial cells. Genes Dev 6:1668–1678

Kotini MP, Mäe MA, Belting H-G, Betsholtz C, Affolter M (2018) Sprouting and anastomosis in the Drosophila trachea and the vertebrate vasculature: similarities and differences in cell behaviour. Vascular pharmacol 112:8–16

Kuijpers M, Hoogenraad CC (2011) Centrosomes, microtubules and neuronal development. Mol Cell Neurosci 48(4):349–358

Kushner EJ, Ferro LS, Liu J-Y, Durrant JR, Rogers SL, Dudley AC, Bautch VL (2014) Excess centrosomes disrupt endothelial cell migration via centrosome scattering. J Cell Biol 206:257–272

Kushner EJ, Ferro LS, Yu Z, Bautch VL (2016) Excess centrosomes perturb dynamic endothelial cell repolarization during blood vessel formation. Mol Biol Cell 27:1911–1920

Lee T, Hacohen N, Krasnow M, Montell DJ (1996) Regulated breathless receptor tyrosine kinase activity required to pattern cell migration and branching in the Drosophila tracheal system. Genes Dev 10:2912–2921

Luxton GWG, Gundersen GG (2011) Orientation and function of the nuclear-centrosomal axis during cell migration. Curr Opin Cell Biol 23:579–588

Madero-Perez J, Fdez E, Fernandez B, Lara Ordonez AJ, Blanca Ramirez M, Gomez-Suaga P, Waschbusch D, Lobbestael E, Baekelandt V, Nairn AC, Ruiz-Martinez J, Aiastui A, Lopez De Munain A, Lis P, Comptdaer T, Taymans JM, Chartier-Harlin MC, Beilina A, Gonnelli A, Cookson MR, Greggio E, Hilfiker S (2018) Parkinson disease-associated mutations in LRRK2 cause centrosomal defects via Rab8a phosphorylation. Mol Neurodegener 13:3

Makanya AN, Hlushchuk R, Djonov VG (2009) Intussusceptive angiogenesis and its role in vascular morphogenesis, patterning, and remodeling. Angiogenesis 12:113–123

Manning G, Krasnow MA (1993) Development of the *Drosophila* tracheal system. In: Bate M, Martínez-Arias A (eds) The development of *Drosophila melanogaster*. Cold Spring Harbor Laboratory Press, Cold Spring Harbor, NY

Martin M, Veloso A, Wu J, Katrukha EA, Akhmanova A (2018) Control of endothelial cell polarity and sprouting angiogenesis by non-centrosomal microtubules. eLife 7:R162

Nguyen MM, Stone MC, Rolls MM (2011) Microtubules are organized independently of the centrosome in Drosophila neurons. Neural Dev 6:38

Nguyen MM, Mccracken CJ, Milner ES, Goetschius DJ, Weiner AT, Long MK, Michael NL, Munro S, Rolls MM (2014) Gamma-tubulin controls neuronal microtubule polarity independently of Golgi outposts. Mol Biol Cell 25:2039–2050

Nishimura M, Inoue Y, Hayashi S (2007) A wave of EGFR signaling determines cell alignment and intercalation in the Drosophila tracheal placode. Development 134:4273–4282

Ochoa-Espinosa A, Affolter M (2012) Branching morphogenesis: from cells to organs and back. Cold Spring Harb Perspect Biol 4(10). https://doi.org/10.1101/cshperspect.a008243

Ogura Y, Wen FL, Sami MM, Shibata T, Hayashi S (2018) A switch-like activation relay of EGFR-ERK signaling regulates a wave of cellular contractility for epithelial invagination. Dev Cell 46:162–172 e5

Ori-Mckenney KM, Jan LY, Jan Y-N (2012) Golgi outposts shape dendrite morphology by functioning as sites of acentrosomal microtubule nucleation in neurons. Neuron 76:921–930

Petry S, Vale RD (2015) Microtubule nucleation at the centrosome and beyond. Nat Cell Biol 17:1089–1093

Puram SV, Kim AH, Park H-Y, Anckar J, Bonni A (2013) The ubiquitin receptor S5a/Rpn10 links centrosomal proteasomes with dendrite development in the mammalian brain. Cell Rep 4 (1):19–30

Ricolo D, Deligiannaki M, Casanova J, Araújo SJ (2016) Centrosome amplification increases single-cell branching in post-mitotic cells. Curr Biol 26:2805–2813

Rodríguez-Fraticelli AE, Auzan M, Alonso MA, Bornens M, Martín-Belmonte F (2012) Cell confinement controls centrosome positioning and lumen initiation during epithelial morphogenesis. J Cell Biol 198:1011–1023

Sanchez-Huertas C, Luders J (2015) The augmin connection in the geometry of microtubule networks. Curr Biol 25:R294–R299

Sanchez-Huertas C, Freixo F, Viais R, Lacasa C, Soriano E, Luders J (2016) Non-centrosomal nucleation mediated by augmin organizes microtubules in post-mitotic neurons and controls axonal microtubule polarity. Nat Commun 7:12187

Sanders AA, Kaverina I (2015) Nucleation and dynamics of Golgi-derived microtubules. Front Neurosci 9:431

Schnatwinkel C, Niswander L (2012) Nubp1 is required for lung branching morphogenesis and distal progenitor cell survival in mice. PLoS One 7:e44871

Schottenfeld-Roames J, Ghabrial AS (2012) Whacked and Rab35 polarize dynein-motor-complex-dependent seamless tube growth. Nat Cell Biol 14(4):386–393

Schottenfeld-Roames J, Rosa JB, Ghabrial AS (2014) Seamless tube shape is constrained by endocytosis-dependent regulation of active moesin. Curr Biol 24:1756–1764

Sharp DJ, Yu W, Baas PW (1995) Transport of dendritic microtubules establishes their nonuniform polarity orientation. J Cell Biol 130:93–103

Sigurbjörnsdóttir S, Mathew R, Leptin M (2014) Molecular mechanisms of de novo lumen formation. Nat Rev Mol Cell Biol 15:665–676

Stiess M, Maghelli N, Kapitein LC, Gomis-Rüth S, Wilsch-Bräuninger M, Hoogenraad CC, Tolić-Nørrelykke IM, Bradke F (2010) Axon extension occurs independently of centrosomal microtubule nucleation. Science 327:704–707

Sutherland D, Samakovlis C, Krasnow MA (1996) Branchless encodes a Drosophila FGF homolog that controls tracheal cell migration and the pattern of branching. Cell 87:1091–1101

Tang N, Marshall WF (2012) Centrosome positioning in vertebrate development. J Cell Sci 125:4951–4961

Tang N, Marshall WF, McMahon M, Metzger RJ, Martin GR (2011) Control of mitotic spindle angle by the RAS-regulated ERK1/2 pathway determines lung tube shape. Science 333 (6040):342–345. https://doi.org/10.1126/science.1204831

Yalgin C, Ebrahimi S, Delandre C, Yoong LF, Akimoto S, Tran H, Amikura R, Spokony R, Torben-Nielsen B, White KP, Moore AW (2015) Centrosomin represses dendrite branching by orienting microtubule nucleation. Nat Neurosci 18:1437–1445

Yu W, Qiang L, Solowska JM, Karabay A, Korulu S, Baas PW (2008) The microtubule-severing proteins spastin and katanin participate differently in the formation of axonal branches. Mol Biol Cell 19:1485–1498

Zeng G, Taylor SM, Mccolm JR, Kappas NC, Kearney JB, Williams LH, Hartnett ME, Bautch VL (2007) Orientation of endothelial cell division is regulated by VEGF signaling during blood vessel formation. Blood 109:1345–1352

Zmuda JF, Rivas RJ (1998) The Golgi apparatus and the centrosome are localized to the sites of newly emerging axons in cerebellar granule neurons in vitro. Cell Motil Cytoskeleton 41:18–38

Chapter 14
MTOC Organization and Competition During Neuron Differentiation

Jason Y. Tann and Adrian W. Moore

Abstract Neurons are polarized cells with long branched axons and dendrites. Microtubule generation and organization machineries are crucial to grow and pattern these complex cellular extensions. Microtubule organizing centers (MTOCs) concentrate the molecular machinery for templating microtubules, stabilizing the nascent polymer, and organizing the resultant microtubules into higher-order structures. MTOC formation and function are well described at the centrosome, in the spindle, and at interphase Golgi; we review these studies and then describe recent results about how the machineries acting at these classic MTOCs are repurposed in the postmitotic neuron for axon and dendrite differentiation. We further discuss a constant tug-of-war interplay between different MTOC activities in the cell and how this process can be used as a substrate for transcription factor-mediated diversification of neuron types.

14.1 Introduction

Neurons transmit electrical signals between functionally associated regions of the nervous system. To do this, they use axons to send information, and dendrites to receive it. These axons and dendrites commonly project over long distances, and they develop complex branching patterns to wire circuits. During neuron differentiation, the polymerization and organization of the microtubule cytoskeleton drive the protrusive growth of axons and dendrites. Further regulation of this process contributes to branch initiation and patterning (Delandre et al. 2016). The microtubule network also supports cargo trafficking along these axons and dendrites in the differentiating and mature neuron (Lewis et al. 2013), and local microtubule polymerization and re-modelling contributes to activity-dependent synaptic remodeling for nervous system plasticity (Bodaleo and Gonzalez-Billault 2016). Progressive loss of microtubule content, accompanied by axon and dendrite recession, is a

J. Y. Tann · A. W. Moore (✉)
Laboratory for Neurodiversity, RIKEN Centre for Brain Science, Saitama, Japan
e-mail: adrian.moore@riken.jp

© Springer Nature Switzerland AG 2019
M. Kloc (ed.), *The Golgi Apparatus and Centriole*, Results and Problems in Cell
Differentiation 67, https://doi.org/10.1007/978-3-030-23173-6_14

hallmark of age-associated neurodegeneration, and is further accelerated in disorders such as Alzheimer's disease and Parkinson's disease (Dubey et al. 2015).

De novo microtubule formation consists of microtubule templating followed by nascent polymer extension. De novo microtubule formation is controlled at cellular microtubule organizing centers (MTOCs). At present, the best understood MTOCs are those in cycling cells; these are at the centrosome, in the mitotic spindle, and at the Golgi. In this article, we overview mechanisms of MTOC function at these structures. We then discuss how the machineries of centrosomal, spindle, and Golgi MTOCs are repurposed to create and pattern microtubule networks in differentiating axons and dendrites.

14.2 Microtubule Polarity Underlies Neuronal Polarity

Microtubules are comprised of 13 protofilaments assembled into a 25-nm-diameter hollow tube. The number of protofilaments may be altered, e.g., in *C. elegans* touch receptor sensory neurons (Chalfie and Thomson 1982). Protofilaments are constructed by head-to-tail binding of heterodimers of α- and β-Tubulin, and then weak lateral associations of these assembling protofilaments create the hollow cylindrical structure (Aldaz et al. 2005). Assembly from heterodimers of Tubulin leads to distinct plus- and minus-microtubule ends. The plus-end is the primary site of microtubule polymerization. The minus-end can also grow, but it is usually stabilized by anchoring at an MTOC or by capping with factors that prevent growth or depolymerisation (Akhmanova and Hoogenraad Casper 2015) (Fig. 14.1).

The distribution of polar microtubules determines the geometry of the microtubule network in a cell. Neuron polarization into axonal and dendritic compartments is critical for information transmission. Setting up differential microtubule polarity organization in axons and dendrites enables selective cargo targeting within the neuron by the precise use of different molecular motors that track along microtubules (Lewis et al. 2013). Through selective cargo targeting, axons and dendrites develop their respective sending and receiving functionalities. In axons, almost all microtubule plus-ends point in the anterograde direction (away from the soma). On the other hand, dendrites have mixed polarity microtubule populations, the majority possessing retrograde polarity (with plus-ends pointing towards from the soma) (Delandre et al. 2016; Lewis et al. 2013) (Fig. 14.2).

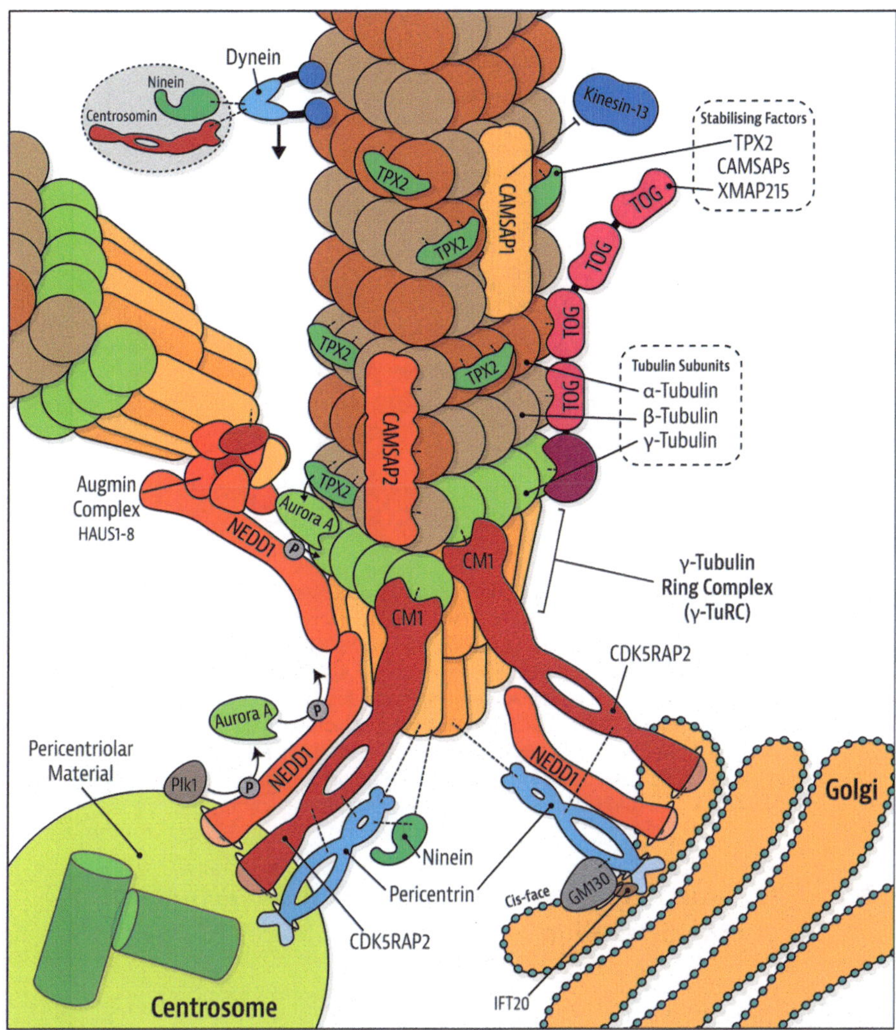

Fig. 14.1 Regulatory factors that control microtubule nucleation and stabilization at the centrosome, Golgi, and pre-existing microtubules. The γ-Tubulin Ring Complex (γ-TuRC) comprises of five Gcp subunits which bind to γ-Tubulin and organize this into a ring. This γ-Tubulin ring in turn templates α-/β-Tubulin heterodimers to initiate microtubule polymerization. Stabilization of the initial microtubule seed occurs through binding of TOG-domain-containing Xmap215 and laterally associating Tpx2. Centrosome-mediated nucleation is through the large coiled-coil proteins Pcnt/Plp and Cdk5rap2/Cnn, as well as Nedd1. This Nedd1 activity is regulated by phosphorylation. Pcnt also recruits the stabilization factor Nin. Many of these factors are brought to the microtubule minus-end through Dynein-mediated transport. Golgi-mediated nucleation is through Cdk5rap2/Myomegalin, Nedd1, and Pcnt. These are recruited by the Golgi proteins Gm130 and Ift20. Augmin and Nedd1 recruit a γ-TuRC complex that initiates microtubule nucleation on the side of existing microtubules

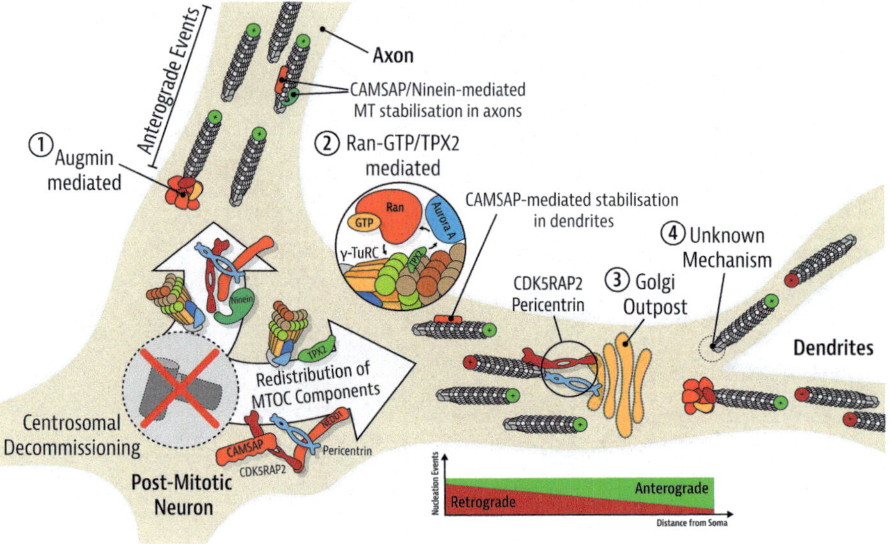

Fig. 14.2 Microtubule regulation in postmitotic neurons occurs at acentrosomal MTOC sites that participate in a tug-of-war-like mechanism. As nascent neurons enter the postmitotic stage, centrosomal decommissioning and the redistribution of MTOC components into the neurites occur. Axons and dendrites possess significantly different microtubule nucleation directionality. In axons, only anterograde events occur; Augmin supports this. In dendrites, both anterograde and retrograde events are present; the retrograde events predominate proximal to the soma. A tug-of-war-like mechanism between (1) Augmin-mediated MTOCs, (2) Ran-GTP-mediated MTOCs, (3) Golgi outpost as an MTOC, and (4) other possible unknown mechanisms may occur in the neurons which allow it to control dendritic arbor patterning and branching

14.3　The Function of MTOCs in Microtubule Templating and Polymerization

γ-Tubulin is key to the initiation of microtubule polymerization. To enable microtubule nucleation, γ-Tubulin is organized into a ring template to act as a platform for the initiation of α/β-Tubulin heterodimer binding (Llanos et al. 1999) (Fig. 14.1). In *Saccharomyces cerevisiae*, two molecules of γ-Tubulin assemble with one γ-Tubulin Complex Protein 2 (Gcp2) and one Gcp3 to form the γ-Tubulin Small Complex (γ-TuSC); then, seven copies of γ-TuSC create a platform for de novo nucleation. In cells of higher organisms, some of the Gcp2 and 3 subunits in the platform are replaced by Gcp4, 5, or 6, to make a γ-Tubulin Ring Complex (γ-TuRC). The γ-TuRC/γ-TuSC assembles into an inverted cone structure, in which the base of the cone acts as a template to form γ-Tubulin subunits into a ring (Kollman et al. 2008, 2011; Tovey and Conduit 2018; Guillet et al. 2011) (Fig. 14.1 and Table 14.1).

The function of MTOCs is to localize and co-concentrate γ-Tubulin with a suite of regulatory factors. γ-TuSC can only form the inverted cone structure capable of

Table 14.1 List of known homologs/orthologs/paralogs of MTOC factors

Human *Homo sapiens*	Fruit fly *Drosophila melanogaster*	Roundworm *Caenorhabditis elegans*∗
γ-TUBULIN 1 γ-TUBULIN 2	γ-Tubulin 23C γ-Tubulin 37C	TBG-1 TBG-2
GCP2 GCP3 GCP4 GCP5 GCP6	Grip84 Grip91 Grip75 Grip128 Grip163	GRIP-1 GRIP-2 –
NEDD1	Grip71	–
CDK5RAP2 MYOMEGALIN	Cnn	SPD-5
PCNT AKAP450	Plp	
XMAP215	Xmap215/Msps	ZYG-9
TPX2	Mei-38	TPXL-1
PLK1 PLK4	Polo SAK	PLK1 ZYG-1
AURORA A	Aurora A	AURORA A/AIR-1
CEP192	Spd-2	SPD-2
NINEIN	Bsg25D	NOCA-1
CLASP1 CLASP2	MAST/Orbit	CLS-1 CLS-2 CLS-3
CAMSAP1 CAMSAP2 CAMSAP3	Patronin	PTRN-1
HAUS1 HAUS2 HAUS3 HAUS4 HAUS5 HAUS6 HAUS7 HAUS8	Dgt2 Dgt3 Dgt4 Dgt5 Dgt6 Msd1 Msd5 Wac	–
GM130	Gm130	GOLGIN-107

templating a microtubule when at an MTOC. On the other hand, incorporation of Gcp4–6 enables γ-TuRCs to assemble in the cytosol. These γ-TuRCs can be subsequently recruited to different MTOCs through a suite of γ-TuRC interacting partners (Tovey and Conduit 2018). γ-TuRC structure is regulated by phosphorylation (Lüders et al. 2005; Haren et al. 2006; Zhang et al. 2009) through kinases including Polo-like kinase 1 (Human: PLK1; Drosophila: Polo). In addition, the WD-40 containing Nedd1 (Neural precursor cell expressed developmentally down-regulated protein 1; Drosophila: Grip71) is a key adaptor that is closely associated with γ-TuRC and regulates its recruitment. Nedd1 is also phosphorylated by different kinases in a context-dependent manner as an additional level of control (Meunier and Vernos 2016).

In addition to the γ-TuRC, MTOCs recruit regulators that promote nucleation activity through stabilizing nascent polymer elongation and preventing catastrophe (sudden rapid microtubule depolymerization). The Xmap215 (Xenopus Microtubule-associated protein 215 kDa) microtubule polymerase, and the anti-catastrophe factor Tpx2 (Targeting protein for Xklp2) are key to this stabilization (Wieczorek et al. 2015; Zhang et al. 2017) (Fig. 14.1). Xmap215 works by binding to the side of one γ-Tubulin molecule, then through a series of TOG domains (Tumor Overexpressed Gene) it also binds along the first few α/β-Tubulin heterodimers of the nascent protofilament (Thawani et al. 2018). The Clasp family (CLIP-associating protein) also contain TOG domains and act as microtubule stabilization and rescue factors (Lindeboom et al. 2019; Al-Bassam et al. 2010). Tpx2 acts in the perpendic-ular direction to Xmap215; it stabilizes the nascent microtubule by binding laterally across neighboring Tubulin dimers cross-bridging protofilaments to strengthen the lateral associations that create the tubular structure (Zhang et al. 2017).

Changing the balance of Gcp subunits, combined with phosphorylation regula-tion, is hypothesized to diversify the nucleation process (Tovey and Conduit 2018). As discussed below, additional factors also specifically recruit γ-TuRC to different sites in the cell. Together, these three levels of control can provide the spatiotem-poral regulation of microtubule nucleation in cells (Teixido-Travesa et al. 2012; Kollman et al. 2011; Petry and Vale 2015; Tovey and Conduit 2018).

Because MTOCs spatially organize γ-TuRC within a cell, there is potential for tug-of-war-like regulatory interactions between different MTOCs, a process that we will argue contributes to patterning the microtubule network in neurons.

14.4 Microtubule Minus-End Stabilization

Once generated, microtubule minus-ends need to be stabilized (Figs. 14.1 and 14.2). At some sites, this occurs through the retention of the γ-TuRC. At other sites, the CAMSAP family (Calmodulin-regulated Spectrin-associated proteins; Drosophila: Patronin) are utilized. CAMSAP binds proximal to the minus-end of microtubules (Hendershott and Vale 2014; Jiang et al. 2014). They prevent depolymerization both by increasing the inherent stability of the microtubule itself and by blocking the access of microtubule destabilizing factors such as Kinesin-13 (Atherton et al. 2017).

In postmitotic epithelial cells, CAMSAP and Ninein (human: NIN; Drosophila: Bsg25D) are used to tether microtubule minus ends to acentrosomal MTOCs (Toya et al. 2016; Tanaka et al. 2012). In neurons, CAMSAPs mark the minus-ends of dendritic and axonal microtubules, and their depletion leads to disruption of micro-tubule growth dynamics and an overall reduction in microtubule mass (Wang et al. 2019; Jiang et al. 2014). Loss of *CAMSAP2* (or the single paralog *Patronin* in *Drosophila*) reduces dendritic branching (Wang et al. 2019; Pongrakhananon et al. 2018; Yau Kah et al. 2014). Intriguingly, CAMSAP3 does not bind to stable acetylated microtubules, and its loss leads to an increase in this population (Pongrakhananon et al. 2018). A slight increase in microtubule stability can lead a neurite that is originally destined to be a dendrite to instead differentiate into an axon; loss of *CAMSAP3* leads to supernumerary axon formation (Pongrakhananon et al. 2018).

14.5 The Centrosome MTOC Machinery

At the centrosome, the centrioles are surrounded by pericentriolar material (PCM). This PCM is the site of MTOC activity; it contains molecular machinery to not only recruit the γ-TuRC, but also to activate it, and then stabilize and attach the minus-ends of the resultant microtubules. The core structural PCM components consist of several large coiled-coil proteins: Pericentrin (Human: PCNT; Drosophila: Pericentrin-like protein, Plp), Cep192 (Centrosomal protein of 192 kDa; Drosophila: Spd-2), and the paralogs Cdk5rap2 (CDK5 regulatory subunit associated protein 2) and Myomegalin (Drosophila: Centrosomin, Cnn). They bind to each other creating a PCM scaffold that supports the recruitment of additional MTOC components. In addition, each of these coiled-coil proteins can directly bind and recruit γ-Tubulin (Conduit et al. 2015) (Fig. 14.1 and Table 14.1).

Pcnt/Plp plays a key role in recruiting and organizing other PCM factors. During interphase, nine radial Pcnt/Plp spokes organize a series of concentric rings of the other PCM factors (Lawo et al. 2012; Mennella et al. 2012). At mitosis, the PCM increases in size primarily through addition of Cdk5rap2/Cnn; this leads to a three- to fivefold increase in γ-Tubulin content to support the formation of the microtubule-dense mitotic spindle (Palazzo et al. 2000). In addition to recruiting γ-Tubulin, Cdk5rap2/Cnn contains a CM1 domain (centrosomin motif 1); this domain is also found in mammalian Pericentrin and Myomegalin. The CM1 domain supports binding to γ-TuRCs, which potentially further helps to activate the γ-TuRCs (Choi et al. 2010; Muroyama et al. 2016).

Following nucleation, microtubule stabilization at the centrosome occurs primarily through capping the minus ends by maintaining the association of the γ-TuRC. Nedd1 binds γ-TuRC at the centrosome to switch γ-TuRC from a nucleation to a minus-end anchoring role (Muroyama et al. 2016). Nin also binds at the centrosome to stabilize the γ-TuRC-microtubule minus-end complex (Delgehyr et al. 2005; Wang et al. 2015). The anchoring of the microtubule minus-ends to the centrosome creates a track for Dynein, a microtubule minus-end directed motor protein (Burakov et al. 2008) (Fig. 14.1). Both microtubule nucleating factors, such as Cdk5rap2 (Jia et al. 2013), and microtubule anchoring factors, such as Nin, bind to Dynein for transport (Redwine et al. 2017); this creates a feedback loop amplifying their recruitment to the centrosome.

14.6 Loss of Centrosomal MTOC Activity Leads to Microcephaly

Mutation of at least 15 centrosome-related genes, including *CDK5RAP2*, *PCNT*, and *NIN*, lead to primary microcephaly and Seckel syndrome. These patients have reduced head and brain size due to reductions in neuron number (Gilmore and Walsh 2013; Saade et al. 2018; O'Neill et al. 2018). In proliferating neural precursor cells, the switch between self-renewal and differentiation is regulated by asymmetric segregation of fate determinants at mitosis, and this segregation requires precise orientation of the mitotic spindle relative to the apicobasal axis of the precursor (Homem et al. 2015). The self-renewal switch is also influenced by whether the mother or daughter centrosome is inherited after division (Wang et al. 2009). Mutant *CDK5RAP2* patient iPS-derived neural precursors, and the precursors in *Cdk5rap2* or *Pcnt* mutant mice, show failed centriole duplication, disorganized PCM, and mis-oriented spindles. These mitotic defects correlate with premature stem cell exit from self-renewal and increased apoptosis (Buchman et al. 2010; Lizarraga et al. 2010; Lancaster et al. 2013; Yigit et al. 2015). Failure in neural precursor cell centrosome MTOC activity is the general causative model of microcephaly (Thornton and Woods 2009; Lasser et al. 2018).

For each specific gene mutation, there may be additional contributions to brain pathology that are due to other functions of each factor (Thornton and Woods 2009; Lasser et al. 2018). For example, Nin is required to maintain a connection between microtubules and the centrosome, and in interphase, this connection is required for interkinetic nuclear migration, a process in which neural precursor nuclei migrate to different positions along the apicobasal cell axis dependent on the cell cycle phase (Shinohara et al. 2013). Disrupting interkinetic nuclear migration reduces neuron output because it is required for proper signaling between precursors to promote cell proliferation and self-renewal (Homem et al. 2015). In postmitotic nascent neurons, this Nin-mediated connection between the centrosome and microtubules coordinates the migration away from the precursor layer (Rao et al. 2016).

In an intriguing converse to microcephaly, expansion of brain size in hominids and nonhuman primates is proposed to have been driven through the evolution of some of these same microcephaly genes, in particular, *CDK5RAP2* (Montgomery et al. 2011; Evans et al. 2006).

14.7 Centrioles Are Repurposed as Dendritic MTOCs in Ciliated Sensory Neurons

Specialized sensory neurons such as olfactory receptor neurons, photoreceptors, chemosensory neurons, and proprioceptive neurons have a single dendrite, which is tipped by a sensory cilium (Burton 1985; Reiter and Leroux 2017; Troutt et al. 1990; Li et al. 2017; Wang and Dynlacht 2018). The basal body, which is derived from a centriole, supports cilia formation across cell types (Wang and Dynlacht 2018). Transport of a centriole from the soma into the dendrite forms the basal body (Li et al. 2017). In ciliated *C. elegans* neurons, the basal body has been shown to be enriched for γ-Tubulin and acts as a distal MTOC (Harterink et al. 2018). Plp is also targeted to the basal body and is required for ciliation and function in a range of sensory neuron types (Jurczyk et al. 2004; Martinez-Campos et al. 2004; Muhlhans et al. 2011; Falk et al. 2018).

14.8 Spindle Microtubule Nucleation Mechanisms Are Reutilized in the Postmitotic Neuron

The centrosome stands out as the major site of MTOC activity during mitosis. Yet two further microtubule nucleation pathways occur within the mitotic spindle itself (Meunier and Vernos 2016); the central mechanisms of these pathways are reutilized during postmitotic neuron differentiation.

During mitosis, a RanGTP (RAs-related nuclear protein) gradient triggers microtubule nucleation free in the cytosol close to the chromosomes; this is through promoting the local interaction of a Nedd1-γTuRC complex with Tpx2. Nedd1 is phosphorylated by Aurora A to regulate this activity (Scrofani et al. 2015). In the early stages of neuron differentiation in culture, Tpx2 is localized to the centrosome and along the neurite shaft. TPX2 facilitates Aurora A activation, and following decommissioning of the centrosome, a new functional MTOC can be generated within the proximal neurite via local activation of Aurora A (Mori et al. 2009). Notably, RanGTP is concentrated specifically at the base and at the tip of neurites; loss of *Tpx2* decreases the frequency of microtubule nucleation events within the neuron, but only at these sites (Chen et al. 2017b) (Fig. 14.2).

Within the mitotic spindle, an eight-part multiprotein Augmin complex (human, HAUS (homologous to Augmin subunits); *Drosophila*, Augmin) drives local amplification of microtubule numbers (Figs. 14.1, 14.2 and Table 14.1). Augmin targets the Nedd1-γTuRC complex to the side of existing microtubules, where it is activated and interacts with Tpx2 for nucleation (Lüders et al. 2005; Zhang et al. 2009). Augmin-generated microtubules then branch from the side of these pre-existing microtubules (Hsia et al. 2014; Petry et al. 2013; Zhu et al. 2008; Teixidó-Travesa et al. 2010; Chen et al. 2017a); electron microscopy data suggest they are quickly released and stabilized by maintaining γ-TuRC as a cap (Kamasaki et al. 2013).

Augmin complexes are localized together with γ-TuRC throughout differentiating axons and dendrites, where they locally amplify microtubule generation (Goshima et al. 2008). Whilst Nedd1 is required for Augmin complex targeting in the spindle, this is not the case in the postmitotic neuron. Nedd1 is lost from differentiating neurons and knockdown of *Nedd1* at this stage does not demonstrate a phenotype (Sánchez-Huertas et al. 2016; Stiess et al. 2010).

Intriguingly, Augmin preferentially supports anterograde microtubule nucleation in the axon. In axons, when individual *Augmin* components are lost, in addition to a reduction of microtubule mass, unipolar anterograde microtubule polarity organization is lost (Sánchez-Huertas et al. 2016; Cunha-Ferreira et al. 2018). Moreover, in dendrites, loss of *Augmin* components alone does not affect polarity (Cunha-Ferreira et al. 2018; Yalgin et al. 2015). Interestingly, however, loss of an *Augmin* component (*Wac; Wee Augmin*) rescues an increase in anterograde polarity caused by loss of *Cnn* in *Drosophila* sensory neurons; this suggests that these factors are competing to shape a balance of microtubule polarities in dendrites (Yalgin et al. 2015; Delandre et al. 2016).

14.9　Golgi and Dendritic Golgi Outpost MTOCs

The Golgi apparatus functions as an alternative MTOC (Fig. 14.1). Because Golgi stacks are polar, they organize directional microtubule networks in the cell. This network directionality regulates cell behavior; for example, it is required for concerted cell migration in a specific direction (Wu et al. 2016; Hurtado et al. 2011; Vinogradova et al. 2009). The network also regulates cargo movement through the cell, and like the dynein-centered feedback loop at the centrosome, Golgi-anchored microtubules are used by dynein to organize the Golgi stacks (Zhu and Kaverina 2013).

During interphase in mammalian cultured cycling cells, the Golgi matrix protein Gm130 recruits Akap450 (the Pcnt paralog) to the Golgi surface (Rivero et al. 2009). Intraflagellar transport protein 20 (IFT20), best described in the organization of cilia, stabilizes this Gm130-Akap450 complex (Nishita et al. 2017). With similarity to the scaffolding role of Pcnt at the centrosome, Akap450 recruits γ-TuRC, Cdk5rap2, and Myomegalin. Cdk5rap2 and Myomegalin also recruit γ-TuRC and activate it (Roubin et al. 2013; Wang et al. 2010). Clasp activity then supports these nascent microtubules (Roubin et al. 2013; Yang et al. 2017). In contrast to centrosome mechanisms, γ-TuRC is not used to stabilize nascent microtubules at the Golgi surface. Instead, the microtubules are coated with CAMSAP2 for release and stabilization; then they are linked back onto the Golgi surface by Gm130 (Jiang et al. 2018). Gm130 can also capture microtubules originating from other places in the cell, including those released from the centrosome or formed in the cytoplasm.

In mitotic cells, the Golgi apparatus is disassembled into clusters of vesicles, and these are linked by microtubule bridges to the spindle. They are required for successful segregation of organelles at cell division. Local cytoplasmic microtubule nucleation creates these bridges. Because the nuclear envelope breaks down in mitosis, Tpx2 escapes into the cytoplasm. Gm130 then triggers Tpx2 activation of Aurora A phosphorylation close to the Golgi surface leading to the local nucleation of microtubules in the cytoplasm, probably involving Nedd1 (Wei et al. 2015).

Dendrites contain fragments of Golgi called Golgi outposts (Bartlett and Banker 1984; Craig and Banker 1994) (Fig. 14.2). In *Drosophila* sensory neurons, Golgi outposts support a substantial minority of microtubule generation events in the arbor (Ori-McKenney et al. 2012). The outposts move through the arbor, then stall, often at dendritic branchpoints; when stalled, they become active as MTOCs (Ori-McKenney et al. 2012). Plp and Cnn couple microtubule generation events to these dendrite Golgi outposts (Ori-McKenney et al. 2012); this is similar to the roles of Pcnt, Cdk5rap2, and Myomegalin at somatic Golgi. On the other hand, in contrast to the multi-compartment structure of somatic Golgi, many Golgi outposts exist in trans or cis single-compartment states. Gm130 regulates the fusion of single compartment Golgi outposts into multi-compartment structures, and this fusion event increases microtubule output (Zhou et al. 2014).

14.10 Tug-of-War Between MTOC Activities at Different Sites Within the Cell

Manipulation of MTOC activities in the spindle, centrosome, and Golgi reveal a constant tug-of-war interplay between these sites. As cells enter mitosis, microtubule nucleation in the centrosome and Golgi are increased and reduced, respectively (Fry et al. 2017; Maia et al. 2013). Even in interphase, γ-Tubulin is difficult to detect at Golgi in normal cells, but it becomes clearly enriched when the centrosomes are removed (Wu et al. 2016). In *Drosophila* syncytial embryos, loss of Augmin-dependent microtubule nucleation in the spindle increases centrosome MTOC activity (Mahoney et al. 2006; Hayward et al. 2014). On the other hand, reducing centrosome MTOC activity leads to a concomitant increase in microtubule nucleation around chromatin, and some nucleation also occurs in the cytoplasm (Mahoney et al. 2006; Hayward et al. 2014). In mammalian cells, simultaneously disrupting both Golgi and centrosome MTOC activity leads Pcnt, Cdk5rap2, and γ-Tubulin to colocalize at active ectopic foci of microtubule nucleation sites within the cytoplasm (Gavilan et al. 2018).

Competition between sites for a limited pool of γ-Tubulin localization factors may be one mechanism underlying these tug-of-war-like interactions. This is seen in *Schizosaccharomyces pombe* where the CM1 domain containing proteins Mto1 and 2 recruit the γ-TuSC and activate it. Competition exists between the spindle pole body, pre-existing microtubules, and the nuclear envelope for a limited Mto1 and 2 protein pool. The spindle pole body and microtubules typically capture this limited pool; however, blocking the targeting to these sites enables Mto1 and Mto2 to move to the nuclear envelope (Lynch et al. 2014). The second level of control may be through differential Nedd1 phosphorylation events. The different MTOC assembly pathways we describe require NEDD1 phosphorylation by different kinases on different distinct residues (Meunier and Vernos 2016); notably, kinase activity regulates interaction between MTOC sites, for example, disruption of both the centrosome-associated kinases *Plk4* and *Aurora A* shifts MTOC function from the centrosome to the Golgi apparatus (Wong et al. 2015).

Competition between different sites of MTOC activity shapes the neuron differentiation process (Fig. 14.2). In nascent neurons, Nedd1 is slowly depleted from the centrosome, the centrosome loses MTOC activity (Stiess et al. 2010). A process of alternative splicing of Nin also couples this decommissioning of the centrosome to the exit from neuronal precursor self-renewal. In neuronal precursors, a precursor-enriched splice isoform of Nin interacts with other centrosome components to orient the mitotic cleavage plane in favor of self-renewal. However, the Nin centrosome targeting domain is spliced out when neuron differentiation is initiated; this new Nin isoform has a dominant negative effect on the activity of the precursor-enriched Nin splice isoform, and thus an expression of this new isoform blocks self-renewal. Moreover, the new splice isoform is not maintained at the centrosome, leading to loss of centrosome MTOC activity (Zhang et al. 2016). As neuron differentiation proceeds further, Nin also localizes to the axon where it is required for proper axonal microtubule dynamics and axonal growth (Srivatsa et al. 2015).

Concomitant with centrosome decommissioning in nascent neurons, Cdk5rap2, Pcnt, and γ-Tubulin redistribute away from the centrosome (Zhang et al. 2016; Baird et al. 2004; Ohama and Hayashi 2009; Yonezawa et al. 2015). However, the differentiating neurons continue to use γ-Tubulin to nucleate microtubules (Sánchez-Huertas et al. 2016; Yau Kah et al. 2014; Nguyen et al. 2014; Ori-McKenney et al. 2012); the position of microtubule generation activity transfers to the outgrowing neurites. Tug-of-war mechanisms between MTOC machinery also control microtubule polarity in the dendrites. In *Drosophila* sensory neuron dendrites, Cnn activity promotes the formation of microtubules that polymerize in the retrograde direction. This retrograde Cnn activity counteracts an Augmin-based activity that supports anterograde events at the growing dendrite tips (Yalgin et al. 2015; Delandre et al. 2016). Intriguingly, disrupting Patronin function (Wang et al. 2019) or γ-Tubulin activity (Nguyen et al. 2014) also regulates the relative levels of anterograde versus retrograde polymerization in dendrites. The reasons remain unclear; we suggest one possibility is that tug-of-war tension between different microtubule generation pathways is exposed when γ-Tubulin activity levels change.

14.11 Transcription Factors Regulate a Tug-of-War Between Neuronal Microtubule Nucleation Mechanisms to Create Diversity in Neuron Branching Patterns

Transcription factors regulate neuron diversification; this includes the branching patterns of specific neuron types (Santiago and Bashaw 2014; Lefebvre et al. 2015). One way this diversification occurs is through changing the balance of MTOC mechanisms; in turn, this shapes branch formation processes. For example, the zinc-finger transcription factor Sip1 (Smad Interacting protein 1) regulates neocortical upper-layer neuron morphology. Sip1 upregulates Nin levels in these postmitotic neurons. *Sip1* or *Nin* loss disrupts axonal microtubule stability and dynamics, reducing axon branching (Srivatsa et al. 2015).

One of the best studied systems of transcription factor-mediated dendritic patterning is in *Drosophila* sensory neurons (Santiago and Bashaw 2014; Lefebvre et al. 2015). The Krüppel-like factor Dar1 (Dendrite arbor reduction 1) supports the generation of complex sensory dendrite arbor shapes. Dar1 promotes the combination of single into multi-component Golgi outposts, upregulating microtubule nucleation at these MTOCs (Zhou et al. 2014). In Class I *Drosophila* sensory neurons, the BTB factor Abrupt suppresses branching to create a simple arbor morphology (Li et al. 2004; Sugimura et al. 2004). Abrupt upregulates Cnn (Yalgin et al. 2015). Changing Cnn levels reduces the frequency of nucleation events at the Golgi outposts and alters a tug-of-war interaction with concomitant Augmin-mediated nucleation in the dendrites (Yalgin et al. 2015).

14.12 Conclusions

Our understanding of the components and organization of MTOC mechanisms in differentiating neurons remains fragmentary. Yet, as we describe in this review, recent studies have revealed how the MTOC activities of growing axons and dendrites re-utilize γ-Tubulin- and γ-TuRC-recruitment and activation machinery from the PCM, spindle, and Golgi surface of cycling cells. The new MTOC activities of the growing axons and dendrites then shape the spatial distribution of microtubules in neurons and organize the polarity of this microtubule network. These features of the neuronal microtubule cytoskeleton are critical for neuronal branch patterning and for regulating cargo trafficking throughout the cell.

While several components of the axon and dendrite microtubule nucleation machinery have now been identified, where are they localized? Augmin-dependent processes occur throughout the neuron (Sánchez-Huertas et al. 2016; Cunha-Ferreira et al. 2018); for other machineries, however, it remains unclear what underlying structures act as the foundations onto which γ-Tubulin and γ-TuRC are recruited. We have described two in this review: the basal body and Golgi outposts. Yet, the basal body is restricted to specialized ciliated dendrites. Similarly, examination of the appearance of polymerizing microtubules at Golgi outposts indicates that most events within the arbor are not associated with these structures (Delandre et al. 2016), and while dendrite branchpoints in *Drosophila* sensory neurons have increased γ-Tubulin levels and are enriched for microtubule generation activity, removing Golgi outposts from these branchpoints using a light-activated transgenic motor did not alter the local concentration of γ-Tubulin (Nguyen et al. 2014). Dendrites contain a variety of membranous organelles which could potentially act as sites to concentrate γ-TuRC recruiting factors. Other membranous organelles, e.g., mitochondria, have also been investigated as potential dendritic MTOCs, but so far none were found to associate with nucleation events (Ori-McKenney et al. 2012). Key sites remain to be discovered.

A further issue is how are the amplitude and polarity of the neuronal microtubule nucleation processes controlled? One mechanism that may play a role in this we have highlighted in this article—MTOCs in cycling cells show tug-of-war like interactions, in which they compete for activity as sites of nucleation. Importantly, in differentiating neurons different MTOC mechanisms are also running simultaneously. For example, Augmin- and Cnn-mediated pathways run concurrently in dendrites, and they interact to control the polarity of microtubule polymerization events (Yalgin et al. 2015; Delandre et al. 2016). It is likely that both positive feedback and tug-of-war antagonistic interactions between different MTOC mechanisms regulate this. We also predict that this interaction network provides a rich substrate for the generation of neuron type branching diversity. Transcription factors that alter the balance of the nucleation machinery activities in a differentiating neuron will alter the balance of alternative MTOC output and amplitude, thus changing axons and dendrite branching and outgrowth dynamics (Yalgin et al. 2015; Delandre et al. 2016; Srivatsa et al. 2015).

Beyond the differentiation mechanisms discussed in this review, how neuronal MTOC networks function has significant implications for the pathophysiology of neuron injury and neurodegenerative disease. Axon injury induces a rapid upregulation in microtubule nucleation dynamics throughout the neuron (Stone et al. 2010; Chen et al. 2012). A similar response occurs upon the induction of a neurodegenerative disease-causing mechanism, such as the expression of a polyQ repeat or a mutated form of SOD1 (Chen et al. 2012; Kleele et al. 2014). The upregulation in microtubule nucleation activity that occurs following injury provides increased stabilization of the neuron against further damage (Chen et al. 2012). Similarly, Regeneration after axonal injury requires the regrowth of the proximal axon for injuries distal from the cell soma, or the repurposing of a dendrite into an axon for injuries that are proximal. Notably, repurposing a dendrite to an axon necessitates a change of microtubule polarity from retrograde to anterograde (Song et al. 2012); we speculate this may involve shifting the tug-of-war between competing MTOC mechanisms in the cell. Overall, further examination of the MTOC mechanisms in differentiating neurons can lead us towards ways to manipulate or reactivate MTOC activity in order to treat neuronal injury and degeneration.

References

Akhmanova A, Hoogenraad Casper C (2015) Microtubule minus-end-targeting proteins. Curr Biol 25(4):R162–R171. https://doi.org/10.1016/j.cub.2014.12.027

Al-Bassam J, Kim H, Brouhard G, van Oijen A, Harrison SC, Chang F (2010) CLASP promotes microtubule rescue by recruiting tubulin dimers to the microtubule. Dev Cell 19(2):245–258. https://doi.org/10.1016/j.devcel.2010.07.016

Aldaz H, Rice LM, Stearns T, Agard DA (2005) Insights into microtubule nucleation from the crystal structure of human γ-tubulin. Nature 435(7041):523. https://doi.org/10.1038/nature03586

Atherton J, Jiang K, Stangier MM, Luo Y, Hua S, Houben K, van Hooff JJE, Joseph AP, Scarabelli G, Grant BJ, Roberts AJ, Topf M, Steinmetz MO, Baldus M, Moores CA, Akhmanova A (2017) A structural model for microtubule minus-end recognition and protection by CAMSAP proteins. Nat Struct Mol Biol 24(11):931–943. https://doi.org/10.1038/nsmb.3483

Baird DH, Myers KA, Mogensen M, Moss D, Baas PW (2004) Distribution of the microtubule-related protein ninein in developing neurons. Neuropharmacology 47(5):677–683. https://doi.org/10.1016/j.neuropharm.2004.07.016

Bartlett WP, Banker GA (1984) An electron microscopic study of the development of axons and dendrites by hippocampal neurons in culture. I. Cells which develop without intercellular contacts. J Neurosci 4(8):1944–1953. https://doi.org/10.1523/jneurosci.04-08-01944.1984

Bodaleo FJ, Gonzalez-Billault C (2016) The presynaptic microtubule cytoskeleton in physiological and pathological conditions: lessons from drosophila fragile X syndrome and hereditary spastic paraplegias. Front Mol Neurosci 9:60. https://doi.org/10.3389/fnmol.2016.00060

Buchman JJ, Tseng HC, Zhou Y, Frank CL, Xie Z, Tsai LH (2010) Cdk5rap2 interacts with pericentrin to maintain the neural progenitor pool in the developing neocortex. Neuron 66 (3):386–402. https://doi.org/10.1016/j.neuron.2010.03.036

Burakov A, Kovalenko O, Semenova I, Zhapparova O, Nadezhdina E, Rodionov V (2008) Cytoplasmic dynein is involved in the retention of microtubules at the centrosome in interphase cells. Traffic 9(4):472–480. https://doi.org/10.1111/j.1600-0854.2007.00698.x

Burton PR (1985) Ultrastructure of the olfactory neuron of the bullfrog: the dendrite and its microtubules. J Comp Neurol 242(2):147–160. https://doi.org/10.1002/cne.902420202

Chalfie M, Thomson JN (1982) Structural and functional diversity in the neuronal microtubules of *Caenorhabditis elegans*. J Cell Biol 93(1):15–23

Chen L, Stone MC, Tao J, Rolls MM (2012) Axon injury and stress trigger a microtubule-based neuroprotective pathway. Proc Natl Acad Sci 109(29):11842–11847. https://doi.org/10.1073/pnas.1121180109

Chen JWC, Chen ZA, Rogala KB, Metz J, Deane CM, Rappsilber J, Wakefield JG (2017a) Cross-linking mass spectrometry identifies new interfaces of Augmin required to localise the gamma-tubulin ring complex to the mitotic spindle. Biol Open 6(5):654–663. https://doi.org/10.1242/bio.022905

Chen WS, Chen YJ, Huang YA, Hsieh BY, Chiu HC, Kao PY, Chao CY, Hwang E (2017b) Ran-dependent TPX2 activation promotes acentrosomal microtubule nucleation in neurons. Sci Rep 7:42297. https://doi.org/10.1038/srep42297

Choi Y-K, Liu P, Sze SK, Dai C, Qi RZ (2010) CDK5RAP2 stimulates microtubule nucleation by the γ-tubulin ring complex. J Cell Biol 191(6):1089–1095. https://doi.org/10.1083/jcb.201007030

Conduit PT, Wainman A, Novak ZA, Weil TT, Raff JW (2015) Re-examining the role of Drosophila Sas-4 in centrosome assembly using two-colour-3D-SIM FRAP. elife 4:15915. https://doi.org/10.7554/elife.08483

Craig AM, Banker G (1994) Neuronal polarity. Annu Rev Neurosci 17(1):267–310. https://doi.org/10.1146/annurev.ne.17.030194.001411

Cunha-Ferreira I, Chazeau A, Buijs RR, Stucchi R, Will L, Pan X, Adolfs Y, Cvd M, Wolthuis JC, Kahn OI, Schätzle P, Altelaar M, Pasterkamp RJ, Kapitein LC, Hoogenraad CC (2018) The HAUS complex is a key regulator of non-centrosomal microtubule organization during neuronal development. Cell Rep 24(4):791–800. https://doi.org/10.1016/j.celrep.2018.06.093

Delandre C, Amikura R, Moore AW (2016) Microtubule nucleation and organization in dendrites. Cell Cycle 15(13):1–8. https://doi.org/10.1080/15384101.2016.1172158

Delgehyr N, Sillibourne J, Bornens M (2005) Microtubule nucleation and anchoring at the centrosome are independent processes linked by ninein function. J Cell Sci 118 (8):1565–1575. https://doi.org/10.1242/jcs.02302

Dubey J, Ratnakaran N, Koushika SP (2015) Neurodegeneration and microtubule dynamics: death by a thousand cuts. Front Cell Neurosci 9:343. https://doi.org/10.3389/fncel.2015.00343

Evans PD, Vallender EJ, Lahn BT (2006) Molecular evolution of the brain size regulator genes CDK5RAP2 and CENPJ. Gene 375:75–79. https://doi.org/10.1016/j.gene.2006.02.019

Falk N, Kessler K, Schramm SF, Boldt K, Becirovic E, Michalakis S, Regus-Leidig H, Noegel AA, Ueffing M, Thiel CT, Roepman R, Brandstatter JH, Giessl A (2018) Functional analyses of Pericentrin and Syne-2 interaction in ciliogenesis. J Cell Sci 131(16):jcs218487. https://doi.org/10.1242/jcs.218487

Fry AM, Sampson J, Shak C, Shackleton S (2017) Recent advances in pericentriolar material organization: ordered layers and scaffolding gels. F1000Research 6:1622. https://doi.org/10.12688/f1000research.11652.1

Gavilan MP, Gandolfo P, Balestra FR, Arias F, Bornens M, Rios RM (2018) The dual role of the centrosome in organizing the microtubule network in interphase. EMBO Rep 19(11):e45942. https://doi.org/10.15252/embr.201845942

Gilmore EC, Walsh CA (2013) Genetic causes of microcephaly and lessons for neuronal development. Wiley Interdiscip Rev Dev Biol 2(4):461–478. https://doi.org/10.1002/wdev.89

Goshima G, Mayer M, Zhang N, Stuurman N, Vale RD (2008) Augmin: a protein complex required for centrosome-independent microtubule generation within the spindle. J Cell Biol 181 (3):421–429. https://doi.org/10.1083/jcb.200711053

Guillet V, Knibiehler M, Gregory-Pauron L, Remy M-H, Chemin C, Raynaud-Messina B, Bon C, Kollman JM, Agard DA, Merdes A, Mourey L (2011) Crystal structure of γ-tubulin complex protein GCP4 provides insight into microtubule nucleation. Nat Struct Mol Biol 18(8):915. https://doi.org/10.1038/nsmb.2083

Haren L, Remy M-H, Bazin I, Callebaut I, Wright M, Merdes A (2006) NEDD1-dependent recruitment of the γ-tubulin ring complex to the centrosome is necessary for centriole duplication and spindle assembly. J Cell Biol 172(4):505–515. https://doi.org/10.1083/jcb.200510028

Harterink M, Edwards SL, de Haan B, Yau KW, van den Heuvel S, Kapitein LC, Miller KG, Hoogenraad CC (2018) Local microtubule organization promotes cargo transport in *C. elegans* dendrites. J Cell Sci 131(20):jcs223107. https://doi.org/10.1242/jcs.223107

Hayward D, Metz J, Pellacani C, Wakefield JG (2014) Synergy between multiple microtubule-generating pathways confers robustness to centrosome-driven mitotic spindle formation. Dev Cell 28(1):81–93. https://doi.org/10.1016/j.devcel.2013.12.001

Hendershott MC, Vale RD (2014) Regulation of microtubule minus-end dynamics by CAMSAPs and Patronin. Proc Natl Acad Sci 111(16):5860–5865. https://doi.org/10.1073/pnas.1404133111

Homem CC, Repic M, Knoblich JA (2015) Proliferation control in neural stem and progenitor cells. Nat Rev Neurosci 16(11):647–659. https://doi.org/10.1038/nrn4021

Hsia KC, Wilson-Kubalek EM, Dottore A, Hao Q, Tsai KL, Forth S, Shimamoto Y, Milligan RA, Kapoor TM (2014) Reconstitution of the augmin complex provides insights into its architecture and function. Nat Cell Biol 16(9):852–863. https://doi.org/10.1038/ncb3030

Hurtado L, Caballero C, Gavilan MP, Cardenas J, Bornens M, Rios RM (2011) Disconnecting the Golgi ribbon from the centrosome prevents directional cell migration and ciliogenesis. J Cell Biol 193(5):917–933. https://doi.org/10.1083/jcb.201011014

Jia Y, Fong KW, Choi YK, See SS, Qi RZ (2013) Dynamic recruitment of CDK5RAP2 to centrosomes requires its association with dynein. PLoS One 8(7):e68523. https://doi.org/10.1371/journal.pone.0068523

Jiang K, Hua S, Mohan R, Grigoriev I, Yau Kah W, Liu Q, Katrukha Eugene A, Altelaar AFM, Heck Albert JR, Hoogenraad Casper C, Akhmanova A (2014) Microtubule minus-end stabilization by polymerization-driven CAMSAP deposition. Dev Cell 28(3):295–309. https://doi.org/10.1016/j.devcel.2014.01.001

Jiang K, Faltova L, Hua S, Capitani G, Prota AE, Landgraf C, Volkmer R, Kammerer RA, Steinmetz MO, Akhmanova A (2018) Structural basis of formation of the microtubule minus-end-regulating CAMSAP-katanin complex. Structure 26(3):375–382.e374. https://doi.org/10.1016/j.str.2017.12.017

Jurczyk A, Gromley A, Redick S, San Agustin J, Witman G, Pazour GJ, Peters DJ, Doxsey S (2004) Pericentrin forms a complex with intraflagellar transport proteins and polycystin-2 and is required for primary cilia assembly. J Cell Biol 166(5):637–643. https://doi.org/10.1083/jcb.200405023

Kamasaki T, O'Toole E, Kita S, Osumi M, Usukura J, McIntosh JR, Goshima G (2013) Augmin-dependent microtubule nucleation at microtubule walls in the spindle. J Cell Biol 202(1):25–33. https://doi.org/10.1083/jcb.201304031

Kleele T, Marinkovic P, Williams PR, Stern S, Weigand EE, Engerer P, Naumann R, Hartmann J, Karl RM, Bradke F, Bishop D, Herms J, Konnerth A, Kerschensteiner M, Godinho L, Misgeld T (2014) An assay to image neuronal microtubule dynamics in mice. Nat Commun 5:4827. https://doi.org/10.1038/ncomms5827

Kollman JM, Zelter A, Muller EGD, Fox B, Rice LM, Davis TN, Agard DA (2008) The structure of the γ-tubulin small complex: implications of its architecture and flexibility for microtubule nucleation. Mol Biol Cell 19(1):207–215. https://doi.org/10.1091/mbc.e07-09-0879

Kollman JM, Merdes A, Mourey L, Agard DA (2011) Microtubule nucleation by γ-tubulin complexes. Nat Rev Mol Cell Biol 12(11):709–721. https://doi.org/10.1038/nrm3209

Lancaster MA, Renner M, Martin CA, Wenzel D, Bicknell LS, Hurles ME, Homfray T, Penninger JM, Jackson AP, Knoblich JA (2013) Cerebral organoids model human brain development and microcephaly. Nature 501(7467):373–379. https://doi.org/10.1038/nature12517

Lasser M, Tiber J, Lowery LA (2018) The role of the microtubule cytoskeleton in neurodevelopmental disorders. Front Cell Neurosci 12:165. https://doi.org/10.3389/fncel.2018.00165

Lawo S, Hasegan M, Gupta GD, Pelletier L (2012) Subdiffraction imaging of centrosomes reveals higher-order organizational features of pericentriolar material. Nat Cell Biol 14(11):1148–1158. https://doi.org/10.1038/ncb2591

Lefebvre JL, Sanes JR, Kay JN (2015) Development of dendritic form and function. Annu Rev Cell Dev Biol 31:741–777. https://doi.org/10.1146/annurev-cellbio-100913-013020

Lewis TL, Courchet J, Polleux F (2013) Cellular and molecular mechanisms underlying axon formation, growth, and branching. J Cell Biol 202(6):837–848. https://doi.org/10.1083/jcb.201305098

Li W, Wang F, Menut L, Gao FB (2004) BTB/POZ-zinc finger protein abrupt suppresses dendritic branching in a neuronal subtype-specific and dosage-dependent manner. Neuron 43(6):823–834. https://doi.org/10.1016/j.neuron.2004.08.040

Li W, Yi P, Zhu Z, Zhang X, Li W, Ou G (2017) Centriole translocation and degeneration during ciliogenesis in *Caenorhabditis elegans* neurons. EMBO J 36(17):2553–2566. https://doi.org/10.15252/embj.201796883

Lindeboom JJ, Nakamura M, Saltini M, Hibbel A, Walia A, Ketelaar T, Emons AMC, Sedbrook JC, Kirik V, Mulder BM, Ehrhardt DW (2019) CLASP stabilization of plus ends created by severing promotes microtubule creation and reorientation. J Cell Biol 218(1):190–205. https://doi.org/10.1083/jcb.201805047

Lizarraga SB, Margossian SP, Harris MH, Campagna DR, Han AP, Blevins S, Mudbhary R, Barker JE, Walsh CA, Fleming MD (2010) Cdk5rap2 regulates centrosome function and chromosome segregation in neuronal progenitors. Development 137(11):1907–1917. https://doi.org/10.1242/dev.040410

Llanos R, Chevrier V, Ronjat M, Meurer-Grob P, Martinez P, Frank R, Bornens M, Wade RH, Wehland J, Job D (1999) Tubulin binding sites on gamma-tubulin: identification and molecular characterization. Biochemistry 38(48):15712–15720. https://doi.org/10.1021/bi990895w

Lüders J, Patel UK, Stearns T (2005) GCP-WD is a γ-tubulin targeting factor required for centrosomal and chromatin-mediated microtubule nucleation. Nat Cell Biol 8(2):137–147. https://doi.org/10.1038/ncb1349

Lynch EM, Groocock LM, Borek WE, Sawin KE (2014) Activation of the gamma-tubulin complex by the Mto1/2 complex. Curr Biol 24(8):896–903. https://doi.org/10.1016/j.cub.2014.03.006

Mahoney NM, Goshima G, Douglass AD, Vale RD (2006) Making microtubules and mitotic spindles in cells without functional centrosomes. Curr Biol 16(6):564–569. https://doi.org/10.1016/j.cub.2006.01.053

Maia ARR, Zhu X, Miller P, Gu G, Maiato H, Kaverina I (2013) Modulation of Golgi-associated microtubule nucleation throughout the cell cycle. Cytoskeleton 70(1):32–43. https://doi.org/10.1002/cm.21079

Martinez-Campos M, Basto R, Baker J, Kernan M, Raff JW (2004) The Drosophila pericentrin-like protein is essential for cilia/flagella function, but appears to be dispensable for mitosis. J Cell Biol. 165(5):673–683. https://doi.org/10.1083/jcb.200402130

Mennella V, Keszthelyi B, McDonald KL, Chhun B, Kan F, Rogers GC, Huang B, Agard DA (2012) Subdiffraction-resolution fluorescence microscopy reveals a domain of the centrosome critical for pericentriolar material organization. Nat Cell Biol 14(11):1159. https://doi.org/10.1038/ncb2597

Meunier S, Vernos I (2016) Acentrosomal microtubule assembly in mitosis: the where, when, and how. Trends Cell Biol 26(2):80–87. https://doi.org/10.1016/j.tcb.2015.09.001

Montgomery SH, Capellini I, Venditti C, Barton RA, Mundy NI (2011) Adaptive evolution of four microcephaly genes and the evolution of brain size in anthropoid primates. Mol Biol Evol 28 (1):625–638. https://doi.org/10.1093/molbev/msq237

Mori D, Yamada M, Mimori-Kiyosue Y, Shirai Y, Suzuki A, Ohno S, Saya H, Wynshaw-Boris A, Hirotsune S (2009) An essential role of the aPKC-Aurora A-NDEL1 pathway in neurite elongation by modulation of microtubule dynamics. Nat Cell Biol 11(9):1057–1068. https://doi.org/10.1038/ncb1919

Muhlhans J, Brandstatter JH, Giessl A (2011) The centrosomal protein pericentrin identified at the basal body complex of the connecting cilium in mouse photoreceptors. PLoS One 6(10):e26496. https://doi.org/10.1371/journal.pone.0026496

Muroyama A, Seldin L, Lechler T (2016) Divergent regulation of functionally distinct γ-tubulin complexes during differentiation. J Cell Biol. 213(6):679–692. https://doi.org/10.1083/jcb.201601099

Nguyen MM, McCracken CJ, Milner ES, Goetschius DJ, Weiner AT, Long MK, Michael NL, Munro S, Rolls MM (2014) γ-Tubulin controls neuronal microtubule polarity independently of Golgi outposts. Mol Biol Cell 25(13):2039–2050. https://doi.org/10.1091/mbc.e13-09-0515

Nishita M, Park SY, Nishio T, Kamizaki K, Wang Z, Tamada K, Takumi T, Hashimoto R, Otani H, Pazour GJ, Hsu VW, Minami Y (2017) Ror2 signaling regulates Golgi structure and transport through IFT20 for tumor invasiveness. Sci Rep 7(1):1. https://doi.org/10.1038/s41598-016-0028-x

Ohama Y, Hayashi K (2009) Relocalization of a microtubule-anchoring protein, ninein, from the centrosome to dendrites during differentiation of mouse neurons. Histochem Cell Biol 132 (5):515–524. https://doi.org/10.1007/s00418-009-0631-z

O'Neill RS, Schoborg TA, Rusan NM (2018) Same but different: pleiotropy in centrosome-related microcephaly. Mol Biol Cell 29(3):241–246. https://doi.org/10.1091/mbc.E17-03-0192

Ori-McKenney KM, Jan LY, Jan Y-N (2012) Golgi outposts shape dendrite morphology by functioning as sites of acentrosomal microtubule nucleation in neurons. Neuron 76 (5):921–930. https://doi.org/10.1016/j.neuron.2012.10.008

Palazzo RE, Vogel JM, Schnackenberg BJ, Hull DR, Wu X (2000) Centrosome maturation. Curr Top Dev Biol 49:449–470

Petry S, Vale RD (2015) Microtubule nucleation at the centrosome and beyond. Nat Cell Biol 17 (9):1089–1093. https://doi.org/10.1038/ncb3220

Petry S, Groen Aaron C, Ishihara K, Mitchison Timothy J, Vale Ronald D (2013) Branching microtubule nucleation in xenopus egg extracts mediated by augmin and TPX2. Cell 152 (4):768–777. https://doi.org/10.1016/j.cell.2012.12.044

Pongrakhananon V, Saito H, Hiver S, Abe T, Shioi G, Meng W, Takeichi M (2018) CAMSAP3 maintains neuronal polarity through regulation of microtubule stability. Proc Natl Acad Sci USA 115(39):9750–9755. https://doi.org/10.1073/pnas.1803875115

Rao AN, Falnikar A, O'Toole ET, Morphew MK, Hoenger A, Davidson MW, Yuan X, Baas PW (2016) Sliding of centrosome-unattached microtubules defines key features of neuronal phenotype. J Cell Biol 213(3):329–341. https://doi.org/10.1083/jcb.201506140

Redwine WB, DeSantis ME, Hollyer I, Htet ZM, Tran PT, Swanson SK, Florens L, Washburn MP, Reck-Peterson SL (2017) The human cytoplasmic dynein interactome reveals novel activators of motility. elife 6:e28257. https://doi.org/10.7554/elife.28257

Reiter JF, Leroux MR (2017) Genes and molecular pathways underpinning ciliopathies. Nat Rev Mol Cell Biol 18(9):533–547. https://doi.org/10.1038/nrm.2017.60

Rivero S, Cardenas J, Bornens M, Rios RM (2009) Microtubule nucleation at the cis-side of the Golgi apparatus requires AKAP450 and GM130. EMBO J 28(8):1016–1028. https://doi.org/10.1038/emboj.2009.47

Roubin R, Acquaviva C, Chevrier V, Sedjaï F, Zyss D, Birnbaum D, Rosnet O (2013) Myomegalin is necessary for the formation of centrosomal and Golgi-derived microtubules. Biol Open 2 (2):238–250. https://doi.org/10.1242/bio.20123392

Saade M, Blanco-Ameijeiras J, Gonzalez-Gobartt E, Marti E (2018) A centrosomal view of CNS growth. Development 145(21). https://doi.org/10.1242/dev.170613

Sánchez-Huertas C, Freixo F, Viais R, Lacasa C, Soriano E, Lüders J (2016) Non-centrosomal nucleation mediated by augmin organizes microtubules in post-mitotic neurons and controls axonal microtubule polarity. Nat Commun 7:12187. https://doi.org/10.1038/ncomms12187

Santiago C, Bashaw GJ (2014) Transcription factors and effectors that regulate neuronal morphology. Development 141(24):4667–4680. https://doi.org/10.1242/dev.110817

Scrofani J, Sardon T, Meunier S, Vernos I (2015) Microtubule nucleation in mitosis by a RanGTP-dependent protein complex. Curr Biol 25(2):131–140. https://doi.org/10.1016/j.cub.2014.11.025

Shinohara H, Sakayori N, Takahashi M, Osumi N (2013) Ninein is essential for the maintenance of the cortical progenitor character by anchoring the centrosome to microtubules. Biol Open 2 (7):739–749. https://doi.org/10.1242/bio.20135231

Song Y, Ori-McKenney KM, Zheng Y, Han C, Jan LY, Jan YN (2012) Regeneration of Drosophila sensory neuron axons and dendrites is regulated by the Akt pathway involving Pten and microRNA bantam. Genes Dev 26(14):1612–1625. https://doi.org/10.1101/gad.193243.112

Srivatsa S, Parthasarathy S, Molnar Z, Tarabykin V (2015) Sip1 downstream effector ninein controls neocortical axonal growth, ipsilateral branching, and microtubule growth and stability. Neuron 85(5):998–1012. https://doi.org/10.1016/j.neuron.2015.01.018

Stiess M, Maghelli N, Kapitein LC, Gomis-Rüth S, Wilsch-Bräuninger M, Hoogenraad CC, Tolić-Nørrelykke IM, Bradke F (2010) Axon extension occurs independently of centrosomal microtubule nucleation. Science 327(5966):704–707. https://doi.org/10.1126/science.1182179

Stone MC, Nguyen MM, Tao J, Allender DL, Rolls MM (2010) Global up-regulation of microtubule dynamics and polarity reversal during regeneration of an axon from a dendrite. Mol Biol Cell 21(5):767–777. https://doi.org/10.1091/mbc.e09-11-0967

Sugimura K, Satoh D, Estes P, Crews S, Uemura T (2004) Development of morphological diversity of dendrites in Drosophila by the BTB-zinc finger protein abrupt. Neuron 43(6):809–822. https://doi.org/10.1016/j.neuron.2004.08.016

Tanaka N, Meng W, Nagae S, Takeichi M (2012) Nezha/CAMSAP3 and CAMSAP2 cooperate in epithelial-specific organization of noncentrosomal microtubules. Proc Natl Acad Sci USA 109 (49):20029–20034. https://doi.org/10.1073/pnas.1218017109

Teixidó-Travesa N, Villén J, Lacasa C, Bertran MT, Archinti M, Gygi SP, Caelles C, Roig J, Lüders J (2010) The γTuRC revisited: a comparative analysis of interphase and mitotic human γTuRC redefines the set of core components and identifies the novel subunit GCP8. Mol Biol Cell 21 (22):3963–3972. https://doi.org/10.1091/mbc.e10-05-0408

Teixido-Travesa N, Roig J, Luders J (2012) The where, when and how of microtubule nucleation – one ring to rule them all. J Cell Sci 125(Pt 19):4445–4456. https://doi.org/10.1242/jcs.106971

Thawani A, Kadzik RS, Petry S (2018) XMAP215 is a microtubule nucleation factor that functions synergistically with the gamma-tubulin ring complex. Nat Cell Biol 20(5):575–585. https://doi.org/10.1038/s41556-018-0091-6

Thornton GK, Woods CG (2009) Primary microcephaly: do all roads lead to Rome? Trends Genet 25(11):501–510. https://doi.org/10.1016/j.tig.2009.09.011

Tovey CA, Conduit PT (2018) Microtubule nucleation by gamma-tubulin complexes and beyond. Essays Biochem 62(6):765–780. https://doi.org/10.1042/EBC20180028

Toya M, Kobayashi S, Kawasaki M, Shioi G, Kaneko M, Ishiuchi T, Misaki K, Meng W, Takeichi M (2016) CAMSAP3 orients the apical-to-basal polarity of microtubule arrays in epithelial cells. Proc Natl Acad Sci USA 113(2):332–337. https://doi.org/10.1073/pnas.1520638113

Troutt LL, Wang E, Pagh-Roehl K, Burnside B (1990) Microtubule nucleation and organization in teleost photoreceptors: microtubule recovery after elimination by cold. J Neurocytol 19 (2):213–223

Vinogradova T, Miller PM, Kaverina I (2009) Microtubule network asymmetry in motile cells: role of Golgi-derived array. Cell Cycle (Georgetown, Tex) 8(14):2168–2174. https://doi.org/10.4161/cc.8.14.9074

Wang L, Dynlacht BD (2018) The regulation of cilium assembly and disassembly in development and disease. Development 145(18):dev151407. https://doi.org/10.1242/dev.151407

Wang X, Tsai JW, Imai JH, Lian WN, Vallee RB, Shi SH (2009) Asymmetric centrosome inheritance maintains neural progenitors in the neocortex. Nature 461(7266):947–955. https://doi.org/10.1038/nature08435

Wang Z, Wu T, Shi L, Zhang L, Zheng W, Qu JY, Niu R, Qi RZ (2010) Conserved motif of CDK5RAP2 mediates its localization to centrosomes and the Golgi complex. J Biol Chem 285 (29):22658–22665. https://doi.org/10.1074/jbc.m110.105965

Wang S, Wu D, Quintin S, Green RA, Cheerambathur DK, Ochoa SD, Desai A, Oegema K (2015) NOCA-1 functions with gamma-tubulin and in parallel to Patronin to assemble non-centrosomal microtubule arrays in C. elegans. elife 4:e08649. https://doi.org/10.7554/eLife.08649

Wang Y, Rui M, Tang Q, Bu S, Yu F (2019) Patronin governs minus-end-out orientation of dendritic microtubules to promote dendrite pruning in Drosophila. elife 8:e39964. https://doi.org/10.7554/eLife.39964

Wei JH, Zhang ZC, Wynn RM, Seemann J (2015) GM130 regulates Golgi-derived spindle assembly by activating TPX2 and capturing microtubules. Cell 162(2):287–299. https://doi.org/10.1016/j.cell.2015.06.014

Wieczorek M, Bechstedt S, Chaaban S, Brouhard GJ (2015) Microtubule-associated proteins control the kinetics of microtubule nucleation. Nat Cell Biol 17(7):907. https://doi.org/10.1038/ncb3188

Wong YL, Anzola JV, Davis RL, Yoon M, Motamedi A, Kroll A, Seo CP, Hsia JE, Kim SK, Mitchell JW, Mitchell BJ, Desai A, Gahman TC, Shiau AK, Oegema K (2015) Reversible centriole depletion with an inhibitor of Polo-like kinase 4. Science 348(6239):1155–1160. https://doi.org/10.1126/science.aaa5111

Wu J, de Heus C, Liu Q, Bouchet Benjamin P, Noordstra I, Jiang K, Hua S, Martin M, Yang C, Grigoriev I, Katrukha Eugene A, Altelaar AFM, Hoogenraad Casper C, Qi Robert Z, Klumperman J, Akhmanova A (2016) Molecular pathway of microtubule organization at the Golgi apparatus. Dev Cell 39(1):44–60. https://doi.org/10.1016/j.devcel.2016.08.009

Yalgin C, Ebrahimi S, Delandre C, Yoong LF, Akimoto S, Tran H, Amikura R, Spokony R, Torben-Nielsen B, White KP, Moore AW (2015) Centrosomin represses dendrite branching by orienting microtubule nucleation. Nat Neurosci 18(10):1437–1445. https://doi.org/10.1038/nn.4099

Yang C, Wu J, Cd H, Grigoriev I, Liv N, Yao Y, Smal I, Meijering E, Klumperman J, Qi RZ, Akhmanova A (2017) EB1 and EB3 regulate microtubule minus end organization and Golgi morphology. J Cell Biol 216(10):3179–3198. https://doi.org/10.1083/jcb.201701024

Yau Kah W, van Beuningen Sam FB, Cunha-Ferreira I, Cloin Bas MC, van Battum Eljo Y, Will L, Schätzle P, Tas Roderick P, van Krugten J, Katrukha Eugene A, Jiang K, Wulf Phebe S,

Mikhaylova M, Harterink M, Pasterkamp RJ, Akhmanova A, Kapitein Lukas C, Hoogenraad Casper C (2014) Microtubule minus-end binding protein CAMSAP2 controls axon specification and dendrite development. Neuron 82(5):1058–1073. https://doi.org/10.1016/j.neuron.2014.04.019

Yigit G, Brown KE, Kayserili H, Pohl E, Caliebe A, Zahnleiter D, Rosser E, Bogershausen N, Uyguner ZO, Altunoglu U, Nurnberg G, Nurnberg P, Rauch A, Li Y, Thiel CT, Wollnik B (2015) Mutations in CDK5RAP2 cause Seckel syndrome. Mol Genet Genomic Med 3 (5):467–480. https://doi.org/10.1002/mgg3.158

Yonezawa S, Shigematsu M, Hirata K, Hayashi K (2015) Loss of gamma-tubulin, GCP-WD/ NEDD1 and CDK5RAP2 from the centrosome of neurons in developing mouse cerebral and cerebellar cortex. Acta Histochem Cytochem 48(5):145–152. https://doi.org/10.1267/ahc.15023

Zhang X, Chen Q, Feng J, Hou J, Yang F, Liu J, Jiang Q, Zhang C (2009) Sequential phosphorylation of Nedd1 by Cdk1 and Plk1 is required for targeting of the gammaTuRC to the centrosome. J Cell Sci 122(Pt 13):2240–2251. https://doi.org/10.1242/jcs.042747

Zhang X, Chen MH, Wu X, Kodani A, Fan J, Doan R, Ozawa M, Ma J, Yoshida N, Reiter JF, Black DL, Kharchenko PV, Sharp PA, Walsh CA (2016) Cell-type-specific alternative splicing governs cell fate in the developing cerebral cortex. Cell 166(5):1147–1162 e1115. https://doi.org/10.1016/j.cell.2016.07.025

Zhang R, Roostalu J, Surrey T, Nogales E (2017) Structural insight into TPX2-stimulated microtubule assembly. elife 6:e30959. https://doi.org/10.7554/eLife.30959

Zhou W, Chang J, Wang X, Savelieff Masha G, Zhao Y, Ke S, Ye B (2014) GM130 is required for compartmental organization of dendritic Golgi outposts. Curr Biol 24(11):1227–1233. https://doi.org/10.1016/j.cub.2014.04.008

Zhu X, Kaverina I (2013) Golgi as an MTOC: making microtubules for its own good. Histochem Cell Biol 140(3):361–367. https://doi.org/10.1007/s00418-013-1119-4

Zhu H, Coppinger JA, Jang C-Y, Yates JR, Fang G (2008) FAM29A promotes microtubule amplification via recruitment of the NEDD1–γ-tubulin complex to the mitotic spindle. J Cell Biol 183(5):835–848. https://doi.org/10.1083/jcb.200807046

Chapter 15
The Golgi Apparatus in Polarized Neuroepithelial Stem Cells and Their Progeny: Canonical and Noncanonical Features

Elena Taverna and Wieland B. Huttner

Abstract Neurons forming the central nervous system are generated by neural stem and progenitor cells, via a process called neurogenesis (Götz and Huttner, Nat Rev Mol Cell Biol, 6:777–788, 2005). In this book chapter, we focus on neurogenesis in the dorsolateral telencephalon, the rostral-most region of the neural tube, which contains the part of the central nervous system that is most expanded in mammals (Borrell and Reillo, Dev Neurobiol, 72:955–971, 2012; Wilsch-Bräuninger et al., Curr Opin Neurobiol 39:122–132, 2016). We will discuss recent advances in the dissection of the cell biological mechanisms of neurogenesis, with particular attention to the organization and function of the Golgi apparatus and its relationship to the centrosome.

15.1 Introduction

15.1.1 The Developing Mammalian Neocortex: Nomenclature and General Organization

The neocortex, which is the evolutionary younger part of the cortex involved in higher-oder cognitive functions, develops around a central cavity, called the ventricle (Borrell and Reillo 2012; Götz and Huttner 2005; Noctor et al. 2007b) (see also Fig. 15.1). The portion of the developing neocortex that occupies the apical-most region of the tissue and is in contact with the ventricle is called ventricular zone (VZ). The VZ contains a class of neural stem and progenitor cells collectively called

E. Taverna (✉)
Max Planck Institute for Evolutionary Anthropology, Leipzig, Germany
e-mail: elena_taverna@eva.mpg.de

W. B. Huttner (✉)
Max Planck Institute of Molecular Cell Biology and Genetics, Dresden, Germany
e-mail: huttner@mpi-cbg.de

© Springer Nature Switzerland AG 2019
M. Kloc (ed.), *The Golgi Apparatus and Centriole*, Results and Problems in Cell Differentiation 67, https://doi.org/10.1007/978-3-030-23173-6_15

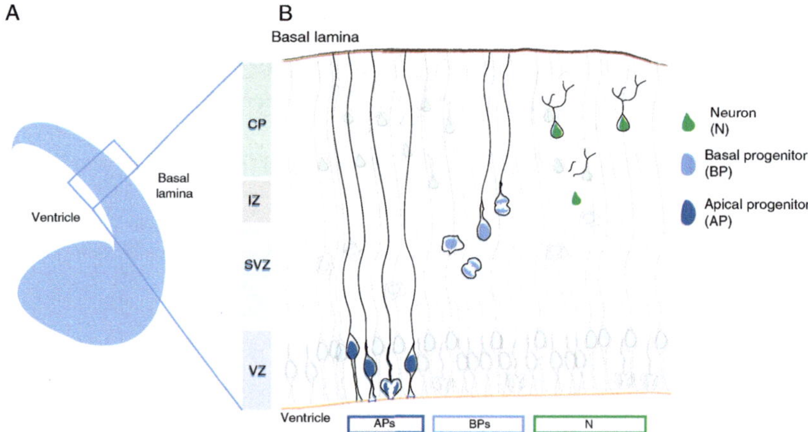

Fig. 15.1 Nomenclature and general organization of the developing mammalian neocortex. (**a**) Cartoon illustrating the general structure of the mammalian developing telencephalon. The neocortex develops around a central cavity, called the ventricle. Opposite to the ventricle, the neocortex is delimited by a basal lamina. (**b**) Schematic representation of the different zones (vertical axis) and cell types (horizontal axis) forming the developing neocortex. From apical to basal: *VZ* ventricular zone, *SVZ* subventricular zone, *IZ* intermediate zone, *CP* cortical plate

apical progenitors (APs). APs give rise, via mitosis, to the second class of progenitor cells, called basal progenitors (BPs). BPs occupy a region basal to the VZ, called subventricular zone (SVZ). Neurons generated in either the VZ (minor source) or SVZ (major source) migrate through the intermediate zone (IZ) and settle in the basal-most region of the developing neocortex called cortical plate (CP). The CP is delimited on the basal side by a basal lamina (see Fig. 15.1 for a summary scheme).

15.1.2 Neural Stem Cell Types and Their Cell Biological Features

We will here review the principal neural cell types found in the developing neocortex (Borrell and Reillo 2012; Götz and Huttner 2005; Noctor et al. 2007b). For every cell type, after a description of the cell's general features (see Sect. 15.2.1 and Fig. 15.2), we will report and discuss what is known about the centrosome and Golgi organization.

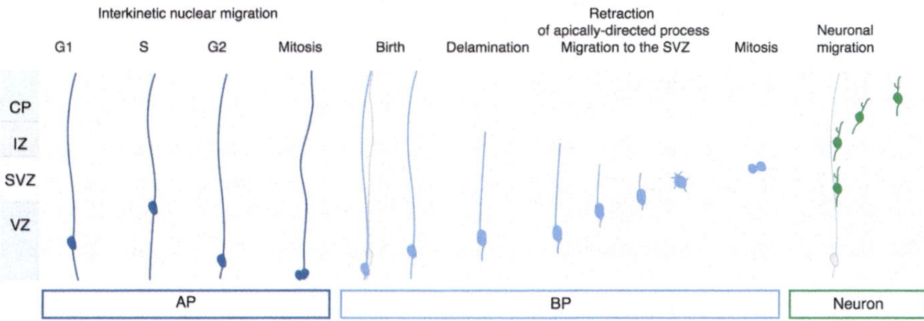

Fig. 15.2 A cell biological perspective on neural stem cell fate transition and neuron generation. Crucial cell biological steps in lineage progression and cell fate transition during neurogenesis. From left to right: apical progenitors (APs) undergo *interkinetic nuclear migration*, moving the nucleus in concert with the different phases of the cell cycle. Apical progenitors via *mitosis* give rise to basal progenitors (BPs). After being generated (*birth*), BPs initially maintain contact with the ventricular surface. BPs then lose their contact with the apical surface via *delamination, migrate* to the SVZ, and eventually undergo *mitosis* to give rise to neurons. The neurons in turn undergo *neuronal migration* toward the basal lamina to reach their correct position in the cortical plate

15.2   Apical Progenitors (APs)

15.2.1   General Remarks

APs are highly polarized epithelial cells, exhibiting apicobasal polarity, and undergo mitosis at the ventricular surface (Fig. 15.2). Their apical plasma membrane represents only a minor fraction (typically 1–2% (Kosodo et al. 2004)) of the total plasma membrane, is delimited by the adherens junction belt, and lines the lumen of ventricle. The basolateral plasma membrane of APs can span the VZ, SVZ, IZ, and CP and reach the basal lamina (Fig. 15.2). Such APs represent an extreme case of cell elongation along the apicobasal axis. This elongation is most prominent in human and other primates, where the thickness of the cortical wall reaches several millimeters (Smart et al. 2002). One characteristic feature of APs is their ability to undergo interkinetic nuclear migration (INM), that is, APs move their nuclei in the VZ in concert with the cell cycle (Norden et al. 2009; Taverna et al. 2014; Taverna and Huttner 2010) (Fig. 15.2). After completing mitosis at the ventricular surface, APs nuclei undergo apical-to-basal migration during G1. After S-phase in the basal part of the VZ, the nuclei undergo basal-to-apical migration. Interestingly, INM is confined to the VZ, and the AP nucleus never moves into the part of the cell spanning the SVZ, IZ, and CP. A key aspect of INM is that AP mitosis typically occurs at the ventricular surface, thereby maximizing the number of AP divisions per apical surface area. An underlying reason why APs mitosis typically occurs at the ventricular surface is that the AP apical plasma membrane bears a primary cilium throughout interphase, as is explained below. AP mitoses can give rise directly to APs, BPs, and/or (rarely) neurons (Figs. 15.1 and 15.2).

15.2.2 Centrosome and Golgi Apparatus in Interphase APs

15.2.2.1 The Centrosome

As for other epithelial cells (Bacallao et al. 1989), the centrosome of APs is associated with the apical plasma membrane. This association has been dissected with regard to centrosome structure. It is known that the centrosome's older centriole (the basal body) is nucleating the microtubules of the primary cilium that, as a specialization of the apical plasma membrane, protrudes into the ventricle. This implies that the cilium's basal body-containing centrosome and the second centrosome that in mitosis builds the mitotic spindle are confined to the apical cell cortex during interphase. Hence, for entry into mitosis, the AP nucleus needs to approach the apical centrosomes (rather than the other way round).

The importance of centrosome function in brain development is highlighted by primary microcephaly, a group of disorders associated with a dramatic reduction in brain size at birth (Bond et al. 2002; Bond and Woods 2005). The mutated genes identified so far in these disorders are associated either with the cytokinesis machinery or with centrosomal proteins.

Cytokinesis Machinery Citron kinase (CRIK) is a RhoA modulator whose ablation results in massive reduction in brain size (Di Cunto et al. 2000; Gai and Di Cunto 2017; Sarkisian et al. 2002). CRIK is expressed by neural stem and progenitor cells and localizes to the midbody bridge, where it is necessary for the final phase of cell division, the abscission (Gai et al. 2011; Naim et al. 2004; Di Cunto et al. 1998). Mutations in CRIK alter abscission, driving neural APs into apoptosis, thereby leading to the observed reduction in brain size. The cytokinesis machinery is also one of the targets of Zika virus infection, which is associated with neurological alterations in both newborn and adult. The Zika protease NS2B-N3, required for virus replication, cleaves Septin-2, a protein that is crucial for the assembly and stability of the midbody bridge (Li et al. 2019). As a consequence, neural stem and progenitor cells are unable to divide, and this drives them into apoptosis. This mechanism might, at least in part, explain the severe reduction in brain size associated with Zika virus infection (Li et al. 2016a, b).

Centrosomal Proteins ASPM (abnormal spindle microcephaly associated) is a centrosomal protein, the gene of which is frequently found to be mutated in microcephaly patients. In mouse brain, *ASPM* mutations cause centrosome amplification, aneuploidy, and tissue degeneration (Marthiens et al. 2013). Cdk5Rap2 represents an additional case of a centrosomal protein implicated in primary microcephaly (Bond et al. 2005). Cdk5Rap2 is recruited to the centrosome via its interaction with pericentrin, and its depletion induces neuronal differentiation by promoting the generation of BPs at the expense of APs. Both Aspm and Cdk5Rap2 have a very well documented role in mitosis and in spindle positioning.

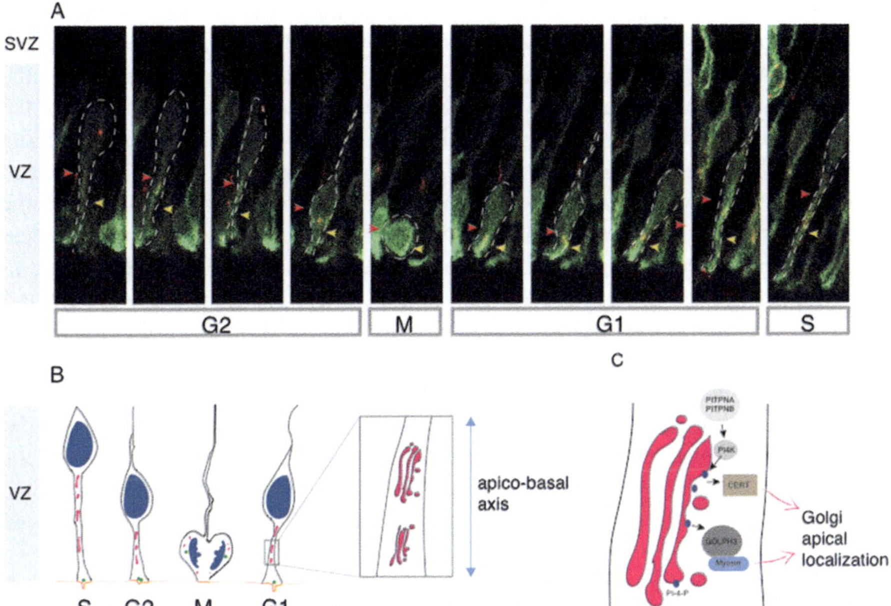

Fig. 15.3 Noncanonical Golgi apparatus organization in apical progenitors. (**a**) Live imaging of the Golgi apparatus. APs were electroporated with plasmids for a Golgi-resident red fluorescent protein (red) and for cytosolic GFP (green) to visualize the cell boundaries (white dashed line). Yellow arrowheads indicate the apical-most Golgi, while orange arrowheads indicate the basal-most Golgi. Note the Golgi apparatus compression in G2-phase, Golgi fragmentation and partitioning during mitosis (M), followed by Golgi extension and stretching in G1-S-phase. (**b**) Cartoon illustrating the Golgi apparatus (magenta) and centrosome (green) organization throughout the cell cycle (Taverna et al. 2016). Orange: apical plasma membrane with primary cilium. Inset on the right: Golgi apparatus orientation in relation with the AP's apicobasal axis (light blue arrow) (Taverna et al. 2016). (**c**) Molecular mechanisms responsible for the apical Golgi confinement (Xie et al. 2018)

15.2.2.2   The Golgi Apparatus. . .

Unlike what has been described for other epithelial cells, the Golgi apparatus in interphase APs shows very specific and noncanonical features, related to (1) the organization and orientation of the Golgi relative to the cell's apicobasal axis, (2) the dynamics during cell cycle progression, and (3) the lack of association with the centrosome (Taverna et al. 2016).

. . .Is Confined to the Apical Process

In APs, the Golgi apparatus is organized in stacks distributed in the apical process, between the apical plasma membrane and the nucleus. Interestingly, the stacks are not forming a typical ribbonlike structure, but appear to be separated entities (Fig. 15.3a, b). No membrane structures identifiable as Golgi cisternae have been observed in the basal process.

Unlike the Golgi apparatus which is confined to the apical process, the endoplasmic reticulum (ER) is evenly distributed in the apical and basal process. The uneven distribution of the Golgi apparatus as a canonical intermediate station within the secretory pathway was found to have consequences regarding the composition of the lateral plasma membrane in the apical vs basal AP process. While proteins carrying ER-derived glycans were found to be evenly distributed along the apicobasal axis of APs, proteins carrying Golgi-modified glycans were found to be confined to the apical process only (Taverna et al. 2016). These data suggest that the apical process relies on the conventional ER-to-Golgi-to-plasma membrane traffic route for membrane supply, while the basal process relies on the unconventional pathway, that is a direct traffic route from the ER to the plasma membrane (Grieve and Rabouille 2011).

It is known that the apical and basal processes perform different functions. For example, the apical process is permissive for INM, while the basal process is heavily involved in cell-to-cell signaling and regulation of cell proliferation vs differentiation. The question arises as to whether the distribution of the Golgi apparatus and the sub-compartmentalization of the basolateral plasma membrane are related to the functional diversity of the apical and basal process. We would like here to propose some speculation that might be interesting to investigate. As to the apical process, the presence of Golgi-derived glycans might render the plasma membrane more fluid, so that it can better adapt to the migration of the nucleus during INM. As to the basal process, work in *Drosophila* has established that integrins are sorted to the basal-most part of epithelial plasma membrane via an unconventional secretory pathway (Schotman et al. 2008). Interestingly, functional manipulations in mouse, ferret, and human developing brain demonstrated that the integrin signaling that regulates progenitor proliferative potential is initiated in the basal process (Fietz et al. 2010; Stenzel et al. 2014). Another hypothesis worth testing is that the ER-derived glycans in the basal process are involved in the radial migration of neurons along the radial fiber. Although highly speculative, this hypothesis is very attractive, considering that a large proportion of disorders of cortical migration are associated with defects in glycosylation (Freeze et al. 2015).

...Is Parallel to the Apicobasal Axis of the Cell

It has been recently shown that in APs the Golgi stacks are oriented with their cis-to-trans axis perpendicular to the apicobasal axis of the cell (Taverna et al. 2016; Xie et al. 2018) (Fig. 15.3b). This organization is somewhat surprising, as in all epithelial cells analyzed so far the cis-to-trans Golgi axis was reported to be parallel to the cell's polarity axis. One can speculate that this orientation minimizes the distance a vesicle has to travel from the trans Golgi network (TGN) to the target plasma membrane. This organization might be instrumental to optimize the membrane supply in APs, as they represent a highly dynamic and fast-elongating cell type.

...Is Reorganized During INM

Since the Golgi apparatus in APs stretches between the nucleus and the apical plasma membrane, what happens during INM, when the nucleus either moves toward or away from the apical plasma membrane? Live imaging experiments have revealed that in APs the Golgi apparatus is reorganized in concert with INM

(Taverna et al. 2016) (Fig. 15.3a, b). In particular, the Golgi apparatus is compressed during the G2-phase basal-to-apical nuclear migration, and then is stretched during the G1-phase apical-to-basal phase of nuclear migration. These observations prompted the authors to propose that the Golgi apparatus in APs is behaving like an accordion, undergoing several phases of compression followed by stretching depending on the cell cycle's phase. The functional consequences of stretching and compressing the Golgi apparatus have not been addressed yet. In this regard, results in tissue culture cells show that mechanical forces applied to the Golgi apparatus perturb the dynamics of Golgi-associated actin and potentially affect membrane trafficking (Guet et al. 2014).

. . .Is Not Pericentrosomal

In mammalian cells, the Golgi apparatus has always been reported to be organized around the centrosome. This organization is called "pericentrosomal" and is considered to be the canonical and well-conserved configuration of the Golgi apparatus in interphase in all mammalian cells, including epithelial cells (Bacallao et al. 1989). This rule does not apply to APs, however, in which the centrosome is strictly located at the apical cell cortex where it nucleates the primary cilium, whereas the Golgi is stretched in the apical process, with no sign of connection with the centrosome (Taverna et al. 2016) (see scheme in Fig. 15.3b).

15.2.2.3   Mechanism of Golgi Apparatus Apical Localization

The confinement of the Golgi apparatus to the apical process of APs is actively maintained via a lipid signaling pathway involving the Golgi-localized pool of phosphatidylinositol-4-phosphate, in addition to GOLPH3 (Golgi phosphoprotein 3) and CERT (ceramide transfer protein) as downstream effectors (Xie et al. 2018) (Fig. 15.3c). In particular, PITPNA and PITPNB, two phosphatidylinositol transfer proteins (PITPs) stimulate the synthesis of phosphatidylinositol-4-phosphate on the Golgi membrane. The phosphatidylinositol-4-phosphate pool in turn recruits GOLPH3, which serves as an adaptor to link the Golgi cisternae to the actin cytoskeleton (Fig. 15.3c). The pathway linking phosphatidylinositol-4-phosphate to actin is necessary for the apical localization of the Golgi apparatus in APs. PITPNA and PITPNB are essential for normal brain development, as their ablation in the embryonic mouse neocortex causes severe developmental defects, characterized by the almost complete absence of the dorsal telencephalon (Xie et al. 2018). The cellular basis for the detrimental effect of PITPNA and PITPNB ablation is the disarrangement of AP structure and architecture, and the subcellular basis for that in turn is the lack of apical confinement of the Golgi apparatus (Xie et al. 2018). Taken together, these data strongly suggest that not only Golgi function is necessary for brain development, but also the Golgi localization and apical confinement (Taverna et al. 2016; Xie et al. 2018) are crucial aspects securing a correct brain development.

15.2.2.4 Centrosome and Golgi Apparatus in Mitotic APs

In mammalian cells, the centrosomes are essential to build the mitotic spindle. In any given cell, the two centrosomes are always asymmetric with respect to centriole age, as one contains the so-called mother centriole and the other the so-called daughter centriole, which were synthesized in different cell cycles (Paridaen and Huttner 2014). For APs undergoing asymmetric cell division, the question arises as to whether the centrosome's asymmetry correlates with the asymmetric fate of the daughter cell. It has been shown (Wang et al. 2009) that the centrosome containing the mother centriole is preferentially inherited by the daughter cell remaining an apical radial glial cell, whereas the centrosome containing the daughter centriole is preferentially inherited by the differentiating daughter cell (a BP, and less frequently a neuron).

Interestingly, also the ciliary membrane is asymmetrically distributed during mitosis, as it—being a single-copy organelle—is inherited by only one of the daughter cells, which tends to be the proliferative daughter cell (Paridaen et al. 2013). Furthermore, the daughter cell inheriting the ciliary membrane tends to re-establish the cilium earlier than the sibling cell (Paridaen et al. 2013). These data suggest that in APs there are most likely two pathways contributing to the biogenesis of the primary cilium: a pathway depending on inheritance of ciliary membrane and a pathway depending on de novo biosynthesis of ciliary membrane (Paridaen et al. 2013). The de novo biosynthesis pathway has been typically linked to Golgi-derived traffic. It would be interesting to understand if and how Golgi traffic contributes to the biogenesis of the primary cilium in newborn BPs.

It is known that in mammalian cells the Golgi undergoes fragmentation during mitosis (Levine et al. 1995). This fragmentation step is necessary for the cell to enter mitosis, and is the cellular basis of the so called "Golgi mitotic checkpoint" (Ayala and Colanzi 2017; Persico et al. 2009, 2010). The distribution of the Golgi in mitotic APs has been analyzed by immunofluorescence and electron microscopy (Taverna et al. 2016). During prophase, the Golgi apparatus of APs appears to undergo fragmentation, and the Golgi remnants are distributed at the cell periphery where they only partially associate with the spindle poles. Interestingly, the Golgi remnants do not form the so called "Golgi haze" (Axelsson and Warren 2004), but rather consist of partially stacked cisternae (as revealed by electron microscopy). In telophase, when the Golgi apparatus is known to re-assemble, the Golgi fragments are still distributed at the cell periphery. Unlike the ciliary membrane, the Golgi apparatus does not seem to be asymmetrically partitioned during mitosis (Taverna et al. 2016).

15.3 Basal Progenitors (BPs)

15.3.1 General Remarks

BPs are cells lacking ventricular contact and apical polarity cues. There are two major subtypes of BPs, the intermediate progenitor cells (IPCs) and the basal radial glial cells (bRGCs) (Fietz et al. 2010; Hansen et al. 2010; Martinez-Cerdeno et al. 2006; Noctor et al. 2004, 2007a; Reillo and Borrell 2012; Reillo et al. 2011; Wang et al. 2011). IPCs are abundant in rodents and are considered to be non-polarized cells, as they lack both apical and basal polarity cues. bRGCs lack apical polarity cues but maintain basal polarity cues, often associated with their attachment to the basal lamina. bRGCs are particularly abundant in gyrencephalic species like human (Reillo et al. 2011). BPs are typically born from an asymmetric division of an AP, in which one daughter cell maintains the AP-fate of the mother cell and the other daughter cell becomes a BP. It has been shown that immediately after being generated, the majority of BPs still retain ventricular contact and feature an apical plasma membrane that is flanked by the apical adherens junction belt (see Fig. 15.2). By the end of G1-phase, BPs have lost their apical attachment in a process called delamination. At the cellular level, delamination is crucial for the basal migration of BPs. At the tissue level, delamination is crucial for the generation of the SVZ, the second germinal zone in the developing brain, the size of which underwent dramatic expansion during human evolution. Reflecting the relevance of delamination for brain development and evolution, considerable efforts have recently been made to dissect the cellular and molecular mechanisms responsible for it. We will here describe this crucial step focusing mainly on the centrosome and primary cilium (see Fig. 15.2 for a general summary of the above-described steps).

15.3.2 Centrosome and Golgi Apparatus in Nascent BPs

Very much like an AP, a newborn BP retains ventricular contact and its apical plasma membrane is delimited by the apical adherens junction belt (Wilsch-Bräuninger et al. 2012), as mentioned above. Which then are the cell biological features that set a newborn BP apart from an AP, allowing a BP to eventually lose apical polarity cues? The disengagement of the BP from the apical adherens junction belt is preceded by the relocation of the centrosome-cilium (Wilsch-Bräuninger et al. 2012). Using electron microscopy, researchers have shown that the first cell biological signature in the generation of a BP is the relocation of the centrosome-cilium from an apical location (which is typical for epithelial cells such as APs) to a basolateral location (Wilsch-Bräuninger et al. 2012). Functional manipulations suggested that the relocation of the centrosome-cilium to the basolateral plasma membrane is a necessary step in delamination. Overexpression of *Insm1* (insulinoma-associated 1), a transcription factor known to promote the generation of BPs (Farkas et al. 2008; Tavano et al. 2018), increases the proportion of basolateral cilia (Wilsch-Bräuninger et al. 2012). Furthermore, microinjection of recombinant dominant-negative Cdc42 into single APs in organotypic slice culture of developing mouse hindbrain leads to their delamination (Taverna et al. 2012). Interestingly, the microinjected cells that were not yet delaminated featured a basolateral (rather than apical) centrosome (Taverna et al. 2012), strongly suggesting that the relocation of the centrosome (and most likely the primary cilium) is a necessary step for the complete disengagement of a delaminating cell from the apical adherens junction belt.

The organization of the Golgi apparatus in nascent BPs is very similar to the one present in G1-phase APs (Taverna et al. 2016), that is, it exhibits multiple non-interconnected stacks distributed between the apical plasma membrane and the nucleus.

Taken together these data suggest that from a cell biological point of view, the organelles orchestrating cell fate transition are most likely the centrosome-cilium (Wilsch-Bräuninger et al. 2012, 2016).

15.3.3 Golgi and Centrosome in Delaminated BPs

In delaminated BPs, the Golgi appears to be localized around the centrosome (Taverna et al. 2016). This aspect is interesting, as it means that during delamination, concomitant with the loss of apical contact, the Golgi shifts from a noncanonical to a canonical, pericentrosomal location. One can wonder how and why cell fate transition and delamination are accompanied by a reorganization of these two organelles. Although there is not a definitive answer, a tempting speculation is that the reorganization at the centrosome-Golgi axis has to do with the different migration characteristics of APs and BPs. As described above, APs undergo INM, which is a very specific type of migration, as it consists of nucleokinesis (movement of the nucleus) without a net translocation of the cell body. Indeed, the AP remains anchored to the apical and basal side while the nucleus is moving. In contrast, upon delamination the newborn BPs translocate both their nucleus and cell body from the VZ to the SVZ. Seminal work performed on neurons in 2D tissue culture has shown that the proximity of the centrosome and the Golgi apparatus is crucial for their directed cell migration (Hurtado et al. 2011). One can therefore speculate that the reorganization of the centrosome and Golgi apparatus helps the directed migration of BPs to the SVZ. It follows that the reorganization at the Golgi-centrosome interface should be regarded as a crucial cell biological step for the generation of the basal germinal zone, the SVZ.

15.3.4 Golgi and Centrosome in bRGCs

As mentioned, all BPs lack apical polarity cues and apical attachment (Fig. 15.1). However, while iPCs lack also basal polarity cues, bRGCs maintain basal polarity cues. Two questions are of interest regarding the cell biology of bRGCs: (1) which are the mechanisms responsible for the selective loss of apical contact, and (2) which is the organization of their intracellular compartments? A recent study suggests that PLEKHA7 acts downstream of INSM1 to promote the generation of bRGCs. PLEKHA7 is an adherens junction belt-specific protein that, when down-regulated, results in the selective loss of apical contact. Hence, the PLEKHA7-manipulated cells maintain their basal attachment and are therefore bRGCs (Tavano et al. 2018). This is the first reported case of a polarity protein selectively regulating apical polarity cues, independently of the basal ones. As for the organization of the intracellular compartments, electron microscopy studies have shown that in bRGCs the centrosome is close to the nucleus (Fietz et al. 2010) and presumably is associated with the Golgi apparatus.

15.3.5 Centrosome and Golgi Apparatus in Mitotic BP

While the Golgi apparatus in mitotic APs is found at the cell periphery, far away from the centrosomes/spindle poles, the Golgi apparatus in mitotic BPs appears to be close to the centrosomes (Taverna et al. 2016). In prophase and metaphase, the Golgi apparatus is compact and mostly founds in close proximity to the centrosomes/ spindle poles. In telophase BPs, the Golgi structures were largely observed near the midbody bridge, which harbors the remnants of the mitotic spindle. Taken together, these data show that the two main classes of neural progenitor cells in the developing mouse neocortex, APs and BPs, differ with regard to the Golgi apparatus–centrosome relationship not only in interphase, but also during mitosis.

15.4 Neurons

15.4.1 General Remarks

Neurons are the final cell output of neurogenesis (Fig. 15.1). Their structural, biochemical, and functional polarization secures the unique ability of neurons of exchanging and storing information. As mentioned above, neurons are mainly produced via divisions of BPs in the SVZ. After being generated, a newborn neuron starts the long journey that will allow it to settle in the forming CP (Fig. 15.2). The CP in turn will give rise to the six-layered cortex, a hallmark of mammals. A newborn neuron engages in two processes that underlie its function and connectivity: the establishment of polarity and the migration to the correct cortical layer. Seminal studies using primary neurons in 2D culture established that the centrosome and the Golgi apparatus play crucial roles in both polarization and migration (Bradke and Dotti 1998, 2000; de Anda et al. 2005; Huttner and Dotti 1991). We will here review some of the main findings in the field. Given the breadth of the subject, we invite the reader to refer to other reviews for a more detailed description of the topic. We also apologize to those colleagues the work of which could not be mentioned for space issues.

15.4.2 Golgi and Centrosome Function in Neuronal Migration

Neuronal migration consists of different phases (Hatten 1999; Marin et al. 2010; Martinez-Martinez et al. 2018; Valiente and Marin 2010). After being generated, a newborn neuron polarizes and starts migrating toward the basal lamina in a process called radial migration (Figs. 15.1 and 15.2). During radial migration, neurons are bipolar and move along the radial fibers (or basal processes) of APs and bRGCs (Martinez-Martinez et al. 2018; Reillo and Borrell 2012; Reillo et al. 2011). While migrating through the IZ, neurons change their morphology from bipolar to multipolar, forming multiple processes and reducing their migration speed (Cooper 2014). During this phase, they can undergo tangential displacement, a crucial process affecting the dispersion of neurons in the tangential dimension and possibly the establishment of connectivity. Eventually neurons acquire a clear bipolar morphology while migrating through the CP (Ayala et al. 2007). Studies from several labs have clearly demonstrated that the centrosome and the Golgi apparatus (along with the actin and microtubule cytoskeleton) work in concert to allow neuronal migration (Bellion et al. 2005; Solecki et al. 2009; Tsai and Gleeson 2005). In particular, the movement of centrosome and Golgi into the leading process precedes and allows nucleokinesis. This is because the centrosome works as a microtubule organizing center, nucleating microtubules that in turn are responsible for force generation on the nucleus. Interestingly, genes associated with disorders of cortical development often perturb neuronal migration by influencing either cytoskeletal elements, the centrosome and/or the Golgi apparatus (Tanaka et al. 2004; Valiente and Marin 2010).

15.4.3 Golgi and Centrosome Function in Neuronal Polarity

15.4.3.1 Centrosome

The centrosome is essential not only for neuronal migration, but also for polarity establishment and neurite outgrowth (Bradke and Dotti 1998, 2000), as demonstrated by studies with 2D neuronal cultures in which the orientation of the centrosome was found to provide spatial cues for axon emergence from the cell body (de Anda et al. 2005; Gärtner et al. 2012). Interestingly, a study suggests that after having set the initial axonal polarity, the centrosome is dispensable for axon extension (Stiess et al. 2010).

15.4.3.2 Golgi Apparatus

The Golgi apparatus plays a key role in establishing and maintaining the highly polarized structure of neurons. A prime example are pyramidal neurons, which feature a prominent dendrite (called the apical dendrite) extending hundreds of microns in the direction of the basal lamina. Several studies have shown that the Golgi apparatus extends into the apical dendrite (Huang et al. 2014; Matsuki et al. 2010; O'Dell et al. 2012). This extension (called dendritic Golgi deployment) depends on the Reelin-Dab1-GM130 pathway (Huang et al. 2014; Matsuki et al. 2010) and is crucial for cell polarization, axon specification, and dendrite growth (Huang et al. 2014; Matsuki et al. 2010). Of note, ubiquitin-protein ligase E3A (Ube3a), a gene mutated in Angelman syndrome, was found to inhibit apical dendrite outgrowth and polarity by disrupting the polarized distribution of the Golgi apparatus (Miao et al. 2013). This finding is interesting, as it illustrates the relevance of the localization of the Golgi apparatus in neurodevelopmental disorders.

Neurons rely not only on a perinuclear or cell body-localized Golgi apparatus but also on the so-called Golgi outposts (Horton and Ehlers 2003; Horton et al. 2005; Ye et al. 2007). These Golgi outposts are small Golgi cisternae often found in dendritic spines, far away from the cell body, where they represent a way to secure the local processing of newly synthesized proteins (Valenzuela and Perez 2015). Golgi outposts influence dendrite architecture by functioning as sites of acentrosomal microtubule nucleation in neurons (Ori-McKenney et al. 2012). This independent traffic route is important as it allows distal compartments such as dendritic spines to become independent from the cell body with regard to protein processing (Quassollo et al. 2015). Furthermore, electrical activity of neurons has been shown to influence protein translation (Evans et al. 2017) and may affect Golgi outpost function as well. This possibility would tremendously increase the level of regulation and fine-tuning at individual synapses, the subcellular structures responsible for neuron-to-neuron communication in the brain.

15.5 Concluding Remarks

Regarding the study of cell fate transition in the developing mammalian brain, classical studies in developmental neurobiology have concentrated on the identification of the transcription factors involved. More recently, thanks to the availability of more sophisticated techniques of manipulation and analysis, we have witnessed an increasing interest in the dissection of the cell biological mechanisms responsible for cell fate transition in brain development and evolution. We have here discussed the main findings linking the centrosome and the Golgi apparatus to neural stem and progenitor cells and brain development.

References

Axelsson MA, Warren G (2004) Rapid, endoplasmic reticulum-independent diffusion of the mitotic Golgi haze. Mol Biol Cell 15:1843–1852

Ayala I, Colanzi A (2017) Mitotic inheritance of the Golgi complex and its role in cell division. Biol Cell 109:364–374

Ayala R, Shu T, Tsai LH (2007) Trekking across the brain: the journey of neuronal migration. Cell 128:29–43

Bacallao R, Antony C, Dotti C, Karsenti E, Stelzer EHK, Simons K (1989) The subcellular organization of Madin-Darby canine kidney cells during the formation of a polarized epithelium. J Cell Biol 109:2817–2832

Bellion A, Baudoin JP, Alvarez C, Bornens M, Metin C (2005) Nucleokinesis in tangentially migrating neurons comprises two alternating phases: forward migration of the Golgi/centrosome associated with centrosome splitting and myosin contraction at the rear. J Neurosci 25:5691–5699

Bond J, Woods CG (2005) Cytoskeletal genes regulating brain size. Curr Opin Cell Biol 18:95–101

Bond J, Roberts E, Mochida GH, Hampshire DJ, Scott S et al (2002) ASPM is a major determinant of cerebral cortical size. Nat Genet 32:316–320

Bond J, Roberts E, Springell K, Lizarraga S, Scott S et al (2005) A centrosomal mechanism involving CDK5RAP2 and CENPJ controls brain size. Nat Genet 37:353–355

Borrell V, Reillo I (2012) Emerging roles of neural stem cells in cerebral cortex development and evolution. Dev Neurobiol 72:955–971

Bradke F, Dotti CG (1998) Membrane traffic in polarized neurons. Biochim Biophys Acta 1404:245–258

Bradke F, Dotti CG (2000) Establishment of neuronal polarity: lessons from cultured hippocampal neurons. Curr Opin Neurobiol 10:574–581

Cooper JA (2014) Molecules and mechanisms that regulate multipolar migration in the intermediate zone. Front Cell Neurosci 8:386

de Anda FC, Pollarolo G, Da Silva JS, Camoletto PG, Feiguin F, Dotti CG (2005) Centrosome localization determines neuronal polarity. Nature 436:704–708

Di Cunto F, Calautti E, Hsiao J, Ong L, Topley G et al (1998) Citron rho-interacting kinase, a novel tissue-specific ser/thr kinase encompassing the Rho-Rac-binding protein Citron. J Biol Chem 273:29706–29711

Di Cunto F, Imarisio S, Hirsch E, Broccoli V, Bulfone A et al (2000) Defective neurogenesis in citron kinase knockout mice by altered cytokinesis and massive apoptosis. Neuron 28:115–127

Evans AJ, Gurung S, Wilkinson KA, Stephens DJ, Henley JM (2017) Assembly, secretory pathway trafficking, and surface delivery of kainate receptors is regulated by neuronal activity. Cell Rep 19:2613–2626

Farkas LM, Haffner C, Giger T, Khaitovich P, Nowick K et al (2008) Insulinoma-associated 1 has a panneurogenic role and promotes the generation and expansion of basal progenitors in the developing mouse neocortex. Neuron 60:40–55

Fietz SA, Kelava I, Vogt J, Wilsch-Brauninger M, Stenzel D et al (2010) OSVZ progenitors of human and ferret neocortex are epithelial-like and expand by integrin signaling. Nat Neurosci 13:690–699

Freeze HH, Eklund EA, Ng BG, Patterson MC (2015) Neurological aspects of human glycosylation disorders. Annu Rev Neurosci 38:105–125

Gai M, Di Cunto F (2017) Citron kinase in spindle orientation and primary microcephaly. Cell Cycle 16:245–246

Gai M, Camera P, Dema A, Bianchi F, Berto G et al (2011) Citron kinase controls abscission through RhoA and anillin. Mol Biol Cell 22:3768–3778

Gärtner A, Fornasiero EF, Munck S, Vennekens K, Seuntjens E et al (2012) N-cadherin specifies first asymmetry in developing neurons. EMBO J 31:1893–1903

Götz M, Huttner WB (2005) The cell biology of neurogenesis. Nat Rev Mol Cell Biol 6:777–788

Grieve AG, Rabouille C (2011) Golgi bypass: skirting around the heart of classical secretion. Cold Spring Harb Perspect Biol 3:a005298

Guet D et al (2014) Mechanical role of actin dynamics in the rheology of the Golgi complex and in Golgi-associated trafficking events. Curr Biol 24(15):P1700–P1711. https://doi.org/10.1016/j.cub.2014.06.048

Hansen DV, Lui JH, Parker PR, Kriegstein AR (2010) Neurogenic radial glia in the outer subventricular zone of human neocortex. Nature 464:554–561

Hatten ME (1999) Central nervous system neuronal migration. Annu Rev Neurosci 22:511–539

Horton AC, Ehlers MD (2003) Dual modes of endoplasmic reticulum-to-Golgi transport in dendrites revealed by live-cell imaging. J Neurosci 23:6188–6199

Horton AC, Racz B, Monson EE, Lin AL, Weinberg RJ, Ehlers MD (2005) Polarized secretory trafficking directs cargo for asymmetric dendrite growth and morphogenesis. Neuron 48:757–771

Huang W, She L, Chang XY, Yang RR, Wang L et al (2014) Protein kinase LKB1 regulates polarized dendrite formation of adult hippocampal newborn neurons. Proc Natl Acad Sci USA 111:469–474

Hurtado L, Caballero C, Gavilan MP, Cardenas J, Bornens M, Rios RM (2011) Disconnecting the Golgi ribbon from the centrosome prevents directional cell migration and ciliogenesis. J Cell Biol 193:917–933

Huttner WB, Dotti CG (1991) Exocytotic and endocytotic membrane traffic in neurons. Curr Opin Neurobiol 1:388–392

Kosodo Y, Röper K, Haubensak W, Marzesco A-M, Corbeil D, Huttner WB (2004) Asymmetric distribution of the apical plasma membrane during neurogenic divisions of mammalian neuroepithelial cells. EMBO J 23:2314–2324

Levine TP, Misteli T, Rabouille C, Warren G (1995) Mitotic disassembly and reassembly of the Golgi apparatus. Cold Spring Harb Symp Quant Biol 60:549–557

Li H, Saucedo-Cuevas L, Regla-Nava JA, Chai G, Sheets N et al (2016a) Zika virus infects neural progenitors in the adult mouse brain and alters proliferation. Cell Stem Cell 19:593–598

Li H, Saucedo-Cuevas L, Shresta S, Gleeson JG (2016b) The neurobiology of Zika virus. Neuron 92:949–958

Li H, Saucedo-Cuevas L, Yuan L, Ross D, Johansen A et al (2019) Zika virus protease cleavage of host protein septin-2 mediates mitotic defects in neural progenitors. Neuron 101(6):1089–1098.e4

Marin O, Valiente M, Ge X, Tsai LH (2010) Guiding neuronal cell migrations. Cold Spring Harb Perspect Biol 2:a001834

Marthiens V, Rujano MA, Pennetier C, Tessier S, Paul-Gilloteaux P, Basto R (2013) Centrosome amplification causes microcephaly. Nat Cell Biol 15:731–40

Martinez-Cerdeno V, Noctor SC, Kriegstein AR (2006) The role of intermediate progenitor cells in the evolutionary expansion of the cerebral cortex. Cereb Cortex 16(Suppl 1):i152–i161

Martinez-Martinez MA, Ciceri G, Espinos A, Fernandez V, Marin O, Borrell V (2018) Extensive branching of radially-migrating neurons in the mammalian cerebral cortex. J Comp Neurol 527 (10):1558–1576

Matsuki T, Matthews RT, Cooper JA, van der Brug MP, Cookson MR et al (2010) Reelin and stk25 have opposing roles in neuronal polarization and dendritic Golgi deployment. Cell 143:826–836

Miao S, Chen R, Ye J, Tan GH, Li S et al (2013) The Angelman syndrome protein Ube3a is required for polarized dendrite morphogenesis in pyramidal neurons. J Neurosci 33:327–333

Naim V, Imarisio S, Di Cunto F, Gatti M, Bonaccorsi S (2004) Drosophila citron kinase is required for the final steps of cytokinesis. Mol Biol Cell 15:5053–5063

Noctor SC, Martinez-Cerdeno V, Ivic L, Kriegstein AR (2004) Cortical neurons arise in symmetric and asymmetric division zones and migrate through specific phases. Nat Neurosci 7:136–144

Noctor SC, Martinez-Cerdeno V, Kriegstein AR (2007a) Contribution of intermediate progenitor cells to cortical histogenesis. Arch Neurol 64:639–642

Noctor SC, Martinez-Cerdeno V, Kriegstein AR (2007b) Neural stem and progenitor cells in cortical development. Novartis Found Symp 288: 59–73; discussion 73–78, 96–98

Norden C, Young S, Link BA, Harris WA (2009) Actomyosin is the main driver of interkinetic nuclear migration in the retina. Cell 138:1195–1208

O'Dell RS, Ustine CJ, Cameron DA, Lawless SM, Williams RM et al (2012) Layer 6 cortical neurons require Reelin-Dab1 signaling for cellular orientation, Golgi deployment, and directed neurite growth into the marginal zone. Neural Dev 7:25

Ori-McKenney KM, Jan LY, Jan YN (2012) Golgi outposts shape dendrite morphology by functioning as sites of acentrosomal microtubule nucleation in neurons. Neuron 76:921–930

Paridaen JT, Huttner WB (2014) Neurogenesis during development of the vertebrate central nervous system. EMBO Rep 15:351–364

Paridaen JT, Wilsch-Brauninger M, Huttner WB (2013) Asymmetric inheritance of centrosome-associated primary cilium membrane directs ciliogenesis after cell division. Cell 155:333–344

Persico A, Cervigni RI, Barretta ML, Colanzi A (2009) Mitotic inheritance of the Golgi complex. FEBS Lett 583:3857–3862

Persico A, Cervigni RI, Barretta ML, Corda D, Colanzi A (2010) Golgi partitioning controls mitotic entry through Aurora-A kinase. Mol Biol Cell 21:3708–3721

Quassollo G, Wojnacki J, Salas DA, Gastaldi L, Marzolo MP et al (2015) A RhoA signaling pathway regulates dendritic golgi outpost formation. Curr Biol 25:971–982

Reillo I, Borrell V (2012) Germinal zones in the developing cerebral cortex of ferret: ontogeny, cell cycle kinetics, and diversity of progenitors. Cereb Cortex 22:2039–2054

Reillo I, de Juan Romero C, Garcia-Cabezas MA, Borrell V (2011) A role for intermediate radial glia in the tangential expansion of the mammalian cerebral cortex. Cereb Cortex 21:1674–1694

Sarkisian MR, Li W, Di Cunto F, D'Mello SR, LoTurco JJ (2002) Citron-kinase, a protein essential to cytokinesis in neuronal progenitors, is deleted in the flathead mutant rat. J Neurosci 22: RC217

Schotman H, Karhinen L, Rabouille C (2008) dGRASP-mediated noncanonical integrin secretion is required for Drosophila epithelial remodeling. Dev Cell 14:171–182

Smart IH, Dehay C, Giroud P, Berland M, Kennedy H (2002) Unique morphological features of the proliferative zones and postmitotic compartments of the neural epithelium giving rise to striate and extrastriate cortex in the monkey. Cereb Cortex 12:37–53

Solecki DJ, Trivedi N, Govek EE, Kerekes RA, Gleason SS, Hatten ME (2009) Myosin II motors and F-actin dynamics drive the coordinated movement of the centrosome and soma during CNS glial-guided neuronal migration. Neuron 63:63–80

Stenzel D, Wilsch-Bräuninger M, Wong FK, Heuer H, Huttner WB (2014) Integrin alphavbeta3 and thyroid hormones promote expansion of progenitors in embryonic neocortex. Development 141:795–806

Stiess M, Maghelli N, Kapitein LC, Gomis-Ruth S, Wilsch-Brauninger M et al (2010) Axon extension occurs independently of centrosomal microtubule nucleation. Science 327:704–707

Tanaka T, Serneo FF, Higgins C, Gambello MJ, Wynshaw-Boris A, Gleeson JG (2004) Lis1 and doublecortin function with dynein to mediate coupling of the nucleus to the centrosome in neuronal migration. J Cell Biol 165:709–721

Tavano S, Taverna E, Kalebic N, Haffner C, Namba T et al (2018) Insm1 induces neural progenitor delamination in developing neocortex via downregulation of the adherens junction belt-specific protein Plekha7. Neuron 97:1299–1314 e8

Taverna E, Huttner WB (2010) Neural progenitor nuclei IN motion. Neuron 67:906–914

Taverna E, Haffner C, Pepperkok R, Huttner WB (2012) A new approach to manipulate the fate of single neural stem cells in tissue. Nat Neurosci 15:329–337

Taverna E, Götz M, Huttner WB (2014) The cell biology of neurogenesis: toward an understanding of the development and evolution of the neocortex. Annu Rev Cell Dev Biol 30:465–502

Taverna E, Mora-Bermudez F, Strzyz PJ, Florio M, Icha J et al (2016) Non-canonical features of the Golgi apparatus in bipolar epithelial neural stem cells. Sci Rep 6:21206

Tsai LH, Gleeson JG (2005) Nucleokinesis in neuronal migration. Neuron 46:383–388

Valenzuela JI, Perez F (2015) Diversifying the secretory routes in neurons. Front Neurosci 9:358

Valiente M, Marin O (2010) Neuronal migration mechanisms in development and disease. Curr Opin Neurobiol 20:68–78

Wang X, Tsai JW, Imai JH, Lian WN, Vallee RB, Shi SH (2009) Asymmetric centrosome inheritance maintains neural progenitors in the neocortex. Nature 461:947–955

Wang X, Tsai JW, Lamonica B, Kriegstein AR (2011) A new subtype of progenitor cell in the mouse embryonic neocortex. Nat Neurosci 14:555–561

Wilsch-Bräuninger M, Peters J, Paridaen JTML, Huttner WB (2012) Basolateral rather than apical primary cilia on neuroepithelial cells committed to delamination. Development 139:95–105

Wilsch-Bräuninger M, Florio M, Huttner WB (2016) Neocortex expansion in development and evolution – from cell biology to single genes. Curr Opin Neurobiol 39:122–132

Xie Z, Hur SK, Zhao L, Abrams CS, Bankaitis VA (2018) A Golgi lipid signaling pathway controls apical Golgi distribution and cell polarity during neurogenesis. Dev Cell 44:725–740 e4

Ye B, Zhang Y, Song W, Younger SH, Jan LY, Jan YN (2007) Growing dendrites and axons differ in their reliance on the secretory pathway. Cell 130:717–729

Chapter 16
Communication of the Cell Periphery with the Golgi Apparatus: A Hypothesis

Werner Jaross

Abstract The Golgi apparatus plays a central role in the numerous traffic tasks in cells. Whereas the well-investigated chemical signaling is sufficient to explain the information processes in the secretory output of cells, it is insufficient to do that for the substitution of structural elements in the three-dimensional space of the cell. Here we review recent work (Jaross, Front Biosci 23:940–946, 2018) suggesting that molecular vibration patterns of those macromolecules which have to be exchanged are recognized by molecules in the Golgi via resonance of the electromagnetic fingerprints. That results in the activation of specific molecules and induction of the whole substitution process. For bridging intracellular distances, the IR radiation must be coherent. It is discussed that coherence is achieved by chemical reaction during the changing process of the molecule along with the quasicrystalline structure of the neighboring water molecules. Several aspects of the relevance of that signaling to the direct interactions of molecules during various intracellular processes are discussed.

16.1 Introduction

In the living cell, numerous traffic tasks have permanently to be performed.

From small to large objects, from simple molecules to complex structured macromolecules up to cell organelles have to be moved from the location where they were synthesized to the cell membrane for exocytosis or to the area of the cell where they will substitute the existing components. The actively directed transport and the Golgi apparatus are the principal regulators of this traffic (Alberts 2017).

The diversity of macromolecules, especially the proteins and the complex lipoides, are synthesized in the endoplasmic reticulum (ER) and transported to the Golgi. There the molecules are specifically modified (maturated), stored, and wrapped with signal proteins and motor proteins and prepared for the active direct transport in the form of particular vesicular containers.

W. Jaross (✉)
Institute of Clinical Chemistry and Laboratory Medicine, University Hospital Carl Gustav Carus, Technische Universitaet Dresden, Dresden, Germany

© Springer Nature Switzerland AG 2019

M. Kloc (ed.), *The Golgi Apparatus and Centriole*, Results and Problems in Cell Differentiation 67, https://doi.org/10.1007/978-3-030-23173-6_16

A great many cells are secretory cells. Proteins designed for secretion are packed in those vesicles that are transported to the plasma membrane, and after fusion with it, the cargo is delivered into the extracellular space. These processes are excellently investigated including the multiple signaling processes (Alberts 2017). A lot of signal molecules which take part in these secretory processes are found. However, this chemical signaling is insufficient to explain the maintenance of specific structure and composition of the three-dimensional cell system. Throughout the whole life, many structural components have to be substituted because they had been changed due to specific synthetic processes (e.g., prostanoid formation) or demolished, e.g., by oxidation, splitting, hydrolysis, or other chemical reactions. Specific signals from that area should be sent out and transferred to the Golgi to start the substitution process. That signal could only be of physical nature as such signals should report on spatial aspects besides the chemical characteristics. Presently, the importance of manifold electric and electromagnetic phenomena in cells and tissues are in discussion (Barghouth et al. 2015; Chernet and Levin 2013; De Ninno and Pregnolato 2017; Foletti et al. 2013; Funk 2015; Kobayashi et al. 1999; Lemeshko et al. 2013; Levin 2012; Nuccitelli 2003; Rouleau and Dotta 2014). It was hypothesized that electromagnetic radiation in the IR range created by the molecular vibration of the changed macromolecules is a candidate for such signaling process. These signals should be able to activate the particular enzymes in the Golgi responsible for maturation of the molecules needed and to induce all steps up to forming the transport units and to navigate them to the area where the molecules have to be substituted.

In the former articles (Jaross 2015, 2016, 2018), this vibrational hypothesis was described, and in this contribution those articles are reviewed.

16.2   The Golgi Apparatus

Despite an enormous variety of the specific structure of the Golgi in the different cell types, its basic structure is principally similar: The Golgi apparatus located near the nucleus is made of endomembrane system connected by the microtubules. The Golgi membranes located at the nucleus called the *cis*-Golgi are continuous with the endoplasmic reticulum (ER), while the *trans*-Golgi membranes are connected to the cytoskeletal filaments and microtubules (MT) for intracellular active transport. The basic elements of the Golgi are membrane-enclosed disks known as cisternae. Different numbers of cisternae form stacks. Similar to the ER, the Golgi cisternae consist of different polar lipids and proteins. Different proteins and macromolecules are synthesized in the ER and transported to the Golgi apparatus for further processing and transport to various cellular destinations (Nebenfuhr and Staehelin 2001; Gyoeva 2014). Various molecules are posttranslationally modified in the Golgi by phosphorylation, sulfatation, glycosylation, the addition of *N*-acetylglucosamine, or removal of mannose molecules. Glycosaminoglycans are synthesized in the Golgi. A cluster of enzymes are acting on the molecules while they translocate from the cis-face to the trans-face of the Golgi. The fully formed and modified proteins and other compounds are sorted, covered with a coat, and transported out of the Golgi in the vesicles. The vesicles contain signal sequences, which determine the final destination of the cargo. The coat proteins, such as COPII and COPI (Nebenfuhr and Staehelin 2001), have a crucial role in posttranslational modification of proteins and maturation of the vesicle cargo. The coat proteins also control the attachment, activity of the motor proteins, and directionality of transport (Balabanian et al. 2016; Hoeprich et al. 2015; Nebenfuhr and Staehelin 2001; Saito and Katada 2015; Zhao and Zhan 2012). The movement of the cargo involves multiple motor proteins, which often have opposite directionality. Because of this, the movement is a discontinuous "tug of war." Motor proteins move on the network of roads—the cytoskeleton. It consists of microtubules (MT), actin filaments, and intermediate filaments, which are highly dynamic and constantly polymerize and depolarize. The chemical energy for moving the motors along with the cargos is provided by energy-rich phosphates.

16.3   The Vibrational Hypothesis

It is known that above the temperature of the absolute zero, all molecules vibrate (Brown et al. 2014; Pelletier and Pelletier 2010). The intensity of these vibrations (oscillations) depends on the temperature. The energy for the oscillations in living systems derives from the metabolic enthalpy. The oscillations form various patterns and involve all the atoms of the molecule. For nonlinear molecules with the n number of atoms, the number of different vibrations is 3n-6. The frequency of oscillations is determined by the type and polarity of molecular bonds. The main oscillation pattern is determined by the functional groups in a macromolecule, while the carbon-hydrogen bonds exhibit only uncharacteristic vibration. The details of the molecular vibrations are thoroughly described in the articles and books of spectroscopic theory (Brown et al. 2014; Pelletier and Pelletier 2010). The molecular vibrations are the source of electromagnetic radiation, which contains uncharacteristic heat radiation and the characteristic peaks resulting from the vibration of the functional groups of the molecule. A specific macromolecule, in the specific conditions, has a characteristic frequency pattern. The proteins such as collagen or elastin, which have uncomplicated structure, have 4–6 strong bands and many weak bands in the Raman or in the IR spectrum (Brown et al. 2014; Steiner and Zimmerer 2013). The emitted energy is in the range of 1 meV–3 eV; the frequencies are between 4000 and 400 cm^{-1} (10^{13}–10^{15} Hz). Majority of the functional groups have bounds in the range of 2000–900 cm^{-1} (Brown et al. 2014; Movasaghi et al. 2008; Steiner and Zimmerer 2013). Because the water absorbs radiation in the range of 1300–1900 cm^{-1}, only the frequencies, which are not absorbed by the water, are able to move a certain distance. The polar lipids and proteins have vibrational peaks in the IR range of 1000–1200 cm^{-1}, and therefore they are able to move over nanoscale distances. The specific structure and charge of the cellular membrane prevent the uncontrolled penetration of the electromagnetic signals. Because the chemical structure and spatial distribution of the sources mirror their electromagnetic radiation patterns, they are very good candidates for the intracellular signaling. The newly synthesized small molecules and the modified macromolecules might produce the fingerprint vibrational electromagnetic signal (e.g., the synthesis of prostaglandins from the phospholipids of the cell membrane).

16.4   Resonant Recognition by Proteins

For signaling to be successful, the receiver has to be able to process the signals created by the changed structures of molecules such as enzymes, signaling, structural and transport proteins, and components of the genetic machinery. Cosic and co-workers (Cosic 1994; Cosic et al. 2006, 2015, 2016) have studied the protein-protein and protein-DNA interactions. They showed that, depending on the number of amino acids, the proteins resonate the frequencies in the range of 10^{13}–10^{15} Hz. Using the Fourier transformation, every amino acid can be assigned with a certain value, which correlates with the energy of delocalized electrons of this amino acid. The distance between the amino acids is about 3.8 Å. The velocity of the electric charge on the backbone of the protein chain is equal to 7.87×10^5 m/s. Using these data, Cosic and their group developed the Resonant Recognition Model (Cosic et al. 2015; Movasaghi et al. 2008), which can be used to identify and calculate the frequencies of proteins and other molecules.

The resonance frequencies of proteins agree with the molecular vibration frequencies (depending on the electromagnetic radiation) of macromolecules present in the complex lipids, proteins, glycoproteins, and various cellular structures. When the electromagnetic signal reaches a protein target and the specific frequency pattern of both molecules matches and is in opposite phase, there will be a transfer of photon energy and the activation of the recipient. This will either result in the conformational change of the molecule followed by the activation for a specific chemical reaction or a transfer of the signal to another molecule. In this hypothesis, which mimics the lock and key theory of chemical reactions, the vibration patterns are equivalent to the chemical structures with the exception that the frequency pattern of a molecule depends not only on its reaction center but also on the frequencies of various parts of the molecule. It has been shown experimentally that the certain isolated frequencies are able to induce the function of the molecule (Cosic et al. 2006). This indicates that there are certain dominant frequencies rich in energy. These authors showed that the proteins with the same or similar biological function share a common dominant frequency, which activates the specific target proteins. These findings indicate that the activation of the molecule does not require a complete match in the frequency pattern. They also suggest that the different biomolecule pairs may have different requirements for the degree of frequency pattern consensus for the successful energy transfer.

16.5 The Coherence of Emitted Photons: A Precondition for the Bridging of Greater Intracellular Distances

It has been shown that single structure-forming molecules emit extremely low electromagnetic energy (Havelka et al. 2014; May and Kühn 2011; Murugan et al. 2015). It has been also suggested that because of the low energy and high background noise, the frequencies generated by molecular vibrations can hardly be perceived by other structural elements (Kucera and Cifra 2013) if the signal has to travel a distance of several micrometers. In contrast, the short distances such as molecular dimensions should not prevent the distribution of the signal between the molecules. This is probably how the molecules of the DNA-repair-machinery or the signal transduction networks interact (Jaross 2018; Kucera and Cifra 2013). If a signal comes from the cell membrane and has to reach the Golgi apparatus, then it has to bridge a distance of several µm. There, the signal has to be received for the activation of at least one particular enzyme. It might be assumed that activation of an enzyme is due to a change of the conformation of that molecule. This change might be different in every molecule depending on the binding to neighboring molecules, on the isomeric characteristics of some components of the macromolecule, and on other factors. Some dipole-dipole (Van der Waals forces) and hydrogen bonds have to be solved and rebounded; also some stereoisomeric transformations of molecule groups have to be induced. The bond energy of a dipole-dipole bond is 6×10^{-19} J; the bond energy of hydrogen bond amounts to its tenfold. The energy differences of conformation isomers are very different in the great variety of eligible chemical structures. The conformation energy of the simple stereoisomers of cyclohexane, e.g., differs up to 3 kcal/mol or 5×10^{-19} J/mol. The energy of one photon in the IR range is about $0.1–4 \times 10^{-19}$ J. If we speculate on the energy needed, we could estimate that a certain number of IR photons are necessary to activate an enzyme. Such an amount can be delivered if the emitted photon stream is coherent. After the activation of the primary molecule, the alteration of the vibration pattern of that molecule induces a chain reaction with many steps until the transport unit is being formed.

What conditions are sufficient for the generation of a coherent photon stream? Funk (2018) has reviewed some conditions resulting in coherent photon emission of macromolecules in the cell. It seems to be the most important that the particular molecule is transferred in an excited state. This could happen in the course of chemical reactions of a molecule. During oxidation, hydrolysis, the substitution of molecule groups, or other reactions, the molecule is transformed into an excited state by activation energy and in the course of the reaction by its enthalpy. During this process, a stream of photons is emitted. As the neighboring environment of the reacting molecules consists of quasicrystalline-ordered water molecules, the coherence of the photon stream should be strongly promoted by them (De Ninno 2017; Funk 2018; Pollack and Clegg 2008; De Ninno et al. 2013) and should persist for certain time (Funk 2018). The importance of quasicrystalline-structured water layers has been shown in several recent contributions (Funk 2018; De Ninno 2017). These assumptions mean that the electromagnetic signal, which results in specific activation of certain molecules in the Golgi, is generated in the moment of chemical reaction of that molecule in the cell periphery, which has to be exchanged later on. In the course of the chemical reaction of this molecule, the vibration pattern is changing until the end of the chemical reaction; however, this dynamic process has not been well investigated. After finishing the chemical reaction, the final vibration pattern might be used for navigation of the traffic unit to the destination area. For this purpose, a signal receiver has to be available in the coat of the transport unit.

Whether the quasicrystalline water layers along the polar MTs and filaments play a role for conducting the coherent photon stream from the periphery to the Golgi is a further topic of discussion (Jaross 2018). The diameter of MTs is too small for the wavelengths of the IR fingerprint signals to be directly used as wave guides; however, they could be indirect wave guides because the multilayered crystalline water structure around the microtubule skeleton structures could reach from the cellular periphery to the Golgi without interruption (Fig. 16.1).

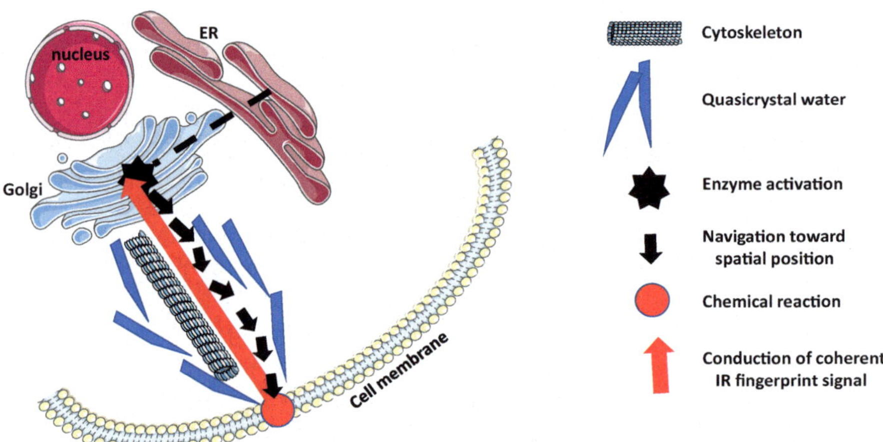

Fig. 16.1 Hypothesis on the interaction of cell periphery with Golgi apparatus via IR-molecular vibration pattern. The molecular vibration pattern of macromolecule changes during a chemical reaction. The coherent photon stream emitted is conducted to the Golgi apparatus, supported by the cytoskeleton microtubule system together with the quasicrystalline water structure. In the Golgi, specific enzymes are activated, which via a chain reaction leading to the formation of the transport unit with the required macromolecule and signal and motor proteins. That motor-driven vehicle is navigated to the location of the request by means of the IR fingerprint signal coming from that location

16.6   Discussion

The basic idea of the hypothesis presented here is that molecules can interact with each other based on electromagnetic radiation created by their molecular vibration. Under specific conditions, the photon stream can be coherent, intracellular distances can be bridged, and specific proteins can be changed by energy transfer, resulting in the activation of enzymes and the induction of a specific chain reaction via emission and recognition of the EM-frequency patterns.

There are many examples in daily life supporting the hypothesis of the interaction of molecules via EM-radiation.

Life is based on photosynthesis. The well-known key process is the transformation of electromagnetic energy for the purpose of synthesis of chemical compounds in plant cells. Or simply, chemical energy in the sun is transferred into electromagnetic energy and again into chemical energy on earth. All the necessary reactions in biochemical systems like activation of enzymes, the transformation of the reactants into excited states, the formation of intermediate products, and the final reaction steps are directly or indirectly promoted or implemented by EM-energy (also via the energy-rich phosphate molecules). Many chemical reactions can be triggered or performed by the direct radiation with specific EM-frequencies of the IR range with sufficient energy in technical applications. They gain importance in green chemistry (Córdova et al. 2011). Light as a specific kind of EM-radiation induces a chain of chemical reactions in specific cells in the eye—the result is the visual ability. Also, the effect of light on plant growth, cell divisions, and the active orientation of plant leaves or flowers (e.g., sunflowers) toward the light supports this mechanism.

The difference to our hypothesis is that the energy amount being transferred forms the central aspect and not the fingerprint frequency pattern as the signaling process in cells and in tissues. However, there are also other reports discussing this aspect:

It has been shown that cells in culture interact, even if they are separated in such a way that only the electromagnetic waves are exchanged (Kucera and Cifra 2013). It has been also shown that the electromagnetic frequencies (calculated using the Cosic's Resonant Recognition Model) influence the differentiation of the tumor and normal cells (Dotta 2016; Murugan et al. 2015). There are studies showing that the specific molecular vibrational patterns are able to differentiate between agonists and antagonists of adenosine receptor ligands, which are the important pharmacotherapy targets (Chee and Oh 2013; Chee et al. 2015). The authors showed features of molecular vibrations of the ligands and the receptor, which "are indispensable for ligand recognition" (Chee and Oh 2013). The findings that "The activation of proteins involves energies of the same order and nature as the electric radiation of light" (Cosic et al. 2006) also support such a hypothesis. Cosic found that certain proteins with the common function, such as the oncogenes, have the same resonance frequency (Cosic et al. 2006).

The biochemical mechanism of smell could also be regarded as the support of the hypothesis that the vibration spectrum of a molecule can lead to an activation of the sensor cells and their specific molecules (Lloyd 2011). Our nose could function as a vibrational spectrometer and not act by chemical binding of odorant molecules to olfactory sensor molecules according to the conventional mechanism of lock and key model. The ability of *Drosophila* fly to differentiate between a certain molecule and its deuterated form confirms the existence of such a mechanism. Both molecules have the same chemical structure and the same chemical affinities to the receptor; however, the deuterium-carbon bounds vibrate differently to hydrogen-carbon bounds, and the fly can only smell the natural molecule showing that the frequency pattern might be the base for realizing the right molecule (Lloyd 2011).

The presented here hypothesis is a generalization of the hypothesis on the precondition of interacting molecules: in analogy to the lock and key theory, the matching frequency pattern between two molecules determines their interaction and energy transfer. The vibration and resonant recognition mechanism might be very important for intracellular processes, such as genetic/nuclear functions, the signal transduction between the nucleus and the cell membrane, the interaction between the cells and the extracellular matrix, and the storage and retrieval of memory units in the brain (Jaross 2015). The elementary living processes, mitosis and apoptosis, run according to the fixed programs; however, all steps have to be strictly coordinated, and intensive feedback is permanently needed. The discussed physical information system might play a role in the multiple feedback signaling and the diverse intra-cellular transportation tasks related to these programs (Funk 2018; Gu 1992). A deeper understanding of these processes requires a comprehensive amount of quantum-physical view, which might be done in the future.

The following aspects should be additionally noted in the discussion: the source of the signal may be only one molecule or many identical molecules present in the same location. Such a situation occurs during the enzymatic catalytic reactions, which continue until terminated by the inhibitory mechanism, such as product inhibition. The turnover frequencies of the enzyme, if they match the vibration frequency of the changed molecule, could also be important for the coherence (Funk 2018; Jaross 2018). In addition, the electromagnetic waves of low frequency and high energy can be used as the carrier waves. Similar to the radio and TV broadcasting, the frequency or amplitude of the carrier waves may be modulated by the frequency of the signal emitted from the changed molecules (Haltiwanger 2010). Such a system has to have a transmitter that sends out the carrier and the signal waves and a receiver that is able to resonate at the carrier-specific frequency. It is possible that the oscillating polarized structures such as microtubules or cell mem-brane rafts (Cifra and Pospisil 2014; Havelka et al. 2014; Barzanjeh et al. 2017) may produce the carrier-specific frequencies able to travel the intracellular distances. For example, the frequency of microtubule oscillations ranges from THz to KHz (De Ninno and Pregnolateo 2017; Foletti et al. 2015).

16.7   Concluding Remarks

The many aspects of presented here hypothesis have to be experimentally proven. This chapter deals with the specific question of how the substitution of specific macromolecules in the three-dimensional structure of a cell is implemented. The basic idea is that the chemical reaction which changes a particular molecule creates a coherent photon stream with characteristic fingerprint-IR-frequency pattern that is able to activate specific enzymes in the Golgi. The excited state of a molecule during a chemical reaction and the quasicrystalline structure of water around the molecule might play a role in creating the coherence of the photon stream as well as its conduction to the Golgi. The generalization of the idea means that molecular vibration has the potency of the interaction of molecules with one another even if they are not in contact, and the intracellular distances have to be bridged. By means of sensible spectrometer, it should be possible to analyze interesting frequency pattern. The application of those in a coherent manner would open many new possibilities to influence a lot of biological as well as technical processes.

References

Alberts B (2017) Molecular biology of the cell. Garland Science, New York. https://doi.org/10.1201/9781315735368

Balabanian L, Berger CL, Hendricks AG (2016) The role of the microtubule cytoskeleton in regulating intracellular *transport*. Biophys J 110(3):466a. https://doi.org/10.1016/j.bpj.2015.11.2494

Barghouth PG, Thiruvalluvan M, Oviedo NJ (2015) Bioelectrical regulation of cell cycle and the planarian model system. Biochim Biophys Acta 1848(10 Pt B):2629–2637. https://doi.org/10.1016/j.bbamem.2015.02.024

Barzanjeh S, Salari V, Tuszynski JA, Cifra M, Simon C (2017) Monitoring microtubule mechanical vibrations via optomechanical coupling. bioRxiv. https://doi.org/10.1101/097725

Brown RW, Cheng YC, Haacke EM, Thompson MR, Venkatesan R (2014) Electromagnetic principles. In: Magnetic resonance imaging. Wiley, London. https://doi.org/10.1002/9781118633953.app1

Chee HK, Oh SJ (2013) Molecular vibration-activity relationship in the agonism of adenosine receptors. Genomics Inform 11(4):282–288. https://doi.org/10.5808/GI.2013.11.4.282

Chee HK, Yang JS, Joung JG, Zhang BT, Oh SJ (2015) Characteristic molecular vibrations of adenosine receptor ligands. FEBS Lett 589(4):548–552. https://doi.org/10.1016/j.febslet.2015.01.024

Chernet BT, Levin M (2013) Transmembrane voltage potential is an essential cellular parameter for the detection and control of tumor development in a Xenopus model. Dis Model Mech 6 (3):595–607. https://doi.org/10.1242/dmm.010835

Cifra M, Pospisil P (2014) Ultra-weak photon emission from biological samples: definition, mechanisms, properties, detection and applications. J Photochem Photobiol B 139:2–10. https://doi.org/10.1016/j.jphotobiol.2014.02.009

Córdova MO, Flores Ramírez CI, Bejarano BV, Arroyo Razo GA, Pérez Flores FJ, Tellez VC, Ruvalcaba RM (2011) Comparative study using different infrared zones of the solventless activation of organic reactions. Int J Mol Sci 12(12):8575–8580

Cosic I (1994) Macromolecular bioactivity: is it resonant interaction between macromolecules? – Theory and applications. IEEE Trans Biomed Eng 41(12):1101–1114. https://doi.org/10.1109/10.33585

Cosic I, Pirogova E, Vojisavljevic V, Fang Q (2006) Electromagnetic properties of biomolecules. FME Trans 34:71–801

Cosic I, Cosic D, Lazar K (2015) Is it possible to predict electromagnetic resonances in proteins, DNA and RNA? EPJ Nonlinear Biomed Phys 3(1):5. https://doi.org/10.1140/epjnbp/s40366-015-0020-6

Cosic I, Cosic D, Lazar K (2016) Environmental light and its relationship with electromagnetic resonances of biomolecular interactions, as predicted by the resonant recognition model. Int J Environ Res Public Health 13(7):647. https://doi.org/10.3390/ijerph130706

De Ninno AD (2017) Dynamics of formation of the exclusion zone near hydrophilic surfaces. Chem Phys Lett 667:322–326. https://doi.org/10.1016/j.cplett.2016.11.015

De Ninno AD, Pregnolato M (2017) Electromagnetic homeostasis and the role of low-amplitude electromagnetic fields on life organization. Electromagn Biol Med 36(2):115–122. https://doi.org/10.1080/15368378.2016.1194293

De Ninno AD, Castellano AC, Giudice ED (2013) The supramolecular structure of liquid water and quantum coherent processes in biology. J Phys Conf Ser 442:012031. https://doi.org/10.1088/1742-6596/442/1/012031

Dotta BT (2016) Ultra-weak photon emissions differentiate malignant cells from non-malignant cells in vitro. Arch Cancer Res 4(2):85. https://doi.org/10.21767/2254-6081.100085

Foletti A, Grimaldi S, Lisi A, Ledda M, Liboff AR (2013) Bioelectromagnetic medicine: the role of resonance signaling. Electromagn Biol Med 32(4):484–499. https://doi.org/10.3109/15368378.2012.743908

Foletti A, Ledda M, Grimaldi S, D'Emilia E, Giuliani L, Liboff A, Lisi A (2015) The trail from quantum electro dynamics to informative medicine. Electromagn Biol Med 34(2):147–150. https://doi.org/10.3109/15368378.2015.1036073

Funk RH (2015) Endogenous electric fields as guiding cue for cell migration. Front Physiol 6:143. https://doi.org/10.3389/fphys.2015.00143

Funk RH (2018) Biophysical mechanisms complementing ldquo classical rdquo cell biology. Front Biosci 23(3):921–939. https://doi.org/10.2741/4625

Gu Q (1992) Quantum theory of biophoton emission. Recent advances in biophoton research and its applications, pp 59–112. https://doi.org/10.1142/9789814439671_0003

Gyoeva FK (2014) The role of motor proteins in signal propagation. Biochemistry (Mosc) 79 (9):849–855. https://doi.org/10.1134/S0006297914090028

Haltiwanger SG (2010) The science of bioenergetic and bioelectric. Technologies formatted full-text article

Havelka D, Cifra M, Kucera O (2014) Multi-mode electro-mechanical vibrations of a microtubule: in silico demonstration of electric pulse moving along a microtubule. Appl Phys Lett 104 (24):243702. https://doi.org/10.1063/1.48841180

Hoeprich G, Hancock W, Berger C (2015) Kinesin-2's role in intracellular cargo transport: navigating the complex microtubule landscape. Biophys J 108(2):135a–136a. https://doi.org/10.1016/j.bpj.2014.11.750

Jaross W (2015) Hypothesis on a signaling system based on molecular vibrations of structure forming macromolecules in cells and tissues. Integr Mol Med 2(5). https://doi.org/10.15761/IMM.1000168

Jaross W (2016) Are molecular vibration patterns of cell structural elements used for intracellular signalling? Open Biochem J 10:12–16. https://doi.org/10.2174/1874091X01610010012

Jaross W (2018) Hypothesis on interactions of macromolecules based on molecular vibration patterns in cells and tissues. Front Biosci 23(3):940–946. https://doi.org/10.2741/4626

Kobayashi M, Takeda M, Sato T, Yamazaki Y, Kaneko K, Ito K, Kato H, Inaba H (1999) In vivo imaging of spontaneous ultraweak photon emission from a rat's brain correlated with cerebral energy metabolism and oxidative stress. Neurosci Res 34(2):103–113. https://doi.org/10.1016/S0168-0102(99)00040-1

Kucera O, Cifra M (2013) Cell-to-cell signaling through light: just a ghost of chance? Cell Commun Signal 11:87. https://doi.org/10.1186/1478-811X-11-87

Lemeshko M, Krems RV, Doyle JM, Kais S (2013) Manipulation of molecules with electromagnetic fields. Mol Phys 111(12–13):1648–1682. https://doi.org/10.1080/00268976.2013.813595

Levin M (2012) Molecular bioelectricity in developmental biology: new tools and recent discoveries: control of cell behavior and pattern formation by transmembrane potential gradients. BioEssays 34(3):205–217. https://doi.org/10.1002/bies.201100136

Lloyd S (2011) Quantum coherence in biological systems. J Phys Conf Ser 302:012037. https://doi.org/10.1088/1742-6596/302/1/012037

May V, Kühn O (2011) Electronic and vibrational molecular states. In: May V, Kühn O (eds) Charge and energy transfer dynamics in molecular systems. Wiley, New York. https://doi.org/10.1002/9783527633791

Movasaghi Z, Rehman S, ur Rehmann DI (2008) Fourier Transform Infrared (FTIR) spectroscopy of biological tissues. Appl Spectrosc Rev 43(2):134–179. https://doi.org/10.1080/0570492070182904

Murugan NJ, Karbowski LM, Persinger MA (2015) Cosic's resonance model for protein sequences and photon emission differentiates lethal and non-lethal ebola strains: implications for treatment. Open J Biophys 05(01):35–43. https://doi.org/10.4236/ojbiphy.2015.51003

Nebenfuhr A, Staehelin LA (2001) Mobile factories: Golgi dynamics in plant cells. Trends Plant Sci 6(4):160–167. https://doi.org/10.1016/S1360-1385(01)01891-X

Nuccitelli R (2003) Endogenous electric fields in embryos during development, regeneration and wound healing. Radiat Prot Dosim 106(4):375–383. https://doi.org/10.1093/oxfordjournals.rpd.a006375

Pelletier J, Pelletier CC (2010) Spectroscopic theory for chemical imaging. In: Raman, infrared, and near-infrared chemical imaging. Wiley, New York. https://doi.org/10.1002/9780470768150.ch1

Pollack GH, Clegg J (2008) Unexpected linkage between unstirred layers, exclusion zones, and water. Phase Trans Cell Biol:143–152. https://doi.org/10.1007/978-1-4020-8651-9_9

Rouleau N, Dotta BT (2014) Electromagnetic fields as structure-function zeitgebers in biological systems: environmental orchestrations of morphogenesis and consciousness. Front Integr Neurosci 8:84. https://doi.org/10.3389/fnint.2014.00084

Saito K, Katada T (2015) Mechanisms for exporting large-sized cargoes from the endoplasmic reticulum. Cell Mol Life Sci 72(19):3709–3720. https://doi.org/10.1007/s00018-015-1952-9

Steiner G, Zimmerer C (2013) IV. Infrared and Raman spectra: datasheet from Landolt-Börnstein – Group VIII advanced materials and technologies Vol 6A1: "Polymer solids and polymer melts – definitions and physical properties I" in Springer materials. In: Arndt KF, Lechner MD (eds) Springer, Berlin. https://doi.org/10.1007/978-3-642-32072-9_22

Zhao Y, Zhan Q (2012) Electric fields generated by synchronized oscillations of microtubules, centrosomes and chromosomes regulate the dynamics of mitosis and meiosis. Theor Biol Med Model 9:26. https://doi.org/10.1186/1742-4682-9-26

Part IV
Golgi- and Centriole-Related Diseases

Chapter 17
Breaking Bad: Uncoupling of Modularity in Centriole Biogenesis and the Generation of Excess Centrioles in Cancer

Harold A. Fisk, Jennifer L. Thomas, and Tan B. Nguyen

Abstract Centrosomes are tiny yet complex cytoplasmic structures that perform a variety of roles related to their ability to act as microtubule-organizing centers. Like the genome, centrosomes are single copy structures that undergo a precise semi-conservative replication once each cell cycle. Precise replication of the centrosome is essential for genome integrity, because the duplicated centrosomes will serve as the poles of a bipolar mitotic spindle, and any number of centrosomes other than two will lead to an aberrant spindle that mis-segregates chromosomes. Indeed, excess centrosomes are observed in a variety of human tumors where they generate abnormal spindles in situ that are thought to participate in tumorigenesis by driving genomic instability. At the heart of the centrosome is a pair of centrioles, and at the heart of centrosome duplication is the replication of this centriole pair. Centriole replication proceeds through a complex macromolecular assembly process. However, while centrosomes may contain as many as 500 proteins, only a handful of proteins have been shown to be essential for centriole replication. Our observations suggest that centriole replication is a modular, bottom-up process that we envision akin to building a house; the proper site of assembly is identified, a foundation is assembled at that site, and subsequent modules are added on top of the foundation. Here, we discuss the data underlying our view of modularity in the centriole assembly process, and suggest that non-essential centriole assembly factors take on greater importance in cancer cells due to their function in coordination between centriole modules, using the Monopolar spindles 1 protein kinase and its substrate Centrin 2 to illustrate our model.

17.1 Centriole Replication

Centrosomes consist of a pair of centrioles surrounded by a matrix containing microtubule nucleation machinery. This microtubule nucleation capacity is responsible for the best known function of centrosomes, namely their role in mitotic spindle

H. A. Fisk (✉) · J. L. Thomas · T. B. Nguyen
Department of Molecular Genetics, The Ohio State University, Columbus, OH, USA
e-mail: fisk.13@osu.edu

© Springer Nature Switzerland AG 2019

391

M. Kloc (ed.), *The Golgi Apparatus and Centriole*, Results and Problems in Cell Differentiation 67, https://doi.org/10.1007/978-3-030-23173-6_17

assembly (Hinchcliffe and Sluder 2001; Rieder et al. 2001). However, centrosomes also regulate cytokinesis (Piel et al. 2001; Pereira and Schiebel 2001) and the assembly of cilia and flagella (Hinchcliffe et al. 2001; Khodjakov and Rieder 2001). In quiescent cells, the older of the two centrioles, which is referred to as the mother centriole and is identified by specialized appendages found at its distal end (Bornens 2002), is converted to a basal body that can dock with the plasma membrane, and coordinate the targeting of incoming Golgi- and endosome-derived vesicles that contain the machinery responsible for building a primary cilium (Hoyer-Fender 2010; Dawe et al. 2007). However, in cells that re-enter the cell cycle, the ciliary axoneme is disassembled (Pugacheva et al. 2007) so that the centrosome can be utilized for centriole replication in preparation for mitotic spindle assembly.

The centriole replication pathway was elucidated through pioneering studies in worm and fly embryos that led to the identification of several of the major players: the Asterless protein identified in the fruit fly *Drosophila melanogaster* and the related Spindles Defective 2 (SPD-2) protein identified in the nematode *Caenorhabditis elegans*, as well as the *Caenorhabditis elegans* proteins Zygote Defective 1 (ZYG-1), and Spindle Assembly Abnormal proteins 4, 5, and 6 (SAS-4, SAS-5, and SAS-6) (Peel et al. 2007; Leidel and Gonczy 2005; Delattre et al. 2006; Pelletier et al. 2006). The human counterparts of SPD-2 and Asterless, ZYG-1, SAS-4, SAS-5, and SAS-6 are Centrosomal protein of 152 kDa (Cep152) and Centrosomal protein of 192 kDa (Cep192), Polo-like Kinase 4 (Plk4), Centrosomal P4.1-associated protein (CPAP), Stem Cell Leukemia/T-cell Acute Lymphocytic Leukemia Protein 1 Interrupting Locus (STIL), and Spindle Assembly Abnormal Protein 6 Homolog (which we will refer to herein simply as Sas6), respectively (see Fig. 17.1). Because this pathway has been the subject of several wonderful reviews (Avidor-Reiss and Fishman 2019; Banterle and Gonczy 2017; Breslow and Holland 2019; Schatten and Sun 2018), we will only provide a sketch of the pathway here. In human cells, Cep152 and Cep192 recruit Plk4 to centrioles (Kim et al. 2013) where it phosphorylates STIL to promote binding of STIL and Sas6 (Moyer et al. 2015; Ohta et al. 2014), and CPAP regulates centriole length (Tang et al. 2009). The initiating event in centriole biogenesis is the binding of Plk4 to STIL. This binding both activates Plk4 and locally suppresses the auto-stimulated degradation of Plk4 at one specific site on the surface of each centriole (Holland et al. 2010; Sillibourne et al. 2010). The resulting localized accumulation of Plk4 defines the site at which a new centriole will be assembled, and promotes STIL-Sas6 binding to initiate centriole assembly at that site (Moyer et al. 2015; Ohta et al. 2014).

Like DNA replication, centriole replication is a semi-conservative process (see Fig. 17.2), but the Watson-Crick template analogy does not apply. Rather, a new centriole is assembled around a precursor called the cartwheel (Figs. 17.1, 17.2), which consists of nine radially arranged Sas6 dimers and will enforce the ninefold symmetry of new centrioles (Hirono 2014). In woodworking terms, the cartwheel can be seen as a jig that guides the assembly of a ninefold symmetrical centriole around it. One exciting model for cartwheel assembly suggests that cartwheels may form inside the lumen of an existing centriole, using the centriole's ninefold

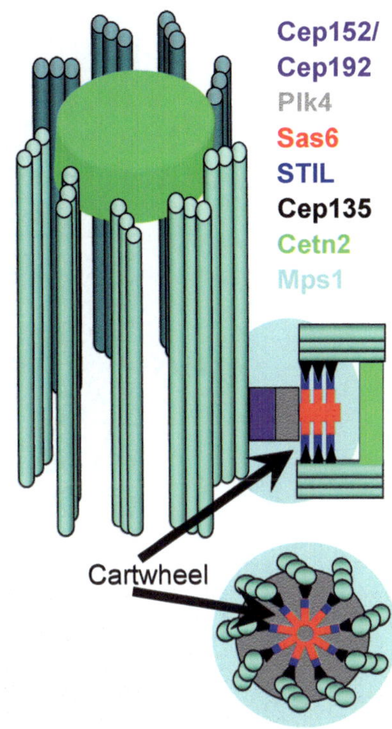

Fig. 17.1 The major players in centriole biogenesis. Depicted are the human orthologs of the major players in the canonical centriole biogenesis pathway

symmetry as a template for the ninefold symmetrical cartwheel (Fong et al. 2014). This model is consistent with the degradation of cartwheels that occurs during mitosis, which would leave the centriole lumen open for cartwheel assembly in the following cell cycle, and suggests that once formed the cartwheel must be released from the centriole lumen and transferred to a site on the surface of each existing centriole (predetermined by Cep152-Cep192/Plk4/STIL). Whether it is assembled in the centriole lumen or on the centriole surface, once a cartwheel is present on the surface of an existing centriole, it serves as the foundation onto which a new ninefold symmetrical centriole is assembled (Hirono 2014).

17.1.1 Centrosomes and Cancer

Due to this semi-conservative replication process (Fig. 17.2), cells typically have either one or two centrosomes, depending on their position in the cell cycle. Through mitosis and cytokinesis, cells inherit a single centrosome that is replicated around the G1/S transition, so that cells in G1 have one centrosome and cells in S, G2, or mitosis have two centrosomes. This precise duplication is critical, because as microtubule-organizing centers centrosomes organize the bipolar mitotic spindle apparatus responsible for segregation of duplicated chromosomes into two identical genomes

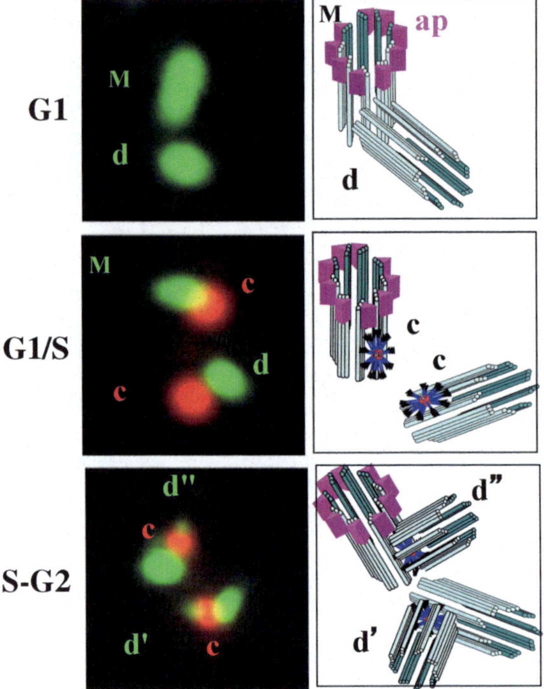

Fig. 17.2 Semi-conservative centriole replication is achieved by building a new centriole on top of a cartwheel template. Panels on the right are diagrams of centrioles before replication in G1, at the onset of replication at the G1/S transition, and after replication in S and G2. Shown are the mother (M) centriole with its appendages (ap), its daughter centriole (d) that was assembled in the previous cell cycle, the two cartwheels (c) that are assembled on the surface of the mother and daughter at the G1/S transition, and the two new daughter centrioles (d′ and d″) that will be assembled on top of each cartwheel. Panels on the left are indirect immunofluorescence of centrosomes from the indicated cell cycle stage showing the centriole protein Cetn2 in green and the cartwheel protein Sas6 in red, highlighting that centrosomes contain either zero Sas6 foci in G1 prior to the initiation of centriole biogenesis, or two cartwheels in S and G2 after initiation of centriole biogenesis

during mitosis. The presence of anything other than two centrosomes leads to the assembly of aberrant spindles that cannot properly segregate chromosomes and lead to the production of aneuploid daughter cells (Ganem et al. 2009). Because the majority of human tumors are aneuploid, this makes centrosomes of potential relevance to tumorigenesis, and indeed a variety of human tumors have excess centrosomes that can generate aberrant mitotic spindles in situ. For example, as many as 80% of invasive breast tumors have extra centrosomes (Lingle et al. 2002). Because these extra centrosomes appear prior to aneuploidy (Lingle et al. 2002), it is thought that they may drive the chromosomal instability that is critical in tumorigenesis (Tsikitis and Chung 2006; Ellsworth et al. 2004a, b; Lengauer et al. 1998). Indeed, excess centrosomes generate aberrant mitotic spindles within tumors in situ

(Lingle and Salisbury 1999), and their presence strongly correlates with aneuploidy in invasive breast tumors (Lingle et al. 1998, 2002).

17.2 Non-essential Centriole Factors

The centriole biogenesis factors identified in worms and flies are deeply conserved (Carvalho-Santos et al. 2010), and orthologs of SPD-2/Asterless, ZYG-1, SAS-4, SAS-5, and SAS-6 appear to be essential for centriole assembly in all eukaryotes that retain centrioles and basal bodies (see Avidor-Reiss and Fishman 2019; Banterle and Gonczy 2017; Breslow and Holland 2019; Schatten and Sun 2018). However, there are several widely conserved centriole proteins found in humans that are missing in worms. These include centrins, the Mps1 protein kinase, and the delta and epsilon tubulins (Fisk et al. 2002; Pike and Fisk 2011), as well as Centrosomal protein of 135 kDa (Cep135) that is associated with microcephaly and bridges Sas6 and CPAP (Lin et al. 2013). In addition, proteomic studies suggest that the number of proteins at human centrioles could be on the order of several hundreds (Andersen et al. 2003; Jakobsen et al. 2011). Clearly, not all of these human proteins have been shown to be essential for building centrioles. As examples of human centriole biogenesis factors that may be non-essential for building centrioles, we will discuss Mps1 and the centrins.

17.2.1 The Mps1 Protein Kinase

Mps1 is a dual-specificity protein kinase discovered based on its requirement in duplication of the budding yeast centrosome equivalent, called the spindle pole body (Winey et al. 1991; Schutz and Winey 1998). Mps1 was subsequently shown to regulate the spindle assembly checkpoint (SAC) (Weiss and Winey 1996; Hardwick et al. 1996), and as such vertebrate Mps1 proteins can be found at both centrosomes and kinetochores (Fisk and Winey 2001; Fisk et al. 2003). The SAC functions of Mps1 will not be discussed here. It appears that centrosomes in human cells are very sensitive to the dosage of Mps1. The total abundance of Mps1 is low in G1 and S phases, and only roughly 10% is found at the centrosomes (Fisk et al. 2003; Kasbek et al. 2007). Subtle increases in this small centrosomal pool of Mps1 cause centrosome amplification (Srinivas et al. 2015; Sawant et al. 2015; Liu et al. 2013; Kasbek et al. 2007, 2009, 2010), but because very little Mps1 is needed to support its function at centrosomes, it requires highly efficient depletion of Mps1 to observe effects on centrosome duplication (Fisk et al. 2003). Less efficient depletion of Mps1 causes SAC defects with no effect on centrosomes (Fisk et al. 2003; Stucke et al. 2002), which led one group to conclude that Mps1 had no centrosomal function (Stucke et al. 2002). While that study remains in the psyche of the centrosome field, several other groups have since validated a role for Mps1 in controlling centrosome number: wild-type Mps1 accelerates centrosome re-duplication in certain tumor-derived cells (Kanai et al. 2007); the protein phosphatase Cell Division Cycle 25B stabilizes a pool of Mps1 that leads to the production of excess centrin-containing foci (Boutros et al. 2013); phosphorylation of Centrin 2 (Cetn2) at the Mps1 phosphorylation site Threonine 118 (T118) modulates incorporation of Cetn2 into centrioles (Dantas et al. 2013); and the phosphatase and tumor suppressor Cyclin-dependent Kinase Inhibitor 3 (CDKN3) restricts centrosome amplification by promoting the degradation of Mps1 at centrosomes (Srinivas et al. 2015). Moreover, BioID experiments show that Mps1 lies in close proximity to Plk4 at centrosomes (Firat-Karalar et al. 2014), consistent with a role for Mps1 in the canonical centriole biogenesis pathway, and our recent data conclusively tie the centrosomal phenotypes associated with manipulation of Mps1 to the pool of Mps1 found at centrosomes (Marquardt et al. 2016). In contrast, an analog-sensitive version of Mps1 (Maciejowski et al. 2010) and several recently developed Mps1 inhibitors (Hewitt et al. 2010; Kwiatkowski et al. 2010; Santaguida et al. 2010; Tardif et al. 2011; Tannous et al. 2013; D'Alise et al. 2008; Wengner et al. 2016) have no reported effects on centrosome duplication, suggesting the possibility that centrioles can be replicated in the absence of Mps1 activity. However, we have developed an Mps1 inhibitor that does perturb centrosome duplication (Sugimoto et al. 2017), and inhibitor studies should be interpreted with caution due to possible off-target effects. As an example, one commonly used Mps1 inhibitor, reversine, was originally developed as an Aurora kinase inhibitor (D'Alise et al. 2008; Santaguida et al. 2010).

17.2.2 The Centrin Family

First identified in green algae, centrins are members of a superfamily of small calcium-binding proteins defined by the presence of a helix-loop-helix calcium-binding motif called an EF-hand. Humans express three centrins, Centrin 1 (Cetn1) that is expressed in male germ cells, neurons, and other ciliated cells and Centrin 2 (Cetn2) and Centrin 3 (Cetn3) that are more ubiquitously expressed (Gavet et al. 2003; Hart et al. 1999; Middendorp et al. 1997; Wolfrum and Salisbury 1998). Centrins fall into two main subfamilies, one represented by budding yeast Cell Division Cycle 31 protein (Cdc31p) that contains human Cetn3 and the other represented by *Chlamydomonas* centrin that contains human Cetn1 and Cetn2 (Hodges et al. 2010; Vonderfecht et al. 2012). Centrins can be found associated with centriolar structures in organisms as diverse as fungi (Baum et al. 1986), ciliates including *Tetrahymena thermophila* (Vonderfecht et al. 2012) and *Paramecium tetraurelia* (Jerka-Dziadosz et al. 2013), and vertebrates (Errabolu et al. 1994; Middendorp et al. 2000; Paoletti et al. 1996), but are also found in the cytoplasm and the nucleus, and have a variety of functions including roles in DNA repair (Dantas et al. 2012; Klein and Nigg 2009; Thompson et al. 2006). While the bulk of centrin in human somatic cells is cytoplasmic, Cetn2 is found in the distal lumen of the centriole (Errabolu et al. 1994; Paoletti et al. 1996) and Cetn3 is found both at centrioles (Middendorp et al. 2000) and in the pericentriolar matrix (Baron et al. 1992). The two centrin families appear to play non-redundant roles at centrioles. In *Tetrahymena*, the Cetn2 ortholog fails to complement a null allele of the Cetn3 ortholog (Vonderfecht et al. 2012), and overexpression of human Cetn2 can generate excess centrioles (Yang et al. 2010), while Cetn3 antagonizes centriole over production (Sawant et al. 2015). However, while early RNA interference approaches suggested that Cetn2 was essential for centriole assembly (Salisbury et al. 2002), several subsequent studies have shown that Cetn2 is dispensable for cartwheel assembly (Kleylein-Sohn et al. 2007; Yang et al. 2010).

Interestingly, both Cetn2 and Cetn3 bind to the Mps1 protein kinase, and both are Mps1 substrates in vitro (Yang et al. 2010). Three sites of in vitro phosphorylation were identified in Cetn2: Threonine 45 (T45) and Threonine 47 (T47) lying in the first EF hand and Threonine 118 (T118) lying in the third EF hand (Yang et al. 2010). Consistent with non-overlapping functions for Cetn2 and Cetn3, phosphorylation of centrins by Mps1 is highly context dependent; Mps1 cannot phosphorylate the third EF hand of Cetn3 even if it is mutated to a sequence identical to the third EF hand in Cetn2 (Yang et al. 2010). Overexpression of Cetn2 can drive the production of excess centrioles in a cell type-specific fashion, and mutation of any of three Mps1 phosphorylation sites in Cetn2 modulates the centriole overproduction phenotype (Yang et al. 2010). However, while Mps1 phosphorylation at T118 is required for incorporation of Cetn2 into centrioles (Dantas et al. 2013; Yang et al. 2010), we found that depletion of Cetn2 caused at most a delay in centriole biogenesis (Yang et al. 2010), supporting the suggestion that Cetn2 is not required for centriole assembly.

17.3 Mps1: The David Banner of Canonical Centriole Biogenesis, or The Hulk of Centrosome Amplification?

As mentioned above, centrosome duplication is exquisitely sensitive to the dosage of centrosomal Mps1. The dosage of centrosomal Mps1 is controlled by the opposing activities of oncogenes and tumor suppressors on a centrosome-specific Mps1 Degradation Signal (MDS) (Kasbek et al. 2007, 2009, 2010). The MDS is a binding site for Ornithine Decarboxylase Antizyme (OAZ) (Kasbek et al. 2010), a tumor suppressor that targets the centrosomal pool of Mps1 to the proteasome for degradation. Cyclin A is an oncogene that is frequently overexpressed in breast cancers (Coletta et al. 2004), and the Cyclin-dependent Kinase 2 (Cdk2)–Cyclin A complex phosphorylates Mps1 at Threonine 468 (T468) within the MDS (Kasbek et al. 2007). T468 phosphorylation blocks binding of OAZ to the MDS to allow accumulation of a centrosomal pool of Mps1 (Kasbek et al. 2010). The tumor suppressor CDKN3 reverses this phosphorylation (Srinivas et al. 2015) to restore OAZ binding and degradation of Mps1 specifically at centrosomes, thus restricting the level and duration of this pool. Mps1 also contains a D-box that controls its Anaphase

Promoting Complex/Cyclosome (APC/C)-dependent degradation. Signaling through the Mitogen-activated Protein Kinase (MAPK) pathway leads to phosphorylation of Mps1 at Serine 281 (S281), a site adjacent to the Mps1 D-box, and S281 phosphorylation prevents APC/C-dependent degradation of Mps1 in G1, leading to elevated centrosomal Mps1 (Liu et al. 2013).

Initial studies of mouse Mps1 found that overexpression of the wild-type Mps1 protein was sufficient to cause centrosome re-duplication in NIH 3T3 cells (Fisk and Winey 2001), but this same perturbation leads to a different result in human cells: simply overexpressing wild-type Mps1 in human cells is not sufficient to increase the centrosomal pool of Mps1 or cause centrosome re-duplication (Kasbek et al. 2007, 2009, 2010; Fisk et al. 2003), perhaps reflecting a tighter regulation of Mps1 degradation at centrosomes in human cells. However, while there are data suggesting centrioles can be built in the absence of the reclusive and mild mannered Mps1, there are consequences for making Mps1 angry: mimicking phosphorylation of Mps1 at T468 or removing the MDS generates excess centrioles at physiological expression levels in all human cells tested (Kasbek et al. 2009). Similarly, mimicking phosphorylation at S281 prevents the D-box dependent-degradation of Mps1, leading to elevated centrosomal Mps1 and also causes centrosome re-duplication (Liu et al. 2013). Accordingly, a modest increase in the levels of Mps1 at centrosomes is sufficient to cause the production of excess centrosomes in human cells.

The consequences of preventing Mps1 degradation are not limited to cultured cells manipulated to express mutant proteins, as defects in Mps1 degradation are also found in both tumor-derived cells and tumors (Kasbek et al. 2009; Liu et al. 2013). The U2OS osteosarcoma-derived cell line expresses Mps1$^{\Delta 12/13}$, a version of Mps1 that lacks exons 12 and 13 that include the MDS (Kasbek et al. 2009). U2OS cells retain both the expression of wild-type Mps1 and the ability to degrade endogenous and exogenous wild-type Mps1. However, Mps1$^{\Delta 12/13}$ cannot bind to OAZ (Kasbek et al. 2010), accumulates at centrosomes even in the absence of Cdk2 activity (Kasbek et al. 2007), and causes centrosome re-duplication when expressed at physiological levels in other human cell lines (Kasbek et al. 2009). Mps1 degradation is also defective in the 21T series of breast cell lines (Kasbek et al. 2009), a series of several cell lines derived from the normal and malignant breast tissue of the same patient (Band et al. 1990; Band and Sager 1991). In contrast to U2OS cells, exogenously expressed wild-type Mps1 cannot be appropriately degraded in the absence of Cdk2 activity in the tumor-derived cells of the 21T series (Kasbek et al. 2009), suggesting that these cells harbor a defect in one of the proteins that control Mps1 degradation (several of which are tumor suppressors, as discussed above). Moreover, Mps1^{T468A} (which cannot be phosphorylated by Cdk2 and is constitutively degraded at centrosomes in other cell lines (Kasbek et al. 2007)) accumulates at centrosomes and enhances centrosome re-duplication in the tumor-derived cells of the 21T series (Kasbek et al. 2009). Because Mps1 is degraded appropriately in a 21T series cell line isolated from normal breast tissue, the defect in Mps1 degradation in the tumor-derived cells in this series correlates not only with their ability to generate excess centrosomes, but also with tumorigenesis. While the observations discussed above were made in tumor-derived cells, and not directly in tumors, expression of the B-RAF^{V600E} oncogene removes Mps1 from its control by the MAPK signaling pathway in tumors, as evidenced by increased S281 phosphorylation in melanoma tumor samples (Liu et al. 2013).

17.3.1 Does Dispensable Mean Unimportant?

Centrosomal Mps1 substrates include the centrins Cetn2 and Cetn3 (Yang et al. 2010; Sawant et al. 2015), and Mps1 promotes the incorporation of Cetn2 into centrioles. Cetn3 inhibits Mps1 activity at centrosomes, preventing the recruitment of Cetn2 to centrioles (Sawant et al. 2015), and this effect is bypassed by mimicking phosphorylation of Cetn2 by Mps1 (Sawant et al. 2015). So far, so dispensable; a kinase that is potentially dispensable for centriole biogenesis regulates two centrins that are clearly dispensable for centriole biogenesis. However, while this Mps1– centrin network may prove dispensable for building a centriole, it is nonetheless of critical importance, because perturbing the network generates excess centrioles: preventing Mps1 degradation generates excess centrioles, as does depletion of Cetn3 (Sawant et al. 2015), overexpression of Cetn2, or expression of Cetn2^{T118D} (Yang et al. 2010), and in each case the centrin-dependent production of excess centrioles requires Mps1. Moreover, the impact of Mps1 on centriole overproduction is not limited to its influence on centrins. Mps1$^{\Delta 12/13}$ can generate excess Sas6-containing structures in Cetn2-depleted cells (Yang et al. 2010), suggesting that Mps1 has a centrin-independent function at centrosomes. Consistent with this suggestion, overexpression of the catalytically inactive Mps1 kinase dead (Mps1KD) leads to a novel defect in Sas6 assembly wherein cells possess a single Sas6 focus (Kasbek et al. 2010) (as opposed to the zero or two foci normally observed). To us, when perturbing one player (Mps1) makes consequential impacts on an essential factor (Sas6) and universally generates excess centrioles, and that player is misregulated in both tumor-derived cells and tumors, it seems unproductive to argue about whether that player is essential. Such a player must be doing something interesting, and it seems more important to understand what it does when all players are present and accounted for in order to understand how and why it generates excess centrosomes in tumor cells. Why might Mps1 be dispensable for making centrioles when its misregulation wreaks havoc with centrosomes? We do not know—but we believe the centrosomal function of Mps1 is worthy of study, whether that function is essential or not.

17.4   Cell Signaling, Centrosomes, Dispensable Factors, and Cancer

Regardless of the importance of Mps1 in the canonical centriole replication pathway, defects in the control of centrosomal Mps1 unequivocally generate excess centrioles, a phenotype that is associated with many tumors. The signaling pathways that promote proliferation ultimately activate Cdk2. Cdk2 activation has been linked to centrosome amplification (Godinho and Pellman 2014), and can result from loss of inhibitory Cdk2 phosphorylation (Zhao et al. 2012), expression of a non-degradable mutant of the Cdk activator Cdc25A (Shreeram et al. 2008), or loss of the Cyclin-dependent Kinase Inhibitor (CKI) p21Cip1 (Mantel et al. 1999). Cdk2 activation can also be achieved by misregulation of its partner cyclins. Overexpression of Cyclin E leads to premature initiation of centriole replication and centrosome amplification (Kawamura et al. 2004; Saavedra et al. 2003), at least in part through its ability to direct Cdk2-dependent phosphorylation of the centrosome duplication inhibitor Nucleophosmin/B23 (Tarapore et al. 2001, 2006). Cyclin A levels are often increased in breast tumors (Coletta et al. 2004) where excess centrosomes are prevalent (Lingle et al. 2002), and Cyclin A overexpression promotes Cdk2-dependent phosphorylation of Mps1 that prevents the degradation of Mps1 at centrosomes and causes centrosome re-duplication (Kasbek et al. 2007).

Because centrosomal Mps1 is controlled by Cdk2, Mps1 must also be responsive to the signaling pathways that are defective in cancers, and signaling events that promote passage through the restriction point should also lead to stabilization of the centrosomal pool of Mps1. The following discussion is not meant to be inclusive of the signaling events that can result in centrosome amplification, much less those that are defective in cancer. Rather, our aim is to illustrate that a centrosomal factor can profoundly impact centrosomes in cancer cells even if an essential role for that factor has not been established.

17.4.1   Mps1 and MAPK Signaling

The MAPK signaling pathway controls various processes, notably including proliferation, differentiation, growth, and apoptosis. Canonical MAPK signaling is initiated by mitogens, which bind to and activate transmembrane receptors that stimulate Rat Sarcoma Viral Oncogene Homolog (RAS), and activated RAS stimulates one of several RAS Effector (RAF) proteins. In the MAPK pathway, a RAF protein kinase activates a Mitogen-activated Protein Kinase Kinase (MAPK2, more commonly known as MEK); MEK in turn activates the Mitogen-activated Protein Kinase (MAPK1, more commonly known as ERK); and activated ERK can translocate to the nucleus where it can stimulate transcription of cell cycle regulators. The MAPK signaling cascade is subject to a high rate of mutation in many cancers, and is perhaps most heavily mutated in melanomas. Mutations can be found in nearly all components of the pathway, but RAS and B-RAF mutations are the most common, with RAS being mutated in 15–20% of melanomas, and B-RAF being mutated in 60–70% of melanomas (Davies et al. 2002). B-RAF mutations almost exclusively occur in the kinase domain and increase the activity of the protein. One such mutation, B-RAF^{V600E}, renders B-RAF constitutively active in the absence of RAS activity. Because B-RAF^{V600E} is activated independently of RAS, it also escapes feedback inhibition that normally occurs upon activation of ERK (Freeman et al. 2013; Davies et al. 2002). Of the human melanomas that contain B-RAF mutations, 90% harbor B-RAF^{V600E} (Davies et al. 2002).

Several B-RAF^{V600E}-responsive phosphorylation sites were identified within Mps1, two of which (S281 and T436) regulate the centrosomal pool of Mps1 (Liu et al. 2013). S281 lies next to the Mps1 D-box, and S281 phosphorylation prevents the APC/C-dependent degradation of Mps1 (Liu et al. 2013). T436 lies within exons 12 and 13 that encode the MDS. A mutant Mps1 where both S281 and T436 have been mutated to glutamic acid prevents the OAZ-dependent degradation of Mps1 at centrosomes, increasing centrosomal Mps1 levels, and causing centrosome re-duplication (Liu et al. 2013). Importantly, expression of B-RAF^{V600E} leads to increased phosphorylation of Mps1, increased centrosomal Mps1 levels, and Mps1-dependent centrosome re-duplication. Moreover, S281 phosphorylation is increased in B-RAF^{V600E}-expressing human melanomas where it correlates with both excess centrosomes and poor prognosis (Liu et al. 2013). Given that S281 and T436 were identified as sites of Mps1 phosphorylation that were lost upon treatment with a MEK inhibitor (Liu et al. 2013), an interesting open question is whether they are phosphorylated by B-RAF or ERK. Their loss in response to inhibition of MEK, the kinase immediately downstream of B-RAF, suggests that activated ERK may be responsible for phosphorylating Mps1.

17.4.2 MAPK Signaling and Cdk2

The canonical view of MAPK signaling is that ERK activation promotes expression of cell cycle genes, ultimately leading to Cdk2 activation. However, the MAPK pathway makes several additional inputs into Cdk2 activation. B-RAF^{V600E} also contributes to activation of Cdk2 by negatively regulating Cyclin-dependent Kinase Inhibitor 1A (CDKN1A) (Jalili et al. 2012) and Cyclin-dependent Kinase Inhibitor 1B (CDKN1B) (Bhatt et al. 2005, 2007). CDKN1A expression is decreased in many melanomas, and levels of the CDKN1A protein (also know as p21Cip1) are reduced in metastatic tumors compared to primary tumors (Maelandsmo et al. 1996; Gray-Schopfer et al. 2006). B-RAF^{V600E} also suppresses expression of CDKN1B (also known as p27Kip1), which can be rescued by B-RAF^{V600E} knockdown or MAPK inhibitors (Bhatt et al. 2005, 2007), and Cdk2 activity is decreased upon treatment of melanoma cells with a MEK inhibitor (Kortylewski et al. 2001). This suggests that the MAPK signaling pathway makes several inputs into Mps1 stability; it has a primary role in promoting phosphorylation of Mps1 at S281 and T436; it has a secondary role in stimulating transcription of genes that drive the cell through the restriction point, leading to activation of Cdk2 that phosphorylates Mps1 at T468; and it plays a tertiary role by removing activity of important CKIs to increase the duration and extent of Cdk2 activation. Thus, while we may not fully understand the role of Mps1 in the canonical centriole biogenesis pathway, Mps1 is a target of the MAPK signaling pathway, and the defects in this signaling and Mps1 degradation are likely to contribute to the production of excess centrosomes in tumor cells. This centrosome amplification is thus likely to involve the Mps1 substrate Cetn2, whose overexpression or mutation can also generate excess centrosomes. Accordingly, while the Mps1–centrin network may be dispensable for the canonical centriole biogenesis pathway, these data suggest proteins that seem unimportant in normal cells may be the very proteins that can break the regulation of centrosome duplication to generate the extra centrosomes in a variety of cancers.

17.5   Modularity in the Centriole Biogenesis Pathway

Despite data suggesting that Mps1 and centrin may not be required to assemble centrioles, perturbations of Mps1, Cetn2, or Cetn3 generate excess centrioles. The precise details of Mps1- and centrin-dependent centriole overproduction previously led us to hypothesize that centriole biogenesis is a modular process (Pike and Fisk 2011). Preventing the degradation of centrosomal Mps1 leads to excess Sas6-containing centrioles, even in the absence of Cetn2, while Mps1KD produces a novel phenotype wherein some cells have a single focus of Sas6 rather than the two Sas6 foci indicative of proper cartwheel assembly. Expression of Cetn2^{T118D} also generates excess Sas6-containing centrioles, and their appearance requires both Mps1 and Sas6. However, overexpression of wild-type Cetn2 can generate excess centrioles whose appearance is dependent on Mps1 but does not require Sas6. These observations suggest that Mps1 acts at multiple points in the assembly process, performing a centrin-independent function that promotes cartwheel assembly and a centrin-dependent function that promotes assembly of distal centriole elements even in the absence of cartwheels.

The centriolar phenotype of cells overexpressing wild-type Cetn2 is key to our suggestion of modularity in centriole biogenesis. Overexpressing wild-type Cetn2 results in the production of just one or two extra centrioles in roughly 50% of cells. However, while this effect is somewhat modest, it is remarkable that it is unaffected by depletion of the cartwheel protein Sas6 (Yang et al. 2010). Yet these centrioles are functional in that they can recruit gamma-Tubulin and serve as mitotic spindle poles, and contain many other centriole proteins—often including Sas6. Moreover, EM analysis showed that many of the excess centrioles are structurally aberrant, lacking the cylindrical organization and radial symmetry typical of centrioles, consistent with having been assembled at sites other than cartwheels (Yang et al. 2010). This observation forms the basis of our hypothesis that centriole biogenesis is a modular process (Fig. 17.3). In essence, we believe our results with Cetn2

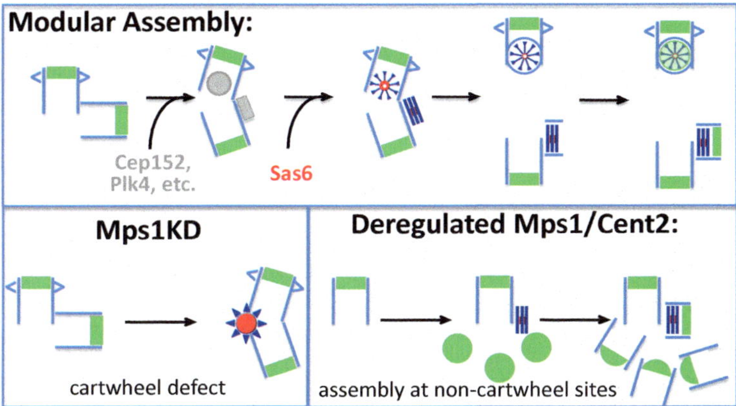

Fig. 17.3 Modularity in centriole biogenesis. The top panel illustrates hypothetical elements in the proposed modular centriole biogenesis pathway, including definition of the assembly site by Cep152-Cep192/Plk4/STIL (gray), assembly of cartwheels driven by Sas6 (red), formation of centriolar microtubules, and incorporation of Cetn2 (green). The catalytically inactive Mps1KD generates an interesting defect in cartwheel assembly where some cells have a single Sas6 focus rather than two that are normally present after initiation of centriole biogenesis. We propose that defective signaling in cancer cells over stimulates the Mps1–centrin network, which in turn uncouples the events of centriole assembly so that downstream centriole modules are assembled in the absence of cartwheels

overexpression support the notion that centrioles are normally assembled by addition of modules from the bottom up, but that perturbations at specific points in the pathway can disrupt the coordination between modules, allowing modules normally added late in the process to be assembled out of order and/or without assembly of previous modules.

Our vision of the canonical centriole biogenesis pathway is that centrioles are normally assembled from the bottom up in a modular fashion akin to assembly of a prefabricated building. The construction site is prepared first, a foundation is built on that site, walls are erected and affixed to the foundation, and a roof is affixed to the walls. In centriole biogenesis, Cep152/Plk4/STIL prepare the site of assembly, the cartwheel is the foundation onto which the rest of the centriole will be assembled, centriolar microtubules represent the walls, and components in the distal lumen such as Cetn2 represent the roof. This view is of course overly simplistic, and there are clearly additional modules between each of these steps. However, the analogy seems useful, particularly when considering dysfunction. Specifically, modularity suggests that modules can be assembled independently. Failure to properly coordinate assembly events for a prefabricated building might produce a building that maintains some aspects of functionality, yet is assembled in the wrong place and is defective in form. We suggest that overexpression of Cetn2 does just that by disrupting coordination between modules. Given that Cetn2 is normally found in the distal lumen of a centriole, we suggest that overexpression of Cetn2 could stabilize distal centriole modules in the absence of proximal modules in a top-down assembly process. The resulting centrioles retain some aspects of functionality but are defective in form because they were not assembled using a cartwheel as a guide. While we have used Mps1 and centrins to illustrate these ideas, it seems likely that there are many other perturbations that could disrupt coordination between modules in centriole biogenesis.

17.6 Non-essential Factors May Break the Canonical Process in Interesting Ways

The semi-conservative replication of centrioles is critical, because the presence of anything other than two centrosomes leads to the assembly of aberrant mitotic spindles. However, how the various centriole biogenesis mechanisms contribute to the production of excess centrosomes has not been established. Based on our observations, we suggest that the assembly of excess centrosomes in tumors may not be the result of simple over-execution or inappropriate re-execution of the canonical centriole biogenesis pathway. Rather, we suggest that a breakdown in the coordination of the events in a modular centriole assembly pathway might lead to the aberrant assembly of downstream modules at inappropriate sites. Such a mechanism could result in centrioles that are structurally aberrant and/or incomplete, but can nonetheless function in mitotic spindle organization—such as those seen in cells overexpressing Cetn2. Because cartwheels may form in the centriole lumen (Fong et al. 2014), these structurally aberrant centrioles may themselves further exacerbate the uncoupling if they lack the characteristic cylindrical structure and/or radial symmetry required to serve as templates for cartwheel assembly in subsequent cell cycles.

Finally, we suggest that factors for which an essential role in the canonical centriole biogenesis pathway is difficult to establish may take on a greater importance in cancer cells, specifically because they regulate coordination between centriole modules. This may be particularly true for factors like Mps1 that are directly responsive to upstream signaling events. Perhaps such factors are tied to signaling events precisely to instill order and dependency in this critical process. However, when it is the signaling events meant to enforce order and dependency that are broken, events that are meant to be staged may occur at the wrong time and at the wrong place, resulting in structures that are defective in some aspects of form, yet retain some aspects of function. For example, constitutive MAPK signaling and/or constitutive BRAF activity in cells expressing B-RAF^{V600E} leads to continuous suppression of Mps1 degradation by both the APC/C- and OAZ-dependent pathways. This could lead to a change in the timing of Mps1-dependent events relative to other events in centriole biogenesis, which could in turn promote the iniation of downstream centriole modules before earlier modules have been completed.

While we maintain that the impact of Mps1 on the essential cartwheel protein Sas6 strongly suggests that Mps1 has some role in canonical centriole biogenesis, perhaps other factors are dispensable for building a centriole because they contribute not to the structure of the new centriole but to the coordination of the events in its assembly. In some cases the uncoupling of centriole modules may be a neomorphic effect of a protein that normally has no role at centrosomes. However, it seems more likely that factors that can uncouple the process are those that are normally involved in its coordination. For example, Mps1 is found at centrosomes during canonical centriole biogenesis (Fisk et al. 2003; Kasbek et al. 2007; Majumder et al. 2012; Sawant et al. 2015; Marquardt et al. 2016), and non-degradable Mps1 mutants generate excess centrioles at physiological expression levels (Kasbek et al. 2009). This suggests that the effects observed in cancer cells reflect the function of Mps1 in a normal cell cycle, even if centrioles can be assembled in the absence of Mps1. Similarly, Cetn2 is targeted to the lumen of the distal end of the centriole (Errabolu et al. 1994; Paoletti et al. 1996) in a late step in the canonical centriole biogenesis pathway. Accordingly, it seems reasonable that the ability of Cetn2 to promote assembly of centrioles at sites other than cartwheels is a reflection of its normal structural contribution to centrioles, even if that contribution is not well understood and is not required to build a centriole. Moreover, the interactions between Mps1 and centrins suggest that the broken signaling events in cancer cells may have synergistic effects on the uncoupling of centriole modules. Defective MAPK signaling impacts Mps1 directly through S281 phosphorylation and indirectly by activating Cdk2 that promotes T468 phosphorylation. Thus, by stabilizing Mps1 signaling defects will also indirectly impact Cetn2, which itself is capable of disrupting coordination of centriole biogenesis. Accordingly, the effects of altered signaling may disrupt the coordination at multiple points that synergize to produce robust overproduction of centrioles that does not proceed by the normally staged bottom-up assembly process.

17.7   Conclusions

Centriole biogenesis is a tightly regulated process that is critical for the maintenance of genomic integrity. We suggest that the canonical centriole biogenesis pathway is a modular process akin to assembly of a building; the site of assembly is first identified and prepared, a foundation is assembled on that site, and sequential modules (floors, walls, ceilings, roof) are added in a specific order to yield a completed structure with a specific form and function. In centriole biogenesis, the corresponding events are definition of the site of assembly on the surface of an existing centriole by Cep152/ Plk4/STIL, assembly or tethering of a cartwheel at this site, construction and elongation of centriolar microtubules, and addition of distal centriole elements such as Cetn2 that reside in the centriole lumen. The data that led us to this notion of a modular assembly process come from observations on perturbations of Mps1 and centrins that disrupt the process in interesting ways. Here, we extend our thoughts on modular centriole biogenesis to suggest that the extra centrosomes observed in cancer are the result of the failure to properly couple assembly of late modules to completion of early modules. Walls can be attached to a roof hanging from a crane in a reversal of the normal building process. Without a foundation, such a building will surely be defective in its overall structure, even though it will likely keep its contents dry. Similarly, assembly and/or stabilization of later centriole modules independently of earlier modules could generate centrioles that are structurally aberrant yet retain their ability to function in microtubule nucleation and thus disrupt spindle assembly, as is seen in cells overexpressing Cetn2. In addition, we suggest that factors that appear to have non-essential roles in canonical centriole biogenesis may become important in cancer cells because they can act to disrupt the normal coordination between centriole modules. For example, Cetn2 is added to centrioles late in the canonical pathway, yet its overexpression can promote the assembly of structurally aberrant centrioles whose assembly does not require cartwheels. Finally, we suggest that factors like Mps1 that are responsive to cell signaling events take on greater importance in cancer cells where those signaling events are misregulated.

References

Andersen JS, Wilkinson CJ, Mayor T, Mortensen P, Nigg EA, Mann M (2003) Proteomic characterization of the human centrosome by protein correlation profiling. Nature 426 (6966):570–574

Avidor-Reiss T, Fishman EL (2019) It takes two (centrioles) to tango. Reproduction 157:R33–R51. https://doi.org/10.1530/REP-18-0350

Band V, Sager R (1991). Tumor progression in breast cancer) In: Rhim JS, Dritschilo A (eds) Neoplastic transformation in human cell systems in vitro. Humana Press, New Jersey, pp 169–178

Band V, Zajchowski D, Swisshelm K, Trask D, Kulesa V, Cohen C, Connolly J, Sager R (1990) Tumor progression in four mammary epithelial cell lines derived from the same patient. Cancer Res 50(22):7351–7357

Banterle N, Gonczy P (2017) Centriole biogenesis: from identifying the characters to understanding the plot. Annu Rev Cell Dev Biol 33:23–49. https://doi.org/10.1146/annurev-cellbio-100616-060454

Baron AT, Greenwood TM, Bazinet CW, Salisbury JL (1992) Centrin is a component of the pericentriolar lattice. Biol Cell 76(3):383–388

Baum P, Furlong C, Byers B (1986) Yeast gene required for spindle pole body duplication: homology of its product with Ca2+-binding proteins. Proc Natl Acad Sci 83:5512–5516

Bhatt KV, Spofford LS, Aram G, McMullen M, Pumiglia K, Aplin AE (2005) Adhesion control of cyclin D1 and p27Kip1 levels is deregulated in melanoma cells through BRAF-MEK-ERK signaling. Oncogene 24(21):3459–3471. https://doi.org/10.1038/sj.onc.1208544

Bhatt KV, Hu R, Spofford LS, Aplin AE (2007) Mutant B-RAF signaling and cyclin D1 regulate Cks1/S-phase kinase-associated protein 2-mediated degradation of p27Kip1 in human melanoma cells. Oncogene 26(7):1056–1066. https://doi.org/10.1038/sj.onc.1209861

Bornens M (2002) Centrosome composition and microtubule anchoring mechanisms. Curr Opin Cell Biol 14(1):25–34

Boutros R, Mondesert O, Lorenzo C, Astuti P, McArthur G, Chircop M, Ducommun B, Gabrielli B (2013) CDC25B overexpression stabilises centrin 2 and promotes the formation of excess centriolar foci. PLoS One 8(7):e67822

Breslow DK, Holland AJ (2019) Mechanism and regulation of centriole and cilium biogenesis. Annu Rev Biochem 88:691–724. https://doi.org/10.1146/annurev-biochem-013118-111153

Carvalho-Santos Z, Machado P, Branco P, Tavares-Cadete F, Rodrigues-Martins A, Pereira-Leal JB, Bettencourt-Dias M (2010) Stepwise evolution of the centriole-assembly pathway. J Cell Sci 123(Pt 9):1414–1426

Coletta RD, Christensen K, Reichenberger KJ, Lamb J, Micomonaco D, Huang L, Wolf DM, Muller-Tidow C, Golub TR, Kawakamia K, Ford HL (2004) The Six1 homeoprotein stimulates tumorigenesis by reactivation of cyclin A1. Proc Natl Acad Sci USA 101(17):6478–6483

D'Alise AM, Amabile G, Iovino M, Di Giorgio FP, Bartiromo M, Sessa F, Villa F, Musacchio A, Cortese R (2008) Reversine, a novel Aurora kinases inhibitor, inhibits colony formation of human acute myeloid leukemia cells. Mol Cancer Ther 7(5):1140–1149. https://doi.org/10.1158/1535-7163.MCT-07-2051

Dantas TJ, Wang Y, Lalor P, Dockery P, Morrison CG (2012) Defective nucleotide excision repair with normal centrosome structures and functions in the absence of all vertebrate centrins. J Cell Biol 193(2):307–318

Dantas TJ, Daly OM, Conroy PC, Tomas M, Wang Y, Lalor P, Dockery P, Ferrando-May E, Morrison CG (2013) Calcium-binding capacity of centrin2 is required for linear POC5 assembly but not for nucleotide excision repair. PLoS One 8(7):e68487

Davies H, Bignell GR, Cox C, Stephens P, Edkins S, Clegg S, Teague J, Woffendin H, Garnett MJ, Bottomley W, Davis N, Dicks E, Ewing R, Floyd Y, Gray K, Hall S, Hawes R, Hughes J, Kosmidou V, Menzies A, Mould C, Parker A, Stevens C, Watt S, Hooper S, Wilson R, Jayatilake H, Gusterson BA, Cooper C, Shipley J, Hargrave D, Pritchard-Jones K, Maitland N, Chenevix-Trench G, Riggins GJ, Bigner DD, Palmieri G, Cossu A, Flanagan A, Nicholson A, Ho JW, Leung SY, Yuen ST, Weber BL, Seigler HF, Darrow TL, Paterson H, Marais R, Marshall CJ, Wooster R, Stratton MR, Futreal PA (2002) Mutations of the BRAF gene in human cancer. Nature 417(6892):949–954. https://doi.org/10.1038/nature00766

Dawe HR, Farr H, Gull K (2007) Centriole/basal body morphogenesis and migration during ciliogenesis in animal cells. J Cell Sci 120(Pt 1):7–15

Delattre M, Canard C, Gonczy P (2006) Sequential protein recruitment in C. elegans centriole formation. Curr Biol 16(18):1844–1849

Ellsworth DL, Ellsworth RE, Liebman MN, Hooke JA, Shriver CD (2004a) Genomic instability in histologically normal breast tissues: implications for carcinogenesis. Lancet Oncol 5(12):753–758

Ellsworth DL, Ellsworth RE, Love B, Deyarmin B, Lubert SM, Mittal V, Shriver CD (2004b) Genomic patterns of allelic imbalance in disease free tissue adjacent to primary breast carcinomas. Breast Cancer Res Treat 88(2):131–139

Errabolu R, Sanders MA, Salisbury JL (1994) Cloning of a cDNA encoding human centrin, an EF-hand protein of centrosomes and mitotic spindle poles. J Cell Sci 107(Pt 1):9–16

Firat-Karalar EN, Rauniyar N, Yates JR, Stearns T (2014) Proximity interactions among centrosome components identify regulators of centriole duplication. Curr Biol 24(6):664

Fisk HA, Winey M (2001) The mouse mps1p-like kinase regulates centrosome duplication. Cell 106(1):95–104

Fisk HA, Mattison CP, Winey M (2002) Centrosomes and tumour suppressors. Curr Opin Cell Biol 14(6):700–705

Fisk HA, Mattison CP, Winey M (2003) Human Mps1 protein kinase is required for centrosome duplication and normal mitotic progression. Proc Natl Acad Sci USA 100(25):14875–14880

Fong CS, Kim M, Yang TT, Liao JC, Tsou MF (2014) SAS-6 assembly templated by the lumen of cartwheel-less centrioles precedes centriole duplication. Dev Cell 30(2):238–245. https://doi.org/10.1016/j.devcel.2014.05.008

Freeman AK, Ritt DA, Morrison DK (2013) Effects of Raf dimerization and its inhibition on normal and disease-associated Raf signaling. Mol Cell 49(4):751–758. https://doi.org/10.1016/j.molcel.2012.12.018

Ganem NJ, Godinho SA, Pellman D (2009) A mechanism linking extra centrosomes to chromosomal instability. Nature 460(7252):278–282

Gavet O, Alvarez C, Gaspar P, Bornens M (2003) Centrin4p, a novel mammalian centrin specifically expressed in ciliated cells. Mol Biol Cell 14(5):1818–1834. https://doi.org/10.1091/mbc. e02-11-0709

Godinho SA, Pellman D (2014) Causes and consequences of centrosome abnormalities in cancer. Philos Trans R Soc Lond Ser B Biol Sci 369(1650):20130467. https://doi.org/10.1098/rstb. 2013.0467

Gray-Schopfer VC, Cheong SC, Chong H, Chow J, Moss T, Abdel-Malek ZA, Marais R, Wynford-Thomas D, Bennett DC (2006) Cellular senescence in naevi and immortalisation in melanoma: a role for p16? Br J Cancer 95(4):496–505. https://doi.org/10.1038/sj.bjc.6603283

Hardwick K, Weiss E, Luca FC, Winey M, Murray A (1996) Activation of the budding yeast spindle assembly checkpoint without mitotic spindle disruption. Science 273:953–956

Hart PE, Glantz JN, Orth JD, Poynter GM, Salisbury JL (1999) Testis-specific murine centrin, Cetn1: genomic characterization and evidence for retroposition of a gene encoding a centrosome protein. Genomics 60(2):111–120. https://doi.org/10.1006/geno.1999.5880

Hewitt L, Tighe A, Santaguida S, White AM, Jones CD, Musacchio A, Green S, Taylor SS (2010) Sustained Mps1 activity is required in mitosis to recruit O-Mad2 to the Mad1-C-Mad2 core complex. J Cell Biol 190(1):25–34. https://doi.org/10.1083/jcb.201002133

Hinchcliffe EH, Sluder G (2001) "It Takes Two to Tango": understanding how centrosome duplication is regulated throughout the cell cycle. Genes Dev 15(10):1167–1181

Hinchcliffe EH, Miller FJ, Cham M, Khodjakov A, Sluder G (2001) Requirement of a centrosomal activity for cell cycle progression through G1 into S phase. Science 291(5508):1547–1550

Hirono M (2014) Cartwheel assembly. Philos Trans R Soc Lond Ser B Biol Sci 369 (1650):20130458. https://doi.org/10.1098/rstb.2013.0458

Hodges ME, Scheumann N, Wickstead B, Langdale JA, Gull K (2010) Reconstructing the evolutionary history of the centriole from protein components. J Cell Sci 123 (Pt 9):1407–1413. https://doi.org/10.1242/jcs.064873

Holland AJ, Lan W, Niessen S, Hoover H, Cleveland DW (2010) Polo-like kinase 4 kinase activity limits centrosome overduplication by autoregulating its own stability. J Cell Biol 188 (2):191–198

Hoyer-Fender S (2010) Centriole maturation and transformation to basal body. Semin Cell Dev Biol 21(2):142–147

Jakobsen L, Vanselow K, Skogs M, Toyoda Y, Lundberg E, Poser I, Falkenby LG, Bennetzen M, Westendorf J, Nigg EA, Uhlen M, Hyman AA, Andersen JS (2011) Novel asymmetrically localizing components of human centrosomes identified by complementary proteomics methods. EMBO J 30(8):1520–1535

Jalili A, Wagner C, Pashenkov M, Pathria G, Mertz KD, Widlund HR, Lupien M, Brunet JP, Golub TR, Stingl G, Fisher DE, Ramaswamy S, Wagner SN (2012) Dual suppression of the cyclin-dependent kinase inhibitors CDKN2C and CDKN1A in human melanoma. J Natl Cancer Inst 104(21):1673–1679. https://doi.org/10.1093/jnci/djs373

Jerka-Dziadosz M, Koll F, Wloga D, Gogendeau D, Garreau de Loubresse N, Ruiz F, Fabczak S, Beisson J (2013) A Centrin3-dependent, transient, appendage of the mother basal body guides the positioning of the daughter basal body in Paramecium. Protist 164(3):352–368. https://doi. org/10.1016/j.protis.2012.11.003

Kanai M, Ma Z, Izumi H, Kim SH, Mattison CP, Winey M, Fukasawa K (2007) Physical and functional interaction between mortalin and Mps1 kinase. Genes Cells 12(6):797–810

Kasbek C, Yang CH, Yusof AM, Chapman HM, Winey M, Fisk HA (2007) Preventing the degradation of mps1 at centrosomes is sufficient to cause centrosome reduplication in human cells. Mol Biol Cell 18(11):4457–4469

Kasbek C, Yang C-H, Fisk HA (2009) Mps1 as a link between centrosomes and genetic instability. Environ Mol Mutagen 50(8):654–665

Kasbek C, Yang CH, Fisk HA (2010) Antizyme restrains centrosome amplification by regulating the accumulation of Mps1 at centrosomes. Mol Biol Cell 21(22):3879–3889

Kawamura K, Izumi H, Ma Z, Ikeda R, Moriyama M, Tanaka T, Nojima T, Levin LS, Fujikawa-Yamamoto K, Suzuki K, Fukasawa K (2004) Induction of centrosome amplification and chromosome instability in human bladder cancer cells by p53 mutation and cyclin E overexpression. Cancer Res 64(14):4800–4809

Khodjakov A, Rieder CL (2001) Centrosomes enhance the fidelity of cytokinesis in vertebrates and are required for cell cycle progression. J Cell Biol 153(1):237–242

Kim T-S, Park J-E, Shukla A, Choi S, Murugan RN, Lee JH, Ahn M, Rhee K, Bang JK, Kim BY, Loncarek J, Erikson RL, Lee KS (2013) Hierarchical recruitment of Plk4 and regulation of centriole biogenesis by two centrosomal scaffolds, Cep192 and Cep152. Proc Natl Acad Sci USA 110(50):E4849

Klein UR, Nigg EA (2009) SUMO-dependent regulation of centrin-2. J Cell Sci 122 (Pt 18):3312–3321

Kleylein-Sohn J, Westendorf J, Le Clech M, Habedanck R, Stierhof YD, Nigg EA (2007) Plk4-induced centriole biogenesis in human cells. Dev Cell 13(2):190–202

Kortylewski M, Heinrich PC, Kauffmann ME, Bohm M, MacKiewicz A, Behrmann I (2001) Mitogen-activated protein kinases control p27/Kip1 expression and growth of human melanoma cells. Biochem J 357(Pt 1):297–303

Kwiatkowski N, Jelluma N, Filippakopoulos P, Soundararajan M, Manak MS, Kwon M, Choi HG, Sim T, Deveraux QL, Rottmann S, Pellman D, Shah JV, Kops GJPL, Knapp S, Gray NS (2010) Small-molecule kinase inhibitors provide insight into Mps1 cell cycle function. Nat Chem Biol 6(5):359–368. https://doi.org/10.1038/Nchembio.345

Leidel S, Gonczy P (2005) Centrosome duplication and nematodes: recent insights from an old relationship. Dev Cell 9(3):317–325

Lengauer C, Kinzler KW, Vogelstein B (1998) Genetic instabilities in human cancers. Nature 396:643–649

Lin YC, Chang CW, Hsu WB, Tang CJ, Lin YN, Chou EJ, Wu CT, Tang TK (2013) Human microcephaly protein CEP135 binds to hSAS-6 and CPAP, and is required for centriole assembly. EMBO J 32(8):1141–1154. https://doi.org/10.1038/emboj.2013.56

Lingle WL, Salisbury JL (1999) Altered centrosome structure is associated with abnormal mitoses in human breast tumors. Am J Pathol 155(6):1941–1951

Lingle WL, Lutz WH, Ingle JN, Maihle NJ, Salisbury JL (1998) Centrosome hypertrophy in human breast tumors: implications for genomic stability and cell polarity. Proc Natl Acad Sci USA 95 (6):2950–2955

Lingle WL, Barrett SL, Negron VC, D'Assoro AB, Boeneman K, Liu W, Whitehead CM, Reynolds C, Salisbury JL (2002) Centrosome amplification drives chromosomal instability in breast tumor development. Proc Natl Acad Sci USA 99(4):1978–1983

Liu J, Cheng X, Zhang Y, Li S, Cui H, Zhang L, Shi R, Zhao Z, He C, Wang C, Zhao H, Zhang C, Fisk HA, Guadagno TM, Cui Y (2013) Phosphorylation of Mps1 by BRAFV600E prevents Mps1 degradation and contributes to chromosome instability in melanoma. Oncogene 32 (6):713–723. https://doi.org/10.1038/onc.2012.94

Maciejowski J, George KA, Terret ME, Zhang C, Shokat KM, Jallepalli PV (2010) Mps1 directs the assembly of Cdc20 inhibitory complexes during interphase and mitosis to control M phase timing and spindle checkpoint signaling. J Cell Biol 190(1):89–100

Maelandsmo GM, Holm R, Fodstad O, Kerbel RS, Florenes VA (1996) Cyclin kinase inhibitor p21WAF1/CIP1 in malignant melanoma: reduced expression in metastatic lesions. Am J Pathol 149(6):1813–1822

Majumder S, Slabodnick M, Pike A, Marquardt J, Fisk HA (2012) VDAC3 regulates centriole assembly by targeting Mps1 to centrosomes. Cell Cycle 11(19):3666–3678. https://doi.org/10. 4161/cc.21927

Mantel C, Braun SE, Reid S, Henegariu O, Liu L, Hangoc G, Broxmeyer HE (1999) p21(cip-1/waf-1) deficiency causes deformed nuclear architecture, centriole overduplication, polyploidy, and relaxed microtubule damage checkpoints in human hematopoietic cells. Blood 93 (4):1390–1398

Marquardt JR, Perkins JL, Beuoy KJ, Fisk HA (2016) Modular elements of the TPR domain in the Mps1 N terminus differentially target Mps1 to the centrosome and kinetochore. Proc Natl Acad Sci USA 113(28):7828–7833. https://doi.org/10.1073/pnas.1607421113

Middendorp S, Paoletti A, Schiebel E, Bornens M (1997) Identification of a new mammalian centrin gene, more closely related to *Saccharomyces cerevisiae* CDC31 gene. Proc Natl Acad Sci USA 94(17):9141–9146

Middendorp S, Kuntziger T, Abraham Y, Holmes S, Bordes N, Paintrand M, Paoletti A, Bornens M (2000) A role for Centrin 3 in centrosome reproduction. J Cell Biol 148(3):405–415

Moyer TC, Clutario KM, Lambrus BG, Daggubati V, Holland AJ (2015) Binding of STIL to Plk4 activates kinase activity to promote centriole assembly. J Cell Biol 209(6):863–878. https://doi.org/10.1083/jcb.201502088

Ohta M, Ashikawa T, Nozaki Y, Kozuka-Hata H, Goto H, Inagaki M, Oyama M, Kitagawa D (2014) Direct interaction of Plk4 with STIL ensures formation of a single procentriole per parental centriole. Nat Commun 5:5267. https://doi.org/10.1038/ncomms6267

Paoletti A, Moudjou M, Paintrand M, Salisbury JL, Bornens M (1996) Most of centrin in animal cells is not centrosome-associated and centrosomal centrin is confined to the distal lumen of centrioles. J Cell Sci 109(Pt 13):3089–3102

Peel N, Stevens NR, Basto R, Raff JW (2007) Overexpressing centriole-replication proteins in vivo induces centriole overduplication and de novo formation. Curr Biol 17(10):834–843

Pelletier L, O'Toole E, Schwager A, Hyman AA, Muller-Reichert T (2006) Centriole assembly in *Caenorhabditis elegans*. Nature 444(7119):619–623

Pereira G, Schiebel E (2001) The role of the yeast spindle pole body and the mammalian centrosome in regulating late mitotic events. Curr Opin Cell Biol 13(6):762–769

Piel M, Nordberg J, Euteneuer U, Bornens M (2001) Centrosome-dependent exit of cytokinesis in animal cells. Science 291(5508):1550–1553

Pike AN, Fisk HA (2011) Centriole assembly and the role of Mps1: defensible or dispensable? Cell Div 6:9

Pugacheva EN, Jablonski SA, Hartman TR, Henske EP, Golemis EA (2007) HEF1-dependent Aurora A activation induces disassembly of the primary cilium. Cell 129(7):1351–1363

Rieder CL, Faruki S, Khodjakov A (2001) The centrosome in vertebrates: more than a microtubule-organizing center. Trends Cell Biol 11(10):413–419

Saavedra HI, Maiti B, Timmers C, Altura R, Tokuyama Y, Fukasawa K, Leone G (2003) Inactivation of E2F3 results in centrosome amplification. Cancer Cell 3(4):333–346

Salisbury J, Suino K, Busby R, Springett M (2002) Centrin-2 is required for centriole duplication in Mammalian cells. Curr Biol 12(15):1287

Santaguida S, Tighe A, D'Alise AM, Taylor SS, Musacchio A (2010) Dissecting the role of MPS1 in chromosome biorientation and the spindle checkpoint through the small molecule inhibitor reversine. J Cell Biol 190(1):73–87. https://doi.org/10.1083/jcb.201001036

Sawant DB, Majumder S, Perkins JL, Yang CH, Eyers PA, Fisk HA (2015) Centrin 3 is an inhibitor of centrosomal Mps1 and antagonizes centrin 2 function. Mol Biol Cell 26(21):3741–3753. https://doi.org/10.1091/mbc.E14-07-1248

Schatten H, Sun QY (2018) Functions and dysfunctions of the mammalian centrosome in health, disorders, disease, and aging. Histochem Cell Biol 150(4):303–325. https://doi.org/10.1007/s00418-018-1698-1

Schutz AR, Winey M (1998) New alleles of the yeast MPS1 gene reveal multiple requirements in spindle pole body duplication. Mol Biol Cell 9(4):759–774

Shreeram S, Hee WK, Bulavin DV (2008) Cdc25A serine 123 phosphorylation couples centrosome duplication with DNA replication and regulates tumorigenesis. Mol Cell Biol 28(24):7442–7450. https://doi.org/10.1128/MCB.00138-08

Sillibourne JE, Tack F, Vloemans N, Boeckx A, Thambirajah S, Bonnet P, Ramaekers FC, Bornens M, Grand-Perret T (2010) Autophosphorylation of polo-like kinase 4 and its role in centriole duplication. Mol Biol Cell 21(4):547–561

Srinivas V, Kitagawa M, Wong J, Liao PJ, Lee SH (2015) The tumor suppressor Cdkn3 is required for maintaining the proper number of centrosomes by regulating the centrosomal stability of Mps1. Cell Rep 13(8):1569–1577. https://doi.org/10.1016/j.celrep.2015.10.039

Stucke VM, Sillje HH, Arnaud L, Nigg EA (2002) Human Mps1 kinase is required for the spindle assembly checkpoint but not for centrosome duplication. EMBO J 21(7):1723–1732

Sugimoto Y, Sawant DB, Fisk HA, Mao L, Li C, Chettiar S, Li PK, Darby MV, Brueggemeier RW (2017) Novel pyrrolopyrimidines as Mps1/TTK kinase inhibitors for breast cancer. Bioorg Med Chem 25(7):2156–2166. https://doi.org/10.1016/j.bmc.2017.02.030

Tang CJ, Fu RH, Wu KS, Hsu WB, Tang TK (2009) CPAP is a cell-cycle regulated protein that controls centriole length. Nat Cell Biol 11(7):825–831

Tannous BA, Kerami M, Van der Stoop PM, Kwiatkowski N, Wang J, Zhou W, Kessler AF, Lewandrowski G, Hiddingh L, Sol N, Lagerweij T, Wedekind L, Niers JM, Barazas M, Nilsson RJ, Geerts D, De Witt Hamer PC, Hagemann C, Vandertop WP, Van Tellingen O, Noske DP, Gray NS, Wurdinger T (2013) Effects of the selective MPS1 inhibitor MPS1-IN-3 on glioblastoma sensitivity to antimitotic drugs. J Natl Cancer Inst 105(17):1322–1331. https://doi.org/10.1093/jnci/djt168

Tarapore P, Tokuyama Y, Horn HF, Fukasawa K (2001) Difference in the centrosome duplication regulatory activity among p53 'hot spot' mutants: potential role of Ser 315 phosphorylation-dependent centrosome binding of p53. Oncogene 20(47):6851–6863

Tarapore P, Shinmura K, Suzuki H, Tokuyama Y, Kim SH, Mayeda A, Fukasawa K (2006) Thr199 phosphorylation targets nucleophosmin to nuclear speckles and represses pre-mRNA processing. FEBS Lett 580(2):399–409

Tardif KD, Rogers A, Cassiano J, Roth BL, Cimbora DM, McKinnon R, Peterson A, Douce TB, Robinson R, Dorweiler I, Davis T, Hess MA, Ostanin K, Papac DI, Baichwal V, McAlexander I, Willardsen JA, Saunders M, Christophe H, Kumar DV, Wettstein DA, Carlson RO, Williams BL (2011) Characterization of the cellular and antitumor effects of MPI-0479605, a small-molecule inhibitor of the mitotic kinase Mps1. Mol Cancer Ther 10(12):2267–2275. https://doi.org/10.1158/1535-7163.Mct-11-0453

Thompson JR, Ryan ZC, Salisbury JL, Kumar R (2006) The structure of the human centrin 2-xeroderma pigmentosum group C protein complex. J Biol Chem 281(27):18746–18752

Tsikitis VL, Chung MA (2006) Biology of ductal carcinoma in situ classification based on biologic potential. Am J Clin Oncol 29(3):305–310

Vonderfecht T, Cookson MW, Giddings TH Jr, Clarissa C, Winey M (2012) The two human centrin homologues have similar but distinct functions at Tetrahymena basal bodies. Mol Biol Cell 23 (24):4766–4777. https://doi.org/10.1091/mbc.E12-06-0454

Weiss E, Winey M (1996) The *Saccharomyces cerevisiae* spindle pole body duplication gene *MPS1* is part of a mitotic checkpoint. J Cell Biol 132(1 & 2):111–123

Wengner AM, Siemeister G, Koppitz M, Schulze V, Kosemund D, Klar U, Stoeckigt D, Neuhaus R, Lienau P, Bader B, Prechtl S, Raschke M, Frisk AL, von Ahsen O, Michels M, Kreft B, von Nussbaum F, Brands M, Mumberg D, Ziegelbauer K (2016) Novel Mps1 kinase inhibitors with potent anti-tumor activity. Mol Cancer Ther 15(4):583–592. https://doi.org/10.1158/1535-7163.MCT-15-0500

Winey M, Goetsch L, Baum P, Byers B (1991) *MPS1* and *MPS2*: novel yeast genes defining distinct steps of spindle pole body duplication. J Cell Biol 114(4):745–754

Wolfrum U, Salisbury JL (1998) Expression of centrin isoforms in the mammalian retina. Exp Cell Res 242(1):10–17. https://doi.org/10.1006/excr.1998.4038

Yang CH, Kasbek C, Majumder S, Mohd Yusof A, Fisk HA (2010) Mps1 phosphorylation sites regulate the function of Centrin 2 in centriole assembly. Mol Biol Cell 21(24):4361–4372

Zhao H, Chen X, Gurian-West M, Roberts JM (2012) Loss of cyclin-dependent kinase 2 (CDK2) inhibitory phosphorylation in a CDK2AF knock-in mouse causes misregulation of DNA replication and centrosome duplication. Mol Cell Biol 32(8):1421–1432. https://doi.org/10.1128/MCB.06721-11

Chapter 18
Centrosome Amplification and Tumorigenesis: Cause or Effect?

Arunabha Bose and Sorab N. Dalal

Abstract Centrosome amplification is a feature of multiple tumour types and has been postulated to contribute to both tumour initiation and tumour progression. This chapter focuses on the mechanisms by which an increase in centrosome number might lead to an increase or decrease in tumour progression and the role of proteins that regulate centrosome number in driving tumorigenesis.

Abbreviations

AKAP450	A-kinase-anchoring protein 450
cdc25C	Cell division cycle 25C
CDK	Cyclin-dependent kinase
CDK5RAP2	CDK5 regulatory Subunit-associated protein 2
CENP-E	CENtrosome-associated Protein E
Cep	Centrosomal protein
CLIP-70	Cytoplasmic LInker Protein 170
C-NAP1	Centrosomal Nek2-associated Protein 1
CP110	Centriolar coiled-coil Protein of 110 kDa
CPAP	Centrosomal P4.1-associated Protein
EGFR	Epidermal growth factor receptor
FBF1	Fas-binding factor 1
GCP	γ-tubulin complex protein
hPOC5	human Proteome of Centriole 5
INCENP	Inner CENtromere Protein
LRRC45	Leucine-rich repeat containing 45
NEDD1	Neural precursor cell expressed developmentally down-regulated protein 1

A. Bose · S. N. Dalal (✉)
KS215, Advanced Centre for Treatment Research and Education in Cancer (ACTREC), Tata Memorial Centre, Navi Mumbai, Maharashtra, India

Homi Bhabha National Institute, Mumbai, Maharashtra, India
e-mail: sdalal@actrec.gov.in

© Springer Nature Switzerland AG 2019
M. Kloc (ed.), *The Golgi Apparatus and Centriole*, Results and Problems in Cell Differentiation 67, https://doi.org/10.1007/978-3-030-23173-6_18

ODF2	Outer dense fiber protein 2
SAS-6	Spindle assembly abnormal protein 6
SCLT1	Sodium channel and clathrin linker 1
SMC3	Structural maintenance of chromosomes protein 3
STIL	SCL interrupting locus protein
TACC2	Transforming acidic coiled-coil-containing protein 2
TRF1	Telomeric repeat-binding factor 1

18.1 Introduction

The cell cycle is a carefully orchestrated process that leads to the accurate segregation of chromosomes into two daughter cells (Malumbres and Barbacid 2007). Accurate genome segregation is accomplished by aligning the paired sister chromatids along the metaphase plate and attachment of the sister chromatids to the mitotic spindle (Reber and Hyman 2015; Walczak and Heald 2008). Following attachment to the spindle, the sister chromatids are 'pulled' apart by forces generated by spindle microtubules that originate from the two poles. Nucleation of the microtubules at the two poles, such that they are organized into a spindle, is mediated by a cellular organelle, called the centrosome (Fu et al. 2016).

The centrosome is a membrane-less organelle, which is the primary microtubule organizing centre (MTOC) in most eukaryotic cells. Its ability to organize microtubules results in it being essential for multiple cellular functions including the generation of the mitotic spindle, regulation of cell cycle progression, the biogenesis of and signalling from cilia, the determination of cell fate, cellular trafficking and the generation of an effective immune response (Arquint et al. 2014; Reina and Gonzalez 2014; Rios 2014; Stinchcombe and Griffiths 2014). Each centrosome consists of a pair of centrioles surrounded by an amorphous structure called the pericentriolar matrix (Luders 2012) (Fig. 18.1). The two centrioles are arranged orthogonally, and each centriole is organized in a typical 9+3 structure, which alludes to the nine sets of triplet microtubules arranged around a central cartwheel (Gonczy 2012). This cartwheel consists of a hub and radial spokes which are made of nine homodimers of Sas-6 (Kitagawa et al. 2011). The spokes emanating from the hub bind to the first microtubule of each triplet through an interaction between Cep135 and the first microtubule of the triplet (Guichard et al. 2017).

In human cells, a mature centriole is a cylinder, ~450 nm in length, with inner and outer diameters of ~130 nm and ~250 nm, respectively (Winey and O'Toole 2014). The centriole is said to have a polarized structure, with a proximal end (the base), and a distal end (the tip). The two centrioles in the centrosome are a mother centriole, which is inherited from the previous cell cycle, and a daughter centriole, whose synthesis is initiated in the current cell cycle during S phase (Fu et al. 2015). The mother centriole can be distinguished from the daughter centriole, by the presence of distal and sub-distal appendages (Fig. 18.1). The distal appendage proteins such as

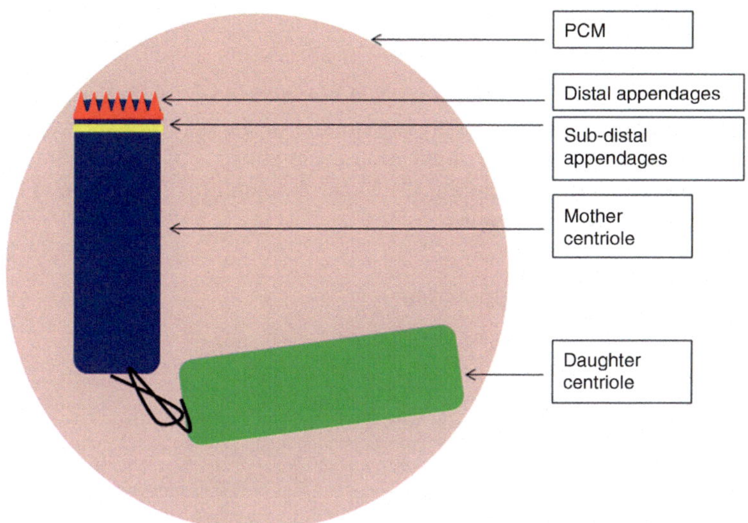

Fig. 18.1 The centrosome. A centrosome consists of a pair of centrioles (represented here in blue and green). The two centrioles differ in age, with one being the mother centriole (in blue) and the other, the daughter (in green). The mother centriole is identified by the presence of distal and sub-distal appendages (red). This centriolar pair is surrounded by an ordered matrix of proteins called the pericentriolar matrix (PCM) (in pink). The PCM contributes towards the nucleation of microtubules and spindle assembly

Cep164, Cep83 and SCLT1 are required to help dock the centrioles at the cell membrane during the formation of cilia (Tanos et al. 2013). Sub-distal appendages are made of proteins such as ninein, Cep170 and centriolin (Jana et al. 2014), and these are associated with nucleating and anchoring microtubules, which contribute to the organization of the mitotic spindle (Piel et al. 2000).

The microtubule organization function of the centrosome is dependent on the presence of the pericentriolar matrix (PCM). The PCM is a multi-protein amorphous structure and contains proteins such as γ-tubulin, Pericentrin, CDK5RAP2, CPAP, AKAP450, TACC2, Cep192 and Cep152 (Dictenberg et al. 1998; Gergely et al. 2000; Gomez-Ferreria et al. 2007; Keryer et al. 2003; Woodruff et al. 2014). On the basis of experiments in different model systems, the current consensus is that initially, a scaffold of proteins such as Cep192, Cep152, Pericentrin and CDK5RAP2 is formed around the paired centrioles (Hatch et al. 2010; Keryer et al. 2003). The activity of kinases such as polo-like kinase-1 (Plk1) and Aurora A stimulates the recruitment of γ-tubulin, TACC2 and other effector proteins to this ring (Gergely et al. 2000; Hannak et al. 2001; Kong et al. 2014). γ-tubulin forms a complex with GCP proteins to form γ-tubulin ring complexes (γ-TURCs), which are required for nucleating microtubules.

18.2   The Centrosome Cycle

The canonical centrosome duplication cycle is synchronized with the DNA replication cycle, thereby ensuring accurate segregation of the genetic material to the two daughter cells. The centrosome duplication cycle consists of four different phases: (1) Disengagement, (2) Duplication, (3) Maturation and (4) Separation (Fig. 18.2). The four steps are discussed below.

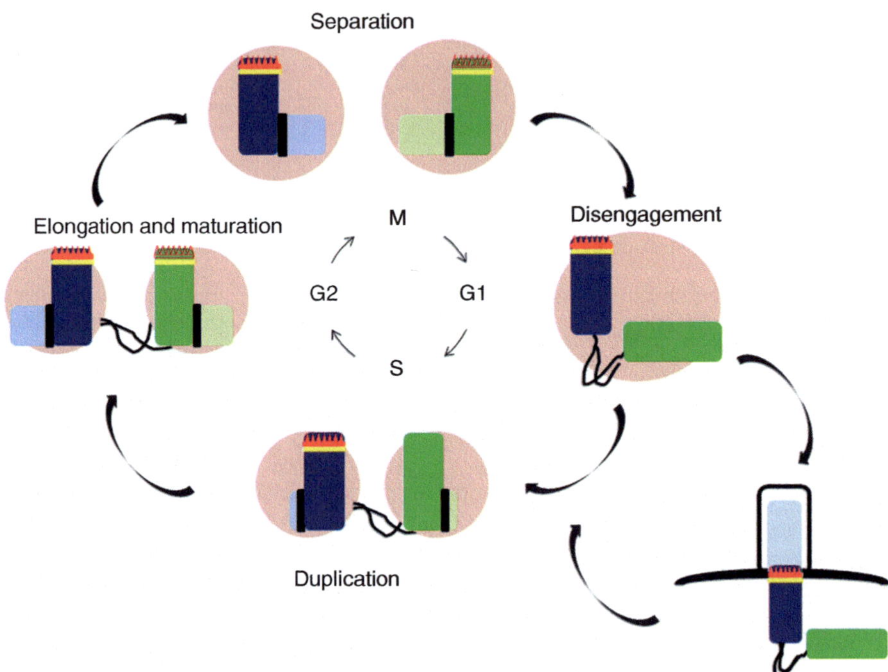

Fig. 18.2 The centrosome cycle. Each daughter cell inherits one centrosome post-mitosis from the mother cell, which consists of two orthogonally arranged centrioles, a mother and daughter centriole. In cycling cells during G1, a new centrosome duplication cycle is initiated by the disengagement of the two centrioles, during G1, which results in a loss of their orthogonal conformation and triggers pro-centriole biogenesis. The two centrioles are now held together by a proteinaceous linker, called the G1-G2 tether. During S phase, the newly formed pro-centrioles are attached to their respective mother centriole, at their proximal end, by the S-M linker. As the cell enters G2, the newly formed pro-centrioles elongate and the new mother centriole matures, so as to be able to nucleate microtubules. As the cell enters mitosis, the G1-G2 tether is degraded and the two centrosomes separate to organize each end of the mitotic spindle. If a cell withdraws from cycle and enters G0, the disengaged centrioles can also participate in the formation of cilia, where the mother centriole is attached to the plasma membrane via distal appendages

18.2.1 Disengagement

From the beginning of the S phase to the onset of mitosis, each mother centriole and its corresponding daughter centriole are orthogonally attached to each other via an S-M linker (Nigg and Stearns 2011). This orthogonal arrangement of the two centrioles is referred to as centriole engagement. The exact composition of the S-M linker has not yet been elucidated; however, studies from *Drosophila* spermatocytes suggest that Sas proteins might be a part of this linker (Stevens et al. 2010). As a cell exits mitosis, the orthogonal arrangement of centrioles in each mature centrosome in the daughter cells is disrupted by proteins such as Separase and Plk1, resulting in the degradation of the S-M linker (Tsou et al. 2009). Loss of the S-M linker and establishment of a G1-G2 tether between the two centrioles serves as a licensing event for the initiation of centrosome duplication. Inhibition of Plk1 activity using the small-molecule BI2536 inhibits centriole disengagement during late G2 or early mitosis (Tsou et al. 2009). Knockdown experiments in HeLa cells have demonstrated that cleavage of Pericentrin by Separase, during anaphase onset, leads to its dissociation from the centrosome, and thus, disengagement (Matsuo et al. 2012). Separase is active only during anaphase onset, and this ensures that centriole duplication occurs only once during the cell cycle (Tsou et al. 2009). Once the centrioles are 'disengaged', they are licensed for the initiation of centriole duplication.

An additional event required for disengagement is the phosphorylation of Nucleophosmin 1 (NPM1) by cyclin-dependent kinases. Phosphorylation of NPM1 at a threonine residue at position 199 by CDK2 or CDK1 results in dissociation of NPM1 from the centrosome, thus licensing the centrosome for duplication (Okuda et al. 2000; Peter et al. 1990). Expression of a phospho-deficient mutant of NPM1 (T199A) has been demonstrated to inhibit centrosome duplication (Tokuyama et al. 2001) while a phospho-mimetic mutant (T199D) promotes centrosome amplification (Mukhopadhyay et al. 2016). Further, NPM1 has been demonstrated to localize between the paired centrioles of unduplicated centrosomes (Shinmura et al. 2005). Plk1 phosphorylates NPM1 at Ser-4, and inhibition of this phosphorylation results in mitotic defects such as abnormal centrosome number and the presence of fragmented nuclei in these cells (Zhang et al. 2004b).

Loss of the orthogonal configuration is not the only licensing event for centriolar duplication (Engle et al. 2008; Gottardo et al. 2014; Shukla et al. 2015), and another factor that governs competency for duplication is the distance between the mother and the daughter centrioles. Plk1 can promote maturation and distancing of the orthogonally arranged daughter centriole, leading to reduplication of the mother centriole (Shukla et al. 2015). Further, another event that occurs during the disengagement process is termed 'centriole to centrosome conversion'. It is essential for the newly formed daughter centriole to become duplication competent and for it to able to function as an MTOC. In vertebrates, this begins with the initial loss of the central cartwheel, which is mediated by CDK1 activity (Arquint and Nigg 2014). An increase in Plk1 activity 'modifies' the daughter centriole inherited from the mother cell and stabilization of the cartwheel-less centriole occurs via its recruitment of Cep295 (Izquierdo et al. 2014). Cep152 and Cep192 are further acquired by the daughter centriole, and the centriole is now competent for duplication and can mature to form the new mother centriole (Hatch et al. 2010; Kim et al. 2013; Wang et al. 2011).

18.2.2 Duplication

Following disengagement, in each cell, there is exactly one mother centriole linked to a daughter centriole via a proteinaceous linker called the G1-G2 tether. The disengaged centrioles are now licensed for duplication (Wang et al. 2011). When cells enter S phase, exactly one pro-centriole must form adjacent to each of the pre-existing centrioles. This process is regulated by proteins such as Plk4 (polo-like kinase 4), Cep192, Cep152, STIL, SAS6 and CDK2. Initially, Cep152 and Cep192 act as scaffolding proteins that recruit Plk4 to the mother centriole (Kim et al. 2013). This results in the ring-like organization of Plk4 around the mother centriole as observed by super-resolution microscopy (Dzhindzhev et al. 2017; Ohta et al. 2018). Plk4 initiates centriole duplication by first marking the site of daughter centriole assembly which is dependent on the interaction between Plk4 and STIL (Ohta et al. 2018; Sonnen et al. 2012). STIL is recruited to the centriole by Cep85 (Liu et al. 2018) and upon recruitment to the mother centriole, STIL further activates and stabilizes Plk4 (Ohta et al. 2018). After recruitment of Plk4 and STIL marks the site of pro-centriole formation, SAS6 localizes to the same site and initiates pro-centriole assembly by assembling into higher order oligomers (Nakazawa et al. 2007). SAS6 oligomerization is the building block for the ninefold symmetrical centriole. Gorab, a trans-golgi protein, interacts with Sas6 and contributes to the establishment of the ninefold symmetry of the centriole and to centriole duplication in *Drosophila* (Kovacs et al. 2018).

Plk4 is initially observed as a ring around the mother centriole, and this ring coalesces into a single spot (Dzhindzhev et al. 2017; Ohta et al. 2018). Several attempts have been made to understand how this transition might occur and how it might regulate centrosome duplication. Currently, there are two proposed models for how the transition occurs. According to a biophysical model proposed by Leda et al. (2018), the activity of Plk4 is dynamic and transiently peaks at several points around the mother centriole. It has been demonstrated previously that complex formation between Plk4 and its substrate, STIL, stabilizes active Plk4 (Moyer et al. 2015). The Plk4–STIL complex could be retained at the centriole due to the binding of STIL to other components of the centriole. This model predicts that there could be multiple points within the ring surrounding the mother centriole, at which this Plk4–STIL complex is stabilized. An exchange of the Plk4–STIL complex occurs between the mother centriole and the cytoplasm, which determines the concentration of the complex at any point. Initially, there is a concentration-based competition between all the Plk4–STIL complexes. The cluster with the highest concentration of Plk4–

STIL is predicted to mark the site of new centriole formation (Leda et al. 2018). The second model suggests that first, Plk4 self-assembles into a ring around the mother centriole. Both inactive and active forms of Plk4 coexist at the centriole. Once it is recruited to the centriole, due to the ability of Plk4 to both laterally inhibit neighbouring molecules of Plk4 and undergo auto-activation, there is a localized increase in Plk4 activity at some points in the ring around the mother centriole. This localized activity is enhanced by complex formation of Plk4 with STIL and SAS6, thus leading to the formation of a single pro-centriole (Takao et al. 2018; Yamamoto and Kitagawa 2018). The formation of the new pro-centriole adjacent to each mother centriole results in the re-establishment of the S-M linker (Nigg and Stearns 2011).

18.2.3 Elongation and Maturation

During late S to G2, once Plk4 has marked the site of pro-centriole biogenesis, centriole elongation is stimulated by proteins such as CPAP, CP110 and SAS6, all of whom play important roles in this process (Kleylein-Sohn et al. 2007). SAS6, the ninefold symmetrical cartwheel protein, acts as a scaffold around which microtubules are assembled (Nakazawa et al. 2007). CPAP is a tubulin dimer-binding capping protein that, along with CP110, promotes microtubule assembly and controls centriolar length (Tang et al. 2009). It has been postulated that Cep135 links SAS6 with CPAP and the microtubule triplet (Lin et al. 2013). Centrin and hPOC5 localize to the distal lumen of centrioles and are essential for elongation at the distal end (Azimzadeh et al. 2009).

Elongation of the pro-centrioles is accompanied by expansion of the PCM. According to one model, Pericentrin recruits the Cep192-Plk1-Aurora A kinase complex to the centrosome. Plk1 has been demonstrated to trigger the accumulation of the scaffolding protein, Cep192 (Joukov et al. 2014). Plk1 phosphorylates Cep192, which creates attachment sites for γ-TURCs (Joukov et al. 2014). Plk1 activity also recruits Cep215, a PCM protein, to the two centrosomes (Haren et al. 2009). It is postulated that Aurora A kinase, in conjunction with Plk1, phosphorylates components of the γ-TURCs, thus making them competent for the recruitment of microtubules. Experiments in *C. elegans* and *D. melanogaster* have demonstrated that loss of Aurora A leads to defects in centrosome maturation, with a reduction in the accumulation of α-tubulin and γ-tubulin at the centrosome (Berdnik and Knoblich 2002; Hannak et al. 2001). This step is essential to ensure that the four centrioles are able to function as two competent MTOCs. Centriole maturation also involves the acquisition of distal and sub-distal appendages by the newly formed mother centriole. Centriole maturation begins with the recruitment of ODF2 as a sub-distal appendage (Ishikawa et al. 2005). Ninein and Cep170 also function as sub-distal appendages and bind microtubules (Mogensen et al. 2000). Distal appendages such as Cep164, Cep89, Cep83, FBF1 and SCLT1 are essential for membrane docking and ciliogenesis (Tanos et al. 2013).

18.2.4 Separation

Before entry into mitosis, the G1-G2 tether that connects the two centrosomes is targeted for degradation. This tether is composed of proteins like C-NAP1 (Cep250), rootletin, Cep68 and LRRC45 (Bahe et al. 2005; Faragher and Fry 2003; Fry et al. 1998a, b; Pagan et al. 2015). C-NAP1 is connected to the mother centriole via Cep135 (Kim et al. 2008). C-NAP1 is present at the proximal end of the mother centrioles and the two pools of C-NAP1 are connected by rootletin fibres (Bahe et al. 2005; Vlijm et al. 2018). Super-resolution microscopy experiments have demonstrated that Cep68 binds to rootletin and increases the thickness of rootletin fibres (Vlijm et al. 2018). LRRC45 might link the rootletin fibres with C-NAP1 (He et al. 2013). The degradation of the tether is initiated by the activity of the NIMA-related kinase, Nek2 (Fry et al. 1998b). Nek2 phosphorylates C-NAP1 and rootletin, leading to their displacement from the proximal end of the mother centrioles (Bahe et al. 2005; Fry et al. 1998a). Plk1-mediated degradation of Cep68 further displaces Cep215, a PCM protein, from the centrosome, contributing to centrosome separation (Pagan et al. 2015).

Once the G1-G2 tether has undergone dissolution, the two centrosomes have to be separated so as to form the mitotic spindle. This occurs primarily via the antiparallel sliding action of Eg5, a motor protein (Cole et al. 1994; Sawin et al. 1992). Additionally, proteins such as dynein, Lis1 and CLIP-70 contribute to spindle formation by inducing microtubule sliding in the opposite direction to Eg5. The combined action of these motor proteins and their binding partners promotes migration of the centrosomes to the two poles (Gonczy et al. 1999; Tanenbaum et al. 2008). Each centrosome migrates towards a different pole in mitotic cells, leading to the formation of a mitotic spindle. The bipolar spindle further contributes towards the accurate segregation of DNA into two daughter cells, thus maintaining ploidy.

18.3 Centrosome Defects and Tumour Progression

Two rules govern the centrosome duplication cycle: cell cycle control and copy number control.

- Cell cycle control ensures that the centrosome replicates only once per cell cycle.
- Copy number control ascertains that only one pro-centriole forms adjacent to each mother centriole.

Therefore, deregulation of any step of the centrosome cycle can give rise to centrosome abnormalities. These can be classified into two types: (a) structural and (b) numerical.

(a) **Structural abnormalities**. Structural abnormalities in centrosomes are errors in the organization of the centrosome. This leads to aberrations in centrosome size or shape and these structural alterations are present at high levels in tumour tissues and have been postulated to contribute to tumour progression (Lingle et al. 1998; Lingle and Salisbury 1999). Recently, it has been demonstrated that overexpression of Ninein-like protein (Nlp) in MCF10A-derived 3D acini leads to the formation of centrosome-related bodies (CRB) harbouring large patches of Nlp, resulting in excessive cell proliferation. The cells in the lumen, which are normally cleared by apoptosis, are not cleared, microtubule organization is altered and this phenotype is associated with cellular transformation and neoplastic progression (Schnerch and Nigg 2016). Further, it has been demonstrated that overexpression of Nlp weakens E-cadherin-based adherens junctions in the epithelium. The weakening of adherens junctions results in increased mechanical stress when a cell in the epithelial tissue enters mitosis. The mitotic programme results in 'budding' of the new daughter cell from the original site and could contribute to metastatic progression in tumour cells (Ganier et al. 2018). In addition to the changes in invasiveness, alterations in centrosome structure could lead to aneuploidy which often contributes to tumour progression (Nigg 2006). However, further studies are needed to investigate the extent of aneuploidy that is induced by structural centrosome aberrations. Moreover, given that the centrosome is the hub of several cellular signalling events, it would be interesting to study how signalling pathways are affected by defects in centrosome structure.

(b) **Numerical abnormalities**. Errors in cell cycle control of centrosome duplication contribute to numerical centrosome abnormalities. They can occur due to centrosome over-duplication, defects in cytokinesis and the de novo formation of centrosomes. These result in the presence of more than two centrosomes in a mitotic cell which is also referred to as centrosome amplification. They are the more extensively studied of the two types of centrosome abnormalities, especially in the context of tumour progression. Given that centrosomes organize microtubules and form the mitotic spindle leading to DNA segregation, any numerical errors in centrosome organization are strongly correlated with chromosomal instability and aneuploidy, which is a hallmark of most tumours. Most tumour cells show an increase in the number of centrosomes, suggesting that the change in centrosome number could drive aneuploidy and thus, tumour progression (Godinho and Pellman 2014).

Cells with multiple centrosomes form a multipolar spindle leading to a multipolar mitosis resulting in aneuploidy (Ring et al. 1982). However, a multipolar mitosis in most tumour cells leads to cell death, probably due to loss of genes that are absolutely required for cell viability. Experiments in Zebrafish demonstrate that neuroepithelial cells harbouring centrosome amplification undergo apoptosis (Dzafic et al. 2015). Similarly, experiments in mice have shown that centrosome amplification leads to microcephaly, due to increased apoptosis (Marthiens et al. 2013). These data suggest that cells harbouring multiple centrosomes must avoid the massive aneuploidy resulting from multipolar mitoses, which raises the interesting question of how tumour cells manage to survive and thrive in the presence of multiple centrosomes. In order to tolerate the burden of extra centrosomes, transformed cells have developed a mechanism called clustering (Ring et al. 1982). In transformed cells with multiple centrosomes, at prophase, centrosomes are not present at only two distinct poles in tumour cells, thus forming a multipolar spindle. During prophase, the centrosomes migrate to and cluster at two different poles leading to the formation of a pseudo-bipolar spindle during metaphase (Basto et al. 2008; Ganem et al. 2009; Kwon et al. 2008; Quintyne et al. 2005). This formation of pseudo-bipolar spindles leads to errors in kinetochore–microtubule attachments, especially the formation of merotelic attachments, which contributes to the presence of lagging chromosomes and an increase in chromosome instability (CIN) (Cosenza et al. 2017; Ganem et al. 2009). The increase in CIN permits the generation of clones with a growth advantage leading to tumour progression.

Multiple studies have attempted to identify gene products/molecular pathways that are required for centrosome clustering. This could lead to the identification of small molecules that inhibit clustering, which could lead to multipolar mitoses resulting in killing of tumour cells. A screen developed by Kwon et al. attempted to identify genes required for clustering in *Drosophila* S2 cells (Kwon et al. 2008). Depletion of HSET, a motor protein, has been demonstrated to induce the formation of multipolar spindles in human cancer cells harbouring multiple centrosomes (Kwon et al. 2008). Depletion of components of the spindle assembly checkpoint (SAC), Mad2, BubR1 (human Bub1) and CENP-E, also leads to the generation of multipolar spindles in S2 cells (Kwon et al. 2008). Several actin-binding proteins were also identified in these screens suggesting that disruption of actin dynamics could inhibit centrosome clustering (Kwon et al. 2008; Leber et al. 2010). This is consistent with recent data from the Godinho laboratory, which suggests that adherens junction functions that are required for maintaining cortical actin organization and cell stiffness can inhibit centrosome clustering (Rhys et al. 2018).

A genome-wide RNAi screen in UPCI:SCC114, a human oral squamous cell carcinoma (OSCC) cell line, has demonstrated that proteins that are a part of the chromosomal passenger complex (CPC), proteins of the Ndc80 complex, Cep164 and Aurora B, which all contribute to spindle tension, are also required for centrosome clustering (Leber et al. 2010). Depletion of Aurora A in a panel of acute myeloid leukaemia (AML) cell lines also results in formation of a multipolar spindle and non-proliferation (Navarro-Serer et al. 2019). Drugs such as Griseofulvin, CP-673451 and Crenolanib that bind to cytoskeletal elements and inhibit clustering demonstrate the contribution of the cytoskeleton towards clustering (Konotop et al. 2016; Rebacz et al. 2007).

Previous results have suggested that loss of 14-3-3γ leads to premature cdc25C, and hence CDK1 activation, which leads to an increase in centrosome number. This is accompanied by an increase in the number of cells with pseudo-bipolar spindles with passage, an increase in aneuploidy and tumour formation (Mukhopadhyay et al. 2016). However, expression of a 14-3-3γ binding defective mutant of cdc25C in 14-3-3γ knockdown cells leads to a reversal of the clustering phenotype and a decrease in tumour growth and cell viability, presumably due to prematurely high levels of CDK1 activity in interphase and mitosis (Mukhopadhyay et al. 2016). These results are consistent with the hypothesis outlined above which suggests that small molecules that might disrupt centrosome clustering in tumour cells could lead to tumour cell killing and cell death. All of these results suggest that centrosome clustering is a complex phenotype that requires multiple cellular pathways.

Given the strong correlation between centrosome amplification, CIN and cancer, it has been a long-standing question of whether centrosome number dysregulation is sufficient to promote tumorigenesis. One of the mechanisms leading to cells acquiring extra centrosomes is a failure in cytokinesis. Therefore, it has been difficult to assess the causal link between centrosome amplification and tumorigenesis. However, several recent studies have addressed this question and have offered multiple solutions of how centrosome amplification might lead to tumour progression. In this chapter, we aim to highlight several recent observations that shed some light on the correlation between the presence of multiple centrosomes and transformation. We have focused on proteins that are essential for different steps of the centrosome cycle and how their aberrant expression has been demonstrated to contribute to centrosome amplification and tumorigenesis.

18.3.1   Polo-Like Kinase 1

Polo-like kinase 1 (Plk1) is a member of the polo-like kinase family of proteins that are serine/threonine kinases. It was initially identified in *Drosophila* embryos as a kinase, active during the late anaphase–telophase transition (Llamazares et al. 1991). Plk1 localization changes throughout the cell cycle, with it localizing at the centrosome throughout interphase, and then moving to the kinetochore, the spindle and the spindle mid-body during mitosis (Golsteyn et al. 1995; Kishi et al. 2009). Plk1 performs multiple functions that contribute to mitotic progression. Plk1 phosphorylates multiple residues in the N-terminus of cdc25C resulting in an increase in cdc25C activity and mitotic progression (Toyoshima-Morimoto et al. 2002). Plk1 phosphorylates BubR1 and INCENP, thus stimulating kinetochore assembly (Arnaud et al. 1998; Elowe et al. 2007; Goto et al. 2006). Inhibition of Plk1 in U-2OS cells decreases the robustness of the SAC, which decreases accurate chromosome segregation and could, hence, contribute to CIN (O'Connor et al. 2015). Plk1 contributes to centrosome maturation by phosphorylating NEDD1, which leads to recruitment of γ-TURCs to centrosomes during centrosome maturation (Haren et al. 2009; Zhang et al. 2009). Depletion of Plk1 using siRNA in HeLa cells leads to a lack of phosphorylation of NPM1, which has been demonstrated to lead to defects in nuclear size, cytokinesis and centrosome amplification (Zhang et al. 2004b). The activity of Plk1 is required for centrosome disengagement, as inhibition of Plk1 using a small molecule prevents centrosome disengagement (Tsou et al. 2009).

Protein levels of Plk1 are elevated in a number of tumour types, including gliomas, breast cancers, oesophageal squamous cell carcinomas, non-small cell lung carcinomas, melanoma, renal cancer, prostate and colorectal cancer (Feng et al. 2009; Liu et al. 2017; Ramani et al. 2015). Plk1 levels are high in Tamoxifen-resistant MCF-7 cells, and inhibition of Plk1 leads to a decrease in cell proliferation (Jeong et al. 2018). A decrease in Plk1 levels confers sensitivity to Gemcitabine in pancreatic cancer cell lines (Jimeno et al. 2010). Plk1 silencing can also enhance the sensitivity of rectal cancer and medulloblastoma cell lines to radiotherapy (Harris et al. 2012; Rodel et al. 2010). Interestingly, several missense and truncation mutations in the C-terminus of Plk1 have been identified in tumour cell lines such as HepG2, A431, MKN74 and A549 (Simizu and Osada 2000). Given that levels of Plk1 are found to be both overexpressed and reduced in tumour cell lines, it is difficult to actually assess the role of Plk1 in neoplastic progression or cellular transformation.

Recent experiments performed in mouse models of Plk1 overexpression suggest that Plk1 may have different effects on neoplastic transformation and tumour progression depending on the presence of other genetic alterations or the tissue type in which Plk1 expression is elevated. Mouse embryonic fibroblasts (MEFs) isolated from mice carrying a doxycycline-inducible transgene for Plk1 exhibit a variety of mitotic defects such as multiple centrosomes, monopolar and multipolar spindles in pro-metaphase, as well as lagging chromosomes and anaphase bridges resulting in an increased duration of mitosis (de Carcer et al. 2018). Despite the defects in mitosis, when Plk1 overexpressing mice were crossed with mice containing a constitutively active K-Ras allele, K-Ras^{G12D}, which is only expressed in the mammary gland, a decrease in tumour progression was observed in the presence of doxycycline (de Carcer et al. 2018). Another study by Li et al. also found chromosomal instability, lagging and misaligned chromosomes and apoptosis in MEFs isolated from mice conditionally overexpressing Plk1 (Li et al. 2017). No malignant transformation was observed in these mice; however, hypersensitivity to ionizing radiation was observed in the liver. Treatment with ionizing radiation led to an increase in the number of liver tumours and the presence of lymphomas, which was accompanied by inhibition in the expression of genes required for DNA repair (Li et al. 2017). In both cases, despite defects in mitosis that lead to aneuploidy, overexpression of Plk1 alone did not lead to tumorigenesis in mice, suggesting that other mechanisms might contribute to tumour formation upon Plk1 overexpression. It is possible that Plk1 overexpression affects other proteins that influence transformation, such as p53 (Smith et al. 2017).

In contrast to its expression levels in most transformed cell lines, loss of Plk1 has been demonstrated to accelerate tumour formation in mice (Lu et al. 2008). Plk1 null mice show embryonic lethality due to the inability of the embryonic cells to divide. However, a significant proportion of Plk1 heterozygous mice develop tumours in various organs suggesting that haploinsufficiency of Plk1 could drive tumour progression. Splenocytes isolated from these mice harboured a higher incidence of aneuploidy, which could contribute to CIN (Lu et al. 2008). Given all of the data on Plk1, it seems likely that Plk1 can act as either an oncogene or a tumour suppressor in a context-dependent manner.

18.3.2   *Separase*

Separase or Separin is a cysteine protease that promotes anaphase entry in mitotic cells by cleaving Ssc1, which is part of the synaptonemal complex. This relaxes the tension in the spindle and leads to anaphase progression (Uhlmann et al. 1999). Inhibiting Separase expression using RNAi in HeLa cells leads to genomic instability (Cucco et al. 2018). Separase forms a complex with the MCM2-7 helicase and loss of Separase leads to an increase in replication fork speed, probably due to an increase in the levels of acetylated SMC3 (Cucco et al. 2018). Antisense oligonuleotide-mediated reduction of Separase gives rise to aberrant mitoses with lagging chromosomes (Chestukhin et al. 2003). Apart from its function in relieving sister chromatid cohesion, Separase also plays a role in ensuring centrosome disengagement (Tsou and Stearns 2006; Tsou et al. 2009). Knockout of Separase in HCT116 and HeLa cell lines inhibits centriole disengagement and subsequent duplication (Tsou et al. 2009). Pericentrin is a potential substrate for Separase and expression of a Pericentrin mutant that cannot be cleaved by Separase suppresses centriole disengagement and duplication (Matsuo et al. 2012). Aki1 and cohesin have also been implicated as substrates of Separase at the centrosome (Nakamura et al. 2009). Separase also functions in double-strand break repair, where it cleaves cohesin (Hellmuth et al. 2018). Activation of Separase in interphase has been demonstrated to aid in DNA double-strand break repair in HEK293 cells (Hellmuth et al. 2018). All of these results suggest that Separase is required for maintaining genomic integrity either by regulating genome organization and duplication or by regulating centrosome duplication.

Separase is significantly overexpressed in osteosarcoma, breast and prostate tumour samples as per tissue immunofluorescence analysis (Meyer et al. 2009). An increased nuclear localization of Separase was also observed in these tissue samples (Meyer et al. 2009). This unusual localization of Separase shows a correlation with tumour status, although the mechanism underlying the correlation between Separase levels and tumour progression remains to be elucidated (Meyer et al. 2009). Increased Separase activity with aberrant centrosome numbers has been observed in bone marrow samples of patients with myelodysplastic syndrome (MDS) (Ruppenthal et al. 2018). IHC staining of breast tumours samples from patients has demonstrated a strong correlation between abnormal Separase expression and impaired survival (Gurvits et al. 2017). These results suggest that an increase in Separase levels can drive tumour progression.

A mutation in the *Zebrafish* Separase gene results in embryos with abnormal mitotic spindles, aneuploidy, polyploidy and multiple centrosomes (Shepard et al. 2007). Adult zebrafish heterozygous for Separase treated with the carcinogen *N*-methyl-*N*-nitro-*N*-nitrosoguanidine (MNNG) were observed to be more susceptible to developing tumours (Shepard et al. 2007). This implies that Separase could act as a tumour suppressor. A knockout of Separase leads to embryonic lethality in mice (Kumada et al. 2006). Experiments in mouse models suggest that Separase haploinsufficiency can lead to tumorigenesis in a p53 mutant background. MEFs isolated from mice heterozygous for Separase were found to exhibit a compromised response to damage induced DNA repair (Hellmuth et al. 2018). Mice with a hypomorphic Separase allele display an increased rate of tumorigenesis with a decrease in survival in a p53 mutant background (Mukherjee et al. 2011). Normal splenocytes isolated from these mice exhibited aneuploidy which might contribute to genomic instability (Mukherjee et al. 2011). In contrast to the results described above, conditional overexpression of Separase in a mouse mammary epithelial cell line leads to increased aneuploidy and tumorigenesis in a p53 mutant background (Zhang et al. 2008). These data suggest that either an increase or a decrease in Separase levels might not be enough to initiate tumour formation, though it might lead to aneuploidy, followed by a second genetic event, which could drive tumorigenesis. It might be interesting to observe the effect of Separase knockout or overexpression in tissue-specific, conditional mouse models.

18.3.3 Polo-Like Kinase 4

Polo-like kinase 4 (Plk4), a member of the polo family of kinases, is a master regulator of centriole biogenesis (Swallow et al. 2005). Plk4 was initially identified in mice, because of its sequence homology to polo in *Drosophila* (Fode et al. 1994). Further experiments in NIH3T3 cells ascertained that Plk4 is associated with the centrosome throughout the cell cycle (Hudson et al. 2001). Experiments in both *Drosophila* and mammalian cell lines using siRNA demonstrated that decrease in Plk4 levels leads to a loss of centrioles (Bettencourt-Dias et al. 2005; Habedanck et al. 2005). Overexpression of Plk4 increases centriole numbers, with multiple pro-centrioles forming from a single mother centriole (Habedanck et al. 2005; Kleylein-Sohn et al. 2007). These results indicate the importance of Plk4 levels in controlling centrosome numbers. Recent experiments in *Drosophila* have demonstrated that Plk4 also plays a role in controlling centriolar length (Aydogan et al. 2018). Plk4 overexpression can also induce de novo centriole and MTOC formation in *Drosophila* embryos and *Xenopus* egg extracts, respectively (Eckerdt et al. 2011; Rodrigues-Martins et al. 2007). This occurs due to the ability of Plk4 condensates to recruit STIL, α-, β-tubulin and γ-tubulin, leading to the formation of acentriolar MTOCs (Gouveia et al. 2018).

Plk4 levels are elevated in several cancers, such as breast, acute lymphoblastic leukaemia, prostate, pediatric medulloblastomas and embryonal brain tumours (Korzeniewski et al. 2012; Li et al. 2016, 2018; Pezuk et al. 2017). An increase in Plk4 levels has also been observed in human lung cancer and gastric cancer cell lines (Shinmura et al. 2014). Inhibition of Plk4 expression in neuroblastoma cell lines has been shown to suppress invasion and migration (Tian et al. 2018). Further, inhibition of Plk4 using CFI-400945 induces aneuploidy in lung cancer cell lines (Kawakami et al. 2018). In contrast to the results described above, Plk4 levels are decreased in colorectal cancer cell lines and hepatocellular cancers suggesting that a decrease in Plk4 levels might contribute to tumour progression (Kuriyama et al. 2009; Liu et al. 2012; Rosario et al. 2010). Mutations in Plk4, including loss of function mutations, cause microcephaly and growth failure with impaired centriole biogenesis (Martin et al. 2014). These data suggest that changes in Plk4 expression lead to defects in centrosome biogenesis leading to neoplastic progression.

As overexpression of Plk4 leads to an increase in centrosome number, it has been the focus of multiple experiments to help understand the role of centrosome amplification in tumour formation and progression. Basto et al. demonstrated for the first time that centrosome amplification could drive tumorigenesis (Basto et al. 2008). Using an assay wherein brain tissue from fruit flies overexpressing Sak (the *Drosophila* homologue of Plk4) was transplanted into the abdomen of WT hosts, they demonstrated that a significant percentage of the hosts developed tumours with multiple centrosomes and a number of the tumours formed metastatic colonies (Basto et al. 2008). Using the same transplantation assay, Castellanos et al. screened for the ability of mutants of different centrosomal proteins to affect tumour formation, in *Drosophila* (Castellanos et al. 2008). They were able to demonstrate that 2% of the WT hosts with wing discs from flies overexpressing Sak transplanted into their abdomen could develop tumours (Castellanos et al. 2008). They were unable to assess the degree of CIN due to the small size of the tumours.

Mice with a homozygous deletion of Plk4 displayed embryonic lethality at E7.5 (Hudson et al. 2001). Plk4+/− MEFs had increased centrosomal amplification, multipolar spindle formation and aneuploidy when compared with WT cells (Ko et al. 2005). Nestin-Cre-driven conditional overexpression of Plk4 in the developing mouse brain led to centrosome amplification, aneuploidy and microcephaly; however, no tumours were observed in the brains of these mice (Marthiens et al. 2013). MEFs isolated from mice conditionally overexpressing Plk4, where a lox stop cassette for Plk4 expression was driven by a chicken β-actin promoter, were found to harbour multiple functional centrosomes which contribute to mitotic errors such as lagging chromosomes and multipolar mitosis (Vitre et al. 2015). Skin fibroblasts derived from mice overexpressing Plk4 had multiple centrosomes, a reduced proliferation rate and poor long-term survival (Vitre et al. 2015). Even when mice overexpressing Plk4 were crossed with p53−/− mice, an increase in the rate of tumour development was not observed, suggesting that Plk4 overexpression in p53−/− mice did not accelerate tumour progression (Vitre et al. 2015). Similar results were obtained in a skin carcinogenesis model, in which mice overexpressing Plk4 were treated with 7,12-dimethylbenz(*a*)anthracene (DMBA), followed by application of the tumour promoter 12-*O*-tetradecanoyl-phorbol-13-acetate (TPA). There was no significant difference in tumour burden between the control and treated mice (Vitre et al. 2015). However, mice carrying a similar Plk4 overexpression allele, where expression was epidermis specific, showed differentiation and barrier defects in the epidermis leading to death in a number of mice at postnatal day one (P1) (Sercin et al. 2016). Mice overexpressing Plk4 that survived past P1 showed a decrease in Plk4 levels suggesting a strong selection pressure against Plk4 overexpression. When these mice were crossed with p53 null mice, the barrier defects were abrogated; however, cells overexpressing Plk4 were not detected post-P1, as in the wild-type mice. Surprisingly, the transient overexpression of Plk4 in the p53−/− mice led to the development of spontaneous skin tumours in these mice with complete penetrance (Sercin et al. 2016). Thus, using essentially the same overexpression model for Plk4 leads to different outcomes for tumorigenesis in the same tissue. One possible difference is that in the first report (Vitre et al. 2015), Plk4 expression is not turned off in the tissues, whereas in the second report (Sercin et al. 2016), Plk4 overexpression is transient and is selected against in mice after P1. This suggests that constant centrosome amplification might lead to cell death due to the presence of multipolar mitoses, while a transient increase in Plk4 levels might allow for the selection of cells that tolerate the initial aneuploidy and then the loss of Plk4 leads to the stable inheritance of the aneuploid genome. This might also depend on the strain of mice used in these experiments, as the first set of experiments were done in C57/Bl6 mice while the second set were performed in a mixed background as the three transgenes were all in different strains. In another set of experiments, a doxycycline-inducible allele of Plk4 was introduced into the Rosa26 locus (Coelho et al. 2015). When mice homozygous for the inducible Plk4 allele were crossed with p53 null mice, hyperproliferation and defects in developmental programmes were observed in tissues such as the pancreas and the epidermis, as compared to mice which were p53 null alone (Coelho et al. 2015). The hyperproliferation could be a precursor to tumour development. The results described herein suggest that the experimental protocols used to generate mice with overexpression of Plk4 and the strain of mice in which these experiments are performed could lead to differences in tumour initiation and development.

In contrast to the results described above, overexpression of Plk4 can initiate tumorigenesis in a mouse intestinal neoplasia model (Levine et al. 2017). These mice harbour an APC^{min} mutation, which causes them to develop multiple adenomas throughout the intestinal tract (Moser et al. 1990). These mice displayed an increase in tumour initiation, with the tumours showing an increase in centrosome number, suggesting that centrosome amplification led to an increase in tumour initiation. MEFs isolated from mice overexpressing Plk4, in combination with an APC^{min} allele, displayed increased centrosome numbers. In the absence of p53, these cells could proliferate in spite of harbouring multiple centrosomes, even after 12 days of doxycycline treatment. After long-term treatment with doxycycline (8 months), they were able to observe severe aneuploidy in splenocytes isolated from mice overexpressing Plk4, due to increased centrosome amplification. In addition to an increase in the intestinal tumour burden, these mice also developed spontaneous lymphomas, squamous cell carcinomas and sarcomas starting at 36 weeks of age. When assayed for centrosome number, it was observed that these tumours exhibited centrosome amplification. It was also observed that these tumours harboured varied levels of p53 target genes, indicating that the p53 pathway might be at least partially compromised in these tissues (Levine et al. 2017).

Driving genomic instability is not the only way centrosome amplification via Plk4 can drive tumorigenesis. Transient overexpression of Plk4 in a three-dimensional organoid breast cancer model leads to centrosome amplification that promotes invasion (Godinho et al. 2014). This was not due to an increase in aneuploidy, but rather the increase in invasion was due to increased Rac1 activation which contributes to greater microtubule nucleation (Godinho et al. 2014). Plk4 has also been demonstrated to phosphorylate Arp2 in the Arp2/3 complex in tumour cell lines, leading to activation of the complex and an increase in migration (Kazazian et al. 2017). Phosphorylation of Plk4 at S305 leads to an increase in migration and cell spreading and increased co-localization with RhoA at cell protrusions, while a decrease in Plk4 expression in MEFs or tumour cells leads to a decrease in migration and invasion, respectively (Rosario et al. 2015). These data suggest that Plk4 expression might lead to tumour progression by regulating cell migration and invasion, in addition to its ability to stimulate centrosome duplication and thus genomic instability.

18.3.4   Aurora A kinase

The Aurora family of kinases consists of three members, A, B and C (Nigg 2001). Aurora A localizes to the centrosome during G2 and the spindle during mitosis (Kimura et al. 1997). Its functions include centrosome maturation, centrosome separation, organization of spindle assembly, chromatin protein modification, chromatid separation and cytokinesis (Fu et al. 2007). Aurora A also localizes to the mitochondria in several cancer cell lines, where it affects mitochondrial dynamics and energy production (Bertolin et al. 2018). Aurora A is recruited to the centrosome by Cep192, which results in its autophosphorylation and activation (Joukov et al. 2014). Aurora A has been demonstrated to contribute to the centrosomal accumulation of γ-tubulin (Hannak et al. 2001). Its centrosomal substrates include γ-tubulin, Centrin1 and D-TACC (Giet et al. 2002; Sardon et al. 2010). Aurora A regulates microtubule growth by organizing the acentriolar spindle in mammalian oocytes during meiosis I (Bury et al. 2017).

Experiments in HeLa cells have shown that increased Aurora A levels can induce centrosome amplification by inhibiting cytokinesis, leading to the formation of tetraploid cells (Meraldi et al. 2002). This effect is augmented in the absence of p53, with 80% of p53–/– MEFs overexpressing Aurora A harbouring extra centrosomes (Meraldi et al. 2002). Knockdown of Aurora A in HeLa cells gives rise to cells with misaligned chromosomes, defective spindle organization, lagging chromosomes, defects in centriole separation and delay in mitotic entry (Hirota et al. 2003; Marumoto et al. 2003). Aurora A also plays a role in the cell cycle; knockdown of Aurora A in HeLa cells leads to arrest of the cells in G2-M and eventually, apoptosis (Du and Hannon 2004).

Aurora A is overexpressed in several cancers, including breast (where it was first identified), ovarian, human gliomas, colon, pancreatic and lung (Kollareddy et al. 2008). The expression of Aurora A results in an inhibition of apoptosis in lung cancer cell lines treated with the EGFR inhibitor, thus ensuring their survival (Shah et al. 2019). Further, a loss of Aurora A in gastrointestinal cell lines with activated KRAS leads to increased cell death due to an increase in the levels of proteins required for apoptosis (Wang-Bishop et al. 2018). Aurora A might also affect epithelial to mesenchymal transition, as demonstrated in OSCC cell lines (Dawei et al. 2018). Silencing of Aurora A leads to an increase in the expression of E-cadherin and a decrease in the levels of Vimentin in OSCC cells accompanied by an increase in the levels of reactive oxygen species (ROS) (Dawei et al. 2018).

MEFs from Aurora A null mice display monopolar spindles, with a single γ-tubulin focus (Cowley et al. 2009). A knockdown of Aurora A in OSCC cell lines led to a decrease in tumour volume in xenograft assays in nude mice. The decrease in tumour volume was accompanied by pronounced apoptosis in tumour tissues (Dawei et al. 2018). However, Aurora A overexpression in primary MEFs is not enough to induce transformation (Anand et al. 2003). Overexpression of Aurora A in a rat mammary carcinogenesis model, where the mice are treated with methylnitrosourea, results in centrosome amplification, but tumours developed at a point much after centrosome amplification was first observed (Goepfert et al. 2002). In contrast to some of the results described above, conditional overexpression of Aurora A in the mammary gland led to increased apoptosis which was reversed upon depletion of p53 (Zhang et al. 2004a). These results indicate that dysregulation of Aurora A levels alone is insufficient to drive transformation. Given the many defects associated with dysregulation of Aurora A kinase, several attempts have been made to target it in cancers. These inhibitors are being used extensively to gain a better understanding of Aurora A biology.

18.3.5   Nek2A

Nek2 kinase is the human homologue of the *Aspergillus nidulans* protein NIMA (never in mitosis). Its expression varies through the cell cycle, with its levels being the highest during the G2 and M phases (Fry et al. 1998b; Schultz et al. 1994). It is a Serine/Threonine kinase that phosphorylates inter-centrosomal linker proteins such as CNAP-1 and Rootletin (Bahe et al. 2005; Fry et al. 1998a, b). Phosphorylation of these proteins leads to the dissolution of the inter-centrosomal linker and thus, initiates centrosomal separation. The centrosomes can now migrate to the two ends of the cell, form a spindle and participate in cell division (Bahe et al. 2005; Fry et al. 1998a, b). Overexpression of Nek2 leads to premature centrosome splitting and a loss of focused microtubules, as demonstrated in U-2OS cells (Fry et al. 1998b). However, these cells are able to enter mitosis despite the presence of unfocused microtubules (Fry et al. 1998b). Nek2 overexpression in MDA-MB 231 and MCF-7 leads to centrosome amplification. However, CIN occurs only upon a concurrent depletion of its interacting partner, TRF1 (Lee and Gollahon 2013).

Several reports suggest an increase in Nek2 levels in different cancers. These include tumours of the breast, B-cell lymphoma, multiple myeloma, bladder cancer and glioblastoma (Fang and Zhang 2016). This change in expression levels is associated with a poor prognosis, drug resistance and tumour progression (Hayward et al. 2004; Lee and Gollahon 2013; Zhou et al. 2013). However, it is not known exactly what leads to the change in Nek2 levels in these multiple cancer types. Nek2 protein levels decrease when CDK4 expression is inhibited using RNA interference

(Pitner et al. 2013). Knockdown of Nek2 in a Her2+ breast cancer model reduced centrosome amplification (Pitner et al. 2013). There is no direct evidence that demonstrates that the centrosome amplification observed upon Nek2 overexpression alone actually contributes to tumorigenesis. Given the large number of cancers that harbour high levels of Nek2, and the poor prognosis associated with it, several attempts have been made to develop Nek2 as a potential therapeutic target in cancer. These drugs include ATP-binding site blockers, siRNA-mediated approaches and drugs that affect the binding of Nek2 to its substrates (Hayward et al. 2010; Suzuki et al. 2010). Nek2 inhibition, in combination with chemotherapeutic drugs, might provide a better treatment approach to target cancer (Meng et al. 2014; Suzuki et al. 2010).

18.4   Conclusion

Centrosome amplification is a major feature of most cancer cells. Recent evidence in different model systems has finally established it to be one of the causes and a potential initiator of tumorigenesis. However, centrosome amplification doesn't always lead to tumour formation and might inhibit or promote tumorigenesis depending on the cell/tissue of origin or the presence of other genetic changes that might contribute to tumour progression. A distinct possibility for why centrosome amplification doesn't always lead to tumorigenesis might be the requirement for the activation or inhibition of pathways that promote or prevent centrosome clustering. While most research has focused on the effects of numerical centrosomal abnormalities on tumour progression, it will be interesting to study the contribution of structural centrosomal defects to neoplastic progression.

References

Anand S, Penrhyn-Lowe S, Venkitaraman AR (2003) AURORA-A amplification overrides the mitotic spindle assembly checkpoint, inducing resistance to Taxol. Cancer Cell 3:51–62

Arnaud L, Pines J, Nigg EA (1998) GFP tagging reveals human Polo-like kinase 1 at the kinetochore/centromere region of mitotic chromosomes. Chromosoma 107:424–429

Arquint C, Nigg EA (2014) STIL microcephaly mutations interfere with APC/C-mediated degradation and cause centriole amplification. Curr Biol 24:351–360. https://doi.org/10.1016/j.cub.2013.12.016

Arquint C, Gabryjonczyk AM, Nigg EA (2014) Centrosomes as signalling centres. Philos Trans R Soc Lond B Biol Sci 369 (1650). pii: 20130464. https://doi.org/10.1098/rstb.2013.0464

Aydogan MG et al (2018) A homeostatic clock sets daughter centriole size in flies. J Cell Biol 217:1233–1248. https://doi.org/10.1083/jcb.201801014

Azimzadeh J, Hergert P, Delouvee A, Euteneuer U, Formstecher E, Khodjakov A, Bornens M (2009) hPOC5 is a centrin-binding protein required for assembly of full-length centrioles. J Cell Biol 185:101–114. https://doi.org/10.1083/jcb.200808082

Bahe S, Stierhof YD, Wilkinson CJ, Leiss F, Nigg EA (2005) Rootletin forms centriole-associated filaments and functions in centrosome cohesion. J Cell Biol 171:27–33. https://doi.org/10.1083/jcb.200504107

Basto R, Brunk K, Vinadogrova T, Peel N, Franz A, Khodjakov A, Raff JW (2008) Centrosome amplification can initiate tumorigenesis in flies. Cell Rep 133:11

Berdnik D, Knoblich JA (2002) Drosophila Aurora-A is required for centrosome maturation and actin-dependent asymmetric protein localization during mitosis. Curr Biol 12:640–647

Bertolin G et al (2018) Aurora kinase A localises to mitochondria to control organelle dynamics and energy production. Elife 7. https://doi.org/10.7554/eLife.38111

Bettencourt-Dias M et al (2005) SAK/PLK4 is required for centriole duplication and flagella development. Curr Biol 15:9

Bury L et al (2017) Plk4 and Aurora A cooperate in the initiation of acentriolar spindle assembly in mammalian oocytes. J Cell Biol 216:3571–3590. https://doi.org/10.1083/jcb.201606077

Castellanos E, Dominguez P, Gonzalez C (2008) Centrosome dysfunction in Drosophila neural stem cells causes tumors that are not due to genome instability. Curr Biol 18:1209–1214. https://doi.org/10.1016/j.cub.2008.07.029

Chestukhin A, Pfeffer C, Milligan S, DeCaprio JA, Pellman D (2003) Processing, localization, and requirement of human separase for normal anaphase progression. Proc Natl Acad Sci USA 100:4574–4579. https://doi.org/10.1073/pnas.0730733100

Coelho PA et al (2015) Over-expression of Plk4 induces centrosome amplification, loss of primary cilia and associated tissue hyperplasia in the mouse. Open Biol 5:150209. https://doi.org/10.1098/rsob.150209

Cole DG, Saxton WM, Sheehan KB, Scholey JM (1994) A "slow" homotetrameric kinesin-related motor protein purified from Drosophila embryos. J Biol Chem 269:22913–22916

Cosenza MR et al (2017) Asymmetric centriole numbers at spindle poles cause chromosome missegregation in cancer. Cell Rep 20:1906–1920. https://doi.org/10.1016/j.celrep.2017.08.005

Cowley DO et al (2009) Aurora-A kinase is essential for bipolar spindle formation and early development. Mol Cell Biol 29:1059–1071. https://doi.org/10.1128/mcb.01062-08

Cucco F et al (2018) Separase prevents genomic instability by controlling replication fork speed. Nucleic Acids Res 46:267–278. https://doi.org/10.1093/nar/gkx1172

Dawei H, Honggang D, Qian W (2018) AURKA contributes to the progression of oral squamous cell carcinoma (OSCC) through modulating epithelial-to-mesenchymal transition (EMT) and apoptosis via the regulation of ROS. Biochem Biophys Res Commun 507:83–90. https://doi.org/10.1016/j.bbrc.2018.10.170

de Carcer G et al (2018) Plk1 overexpression induces chromosomal instability and suppresses tumor development. Nat Commun 9:3012. https://doi.org/10.1038/s41467-018-05429-5

Dictenberg JB et al (1998) Pericentrin and gamma-tubulin form a protein complex and are organized into a novel lattice at the centrosome. J Cell Biol 141:163–174

Du J, Hannon GJ (2004) Suppression of p160ROCK bypasses cell cycle arrest after Aurora-A/STK15 depletion. Proc Natl Acad Sci USA 101:8975–8980. https://doi.org/10.1073/pnas.0308484101

Dzafic E, Strzyz PJ, Wilsch-Brauninger M, Norden C (2015) Centriole amplification in zebrafish affects proliferation and survival but not differentiation of neural progenitor cells. Cell Rep 13:168–182. https://doi.org/10.1016/j.celrep.2015.08.062

Dzhindzhev NS, Tzolovsky G, Lipinszki Z, Abdelaziz M, Debski J, Dadlez M, Glover DM (2017) Two-step phosphorylation of Ana2 by Plk4 is required for the sequential loading of Ana2 and Sas6 to initiate procentriole formation. Open Biol 7(12). pii: 170247. https://doi.org/10.1098/rsob.170247

Eckerdt F, Yamamoto TM, Lewellyn AL, Maller JL (2011) Identification of a polo-like kinase 4-dependent pathway for de novo centriole formation. Curr Biol 21:428–432. https://doi.org/10.1016/j.cub.2011.01.072

Elowe S, Hummer S, Uldschmid A, Li X, Nigg EA (2007) Tension-sensitive Plk1 phosphorylation on BubR1 regulates the stability of kinetochore microtubule interactions. Genes Dev 21:2205–2219. https://doi.org/10.1101/gad.436007

Engle KM, Mei T-S; Wasa M, Yu J-Q (2008) Anomalous centriole configurations are detected in Drosophila wing disc cells upon Cdk1 inactivation Acc Chem Res 45:5

Fang Y, Zhang X (2016) Targeting NEK2 as a promising therapeutic approach for cancer treatment. Cell Cycle 15:895–907. https://doi.org/10.1080/15384101.2016.1152430

Faragher AJ, Fry AM (2003) Nek2A kinase stimulates centrosome disjunction and is required for formation of bipolar mitotic spindles. Mol Biol Cell 14:2876–2889. https://doi.org/10.1091/mbc.e03-02-0108

Feng YB et al (2009) Overexpression of PLK1 is associated with poor survival by inhibiting apoptosis via enhancement of survivin level in esophageal squamous cell carcinoma. Int J Cancer 124:578–588. https://doi.org/10.1002/ijc.23990

Fode C, Motro B, Yousefi S, Heffernan M, Dennis JW (1994) Sak, a murine protein-serine/threonine kinase that is related to the Drosophila polo kinase and involved in cell proliferation. Proc Natl Acad Sci USA 91:6388–6392

Fry AM, Mayor T, Meraldi P, Stierhof YD, Tanaka K, Nigg EA (1998a) C-Nap1, a novel centrosomal coiled-coil protein and candidate substrate of the cell cycle-regulated protein kinase Nek2. J Cell Biol 141:1563–1574

Fry AM, Meraldi P, Nigg EA (1998b) A centrosomal function for the human Nek2 protein kinase, a member of the NIMA family of cell cycle regulators. EMBO J 17:470–481. https://doi.org/10.1093/emboj/17.2.470

Fu J, Bian M, Jiang Q, Zhang C (2007) Roles of Aurora kinases in mitosis and tumorigenesis. Mol Cancer Res 5:1–10. https://doi.org/10.1158/1541-7786.mcr-06-0208

Fu J, Hagan IM, Glover DM (2015) The centrosome and its duplication cycle. Cold Spring Harb Perspect Biol 7:a015800. https://doi.org/10.1101/cshperspect.a015800

Fu J et al (2016) Conserved molecular interactions in centriole-to-centrosome conversion. Nat Cell Biol 18:87–99. https://doi.org/10.1038/ncb3274

Ganem NJ, Godinho SA, Pellman D (2009) A mechanism linking extra centrosomes to chromosomal instability. Nature 460:5

Ganier O, Schnerch D, Oertle P, Lim RY, Plodinec M, Nigg EA (2018) Structural centrosome aberrations promote non-cell-autonomous invasiveness. EMBO J 37. https://doi.org/10.15252/embj.201798576

Gergely F, Karlsson C, Still I, Cowell J, Kilmartin J, Raff JW (2000) The TACC domain identifies a family of centrosomal proteins that can interact with microtubules. Proc Natl Acad Sci USA 97:14352–14357. https://doi.org/10.1073/pnas.97.26.14352

Giet R, McLean D, Descamps S, Lee MJ, Raff JW, Prigent C, Glover DM (2002) Drosophila Aurora A kinase is required to localize D-TACC to centrosomes and to regulate astral microtubules. J Cell Biol 156:437–451. https://doi.org/10.1083/jcb.200108135

Godinho SA, Pellman D (2014) Causes and consequences of centrosome abnormalities in cancer. Philos Trans R Soc Lond B Biol Sci 369. pii: 20130467. https://doi.org/10.1098/rstb.2013.0467

Godinho SA et al (2014) Oncogene-like induction of cellular invasion from centrosome amplification. Nature 510:18

Goepfert TM, Adigun YE, Zhong L, Gay J, Medina D, Brinkley WR (2002) Centrosome amplification and overexpression of aurora A are early events in rat mammary carcinogenesis. Cancer Res 62:4115–4122

Golsteyn RM, Mundt KE, Fry AM, Nigg EA (1995) Cell cycle regulation of the activity and subcellular localization of Plk1, a human protein kinase implicated in mitotic spindle function. J Cell Biol 129:1617–1628

Gomez-Ferreria MA, Rath U, Buster DW, Chanda SK, Caldwell JS, Rines DR, Sharp DJ (2007) Human Cep192 is required for mitotic centrosome and spindle assembly. Curr Biol 17:1960–1966. https://doi.org/10.1016/j.cub.2007.10.019

Gonczy P (2012) Towards a molecular architecture of centriole assembly. Nat Rev Mol Cell Biol 13:425–435. https://doi.org/10.1038/nrm3373

Gonczy P, Pichler S, Kirkham M, Hyman AA (1999) Cytoplasmic dynein is required for distinct aspects of MTOC positioning, including centrosome separation, in the one cell stage Caenorhabditis elegans embryo. J Cell Biol 147:135–150

Goto H et al (2006) Complex formation of Plk1 and INCENP required for metaphase-anaphase transition. Nat Cell Biol 8:180–187. https://doi.org/10.1038/ncb1350

Gottardo M, Callaini G, Riparbelli MG (2014) Procentriole assembly without centriole disengagement—a paradox of male gametogenesis. J Cell Sci 127:3434–3439. https://doi.org/10.1242/jcs.152843

Gouveia SM et al (2018) PLK4 is a microtubule-associated protein that self assembles promoting *de novo* MTOC formation. J Cell Sci. https://doi.org/10.1242/jcs.219501

Guichard P et al (2017) Cell-free reconstitution reveals centriole cartwheel assembly mechanisms. Nat Commun 8:14813. https://doi.org/10.1038/ncomms14813

Gurvits N, Loyttyniemi E, Nykanen M, Kuopio T, Kronqvist P, Talvinen K (2017) Separase is a marker for prognosis and mitotic activity in breast cancer. Br J Cancer 117:1383–1391. https://doi.org/10.1038/bjc.2017.301

Habedanck R, Stierhof YD, Wilkinson CJ, Nigg EA (2005) The Polo kinase Plk4 functions in centriole duplication. Nat Cell Biol 7:1140–1146. https://doi.org/10.1038/ncb1320

Hannak E, Kirkham M, Hyman AA, Oegema K (2001) Aurora-A kinase is required for centrosome maturation in Caenorhabditis elegans. J Cell Biol 155:1109–1116. https://doi.org/10.1083/jcb.200108051

Haren L, Stearns T, Luders J (2009) Plk1-dependent recruitment of gamma-tubulin complexes to mitotic centrosomes involves multiple PCM components. PLoS One 4:e5976. https://doi.org/10.1371/journal.pone.0005976

Harris PS et al (2012) Polo-like kinase 1 (PLK1) inhibition suppresses cell growth and enhances radiation sensitivity in medulloblastoma cells. BMC Cancer 12:80. https://doi.org/10.1186/1471-2407-12-80

Hatch EM, Kulukian A, Holland AJ, Cleveland DW, Stearns T (2010) Cep152 interacts with Plk4 and is required for centriole duplication. J Cell Biol 191:721–729. https://doi.org/10.1083/jcb.201006049

Hayward DG, Clarke RB, Faragher AJ, Pillai MR, Hagan IM, Fry AM (2004) The centrosomal kinase Nek2 displays elevated levels of protein expression in human breast cancer. Cancer Res 64:7370–7376. https://doi.org/10.1158/0008-5472.can-04-0960

Hayward DG et al (2010) Identification by high-throughput screening of viridin analogs as biochemical and cell-based inhibitors of the cell cycle-regulated nek2 kinase. J Biomol Screen 15:918–927. https://doi.org/10.1177/1087057110376537

He R, Huang N, Bao Y, Zhou H, Teng J, Chen J (2013) LRRC45 is a centrosome linker component required for centrosome cohesion. Cell Rep 4:1100–1107. https://doi.org/10.1016/j.celrep.2013.08.005

Hellmuth S, Gutierrez-Caballero C, Llano E, Pendas AM, Stemmann O (2018) Local activation of mammalian separase in interphase promotes double-strand break repair and prevents oncogenic transformation. EMBO J 37. pii: e99184. https://doi.org/10.15252/embj.201899184

Hirota T et al (2003) Aurora-A and an interacting activator, the LIM protein Ajuba, are required for mitotic commitment in human cells. Cell 114:585–598

Hudson JW, Kozarova A, Cheung P, Macmillan JC, Swallow CJ, Cross JC, Dennis JW (2001) Late mitotic failure in mice lacking Sak, a polo-like kinase. Curr Biol 11:441–446

Ishikawa H, Kubo A, Tsukita S, Tsukita S (2005) Odf2-deficient mother centrioles lack distal/subdistal appendages and the ability to generate primary cilia. Nat Cell Biol 7:517–524. https://doi.org/10.1038/ncb1251

Izquierdo D, Wang WJ, Uryu K, Tsou MF (2014) Stabilization of cartwheel-less centrioles for duplication requires CEP295-mediated centriole-to-centrosome conversion. Cell Rep 8:957–965. https://doi.org/10.1016/j.celrep.2014.07.022

Jana SC, Marteil G, Bettencourt-Dias M (2014) Mapping molecules to structure: unveiling secrets of centriole and cilia assembly with near-atomic resolution. Curr Opin Cell Biol 26:96–106. https://doi.org/10.1016/j.ceb.2013.12.001

Jeong SB et al (2018) Essential role of polo-like kinase 1 (Plk1) oncogene in tumor growth and metastasis of tamoxifen-resistant breast cancer. Mol Cancer Ther 17:825–837. https://doi.org/10.1158/1535-7163.mct-17-0545

Jimeno A, Rubio-Viqueira B, Rajeshkumar NV, Chan A, Solomon A, Hidalgo M (2010) A fine-needle aspirate-based vulnerability assay identifies polo-like kinase 1 as a mediator of gemcitabine resistance in pancreatic cancer. Mol Cancer Ther 9:311–318. https://doi.org/10.1158/1535-7163.mct-09-0693

Joukov V, Walter JC, De Nicolo A (2014) The Cep192-organized aurora A-Plk1 cascade is essential for centrosome cycle and bipolar spindle assembly. Mol Cell 55:578–591. https://doi.org/10.1016/j.molcel.2014.06.016

Kawakami M et al (2018) Polo-like kinase 4 inhibition produces polyploidy and apoptotic death of lung cancers. Proc Natl Acad Sci USA 115:1913–1918. https://doi.org/10.1073/pnas.1719760115

Kazazian K et al (2017) Plk4 promotes cancer invasion and metastasis through Arp2/3 complex regulation of the actin cytoskeleton. Cancer Res 77:434–447. https://doi.org/10.1158/0008-5472.can-16-2060

Keryer G, Witczak O, Delouvée A, Kemmner WA, Rouillard D, Tasken K, Bornens M (2003) Dissociating the centrosomal matrix protein AKAP450 from centrioles impairs centriole duplication and cell cycle progression. Mol Biol Cell 14:2436–2446. https://doi.org/10.1091/mbc.e02-09-0614

Kim K, Lee S, Chang J, Rhee K (2008) A novel function of CEP135 as a platform protein of C-NAP1 for its centriolar localization. Exp Cell Res 314:3692–3700. https://doi.org/10.1016/j.yexcr.2008.09.016

Kim TS et al (2013) Hierarchical recruitment of Plk4 and regulation of centriole biogenesis by two centrosomal scaffolds, Cep192 and Cep152. Proc Natl Acad Sci USA 110:E4849–E4857. https://doi.org/10.1073/pnas.1319656110

Kimura M, Kotani S, Hattori T, Sumi N, Yoshioka T, Todokoro K, Okano Y (1997) Cell cycle-dependent expression and spindle pole localization of a novel human protein kinase, Aik, related to Aurora of Drosophila and yeast Ipl1. J Biol Chem 272:13766–13771

Kishi K, van Vugt MA, Okamoto K, Hayashi Y, Yaffe MB (2009) Functional dynamics of Polo-like kinase 1 at the centrosome. Mol Cell Biol 29:3134–3150. https://doi.org/10.1128/mcb.01663-08

Kitagawa D et al (2011) Structural basis of the 9-fold symmetry of centrioles. Cell 144:364–375. https://doi.org/10.1016/j.cell.2011.01.008

Kleylein-Sohn J, Westendorf J, Le Clech M, Habedanck R, Stierhof YD, Nigg EA (2007) Plk4-induced centriole biogenesis in human cells. Dev Cell 13:190–202. https://doi.org/10.1016/j.devcel.2007.07.002

Ko MA, Rosario CO, Hudson JW, Kulkarni S, Pollett A, Dennis JW, Swallow CJ (2005) Plk4 haploinsufficiency causes mitotic infidelity and carcinogenesis. Nat Genet 37:883–888. https://doi.org/10.1038/ng1605

Kollareddy M, Dzubak P, Zheleva D, Hajduch M (2008) Aurora kinases: structure, functions and their association with cancer. Biomed Pap Med Fac Univ Palacky Olomouc Czech Repub 152:27–33

Kong D, Farmer V, Shukla A, James J, Gruskin R, Kiriyama S, Loncarek J (2014) Centriole maturation requires regulated Plk1 activity during two consecutive cell cycles. J Cell Biol 206:855–865. https://doi.org/10.1083/jcb.201407087

Konotop G et al (2016) Pharmacological inhibition of centrosome clustering by slingshot-mediated cofilin activation and actin cortex destabilization. Cancer Res 76:6690–6700. https://doi.org/10.1158/0008-5472.can-16-1144

Korzeniewski N, Hohenfellner M, Duensing S (2012) CAND1 promotes PLK4-mediated centriole overduplication and is frequently disrupted in prostate cancer. Neoplasia (New York, NY) 14:799–806

Kovacs L et al (2018) Gorab is a Golgi protein required for structure and duplication of Drosophila centrioles. Nat Genet 50:1021–1031. https://doi.org/10.1038/s41588-018-0149-1

Kumada K et al (2006) The selective continued linkage of centromeres from mitosis to interphase in the absence of mammalian separase. J Cell Biol 172:835–846. https://doi.org/10.1083/jcb. 200511126

Kuriyama R, Bettencourt-Dias M, Hoffmann I, Arnold M, Sandvig L (2009) Gamma-tubulin-containing abnormal centrioles are induced by insufficient Plk4 in human HCT116 colorectal cancer cells. J Cell Sci 122:2014–2023. https://doi.org/10.1242/jcs.036715

Kwon M, Godinho SA, Chandhok NS, Ganem NJ, Azioune A, Thery M, Pellman D (2008) Mechanisms to suppress multipolar divisions in cancer cells with extra centrosomes. Genes Dev 22:2189–2203. https://doi.org/10.1101/gad.1700908

Leber B et al (2010) Proteins required for centrosome clustering in cancer cells. Sci Transl Med 2:33ra38. https://doi.org/10.1126/scitranslmed.3000915

Leda M, Holland AJ, Goryachev AB (2018) Autoamplification and competition drive symmetry breaking: initiation of centriole duplication by the PLK4-STIL network. iScience 8:222–235. https://doi.org/10.1016/j.isci.2018.10.003

Lee J, Gollahon L (2013) Mitotic perturbations induced by Nek2 overexpression require interaction with TRF1 in breast cancer cells. Cell Cycle 12:3599–3614. https://doi.org/10.4161/cc.26589

Levine MS et al (2017) Centrosome amplification is sufficient to promote spontaneous tumorigenesis in mammals. Dev Cell 40:313–322.e315. https://doi.org/10.1016/j.devcel.2016.12.022

Li Z, Dai K, Wang C, Song Y, Gu F, Liu F, Fu L (2016) Expression of polo-like kinase 4(PLK4) in breast cancer and its response to taxane-based neoadjuvant chemotherapy. J Cancer 7:1125–1132. https://doi.org/10.7150/jca.14307

Li Z et al (2017) Polo-like kinase 1 (Plk1) overexpression enhances ionizing radiation-induced cancer formation in mice. J Biol Chem 292:17461–17472. https://doi.org/10.1074/jbc.M117. 810960

Li S, Wang C, Wang W, Liu W, Zhang G (2018) Abnormally high expression of POLD1, MCM2, and PLK4 promotes relapse of acute lymphoblastic leukemia. Medicine 97:e10734. https://doi. org/10.1097/md.0000000000010734

Lin YC et al (2013) Human microcephaly protein CEP135 binds to hSAS-6 and CPAP, and is required for centriole assembly. EMBO J 32:1141–1154. https://doi.org/10.1038/emboj.2013. 56

Lingle WL, Salisbury JL (1999) Altered centrosome structure is associated with abnormal mitoses in human breast tumors. Am J Pathol 155:1941–1951. https://doi.org/10.1016/s0002-9440(10) 65513-7

Lingle WL, Lutz WH, Ingle JN, Maihle NJ, Salisbury JL (1998) Centrosome hypertrophy in human breast tumors: implications for genomic stability and cell polarity. Proc Natl Acad Sci USA 95:2950–2955

Liu L, Zhang CZ, Cai M, Fu J, Chen GG, Yun J (2012) Downregulation of polo-like kinase 4 in hepatocellular carcinoma associates with poor prognosis. PLoS One 7:e41293. https://doi.org/ 10.1371/journal.pone.0041293

Liu Z, Sun Q, Wang X (2017) PLK1, a potential target for cancer therapy. Transl Oncol 10:22–32. https://doi.org/10.1016/j.tranon.2016.10.003

Liu Y et al (2018) Direct binding of CEP85 to STIL ensures robust PLK4 activation and efficient centriole assembly. Nat Commun 9:15

Llamazares S et al (1991) Polo encodes a protein kinase homolog required for mitosis in Drosophila. Genes Dev 5:2153–2165

Lu LY, Wood JL, Minter-Dykhouse K, Ye L, Saunders TL, Yu X, Chen J (2008) Polo-like kinase 1 is essential for early embryonic development and tumor suppression. Mol Cell Biol 28:6870–6876. https://doi.org/10.1128/mcb.00392-08

Luders J (2012) The amorphous pericentriolar cloud takes shape. Nat Cell Biol 14:1126–1128. https://doi.org/10.1038/ncb2617

Malumbres M, Barbacid M (2007) Cell cycle kinases in cancer. Curr Opin Genet Dev 17:60–65. https://doi.org/10.1016/j.gde.2006.12.008

Marthiens V, Rujano MA, Pennetier C, Tessier S, Paul-Gilloteaux P, Basto R (2013) Centrosome amplification causes microcephaly. Nat Cell Biol 15:731–740. https://doi.org/10.1038/ncb2746

Martin CA et al (2014) Mutations in PLK4, encoding a master regulator of centriole biogenesis, cause microcephaly, growth failure and retinopathy. Nat Genet 46:1283–1292. https://doi.org/10.1038/ng.3122

Marumoto T, Honda S, Hara T, Nitta M, Hirota T, Kohmura E, Saya H (2003) Aurora-A kinase maintains the fidelity of early and late mitotic events in HeLa cells. J Biol Chem 278:51786–51795. https://doi.org/10.1074/jbc.M306275200

Matsuo K, Ohsumi K, Iwabuchi M, Kawamata T, Ono Y, Takahashi M (2012) Kendrin is a novel substrate for separase involved in the licensing of centriole duplication. Curr Biol 22:915–921. https://doi.org/10.1016/j.cub.2012.03.048

Meng L et al (2014) Inhibition of Nek2 by small molecules affects proteasome activity. BioMed Res Int 2014:13. https://doi.org/10.1155/2014/273180

Meraldi P, Honda R, Nigg EA (2002) Aurora-A overexpression reveals tetraploidization as a major route to centrosome amplification in p53-/- cells. EMBO J 21:483–492

Meyer R, Fofanov V, Panigrahi A, Merchant F, Zhang N, Pati D (2009) Overexpression and mislocalization of the chromosomal segregation protein separase in multiple human cancers. Clin Cancer Res 15:2703–2710. https://doi.org/10.1158/1078-0432.ccr-08-2454

Mogensen MM, Malik A, Piel M, Bouckson-Castaing V, Bornens M (2000) Microtubule minus-end anchorage at centrosomal and non-centrosomal sites: the role of ninein. J Cell Sci 113 (Pt 17):3013–3023

Moser AR, Pitot HC, Dove WF (1990) A dominant mutation that predisposes to multiple intestinal neoplasia in the mouse. Science 247:322–324

Moyer TC, Clutario KM, Lambrus BG, Daggubati V, Holland AJ (2015) Binding of STIL to Plk4 activates kinase activity to promote centriole assembly. J Cell Biol 209:863–878. https://doi.org/10.1083/jcb.201502088

Mukherjee M et al (2011) Separase loss of function cooperates with the loss of p53 in the initiation and progression of T- and B-cell lymphoma, leukemia and aneuploidy in mice. PLoS One 6: e22167. https://doi.org/10.1371/journal.pone.0022167

Mukhopadhyay A et al (2016) 14-3-3gamma prevents centrosome amplification and neoplastic progression. Sci Rep 6:26580. https://doi.org/10.1038/srep26580

Nakamura A, Arai H, Fujita N (2009) Centrosomal Aki1 and cohesin function in separase-regulated centriole disengagement. J Cell Biol 187:607–614. https://doi.org/10.1083/jcb.200906019

Nakazawa Y, Hiraki M, Kamiya R, Hirono M (2007) SAS-6 is a cartwheel protein that establishes the 9-fold symmetry of the centriole. Curr Biol 17:2169–2174. https://doi.org/10.1016/j.cub.2007.11.046

Navarro-Serer B, Childers EP, Hermance NM, Mercadante D, Manning AL (2019) Aurora A inhibition limits centrosome clustering and promotes mitotic catastrophe in cells with supernumerary centrosomes. Oncotarget 10(17):1649–1659

Nigg EA (2001) Mitotic kinases as regulators of cell division and its checkpoints. Nat Rev Mol Cell Biol 2:21–32. https://doi.org/10.1038/35048096

Nigg EA (2006) Origins and consequences of centrosome aberrations in human cancers. Int J Cancer 119:2717–2723. https://doi.org/10.1002/ijc.22245

Nigg EA, Stearns T (2011) The centrosome cycle: centriole biogenesis, duplication and inherent asymmetries. Nat Cell Biol 13:1154–1160. https://doi.org/10.1038/ncb2345

O'Connor A, Maffini S, Rainey MD, Kaczmarczyk A, Gaboriau D, Musacchio A, Santocanale C (2015) Requirement for PLK1 kinase activity in the maintenance of a robust spindle assembly checkpoint. Biol Open 5:11–19. https://doi.org/10.1242/bio.014969

Ohta M, Watanabe K, Ashikawa T, Nozaki Y, Yoshiba S, Kimura A, Kitagawa D (2018) Bimodal binding of STIL to Plk4 controls proper centriole copy number. Cell Rep 23:3160–3169.e3164. https://doi.org/10.1016/j.celrep.2018.05.030

Okuda M et al (2000) Nucleophosmin/B23 is a target of CDK2/cyclin E in centrosome duplication. Cell 103:127–140

Pagan JK et al (2015) Degradation of Cep68 and PCNT cleavage mediate Cep215 removal from the PCM to allow centriole separation, disengagement and licensing. Nat Cell Biol 17:31–43. https://doi.org/10.1038/ncb3076

Peter M, Nakagawa J, Doree M, Labbe JC, Nigg EA (1990) Identification of major nucleolar proteins as candidate mitotic substrates of cdc2 kinase. Cell 60:791–801

Pezuk JA, Brassesco MS, de Oliveira RS, Machado HR, Neder L, Scrideli CA, Tone LG (2017) PLK1-associated microRNAs are correlated with pediatric medulloblastoma prognosis. Childs Nerv Syst 33:609–615. https://doi.org/10.1007/s00381-017-3366-5

Piel M, Meyer P, Khodjakov A, Rieder CL, Bornens M (2000) The respective contributions of the mother and daughter centrioles to centrosome activity and behavior in vertebrate cells. J Cell Biol 149:317–330

Pitner H, Kathryn M, Saavedra HI (2013) Cdk4 and Nek2 signal binucleation and centrosome amplification in a Her2+ breast cancer model. PLoS One 8:4

Quintyne NJ, Reing JE, Hoffelder DR, Gollin SM, Saunders WS (2005) Spindle multipolarity is prevented by centrosomal clustering. Science 307:3

Ramani P, Nash R, Sowa-Avugrah E, Rogers C (2015) High levels of polo-like kinase 1 and phosphorylated translationally controlled tumor protein indicate poor prognosis in neuroblastomas. J Neurooncol 125:103–111. https://doi.org/10.1007/s11060-015-1900-4

Rebacz B, Larsen TO, Clausen MH, Ronnest MH, Loffler H, Ho AD, Kramer A (2007) Identification of griseofulvin as an inhibitor of centrosomal clustering in a phenotype-based screen. Cancer Res 67:6342–6350. https://doi.org/10.1158/0008-5472.can-07-0663

Reber S, Hyman AA (2015) Emergent properties of the metaphase spindle. Cold Spring Harb Perspect Biol 7:a015784. https://doi.org/10.1101/cshperspect.a015784

Reina J, Gonzalez C (2014) When fate follows age: unequal centrosomes in asymmetric cell division. Philos Trans R Soc Lond B Biol Sci 369. https://doi.org/10.1098/rstb.2013.0466

Rhys AD et al (2018) Loss of E-cadherin provides tolerance to centrosome amplification in epithelial cancer cells. J Cell Biol 217:195–209. https://doi.org/10.1083/jcb.201704102

Ring D, Hubble R, Kirschner M (1982) Mitosis in a cell with multiple centrioles. J Cell Biol 94:549–556

Rios RM (2014) The centrosome-Golgi apparatus nexus. Philos Trans R Soc Lond B Biol Sci 369. https://doi.org/10.1098/rstb.2013.0462

Rodel F et al (2010) Polo-like kinase 1 as predictive marker and therapeutic target for radiotherapy in rectal cancer. Am J Pathol 177:918–929. https://doi.org/10.2353/ajpath.2010.100040

Rodrigues-Martins A, Riparbelli M, Callaini G, Glover DM, Bettencourt-Dias M (2007) Revisiting the role of the mother centriole in centriole biogenesis. Science 316:1046–1050. https://doi.org/10.1126/science.1142950

Rosario CO et al (2010) Plk4 is required for cytokinesis and maintenance of chromosomal stability. Proc Natl Acad Sci USA 107:6888–6893. https://doi.org/10.1073/pnas.0910941107

Rosario CO et al (2015) A novel role for Plk4 in regulating cell spreading and motility. Oncogene 34:3441–3451. https://doi.org/10.1038/onc.2014.275

Ruppenthal S, Kleiner H, Nolte F, Fabarius A, Hofmann WK, Nowak D, Seifarth W (2018) Increased separase activity and occurrence of centrosome aberrations concur with transformation of MDS. PLoS One 13:e0191734. https://doi.org/10.1371/journal.pone.0191734

Sardon T, Pache RA, Stein A, Molina H, Vernos I, Aloy P (2010) Uncovering new substrates for Aurora A kinase. EMBO Rep 11:977–984. https://doi.org/10.1038/embor.2010.171

Sawin KE, LeGuellec K, Philippe M, Mitchison TJ (1992) Mitotic spindle organization by a plus-end-directed microtubule motor. Nature 359:540–543. https://doi.org/10.1038/359540a0

Schnerch D, Nigg EA (2016) Structural centrosome aberrations favor proliferation by abrogating microtubule-dependent tissue integrity of breast epithelial mammospheres. Oncogene 35:2711–2722. https://doi.org/10.1038/onc.2015.332

Schultz SJ, Fry AM, Sutterlin C, Ried T, Nigg EA (1994) Cell cycle-dependent expression of Nek2, a novel human protein kinase related to the NIMA mitotic regulator of Aspergillus nidulans. Cell Growth Differ 5:625–635

Sercin O et al (2016) Transient PLK4 overexpression accelerates tumorigenesis in p53-deficient epidermis. Nat Cell Biol 18:100–110. https://doi.org/10.1038/ncb3270

Shah KN et al (2019) Aurora kinase A drives the evolution of resistance to third-generation EGFR inhibitors in lung cancer. Nat Med 25:111–118. https://doi.org/10.1038/s41591-018-0264-7

Shepard JL et al (2007) A mutation in separase causes genome instability and increased susceptibility to epithelial cancer. Genes Dev 21:55–59. https://doi.org/10.1101/gad.1470407

Shinmura K, Tarapore P, Tokuyama Y, George KR, Fukasawa K (2005) Characterization of centrosomal association of nucleophosmin/B23 linked to Crm1 activity. FEBS Lett 579:6621–6634. https://doi.org/10.1016/j.febslet.2005.10.057

Shinmura K, Kurabe N, Goto M, Yamada H, Natsume H, Konno H, Sugimura H (2014) PLK4 overexpression and its effect on centrosome regulation and chromosome stability in human gastric cancer. Mol Biol Rep 41:6635–6644. https://doi.org/10.1007/s11033 014-3546-2

Shukla A, Kong D, Sharma M, Magidson V, Loncarek J (2015) Plk1 relieves centriole block to reduplication by promoting daughter centriole maturation. Nat Commun 6:8077. https://doi.org/10.1038/ncomms9077

Simizu S, Osada H (2000) Mutations in the Plk gene lead to instability of Plk protein in human tumour cell lines. Nat Cell Biol 2:852–854. https://doi.org/10.1038/35041102

Smith L, Farzan R, Ali S, Buluwela L, Saurin AT, Meek DW (2017) The responses of cancer cells to PLK1 inhibitors reveal a novel protective role for p53 in maintaining centrosome separation. Sci Rep 7:16115. https://doi.org/10.1038/s41598-017-16394-2

Sonnen KF, Schermelleh L, Leonhardt H, Nigg EA (2012) 3D-structured illumination microscopy provides novel insight into architecture of human centrosomes. Biol Open 1:965–976. https://doi.org/10.1242/bio.20122337

Stevens NR, Roque H, Raff JW (2010) DSas-6 and Ana2 coassemble into tubules to promote centriole duplication and engagement. Dev Cell 19:913–919. https://doi.org/10.1016/j.devcel.2010.11.010

Stinchcombe JC, Griffiths GM (2014) Communication, the centrosome and the immunological synapse. Philos Trans R Soc Lond B Biol Sci 369. https://doi.org/10.1098/rstb.2013.0463

Suzuki K, Kokuryo T, Senga T, Yokoyama Y, Nagino M, Hamaguchi M (2010) Novel combination treatment for colorectal cancer using Nek2 siRNA and cisplatin. Cancer Sci 101:1163–1169. https://doi.org/10.1111/j.1349-7006.2010.01504.x

Swallow CJ, Ko MA, Siddiqui NU, Hudson JW, Dennis JW (2005) Sak/Plk4 and mitotic fidelity. Oncogene 24:306–312. https://doi.org/10.1038/sj.onc.1208275

Takao D, Yamamoto S, Kitagawa D (2018) A theory of centriole duplication based on self-organized spatial pattern formation. bioRxiv 424754. https://doi.org/10.1101/424754

Tanenbaum ME, Macurek L, Galjart N, Medema RH (2008) Dynein, Lis1 and CLIP-170 counteract Eg5-dependent centrosome separation during bipolar spindle assembly. EMBO J 27:3235–3245. https://doi.org/10.1038/emboj.2008.242

Tang CJ, Fu RH, Wu KS, Hsu WB, Tang TK (2009) CPAP is a cell-cycle regulated protein that controls centriole length. Nat Cell Biol 11:825–831. https://doi.org/10.1038/ncb1889

Tanos BE, Yang HJ, Soni R, Wang WJ, Macaluso FP, Asara JM, Tsou MF (2013) Centriole distal appendages promote membrane docking, leading to cilia initiation. Genes Dev 27:163–168. https://doi.org/10.1101/gad.207043.112

Tian X et al (2018) Polo-like kinase 4 mediates epithelial-mesenchymal transition in neuroblastoma via PI3K/Akt signaling pathway. Cell Death Dis 9:54. https://doi.org/10.1038/s41419-017-0088-2

Tokuyama Y, Horn HF, Kawamura K, Tarapore P, Fukasawa K (2001) Specific phosphorylation of nucleophosmin on Thr(199) by cyclin-dependent kinase 2-cyclin E and its role in centrosome duplication. J Biol Chem 276:21529–21537. https://doi.org/10.1074/jbc.M100014200

Toyoshima-Morimoto F, Taniguchi E, Nishida E (2002) Plk1 promotes nuclear translocation of human Cdc25C during prophase. EMBO Rep 3:341–348. https://doi.org/10.1093/embo-reports/kvf069

Tsou MF, Stearns T (2006) Mechanism limiting centrosome duplication to once per cell cycle. Nature 442:947–951. https://doi.org/10.1038/nature04985

Tsou MF, Wang WJ, George KA, Uryu K, Stearns T, Jallepalli PV (2009) Polo kinase and separase regulate the mitotic licensing of centriole duplication in human cells. Dev Cell 17:344–354. https://doi.org/10.1016/j.devcel.2009.07.015

Uhlmann F, Lottspelch F, Nasmyth K (1999) Sister-chromatid separation at anaphase onset is promoted by cleavage of the cohesin subunit Scc1. Nature 400:37–42. https://doi.org/10.1038/21831

Vitre B et al (2015) Chronic centrosome amplification without tumorigenesis. Proc Natl Acad Sci USA 112:11

Vlijm R et al (2018) STED nanoscopy of the centrosome linker reveals a CEP68-organized, periodic rootletin network anchored to a C-Nap1 ring at centrioles. Proc Natl Acad Sci USA 115:8

Walczak CE, Heald R (2008) Mechanisms of mitotic spindle assembly and function. Int Rev Cytol 265:111–158. https://doi.org/10.1016/s0074-7696(07)65003-7

Wang WJ, Soni RK, Uryu K, Tsou MF (2011) The conversion of centrioles to centrosomes: essential coupling of duplication with segregation. J Cell Biol 193:727–739. https://doi.org/10.1083/jcb.201101109

Wang-Bishop L et al (2018) Inhibition of AURKA reduces proliferation and survival of gastrointestinal cancer cells with activated KRAS by preventing activation of RPS6KB1. Gastroenterology. https://doi.org/10.1053/j.gastro.2018.10.030

Winey M, O'Toole E (2014) Centriole structure. Philos Trans R Soc Lond B Biol Sci 369. https://doi.org/10.1098/rstb.2013.0457

Woodruff JB, Wueseke O, Hyman AA (2014) Pericentriolar material structure and dynamics. Philos Trans R Soc Lond B Biol Sci 369. https://doi.org/10.1098/rstb.2013.0459

Yamamoto S, Kitagawa D (2018) Self-organization of Plk4 regulates symmetry breaking in centriole duplication. bioRxiv 313635. https://doi.org/10.1101/313635

Zhang D et al (2004a) Cre-loxP-controlled periodic Aurora-A overexpression induces mitotic abnormalities and hyperplasia in mammary glands of mouse models. Oncogene 23:8720–8730. https://doi.org/10.1038/sj.onc.1208153

Zhang H, Shi X, Paddon H, Hampong M, Dai W, Pelech S (2004b) B23/nucleophosmin serine 4 phosphorylation mediates mitotic functions of polo-like kinase 1. J Biol Chem 279:35726–35734. https://doi.org/10.1074/jbc.M403264200

Zhang N et al (2008) Overexpression of Separase induces aneuploidy and mammary tumorigenesis. Proc Natl Acad Sci USA 105:13033–13038. https://doi.org/10.1073/pnas.0801610105

Zhang X et al (2009) Sequential phosphorylation of Nedd1 by Cdk1 and Plk1 is required for targeting of the gammaTuRC to the centrosome. J Cell Sci 122:2240–2251. https://doi.org/10.1242/jcs.042747

Zhou W et al (2013) NEK2 induces drug resistance mainly through activation of efflux drug pumps and is associated with poor prognosis in myeloma and other cancers. Cancer Cell 23:48–62. https://doi.org/10.1016/j.ccr.2012.12.001

Chapter 19
Golgi Structure and Function in Health, Stress, and Diseases

Jie Li, Erpan Ahat, and Yanzhuang Wang

Abstract The Golgi apparatus is a central intracellular membrane-bound organelle with key functions in trafficking, processing, and sorting of newly synthesized membrane and secretory proteins and lipids. To best perform these functions, Golgi membranes form a unique stacked structure. The Golgi structure is dynamic but tightly regulated; it undergoes rapid disassembly and reassembly during the cell cycle of mammalian cells and is disrupted under certain stress and pathological conditions. In the past decade, significant amount of effort has been made to reveal the molecular mechanisms that regulate the Golgi membrane architecture and function. Here we review the major discoveries in the mechanisms of Golgi structure formation, regulation, and alteration in relation to its functions in physiological and pathological conditions to further our understanding of Golgi structure and function in health and diseases.

19.1 Golgi Architecture and Its Maintenance

The Golgi apparatus is a central intracellular membrane-bound organelle often located adjacent to the nucleus in mammalian cells. Electron microscope (EM) images revealed its unique feature as stacks of five to seven flattened cisternae overlaying one another, with multiple stacks often lined up and interconnected by tubular structures to form a ribbon (Shorter and Warren 2002; Rabouille and Kondylis 2007; Wei and Seemann 2010). The Golgi stacks are polarized; they

J. Li · E. Ahat
Department of Molecular, Cellular and Developmental Biology, University of Michigan, Ann Arbor, MI, USA
e-mail: jieltian@umich.edu; erpan@umich.edu

Y. Wang (✉)
Department of Molecular, Cellular and Developmental Biology, University of Michigan, Ann Arbor, MI, USA

Department of Neurology, University of Michigan School of Medicine, Ann Arbor, MI, USA
e-mail: yzwang@umich.edu

© Springer Nature Switzerland AG 2019
M. Kloc (ed.), *The Golgi Apparatus and Centriole*, Results and Problems in Cell Differentiation 67, https://doi.org/10.1007/978-3-030-23173-6_19

receive proteins and lipids from the endoplasmic reticulum (ER) by the *cis* cisternae and export them from the *trans* cisternae and the *trans*-Golgi network (TGN) to other intracellular membranes such as the endosomes, lysosomes, plasma membrane, and outside of the cell (Tang and Wang 2013; Wang and Seemann 2011). While traversing the Golgi stack, cargo molecules are modified and processed. The sub-compartments of the Golgi stacks house a set of glycosidases and glycosyltransferases responsible for the synthesis of glycoproteins and glycolipids. In the TGN, many secretory proteins are proteolytically cleaved by lumenal proteinases (Huang and Wang 2017; Zhang and Wang 2016; Huttner et al. 1995).

Many efforts have been made to understand the mechanism of Golgi structure formation. The formation of the Golgi ribbon depends on Golgi matrix proteins and an intact microtubule organization. Cytosolic dynein moves Golgi membranes along centrosome-derived microtubules toward the (–) end of the microtubules (Matteis et al. 2008; Rabouille and Kondylis 2007; Wei and Seemann 2010). Subsequently, Golgi-oriented microtubules maintain Golgi stacks in the proximity and facilitate tubular connections between them (Zhu and Kaverina 2013). While dynein and microtubules are required for the concentration of Golgi stacks in the pericentrosomal region and Golgi ribbon formation, they are not essential for the generation and maintenance of the stacked structure, as depolymerization of microtubules disrupts the Golgi ribbon but not the stacks (Thyberg and Moskalewski 1999). From the 1960s, morphological studies have shown connections in the space between cisternae that might be involved in the adhesion of cisternae into stacks (Franke et al. 1972; Mollenhauer 1965; Cluett and Brown 1992), which were later identified as Golgi matrix proteins. These include Golgi stacking proteins and membrane tethers, as discussed below.

19.1.1 *Golgi Matrix Proteins and Golgi Structure Formation*

In 1994, the concept of "Golgi matrix" proteins was first introduced (Slusarewicz et al. 1994). Since then, several Golgi matrix proteins have been identified to be responsible for maintaining the unique architecture and function of the Golgi apparatus. Major components of the Golgi matrix are summarized in Table 19.1. Key proteins involved in Golgi structure formation are discussed below.

19.1.1.1 Golgi *ReA*ssembly Stacking Proteins (GRASPs)

Among all Golgi matrix proteins, the Golgi ReAssembly Stacking Proteins GRASP65 and GRASP55 (GRASPs, also called GORASP1 and GORASP2, respectively) are best characterized for their roles in Golgi structure formation, including stacking (Wang et al. 2003, Shorter et al. 1999, Xiang and Wang 2010, Bekier et al. 2017, Shin et al. 2018, Barr et al. 1997), ribbon-linking (Puthenveedu et al. 2006; Tang et al. 2016; Feinstein and Linstedt 2008), cargo transportation (Kuo et al. 2000;

Table 19.1 Major components of the Golgi matrix

Names	Golgi localization	Membrane anchor	Associated GTPases	Interactions	Functions
GRASPs					
GRASP55/ GORASP2	*medial/ trans*	N-myr (Shorter et al. 1999)	Rab2 (Barr 2005; Short et al. 2001)	– Golgin-45 (Short et al. 2001; Zhao et al. 2017) – p24 (Barr et al. 2001) – TGF-α (Kuo et al. 2000) – LC3, LAMP2 (Zhang et al. 2018) – Sec16 (Piao et al. 2017) – CFTR (Gee et al. 2011) – JAM-B, JAM-C (Cartier-Michaud et al. 2017) – KCTD5 (Dementieva et al. 2009)	• Golgi stacking (Shorter et al. 1999; Xiang and Wang 2010; Bekier et al. 2017) • Golgi ribbon formation (Feinstein and Linstedt 2008) • Cell cycle control (Duran et al. 2008) • Transport of specific cargo (Kuo et al. 2000; D'Angelo et al. 2009) • p24 cargo receptor retention (Barr et al. 2001) • Autophagy (Zhang et al. 2018; Zhang and Wang 2018a, b) • Unconventional secretion (Dupont et al. 2011; Rabouille and Linstedt 2016; Vinke et al. 2011; Gee et al. 2011; Piao et al. 2017) • Spermatogenesis (Cartier-Michaud et al. 2017)
GRASP65/ GORASP1	*cis*	N-myr (Barr et al. 1997)		– GM130 (Barr et al. 1998) – Mena (Tang et al. 2016) – DjA1 (Li et al. 2019) – p24 (Barr et al. 2001) – CD8a, Frizzled 4 (D'Angelo et al. 2009) – Bcl-X$_L$ (Cheng et al. 2010)	• Stacking (Barr et al. 1997; Xiang and Wang 2010; Bekier et al. 2017; Shin et al. 2018) • Golgi ribbon formation (Puthenveedu et al. 2006; Tang et al. 2016) • Cell cycle progression (Preisinger et al. 2005; Sutterlin et al. 2005; Yoshimura et al. 2005; Duran et al. 2008; Tang et al. 2010b) • Mitotic spindle formation (Sutterlin et al. 2005)

(continued)

Table 19.1 (continued)

Names	Golgi localization	Membrane anchor	Associated GTPases	Interactions	Functions
					• Transport of specific cargo (D'Angelo et al. 2009) • p24 cargo receptor retention (Barr et al. 2001) • Unconventional secretion (Gee et al. 2011) • Apoptosis (Lane et al. 2002)
Golgins					
GM130/ GOLGA2	*cis*	P	Rab1 (Weide et al. 2001)	– p115 (Nakamura et al. 1997) – GRASP65 (Hu et al. 2015) – Syntaxin 5 (Diao et al. 2008) – AKAP450 (Rivero et al. 2009) – ZFPL1 (Chiu et al. 2008) – Tuba (Kodani et al. 2009)	• Golgi ribbon formation (Puthenveedu et al. 2006) • COPII vesicle tethering (Moyer et al. 2001; Alvarez et al. 2001) • Non-centrosomal microtubule organization (Rivero et al. 2009) • Centrosome regulation (Kodani and Sutterlin 2008; Kodani et al. 2009) • Spindle formation (Kodani and Sutterlin 2008) • Cell migration (Preisinger et al. 2004) • Apoptosis (Walker et al. 2004) • Purkinje neuron development (Liu et al. 2017)
p115	*cis*	P	Rab1 (Allan et al. 2000)	– GM130 (Nakamura et al. 1997) – Giantin (Sonnichsen et al. 1998) – syntaxin 5, GOS-28 (Shorter et al. 2002)	• Post-mitotic Golgi reassembly (Shorter and Warren 1999; Brandon et al. 2003; Dirac-Svejstrup et al. 2000) • SNARE assembly (Wang et al. 2015; Shorter et al. 2002) • Membrane tethering (Shorter and Warren 1999; Nakamura et al. 1997;

					Alvarez et al. 2001; Seemann et al. 2000; Shorter et al. 2002; Allan et al. 2000) • Apoptosis (Chiu et al. 2002) • Nuclear import (Mukherjee and Shields 2009)
Giantin/ GOLGB1	*cis/medial*	TMD	Rab1/6 (Rosing et al. 2007)	– p115 (Sonnichsen et al. 1998) – GM130 (Sonnichsen et al. 1998) – GCP60/ACBD3 (Sohda et al. 2001)	• Golgi stack organization (Rosing et al. 2007; Koreishi et al. 2013) • Ribbon organization (Koreishi et al. 2013) • Membrane tethering (Sonnichsen et al. 1998; Linstedt et al. 2000; Lowe et al. 2004; Misumi et al. 2001; Derby and Gleeson 2007; Alvarez et al. 2001; Munro 2011) • ER-to-Golgi trafficking (Sohda et al. 2001) • Apoptosis (Lowe et al. 2004) • Cilia formation (Bergen et al. 2017)
Golgin-45/ BLZF1	*medial/ trans*	P	Rab2 (Short et al. 2001)	– GRASP55 (Short et al. 2001, Zhao et al. 2017)	• Golgi stacking (Short et al. 2001, Zhao et al. 2017) • Membrane tethering (Short et al. 2001)
Golgin-67/ GOLGA8B		TMD[a]			• Uncharacterized (Jakymiw et al. 2000, Eystathioy et al. 2000)
Golgin-84/ GOLGA5	*cis*	TMD	Rab1(Satoh et al. 2003)	– CASP (Osterrieder et al. 2017) – COG complex (Sohda et al. 2010)	• Golgi stacking and reassembly (Satoh et al. 2003) • Golgi ribbon formation (Diao et al. 2003) • ER-Golgi tethering and protein transport (Osterrieder et al. 2017) • Membrane tethering (Malsam et al. 2005)

(continued)

Table 19.1 (continued)

Names	Golgi localization	Membrane anchor	Associated GTPases	Interactions	Functions
					• Intra-Golgi retrograde transport (Sohda et al. 2010) • Proinsulin conversion (Liu et al. 2016) • Tau phosphorylation in Alzheimer's disease (Jiang et al. 2014) • Bacterial infection (Rejman Lipinski et al. 2009)
Golgin-97/ GOLGA1	*trans*	GRIP	ARL1/3 (Lu and Hong 2003) Rab6/19/30 (Sinka et al. 2008)	– TBC1D23 (Shin et al. 2017, 2018)	• TGN-to-PM trafficking of E-cadherin (Lock et al. 2005) • Endosome-to-TGN trafficking (Lu and Hong 2003; Lu et al. 2004) • Retrograde transport from recycling endosomes to the TGN (Jing et al. 2010) • Poxvirus morphogenesis (Alzhanova and Hruby 2007)
Golgin-160/ GOLGA3/ GCP170	*cis*	P		– GCP60/ACBD3 (Sbodio et al. 2006; Sbodio and Machamer 2007) – GCP16 (Ohta et al. 2003) – β1AR (Gilbert et al. 2014) – ROMK, PIST (Bundis et al. 2006)	• Post-Golgi trafficking (Bundis et al. 2006; Gilbert et al. 2014) • Apoptosis (Mancini et al. 2000, Maag et al. 2005, Sbodio et al. 2006)
Golgin-245/ GOLGA4/ p230	*trans*	GRIP	ARL1/3 (Wu et al. 2004; Sinka et al. 2008) Rab2/6/19/30 (Sinka et al. 2008)	– TBC1D23 (Shin et al. 2017, 2018)	• TGN-to-PM traffic (Lu and Hong 2003) • Phagophore formation (Sohda et al. 2015)
GCP16		Acylation		– Golgin-160/GOLGA1/ GCP170 (Ohta et al. 2003)	• Golgi-to-PM trafficking (Ohta et al. 2003)

GCC88	*trans*	GRIP	ARL1/3 (Sinka et al. 2008), Rab6/19/30 (Sinka et al. 2008)		• TGN organization and endosome-to-TGN trafficking (Luke et al. 2003; Lieu et al. 2007) • Golgi ribbon formation (Gosavi et al. 2018) • Actin organization (Makhoul et al. 2019)
GCC185	*trans*	GRIP	ARL1/3 (Sinka et al. 2008) ARL4A (Lin et al. 2011) Rab1/2/6/9/30 (Burguete et al. 2008, Sinka et al. 2008, Hayes et al. 2009)	– Syntaxin 16 (Ganley et al. 2008) – CLASP (Efimov et al. 2007)	• Golgi ribbon formation (Derby et al. 2007) • Golgi integrity (Lin et al. 2011) • Membrane tethering (Derby et al. 2007; Reddy et al. 2006; Ganley et al. 2008) • MPR recycling (Reddy et al. 2006) • Attachment of non-centrosomal microtubules (Efimov et al. 2007)
ACBD3/ GCP60		P		– Golgin-45 (Yue et al. 2017) – giantin (Sohda et al. 2001) – Golgin-160 (Sbodio and Machamer 2007; Sbodio et al. 2006) – PI4KIIIβ (Greninger et al. 2012) – TBC1D22 (Greninger et al. 2013) – Aichi virus (Sasaki et al. 2012) – Poliovirus 3A protein (Greninger et al. 2013) – Numb (Zhou et al. 2007)	• Golgi integrity and ER-to-Golgi trafficking (Sohda et al. 2001) • Nuclear translocation of caspase-generated Golgin-160 fragments (residues 140-311) (Sbodio and Machamer 2007, Sbodio et al. 2006) • Asymmetric cell division (Zhou et al. 2007) • Aichi virus and poliovirus replication (Greninger et al. 2012)
Bicaudal-D/ BICD2		P	Rab6	– Dynactin, p50-dynamitin (Hoogenraad et al. 2001)	• Recruitment of the dynein-dynactin complex (Hoogenraad et al. 2001)

(continued)

Table 19.1 (continued)

Names	Golgi localization	Membrane anchor	Associated GTPases	Interactions	Functions
					• COPI-independent Golgi-to-ER transport (Matanis et al. 2002) • Endosome-to-Golgi transport (Wanschers et al. 2007) • Autosomal-dominant spinal muscular atrophy (Neveling et al. 2013, Martinez-Carrera and Wirth 2015)
CASP/Coy1	*cis*	TMD		– Golgin-84 (Osterrieder et al. 2017) – Cog3, Sed5, Sly1, Gos1, Sft1/GS15 (Anderson et al. 2017)	• ER-Golgi tethering and protein transport (Osterrieder et al. 2017) • Retrograde-directed COPI vesicles (Anderson et al. 2017)
CG-NAP				– PKN, RIIα, PP2A, protein phosphatase 1 (Takahashi et al. 1999) – Dynein (Kim et al. 2007) – Cyclin E (Nishimura et al. 2005) – Kendrin, GCP2, γ-tubulin (Takahashi et al. 2002) – CK1δ (Sillibourne et al. 2002)	• Scaffold protein for kinases/phosphatases (Takahashi et al. 1999) • Centrosome amplification (Nishimura et al. 2005) • Microtubule nucleation (Takahashi et al. 2002)
COH1/ VPS13B	*cis*	P			• Golgi ribbon formation (Seifert et al. 2011)
GCP364		C-terminal hydrophobic domain			• Golgi ribbon formation and perinuclear localization (Toki et al. 1997)

GMAP-210/ Trip11/ Trip230	cis	GRAB (Gillingham et al. 2004)	ARF1 (Gillingham et al. 2004)	– ITF20, γ-tubulin (Rios et al. 2004) – Thyroid receptor (Chen et al. 1999)	• Golgi ribbon formation (Rios et al. 2004) • Membrane tethering (Drin et al. 2008) • ER-to-Golgi trafficking (Gillingham et al. 2004) • γ-tubulin recruitment (Rios et al. 2004) • Sorting to primary cilia (Follit et al. 2008) • Interacts with thyroid hormone receptor beta (Chen et al. 1999)
IIGP165		P			• Anti-apoptosis (Ran et al. 2007)
Imh1/SYS3	trans	GRIP	ARL1/Arl1p (Panic et al. 2003)	– Env7 (Rao et al. 2016)	• Endosome-to-Golgi trafficking (Tsukada et al. 1999)
Lava lamp/ Lva				– Dynein heavy chain, Lrrk (Lin et al. 2015)	• Dynein-based Golgi movements (Papoulas et al. 2005)
NECC1/2		TMD[a]			• Uncharacterized (Cruz-Garcia et al. 2007)
SCOCO		P			• Uncharacterized (Van Valkenburgh et al. 2001)
SCYL1BP1/ GORAB		P	Rab6 (Hennies et al. 2008)	– Scyl1	• Gerodermia osteodysplastica (Hennies et al. 2008; Al-Dosari and Alkuraya 2009) • Suppresses Schwann cell (SC) differentiation and neurite outgrowth by enhancing the Rhoa Pathway (Zhang et al. 2016a) • Type 2 diabetes (Peng et al. 2015) • Suppression of hepatocellular carcinoma development (Hu et al. 2012)
TMP/ ARA160		P	Rab6 (Fridmann-Sirkis et al. 2004)	– hSNF2a/b (Mori and Kato 2002) – Fer, AR (Hsiao and Chang 1999) – Stat3 (Perry et al. 2004)	• Golgi integrity (Fridmann-Sirkis et al. 2004) • Early/recycling endosomes-to-TGN trafficking, Golgi enzyme retention (Yamane et al. 2007, Ramirez and Lowe 2009)

(continued)

Table 19.1 (continued)

Names	Golgi localization	Membrane anchor	Associated GTPases	Interactions	Functions
Other Golgi structural proteins					
p24/TMED2	*cis*	TMD		– GRASP55 (Barr et al. 2001) – GRASP65 (Barr et al. 2001)	• Post-Golgi trafficking (Luo et al. 2007) • GPI-anchored proteins export from the ER (Bonnon et al. 2010)
p28	*cis*				• Golgi ribbon formation (Koegler et al. 2010)
p37	*trans*	P		– p97 (Uchiyama et al. 2006)	• Membrane fusion (Uchiyama et al. 2006)
p47	*trans*	P		– VCIP135 (Zhang and Wang 2015a) – p97 (Kondo et al. 1997)	• Post-mitotic Golgi reassembly (Zhang and Wang 2015a)
p97/VCP	*trans*	P		– VCIP135 (Zhang and Wang 2015a) – p47 (Kondo et al. 1997) – p37 (Uchiyama et al. 2006) – YOD1, UBXD1 and PLAA (Papadopoulos et al. 2017)	• Post-mitotic Golgi reassembly (Zhang and Wang 2015a) • Membrane reassembly (Kondo et al. 1997; Ramadan et al. 2007; Hetzer et al. 2001) • Cell cycle progression (Cao et al. 2003; Fu et al. 2003; Parisi et al. 2018) • Protein aggregation (Ghosh et al. 2018; Ristic et al. 2018; Yi and Yuan 2017; Kobayashi et al. 2007; Song et al. 2007; Nishikori et al. 2008) • Autophagy (Papadopoulos et al. 2017) • DNA repair (Van Den Boom et al. 2016; Fujita et al. 2013; Davis et al. 2012; Meerang et al. 2011; Partridge et al. 2003) • ER-stress-induced gene transcription (Marza et al. 2015) • Antiviral signaling (Hauler et al. 2012, Hao et al. 2015)

ACBD Acyl-CoA-binding domain-containing protein, *AKAP* A-kinase anchor protein, *AR* adrenergic receptor, *ARL* ADP-ribosylation factor-like protein, *BICD2* protein bicaudal D homolog 2, *CFTR* cystic fibrosis transmembrane conductance regulator, *CG-NAP* centrosome- and Golgi-localized PKN-associated protein, *CK* casein kinase, *COG complex* conserved oligomeric Golgi complex, *COH* Cohen syndrome protein, *DjA1* DnaJ homolog subfamily A member 1, *GCP* Golgi complex-associated protein, *GMAP* Golgi-associated microtubule-binding protein, *GORAB* RAB6-interacting golgin, *GPI* glycosylphosphatidylinositol, *GRAB* GRIP-related ARF-binding domain, *GRIP* Arl-binding domain, *KCTD5* potassium channel tetramerization domain-containing protein 5, *Lrrk* leucine-rich repeat serine/threonine-protein kinase, *Mena* mammalian enabled homologue, *MPR* mannose 6-phosphate receptor, *NECC* neuroendocrine long coiled-coil protein, *N-Myr* amino-terminal myristoylation, *P* peripheral membrane protein, *PIST* PDZ protein interacting specifically with TC10, *PM* plasma membrane, *ROMK* renal outer medullary potassium, *SCOCO* short coiled-coil protein, *SNARE* soluble NSF attachment protein receptor, *TBC1D23* TBC1 domain family member 23, *TGN* *trans*-Golgi network, *TGF* transforming growth factor, *TMD* transmembrane domain, *TMED* transmembrane emp24 domain-containing protein, *VCIP* valosin-containing protein p97/p47 complex-interacting protein, *VCP* valosin-containing protein, *VPS* vacuolar protein sorting-associated protein, *ZFPL1* zinc finger protein-like 1
This table is updated from Xiang and Wang (2011)
[a]Predicted

D'Angelo et al. 2009; Barr et al. 2001), unconventional secretion (Dupont et al. 2011; Rabouille and Linstedt 2016; Vinke et al. 2011; Gee et al. 2011; Piao et al. 2017), cell cycle regulation (Preisinger et al. 2005; Sutterlin et al. 2005; Yoshimura et al. 2005; Duran et al. 2008; Tang et al. 2010b), apoptosis (Lane et al. 2002), and autophagy (Zhang et al. 2018; Zhang and Wang 2018a, b), although the mechanisms are less well understood.

Both GRASPs share a similar structure: a conserved N-terminal GRASP domain consisting of two PDZ domains (PDZ1 and PDZ2) and an intrinsically disordered C-terminal Serine/Proline-Rich (SPR) domain with multiple phosphorylation sites (Zhang and Wang 2015b) (Fig. 19.1). Both GRASP65 and GRASP55 are peripheral membrane proteins that are attached to the Golgi membranes via an N-terminal myristic acid modification and the interaction with their membrane-bound partner proteins (GM130 and Golgin-45, respectively) and therefore are concentrated at the interface between the cisternae where stacking occurs (Short et al. 2001; Barr et al. 1998). GRASP65 is concentrated on the *cis*-Golgi cisternae, whereas GRASP55 localizes to the *medial/trans*-Golgi cisternae (Barr et al. 1997; Shorter et al. 1999). Both play complementary roles in Golgi stack formation (Xiang and Wang 2010).

Mechanistically, GRASP proteins form homodimers via the N-terminal PDZ domains, and dimers from adjacent Golgi cisternae further oligomerize in *trans* and function as the "glue" that tethers the cisternae into a stack (Wang et al. 2003, 2005). An in vitro study using modified GRASP domain peptides indicated that insertion of the myristic acid moiety is required for the oriented association to Golgi membranes, which ensures the protein-protein interaction in *trans*. Furthermore, the conformational change caused by myristoylation affects the tendency of GRASP domain for self-interaction (Heinrich et al. 2014). Depletion of either GRASP65 or GRASP55 reduces the number of cisternae per Golgi stack, whereas depletion of both GRASPs leads to disassembly of the entire Golgi stack (Sutterlin et al. 2005, Xiang and Wang 2010, Bekier et al. 2017). The GRASPs are tightly modulated by a phosphorylation and dephosphorylation cycle during cell division, resulting in mitotic disassembly and post-mitotic reassembly of the Golgi (Feinstein and Linstedt 2008; Cervigni et al. 2015; Lin et al. 2000; Wang et al. 2003; Preisinger et al. 2005; Tang et al. 2012; Xiang and Wang 2010; Tang et al. 2008; Truschel et al. 2012).

Recent studies have identified novel GRASP-binding proteins involved in Golgi biogenesis and morphology modulation. Recent research of the crystal structure of GRASP55 bound to the Golgin-45 C-terminal peptide revealed that Golgin-45 promotes the oligomerization of GRASP55 by forming a new interaction between two neighboring PDZ2 molecules to play an important role in Golgi stacking (Zhao et al. 2017). Meanwhile, using an optimized in vitro system, mammalian enabled homologue (Mena) and DnaJ homolog subfamily A member 1 (DjA1) were identified as GRASP65 binding partners with potential functions on Golgi structure maintenance (Tang et al. 2016; Li et al. 2019). Mena is an actin elongation factor recruited to the Golgi membranes by GRASP65 to facilitate actin polymerization and GRASP65 oligomerization and thus functions with actin as bridging proteins of GRASP65 in Golgi ribbon linking. DjA1 is a co-chaperone of Heat shock cognate 71 kDa protein (Hsc70), but the activity of DjA1 in Golgi structure formation is independent of its

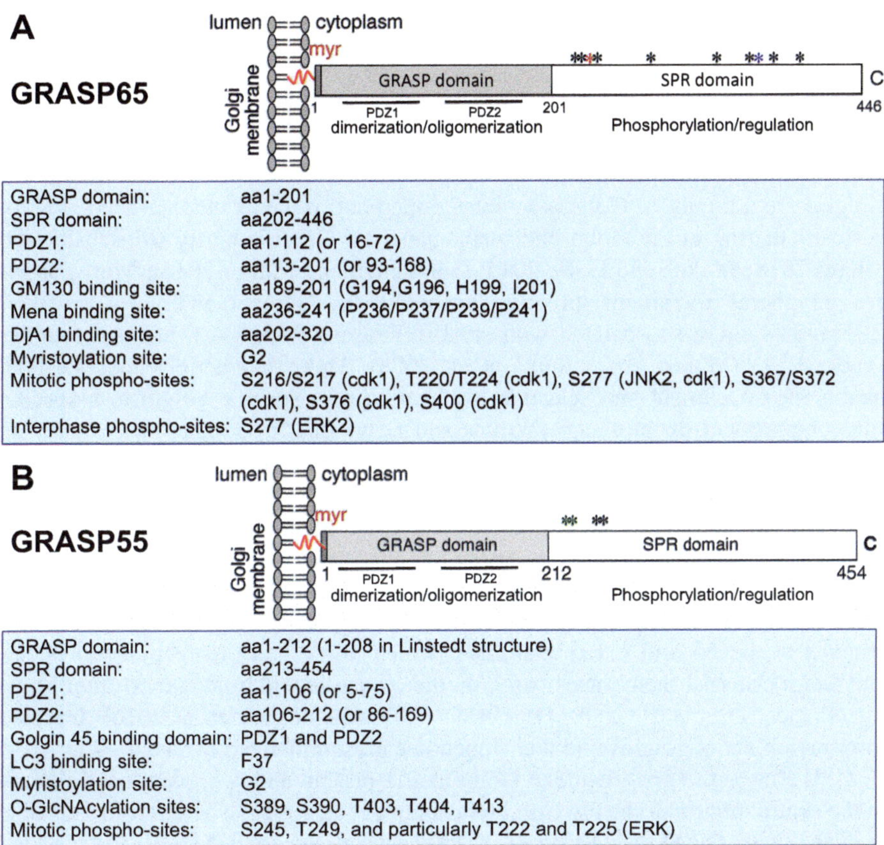

GRASP domain:	aa1-201
SPR domain:	aa202-446
PDZ1:	aa1-112 (or 16-72)
PDZ2:	aa113-201 (or 93-168)
GM130 binding site:	aa189-201 (G194,G196, H199, I201)
Mena binding site:	aa236-241 (P236/P237/P239/P241)
DjA1 binding site:	aa202-320
Myristoylation site:	G2
Mitotic phospho-sites:	S216/S217 (cdk1), T220/T224 (cdk1), S277 (JNK2, cdk1), S367/S372 (cdk1), S376 (cdk1), S400 (cdk1)
Interphase phospho-sites:	S277 (ERK2)

GRASP domain:	aa1-212 (1-208 in Linstedt structure)
SPR domain:	aa213-454
PDZ1:	aa1-106 (or 5-75)
PDZ2:	aa106-212 (or 86-169)
Golgin 45 binding domain:	PDZ1 and PDZ2
LC3 binding site:	F37
Myristoylation site:	G2
O-GlcNAcylation sites:	S389, S390, T403, T404, T413
Mitotic phospho-sites:	S245, T249, and particularly T222 and T225 (ERK)

Fig. 19.1 Structure, modification, and binding sites on GRASP65 (**a**) and GRASP55 (**b**). Rat GRASP65 and GRASP55 sequences are used for illustration. Both GRASPs share a similar structure: a conserved N-terminal GRASP domain consisting of two PDZ domains (PDZ1 and PDZ2) and a C-terminal Serine/Proline-Rich (SPR) domain with multiple phosphorylation sites (indicated by asterisks) that are involved in GRASP modulation during the cell cycle. Both GRASP65 and GRASP55 are peripheral membrane proteins attached to the Golgi membranes via N-terminal myristoylation and the interaction with their membrane-bound partner proteins (GM130 and Golgin-45, respectively). GRASP65-binding proteins Mena and DjA1 have been identified to enhance Golgi ribbon linking and stacking, respectively. GRASP55 is regulated by O-GlcNAcylation depending on the glucose level and interacts with LC3 and LAMP2 to facilitate glucose starvation-induced autophagy

co-chaperone activity or Hsc70, rather, through DjA1-GRASP65 interaction to promote GRASP65 oligomerization. Thus, DjA1 facilitates Golgi structure formation through an unconventional Hsc70-independent pathway. These studies further confirmed GRASP65 as a multifaceted protein in Golgi structure formation and indicated that an array of GRASP binding proteins could play important roles in Golgi morphology maintenance (Fig. 19.1). In addition, GRASP55 was reported to be involved in glucose starvation-induced autophagy (Zhang et al. 2018), where it interacts with LC3 on autophagosomes and LAMP2 on lysosomes to facilitate autophagosome maturation, which will be discussed in later sections.

19.1.1.2 Golgins

Golgins are a family of Golgi-associated coiled-coil proteins that are necessary for vesicle tethering at the Golgi and maintenance of Golgi integrity (Muschalik and Munro 2018; Witkos and Lowe 2015; Gillingham and Munro 2016). Most golgins are peripheral membrane proteins anchored on Golgi membranes via their C-terminus and are associated with small GTPases of Rab, Arf, and Arl families (Table 19.1) (Munro 2011; Sinka et al. 2008). These interactions mediate both membrane attachment and selective localization of a specific golgin to a specific sub-compartment of the Golgi (Witkos and Lowe 2015). Golgins lack significant sequence homology between the family members and localize to different regions of the Golgi to play distinct roles in tethering events, membrane traffic, and Golgi organization. The coiled-coil regions provide the golgins with an extended structure required for the tethering function, while the interactions with Rab GTPases control these molecules in their open (extended) or closed (folded) confirmation (Cheung et al. 2015). In addition, golgins often contain specific sequence and structural features at the N- and C-terminal ends, which allow them to recognize vesicles and Golgi cisternal membranes based on the curvature and lipid composition of the membranes (Drin et al. 2008; Drin et al. 2007; Magdeleine et al. 2016). Detailed information about golgins and their functions are summarized in Table 19.1.

GM130 was the first identified Golgi matrix protein and is predominantly found in the central region of the *cis*-Golgi (Nakamura et al. 1995), where it forms a stable complex with GRASP65 (Barr et al. 1998). Depletion of GM130 results in the disruption of the Golgi ribbon and causes protein glycosylation defects (Puthenveedu et al. 2006). There are two possible ways GM130 contributes to Golgi ribbon formation. First, GM130 targets GRASP65 to the rim of the cisternae where GRASP65 promotes lateral linking of cisternae via oligomerization (Puthenveedu et al. 2006). Mena and actin cytoskeleton may facilitate GRASP65 in this action (Tang et al. 2016). Second, GM130 recruits A-kinase anchoring protein 450 (AKAP450) onto the *cis*-Golgi and allows Golgi-associated nucleation of microtubules, which arranges Golgi stacks in close proximity to form a ribbon (Rivero et al. 2009). Similarly, other golgins may also work with the microtubule cytoskeleton in a similar way to facilitate Golgi structure organization. For example, GMAP-210, another *cis*-Golgi-localized golgin, recruits the γ-tubulin-containing complexes to the Golgi membranes and promotes the formation of tubulin oligomers on Golgi membranes. Depletion of GMAP-210 results in extensive Golgi fragmentation, suggesting a role in Golgi ribbon formation (Rios et al. 2004).

In addition to GRASP65, GM130 also interacts with p115 and giantin to form the GM130-p115-giantin tethering complex (Sonnichsen et al. 1998). The Rab1 effector p115, when recruited to coat protein complex (COP) II vesicles during the budding from the ER, interacts with the soluble ATPases *N*-ethylmaleimide-sensitive factor

(NSF) attachment protein receptors (SNAREs), a specialized set of COPII v-SNAREs, to form a *cis*-SNARE complex that promotes vesicle targeting to the Golgi apparatus (Allan et al. 2000). Meanwhile, p115 also binds giantin on COPI vesicles and works with GM130 on Golgi membranes to provide a bridging role for vesicle tethering (Sonnichsen et al. 1998). Thus, GM130 and p115 are two major tethering factors in ER-to-Golgi trafficking (Munro 2011).

Other than the well-studied GRASP65-GM130 and GM130-p115-giantin complexes, the GRIP domain containing golgins are another group of proteins associated with the Golgi structure. Most GRIP domain-containing golgins localize to the TGN via their GRIP domains and are involved in Golgi organization (Luke et al. 2005). GCC185 is reported to localize independently of Arl1 on TGN and plays an essential role in Golgi structure formation. Depletion of GCC185 results in fragmentation of both *cis*- and *trans*-Golgi that are dispersed throughout the cytoplasm (Derby et al. 2007). On the other hand, another TGN golgin GCC88 is reported to play a role in TGN organization and ribbon-linking. Overexpression of GCC88 causes a loss of the compact Golgi ribbon and dispersal of mini-stacks throughout the cytoplasm, while knockdown of GCC88 results in a longer ribbon structure (Gosavi et al. 2018). A recent report suggests that GCC88-induces Golgi ribbon dispersal via actin and non-muscle myosin IIA. In addition, a novel GCC88-binding partner, the long isoform of intersectin-1 (ITSN-1), a guanine nucleotide exchange factor for Cdc42, is identified to be involved in this process (Makhoul et al. 2019).

19.1.2   *Other Golgi Structure-Related Proteins*

Besides the Golgi matrix proteins and their cofactors described above, other proteins including SNAREs, kinases, methyltransferases, and GTPases have also been reported to be related to Golgi structural organization and function. A few examples are discussed below.

Vesicle-associated membrane protein 4 (VAMP4), a v-SNARE protein located on the TGN, was first shown to play a role in retrograde trafficking from early endosomes to the TGN (Steegmaier et al. 1999). It was later reported that depletion of VAMP4 led to fragmentation of the Golgi ribbon, although Golgi membranes remained in the juxtanuclear area. EM studies revealed shortened Golgi stacks with a normal arrangement. Depletion of the cognate SNAREs of VAMP4, syntaxin 6, syntaxin 16, and Vti1a also disrupted the Golgi ribbon. These findings suggest that the maintenance of the Golgi ribbon structure requires normal retrograde trafficking, which is likely mediated by the formation of VAMP4-containing SNARE complexes (Shitara et al. 2013).

Serine/threonine-protein kinase H1 (PSKH1) was primarily characterized with multiple intracellular localizations, including Brefeldin A-sensitive Golgi compartment, centrosomes, nucleus, and cytoplasm (Brede et al. 2000). PSKH1 targeting to Golgi depends on dual N-terminal acylation, myristoylation on glycine 2, and palmitoylation on cysteine 3. Expression of palmitoylation site mutant PSKH1 results in the disassembly of the Golgi apparatus to a diffused cytoplasmic pattern without interrupting the microtubule cytoskeleton (Brede et al. 2003). The substrates of this kinase on the Golgi are so far unidentified.

Protein arginine methyltransferase 5 (PRMT5) localizes to the Golgi apparatus and forms complexes with several components, including GM130, which was later identified as a substrate of PRMT5. N-terminal methylation of GM130 does not affect its Golgi localization but is critical for Golgi ribbon formation. Depletion of PRMT5 and expression of methylation-defective GM130 mutants result in fragmentation and dispersal of the Golgi ribbon (Zhou et al. 2010).

In addition to the proteins mentioned above, Rab small GTPases are another group of key regulators of mammalian Golgi organization. Many Rab proteins, including but not limited to Rab1, Rab2 and Rab8 (Aizawa and Fukuda 2015), Rab18 and Rab43 (Dejgaard et al. 2008), Rab6/41 (Goud et al. 1990; Martinez et al. 1997; Liu et al. 2013), and Rab30 (Kelly et al. 2012), have been shown to play a role in Golgi structure organization and reviewed previously in detail (Goud et al. 2018). Considering that Rab proteins switch between inactive GDP-bound and active GTP-bound forms, it has been proposed that Golgi organization-related Rab proteins are divided into two categories. With Class 1 Rabs, the Golgi ribbon is disrupted by Rab inactivation but appears normal with overexpression, whereas with Class 2 Rabs, Rab inactivation has little effect on Golgi ribbon organization, while overexpression leads to the redistribution of Golgi enzymes to the ER (Liu and Storrie 2015). These results indicate that Rabs control the Golgi structure through modulating membrane tethering and trafficking.

19.2   Golgi Dynamics in the Mammalian Cell Cycle

The Golgi undergoes a series of sophisticated cell cycle-dependent disassembly and reassembly processes, including the deformation and reformation of Golgi ribbon, stacks, and cisternae. At the onset of mitosis, the Golgi ribbon unlinks into ministacks, which further undergo unstacking and vesiculation. These processes ensure the equal distribution of Golgi compartments into the two daughter cells (Wang 2008; Wei and Seemann 2010). In telophase, the Golgi vesicles fuse into cisternae and form stacks. The new stacks then accumulate in the perinuclear region and further link into a ribbon. The molecular factors that control these processes include Golgi matrix proteins, kinases and phosphatases, ubiquitin ligases and deubiquitinating enzymes, vesicle budding and fusion factors, and actin and microtubule cytoskeleton. An in vitro system has been developed to replicate the Golgi disassembly and reassembly process through sequential treatments of purified Golgi membranes with mitotic (MC) and interphase (IC) cytosol, or with purified proteins (Tang et al. 2008, 2010a). This system provides a powerful tool for testing key proteins in Golgi structure formation, which makes it possible to identify the minimal machinery and key components that control mitotic Golgi disassembly and post-mitotic reassembly (Huang and Wang 2017).

19.2.1 Mechanisms of Golgi Disassembly and Reassembly in the Mammalian Cell Cycle

The first step of Golgi disassembly at the onset of mitosis is Golgi ribbon unlinking. At this step, Golgi stacks in the ribbon are disconnected and dispersed. This step involves disconnecting the tubules between the stacks by the membrane fission protein CtBP/BARS, which is crucial for G2/M transition (Hidalgo Carcedo et al. 2004; Colanzi et al. 2007). Further, GRASPs undergo mitotic phosphorylation which are also required for ribbon-unlinking. The extracellular-signal-regulated kinase (ERK) directly phosphorylates GRASP55 and blocks its activity in both Golgi ribbon formation and *trans*-oligomerization (Feinstein and Linstedt 2008), while GRASP65 is phosphorylated by c-Jun N-terminal kinase (JNK) on Serine 277 (S277), which causes the separation of the Golgi stacks (Cervigni et al. 2015).

Sequential phosphorylation of GRASP65 on multiple sites by cyclin-dependent kinase 1 (Cdk1) and polo-like kinase 1 (Plk1) results in its conformational changes and subsequent de-oligomerization (Lin et al. 2000; Preisinger et al. 2005; Wang et al. 2003; Tang et al. 2012; Vielemeyer et al. 2009). On the other hand, GRASP55 is phosphorylated by ERK and partially by Cdk1 (Xiang and Wang 2010). In addition to the unique phosphorylation sites in the SPR domains of GRASPs, S189 within the GRASP domain of GRASP65 is modified by Plk1, which causes conformational change and impaired self-association (Sengupta and Linstedt 2010). An in vivo membrane-tethering activity assay using a construct of full-length GRASP55 fused to the C-terminal mitochondrial anchoring sequence shows that the tethering activity is diminished by introducing a phosphomimic S189D mutant. This result suggests that S189 might be a Plk1 target site on both GRASPs, although direct evidence remains to be provided (Truschel et al. 2012). GRASP65 is dephosphorylated by PP2A in late mitosis and the *trans*-oligomer reformation is therefore rehabilitated to promote cisternae stacking (Tang et al. 2008).

The unstacked cisternae further disassemble into vesicles, which depends on COPI vesicle formation and blockage of vesicle docking and membrane fusion. As mentioned above, the GM130-p115-giantin complex promotes COPII vesicle docking to the Golgi. Cdk1 phosphorylation of GM130 on S25 during mitosis inhibits p115-interaction and therefore blocks vesicle docking (Lowe et al. 1998). Inhibition of Cdk1 causes Golgi vesiculation failure, suggesting an essential role of Cdk1 activity in mitotic Golgi vesiculation. However, expression of the GM130 S25A non-phosphorylatable mutant in GM130-depleted cells causes no apparent defects in Golgi vesiculation and mitotic progression, indicating the existence of GM130 S25 phosphorylation-independent pathways that ensure Golgi vesiculation and mitotic progression in mammalian cells (Sundaramoorthy et al. 2010).

Recently, the recruitment and activation of Aurora kinase family member Aurora A at the centrosomes in M phase was reported to depend on Golgi ribbon unlinking in G2 phase (Barretta et al. 2016). Aurora A functions in centrosome maturation, mitotic entry, and bipolar spindle formation during mitosis (Nikonova et al. 2013; Carmena et al. 2009; Kimura et al. 2013). This finding indicates a potential link between Aurora A activity and cell cycle-associated Golgi structure modulation. Indeed, it was later confirmed that knockdown or inhibition of Aurora A induces Golgi dispersal without affecting the GM130 protein level in interphase. Further investigation revealed that interference of Aurora A causes Golgi dispersal only after mitosis via the dissociation of the Golgi and centrosome (Kimura et al. 2018). These studies revealed a novel relationship between G2 phase Golgi unlinking, M phase Aurora A activation, and interphase Golgi structure formation.

19.2.2 Post-mitotic Golgi Membrane Fusion and Its Regulation

Two AAA ATPases, NSF and p97/VCP, are involved in membrane fusion during post-mitotic Golgi reassembly, and their activities are regulated by phosphorylation during mitosis (Rabouille et al. 1995). For NSF-catalyzed fusion, the p115-GM130 tethering complex is disrupted by GM130 phosphorylation during mitosis (Lowe et al. 1998), while post-mitotic phosphorylation of p115 by a casein kinase II (CKII)-like enzyme is required for cisterna reassembly (Dirac-Svejstrup et al. 2000). Contemporarily, homotypic fusion of Golgi membranes mediated by p97 is also blocked upon phosphorylation of p47 and p37. p97 uses these two distinct cofactors for its membrane fusion function: p47 is essential for the regrowth of Golgi cisternae from mitotic Golgi fragments (Kondo et al. 1997), while p37 is required for the maintenance of the Golgi structure in interphase as well as for its reassembly in late mitosis (Uchiyama et al. 2006). Both pathways are regulated by Cdk1-mediated phosphorylation. Phosphorylation of p47 on S140 abolishes its binding to Golgi membranes, resulting in mitotic inhibition of the p97/p47 pathway (Uchiyama et al. 2003). Phosphorylation on S56 and Threonine 59 (T59) disables p37 from binding to Golgi membranes and consequently blocks p97/p37-mediated Golgi membrane fusion at late mitosis (Kaneko et al. 2010).

In addition to phosphorylation, p97/p47-mediated Golgi membrane fusion is also regulated by ubiquitination (Tang and Wang 2013). Tang et al. discovered that the Homologous to the E6-AP Carboxyl Terminus (HECT) domain containing ubiquitin ligase HACE1 is targeted to the Golgi membrane through the interaction with Rab1 and participates in post-mitotic Golgi biogenesis (Tang et al. 2011). Depletion of HACE1 or expression of an inactive mutant impairs post-mitotic Golgi membrane fusion. The identification of HACE1 as a Golgi-localized ubiquitin ligase provides evidence that ubiquitin has a critical role in Golgi biogenesis during the cell cycle (Tang et al. 2011). Later, the Golgi t-SNARE syntaxin 5 was identified as a ubiquitination substrate (Huang et al. 2016). Syntaxin 5 is monoubiquitinated by HACE1 in early mitosis and deubiquitinated by the de-ubiquitinase VCIP135 in late mitosis (Wang et al. 2004). The monoubiquitination of syntaxin 5 at Lysine 270 (K270) in the SNARE domain impairs the interaction between syntaxin 5 and the cognate v-SNARE Bet1 but increases its binding to the p97 adaptor p47 through the UBA domain of p47, which is required for post-mitotic Golgi membrane fusion (Meyer et al. 2002). Expression of the syntaxin 5 K270R mutant in cells impairs post-mitotic Golgi reassembly. Therefore, monoubiquitinated syntaxin 5 recruits p97/p47 to the mitotic Golgi fragments and promotes post-mitotic Golgi reassembly upon ubiquitin removal by VCIP135 (Huang et al. 2016). VCIP135 was originally identified as a p97 interacting protein (Kano et al. 2005). It was later shown to be a de-ubiquitinating (DUB) enzyme involved in p97/p47-mediated membrane fusion (Wang et al. 2004). VCIP135 DUB activity, as well as its interaction with p97 and association with Golgi membranes, is regulated by phosphorylation (Zhang and Wang 2015a; Zhang et al. 2014). In mitosis, VCIP135 is phosphorylated at S130 by Cdk1 and thus is inactivated, allowing syntaxin 5 to be ubiquitinated by HACE1; in telophase, VCIP135 is dephosphorylated and reactivated, removing ubiquitin from syntaxin 5 to allow p97-mediated membrane fusion (Huang and Wang 2017; Wang 2008). These studies revealed a novel mechanism that monoubiquitination regulates Golgi membrane dynamics during the mammalian cell cycle.

19.3 Golgi Stress Response

As stated above, the Golgi apparatus in mammalian cells forms a unique stacked structure under normal growth conditions, which undergoes a regulated disassembly and reassembly process during the cell cycle. However, the Golgi structure and function could be impaired under stress conditions, such as DNA damage, energy and nutrient deprivation, and pro-apoptotic conditions. This could be attributed to perturbation of microtubule organization or phosphorylation, degradation, or cleavage of Golgi structural proteins. Additionally, many signaling molecules have been identified to be associated with the Golgi. Thus, it has been proposed that the Golgi could sense and transduce stress signals and therefore serves as a hub in the cellular signaling network (Farhan and Rabouille 2011; Mayinger 2011; Makhoul et al. 2019).

19.3.1 Apoptotic Stress and Golgi Fragmentation

Apoptosis, also known as programmed cell death, is a cell suicide mechanism carried out by organelle-directed regulators such as the Bcl-2 proteins and ultimately executed by the caspase family proteases (Nicholson and Thornberry 1997). Organellar response to apoptotic initiation includes death receptor endocytosis, mitochondrial and lysosomal permeabilization, ER calcium release, and Golgi fragmentation. The Golgi is one of the first organelles to be affected during apoptosis (Mukherjee et al. 2007; Aslan and Thomas 2009). During apoptosis, several Golgi matrix proteins related to Golgi structure maintenance are cleaved by caspases, leading to Golgi fragmentation (Hicks and Machamer 2005). Apoptotic Golgi

Table 19.2 Apoptotic cleavage of Golgi proteins

Names	Apoptosis inducer	Caspases	Cleavage site	Golgi structural change
Golgin-160	STS; CH11	Caspase-2, 3 and 7	D59, D139, D311	Golgi fragmentation (Mancini et al. 2000; Mukherjee et al. 2007; Nozawa et al. 2002; Hicks and Machamer 2002; Maag et al. 2005)
GRASP65	Anisomycin; STS	Caspase-3	D320, D375, D393	Golgi fragmentation (Lane et al. 2002; Cheng et al. 2010)
p115	STS; 4-hydroxytamoxifen; CH11; CA	Caspase-3 and 8	D757	Golgi fragmentation (Chiu et al. 2002; How and Shields 2011; Mukherjee et al. 2007; Mukherjee and Shields 2009; Woldemichael et al. 2011)
GM130	CH11	Caspase-3	–	Golgi fragmentation (Walker et al. 2004; Lowe et al. 2004; Mukherjee et al. 2007)
Syntaxin 5	STS; anisomycin	Caspase-3	D1882, D1083	Secretion inhibition (Lowe et al. 2004)
Giantin	STS; anisomycin	Caspase-3	D263	Secretion inhibition (Lowe et al. 2004; Nozawa et al. 2002)

CA carminomycin I, *CH11* an anti-Fas monoclonal antibody, *D* aspartic acid, *STS* staurosporine

fragmentation is one of the most extensively studied Golgi stress responses. Reported caspases-cleaved Golgi proteins include GRASP65, golgin-160, GM130, p115, syntaxin 5, and giantin (Lane et al. 2002; Mancini et al. 2000; Walker et al. 2004; Chiu et al. 2002; Lowe et al. 2004; Machamer 2015), as summarized in Table 19.2 and discussed below.

19.3.1.1 Golgin-160

Golgin-160 is a golgin that plays a role in vesicle tethering and trafficking (Misumi et al. 1997). It is cleaved by caspase-2, caspase-3, and caspase-7 during apoptosis. Under pro-apoptotic conditions stimulated by staurosporine, the Golgi senses and transduces apoptotic signals using a local caspase, caspase-2. Caspase-2 is special in a way that it has both the property of initiator caspases and the substrate specificity of executioner caspases (Mancini et al. 2000). Although it is unclear how caspase-2 is activated by pro-apoptotic signals, in vitro and in vivo caspase cleavage assays showed that caspase-2 cleavage of golgin-160 at aspartate 59 (D59) happens prior to golgin-160 cleavage by caspase-3 and 7 at D139 and D311 (Mancini et al. 2000). Expression of the D59A cleavage-defective mutant of golgin-160 delays Golgi disintegration under staurosporine treatment (Machamer 2003; Hicks and Machamer 2005).

Subsequently, it was shown that an N-terminal 85 amino acid fragment of golgin-160 contains both a Golgi localization signal and a nuclear localization signal (Hicks and Machamer 2002). Expression of a non-cleavable golgin-160 mutant inhibits ER stress or ligation of death receptor-induced apoptosis (Maag et al. 2005). Latterly, yeast two hybrid screening revealed that GCP60 preferentially binds to one of the caspase cleavage products of golgin-160, aa 140-311, to inhibit its nuclear localization (Sbodio et al. 2006). Overexpression of GCP60 sensitizes cells to staurosporine-induced apoptosis, while nuclear localization of a golgin-160 apoptotic cleavage fragment (aa 140-311) protects cells from apoptosis. However, another report indicates that golgin-160 depletion does not affect the Golgi morphology nor constitutive secretion (Williams et al. 2006). Therefore, the mechanism of how golgin-160 transduces apoptotic signals and regulates the apoptotic response needs to be further studied.

19.3.1.2   GRASP65

GRASP65 is cleaved in apoptosis induced by oxygen- and glucose-deprivation (OGD) as in ischemia-induced cerebral vascular endothelial injury (Yin et al. 2010) and in staurosporine- or Fas ligand-induced apoptosis (Lane et al. 2002, Cheng et al. 2010). In apoptosis, GRASP65 is cleaved by caspase-3 on D320, D375, and D393. Expression of a cleavage-resistant form of GRASP65 delays Golgi fragmentation in apoptosis and protects cells from Fas/CD95-mediated apoptosis, whereas expression of an N-terminal caspase-cleaved fragment dramatically sensitizes cells to Fas/CD95-mediated apoptosis (Lane et al. 2002, Cheng et al. 2010). Further results revealed that the C-terminal fragments of GRASP65 produced by caspase cleavage promotes Fas/CD95-mediated apoptosis via being targeted to mitochondria by binding to Bcl-X_L (Cheng et al. 2010). However, the mechanism of how the C-terminal cleavage fragment of GRASP65 regulates apoptosis at the mitochondria and the role of Bcl-X_L in this process are still unknown. The Golgi fragmentation phenotype induced by apoptotic GRASP65 cleavage is similar to that of GRASP65 phosphorylation in mitosis (Warren 1995). There is evidence that several kinases involved in mitotic GRASP65 phosphorylation such as Cdk1 and ERK are activated during apoptosis and regulate apoptosis by phosphorylating caspase and Bcl-2 family proteins (Terrano et al. 2010; Yamaguchi et al. 2008; Lu et al. 2011). Whether these kinases directly regulate apoptotic Golgi fragmentation by phosphorylating GRASP65 or other Golgi proteins remains unclear (Ji et al. 2013).

19.3.1.3   p115

The ER-to-Golgi membrane tether, p115, is cleaved by caspase-3 and caspase-8 during apoptosis. Expression of a caspase-resistant form of p115 delays Golgi fragmentation in apoptosis. Exogenous expression of a p115 C-terminal apoptotic fragment leads to apoptosis and Golgi fragmentation (Chiu et al. 2002). The extreme C-terminal fragment, generated by caspase cleavage during apoptosis, translocates into the nucleus and further activates the apoptosis machinery. Interestingly, translocation of the p115 C-terminal fragment happens prior to major Golgi structural changes, indicating it as an early event (Mukherjee and Shields 2009). The p115 C-terminus is SUMOylated, which regulates its nuclear translocation and amplification of apoptosis signals in a p53-dependent manner (How and Shields 2011; Mukherjee and Shields 2009). In a high-throughput screen, Carminomycin I (CA) was discovered to inhibit cell proliferation of Von Hippel-Lindau (VHL) defective Clear Cell Renal Cell Carcinoma (*VHL−/−* CCRCC) (Woldemichael et al. 2011). CA activates caspase-2 and caspase-3 to cleave p115, which inhibits CCRCC proliferation (Woldemichael et al. 2011).

19.3.1.4 Other Proteins

Several other Golgi structural proteins are also involved in apoptosis or cleaved by caspases. The level of GM130 is reduced during Fas-mediated apoptosis but not in staurosporine-induced apoptosis (Walker et al. 2004). However, it is not clear whether the reduction is due to GM130 cleavage or degradation. Syntaxin 5 and giantin are also cleaved by caspase-3 during apoptosis, which inhibits ER-to-Golgi transport (Lowe et al. 2004). Golgin-95 and golgin-97 are cleaved during necrosis but not apoptosis (Nozawa et al. 2002). Cleavage of golgin-95 and golgin-97 during necrosis is also caspase-dependent, since pretreatment of the cells with pan-caspase inhibitor, zVAD-fmk, abolished the cleavage of these two proteins (Nozawa et al. 2002).

Some of the Golgi proteins are also reported to regulate apoptosis. The Golgi SNARE GS28 is involved in cisplatin-induced apoptosis in a p53-dependent manner (Sun et al. 2012). Overexpression of GS28 sensitizes HEK293 cells to the apoptosis-inducer cisplatin via the accumulation of p53 and Bax and the stimulation of p53 pro-apoptotic phosphorylation at S46. It was also shown that GS28 forms a complex with the p53 ubiquitin E3 ligase Murine Double Minute 2 (MDM2) to inhibit its function and consequential p53 ubiquitination and degradation (Sun et al. 2012). Therefore, GS28 promotes cisplatin-induced apoptosis by stabilizing and regulating pro-apoptotic phosphorylation of p53.

A well-studied mediator of intracellular vesicle fusion, NSF attachment protein α (αSNAP), has been reported to have pro-survival functions (Naydenov et al. 2012). Depletion of αSNAP triggers apoptosis in epithelial cells by reducing the anti-apoptotic protein Bcl-2. Depletion of αSNAP in p53 null or Bax null cells still results in apoptosis, indicating that the anti-apoptotic function is independent of p53 and Bax. Interestingly, αSNAP depletion induces apoptosis independent of the cleavage of Golgi proteins such as GRASP65, golgin-160, and p115 but rather by dysregulation of ER-Golgi vesicle cycling and possibly through ER stress (Naydenov et al. 2012). Some other Golgi proteins, including human Golgi anti-apoptotic protein (h-GAAP) (Gubser et al. 2007; Saraiva et al. 2013) and Golgi integral membrane protein 4 (GOLIM4) (Bai et al. 2018), are also reported to have anti-apoptotic functions. GOLIM4 is overexpressed in some head and neck cancers, and depletion of GOLIM4 reduces cell proliferation and cell viability by inducing apoptosis (Bai et al. 2018).

Although microtubule and actin filaments play important roles in Golgi orientation and structure, Golgi fragmentation in apoptosis occurs prior to cytoskeleton disorganization (Mukherjee et al. 2007; Yadav and Linstedt 2011). Furthermore, the level of actin and tubulin did not change during apoptosis, while Golgi structural proteins are cleaved as discussed above. Therefore, Golgi fragmentation in early apoptosis is independent of microtubule and actin filament disorganization.

19.3.2 *GOLPH3 and DNA Damage-Induced Golgi Fragmentation*

The Golgi phosphoprotein 3 (GOLPH3) is a peripheral membrane protein that regulates vesicle budding and TGN-to-plasma membrane trafficking (Dippold et al. 2009). GOLPH3 is localized to the TGN by binding to phosphatidylinositol 4-phosphate (PI4P). Depletion of PI4P leads to GOLPH3 dissociation from the TGN. GOLPH3 also binds to the actin-based motor protein MYO18A to link Golgi membranes with the actin cytoskeleton. This bridging effect creates a tension required for vesicle budding, trafficking, and maintenance of the Golgi ribbon. Depletion of GOLPH3 or MYO18A leads to the loss of the tensile force, resulting in the shrinkage of the Golgi ribbon and a reduction of vesicles formed at the TGN (Dippold et al. 2009; Bishe et al. 2012; Ng et al. 2013). GOLPH3 is an oncogene known to be overexpressed in some solid tumors, including lung cancer and breast cancer (Scott et al. 2009; Zeng et al. 2012). It is also reported that GOLPH3 increases cell proliferation and cell size by regulating cell proliferation through the interaction with the retromer complex and activation of the mammalian target of rapamycin mTOR (Scott et al. 2009). GOLPH3 induces cell proliferation in breast cancer cells by inhibiting the tumor suppressor transcription factor FOXO1 through activating AKT (Zeng et al. 2012). These findings demonstrate that a *trans*-Golgi protein can serve as an oncogene (Scott et al. 2009; Buschman et al. 2015; Kuna and Field 2018).

Interestingly, DNA damage causes Golgi dispersal in a GOLPH3-dependent manner. DNA damage activates the DNA-PK kinase to phosphorylate GOLPH3 on T143/T148, which aberrantly increases the tensile force for the Golgi to fragment. Golgi fragmentation in this scenario increases cell survival with an unknown mechanism (Farber-Katz et al. 2014). Depletion of GOLPH3 or MYO18A increases cancer cells' sensitivity to DNA damage inducing agents, suggesting that GOLPH3 phosphorylation-induced Golgi fragmentation may serve as a protective mechanism (Farber-Katz et al. 2014). Considering that GOLPH3 is overexpressed in many solid tumors, it is reasonable to speculate that this may be a mechanism of how cancer cells escape DNA damage-induced apoptosis (Farber-Katz et al. 2014; Buschman et al. 2015; Li et al. 2016b).

In addition to its high expression level in some cancer cells, GOLPH3 overexpression is also reported in mouse N2A cells under oxygen-glucose deprivation and reoxygenation (OGD/R), a model mimicking severe oxidative injury (Li et al. 2016a). In this OGD/R model, GOLPH3 is overexpressed and forms puncta in the cytosol, which induces the formation of reactive oxygen species (ROS) and lipidation of LC3. Opposed to its anti-apoptotic role in cancer cells, depletion of GOLPH3 in OGD/R desensitizes the cells to apoptosis (Li et al. 2016a).

19.3.3   *Golgi in Autophagy Regulation*

Most recently, the Golgi stacking protein GRASP55 was reported to regulate autophagy upon energy deprivation (Zhang et al. 2018; Zhang and Wang 2018a, b). Under normal growth condition, GRASP55 is O-GlcNAcylated and localizes in the *medial-* and *trans-*Golgi for stacking. However, under glucose starvation, a pool of de-O-GlcNAcylated GRASP55 translocates to the interface between autophagosomes and lysosomes to facilitate autophagosome-lysosome fusion. After a short-term energy deprivation, the Golgi structure is only mildly affected, possibly due to sufficient GRASP55 molecules remaining in the Golgi to maintain its structure. Among over a dozen Golgi proteins tested, only GRASP55, but not GRASP65, GM130, or golgin-45, is O-GlcNAcylated under growth conditions and targets to autophagosomes upon energy deprivation, indicating that GRASP55 serves as an energy sensor on the Golgi to regulate both intracellular trafficking and autophagy. Significantly, the same scenario may be seen in autophagy induced by amino acid starvation and inhibition of mTOR (Zhang et al. 2018).

In addition to Golgi fragmentation, GCC88 overexpression also induces autophagy via reducing the activity of mTOR. A considerable pool of mTOR is localized and activated on the Golgi, which is dependent on the ribbon structure for recruitment but independent of lysosomal mTOR activation (Gosavi et al. 2018). These findings indicate the Golgi ribbon as an important location for the functional regulation of mTOR activity.

Additionally, autophagosomes may directly form on Golgi membranes (Guo et al. 2012) or obtain membranes from the Golgi (Geng et al. 2010; Geng and Klionsky 2010). The only recognized transmembrane ATG protein, ATG9, localizes at the *trans-*Golgi network and late endosomes and is essential for autophagosome formation (Yamamoto et al. 2012), although the detailed mechanism awaits further investigation (Orsi et al. 2012). Recently, the endoplasmic reticulum-Golgi intermediate compartment (ERGIC) is proposed to serve as a key membrane source for autophagosome formation (Ge et al. 2013). Under normal condition, COPII vesicles are generated from the ER-exit sites (ERES) for ER-Golgi membrane trafficking, while upon starvation, the COPII assembly activator Prolactin Regulatory Element-

Binding protein (PREB)/SEC12 relocates to the ERGIC and triggers ERGIC-COPII vesicle formation as membrane templates for LC3 lipidation.

19.4 Alteration of Golgi Structure and Function in Diseases

Golgi structure defects and dysfunctions have been observed in many diseases, including pathogen infection, neurodegenerative diseases, and cancer (Aridor and Hannan 2000). Generally, the mechanisms of Golgi fragmentation include imbalanced membrane flux, altered microtubule dynamics, and posttranslational modifications or proteolytic cleavage of Golgi structural proteins (Wei and Seemann 2017). In many cases, the correlation between Golgi defect and disease progression is unclear. A few interesting cases reported recently are discussed below.

19.4.1 Alzheimer's Disease (AD)

AD is an age-related neurodegenerative disease of the central nerve system characterized by progressive loss of cognition and memory. Golgi fragmentation occurs in neurons of patients with AD since the earliest stages of disease development (Sundaramoorthy et al. 2015). Some cases of early-onset AD are related to mutations in the Amyloid Precursor Protein (APP) or Presenilin 1 and 2 (PSN1 and 2). The amyloid-beta (Aβ) peptide is a proteolytic product of APP, which is considered to be the major inducer of AD (Selkoe and Hardy 2016). Aβ accumulation is likely the direct cause of Golgi fragmentation, as Aβ-treatment causes reversible Golgi fragmentation in cultured neurons (Joshi et al. 2014).

Present research supports that activation of Cyclin-dependent kinase 5 (Cdk5) by Aβ accumulation via the $[Ca^{2+}]$-calpain-p25 pathway may be the major trigger of Golgi fragmentation in AD (Lee and Linstedt 2000; Joshi et al. 2015; Joshi and Wang 2015; Evin 2015; Ayala and Colanzi 2017). Cdk5 may function in two ways. First, Cdk5 phosphorylates GM130 at S25 and inhibits its interaction with the Golgi tethering protein p115 (Sun et al. 2008). Second, Cdk5 phosphorylates GRASP65 at T220/T224, which inhibits GRASP65 function in Golgi stack formation and ribbon linking (Joshi et al. 2014). Furthermore, inhibiting Cdk5 or expressing non-phosphorylatable GRASP65 mutants both rescued the Golgi structure and reduced Aβ secretion by elevating the α-cleavage of APP (Joshi and Wang 2015; Joshi et al. 2014), indicating that GRASP65 phosphorylation may be the main reason for AD-induced Golgi fragmentation. These studies not only provide molecular mechanisms for Golgi fragmentation but also suggest Golgi as a potential drug target for AD treatment (Joshi et al. 2015; Joshi and Wang 2015; Ayala and Colanzi 2017).

19.4.2 Amyotrophic Lateral Sclerosis (ALS)

ALS is a fatal neurodegenerative disorder specifically targeted to motor neurons. Fragmented Golgi has been observed in numerous models of superoxide dismutase 1 (SOD1), TDP-43, FUS, and optineurin-associated ALS (Fujita et al. 2008; Soo et al. 2015; Van Dis et al. 2014; Wallis et al. 2018). SOD1 inhibits ER-to-Golgi transport and causes Golgi fragmentation (Atkin et al. 2014) by reducing the β-COP protein level, accumulating the ER-Golgi v-SNAREs GS15 and GS28, and destabilizing microtubules by the upregulation of Stathmins 1 and 2 (Bellouze et al. 2016; Atkin et al. 2014). FUS and TDP-43 impair the incorporation of secretory cargo into COPII vesicles (Soo et al. 2015). Furthermore, expression of ALS-related optineurin mutants impairs myosin VI-mediated protein trafficking from Golgi to plasma membrane, which also induces Golgi fragmentation (Sundaramoorthy et al. 2015). Conclusively, impairment of distinct protein trafficking pathways by different ALS-linked proteins are specific triggers for Golgi fragmentation in ALS.

19.4.3 Parkinson's Disease (PD)

PD is pathologically characterized by the loss of dopamine-containing neurons and by the formation of intracellular protein aggregates known as Lewy bodies in which α-synuclein has been recognized as a major constituent (Forno 1996; Wakabayashi et al. 1998). Golgi fragmentation can be detected in early-stage PD brains (Fujita et al. 2006) and is strongly correlated to the presence of prefibrillar α-synuclein (Gosavi et al. 2002). Since then, emerging studies have provided important insights into the mechanisms of how α-synuclein causes pathological Golgi fragmentation and neuronal degeneration. The primary effect of α-synuclein aggregation is the inhibition of ER-to-Golgi transport (Lashuel and Hirling 2006), which can be rescued by the overexpression of Rab1 and Rab8 and depletion of Rab2 and syntaxin 5 (Rendon et al. 2013; Coune et al. 2011). In a most recent report, mutations in the leucine-rich repeat kinase 2 (LRRK2), a major genetic cause of autosomal-dominantly inherited PD, markedly enhance Rab7L1 phosphorylation on S72, resulting in TGN fragmentation (Fujimoto et al. 2018).

19.4.4 Cancer

Golgi disorganization may be related to cancer progression and metastasis in the following aspects: aberrant glycosylation, abnormal expression of Ras GTPase, dysregulation of kinases, and hyperactivation of myosin motor proteins (Petrosyan 2015). Perturbation of the Golgi morphology in cancers results in an increase of sialylation which is associated with a metastatic cell phenotype (Schultz et al. 2012).

Overexpression of sialylated antigens is significantly correlated with tumor progression and therapy resistance due to an anti-apoptotic effect (Lee et al. 2008; Park et al. 2012; Petrosyan et al. 2014). Over-activation of Rabs, which coordinate with golgins in protein transportation and Golgi structure maintenance, has been reported in different types of cancers (Goldenring 2013). Furthermore, Golgi disorganization-related kinases, including Src, ERK9, and P21-activated protein kinase (Pak1), are found elevated in tumor cells (Chia et al. 2014; Ching et al. 2007; Weller et al. 2010). The ubiquitin ligase HACE1, which regulates p97-mediated Golgi membrane fusion as discussed above, is reported as a tumor suppressor downregulated in multiple tumors including Wilms' tumor (Anglesio et al. 2004), resulting in Golgi fragmentation (Tang et al. 2011; Cui and Wang 2012). Additionally, hyperactivation of Golgi-associated myosins, including MYO18A that directly binds to GOLPH3 to promote Golgi dispersal (Dippold et al. 2009; Allan et al. 2002), is detected in many aggressive cancers. Golgi fragmentation is one of the essential and earliest events in apoptosis, where several golgins and GRASP65 are cleaved by activated caspases, as described in previous sections.

19.4.5 Viral Infection

Several membrane structures including the Golgi are used by viruses as viral factories to replicate, concentrate, and assemble the viral genome and proteins into viral particles (Miller and Krijnse-Locker 2008; Netherton et al. 2007; Salonen et al. 2005). As a highly dynamic organelle, Golgi serves as a membrane scaffold for multiple viruses, including infectious hepatitis C virus, enteroviruses, poliovirus, foot-and-mouth-disease virus, dengue virus, coronavirus, Kunjin virus, tick-borne encephalitis virus, rubella virus, and bunyamwera virus (Miller and Krijnse-Locker 2008; Harak and Lohmann 2015; Risco et al. 2003; Salanueva et al. 2003; Delgui et al. 2013; Westerbeck and Machamer 2015), and is frequently fragmented after infection (Campadelli et al. 1993; Salanueva et al. 2003; Yadav et al. 2016; Avitabile et al. 1995; Lavi et al. 1996; Hansen et al. 2017; Rebmann et al. 2016). Viruses use Golgi membranes directly and/or hijack master controllers of Golgi biogenesis and trafficking to generate vesicles that are used as the site of viral RNA replication (Quiner and Jackson 2010; Hansen et al. 2017; Short et al. 2013), wrapping (Sivan et al. 2016; Alzhanova and Hruby 2007; Alzhanova and Hruby 2006; Nanbo et al. 2018; Lundu et al. 2018; Procter et al. 2018), intracellular transduction (Nonnenmacher et al. 2015), and secretion (Zhang et al. 2016b). Viral infection triggers Golgi fragmentation via diverse mechanisms, ranging from phosphorylating key Golgi structural proteins such as GRASP65 (Rebmann et al. 2016), activating the Src kinase to phosphorylate the Dynamin 2 GTPase (Martin et al. 2017), targeting the immunity-related GTPase M (IRGM) to the Golgi to induce GBF1 phosphorylation (Hansen et al. 2017), modulating vesicular trafficking (Yadav et al. 2016; Johns et al. 2014), to impeding the major histocompatibility complex (MHC) class I trafficking, antigen presentation, and/or cytokine secretion (Moffat et al. 2007; Rohde et al. 2012).

19.5 Conclusions and Perspectives

The Golgi is the central hub in the secretory pathway, where proteins and lipids are processed, sorted, and dispatched to distinct destinations. As a well-organized polarized membrane structure, Golgi function is tightly related to its structural integrity. Thus, the first key question in Golgi biology concerns how the stacked Golgi structure forms. During the past decades, proteins with a variety of functions have been identified in the maintenance of Golgi structure and regulation of Golgi function, including but not limited to GRASPs, golgins, kinases, phosphatases, ubiquitin E3 ligases, and deubiquitinases, as summarized above. More detailed investigations need to be done to investigate how Golgi structural proteins and their interacting molecules cooperate together to form the stacked Golgi structure.

Golgi structural and functional defects have been increasingly reported in stress and disease conditions. In addition to its central role in protein sorting and trafficking, the Golgi has been more recently recognized as a hub of signaling pathways, which facilitates Golgi reaction upon stresses and diseases. Thus, the second key question in Golgi biology concerns how the Golgi structure becomes defective in stress and disease conditions. In this regard, much effort and some progress have been made, such as Golgi fragmentation in AD by Cdk5-mediated GRASP65 phosphorylation as discussed above. However, many questions remain. For example, is there an unfolded protein response (UPR)-like mechanism at the Golgi to cope with different stresses? Is there a common stress sensor on the Golgi? How do the signaling pathways on the Golgi sense and transduce stress signals? Apparently, a more systematic analysis of Golgi response to different stressors is necessary. Gene expression profile and posttranslational modification analysis of Golgi structural proteins, Golgi enzymes, and signaling molecules will fast forward the field and shed light on new directions. Considering lipid organization and modification are important for Golgi function and certain lipids can work as signaling molecules, more attention may be put on lipids at the Golgi, in addition to proteins.

The third key question in Golgi biology concerns how Golgi structure alteration affects its function in trafficking, glycosylation, and sorting. The consequence of Golgi fragmentation in different diseases is likely different, but it has been reported that Golgi cisternal unstacking by depleting GRASP proteins enhances protein trafficking. This, however, impairs accurate glycosylation and causes missorting of lysosomal enzymes to the extracellular space (Xiang et al. 2013; Zhang and Wang 2016; Bekier et al. 2017; Wang et al. 2008). Consistently, Golgi fragmentation in AD enhances APP trafficking and Aβ production, while rescue of the Golgi causes APP accumulation in the Golgi and reduces Aβ secretion (Joshi et al. 2014). Most recently, we have obtained evidence that Golgi structure disassembly by GRASP depletion reduces cell attachment and migration while accelerating cell growth and cell cycle progression (Ahat et al. 2019). It will be interesting to investigate how Golgi fragmentation enhances cancer cell proliferation and metastasis in the future. Future efforts may also aim at developing small chemicals or molecular tools to rescue the Golgi structure in diseases, which may delay the disease development.

Acknowledgment We thank members of the Wang lab for stimulating discussions. This work was supported by the National Institutes of Health (Grants GM112786 and GM105920), MCubed, and the Fast Forward Protein Folding Disease Initiative of the University of Michigan to Y. Wang, and a University of Michigan Rackham Predoctoral fellowship to E. Ahat.

References

Ahat E, Xiang Y, Zhang X, Bekier M, Wang Y (2019) GRASP depletion-mediated Golgi destruction decreases cell adhesion and migration via the reduction of $\alpha5\beta1$ integrin. Mol Biol Cell 30 (6):766–777

Aizawa M, Fukuda M (2015) Small GTPase Rab2B and its specific binding protein Golgi-associated Rab2B interactor-like 4 (GARI-L4) regulate Golgi morphology. J Biol Chem 290:22250–22261

Al-Dosari M, Alkuraya FS (2009) A novel missense mutation in SCYL1BP1 produces geroderma osteodysplastica phenotype indistinguishable from that caused by nullimorphic mutations. Am J Med Genet A 149A:2093–2098

Allan BB, Moyer BD, Balch WE (2000) Rab1 recruitment of p115 into a cis-SNARE complex: programming budding COPII vesicles for fusion. Science 289:444–448

Allan VJ, Thompson HM, Mcniven MA (2002) Motoring around the Golgi. Nat Cell Biol 4:E236–E242

Alvarez C, Garcia-Mata R, Hauri HP, Sztul E (2001) The p115-interactive proteins GM130 and giantin participate in endoplasmic reticulum-Golgi traffic. J Biol Chem 276:2693–2700

Alzhanova D, Hruby DE (2006) A trans-Golgi network resident protein, golgin-97, accumulates in viral factories and incorporates into virions during poxvirus infection. J Virol 80:11520–11527

Alzhanova D, Hruby DE (2007) A host cell membrane protein, golgin-97, is essential for poxvirus morphogenesis. Virology 362:421–427

Anderson NS, Mukherjee I, Bentivoglio CM, Barlowe C (2017) The Golgin protein Coy1 functions in intra-Golgi retrograde transport and interacts with the COG complex and Golgi SNAREs. Mol Biol Cell. https://doi.org/10.1091/mbc.E17-03-0137

Anglesio MS, Evdokimova V, Melnyk N, Zhang L, Fernandez CV, Grundy PE, Leach S, Marra MA, Brooks-Wilson AR, Penninger J, Sorensen PH (2004) Differential expression of a novel ankyrin containing E3 ubiquitin-protein ligase, Hace1, in sporadic Wilms' tumor versus normal kidney. Hum Mol Genet 13:2061–2074

Aridor M, Hannan LA (2000) Traffic jam: a compendium of human diseases that affect intracellular transport processes. Traffic 1:836–851

Aslan JE, Thomas G (2009) Death by committee: organellar trafficking and communication in apoptosis. Traffic 10:1390–1404

Atkin JD, Farg MA, Soo KY, Walker AK, Halloran M, Turner BJ, Nagley P, Horne MK (2014) Mutant SOD1 inhibits ER-Golgi transport in amyotrophic lateral sclerosis. J Neurochem 129:190–204

Avitabile E, Di Gaeta S, Torrisi MR, Ward PL, Roizman B, Campadelli-Fiume G (1995) Redistribution of microtubules and Golgi apparatus in herpes simplex virus-infected cells and their role in viral exocytosis. J Virol 69:7472–7482

Ayala I, Colanzi A (2017) Alterations of Golgi organization in Alzheimer's disease: A cause or a consequence? Tissue Cell 49:133–140

Bai Y, Cui X, Gao D, Wang Y, Wang B, Wang W (2018) Golgi integral membrane protein 4 manipulates cellular proliferation, apoptosis, and cell cycle in human head and neck cancer. Biosci Rep 38. https://doi.org/10.1042/BSR20180454

Barr FA (2005) Purification and functional interactions of GRASP55 with Rab2. Methods Enzymol 403:391–401

Barr FA, Puype M, Vandekerckhove J, Warren G (1997) GRASP65, a protein involved in the stacking of Golgi cisternae. Cell 91:253–262

Barr FA, Nakamura N, Warren G (1998) Mapping the interaction between GRASP65 and GM130, components of a protein complex involved in the stacking of Golgi cisternae. EMBO J 17:3258–3268

Barr FA, Preisinger C, Kopajtich R, Korner R (2001) Golgi matrix proteins interact with p24 cargo receptors and aid their efficient retention in the Golgi apparatus. J Cell Biol 155:885–891

Barretta ML, Spano D, D'Ambrosio C, Cervigni RI, Scaloni A, Corda D, Colanzi A (2016) Aurora-A recruitment and centrosomal maturation are regulated by a Golgi-activated pool of Src during G2. Nat Commun 7:11727

Bekier ME 2nd, Wang L, Li J, Huang H, Tang D, Zhang X, Wang Y (2017) Knockout of the Golgi stacking proteins GRASP55 and GRASP65 impairs Golgi structure and function. Mol Biol Cell 28:2833–2842

Bellouze S, Baillat G, Buttigieg D, De La Grange P, Rabouille C, Haase G (2016) Stathmin 1/2-triggered microtubule loss mediates Golgi fragmentation in mutant SOD1 motor neurons. Mol Neurodegener 11:43

Bergen DJM, Stevenson NL, Skinner REH, Stephens DJ, Hammond CL (2017) The Golgi matrix protein giantin is required for normal cilia function in zebrafish. Biol Open 6:1180–1189

Bishe B, Syed GH, Field SJ, Siddiqui A (2012) Role of phosphatidylinositol 4-phosphate (PI4P) and its binding protein GOLPH3 in hepatitis C virus secretion. J Biol Chem 287:27637–27647

Bonnon C, Wendeler MW, Paccaud JP, Hauri HP (2010) Selective export of human GPI-anchored proteins from the endoplasmic reticulum. J Cell Sci 123:1705–1715

Brandon E, Gao Y, Garcia-Mata R, Alvarez C, Sztul E (2003) Membrane targeting of p115 phosphorylation mutants and their effects on Golgi integrity and secretory traffic. Eur J Cell Biol 82:411–420

Brede G, Solheim J, Troen G, Prydz H (2000) Characterization of PSKH1, a novel human protein serine kinase with centrosomal, golgi, and nuclear localization. Genomics 70:82–92

Brede G, Solheim J, Stang E, Prydz H (2003) Mutants of the protein serine kinase PSKH1 disassemble the Golgi apparatus. Exp Cell Res 291:299–312

Bundis F, Neagoe I, Schwappach B, Steinmeyer K (2006) Involvement of Golgin-160 in cell surface transport of renal ROMK channel: co-expression of Golgin-160 increases ROMK currents. Cell Physiol Biochem 17:1–12

Burguete AS, Fenn TD, Brunger AT, Pfeffer SR (2008) Rab and Arl GTPase family members cooperate in the localization of the golgin GCC185. Cell 132:286–298

Buschman MD, Xing M, Field SJ (2015) The GOLPH3 pathway regulates Golgi shape and function and is activated by DNA damage. Front Neurosci 9:362

Campadelli G, Brandimarti R, Di Lazzaro C, Ward PL, Roizman B, Torrisi MR (1993) Fragmentation and dispersal of Golgi proteins and redistribution of glycoproteins and glycolipids processed through the Golgi apparatus after infection with herpes simplex virus 1. Proc Natl Acad Sci USA 90:2798–2802

Cao K, Nakajima R, Meyer HH, Zheng Y (2003) The AAA-ATPase Cdc48/p97 regulates spindle disassembly at the end of mitosis. Cell 115:355–367

Carmena M, Ruchaud S, Earnshaw WC (2009) Making the Auroras glow: regulation of Aurora A and B kinase function by interacting proteins. Curr Opin Cell Biol 21:796–805

Cartier-Michaud A, Bailly AL, Betzi S, Shi X, Lissitzky JC, Zarubica A, Serge A, Roche P, Lugari A, Hamon V, Bardin F, Derviaux C, Lembo F, Audebert S, Marchetto S, Durand B, Borg JP, Shi N, Morelli X, Aurrand-Lions M (2017) Genetic, structural, and chemical insights into the dual function of GRASP55 in germ cell Golgi remodeling and JAM-C polarized localization during spermatogenesis. PLoS Genet 13:e1006803

Cervigni RI, Bonavita R, Barretta ML, Spano D, Ayala I, Nakamura N, Corda D, Colanzi A (2015) JNK2 controls fragmentation of the Golgi complex and the G2/M transition through phosphorylation of GRASP65. J Cell Sci 128:2249–2260

Chen Y, Chen PL, Chen CF, Sharp ZD, Lee WH (1999) Thyroid hormone, T3-dependent phosphorylation and translocation of Trip230 from the Golgi complex to the nucleus. Proc Natl Acad Sci USA 96:4443–4448

Cheng JP, Betin VM, Weir H, Shelmani GM, Moss DK, Lane JD (2010) Caspase cleavage of the Golgi stacking factor GRASP65 is required for Fas/CD95-mediated apoptosis. Cell Death Dis 1: e82

Cheung PY, Limouse C, Mabuchi H, Pfeffer SR (2015) Protein flexibility is required for vesicle tethering at the Golgi. Elife 4. https://doi.org/10.7554/eLife.12790

Chia J, Tham KM, Gill DJ, Bard-Chapeau EA, Bard FA (2014) ERK8 is a negative regulator of O-GalNAc glycosylation and cell migration. Elife 3:e01828

Ching YP, Leong VY, Lee MF, Xu HT, Jin DY, Ng IO (2007) P21-activated protein kinase is overexpressed in hepatocellular carcinoma and enhances cancer metastasis involving c-Jun NH2-terminal kinase activation and paxillin phosphorylation. Cancer Res 67:3601–3608

Chiu R, Novikov L, Mukherjee S, Shields D (2002) A caspase cleavage fragment of p115 induces fragmentation of the Golgi apparatus and apoptosis. J Cell Biol 159:637–648

Chiu CF, Ghanekar Y, Frost L, Diao A, Morrison D, Mckenzie E, Lowe M (2008) ZFPL1, a novel ring finger protein required for cis-Golgi integrity and efficient ER-to-Golgi transport. EMBO J 27:934–947

Cluett EB, Brown WJ (1992) Adhesion of Golgi cisternae by proteinaceous interactions: intercisternal bridges as putative adhesive structures. J Cell Sci 103:773–784

Colanzi A, Hidalgo Carcedo C, Persico A, Cericola C, Turacchio G, Bonazzi M, Luini A, Corda D (2007) The Golgi mitotic checkpoint is controlled by BARS-dependent fission of the Golgi ribbon into separate stacks in G2. EMBO J 26:2465–2476

Coune PG, Bensadoun JC, Aebischer P, Schneider BL (2011) Rab1A over-expression prevents Golgi apparatus fragmentation and partially corrects motor deficits in an alpha-synuclein based rat model of Parkinson's disease. J Parkinsons Dis 1:373–387

Cruz-Garcia D, Vazquez-Martinez R, Peinado JR, Anouar Y, Tonon MC, Vaudry H, Castano JP, Malagon MM (2007) Identification and characterization of two novel (neuro)endocrine long coiled-coil proteins. FEBS Lett 581:3149–3156

Cui F, Wang Y (2012) HACE1 (HECT domain and ankyrin repeat containing E3 ubiquitin protein ligase 1). Atlas Genet Cytogenet Oncol Haematol 17:333–336

D'Angelo G, Prencipe L, Iodice L, Beznoussenko G, Savarese M, Marra P, Di Tullio G, Martire G, De Matteis MA, Bonatti S (2009) GRASP65 and GRASP55 sequentially promote the transport of C-terminal valine-bearing cargos to and through the Golgi complex. J Biol Chem 284:34849–34860

Davis EJ, Lachaud C, Appleton P, Macartney TJ, Nathke I, Rouse J (2012) DVC1 (C1orf124) recruits the p97 protein segregase to sites of DNA damage. Nat Struct Mol Biol 19:1093–1100

Dejgaard SY, Murshid A, Erman A, Kizilay O, Verbich D, Lodge R, Dejgaard K, Ly-Hartig TB, Pepperkok R, Simpson JC, Presley JF (2008) Rab18 and Rab43 have key roles in ER-Golgi trafficking. J Cell Sci 121:2768–2781

Delgui LR, Rodriguez JF, Colombo MI (2013) The endosomal pathway and the Golgi complex are involved in the infectious bursal disease virus life cycle. J Virol 87:8993–9007

Dementieva IS, Tereshko V, McCrossan ZA, Solomaha E, Araki D, Xu C, Grigorieff N, Goldstein SA (2009) Pentameric assembly of potassium channel tetramerization domain-containing protein 5. J Mol Biol 387(1):175–191

Derby MC, Gleeson PA (2007) New insights into membrane trafficking and protein sorting. Int Rev Cytol 261:47–116

Derby MC, Lieu ZZ, Brown D, Stow JL, Goud B, Gleeson PA (2007) The trans-Golgi network golgin, GCC185, is required for endosome-to-Golgi transport and maintenance of Golgi structure. Traffic 8:758–773

Diao A, Rahman D, Pappin DJ, Lucocq J, Lowe M (2003) The coiled-coil membrane protein golgin-84 is a novel rab effector required for Golgi ribbon formation. J Cell Biol 160:201–212

Diao A, Frost L, Morohashi Y, Lowe M (2008) Coordination of golgin tethering and SNARE assembly: GM130 binds syntaxin 5 in a p115-regulated manner. J Biol Chem 283:6957–6967

Dippold HC, Ng MM, Farber-Katz SE, Lee SK, Kerr ML, Peterman MC, Sim R, Wiharto PA, Galbraith KA, Madhavarapu S, Fuchs GJ, Meerloo T, Farquhar MG, Zhou H, Field SJ (2009) GOLPH3 bridges phosphatidylinositol-4-phosphate and actomyosin to stretch and shape the Golgi to promote budding. Cell 139:337–351

Dirac-Svejstrup AB, Shorter J, Waters MG, Warren G (2000) Phosphorylation of the vesicle-tethering protein p115 by a casein kinase II-like enzyme is required for Golgi reassembly from isolated mitotic fragments. J Cell Biol 150:475–488

Drin G, Casella JF, Gautier R, Boehmer T, Schwartz TU, Antonny B (2007) A general amphipathic alpha-helical motif for sensing membrane curvature. Nat Struct Mol Biol 14:138–146

Drin G, Morello V, Casella JF, Gounon P, Antonny B (2008) Asymmetric tethering of flat and curved lipid membranes by a golgin. Science 320:670–673

Dupont N, Jiang S, Pilli M, Ornatowski W, Bhattacharya D, Deretic V (2011) Autophagy-based unconventional secretory pathway for extracellular delivery of IL-1beta. EMBO J 30:4701–4711

Duran JM, Kinseth M, Bossard C, Rose DW, Polishchuk R, Wu CC, Yates J, Zimmerman T, Malhotra V (2008) The role of GRASP55 in Golgi fragmentation and entry of cells into mitosis. Mol Biol Cell 19:2579–2587

Efimov A, Kharitonov A, Efimova N, Loncarek J, Miller PM, Andreyeva N, Gleeson P, Galjart N, Maia AR, Mcleod IX, Yates JR 3rd, Maiato H, Khodjakov A, Akhmanova A, Kaverina I (2007) Asymmetric CLASP-dependent nucleation of noncentrosomal microtubules at the trans-Golgi network. Dev Cell 12:917–930

Evin G (2015) How accelerated Golgi trafficking may drive Alzheimer's disease (comment on DOI 10.1002/bies.201400116). Bioessays 37:232–233

Eystathioy T, Jakymiw A, Fujita DJ, Fritzler MJ, Chan EK (2000) Human autoantibodies to a novel Golgi protein golgin-67: high similarity with golgin-95/gm 130 autoantigen. J Autoimmun 14:179–187

Farber-Katz SE, Dippold HC, Buschman MD, Peterman MC, Xing M, Noakes CJ, Tat J, Ng MM, Rahajeng J, Cowan DM, Fuchs GJ, Zhou H, Field SJ (2014) DNA damage triggers Golgi dispersal via DNA-PK and GOLPH3. Cell 156:413–427

Farhan H, Rabouille C (2011) Signalling to and from the secretory pathway. J Cell Sci 124:171–180

Feinstein TN, Linstedt AD (2008) GRASP55 regulates Golgi ribbon formation. Mol Biol Cell 19:2696–2707

Follit JA, San Agustin JT, Xu F, Jonassen JA, Samtani R, Lo CW, Pazour GJ (2008) The Golgin GMAP210/TRIP11 anchors IFT20 to the Golgi complex. PLoS Genet 4:e1000315

Forno LS (1996) Neuropathology of Parkinson's disease. J Neuropathol Exp Neurol 55:259–272

Franke WW, Kartenbeck J, Krien S, Vanderwoude WJ, Scheer U, Morre DJ (1972) Inter- and intracisternal elements of the Golgi apparatus. A system of membrane-to-membrane cross-links. Z Zellforsch Mikrosk Anat 132:365–380

Fridmann-Sirkis Y, Siniossoglou S, Pelham HR (2004) TMF is a golgin that binds Rab6 and influences Golgi morphology. BMC Cell Biol 5:18

Fu X, Ng C, Feng D, Liang C (2003) Cdc48p is required for the cell cycle commitment point at Start via degradation of the G1-CDK inhibitor Far1p. J Cell Biol 163:21–26

Fujimoto T, Kuwahara T, Eguchi T, Sakurai M, Komori T, Iwatsubo T (2018) Parkinson's disease-associated mutant LRRK2 phosphorylates Rab7L1 and modifies trans-Golgi morphology. Biochem Biophys Res Commun 495:1708–1715

Fujita Y, Ohama E, Takatama M, Al-Sarraj S, Okamoto K (2006) Fragmentation of Golgi apparatus of nigral neurons with alpha-synuclein-positive inclusions in patients with Parkinson's disease. Acta Neuropathol 112:261–265

Fujita Y, Mizuno Y, Takatama M, Okamoto K (2008) Anterior horn cells with abnormal TDP-43 immunoreactivities show fragmentation of the Golgi apparatus in ALS. J Neurol Sci 269:30–34

Fujita K, Nakamura Y, Oka T, Ito H, Tamura T, Tagawa K, Sasabe T, Katsuta A, Motoki K, Shiwaku H, Sone M, Yoshida C, Katsuno M, Eishi Y, Murata M, Taylor JP, Wanker EE, Kono K, Tashiro S, Sobue G, La Spada AR, Okazawa H (2013) A functional deficiency of TERA/VCP/p97 contributes to impaired DNA repair in multiple polyglutamine diseases. Nat Commun 4:1816

Ganley IG, Espinosa E, Pfeffer SR (2008) A syntaxin 10-SNARE complex distinguishes two distinct transport routes from endosomes to the trans-Golgi in human cells. J Cell Biol 180:159–172

Ge L, Melville D, Zhang M, Schekman R (2013) The ER-Golgi intermediate compartment is a key membrane source for the LC3 lipidation step of autophagosome biogenesis. Elife 2:e00947

Gee HY, Noh SH, Tang BL, Kim KH, Lee MG (2011) Rescue of DeltaF508-CFTR trafficking via a GRASP-dependent unconventional secretion pathway. Cell 146:746–760

Geng J, Klionsky DJ (2010) The Golgi as a potential membrane source for autophagy. Autophagy 6:950–951

Geng J, Nair U, Yasumura-Yorimitsu K, Klionsky DJ (2010) Post-Golgi Sec proteins are required for autophagy in Saccharomyces cerevisiae. Mol Biol Cell 21:2257–2269

Ghosh DK, Roy A, Ranjan A (2018) The ATPase VCP/p97 functions as a disaggregase against toxic Huntingtin-exon1 aggregates. FEBS Lett 592:2680–2692

Gilbert CE, Zuckerman DM, Currier PL, Machamer CE (2014) Three basic residues of intracellular loop 3 of the beta-1 adrenergic receptor are required for golgin-160-dependent trafficking. Int J Mol Sci 15:2929–2945

Gillingham AK, Munro S (2016) Finding the Golgi: Golgin coiled-coil proteins show the way. Trends Cell Biol 26:399–408

Gillingham AK, Tong AH, Boone C, Munro S (2004) The GTPase Arf1p and the ER to Golgi cargo receptor Erv14p cooperate to recruit the golgin Rud3p to the cis-Golgi. J Cell Biol 167:281–292

Goldenring JR (2013) A central role for vesicle trafficking in epithelial neoplasia: intracellular highways to carcinogenesis. Nat Rev Cancer 13:813–820

Gosavi N, Lee HJ, Lee JS, Patel S, Lee SJ (2002) Golgi fragmentation occurs in the cells with prefibrillar alpha-synuclein aggregates and precedes the formation of fibrillar inclusion. J Biol Chem 277:48984–48992

Gosavi P, Houghton FJ, Mcmillan PJ, Hanssen E, Gleeson PA (2018) The Golgi ribbon in mammalian cells negatively regulates autophagy by modulating mTOR activity. J Cell Sci 131. https://doi.org/10.1242/jcs.211987

Goud B, Zahraoui A, Tavitian A, Saraste J (1990) Small GTP-binding protein associated with Golgi cisternae. Nature 345:553–556

Goud B, Liu S, Storrie B (2018) Rab proteins as major determinants of the Golgi complex structure. Small GTPases 9:66–75

Greninger AL, Knudsen GM, Betegon M, Burlingame AL, Derisi JL (2012) The 3A protein from multiple picornaviruses utilizes the golgi adaptor protein ACBD3 to recruit PI4KIIIbeta. J Virol 86:3605–3616

Greninger AL, Knudsen GM, Betegon M, Burlingame AL, Derisi JL (2013) ACBD3 interaction with TBC1 domain 22 protein is differentially affected by enteroviral and kobuviral 3A protein binding. MBio 4:e00098-13

Gubser C, Bergamaschi D, Hollinshead M, Lu X, Van Kuppeveld FJ, Smith GL (2007) A new inhibitor of apoptosis from vaccinia virus and eukaryotes. PLoS Pathog 3:e17

Guo Y, Chang C, Huang R, Liu B, Bao L, Liu W (2012) AP1 is essential for generation of autophagosomes from the trans-Golgi network. J Cell Sci 125:1706–1715

Hansen MD, Johnsen IB, Stiberg KA, Sherstova T, Wakita T, Richard GM, Kandasamy RK, Meurs EF, Anthonsen MW (2017) Hepatitis C virus triggers Golgi fragmentation and autophagy through the immunity-related GTPase M. Proc Natl Acad Sci USA 114:E3462–E3471

Hao Q, Jiao S, Shi Z, Li C, Meng X, Zhang Z, Wang Y, Song X, Wang W, Zhang R, Zhao Y, Wong CC, Zhou Z (2015) A non-canonical role of the p97 complex in RIG-I antiviral signaling. EMBO J 34:2903–2920

Harak C, Lohmann V (2015) Ultrastructure of the replication sites of positive-strand RNA viruses. Virology 479–480:418–433

Hauler F, Mallery DL, Mcewan WA, Bidgood SR, James LC (2012) AAA ATPase p97/VCP is essential for TRIM21-mediated virus neutralization. Proc Natl Acad Sci USA 109:19733–19738

Hayes GL, Brown FC, Haas AK, Nottingham RM, Barr FA, Pfeffer SR (2009) Multiple Rab GTPase binding sites in GCC185 suggest a model for vesicle tethering at the trans-Golgi. Mol Biol Cell 20:209–217

Heinrich F, Nanda H, Goh HZ, Bachert C, Losche M, Linstedt AD (2014) Myristoylation restricts orientation of the GRASP domain on membranes and promotes membrane tethering. J Biol Chem 289:9683–9691

Hennies HC, Kornak U, Zhang H, Egerer J, Zhang X, Seifert W, Kuhnisch J, Budde B, Natebus M, Brancati F, Wilcox WR, Muller D, Kaplan PB, Rajab A, Zampino G, Fodale V, Dallapiccola B, Newman W, Metcalfe K, Clayton-Smith J, Tassabehji M, Steinmann B, Barr FA, Nurnberg P, Wieacker P, Mundlos S (2008) Gerodermia osteodysplastica is caused by mutations in SCYL1BP1, a Rab-6 interacting golgin. Nat Genet 40:1410–1412

Hetzer M, Meyer HH, Walther TC, Bilbao-Cortes D, Warren G, Mattaj IW (2001) Distinct AAA-ATPase p97 complexes function in discrete steps of nuclear assembly. Nat Cell Biol 3:1086–1091

Hicks SW, Machamer CE (2002) The NH2-terminal domain of Golgin-160 contains both golgi and nuclear targeting information. J Biol Chem 277:35833–35839

Hicks SW, Machamer CE (2005) Golgi structure in stress sensing and apoptosis. Biochim Biophys Acta 1744:406–414

Hidalgo Carcedo C, Bonazzi M, Spano S, Turacchio G, Colanzi A, Luini A, Corda D (2004) Mitotic Golgi partitioning is driven by the membrane-fissioning protein CtBP3/BARS. Science 305:93–96

Hoogenraad CC, Akhmanova A, Howell SA, Dortland BR, De Zeeuw CI, Willemsen R, Visser P, Grosveld F, Galjart N (2001) Mammalian Golgi-associated Bicaudal-D2 functions in the dynein-dynactin pathway by interacting with these complexes. EMBO J 20:4041–4054

How PC, Shields D (2011) Tethering function of the caspase cleavage fragment of Golgi protein p115 promotes apoptosis via a p53-dependent pathway. J Biol Chem 286:8565–8576

Hsiao PW, Chang C (1999) Isolation and characterization of ARA160 as the first androgen receptor N-terminal-associated coactivator in human prostate cells. J Biol Chem 274:22373–22379

Hu L, Liu M, Chen L, Chan TH, Wang J, Huo KK, Zheng BJ, Xie D, Guan XY (2012) SCYL1 binding protein 1 promotes the ubiquitin-dependent degradation of Pirh2 and has tumor-suppressive function in the development of hepatocellular carcinoma. Carcinogenesis 33:1581–1588

Hu F, Shi X, Li B, Huang X, Morelli X, Shi N (2015) Structural basis for the interaction between the Golgi reassembly-stacking protein GRASP65 and the Golgi matrix protein GM130. J Biol Chem 290:26373–26382

Huang S, Wang Y (2017) Golgi structure formation, function, and post-translational modifications in mammalian cells. F1000Res 6:2050

Huang S, Tang D, Wang Y (2016) Monoubiquitination of syntaxin 5 regulates Golgi membrane dynamics during the cell cycle. Dev Cell 38:73–85

Huttner WB, Ohashi M, Kehlenbach RH, Barr FA, Bauerfeind R, Braunling O, Corbeil D, Hannah M, Pasolli HA, Schmidt A et al (1995) Biogenesis of neurosecretory vesicles. Cold Spring Harb Symp Quant Biol 60:315–327

Jakymiw A, Raharjo E, Rattner JB, Eystathioy T, Chan EK, Fujita DJ (2000) Identification and characterization of a novel Golgi protein, golgin-67. J Biol Chem 275:4137–4144

Ji G, Ji H, Mo X, Li T, Yu Y, Hu Z (2013) The role of GRASPs in morphological alterations of Golgi apparatus: mechanisms and effects. Rev Neurosci 24:485–497

Jiang Q, Wang L, Guan Y, Xu H, Niu Y, Han L, Wei YP, Lin L, Chu J, Wang Q, Yang Y, Pei L, Wang JZ, Tian Q (2014) Golgin-84-associated Golgi fragmentation triggers tau hyperphosphorylation by activation of cyclin-dependent kinase-5 and extracellular signal-regulated kinase. Neurobiol Aging 35:1352–1363

Jing J, Junutula JR, Wu C, Burden J, Matern H, Peden AA, Prekeris R (2010) FIP1/RCP binding to Golgin-97 regulates retrograde transport from recycling endosomes to the trans-Golgi network. Mol Biol Cell 21:3041–3053

Johns HL, Gonzalez-Lopez C, Sayers CL, Hollinshead M, Elliott G (2014) Rab6 dependent post-Golgi trafficking of HSV1 envelope proteins to sites of virus envelopment. Traffic 15:157–178

Joshi G, Wang Y (2015) Golgi defects enhance APP amyloidogenic processing in Alzheimer's disease. Bioessays 37:240–247

Joshi G, Chi Y, Huang Z, Wang Y (2014) Abeta-induced Golgi fragmentation in Alzheimer's disease enhances Abeta production. Proc Natl Acad Sci USA 111:E1230–E1239

Joshi G, Bekier ME 2nd, Wang Y (2015) Golgi fragmentation in Alzheimer's disease. Front Neurosci 9:340

Kaneko Y, Tamura K, Totsukawa G, Kondo H (2010) Phosphorylation of p37 is important for Golgi disassembly at mitosis. Biochem Biophys Res Commun 402:37–41

Kano F, Kondo H, Yamamoto A, Kaneko Y, Uchiyama K, Hosokawa N, Nagata K, Murata M (2005) NSF/SNAPs and p97/p47/VCIP135 are sequentially required for cell cycle-dependent reformation of the ER network. Genes Cells 10:989–999

Kelly EE, Giordano F, Horgan CP, Jollivet F, Raposo G, Mccaffrey MW (2012) Rab30 is required for the morphological integrity of the Golgi apparatus. Biol Cell 104:84–101

Kim HS, Takahashi M, Matsuo K, Ono Y (2007) Recruitment of CG-NAP to the Golgi apparatus through interaction with dynein-dynactin complex. Genes Cells 12:421–434

Kimura M, Yoshioka T, Saio M, Banno Y, Nagaoka H, Okano Y (2013) Mitotic catastrophe and cell death induced by depletion of centrosomal proteins. Cell Death Dis 4:e603

Kimura M, Takagi S, Nakashima S (2018) Aurora A regulates the architecture of the Golgi apparatus. Exp Cell Res 367:73–80

Kobayashi T, Manno A, Kakizuka A (2007) Involvement of valosin-containing protein (VCP)/p97 in the formation and clearance of abnormal protein aggregates. Genes Cells 12:889–901

Kodani A, Sutterlin C (2008) The Golgi protein GM130 regulates centrosome morphology and function. Mol Biol Cell 19:745–753

Kodani A, Kristensen I, Huang L, Sutterlin C (2009) GM130-dependent control of Cdc42 activity at the Golgi regulates centrosome organization. Mol Biol Cell 20:1192–1200

Koegler E, Bonnon C, Waldmeier L, Mitrovic S, Halbeisen R, Hauri HP (2010) p28, a novel ERGIC/cis Golgi protein, required for Golgi ribbon formation. Traffic 11:70–89

Kondo H, Rabouille C, Newman R, Levine TP, Pappin D, Freemont P, Warren G (1997) p47 is a cofactor for p97-mediated membrane fusion. Nature 388:75–78

Koreishi M, Gniadek TJ, Yu S, Masuda J, Honjo Y, Satoh A (2013) The golgin tether giantin regulates the secretory pathway by controlling stack organization within Golgi apparatus. PLoS One 8:e59821

Kuna RS, Field SJ (2018) GOLPH3: A Golgi phosphatidylinositol(4)phosphate effector that directs vesicle trafficking and drives cancer. J Lipid Res 60(2):269–275

Kuo A, Zhong C, Lane WS, Derynck R (2000) Transmembrane transforming growth factor-alpha tethers to the PDZ domain-containing, Golgi membrane-associated protein p59/GRASP55. EMBO J 19:6427–6439

Lane JD, Lucocq J, Pryde J, Barr FA, Woodman PG, Allan VJ, Lowe M (2002) Caspase-mediated cleavage of the stacking protein GRASP65 is required for Golgi fragmentation during apoptosis. J Cell Biol 156:495–509

Lashuel HA, Hirling H (2006) Rescuing defective vesicular trafficking protects against alpha-synuclein toxicity in cellular and animal models of Parkinson's disease. ACS Chem Biol 1:420–424

Lavi E, Wang Q, Weiss SR, Gonatas NK (1996) Syncytia formation induced by coronavirus infection is associated with fragmentation and rearrangement of the Golgi apparatus. Virology 221:325–334

Lee TH, Linstedt AD (2000) Potential role for protein kinases in regulation of bidirectional endoplasmic reticulum-to-Golgi transport revealed by protein kinase inhibitor H89. Mol Biol Cell 11:2577–2590

Lee M, Lee HJ, Bae S, Lee YS (2008) Protein sialylation by sialyltransferase involves radiation resistance. Mol Cancer Res 6:1316–1325

Li T, You H, Mo X, He W, Tang X, Jiang Z, Chen S, Chen Y, Zhang J, Hu Z (2016a) GOLPH3 mediated Golgi stress response in modulating N2A cell death upon oxygen-glucose deprivation and reoxygenation injury. Mol Neurobiol 53:1377–1385

Li X, Li M, Tian X, Li Q, Lu Q, Jia Q, Zhang L, Yan J, Li X, Li X (2016b) Golgi phosphoprotein 3 inhibits the apoptosis of human glioma cells in part by downregulating N-myc downstream regulated gene 1. Med Sci Monit 22:3535–3543

Li J, Tang D, Ireland SC, Wang Y (2019) DjA1 maintains Golgi integrity via interaction with GRASP65. Mol Biol Cell 30(4):478–490

Lieu ZZ, Derby MC, Teasdale RD, Hart C, Gunn P, Gleeson PA (2007) The golgin GCC88 is required for efficient retrograde transport of cargo from the early endosomes to the trans-Golgi network. Mol Biol Cell 18:4979–4991

Lin CY, Madsen ML, Yarm FR, Jang YJ, Liu X, Erikson RL (2000) Peripheral Golgi protein GRASP65 is a target of mitotic polo-like kinase (Plk) and Cdc2. Proc Natl Acad Sci USA 97:12589–12594

Lin YC, Chiang TC, Liu YT, Tsai YT, Jang LT, Lee FJ (2011) ARL4A acts with GCC185 to modulate Golgi complex organization. J Cell Sci 124:4014–4026

Lin CH, Li H, Lee YN, Cheng YJ, Wu RM, Chien CT (2015) Lrrk regulates the dynamic profile of dendritic Golgi outposts through the golgin Lava lamp. J Cell Biol 210:471–483

Linstedt AD, Jesch SA, Mehta A, Lee TH, Garcia-Mata R, Nelson DS, Sztul E (2000) Binding relationships of membrane tethering components. The giantin N terminus and the GM130 N terminus compete for binding to the p115 C terminus. J Biol Chem 275:10196–10201

Liu S, Storrie B (2015) How Rab proteins determine Golgi structure. Int Rev Cell Mol Biol 315:1–22

Liu S, Hunt L, Storrie B (2013) Rab41 is a novel regulator of Golgi apparatus organization that is needed for ER-to-Golgi trafficking and cell growth. PLoS One 8:e71886

Liu X, Wang Z, Yang Y, Li Q, Zeng R, Kang J, Wu J (2016) Rab1A mediates proinsulin to insulin conversion in beta-cells by maintaining Golgi stability through interactions with golgin-84. Protein Cell 7:692–696

Liu C, Mei M, Li Q, Roboti P, Pang Q, Ying Z, Gao F, Lowe M, Bao S (2017) Loss of the golgin GM130 causes Golgi disruption, Purkinje neuron loss, and ataxia in mice. Proc Natl Acad Sci USA 114:346–351

Lock JG, Hammond LA, Houghton F, Gleeson PA, Stow JL (2005) E-cadherin transport from the trans-Golgi network in tubulovesicular carriers is selectively regulated by golgin-97. Traffic 6:1142–1156

Lowe M, Rabouille C, Nakamura N, Watson R, Jackman M, Jamsa E, Rahman D, Pappin DJ, Warren G (1998) Cdc2 kinase directly phosphorylates the cis-Golgi matrix protein GM130 and is required for Golgi fragmentation in mitosis. Cell 94:783–793

Lowe M, Lane JD, Woodman PG, Allan VJ (2004) Caspase-mediated cleavage of syntaxin 5 and giantin accompanies inhibition of secretory traffic during apoptosis. J Cell Sci 117:1139–1150

Lu L, Hong W (2003) Interaction of Arl1-GTP with GRIP domains recruits autoantigens Golgin-97 and Golgin-245/p230 onto the Golgi. Mol Biol Cell 14:3767–3781

Lu L, Tai G, Hong W (2004) Autoantigen Golgin-97, an effector of Arl1 GTPase, participates in traffic from the endosome to the trans-golgi network. Mol Biol Cell 15:4426–4443

Lu PH, Yu CC, Chiang PC, Chen YC, Ho YF, Kung FL, Guh JH (2011) Paclitaxel induces apoptosis through activation of nuclear protein kinase C-delta and subsequent activation of Golgi associated Cdk1 in human hormone refractory prostate cancer. J Urol 186:2434–2441

Luke MR, Kjer-Nielsen L, Brown DL, Stow JL, Gleeson PA (2003) GRIP domain-mediated targeting of two new coiled-coil proteins, GCC88 and GCC185, to subcompartments of the trans-Golgi network. J Biol Chem 278:4216–4226

Luke MR, Houghton F, Perugini MA, Gleeson PA (2005) The trans-Golgi network GRIP-domain proteins form alpha-helical homodimers. Biochem J 388:835–841

Lundu T, Tsuda Y, Ito R, Shimizu K, Kobayashi S, Yoshii K, Yoshimatsu K, Arikawa J, Kariwa H (2018) Targeting of severe fever with thrombocytopenia syndrome virus structural proteins to the ERGIC (endoplasmic reticulum Golgi intermediate compartment) and Golgi complex. Biomed Res 39:27–38

Luo W, Wang Y, Reiser G (2007) p24A, a type I transmembrane protein, controls ARF1-dependent resensitization of protease-activated receptor-2 by influence on receptor trafficking. J Biol Chem 282:30246–30255

Maag RS, Mancini M, Rosen A, Machamer CE (2005) Caspase-resistant Golgin-160 disrupts apoptosis induced by secretory pathway stress and ligation of death receptors. Mol Biol Cell 16:3019–3027

Machamer CE (2003) Golgi disassembly in apoptosis: cause or effect? Trends Cell Biol 13:279–281

Machamer CE (2015) The Golgi complex in stress and death. Front Neurosci 9:421

Magdeleine M, Gautier R, Gounon P, Barelli H, Vanni S, Antonny B (2016) A filter at the entrance of the Golgi that selects vesicles according to size and bulk lipid composition. Elife 5. https://doi.org/10.7554/eLife.16988

Makhoul C, Gosavi P, Duffield R, Delbridge B, Williamson NA, Gleeson PA (2019) Intersectin-1 interacts with the golgin, GCC88, to couple the actin network and Golgi architecture. Mol Biol Cell 30(3):370–386. https://doi.org/10.1091/mbc.E18-05-0313

Malsam J, Satoh A, Pelletier L, Warren G (2005) Golgin tethers define subpopulations of COPI vesicles. Science 307:1095–1098

Mancini M, Machamer CE, Roy S, Nicholson DW, Thornberry NA, Casciola-Rosen LA, Rosen A (2000) Caspase-2 is localized at the Golgi complex and cleaves golgin-160 during apoptosis. J Cell Biol 149:603–612

Martin C, Leyton L, Hott M, Arancibia Y, Spichiger C, Mcniven MA, Court FA, Concha MI, Burgos PV, Otth C (2017) Herpes simplex virus type 1 neuronal infection perturbs Golgi apparatus integrity through activation of Src tyrosine kinase and Dyn-2 GTPase. Front Cell Infect Microbiol 7:371

Martinez O, Antony C, Pehau-Arnaudet G, Berger EG, Salamero J, Goud B (1997) GTP-bound forms of rab6 induce the redistribution of Golgi proteins into the endoplasmic reticulum. Proc Natl Acad Sci USA 94:1828–1833

Martinez-Carrera LA, Wirth B (2015) Dominant spinal muscular atrophy is caused by mutations in BICD2, an important golgin protein. Front Neurosci 9:401

Marza E, Taouji S, Barroso K, Raymond AA, Guignard L, Bonneu M, Pallares-Lupon N, Dupuy JW, Fernandez-Zapico ME, Rosenbaum J, Palladino F, Dupuy D, Chevet E (2015) Genome-wide screen identifies a novel p97/CDC-48-dependent pathway regulating ER-stress-induced gene transcription. EMBO Rep 16:332–340

Matanis T, Akhmanova A, Wulf P, Del Nery E, Weide T, Stepanova T, Galjart N, Grosveld F, Goud B, De Zeeuw CI, Barnekow A, Hoogenraad CC (2002) Bicaudal-D regulates COPI-independent Golgi-ER transport by recruiting the dynein-dynactin motor complex. Nat Cell Biol 4:986–992

Matteis M, Mironov A, Beznoussenko G (2008) The Golgi ribbon and the function of the Golgins. In: Mironov A, Pavelka M (eds) The Golgi apparatus. Springer, Vienna

Mayinger P (2011) Signaling at the Golgi. Cold Spring Harb Perspect Biol 3. https://doi.org/10.1101/cshperspect.a005314

Meerang M, Ritz D, Paliwal S, Garajova Z, Bosshard M, Mailand N, Janscak P, Hubscher U, Meyer H, Ramadan K (2011) The ubiquitin-selective segregase VCP/p97 orchestrates the response to DNA double-strand breaks. Nat Cell Biol 13:1376–1382

Meyer HH, Wang Y, Warren G (2002) Direct binding of ubiquitin conjugates by the mammalian p97 adaptor complexes, p47 and Ufd1-Npl4. EMBO J 21:5645–5652

Miller S, Krijnse-Locker J (2008) Modification of intracellular membrane structures for virus replication. Nat Rev Microbiol 6:363–374

Misumi Y, Sohda M, Yano A, Fujiwara T, Ikehara Y (1997) Molecular characterization of GCP170, a 170-kDa protein associated with the cytoplasmic face of the Golgi membrane. J Biol Chem 272:23851–23858

Misumi Y, Sohda M, Tashiro A, Sato H, Ikehara Y (2001) An essential cytoplasmic domain for the Golgi localization of coiled-coil proteins with a COOH-terminal membrane anchor. J Biol Chem 276:6867–6873

Moffat K, Knox C, Howell G, Clark SJ, Yang H, Belsham GJ, Ryan M, Wileman T (2007) Inhibition of the secretory pathway by foot-and-mouth disease virus 2BC protein is reproduced by coexpression of 2B with 2C, and the site of inhibition is determined by the subcellular location of 2C. J Virol 81:1129–1139

Mollenhauer HH (1965) An intercisternal structure in the Golgi apparatus. J Cell Biol 24:504–511

Mori K, Kato H (2002) A putative nuclear receptor coactivator (TMF/ARA160) associates with hbrm/hSNF2 alpha and BRG-1/hSNF2 beta and localizes in the Golgi apparatus. FEBS Lett 520:127–132

Moyer BD, Allan BB, Balch WE (2001) Rab1 interaction with a GM130 effector complex regulates COPII vesicle cis—Golgi tethering. Traffic 2:268–276

Mukherjee S, Shields D (2009) Nuclear import is required for the pro-apoptotic function of the Golgi protein p115. J Biol Chem 284:1709–1717

Mukherjee S, Chiu R, Leung SM, Shields D (2007) Fragmentation of the Golgi apparatus: an early apoptotic event independent of the cytoskeleton. Traffic 8:369–378

Munro S (2011) The golgin coiled-coil proteins of the Golgi apparatus. Cold Spring Harb Perspect Biol 3:1–14

Muschalik N, Munro S (2018) Golgins. Curr Biol 28:R374–R376

Nakamura N, Rabouille C, Watson R, Nilsson T, Hui N, Slusarewicz P, Kreis TE, Warren G (1995) Characterization of a cis-Golgi matrix protein, GM130. J Cell Biol 131:1715–1726

Nakamura N, Lowe M, Levine TP, Rabouille C, Warren G (1997) The vesicle docking protein p115 binds GM130, a cis-Golgi matrix protein, in a mitotically regulated manner. Cell 89:445–455

Nanbo A, Noda T, Ohba Y (2018) Epstein-Barr virus acquires its final envelope on intracellular compartments with Golgi markers. Front Microbiol 9:454

Naydenov NG, Harris G, Brown B, Schaefer KL, Das SK, Fisher PB, Ivanov AI (2012) Loss of soluble N-ethylmaleimide-sensitive factor attachment protein alpha (alphaSNAP) induces epithelial cell apoptosis via down-regulation of Bcl-2 expression and disruption of the Golgi. J Biol Chem 287:5928–5941

Netherton C, Moffat K, Brooks E, Wileman T (2007) A guide to viral inclusions, membrane rearrangements, factories, and viroplasm produced during virus replication. Adv Virus Res 70:101–182

Neveling K, Martinez-Carrera LA, Holker I, Heister A, Verrips A, Hosseini-Barkooie SM, Gilissen C, Vermeer S, Pennings M, Meijer R, Te Riele M, Frijns CJ, Suchowersky O, Maclaren L, Rudnik-Schoneborn S, Sinke RJ, Zerres K, Lowry RB, Lemmink HH, Garbes L, Veltman JA, Schelhaas HJ, Scheffer H, Wirth B (2013) Mutations in BICD2, which encodes a golgin and important motor adaptor, cause congenital autosomal-dominant spinal muscular atrophy. Am J Hum Genet 92:946–954

Ng MM, Dippold HC, Buschman MD, Noakes CJ, Field SJ (2013) GOLPH3L antagonizes GOLPH3 to determine Golgi morphology. Mol Biol Cell 24:796–808

Nicholson DW, Thornberry NA (1997) Caspases: killer proteases. Trends Biochem Sci 22:299–306

Nikonova AS, Astsaturov I, Serebriiskii IG, Dunbrack RL Jr, Golemis EA (2013) Aurora A kinase (AURKA) in normal and pathological cell division. Cell Mol Life Sci 70:661–687

Nishikori S, Yamanaka K, Sakurai T, Esaki M, Ogura T (2008) p97 Homologs from Caenorhabditis elegans, CDC-48.1 and CDC-48.2, suppress the aggregate formation of huntingtin exon1 containing expanded polyQ repeat. Genes Cells 13:827–838

Nishimura T, Takahashi M, Kim HS, Mukai H, Ono Y (2005) Centrosome-targeting region of CG-NAP causes centrosome amplification by recruiting cyclin E-cdk2 complex. Genes Cells 10:75–86

Nonnenmacher ME, Cintrat JC, Gillet D, Weber T (2015) Syntaxin 5-dependent retrograde transport to the trans-Golgi network is required for adeno-associated virus transduction. J Virol 89:1673–1687

Nozawa K, Casiano CA, Hamel JC, Molinaro C, Fritzler MJ, Chan EK (2002) Fragmentation of Golgi complex and Golgi autoantigens during apoptosis and necrosis. Arthritis Res 4:R3

Ohta E, Misumi Y, Sohda M, Fujiwara T, Yano A, Ikehara Y (2003) Identification and characterization of GCP16, a novel acylated Golgi protein that interacts with GCP170. J Biol Chem 278:51957–51967

Orsi A, Razi M, Dooley HC, Robinson D, Weston AE, Collinson LM, Tooze SA (2012) Dynamic and transient interactions of Atg9 with autophagosomes, but not membrane integration, are required for autophagy. Mol Biol Cell 23:1860–1873

Osterrieder A, Sparkes IA, Botchway SW, Ward A, Ketelaar T, De Ruijter N, Hawes C (2017) Stacks off tracks: a role for the golgin AtCASP in plant endoplasmic reticulum-Golgi apparatus tethering. J Exp Bot 68:3339–3350

Panic B, Whyte JR, Munro S (2003) The ARF-like GTPases Arl1p and Arl3p act in a pathway that interacts with vesicle-tethering factors at the Golgi apparatus. Curr Biol 13:405–410

Papadopoulos C, Kirchner P, Bug M, Grum D, Koerver L, Schulze N, Poehler R, Dressler A, Fengler S, Arhzaouy K, Lux V, Ehrmann M, Weihl CC, Meyer H (2017) VCP/p97 cooperates with YOD1, UBXD1 and PLAA to drive clearance of ruptured lysosomes by autophagy. EMBO J 36:135–150

Papoulas O, Hays TS, Sisson JC (2005) The golgin Lava lamp mediates dynein-based Golgi movements during Drosophila cellularization. Nat Cell Biol 7:612–618

Parisi E, Yahya G, Flores A, Aldea M (2018) Cdc48/p97 segregase is modulated by cyclin-dependent kinase to determine cyclin fate during G1 progression. EMBO J 37. https://doi.org/10.15252/embj.201798724

Park JJ, Yi JY, Jin YB, Lee YJ, Lee JS, Lee YS, Ko YG, Lee M (2012) Sialylation of epidermal growth factor receptor regulates receptor activity and chemosensitivity to gefitinib in colon cancer cells. Biochem Pharmacol 83:849–857

Partridge JJ, Lopreiato JO Jr, Latterich M, Indig FE (2003) DNA damage modulates nucleolar interaction of the Werner protein with the AAA ATPase p97/VCP. Mol Biol Cell 14:4221–4229

Peng D, Wang J, Zhang R, Jiang F, Tang S, Chen M, Yan J, Sun X, Wang S, Wang T, Yan D, Bao Y, Hu C, Jia W (2015) Common variants in or near ZNRF1, COLEC12, SCYL1BP1 and API5 are associated with diabetic retinopathy in Chinese patients with type 2 diabetes. Diabetologia 58:1231–1238

Perry E, Tsruya R, Levitsky P, Pomp O, Taller M, Weisberg S, Parris W, Kulkarni S, Malovani H, Pawson T, Shpungin S, Nir U (2004) TMF/ARA160 is a BC-box-containing protein that mediates the degradation of Stat3. Oncogene 23:8908–8919

Petrosyan A (2015) Onco-Golgi: is fragmentation a gate to cancer progression? Biochem Mol Biol J 1:16

Petrosyan A, Holzapfel MS, Muirhead DE, Cheng PW (2014) Restoration of compact Golgi morphology in advanced prostate cancer enhances susceptibility to galectin-1-induced apoptosis by modifying mucin O-glycan synthesis. Mol Cancer Res 12:1704–1716

Piao H, Kim J, Noh SH, Kweon HS, Kim JY, Lee MG (2017) Sec16A is critical for both conventional and unconventional secretion of CFTR. Sci Rep 7:39887

Preisinger C, Short B, De Corte V, Bruyneel E, Haas A, Kopajtich R, Gettemans J, Barr FA (2004) YSK1 is activated by the Golgi matrix protein GM130 and plays a role in cell migration through its substrate 14-3-3{zeta}. J Cell Biol 164:1009–1020

Preisinger C, Korner R, Wind M, Lehmann WD, Kopajtich R, Barr FA (2005) Plk1 docking to GRASP65 phosphorylated by Cdk1 suggests a mechanism for Golgi checkpoint signalling. EMBO J 24:753–765

Procter DJ, Banerjee A, Nukui M, Kruse K, Gaponenko V, Murphy EA, Komarova Y, Walsh D (2018) The HCMV assembly compartment is a dynamic Golgi-derived MTOC that controls nuclear rotation and virus spread. Dev Cell 45:83–100.e7

Puthenveedu MA, Bachert C, Puri S, Lanni F, Linstedt AD (2006) GM130 and GRASP65-dependent lateral cisternal fusion allows uniform Golgi-enzyme distribution. Nat Cell Biol 8:238–248

Quiner CA, Jackson WT (2010) Fragmentation of the Golgi apparatus provides replication membranes for human rhinovirus 1A. Virology 407:185–195

Rabouille C, Kondylis V (2007) Golgi ribbon unlinking: an organelle-based G2/M checkpoint. Cell Cycle 6:2723–2729

Rabouille C, Linstedt AD (2016) GRASP: a multitasking tether. Front Cell Dev Biol 4:1

Rabouille C, Levine TP, Peters JM, Warren G (1995) An NSF-like ATPase, p97, and NSF mediate cisternal regrowth from mitotic Golgi fragments. Cell 82:905–914

Ramadan K, Bruderer R, Spiga FM, Popp O, Baur T, Gotta M, Meyer HH (2007) Cdc48/p97 promotes reformation of the nucleus by extracting the kinase Aurora B from chromatin. Nature 450:1258–1262

Ramirez IB, Lowe M (2009) Golgins and GRASPs: holding the Golgi together. Semin Cell Dev Biol 20:770–779

Ran R, Pan R, Lu A, Xu H, Davis RR, Sharp FR (2007) A novel 165-kDa Golgin protein induced by brain ischemia and phosphorylated by Akt protects against apoptosis. Mol Cell Neurosci 36:392–407

Rao KH, Ghosh S, Datta A (2016) Env7p associates with the Golgin protein Imh1 at the trans-Golgi Network in Candida albicans. mSphere 1. https://doi.org/10.1128/mSphere.00080-16

Rebmann GM, Grabski R, Sanchez V, Britt WJ (2016) Phosphorylation of Golgi peripheral membrane protein Grasp65 is an integral step in the formation of the human cytomegalovirus cytoplasmic assembly compartment. MBio 7. https://doi.org/10.1128/mBio.01554-16

Reddy JV, Burguete AS, Sridevi K, Ganley IG, Nottingham RM, Pfeffer SR (2006) A functional role for the GCC185 golgin in mannose 6-phosphate receptor recycling. Mol Biol Cell 17:4353–4363

Rejman Lipinski A, Heymann J, Meissner C, Karlas A, Brinkmann V, Meyer TF, Heuer D (2009) Rab6 and Rab11 regulate Chlamydia trachomatis development and golgin-84-dependent Golgi fragmentation. PLoS Pathog 5:e1000615

Rendon WO, Martinez-Alonso E, Tomas M, Martinez-Martinez N, Martinez-Menarguez JA (2013) Golgi fragmentation is Rab and SNARE dependent in cellular models of Parkinson's disease. Histochem Cell Biol 139:671–684

Rios RM, Sanchis A, Tassin AM, Fedriani C, Bornens M (2004) GMAP-210 recruits gamma-tubulin complexes to cis-Golgi membranes and is required for Golgi ribbon formation. Cell 118:323–335

Risco C, Carrascosa JL, Frey TK (2003) Structural maturation of rubella virus in the Golgi complex. Virology 312:261–269

Ristic G, Sutton JR, Libohova K, Todi SV (2018) Toxicity and aggregation of the polyglutamine disease protein, ataxin-3 is regulated by its binding to VCP/p97 in Drosophila melanogaster. Neurobiol Dis 116:78–92

Rivero S, Cardenas J, Bornens M, Rios RM (2009) Microtubule nucleation at the cis-side of the Golgi apparatus requires AKAP450 and GM130. EMBO J 28:1016–1028

Rohde J, Emschermann F, Knittler MR, Rziha HJ (2012) Orf virus interferes with MHC class I surface expression by targeting vesicular transport and Golgi. BMC Vet Res 8:114

Rosing M, Ossendorf E, Rak A, Barnekow A (2007) Giantin interacts with both the small GTPase Rab6 and Rab1. Exp Cell Res 313:2318–2325

Salanueva IJ, Novoa RR, Cabezas P, Lopez-Iglesias C, Carrascosa JL, Elliott RM, Risco C (2003) Polymorphism and structural maturation of bunyamwera virus in Golgi and post-Golgi compartments. J Virol 77:1368–1381

Salonen A, Ahola T, Kaariainen L (2005) Viral RNA replication in association with cellular membranes. Curr Top Microbiol Immunol 285:139–173

Saraiva N, Prole DL, Carrara G, Maluquer De Motes C, Johnson BF, Byrne B, Taylor CW, Smith GL (2013) Human and viral Golgi anti-apoptotic proteins (GAAPs) oligomerize via different mechanisms and monomeric GAAP inhibits apoptosis and modulates calcium. J Biol Chem 288:13057–13067

Sasaki J, Ishikawa K, Arita M, Taniguchi K (2012) ACBD3-mediated recruitment of PI4KB to picornavirus RNA replication sites. EMBO J 31:754–766

Satoh A, Wang Y, Malsam J, Beard MB, Warren G (2003) Golgin-84 is a rab1 binding partner involved in Golgi structure. Traffic 4:153–161

Sbodio JI, Machamer CE (2007) Identification of a redox-sensitive cysteine in GCP60 that regulates its interaction with golgin-160. J Biol Chem 282:29874–29881

Sbodio JI, Hicks SW, Simon D, Machamer CE (2006) GCP60 preferentially interacts with a caspase-generated golgin-160 fragment. J Biol Chem 281:27924–27931

Schultz MJ, Swindall AF, Bellis SL (2012) Regulation of the metastatic cell phenotype by sialylated glycans. Cancer Metastasis Rev 31:501–518

Scott KL, Kabbarah O, Liang MC, Ivanova E, Anagnostou V, Wu J, Dhakal S, Wu M, Chen S, Feinberg T, Huang J, Saci A, Widlund HR, Fisher DE, Xiao Y, Rimm DL, Protopopov A, Wong KK, Chin L (2009) GOLPH3 modulates mTOR signalling and rapamycin sensitivity in cancer. Nature 459:1085–1090

Seemann J, Jokitalo EJ, Warren G (2000) The role of the tethering proteins p115 and GM130 in transport through the Golgi apparatus in vivo. Mol Biol Cell 11:635–645

Seifert W, Kuhnisch J, Maritzen T, Horn D, Haucke V, Hennies HC (2011) Cohen syndrome-associated protein, COH1, is a novel, giant Golgi matrix protein required for Golgi integrity. J Biol Chem 286:37665–37675

Selkoe DJ, Hardy J (2016) The amyloid hypothesis of Alzheimer's disease at 25 years. EMBO Mol Med 8:595–608

Sengupta D, Linstedt AD (2010) Mitotic inhibition of GRASP65 organelle tethering involves Polo-like kinase 1 (PLK1) phosphorylation proximate to an internal PDZ ligand. J Biol Chem 285:39994–40003

Shin JJH, Gillingham AK, Begum F, Chadwick J, Munro S (2017) TBC1D23 is a bridging factor for endosomal vesicle capture by golgins at the trans-Golgi. Nat Cell Biol 19:1424–1432

Shin JJH, Gillingham AK, Begum F, Chadwick J, Munro S (2018) Author Correction: TBC1D23 is a bridging factor for endosomal vesicle capture by golgins at the trans-Golgi. Nat Cell Biol 20:222

Shitara A, Shibui T, Okayama M, Arakawa T, Mizoguchi I, Sakakura Y, Takuma T (2013) VAMP4 is required to maintain the ribbon structure of the Golgi apparatus. Mol Cell Biochem 380:11–21

Short B, Preisinger C, Korner R, Kopajtich R, Byron O, Barr FA (2001) A GRASP55-rab2 effector complex linking Golgi structure to membrane traffic. J Cell Biol 155:877–883

Short JR, Nakayinga R, Hughes GE, Walter CT, Dorrington RA (2013) Providence virus (family: Carmotetraviridae) replicates vRNA in association with the Golgi apparatus and secretory vesicles. J Gen Virol 94:1073–1078

Shorter J, Warren G (1999) A role for the vesicle tethering protein, p115, in the post-mitotic stacking of reassembling golgi cisternae in a cell-free system. J Cell Biol 146:57–70

Shorter J, Warren G (2002) Golgi architecture and inheritance. Annu Rev Cell Dev Biol 18:379–420

Shorter J, Watson R, Giannakou ME, Clarke M, Warren G, Barr FA (1999) GRASP55, a second mammalian GRASP protein involved in the stacking of Golgi cisternae in a cell-free system. EMBO J 18:4949–4960

Shorter J, Beard MB, Seemann J, Dirac-Svejstrup AB, Warren G (2002) Sequential tethering of Golgins and catalysis of SNAREpin assembly by the vesicle-tethering protein p115. J Cell Biol 157:45–62

Sillibourne JE, Milne DM, Takahashi M, Ono Y, Meek DW (2002) Centrosomal anchoring of the protein kinase CK1delta mediated by attachment to the large, coiled-coil scaffolding protein CG-NAP/AKAP450. J Mol Biol 322:785–797

Sinka R, Gillingham AK, Kondylis V, Munro S (2008) Golgi coiled-coil proteins contain multiple binding sites for Rab family G proteins. J Cell Biol 183:607–615

Sivan G, Weisberg AS, Americo JL, Moss B (2016) Retrograde transport from early endosomes to the trans-Golgi network enables membrane wrapping and egress of vaccinia virus virions. J Virol 90:8891–8905

Slusarewicz P, Nilsson T, Hui N, Watson R, Warren G (1994) Isolation of a matrix that binds medial Golgi enzymes. J Cell Biol 124:405–413

Sohda M, Misumi Y, Yamamoto A, Yano A, Nakamura N, Ikehara Y (2001) Identification and characterization of a novel Golgi protein, GCP60, that interacts with the integral membrane protein giantin. J Biol Chem 276:45298–45306

Sohda M, Misumi Y, Yamamoto A, Nakamura N, Ogata S, Sakisaka S, Hirose S, Ikehara Y, Oda K (2010) Interaction of Golgin-84 with the COG complex mediates the intra-Golgi retrograde transport. Traffic 11:1552–1566

Sohda M, Misumi Y, Ogata S, Sakisaka S, Hirose S, Ikehara Y, Oda K (2015) Trans-Golgi protein p230/golgin-245 is involved in phagophore formation. Biochem Biophys Res Commun 456:275–281

Song C, Wang Q, Li CC (2007) Characterization of the aggregation-prevention activity of p97/valosin-containing protein. Biochemistry 46:14889–14898

Sonnichsen B, Lowe M, Levine T, Jamsa E, Dirac-Svejstrup B, Warren G (1998) A role for giantin in docking COPI vesicles to Golgi membranes. J Cell Biol 140:1013–1021

Soo KY, Halloran M, Sundaramoorthy V, Parakh S, Toth RP, Southam KA, Mclean CA, Lock P, King A, Farg MA, Atkin JD (2015) Rab1-dependent ER-Golgi transport dysfunction is a common pathogenic mechanism in SOD1, TDP-43 and FUS-associated ALS. Acta Neuropathol 130:679–697

Steegmaier M, Klumperman J, Foletti DL, Yoo JS, Scheller RH (1999) Vesicle-associated membrane protein 4 is implicated in trans-Golgi network vesicle trafficking. Mol Biol Cell 10:1957–1972

Sun KH, De Pablo Y, Vincent F, Johnson EO, Chavers AK, Shah K (2008) Novel genetic tools reveal Cdk5's major role in Golgi fragmentation in Alzheimer's disease. Mol Biol Cell 19:3052–3069

Sun NK, Huang SL, Chien KY, Chao CC (2012) Golgi-SNARE GS28 potentiates cisplatin-induced apoptosis by forming GS28-MDM2-p53 complexes and by preventing the ubiquitination and degradation of p53. Biochem J 444:303–314

Sundaramoorthy S, Goh JB, Rafee S, Murata-Hori M (2010) Mitotic Golgi vesiculation involves mechanisms independent of Ser25 phosphorylation of GM130. Cell Cycle 9:3100–3105

Sundaramoorthy V, Sultana JM, Atkin JD (2015) Golgi fragmentation in amyotrophic lateral sclerosis, an overview of possible triggers and consequences. Front Neurosci 9:400

Sutterlin C, Polishchuk R, Pecot M, Malhotra V (2005) The Golgi-associated protein GRASP65 regulates spindle dynamics and is essential for cell division. Mol Biol Cell 16:3211–3222

Takahashi M, Shibata H, Shimakawa M, Miyamoto M, Mukai H, Ono Y (1999) Characterization of a novel giant scaffolding protein, CG-NAP, that anchors multiple signaling enzymes to centrosome and the golgi apparatus. J Biol Chem 274:17267–17274

Takahashi M, Yamagiwa A, Nishimura T, Mukai H, Ono Y (2002) Centrosomal proteins CG-NAP and kendrin provide microtubule nucleation sites by anchoring gamma-tubulin ring complex. Mol Biol Cell 13:3235–3245

Tang D, Wang Y (2013) Cell cycle regulation of Golgi membrane dynamics. Trends Cell Biol 23:296–304

Tang D, Mar K, Warren G, Wang Y (2008) Molecular mechanism of mitotic Golgi disassembly and reassembly revealed by a defined reconstitution assay. J Biol Chem 283:6085–6094

Tang D, Xiang Y, Wang Y (2010a) Reconstitution of the cell cycle-regulated Golgi disassembly and reassembly in a cell-free system. Nat Protoc 5:758–772

Tang D, Yuan H, Wang Y (2010b) The role of GRASP65 in Golgi cisternal stacking and cell cycle progression. Traffic 11:827–842

Tang D, Xiang Y, De Renzis S, Rink J, Zheng G, Zerial M, Wang Y (2011) The ubiquitin ligase HACE1 regulates Golgi membrane dynamics during the cell cycle. Nat Commun 2:501

Tang D, Yuan H, Vielemeyer O, Perez F, Wang Y (2012) Sequential phosphorylation of GRASP65 during mitotic Golgi disassembly. Biol Open 1:1204–1214

Tang D, Zhang X, Huang S, Yuan H, Li J, Wang Y (2016) Mena-GRASP65 interaction couples actin polymerization to Golgi ribbon linking. Mol Biol Cell 27:137–152

Terrano DT, Upreti M, Chambers TC (2010) Cyclin-dependent kinase 1-mediated Bcl-xL/Bcl-2 phosphorylation acts as a functional link coupling mitotic arrest and apoptosis. Mol Cell Biol 30:640–656

Thyberg J, Moskalewski S (1999) Role of microtubules in the organization of the Golgi complex. Exp Cell Res 246:263–279

Toki C, Fujiwara T, Sohda M, Hong HS, Misumi Y, Ikehara Y (1997) Identification and characterization of rat 364-kDa Golgi-associated protein recognized by autoantibodies from a patient with rheumatoid arthritis. Cell Struct Funct 22:565–577

Truschel ST, Zhang M, Bachert C, Macbeth MR, Linstedt AD (2012) Allosteric regulation of GRASP protein-dependent Golgi membrane tethering by mitotic phosphorylation. J Biol Chem 287:19870–19875

Tsukada M, Will E, Gallwitz D (1999) Structural and functional analysis of a novel coiled-coil protein involved in Ypt6 GTPase-regulated protein transport in yeast. Mol Biol Cell 10:63–75

Uchiyama K, Jokitalo E, Lindman M, Jackman M, Kano F, Murata M, Zhang X, Kondo H (2003) The localization and phosphorylation of p47 are important for Golgi disassembly-assembly during the cell cycle. J Cell Biol 161:1067–1079

Uchiyama K, Totsukawa G, Puhka M, Kaneko Y, Jokitalo E, Dreveny I, Beuron F, Zhang X, Freemont P, Kondo H (2006) p37 is a p97 adaptor required for Golgi and ER biogenesis in interphase and at the end of mitosis. Dev Cell 11:803–816

Van Den Boom J, Wolf M, Weimann L, Schulze N, Li F, Kaschani F, Riemer A, Zierhut C, Kaiser M, Iliakis G, Funabiki H, Meyer H (2016) VCP/p97 extracts sterically trapped Ku70/80 rings from DNA in double-strand break repair. Mol Cell 64:189–198

Van Dis V, Kuijpers M, Haasdijk ED, Teuling E, Oakes SA, Hoogenraad CC, Jaarsma D (2014) Golgi fragmentation precedes neuromuscular denervation and is associated with endosome abnormalities in SOD1-ALS mouse motor neurons. Acta Neuropathol Commun 2:38

Van Valkenburgh H, Shern JF, Sharer JD, Zhu X, Kahn RA (2001) ADP-ribosylation factors (ARFs) and ARF-like 1 (ARL1) have both specific and shared effectors: characterizing ARL1-binding proteins. J Biol Chem 276:22826–22837

Vielemeyer O, Yuan H, Moutel S, Saint-Fort R, Tang D, Nizak C, Goud B, Wang Y, Perez F (2009) Direct selection of monoclonal phosphospecific antibodies without prior phosphoamino acid mapping. J Biol Chem 284:20791–20795

Vinke FP, Grieve AG, Rabouille C (2011) The multiple facets of the Golgi reassembly stacking proteins. Biochem J 433:423–433

Wakabayashi K, Hayashi S, Kakita A, Yamada M, Toyoshima Y, Yoshimoto M, Takahashi H (1998) Accumulation of alpha-synuclein/NACP is a cytopathological feature common to Lewy body disease and multiple system atrophy. Acta Neuropathol 96:445–452

Walker A, Ward C, Sheldrake TA, Dransfield I, Rossi AG, Pryde JG, Haslett C (2004) Golgi fragmentation during Fas-mediated apoptosis is associated with the rapid loss of GM130. Biochem Biophys Res Commun 316:6–11

Wallis N, Lau CL, Farg MA, Atkin JD, Beart PM, O'Shea RD (2018) SOD1 mutations causing familial amyotrophic lateral sclerosis induce toxicity in astrocytes: evidence for bystander effects in a continuum of astrogliosis. Neurochem Res 43:157–170

Wang Y (2008) Golgi apparatus inheritance. In: Mironov A, Pavelka M, Luini A (eds) The Golgi apparatus. State of the art 110 years after Camillo Golgi's discovery. Springer, New York

Wang Y, Seemann J (2011) Golgi biogenesis. Cold Spring Harb Perspect Biol 3:a005330

Wang Y, Seemann J, Pypaert M, Shorter J, Warren G (2003) A direct role for GRASP65 as a mitotically regulated Golgi stacking factor. EMBO J 22:3279–3290

Wang Y, Satoh A, Warren G, Meyer HH (2004) VCIP135 acts as a deubiquitinating enzyme during p97-p47-mediated reassembly of mitotic Golgi fragments. J Cell Biol 164:973–978

Wang Y, Satoh A, Warren G (2005) Mapping the functional domains of the Golgi stacking factor GRASP65. J Biol Chem 280:4921–4928

Wang Y, Wei JH, Bisel B, Tang D, Seemann J (2008) Golgi cisternal unstacking stimulates COPI vesicle budding and protein transport. PLoS One 3:e1647

Wang T, Grabski R, Sztul E, Hay JC (2015) p115-SNARE interactions: a dynamic cycle of p115 binding monomeric SNARE motifs and releasing assembled bundles. Traffic 16:148–171

Wanschers BF, Van De Vorstenbosch R, Schlager MA, Splinter D, Akhmanova A, Hoogenraad CC, Wieringa B, Fransen JA (2007) A role for the Rab6B Bicaudal-D1 interaction in retrograde transport in neuronal cells. Exp Cell Res 313:3408–3420

Warren G (1995) Intracellular membrane morphology. Philos Trans R Soc Lond B Biol Sci 349:291–295

Wei JH, Seemann J (2010) Unraveling the Golgi ribbon. Traffic 11:1391–1400

Wei JH, Seemann J (2017) Golgi ribbon disassembly during mitosis, differentiation and disease progression. Curr Opin Cell Biol 47:43–51

Weide T, Bayer M, Koster M, Siebrasse JP, Peters R, Barnekow A (2001) The Golgi matrix protein GM130: a specific interacting partner of the small GTPase rab1b. EMBO Rep 2:336–341

Weller SG, Capitani M, Cao H, Micaroni M, Luini A, Sallese M, Mcniven MA (2010) Src kinase regulates the integrity and function of the Golgi apparatus via activation of dynamin 2. Proc Natl Acad Sci USA 107:5863–5868

Westerbeck JW, Machamer CE (2015) A coronavirus E protein is present in two distinct pools with different effects on assembly and the secretory pathway. J Virol 89:9313–9323

Williams D, Hicks SW, Machamer CE, Pessin JE (2006) Golgin-160 is required for the Golgi membrane sorting of the insulin-responsive glucose transporter GLUT4 in adipocytes. Mol Biol Cell 17:5346–5355

Witkos TM, Lowe M (2015) The Golgin family of coiled-coil tethering proteins. Front Cell Dev Biol 3:86

Woldemichael GM, Turbyville TJ, Linehan WM, Mcmahon JB (2011) Carminomycin I is an apoptosis inducer that targets the Golgi complex in clear cell renal carcinoma cells. Cancer Res 71:134–142

Wu M, Lu L, Hong W, Song H (2004) Structural basis for recruitment of GRIP domain golgin-245 by small GTPase Arl1. Nat Struct Mol Biol 11:86–94

Xiang Y, Wang Y (2010) GRASP55 and GRASP65 play complementary and essential roles in Golgi cisternal stacking. J Cell Biol 188:237–251

Xiang Y, Wang Y (2011) New components of the Golgi matrix. Cell Tissue Res 344:365–379

Xiang Y, Zhang X, Nix DB, Katoh T, Aoki K, Tiemeyer M, Wang Y (2013) Regulation of protein glycosylation and sorting by the Golgi matrix proteins GRASP55/65. Nat Commun 4:1659

Yadav S, Linstedt AD (2011) Golgi positioning. Cold Spring Harb Perspect Biol 3. https://doi.org/10.1101/cshperspect.a005322

Yadav V, Panganiban AT, Honer Zu Bentrup K, Voss TG (2016) Influenza infection modulates vesicular trafficking and induces Golgi complex disruption. Virusdisease 27:357–368

Yamaguchi Y, Larkin D, Lara-Lemus R, Ramos-Castaneda J, Liu M, Arvan P (2008) Endoplasmic reticulum (ER) chaperone regulation and survival of cells compensating for deficiency in the ER stress response kinase, PERK. J Biol Chem 283:17020–17029

Yamamoto H, Kakuta S, Watanabe TM, Kitamura A, Sekito T, Kondo-Kakuta C, Ichikawa R, Kinjo M, Ohsumi Y (2012) Atg9 vesicles are an important membrane source during early steps of autophagosome formation. J Cell Biol 198:219–233

Yamane J, Kubo A, Nakayama K, Yuba-Kubo A, Katsuno T, Tsukita S (2007) Functional involvement of TMF/ARA160 in Rab6-dependent retrograde membrane traffic. Exp Cell Res 313:3472–3485

Yi Z, Yuan Z (2017) Aggregation of a hepatitis C virus replicase module induced by ablation of p97/VCP. J Gen Virol 98:1667–1678

Yin KJ, Deng Z, Hamblin M, Xiang Y, Huang H, Zhang J, Jiang X, Wang Y, Chen YE (2010) Peroxisome proliferator-activated receptor delta regulation of miR-15a in ischemia-induced cerebral vascular endothelial injury. J Neurosci 30:6398–6408

Yoshimura S, Yoshioka K, Barr FA, Lowe M, Nakayama K, Ohkuma S, Nakamura N (2005) Convergence of cell cycle regulation and growth factor signals on GRASP65. J Biol Chem 280:23048–23056

Yue X, Bao M, Christiano R, Li S, Mei J, Zhu L, Mao F, Yue Q, Zhang P, Jing S, Rothman JE, Qian Y, Lee I (2017) ACBD3 functions as a scaffold to organize the Golgi stacking proteins and a Rab33b-GAP. FEBS Lett 591:2793–2802

Zeng Z, Lin H, Zhao X, Liu G, Wang X, Xu R, Chen K, Li J, Song L (2012) Overexpression of GOLPH3 promotes proliferation and tumorigenicity in breast cancer via suppression of the FOXO1 transcription factor. Clin Cancer Res 18:4059–4069

Zhang X, Wang Y (2015a) Cell cycle regulation of VCIP135 deubiquitinase activity and function in p97/p47-mediated Golgi reassembly. Mol Biol Cell 26:2242–2251

Zhang X, Wang Y (2015b) GRASPs in Golgi structure and function. Front Cell Dev Biol 3:84

Zhang X, Wang Y (2016) Glycosylation quality control by the Golgi structure. J Mol Biol 428:3183–3193

Zhang X, Wang Y (2018a) The Golgi stacking protein GORASP2/GRASP55 serves as an energy sensor to promote autophagosome maturation under glucose starvation. Autophagy 14:1649–1651

Zhang X, Wang Y (2018b) GRASP55 facilitates autophagosome maturation under glucose deprivation. Mol Cell Oncol 5:e1494948

Zhang X, Zhang H, Wang Y (2014) Phosphorylation regulates VCIP135 function in Golgi membrane fusion during the cell cycle. J Cell Sci 127:172–181

Zhang W, Liu Y, Zhu X, Cao Y, Liu Y, Mao X, Yang H, Zhou Z, Wang Y, Shen A (2016a) SCYL1-like 1-binding protein 1 (SCYL1BP1) suppressed sciatic nerve regeneration by enhancing the RhoA pathway. Mol Neurobiol 53:6342–6354

Zhang X, Wang T, Dai X, Zhang Y, Jiang H, Zhang Q, Liu F, Wu K, Liu Y, Zhou H, Wu J (2016b) Golgi protein 73 facilitates the interaction of hepatitis C virus NS5A with apolipoprotein E to promote viral particle secretion. Biochem Biophys Res Commun 479:683–689

Zhang X, Wang L, Lak B, Li J, Jokitalo E, Wang Y (2018) GRASP55 senses glucose deprivation through O-GlcNAcylation to promote autophagosome-lysosome fusion. Dev Cell 45:245–261. e6

Zhao J, Li B, Huang X, Morelli X, Shi N (2017) Structural basis for the interaction between Golgi reassembly-stacking protein GRASP55 and Golgin45. J Biol Chem 292:2956–2965

Zhou Y, Atkins JB, Rompani SB, Bancescu DL, Petersen PH, Tang H, Zou K, Stewart SB, Zhong W (2007) The mammalian Golgi regulates numb signaling in asymmetric cell division by releasing ACBD3 during mitosis. Cell 129:163–178

Zhou Z, Sun X, Zou Z, Sun L, Zhang T, Guo S, Wen Y, Liu L, Wang Y, Qin J, Li L, Gong W, Bao S (2010) PRMT5 regulates Golgi apparatus structure through methylation of the golgin GM130. Cell Res 20:1023–1033

Zhu X, Kaverina I (2013) Golgi as an MTOC: making microtubules for its own good. Histochem Cell Biol 140:361–367

Chapter 20
Selected Golgi-Localized Proteins and Carcinogenesis: What Do We Know?

Piotr Donizy and Jakub Marczuk

Abstract The role of the Golgi apparatus in carcinogenesis still remains unclear. A number of structural and functional cis-, medial-, and trans-Golgi proteins as well as a complexity of metabolic pathways which they mediate may indicate a central role of the Golgi apparatus in the development and progression of cancer. Pleiotropy of cellular function of the Golgi apparatus makes it a "metabolic heart" or a relay station of a cell, which combines multiple signaling pathways involved in carcinogenesis. Therefore, any damage to or structural abnormality of the Golgi apparatus, causing its fragmentation and/or biochemical dysregulation, results in an up- or downregulation of signaling pathways and may in turn promote tumor progression, as well as local nodal and distant metastases. Three alternative or parallel models of spatial and functional Golgi organization within tumor cells were proposed: (1) compacted Golgi structure, (2) normal Golgi structure with its increased activity, and (3) the Golgi fragmentation with ministacks formation. Regardless of the assumed model, the increased activity of oncogenesis initiators and promoters with inhibition of suppressor proteins results in an increased cell motility and migration, increased angiogenesis, significantly activated trafficking kinetics, proliferation, EMT induction, decreased susceptibility to apoptosis-inducing factors, and modulating immune response to tumor cell antigens. Eventually, this will lead to the increased metastatic potential of cancer cells and an increased risk of lymph node and distant metastases. This chapter provided an overview of the current state of knowledge of selected Golgi proteins, their role in cytophysiology as well as potential involvement in tumorigenesis.

P. Donizy (✉)
Department of Pathomorphology and Oncological Cytology, Wroclaw Medical University, Wroclaw, Poland

Jan Mikulicz-Radecki University Teaching Hospital, Wroclaw, Poland

J. Marczuk
Department of Pathomorphology and Oncological Cytology, Wroclaw Medical University, Wroclaw, Poland

© Springer Nature Switzerland AG 2019
M. Kloc (ed.), *The Golgi Apparatus and Centriole*, Results and Problems in Cell Differentiation 67, https://doi.org/10.1007/978-3-030-23173-6_20

20.1   Introduction

The Golgi apparatus (GA) is a strictly organized membrane organelle responsible for trafficking of proteins and lipids in the cell as well as for their post-translational modification (Witkos and Lowe 2016; Howley and Howe 2018). This cellular factory has a form of stacks consisting of five to eight flattened cisternae, laid tightly parallel to each other. It can be divided into three distinct parts: the cis-Golgi network (CGN; close to the endoplasmic reticulum [ER] and receiving its output), the medial part, and the trans-Golgi network (TGN; close to the plasma membrane and sending cargo molecules to different destinations) (Witkos and Lowe 2016). Golgi stacks are able to aggregate to form the so-called Golgi ribbon, comprising of multiple stacks linked laterally by tubular structures. Other postulated cytophysiological functions of the GA and its proteins involve, for instance, regulating apoptosis, cell mobility and migration (Millarte and Farhan 2012).

Pleiotropy of cellular function of the GA makes it a "metabolic heart" of a cell, which combines multiple signaling pathways involved in carcinogenesis. Therefore, any damage to the GA, causing its fragmentation and/or biochemical disruption, results in a dysregulation (up- or downregulation) of cellular metabolic pathways and may in turn promote tumor progression, as well as local nodal and distant metastases. This chapter provides an overview of the current state of knowledge of selected Golgi proteins, their role in cytophysiology as well as potential involvement in tumorigenesis and progression.

20.2   Structural-Functional cis-Golgi Proteins and Their Role in Carcinogenesis

20.2.1   GMAP-210 (Golgi-Microtubule-Associated Protein of 210 kDa, Thyroid Receptor-Interacting Protein 11, TRIP11)

GMAP-210 is a protein involved in vesicle tethering and vesicular traffic located within the CGN (Pranke et al. 2011; Sato et al. 2015) (Fig. 20.1). The important roles of GMAP-210 in cell physiology involve being a membrane tether as well as linking the GA to microtubules and centrosomes (Ríos et al. 2004; Infante et al. 1999). Loss-of-function mutations in GMAP-210 lead to the achondrogenesis type 1A, which may suggest a significant role of secretory trafficking impairment in chondrocytes (Vanegas et al. 2018). Interestingly, GMAP-210 can localize in the nucleus, acting as a transcription coactivator and interacting with thyroid hormone receptors (THR) (Popławski et al. 2017; Chen et al. 1999). In the presence of triiodothyronine, GMAP-210 (TRIP11) binds THRs and enhances THR-dependent transcription. Furthermore, TRIP11 also functions as a coactivator of hypoxia-inducible factor (HIF) (Beischlag et al. 2004). Until now, only one study has been published to discuss the expression of

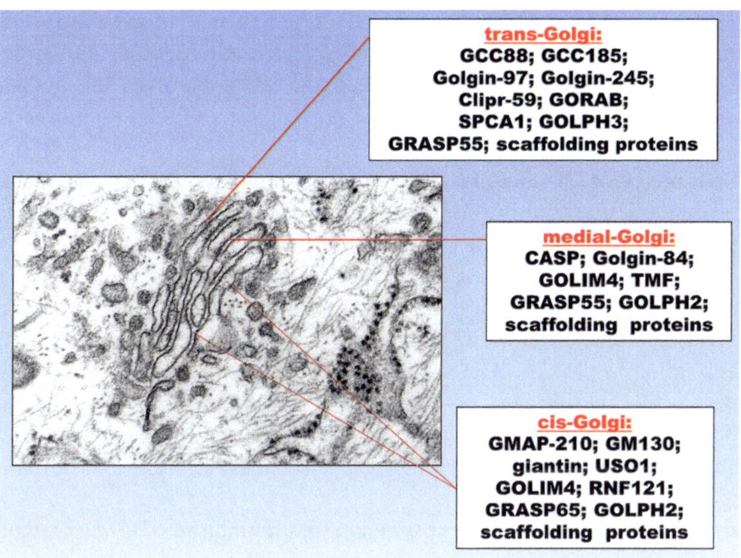

Fig. 20.1 The structure of the Golgi apparatus with a division into the cis-, medial-, and trans-Golgi compartments as well as their selected characteristic proteins (electron microphotograph; ×120,000 magnification)

GMAP-210 (TRIP11) in renal cancer, without specifying its histological subtypes. A significant correlation between its decreased expression and shorter overall survival was demonstrated (Popławski et al. 2017). A potential role of GMAP-210 in human tumors needs further in-depth research using clinical tissue specimens.

20.2.2 *GM130 (Golgin Subfamily A Member 2, 130 kDa cis-Golgi Matrix Protein; SY11 Protein, Golgin Subfamily a2; Golgin-95, Golgin A2)*

GM130 is another cis-Golgi protein of golgin subfamily (Fig. 20.1). It is a pleiotropic intracellular protein, which plays a key role in the maintenance of ER-to-Golgi transport vesicles through direct interaction with vesicle docking protein p115 (Nakamura et al. 1997). Furthermore, GM130 promotes microtubule nucleation and microtubule growth at the cis-Golgi, recruiting the nucleation-promoting factor AKAP450 there (Rivero et al. 2009). GM130 is also involved in GA fragmentation during mitosis and, by regulating microtubular dynamics, it enables equal partitioning into the two daughter cells (Wei and Seemann 2009; Wei and Seemann

2017). At early stages of mitosis, GM130 interacts with importin-α causing its sequestration, and, in turn, TPX2 (microtubule nucleation factor) activation within Golgi membranes (Wei et al. 2015). The main effect involves the nucleation of spindle microtubules at the Golgi and equal segregation of Golgi membranes during mitosis.

Furthermore, GM130 directly affects active, polarized cell migration. Binding and subsequent sequestration of Tuba protein (Ccd42-specific guanine nucleotide exchange factor) in the GA plays the key role in this process. As a result of Tuba protein translocation to GA, Cdc42 redistribution and activation take place (Kodani et al. 2009; Baschieri et al. 2014). Baschieri et al. (2014) demonstrated that interaction between GM130 and RasGRF (inhibitor of Cdc42) also leads to Cdc42 activation in the GA. The second molecular mechanism associated with cell migration is the recruitment of protein kinase Stk25 (serine/threonine kinase 25, YSK1) to the GA and its subsequent activation (Preisinger et al. 2004). Interestingly, GM130 also inhibits autophagy by sequestrating its known inducer, GABARAP (Joachim et al. 2015).

Another important issue is a post-translational regulation of GM130, which may play a role in the maintenance of Golgi structure. Zhou et al. (2010) demonstrated that protein arginine methyltransferase 5 (PRMT5) is essential for the maintenance of the GA architecture through its arginine methylation of GM130. The prognostic role of PRMT5 upregulation and downregulation in human cancers still remains an open question, though.

Recent studies indicate the crucial role of GM130 in carcinogenesis. Zhao et al. (2015) demonstrated that overexpression of GM130 in patients with gastric cancer was associated with increased invasiveness of tumor cells. Kaplan-Meier survival analysis demonstrated a significant correlation between overexpression of GM130 and shorter survival. Furthermore, an association was demonstrated between the downregulation of E-cadherin and overexpression of GM130 as well as of Snail (the epithelial-mesenchymal transition [EMT] regulator), which suggests a role of GM130 as the EMT regulator (Zhao et al. 2015).

A downregulation of GM130 by short hairpin RNA (shRNA) constructs in lung cancer model in vitro inhibits angiogenesis and decreases the invasiveness of lung cancer cells. Furthermore, a downregulation of GM130 was shown to induce autophagy in A549 lung cancer cell line (Chang et al. 2012). Intriguingly, though, in breast cancer cell lines, depletion of GM130 increases cellular velocity and increases the invasiveness of cancer cells (Baschieri et al. 2015).

Contradictory and equivocal conclusions regarding a potential involvement of GM130 in carcinogenesis warrant further in-depth multicenter research in large samples of patients with solid organ tumors.

20.2.3   Giantin (GC; GCP372; GOLIM1; Golgin Subfamily B Member 1; 372 kDa Golgi Complex-Associated Protein; Golgi Autoantigen; Golgin Subfamily b, Macrogolgin (with Transmembrane Signal), 1; Golgi Integral Membrane Protein 1; Golgin B1, Golgi Integral Membrane Protein; Macrogolgin; Golgin B1)

Giantin is the largest golgin subfamily member present in mammals (Koreishi et al. 2013). It participates in vesicle tethering, membrane fusion, and cargo transport processes. Furthermore, it is responsible for spatial organization of the Golgi ribbons (Koreishi et al. 2013).

Alongside GM130, p115, and GRASP65 (Golgi peripheral membrane protein p65), it forms the "cis-golgin tether," which plays a role in vesicle tethering and anterograde cargo. To form cis-golgin tether, giantin is linked to GM130 via p115. GM130 is in turn linked to GRASP65 anchored to the Golgi membrane (Fig. 20.1). Furthermore, giantin is able to bind directly to Rab1 (small GTPase of the Rab family), the role of which in forming cis-golgin tether still remains unclear (Koreishi et al. 2013).

Little is known about the role of giantins in oncogenesis. A study using prostate cancer cell lines demonstrated that the GA in androgen-sensitive prostate cancer cells and normal prostate cells has a compact morphology whereas it is fragmented in androgen-refractory cells and primary prostate tumors. At the same time, mislocalization of C2GnT-L (core 2 N-acetylglucosaminyltransferase-L coded by GCNT1 gene) is found in cells with fragmented GA (Petrosyan et al. 2014). It was observed that giantin is essential for Golgi targeting of C2GnT-L (unlike e.g. sialyl-T antigen produced by β-galactoside α-2,3-sialyltransferase-1 [ST3Gal1], which can also use GM130-GRASP65 for Golgi targeting if giantin is defective) (Petrosyan et al. 2014). Targeting of C2GnT-L is only possible with phosphorylated and dimerized form of giantin. Failure of giantin monomers to be phosphorylated and dimerized prevents them from being used for Golgi targeting and, in turn, prevents Golgi from forming compact morphology (Petrosyan et al. 2014). It was also demonstrated that inhibition or knockdown of non-muscle myosin IIA (MYH9) motor protein enables Rab6a GTPase to promote phosphorylation of giantin by polo-like kinase 3 (PLK3), thereby allowing dimerization of giantin assisted by protein disulfide isomerase A3 (PDIA3), restoration of compact Golgi morphology and targeting of C2GnT-L (Petrosyan et al. 2014). Interestingly, Golgi relocation of C2GnT-L in androgen-refractory cells was shown to result in their increased susceptibility to galectin-1-induced apoptosis by replacing sialyl-T antigen with polylactosamine (Petrosyan et al. 2014). To sum up, giantin (alongside the mediators of its phosphorylation and subsequent dimerization) and C2GnT-L are the key players affecting the susceptibility of prostate cancer cells to galectin-1-mediated apoptosis.

Another study using prostate cancer cell lines demonstrated that by being a mediator of glycosylation, giantin regulates cancer cell phenotype (Bhat et al. 2017). It was shown that defective giantin in androgen-independent prostate cancer cells results in a shift of Golgi targeting of glycosyltransferases and α-mannosidase IA from giantin to GM130-GRASP65. A consequence of shifted enzyme targeting is an abnormal modification of trans-Golgi enzymes and cell surface glycoproteins, which acquire high mannose N-glycans, absent in cells with functional giantin (Bhat et al. 2017). The role of the discussed abnormalities in regulating the phenotype of prostate cancer is the fact that in situ proximity ligation assays of co-localization of α-mannosidase IA with GM130 and GRASP65, and trans-Golgi glycosyltransferases with high mannose *N*-glycans are negative in androgen-sensitive LNCaP C-33 cells, but positive in androgen-independent LNCaP C-81 and DU145 cells, and giantin-devoid LNCaP C-33 cells. It was, therefore, suggested that in situ proximity ligation assay can be used to differentiate between indolent and invasive clinical subtypes of prostate cancer (Bhat et al. 2017).

There is scant data on the role of giantin in eosinophilia-associated myeloproliferative neoplasms (MPN-eo). Naumann et al. reported a case of a 75-year-old male, in whom a new PDGFRB (platelet-derived growth factor receptor beta) fusion gene, GOLGB1-PDGFRB [karyotype 46,XYt(5;17)(q33;p11)], was identified (Naumann et al. 2015). They proposed that fusion protein encoded by the gene is predicted to contain the tyrosine kinase (TK) domain, whereas partner gene contains domains like coiled-coil structures which putatively facilitate dimerization and constitutive activation of TK. The patient achieved rapid and durable complete remission on imatinib (Naumann et al. 2015). Troadec et al. (2017) reported a case of a 71-year-old female with an atypical mixed lymphoid/myeloid neoplasm—MPN-Eo characterized by the coexistence of bone marrow myeloproliferation with circulating hypereosinophilia and T-cell lymphoblastic lymphoma in lymph nodes. Karyotype of tumor cells from lymph nodes and bone marrow revealed a single clonal t(3;13)(q13;q12) translocation resulting in a FLT3 (fms-like tyrosine kinase 3) and GOLGB1 fusion transcript. The GOLGB1-FLT3 protein fused together the three coiled-coil GOLGB1 domains with the split kinase TK domain of FLT3, that could lead to a constitutively multimerized active protein (Troadec et al. 2017). An alternative hypothesis assumes that constitutive TK activation is due to the loss of the inhibitory juxtamembrane domain of FLT3. Regardless of the mechanism of persistent protein activation, the resultant resistance to imatinib developed and the patient died within 3 months, despite conventional CHOP chemotherapy (cyclophosphamide, hydroxydaunorubicin, oncovin, and prednisolone) (Troadec et al. 2017).

20.2.4 *USO1 (p115, USO1 Vesicle Transport Factor; USO1 Vesicle Docking Protein Homolog; TAP; VDP; General Vesicular Transport Factor p115; Transcytosis Associated Protein; Vesicle Docking Protein p115)*

USO1 is located predominantly in cis-Golgi (Fig. 20.1), where it plays an important role in vesicle trafficking between ER and GA (Sui et al. 2015; Barroso et al. 1995). It was demonstrated that by its domains, USO1 interacts with diverse proteins, such as GM130, giantin, Rab1a, N-ethylmaleimide-sensitive factor attachment protein receptor (SNARE) family members, and GBF1 (ARF guanine nucleotide exchange factor) (García-Mata and Sztul 2003). It was demonstrated that mice lacking USO1 exhibit disruption of Golgi structure and early embryonic lethality, indicating that USO1 is a crucial protein for early embryonic development (Kim et al. 2012a).

The importance and actual role of USO1 in carcinogenesis is still unclear. In an in vitro study using the colon cancer cells (HCT116 and HT-29), the expression of USO1 was knockdown by using a special lentivirus shRNA approach. It was shown that knockdown of USO1 inhibits the ability of cell proliferation and migration. Furthermore, the dysregulation of USO1 induces early apoptosis and decreased cells in G2-M phase (Sui et al. 2015). The USO1 expression analysis in multiple myeloma using qRT-PCR and western blot assay demonstrated a significantly higher USO1 expression in cancer cells than in normal plasma cells (Jin and Dai 2016). Additionally, knockdown of USO1 resulted in the inhibited ability of cell proliferation and induced cell apoptosis with reduced expression of cyclin D1, Mcm2 (minichromosome maintenance protein 2), PCNA (proliferating cell nuclear antigen), and p-Erk1/2 (extracellular signal-regulated kinase 1 and 2) (Jin and Dai 2016).

An experimental study using breast cancer cell lines demonstrated that upregulation of ER-Golgi trafficking genes, manifesting as e.g. overexpression of USO1, enhances trafficking kinetics and promotes cell adhesion, migration, and invasion (Howley et al. 2018). Furthermore, high trafficking gene expression significantly correlated with increased risk of distant metastasis as well as significantly shorter overall survival and relapse-free survival in breast cancer patients, suggesting that dysregulation of ER-Golgi trafficking plays an important role in breast cancer progression (Howley et al. 2018).

Other than the studies discussed above, there are no reports to analyze the expression of USO1 in cancer and to determine its potential prognostic role in cancer patients.

20.2.5 GOLIM4 (GIMPC; GOLPH4; P138; Golgi Integral Membrane Protein 4; 130 kDa Golgi-Localized Phosphoprotein; cis Golgi-Localized Calcium-Binding Protein; Golgi Integral Membrane Protein, cis; Golgi Phosphoprotein 4; Golgi Phosphoprotein of 130 kDa; Golgi-Localized Phosphoprotein of 130 kDa; Type II Golgi Membrane Protein)

GOLIM4 is a cis and medial Golgi-localized membrane protein (Fig. 20.1), which actively migrates between the Golgi and endosomes. It is a target protein for the calcium channel protein STIM1 (stromal interaction molecule 1) (Bai et al. 2018a). STIM1 is the endoplasmic reticulum calcium sensor, which participates in calcium ion influx via the store-operated calcium entry (SOCE) (Mo and Yang 2018). STIM1 overexpression and SOCE upregulation may promote metastases by affecting actin cytoskeleton, extracellular matrix distribution and interfering with tumor microenvironment (Mo and Yang 2018).

The role of GOLIM4 in carcinogenesis and its association with STIM1 has been poorly understood. What is known, though, is that GOLIM4 regulates cancer cell cycle, apoptosis and proliferation as demonstrated in studies using head and neck cancer cell lines. GOLIM4 overexpression was demonstrated in two cell lines of head and neck cancer: FaDu (human pharyngeal squamous carcinoma cell) and Tca-8113 (human tongue squamous carcinoma cell) (Bai et al. 2018a). Interestingly, GOLIM4 knockdown leads to inhibited growth of cancer cells, slowing tumor progression by inducing apoptosis, cell cycle arrest at the G1 phase, and preventing the cells from entering the interphase (S phase) (Bai et al. 2018a). These findings suggest that GOLIM4 promotes carcinogenesis. However, due to a low number of studies to analyze it in clinical specimens, a full explanation of its role is yet to be provided and warrants further research.

20.2.6 RNF121 (RING Finger Protein 121)

The RNF121 is a member of a large family of proteins of unclear cytophysiological function, which share a common RING domain. The RNF121, an ER- and cis-Golgi-resident E3-ubiquitin ligase (Zhao et al. 2014) (Fig. 20.1), facilitates the degradation and membrane localization of voltage-gated sodium channels (Ogino et al. 2015). Zhao et al. (2014) demonstrated that RNF121 knockdown inhibited cell growth and induced apoptosis, which was accompanied by caspase-3 activation and the cleavage of poly (adenosine diphosphate-ribose) polymerase (PARP1). Importantly, RNF121 knockdown enhanced etoposide-induced apoptosis, which has significant clinical implications.

By its RING catalytic domain, RNF121 as a Golgi-anchored E3 ubiquitin ligase regulates intracellular signaling pathway leading to NF-κB (nuclear factor kappa-light-chain-enhancer of activated B cells) activation (Zemirli et al. 2014). RNF121 overexpression was also shown to promote NF-κB activity (Zemirli et al. 2014). The detailed role of this regulation requires further research and makes the RNF121 an interesting putative target of molecular therapies, due to pleiotropy of NF-κB family members (they are known to regulate cell proliferation and differentiation, inflammation, immunity, and tumor progression).

Little is known about the role of RNF121 in oncogenesis. It was postulated that RNF121 might be a tumor suppressor. It was demonstrated that RNF121 was less expressed in tumor tissues than adjacent normal tissues of renal cell carcinoma (RCC) patients (Xiang et al. 2018). Furthermore, RNF121 overexpression inhibited the growth of human RCC cells (768-O cell line) and decreased the proliferation, migration, and invasion of human RCC cells. The NF-κB signaling pathway activation, on the other hand, occurs by RNF121-mediated IκBα degradation in RCC cells (Xiang et al. 2018).

Other studies indicate that RNF121 may promote carcinogenesis. The analysis including the real-time quantitative polymerase chain reaction (PCR), tissue microarray assay, and western blot analysis demonstrated over 16-fold increase of RNF121 expression in Barrett esophagus and esophageal adenocarcinoma as compared to the normal esophagus. Furthermore, mRNA expression level of RNF121 was higher in esophageal adenocarcinoma than in Barrett esophagus, which indicates a crucial role of RNF121 in the development and progression of esophageal adenocarcinoma (Wang et al. 2014). Similar conclusions regarding oncogenic potential of RNF121 can be found in a study by Wu et al. (2018), who demonstrated that gene amplification of RNF121 may be associated with a higher risk of progression to invasive cancer in oral verrucous hyperplasia, and Kaplan-Meier survival analysis showed that overexpression of RNF121 was associated with poorer prognosis in patients with oral verrucous hyperplasia.

Interestingly, Choi et al. (2018) analyzed RNA sequencing data of colorectal cancer patients and identified a new oncogenic fusion gene RNF121-FOLR2. They demonstrated that colorectal cancer cells with RNF121-FOLR2 overexpression exhibited increased cell proliferation. Better understanding of clinical and prognostic role of this novel fusion gene requires further research in a larger sample.

20.3 Structural-Functional Medial-Golgi Proteins and Their Role in Carcinogenesis

20.3.1 *CASP (CDP; CDP/Cut; CDP1; COY1; CUTL1; CUX; Clox; Cux/CDP; GOLIM6; Nbla10317; p100; p110; p200; p75; CCAAT Displacement Protein; Cut Homolog; Golgi Integral Membrane Protein 6; Homeobox Protein Cux-1; Cut like Homeobox 1)*

CASP is a coiled-coil protein and alternatively spliced product of the gene encoding the CCAAT-displacement protein transcription factor, which is described as a Golgi membrane protein related to giantin (Gillingham et al. 2002). The Cutl1 protein (CDP, a nuclear transcription factor), which undergoes numerous transcriptional and translational modifications, has an oncogenic potential, promoting migration and invasiveness of cancer cells (Liu et al. 2013). Little, though, is known about its modified form, CASP. As shown in a pilot study, CASP lacks cut-repeat and homeo DNA-binding domains, but instead it has a unique C-terminal region, which may be responsible for its Golgi localization. CASP was also shown to be structurally similar to giantin and golgin-84 (Gillingham et al. 2002). CASP is a structural protein, which participates in forming protein complexes and lacks any enzymatic activity (MacNeil and Pohajdak 2009). CASP was proposed to mediate the intra-Golgi antero- and retrograde axoplasmic transport and potential interactions with giantin and golgin-84 (Oka et al. 2004; Sohda et al. 2010). Other studies demonstrated overexpression of CASP in immune cells (T cells and NK cells) (MacNeil and Pohajdak 2009), which may indicate its involvement as an immune response mediator. The CASP encoding gene has several potential binding sites for several lymphoid-specific transcription factors (e.g., AP-1 and NFAT), and all their interactions may play a role in T-cell activation and proliferation (MacNeil and Pohajdak 2009). Tompkins et al. (2014a) demonstrated that CASP has a direct role in NK-mediated immune response by promoting NK cell motility and ability to kill tumor cells (Tompkins et al. 2014a). The formation of the NK cell immunological synapse targeting tumor cells with their resultant death is a complex process. The research has shown that CASP may be considered a granzyme B substrate, as it contains a conserved granzyme B cleavage site, which can modify its intracellular localization and interaction with sorting nexin 27 (Tompkins et al. 2014b). Granzyme B-dependent biochemical modifications of CASP with their resultant localization and structural changes as well as their role in regulating the NK-mediated immune response require further research.

So far, several protein groups were discovered, which directly interact with CASP. This ever-open list includes (1) cytohesin/ARNO family, mediating cell adhesion and migration, apical endocytosis and plasma membrane signaling, as well as (2) sorting nexin 27 (SNX27), which is implied in immune cell polarization, receptor sorting and recycling (MacNeil and Pohajdak 2009).

A detailed role of CASP in carcinogenesis has not been established yet. To date, there has been no study to analyze its expression in clinical cancer specimens.

20.3.2 Golgin-84 (Golgin-84; GOLIM5; RFG5; ret-II; Golgin Subfamily A Member 5; RET-Fused Gene 5 Protein; Cell Proliferation-Inducing Gene 31 Protein; Golgi Autoantigen, Golgin Subfamily a, 5; Golgi Integral Membrane Protein 5; Golgin A5)

Golgin-84 is an integral membrane protein with a single transmembrane domain close to its C terminus located in the GA (Bascom et al. 1999). Structural studies showed that golgin-84 inserts post-translationally into microsomal membranes and is similar in structure and sequence to giantin, a membrane protein which tethers coatamer complex I vesicles to the Golgi. It was also demonstrated that golgin-84 can form dimers (Bascom et al. 1999; McGee et al. 2017).

The primary cytophysiological function of golgin-84 involves its role in intra-Golgi traffic as a tethering factor for coat protein I (COPI) and for the conserved oligomeric Golgi (COG) complex, through its subunit Cog7 (Sohda et al. 2010). Furthermore, Sohda et al. (2010) demonstrated that depletion of golgin-84 manifests morphologically as Golgi fragmentation and impaired localization of Golgi resident proteins. Satoh et al. (2003) showed that by a direct interaction with Rab1 golgin-84 is involved in Golgi structure. Intriguingly, Diao et al. (2003) found that any change to the expression of golgin-84 (overexpression or low immunoreactivity) leads to Golgi ribbon fragmentation.

Interestingly, golgin-84 plays a role in infections with *Chlamydia trachomatis* (Heuer et al. 2009; Rejman Lipinski et al. 2009). In epithelial cells, infection with *Chlamydia trachomatis* induces Golgi fragmentation with subsequent Golgi ministack formation surrounding the bacterial inclusion. It was demonstrated that ministack formation and infection progression is triggered by the proteolytic cleavage of golgin-84. Golgin-84 knockdown inhibited *Chlamydia trachomatis* development, restricting the spread of infection (Rejman Lipinski et al. 2009).

There are no reports to discuss the expression of golgin-84 in human cancer cells; therefore understanding its prognostic value and role in carcinogenesis warrants further research using clinical specimens.

20.3.3 TMF (ARA160; TATA Element Modulatory Factor; Androgen Receptor Coactivator 160 kDa Protein; Androgen Receptor-Associated Protein of 160 kDa; TATA Element Modulatory Factor 1)

TMF is a coiled-coil protein of golgins family, which was first described as a binding protein to and coactivator for the androgen receptor, which due to its role localizes predominantly in the nucleus. However, the study by Mori and Kato (2002) demonstrated its presence in the GA, as well (Fig. 20.1). They demonstrated that TMF specifically associated in vitro and in vivo with hbrm/hSNF2 alpha and BRG-1/hSNF2 beta, the ATPase subunits of the human SNF/SWI complexes. Furthermore, TMF was shown to regulate chromatin remodeling and PDGFRβ translocation to the nucleus (Papadopoulos et al. 2018). The exact biological role of these interactions has not been fully understood yet.

Importantly, Fridmann-Sirkis et al. (2004) demonstrated that by binding Rab6 GTPase, TMF is involved in the maintenance of Golgi structural organization. Depletion of the protein by RNA interference in rat NRK cells results in Golgi membrane dispersal, suggesting a role for TMF in the movement or adherence of Golgi stacks. By interacting with Rab6, TMF also regulates retrograde membrane traffic from endosomes to the GA and from the Golgi to the ER (Yamane et al. 2007).

In an in vitro study using myogenic C2C12 cells, Perry et al. (2004) demonstrated a direct role of TMF in proteasomal degradation of Stat3, which is the key cell growth regulator. Interestingly, preliminary analyzes using tissue specimens of selected brain tumors (meningioma, anaplastic meningioma, and glioblastoma) demonstrated a significantly decreased level of TMF/ARA160 in malignant brain tumors. It was proposed that low immunoreactivity of TMF in cancer cells prevents Stat3 degradation, which contributes to increased Stat3-mediated cell proliferation (Perry et al. 2004).

Studies using cancer cell lines demonstrated that TMF overexpression in PC3 prostate carcinoma cells significantly attenuated the development and growth of xenograft tumors elicited by these cells in athymic mice. Interestingly, impaired angiogenesis and accelerated onset of apoptosis in the TMF-expressing tumoral cells were also shown. Finally, TMF was shown to mediate ubiquitination and proteasomal degradation of p65/RelA (nuclear factor-kB component), because the level of the p65/RelA in TMF-expressing xenografts was decreased (Abrham et al. 2009).

To date, little is known about the role of TMF expression in human cancer cells. There is no published study to analyze its expression in clinical specimens.

20.4   Structural-Functional trans-Golgi Proteins and Their Potential Role in Carcinogenesis

20.4.1   GCC88 (GCC1P; GRIP and Coiled-Coil Domain-Containing Protein 1; Golgi Coiled-Coil 1; Golgi Coiled-Coil Protein 1; Peripheral Membrane Golgi Protein; GRIP and Coiled-Coil Domain Containing 1)

GCC88 is a membrane tether of the TGN (Fig. 20.1). A GRIP domain present at its C-terminus determines the trans-Golgi localization of GCC88 (Luke et al. 2003). In an in vitro study, Luke et al. (2003) demonstrated that overexpression of GCC88 induces structural changes to the GA, mainly involving the enlargement of Golgi-associated structures and forming cauliflower structures of trans-Golgi fragments containing high levels of GCC88. GCC88 overexpression may trigger drawing TGN membranes from a tightly packed compact ribbon structure of the GA into an open structure as a result of enhanced interactions with the cytoskeleton (Luke et al. 2003). Gosavi et al. (2018) confirmed that GCC88 overexpression results in the loss of the compact Golgi ribbon and dispersal of the organelle as ministacks throughout the cytoplasm. Furthermore, they demonstrated that RNAi-mediated depletion of GCC88 restored the Golgi ribbon and reduced autophagy. It is likely that this Golgi fragmentation increases its effective volume whereas the overexpressed GCC88 may function as a protein required for more efficient membrane sorting and transport. Interestingly, in another study, Luke et al. (2005) demonstrated that GCC88 can interact with themselves forming homodimers, whereas it did not form heterodimers with other trans-Golgi proteins.

GCC88 also mediates retrograde transport pathways from early or recycling endosomes to the TGN. In a study using cell lines, Lieu et al. (2007) found a depletion of GCC88 in HeLa cells by interference RNA which resulted in a block in plasma membrane-TGN recycling of two cargo proteins, TGN38 and a CD8 mannose-6-phosphate receptor cytoplasmic tail fusion protein. GCC88 regulates intracellular transport processes and ensures they fuse only with their correct destination of the transport carriers (such as vesicles) (Wong et al. 2017).

To date, there are no reports to assess GCC88 expression in clinical specimens originating from cancer patients.

20.4.2 GCC185 (Golgi Coiled-Coil Protein 185, RANBP2L4, REN53, GRIP and Coiled-Coil Domain-Containing Protein 2; 185 kDa Golgi Coiled-Coil Protein; CLL-Associated Antigen KW-11; CTCL Tumor Antigen se1-1, 185-kD; Ran-Binding Protein 2-like 4; Renal Carcinoma Antigen NY-REN-53; GRIP and Coiled-Coil Domain Containing 2)

The precise mechanism of trans-Golgi localization of GCC185 still remains a subject of extensive studies (Fig. 20.1). Hayes et al. (2009) and Burguete et al. (2008) demonstrated that for its trans-Golgi localization, GCC185 requires an interaction with Rab6 and Arl1 (ADP-ribosylation factor-like protein 1) GTPases. However, Houghton et al. (2009) who demonstrated that Golgi recruitment of endogenous GCC185 does not involve Rab6A/A' and Arl1, did not confirm this finding.

Maintaining the organization of the GA is one of the known cytophysiological functions of GCC185. It was shown, using both small interfering RNA (siRNA) and microRNA (miRNA), that depletion of GCC185 in HeLa cells frequently resulted in fragmentation of the GA (Derby et al. 2007). GCC185 overexpression resulted in the appearance of small punctate structures dispersed in the cytoplasm, identified by immunoelectron microscopy as membrane tubular structures (Derby et al. 2004). Importantly, three other mammalian GRIP family members were not found in the abovementioned GCC185-positive structures. It should also be noted that in that study, GCC185 overexpression did not affect global Golgi morphology (Derby et al. 2004). Another study found GCC185 to be Rab9 effector required for mannose 6-phosphate receptor (MPR) recycling from endosomes to the TGN. Interestingly, depletion of GCC185 slightly altered the Golgi ribbon but did not interfere with Golgi function in that study (Reddy et al. 2006). GCC185 was shown to participate in retrograde transport of *Shiga* toxin (Lieu and Gleeson 2010). It is also involved in the formation of microtubules at the Golgi through recruiting CLASPs (cytoplasmic linker-associated proteins), microtubule-binding proteins that selectively coat noncentrosomal microtubule seeds (Efimov et al. 2007; Yu et al. 2016).

Furthermore, GCC185 acts as a direct effector for ARL4A (ADP-ribosylation factor-like protein 4A) protein which maintains the Golgi structure (Lin et al. 2011). Depletion of ARL4A leads to Golgi fragmentation and impairs endosome-to-Golgi transport. Interestingly, depletion of ARL4A also impairs the interaction between GCC185 and CLASP1 and CLASP2 proteins, which are of key importance for the Golgi structure (Lin et al. 2011).

To date, there is no study to analyze immune reactivity of GCC185 in human cancer cells. A potential prognostic value of altered expression of GCC185 still remains poorly understood.

20.4.3 Golgin-97 (Golgin Subfamily A Member 1; Gap Junction Protein, Alpha 4, 37 kDa; Golgi Autoantigen, Golgin Subfamily a, 1; Golgin A1)

Golgin-97 is another trans-Golgi protein putatively involved in carcinogenesis (Fig. 20.1). Its detailed role in cell physiology has not been understood yet. It was shown to act as a FIP1/RCP-binding protein, whereas FIP1/RCP is a Rab11a/b-binding protein (Jing et al. 2010). The FIP1/RCP-golgin-97 protein complex is required for tethering and fusion of recycling endosome-derived retrograde transport vesicles to the TGN. Lu et al. (2004) demonstrated that anti-golgin-97 antibody inhibited the transport of STxB (*Shiga* toxin B fragment) in vitro, thus confirming a crucial role of golgin-97 in traffic from the endosome to the TGN.

Furthermore, golgin-97 is identified as a selective and crucial component of the tubulovesicular carriers transporting E-cadherin out of the TGN (Lock et al. 2005). It was demonstrated that golgin-97 was recruited to trans-Golgi by direct interaction of its GRIP domain with Arl1 (Derby et al. 2004). Golgin-97 was confirmed to be a Kir2.1 (inward rectifying potassium channel) binding partner, required for targeting the channel to the TGN. RNA interference-mediated knockdown of golgin-97 prevented exit of Kir2.1 from the Golgi (Taneja et al. 2018).

There is just one study to discuss the involvement of golgin-97 in the development of breast cancer. Hsu et al. (2018) demonstrated a correlation between low expression of golgin-97 and poor overall survival of cancer patients, which was associated with invasiveness of breast cancer cells. Furthermore, golgin-97 knockdown promoted cell migration and invasion. Subsequent analyses demonstrated that golgin-97 knockdown induced the expression of several invasion-promoting genes that were transcriptionally regulated by NF-κB (Hsu et al. 2018).

The function of golgin-97 in cell physiology described above and a single report regarding its expression in breast cancer cells may represent its important role in carcinogenesis. However, to date, there have been no detailed analyzes to determine the prognostic role of altered expression of golgin-97 in human cancer development and progression.

20.4.4 Golgin-245 (p230, Golgin-245; CRPF46; GCP2; GOLG; MU-RMS-40.18; Golgin Subfamily A Member 4; 256 kDa Golgin; 72.1 Protein; Centrosome-Related Protein F46; Golgi Autoantigen, Golgin Subfamily a, 4; Golgin-240; Protein 72.1; trans-Golgi p230; Golgin A4)

First reports describing a trans-Golgi protein referred to as golgin-245 were published in 1995 (Fritzler et al. 1995). The trans-Golgi localization of golgin-245 is mediated by an interaction of Arl1 with GRIP domain of golgin-245 in the C-terminus (Derby et al. 2004; Lu et al. 2005; Wu et al. 2004; Lu and Hong 2003) (Fig. 20.1).

Golgin-245 regulates vesicular transport of proteins from endosomes to the TGN in cooperation with the adaptor protein-1 (AP-1 complex) and WDR11 protein (Navarro Negredo et al. 2018). It was shown that golgin-245 binds directly to TBC1D23, a member of a family of Rab GTPase-activating proteins (GAPs), which is another membrane traffic mediator ensuring target localization during endosome-to-Golgi trafficking (Shin et al. 2017).

The regulatory role golgin-245 in autophagy via an interaction with MACF1 (microtubule actin crosslinking protein 1) was also postulated. Golgin-245 or MACF1 knockdown cells failed to increase the autophagic flow rate, so these results indicate that golgin-245 and MACF1 cooperatively play an important role in phagophore formation (Sohda et al. 2015). Interestingly, golgin-245 can also bind to CD99 and it mediates HLA class I modulation. As shown by Brémond et al. (2009), overexpression of p230 (GRIP) domain can lead to surface and intracellular down-modulation of HLA class I molecules.

It was also demonstrated that golgin-245 is essential for intracellular trafficking and cell surface delivery of tumor necrosis factor-alpha (TNF-α) in activated macrophages. Using live-cell imaging, Lieu et al. (2008) observed that TNF-α transport from the TGN is mediated selectively by carriers marked by golgin-245. This suggests that golgin-245 may be a key regulator of TNF-α secretion in macrophages.

The precise role of reduced or enhanced expression of golgin-245 in carcinogenesis has not been studied extensively. The involvement of this protein in cell physiology described to date may indicate its role initiation and/or progression of human cancers, by e.g. mediating the immune response. In order to confirm it, however, extensive research is required using tissue specimens originating from cancer patients to determine the correlations between the expression of golgin-245 and individual clinical and histological parameters.

20.4.5 *Clipr-59 (Cytoplasmic Linker Protein 170-Related 59 kDa Protein)*

Clipr-59 is not only associated with TGN membranes (Fig. 20.1), but also with plasma membrane (Perez et al. 2002) and lipid rafts (Lallemand-Breitenbach et al. 2004). Multiple roles of CLIPR-59 have been proposed, such as regulating membrane trafficking, microtubule dynamics (Lallemand-Breitenbach et al. 2004) and TNF-α-mediated apoptosis by controlling ubiquitination of RIP1 (receptor-interacting protein 1) (Fujikura et al. 2012).

The role of CLIPR-59 in carcinogenesis has been widely studied. A downregulation of CLIPR-59 was demonstrated in glioblastoma multiforme (GBM) cells and high-grade glioma samples as compared to low-grade glioma and normal tissue samples (Ding et al. 2015). A decreased expression of CLIPR-59 was demonstrated using both western blotting and immunohistochemistry (IHC) staining. In IHC, CLIPR-59 immunoreactivity was predominantly located in the cytoplasm. Importantly, there was a significant correlation between decreased expression of CLIPR-59 and a decreased cancer-free survival (Ding et al. 2015).

Regarding the tumorigenesis, CLIPR-59 interaction with Spy1 (human speedy A1) was demonstrated in GBM cells, which decreased the association of CLIPR-59 with CYLD (de-ubiquitinating enzyme). On the other hand, inhibition of Spy1 reduced GBM cell viability and increased the binding ability of CLIPR-59 and CYLD. As a result, the lysine-63-dependent de-ubiquitinating activity of RIP1 increased alongside upregulation of caspase-8 and caspase-3, which induced TNF-α-mediated apoptosis (Ding et al. 2015). Thus, it can be concluded that an interplay between Spy1 and CLIPR-59 is a key determinant of resistance of GBM cells to TNF-α-induced programmed cell death.

In a study using lymphoblastoid T cells, a role of CLIPR-59 in apoptosis triggered by CD95/Fas was demonstrated (Sorice et al. 2010). The interaction of CLIPR-59 with GD3 (a ganglioside present in abundance in small lipid domains on mitochondrial membrane and being a lipid raft component) occurs at early time points after CD95/Fas ligation, which is followed by the GD3-tubulin association. It was also shown that silencing CLIPR-59 by siRNA impaired the kinetics of GD3-tubulin association, spreading of GD3 toward mitochondria and apoptosis execution (Sorice et al. 2010). It was, therefore, concluded that CLIPR-59 may act as a typical chaperone, which enables an interaction between tubulin and GD3 during CD95/Fas-mediated cell apoptosis. In the same study, it was hypothesized that CLIPR-59 could delay apoptotic execution through its interaction with Akt (Sorice et al. 2010; Ding and Du 2009). Further research is needed to verify this hypothesis.

20.4.6 *GORAB (NTKLBP1; SCYL1BP1; RAB6-Interacting Golgin; N-terminal Kinase-Like-Binding Protein 1; NTKL-Binding Protein 1; SCY1-Like 1-Binding Protein 1; SCYL1-BP1; SCYL1-Binding Protein 1; hNTKL-BP1; Golgin, RAB6 Interacting)*

GORAB is a coiled-coil membrane protein which localizes in trans-Golgi via a specific, GTP-dependent interaction with the small GTPases ARF5 and RAB6 (Egerer et al. 2015) (Fig. 20.1). Mutations of the gene encoding for GORAB are associated with gerodermia osteodysplastica (GO) characterized by skin laxity and early-onset osteoporosis, which suggests the key role of GORAB in early stages of osteoblast differentiation (Hennies et al. 2008; Chan et al. 2018). Furthermore, GORAB inactivation reduces dermatan sulfate levels and proteoglycan glycanation in skin and bone tissue. Interestingly, loss of GORAB in fibroblasts is associated with decorin retention and lower GAG levels in the GA (Chan et al. 2018). GORAB was also implied in hair follicle morphogenesis via Hedgehog signaling pathway (Liu et al. 2016). In their experimental study, Kovacs et al. (2018) demonstrated that GORAB and Sas6 are the key players in centriole duplication.

To date, no study has been published to analyze GORAB expression in human cancer cells. Its newly discovered functions, especially involvement in centriole duplication, make GORAB a protein likely participating in oncogenesis. In order to confirm it, however, immunohistochemistry analyzes using tissue specimens originating from cancer patients are required to determine correlations between the expression of GORAB and individual clinical and histological parameters as well as to verify its prognostic importance.

20.4.7 *SPCA1 (SPCA1; ATP2C1A; BCPM; HHD; PMR1; hSPCA1; Calcium-transporting ATPase Type 2C Member 1; ATP-dependent Ca(2+) Pump PMR1; ATPase 2C1; ATPase, Ca(2+)-Sequestering; ATPase, Ca++ Transporting, Type 2C, Member 1; HUSSY-28; Secretory Pathway Ca^{2+}/Mn^{2+} ATPase 1; ATPase Secretory Pathway Ca^{2+} Transporting 1)*

SPCA1 is a trans-Golgi protein involved in membrane trafficking and intracellular calcium level regulation (Pizzo et al. 2010; Pakdel and von Blume 2018) (Fig. 20.1). Functional variants of ATP2C1 gene encoding for SPCA1 have been associated with Hailey-Hailey disease, a rare familial benign chronic pemphigus (Micaroni et al. 2016; Missiaen et al. 2004). SPCA1 was shown to play a key role in viral maturation and spread (*Paramyxoviridae, Flaviviridae*, and *Togaviridae* families, including measles, dengue, West Nile, Zika, and chikungunya viruses) (Hoffmann et al. 2017). The research using genome-wide knockout screen in human haploid cells demonstrated that SPCA1 regulates proteases within the TGN, which require calcium for their activity and are critical for virus glycoprotein maturation. Interestingly, SPCA1-deficient cells prevent viral spread, which makes the protein an attractive antiviral host target (Hoffmann et al. 2017).

One of the newly discovered roles of SPCA1 is its involvement in the process that links the cytoplasmic actin cytoskeleton to the membrane-anchored Ca^{2+} ATPase SPCA1 and the lumenal Ca^{2+}-binding protein Cab45. Cab45 mediates sorting of a subset of secretory proteins at the TGN and in response to Ca^{2+} influx, it forms oligomers, enabling it to bind a variety of specific cargo molecules (Blank and von Blume 2017). Furthermore, it was shown that cofilin recruits F-actin to SPCA1 and promotes Ca^{2+}-mediated secretory cargo sorting (Kienzle et al. 2014).

A potential role of SPCA1 in carcinogenesis is yet to be understood. In their study of gene expressions of intracellular calcium channels, pumps, and exchangers with epidermal growth factor-induced EMT in a MDA-MB-468 breast cancer cell line,

Davis et al. (2013) found no significant alterations in mRNA levels of the Golgi calcium pump secretory pathway calcium ATPases SPCA1 and SPCA2. Furthermore, SPCA1 was shown to be a key regulator of insulin-like growth factor receptor (IGF1R) processing in basal-like breast cancer cell line MDA-MB-231 (Grice et al. 2010).

Whereas there have been no studies to determine SPCA1 expression in clinical specimens and its correlation with clinical and histological parameters, the published research of SCPA1 interactions with cofilin and IGF1R may suggest its role in carcinogenesis.

20.5 Golgi Scaffold Proteins and Their Role in Carcinogenesis

Scaffold proteins are an umbrella term denoting a group of proteins, which (1) recruit particular proteins to a specific location, (2) organize the proteins into a higher order macromolecular complex, (3) facilitate the interaction of and fine-tune the activity and crosstalk among the proteins within the entire assembly, and (4) coordinate the functions of the different molecular assemblies in different cellular parts or "microdomains." Scaffold proteins specific for GA include Sef, PAQR3, PAQR10, and PAQR11 (Peng et al. 2014).

20.5.1 Sef (Similar Expression to FGF Genes)

Sef is predominantly localized in the Golgi (Fig. 20.1); however, under some conditions it can also translocate to the cell membrane. It has a pleiotropic effect, participating in the development of different structures during embryogenesis as well as acting as tumor suppressor (Peng et al. 2014; Philips 2004). The multiple functions of Sef result from its mediating a number of signaling pathways.

Sef is a known inhibitor of FGF (fibroblast growth factor) signaling, via the FGF1/Ras/MAPK (mitogen-activated protein kinase) pathway (Ren et al. 2006). However, with EGF (epidermal growth factor) signaling, it exerts an opposite effect—by interacting with EGF receptor (EGFR), Sef interferes with EGFR trafficking and attenuates its degradation, thus potentiating EGF-mediated MAPK signaling (Ren et al. 2008). Furthermore, Sef mediates IL-17 signaling (Rong et al. 2009) while inhibiting inflammation via sequestering nuclear factor-kappa B (NF-κB) in the cytoplasm (Fuchs et al. 2012). Sef is also involved in processes leading to apoptosis, interacting with transforming growth factor-β-associated kinase (TAK1) and activating Jun N-terminal kinase (JNK) through the TAK1-MKK4 (mitogen-activated protein kinase kinase 4)-JNK pathway (Yang et al. 2004).

Sef downregulation was demonstrated in selected epithelial tumors (e.g., breast, thyroid, and ovarian cancer) (Zisman-Rozen et al. 2007). A decreased Sef expression was also demonstrated in prostate cancer (Zisman-Rozen et al. 2007), where it was additionally associated with enhanced FGF signaling, high-grade of primary tumor and unfavorable clinical outcome (Darby et al. 2006, 2009; Murphy et al. 2010). Furthermore, in endometrial cancer, an ability of Sef to inhibit growth and proliferation of cancer cells via the FGF2/MAPK/ERK pathway was shown (Zhang et al. 2011). In colorectal cancer, Sef downregulation was associated with Ras-induced nuclear accumulation of activated MEK1/2 (mitogen-activated protein kinase kinase 1 and 2) and the re-expression of Sef was sufficient to restore normal MEK1/2 localization and a reversal of Ras-induced proliferation and tumorigenesis (Duhamel et al. 2012). Interestingly, He et al. (2016) demonstrated that the knockdown of Sef in breast cancer cell lines was directly associated with overexpression of EMT regulatory genes and the resultant increased migration potential and invasiveness of cancer cells. Furthermore, Sef upregulation inhibited EMT in breast cancer cell line.

The above studies suggest the role of Sef as a significant suppressor of carcinogenesis. However, further in-depth clinical and histological studies are warranted.

20.5.2 PAQR3 (RKTG; Progestin and adipoQ Receptor Family Member 3; Raf Kinase Trapping to Golgi; Progestin and adipoQ Receptor Family Member III)

PAQR3 is exclusively localized in the GA (Peng et al. 2014) (Fig. 20.1). One of its functions involves regulating vesicle fission and transport via the Gβγ-PKD (G-protein βγ subunit of protein kinase D) signaling pathway (Hewavitharana and Wedegaertner 2015). It also acts as a tumor suppressor via inhibiting the Ras/Raf/MEK/ERK pathway by binding Raf-1 which results in translocation of Raf-1 into the GA, inhibiting Raf-1 activation and attenuating the association of Raf-1 with Ras and MEK (Feng et al. 2007). It may adversely affect cell proliferation and survival by inhibiting the phosphoinositide 3-kinase (PI3K)/Akt PAQR3 pathway via two mechanisms, i.e., by inhibiting signaling of Gβγ (Jiang et al. 2010) and by inhibiting PI3K via spatial regulation of p110a subunit (Wang et al. 2013). Furthermore, PAQR3 is a negative regulator of angiogenesis by suppressing VEGF (vascular endothelial growth factor) transcription (Zhang et al. 2010). The interplay of PAQR3 and p53 regulates tumorigenesis and EMT (Jiang et al. 2011). Another mechanism associated with the antitumor effect of PAQR3, demonstrated in gastric cancer cells, is its effect on the Twist-related protein 1 (Twist1) (Guo et al. 2016), which is a transcription factor playing the key role in initiating EMT and promotion of tumor metastasis. PAQR3 was shown to form a protein complex with Twist1 and BTRC (beta-transducin repeat containing E3 ubiquitin protein ligase). As a result of this interplay, PAQR3 overexpression results in increased Twist1 degradation, followed by EMT inhibition, which is how metastases develop (Guo et al. 2016). The above findings make PAQR3 an attractive target for novel molecular therapies.

The decreased expression of PAQR3 was observed in different human cancer cells. In colorectal cancer, downregulation of PAQR3 was associated with higher histological tumor grade (Wang et al. 2012). Another study demonstrated the association between decreased expression of PAQR3 mRNA and PAQR3 protein in colorectal cancer tissues with the differentiation degree, lymphatic metastasis, and tumor infiltration depth. Furthermore, the methylation rates of PAQR3 in cancer tissues were significantly higher than in adjacent tissue (Li et al. 2016).

In clear-cell renal carcinoma, a decreased expression of PAQR3 mRNA and PAQR3 protein was demonstrated (Zhang et al. 2010), which inversely correlated with VEGF expression. The mechanism of VEGF inhibition occurs via negative regulation of the Ras/Raf/MEK/ERK signaling cascade, which results in inhibiting the formation of the HIF-1α/p300 complex and suppressing the activity of HIF-1α (Zhang et al. 2010). This, in turn, leads to the suppression of hypoxia-stimulated VEGF transcription and expression.

In patients with gastric cancer, PAQR3 downregulation was associated with greater tumor size, more advanced clinical stage, presence of venous and lymphatic invasion, presence of nodal and distant metastases, as well as with shorter patient survival. Furthermore, reduced expression of PAQR3 was also highly correlated with increased EMT features in gastric cancer (Ling et al. 2014). One of the proposed PAQR3-dependent mechanisms in gastric carcinogenesis is its dysregulation by damage-specific DNA-binding protein 2 (DDB2), involved in DNA repair processes (Qiao et al. 2015). DDB2 is capable of ubiquitination and degradation of PAQR3. Overexpression of DDB2 elevates the degradation and ubiquitination of PAQR3, which is reflected by the progression of tumorigenesis (Qiao et al. 2015). On the other hand, DDB2 knockdown inhibits cell proliferation rate and migration of gastric cancer cells. Interestingly, simultaneous knockdown of DDB2 and PAQR3 leads to tumor progression. This shows that PAQR3 is not primarily responsible for carcinogenesis under any circumstances, and that changes to its expression can be secondary to dysregulation mediated by other proteins (Qiao et al. 2015).

Another example of changes secondarily affecting PAQR3 comes from the study using gastric cancer cell lines, which analyzed the effect of miRNA on PAQR3 expression (Zhao et al. 2017). The upregulation of miR-15b-5p was demonstrated in gastric cancer cell lines, which correlated with the presence of distant metastases. At the same time, it was shown that elevated expression of miR-15b-5p contributed to cell proliferation, migration, invasion, and EMT (Zhao et al. 2017). Importantly, a direct effect of miR-15b-5p on PAQR3 was demonstrated, which resulted in the decreased expression of the latter. Furthermore, re-expression of PAQR3 in miR-15b-5p-overexpressing gastric cancer cells attenuated the effect of miR-15b-5p on cell proliferation, migration, and invasion, which suggests that miR-15b-5p may be a key regulator of PAQR3 expression (Zhao et al. 2017).

Additionally, a decreased expression of PAQR3 mRNA and PAQR3 protein was demonstrated in esophageal squamous cell carcinoma (ESCC) as compared to adjacent tissues (Zhou et al. 2017; Bai et al. 2018b). Overexpression of PAQR3 was found to be associated with inhibition of cell proliferation, migration, invasion, and colony formation in ESCC in vitro, as well as slowing tumor growth in a tumor xenograft model (Zhou et al. 2017; Bai et al. 2017). Furthermore, there was a correlation between PAQR3 and tumor stage, lymph node metastasis and local recurrence (Bai et al. 2018b). Additionally, the Kaplan-Meier analysis revealed that a low level of PAQR3 expression was associated with unfavorable clinical outcome in patients with ESCC. Multivariate analysis indicated that PAQR3 expression was an independent prognostic factor in ESCC (Bai et al. 2018b).

PAQR3 as a tumor suppressor was demonstrated to inhibit EMT leading to an increased expression of E-cadherin in cancer cells (Bai et al. 2018b). Furthermore, PAQR3 attenuated the expression of RAF1, p-MEK1, and p-ERK1/2 in ESCC cells (Zhou et al. 2017). Another study showed that overexpression of PAQR3 was associated with inhibited ERK1/2 phosphorylation, which attenuated the ERK signaling. It was also shown that overexpressed PAQR3 blocked cell cycle transition from G1 to S phase (Bai et al. 2017).

A study assessing the expression of PAQR3 in non-small cell lung cancer (NSCLC) demonstrated decreased expression of both PAQR3 mRNA and PAQR3 protein (Li et al. 2018a). PAQR3 was demonstrated to be a tumor suppressor in NSCLC, since overexpression of PAQR3 led to the inhibition of cell proliferation, induction of apoptosis, and promotion of cell cycle arrest at G0/G1 phase. Furthermore, knockdown of PAQR3 showed a reverse effect on NSCLC cells (Li et al. 2018a). The suppressant activity of PAQR3 demonstrated in NSCLC was shown to occur via suppressing the PI3K/AKT signaling pathway (Li et al. 2018a).

The miR-137-mediated inhibition of PAQR3 expression was shown in bladder cancer cells. Additionally, the upregulation of miR-137 was associated with the presence of distant metastases and more advanced clinical stage as well as increased cell proliferation, migration, and invasion (Xiu et al. 2014), which may indirectly suggest the tumor suppressor role of PAQR3 in bladder cancer.

To date, two studies have been conducted to determine the role of PAQR3 in prostate cancer. The first of them demonstrated that PAQR3 inhibited cell proliferation and migration in vitro, as well as tumor growth in vivo (Huang et al. 2016). PAQR3 was shown to exert its suppressant effect in a dual mechanism of inhibiting the Ras/Raf/MEK/ERK and PI3K/AKT pathways (Huang et al. 2016). Furthermore, PAQR3 inhibited EMT, which was reflected in an upregulation of epithelial markers, E-cadherin and zona occludens protein (ZO-1) as well as a downregulation of vimentin (a mesenchymal phenotype marker) (Huang et al. 2016). The second study demonstrated that PAQR3 deregulation may be associated with promoter methylation of PAQR3 encoding gene (Lounglaithong et al. 2018). Unlike benign prostate hyperplasia cells, the prostate cancer cells showed a hypermethylation of PAQR3 gene promoter. Furthermore, PAQR3 hypermethylation was correlated with perineural invasion, which is an indicator of tumor aggressiveness and poor prognosis (Lounglaithong et al. 2018).

A decreased expression of PAQR3 mRNA and PAQR3 protein was also demonstrated in glioma cells (Tang et al. 2017). PAQR3 was shown to act as a tumor suppressor, as its overexpression inhibited the proliferation of glioma cells in vitro and attenuated tumor xenograft growth in vivo, as well as suppressed the migration and invasion of glioma cells, as well as prevented the EMT process. On the other hand, PAQR3 knockdown enhanced the proliferation and migration of glioma cells (Tang et al. 2017). The mechanism of action of PAQR3 as a tumor suppressor involves downregulation of phosphorylated PI3K and Akt (Tang et al. 2017).

In osteosarcoma, decreased levels of PAQR3 protein and mRNA were also demonstrated (Ma et al. 2015). PAQR3 expression was tumor grade-dependent. Its expression in metastatic osteosarcoma cells was even lower than in non-metastatic ones, with a general attenuation demonstrated in osteosarcoma cells as compared to normal cells (Ma et al. 2015).

Downregulation of PAQR3 secondary to promoter hypermethylation plays an important role in breast cancer development by affecting the proliferation and migration of breast cancer cells (Li et al. 2015; Chen et al. 2016). An inverse association between the expression of PAQR3 mRNA and HER2 (human epidermal growth factor receptor 2) was demonstrated. At the same time, trastuzumab-induced inhibition of HER2 resulted in increased expression of PAQR3, which suggests that HER2 may be a negative regulator of PAQR3 expression (Li et al. 2015). An inverse correlation was also observed between the PAQR3 mRNA expression and disease-free survival. PAQR3 overexpression was shown to inhibit cell proliferation, colony formation and migration in breast cancer (Li et al. 2015).

In hepatocellular carcinoma (HCC), a decreased expression of PAQR3 mRNA and PAQR3 protein was demonstrated (Li et al. 2015; Yu et al. 2014), correlating with an elevated serum α-fetoprotein level, high mitotic index, tumor size, histological grade, recurrence and poor prognosis (Li et al. 2015). A decreased PAQR3 expression was an independent prognostic factor for overall and disease-free survival in patients with HCC. On the other hand, the overexpression of PAQR3 was associated with attenuated cell proliferation and colony formation, which indicates a suppressor role of PAQR3 in HCC (Li et al. 2015). Furthermore, a silenced PAQR3 expression triggered cell growth (Li et al. 2015). The miRNA-dependent PAQR3 activation may play a role in carcinogenesis. A significant overexpression of miR-543 associated with the increased proliferation and invasiveness of tumor cells was demonstrated in clinical HCC tissues and cell lines (Yu et al. 2014). At the same time, PAQR3 was shown to be a direct target gene for miR-543, whereas PAQR3 mRNA levels were inversely correlated with the miR-543 expression level. This, on one hand, suggests the role of miR-543 as a negative regulator of PAQR3 expression. On the other hand, though, it indicates that PAQR3 is a tumorigenesis inhibitor in HCC (Yu et al. 2014).

In laryngeal squamous cell carcinoma (LSCC), a decreased expression of PAQR3 was demonstrated as compared to adjacent non-neoplastic tissues (Wu et al. 2016). On the other hand, an opposite process of PAQR3 overexpression suppressed proliferation and invasion of LSCC cells, which suggests the role of PAQR3 as a tumor suppressor in LSCC (Wu et al. 2016). At the same time, it was shown that PAQR3 overexpression-mediated inhibition of cell proliferation and invasion occurred through inhibited ERK phosphorylation in Hep2 cell line (Wu et al. 2016).

There has been a single report of the role of PAQR3 in the development of acute myeloid leukemia (AML) (Xu et al. 2017). In a study using human leukemia cell line U937, exogenous overexpression of PAQR3 was shown to inhibit cell proliferation causing a cell cycle arrest. An increased proportion of AML cells undergoing apoptosis was also demonstrated (Xu et al. 2017). Other effects of PAQR3 overexpression included increased level of cleaved caspase 3, B-cell lymphoma 2 (Bcl2)-associated X apoptosis regulator, reduced the level of Bcl-2, as well as inhibition of ERK and PI3K/AKT serine/threonine kinase 1 signaling (Xu et al. 2017). Thus, it can be concluded that, just as it is the case with other human cancers, PAQR3 is a tumor suppressor in AML.

The issue of PAQR3 immune reactivity in melanomagenesis still remains unclear. To date, only one study using melanoma cell lines was published. In A375 human melanoma cells harboring an oncogenic BRAF mutation V600E, the most common mutation in melanoma, overexpression of PAQR3 was shown to inhibit proliferation and tumorigenecity of malignant cells (Fan et al. 2008). It can be explained by the likely interaction of PAQR3 with BRAF, causing its sequestration to the GA (Fan et al. 2008).

20.5.3 PAQR10/11

PAQR10 and PAQR11, just as PAQR3, are exclusively localized in the GA (Peng et al. 2014) (Fig. 20.1). In terms of their involvement in human carcinogenesis, these are the least known scaffold proteins. It is currently known that PAQR10/11 may interact with Ras (including Hras, Nras, and Kras4A), markedly elevating their Golgi localization, and in turn enhancing ERK signaling (Jin et al. 2012). These proteins are able to interact with RasGRP1 (a guanine nucleotide exchange protein of Ras), elevating its Golgi levels, which eventually enhances Ras signaling (Jin et al. 2012).

Tan et al. (2017) demonstrated key involvement of PAQR11 overexpression in pro-metastatic vesicular trafficking, which is closely linked to the EMT. It was shown for the first time that increased metastatic potential does not only result from Golgi phosphoprotein 3-dependent (GOLPH3-dependent) Golgi membrane dispersal (via upregulated budding and transport of secretory vesicles), but a completely reverse mechanism may also play a role, as the formation of compact Golgi organelles with improved ribbon linking and cisternal stacking mediated by PAQR11 was associated with overexpression of EMT-encoding genes and increased metastatic potential both in vitro and in vivo. Thus, the above study suggests that EMT initiation results in PAQR11-mediated Golgi compaction process that drives metastasis. Further research using clinical specimens will prove the key to understanding the prognostic role of PAQR10 and PAQR11 expression in human cancer cells.

20.6 Reassembly Stacking Proteins: Their Role in Physiology and Carcinogenesis

The structure of the Golgi depends on the phase of the cell cycle, as it undergoes a disassembly in early mitosis and reassembly in late mitosis and early interphase. During the disassembly, Golgi ribbon links are severed, cisternae unstack and turn into vesicles with tubular structures dispersed in the cytoplasm (Huang and Wang 2017). A reversal of this process, the reassembly, consists of forming new cisternae, which re-stack and re-aggregate to form Golgi ribbons. Whereas the exact role of those processes is so far unclear, they are believed to facilitate equal distribution of Golgi membranes into the two daughter cells. They might also constitute a cell cycle checkpoint before mitosis may progress and enable proper assembly of the mitotic spindle (Huang and Wang 2017).

The two main proteins involved in disassembly, reassembly, and ribbon forming are peripheral membrane molecules, GRASP65 and GRASP55 (Golgi reassembly stacking protein of 55 kDa) (Zhang and Wang 2016) (Fig. 20.1).

20.6.1 GRASP65 (GORASP1, GOLPH5; P65; Golgi Reassembly-Stacking Protein 1; Golgi Peripheral Membrane Protein p65; Golgi Phosphoprotein 5; Golgi Reassembly and Stacking Protein 1; Golgi Reassembly Stacking Protein 1, 65 kDa; Golgi Reassembly-Stacking Protein of 65 kDa; Golgi Reassembly Stacking Protein 1)

GRASP65 can be found mainly in the cis-Golgi (Huang and Wang 2017). The protein contains a C-terminal serine/proline-rich (SPR) domain comprising multiple phosphorylation sites and is able to form dimers and oligomers (Zhang and Wang 2016). At the beginning of mitosis, SPR domains undergo phosphorylation by mitotic kinases Cdk1 (cyclin-dependent kinase 1) and Plk1 (polo-like kinase 1). As a result, GRASP65 molecules deoligomerize and the Golgi stack disassembles. In the late mitotic phase, GRASP65 is dephosphorylated by phosphatase PP2A and reforms oligomers, which results in restacking of Golgi cisternae (Wang et al. 2003; Huang and Wang 2017). Moreover, Bekier et al. (2017) applied the CRISPR-Cas9 technology to knock out GRASP55 and GRASP65 in HeLa and HEK293 cells, and they revealed that double knockout of GRASP proteins dispersed the Golgi stack into single cisternae and tubulovesicular structures and enhanced protein trafficking.

To date, there is no report to discuss the role of GRASP65 in carcinogenesis. There has been no study to analyze the expression of GRASP65 in tissue specimens of patients with cancers and/or to analyze the prognostic value of its expression.

20.6.2 *GRASP55 (GORASP2, GOLPH6; GRASP55; GRS2; p59; Golgi Reassembly-Stacking Protein 2; Golgi Phosphoprotein 6; Golgi Reassembly Stacking Protein 2, 55 kDa; Golgi Reassembly Stacking Protein 2)*

Unlike GRASP65, GRASP55 can be predominantly observed in the medial-trans-cisternae (Huang and Wang 2017). Its structure, chemical properties, and functions are similar to those of GRASP65—the protein also contains SPR domain with multiple phosphorylation sites, is capable of oligomerization and contributes to Golgi de- and reassembly as well as ribbon forming (Huang and Wang 2017; Zhang and Wang 2016). The key difference is that GRASP65 is regulated by different enzymes; during the early stages of mitosis, the protein is phosphorylated by MEK1/ERK kinases, resulting in deoligomerization and Golgi stack disassembly. The reverse process, dephosphorylation, is catalyzed by a still unknown phosphatase. Regarding Golgi ribbon forming, ERK-mediated phosphorylation prevents Golgi from ribbon linking (Huang and Wang 2017).

There is no paper to discuss the expression of GRASP55 in human cancer and its role in oncogenesis. There is, however, a case report of a 60-year-old male with anaplastic lymphoma kinase-positive large B-cell lymphoma (ALK(+) LBCL) (Ise et al. 2018), initially diagnosed with multiple extramedullary plasmocytomas (EMPs), who underwent the standard of care treatment. Two years later, the diagnosis was changed to ALK(+) LBCL with a GORASP2-ALK rearrangement (Ise et al. 2018). Despite the worst prognosis with his non-GCS pattern ALK (ALK was not expressed as a granular cytoplasmic staining pattern), where the overall survival is 20.3 months, and the 5-year survival rate is 25%, the discussed patient survived for 3 years following the initial diagnosis (Ise et al. 2018). It may indicate weak tumorigenicity of GORASP2-ALK. It is the first reported case of malignancy associated with GORASP2 rearrangement.

20.7 Other Golgi-Located Proteins: Cytophysiology and Role in Carcinogenesis

Little academic research has sparingly discussed the role of some Golgi-located proteins, the function of which does not identify them with any of the above-mentioned families. This subsection discusses those Golgi proteins of poorly understood role, which have not been formally classified as belonging to any of the previously discussed groups. These are: TMEM165, LZTR-1, KBTBD8, and STK16.

As there is a plethora of scientific evidence regarding GOLPH2 (Kim et al. 2012b; Ju et al. 2013; Liu et al. 2014; Donizy et al. 2016; Zhang et al. 2017; Duan et al. 2018; Li et al. 2018b) and GOLPH3 (Bergeron et al. 2017; Makowski et al.

2017; Rizzo et al. 2017; Jiang et al. 2015; Buschman et al. 2015a, b; Sechi et al. 2015) we have not covered their pleiotropic function in carcinogenesis in this chapter.

20.7.1   TMEM165 (CDG2K; FT27; GDT1; TMPT27; TPARL; Transmembrane Protein 165; TPA Regulated Locus; Transmembrane Protein PT27; Transmembrane Protein TPARL)

Transmembrane protein 165 (TMEM165) is a Golgi protein which plays a crucial role in manganese homeostasis and vesicular trafficking in the GA (Potelle et al. 2017). Interestingly, TMEM165 deficiency causes a congenital disorder of glycosylation (CDG), a recessive autosomal metabolic disease in which patients exhibit cartilage and bone dysplasia and altered glycosylation of serum glycoproteins (Foulquier et al. 2012; Bammens et al. 2015).

So far, little is known about a potential role of TMEM165 in carcinogenesis. The only paper which focuses on its involvement in carcinogenesis discusses HCC (Lee et al. 2018). Lee et al. demonstrated TMEM165 overexpression in HCC. Furthermore, its depletion reduced the invasive activity of cancer cells through suppression of matrix metalloproteinase 2 (MMP-2) expression. The TMEM165 expression analysis revealed an association between its overexpression and more frequent macroscopic vascular invasion, microscopic serosal invasion and higher α-fetoprotein levels. These findings suggest its important role in HCC progression. However, its role in other cancer types is yet to be explained by further research.

20.7.2   LZTR-1 (BTBD29, NS10, SWNTS2, Leucine Zipper like Transcription Regulator 1)

LZTR-1 is a member of the BTB-kelch superfamily, which play important roles during crucial cellular processes, such as migration and regulation of cell morphology (Stogios et al. 2005). LZTR-1 is deleted in the majority of patients with DiGeorge syndrome (Kurahashi et al. 1995). The study using confocal analysis of the subcellular distribution of LZTR-1 demonstrated that it localizes to the Golgi complex (Nacak et al. 2006). Furthermore, treatment with brefeldin A (BFA) did not lead to redistribution of LZTR-1 but caused its relocation to dispersed punctate structures, also positive for GM130, which supports the role of LZTR-1 as a Golgi structural protein. Finally, it was shown that LZTR-1 is degraded upon induction of apoptosis (Nacak et al. 2006). Little is known about the biological function of LZTR-1. There are no publications to discuss the potential role of its impaired expression in oncogenesis.

20.7.3　KBTBD8 (TA-KRP; TAKRP; Kelch Repeat and BTB Domain-Containing Protein 8; T-Cell Activation Kelch Repeat Protein; Kelch Repeat and BTB (POZ) Domain Containing 8; Kelch Repeat and BTB Domain Containing 8)

KBTBD8 is a second protein of the BTB-kelch superfamily, which localizes to the Golgi complex (Lührig et al. 2013). Little is known about its role in cytophysiology. However, available research indicates its putative role in the regulation of pluripotency (Meyer et al. 2010). Lühring et al. (2013) demonstrated that KBTBD8 localizes within the GA in non-dividing cells and co-localizes with α-tubulin on the spindle apparatus during mitosis, which suggests its role in cell proliferation. Involvement in mitosis may make KBTBD8 a useful prognostic marker. However, there are no studies to analyze its prognostic value in the development and progression of cancer.

20.7.4　STK16 (PKL12, KRCT; MPSK; PSK; TSF1; hPSK; Serine/Threonine-Protein Kinase 16; TGF-beta-Stimulated Factor 1; Myristoylated and Palmitoylated Serine/Threonine-Protein Kinase; Protein Kinase PKL12; Protein Kinase Expressed in Day 12 Fetal Liver; Serine/Threonine Kinase 16)

STK16 is a myristoylated, and palmitoylated serine/threonine protein kinase with underexplored cytophysiological functions which localizes to the Golgi complex (Liu et al. 2017). It was demonstrated that STK16 is involved in cell division and plays important roles in Golgi structure regulation. It also directly binds to actin and regulates actin dynamics. STK16 knockdown or kinase inhibition disrupts actin polymers and causes fragmented Golgi in cells. Furthermore, STK16 knockdown delays mitotic entry and arrests cells at prometaphase and cytokinesis (Liu et al. 2017). Interestingly, the studies using NIH/3T3 and NRK cell lines showed that treatment with BFA or nocodazole, drugs that promote Golgi disorganization, resulted in STK16 translocation to the nuclear compartment. (Guinea et al. 2006). The same study also showed that overexpression of STK16 enhances the capacity of analyzed cell lines to produce and secrete VEGF, which suggests a potential involvement of STK16 in regulating angiogenesis. A potential role in stimulating neoangiogenesis may make STK16 a very promising target of future molecular therapies.

There is no published paper to analyze STK16 expression in tissue specimens collected from patients with cancer and to discuss its potential prognostic role in oncology.

20.8   Conclusions

The role of the Golgi apparatus in carcinogenesis still remains unclear. A number of structural and functional cis-, medial- and trans-Golgi proteins as well as a complexity of metabolic pathways which they mediate, may indicate a central role of the Golgi apparatus in development and progression of cancer (Fig. 20.1). Pleiotropy of cellular function of the Golgi apparatus makes it a "metabolic heart" or a relay station of a cell, which combines multiple signaling pathways involved in carcinogenesis. Therefore, any damage to or structural abnormality of the Golgi apparatus, causing its fragmentation and/or biochemical dysregulation, results in an up- or downregulation of signaling pathways and may in turn promote tumor progression, as well as local nodal and distant metastases. Three alternative or parallel models of spatial and functional Golgi organization within tumor cells were proposed. One of them involves forming the compact Golgi structure, mediated by PARQ11 and GRASPs (Fig. 20.2a). The second Golgi organization model in a tumor cell postulates that its spatial structure is close to normal. However, it is the increased metabolic activity in the Golgi which differs it from normal cells. A number of Golgi proteins (GM130, GOLIM4, USO1, TMEM165, STK16, and GOLPH2) are overexpressed during oncogenesis, whereas the expression of others (GMAP210, PAQR3, golgin-97, Clipr-59, and Sef) is inhibited or significantly decreased (Fig. 20.2b). The third proposed model involves Golgi fragmentation and ministacks formation. It was postulated that cancer-specific Golgi fragmentation is associated with a significant increase of trafficking kinetics and intensification of intracellular signal transduction pathways, which increases the metastatic potential of a tumor cell. Overexpression of GOLPH3 and GCC88 alongside a decreased immunoreactivty of TMF, GCC185, and GRASPs are considered the key mediators of Golgi fragmentation (Fig. 20.2c). The role of giantins and golgin-84 was also postulated. However, the reports on their involvement in Golgi dispersion remain unclear. Regardless of the assumed model, the increased activity of oncogenesis initiators and promoters with inhibition of suppressor proteins results in an increased cell motility and migration, increased angiogenesis, significantly activated trafficking kinetics, proliferation, EMT induction, decreased susceptibility to apoptosis-inducing factors, and modulating immune response to tumor cell antigens. Eventually, this will lead to the increased metastatic potential of cancer cells and an increased risk of lymph node and distant metastases (Fig. 20.3).

This chapter provided an overview of the current state of knowledge of selected Golgi proteins, their role in cytophysiology, as well as potential involvement in tumorigenesis and progression. Many questions still remain unanswered; so further in-depth research of Golgi role in oncogenesis is needed.

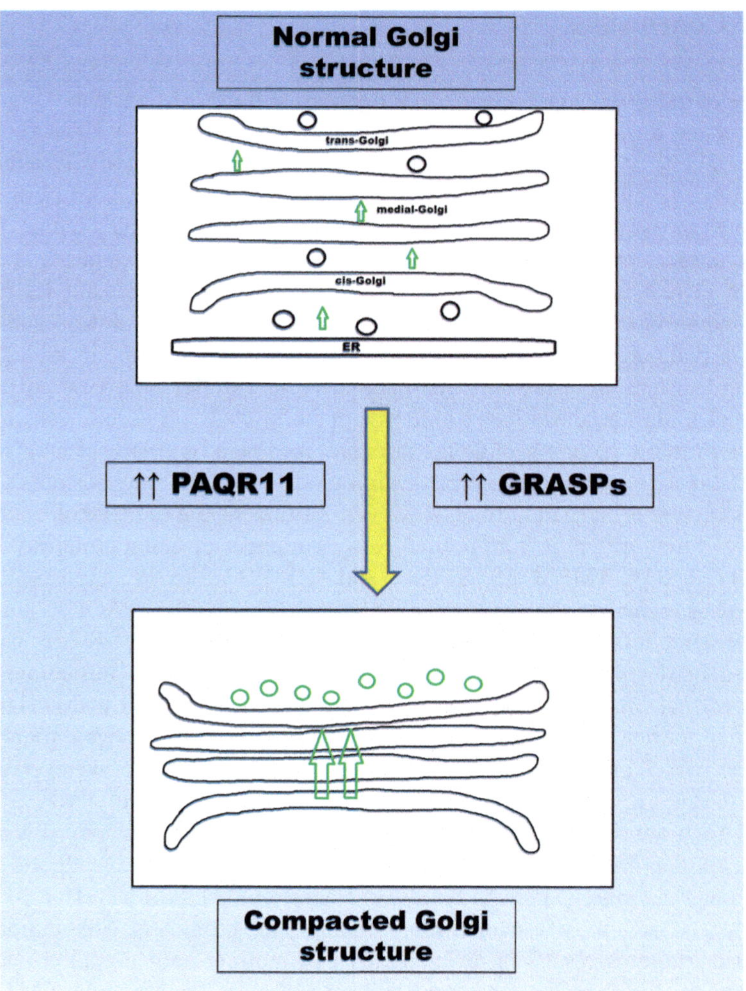

Fig. 20.2 The mechanism of three postulated models of spatial and functional Golgi organization in cancer. (**a**) Formation of compact Golgi apparatus with improved ribbon linking and cisternal stacking mediated by PAQR11 may be associated with an increased metastatic potential of tumor cells by e.g. EMT activation and enhanced trafficking kinetics. The potential involvement of GRASPs in ensuring a compact Golgi structure has also been postulated. (**b**) The second Golgi organization model in a tumor cell postulates that its spatial structure is close to normal. However, it is the increased metabolic activity in the Golgi which differs it from normal cells. A number of Golgi proteins (GM130, GOLIM4, USO1, TMEM165, STK16, and GOLPH2) are overexpressed during oncogenesis, whereas the expression of others (GMAP210, PAQR3, golgin-97, Clipr-59, and Sef) is inhibited or significantly decreased. (**c**) Golgi fragmentation with ministack forming is the third type of its structural and functional organization in a tumor cell. It was postulated that cancer-specific Golgi fragmentation is associated with a significant increase of trafficking kinetics and intensification of intracellular signal transduction pathways, which increases the metastatic potential of a tumor cell. Overexpression of GOLPH3 and GCC88 alongside a decreased immunoreactivty of TMF, GCC185, and GRASPs are considered the key mediators of Golgi fragmentation. The role of giantins and golgin-84 was also postulated. However, the reports on their involvement in Golgi dispersion remain unclear

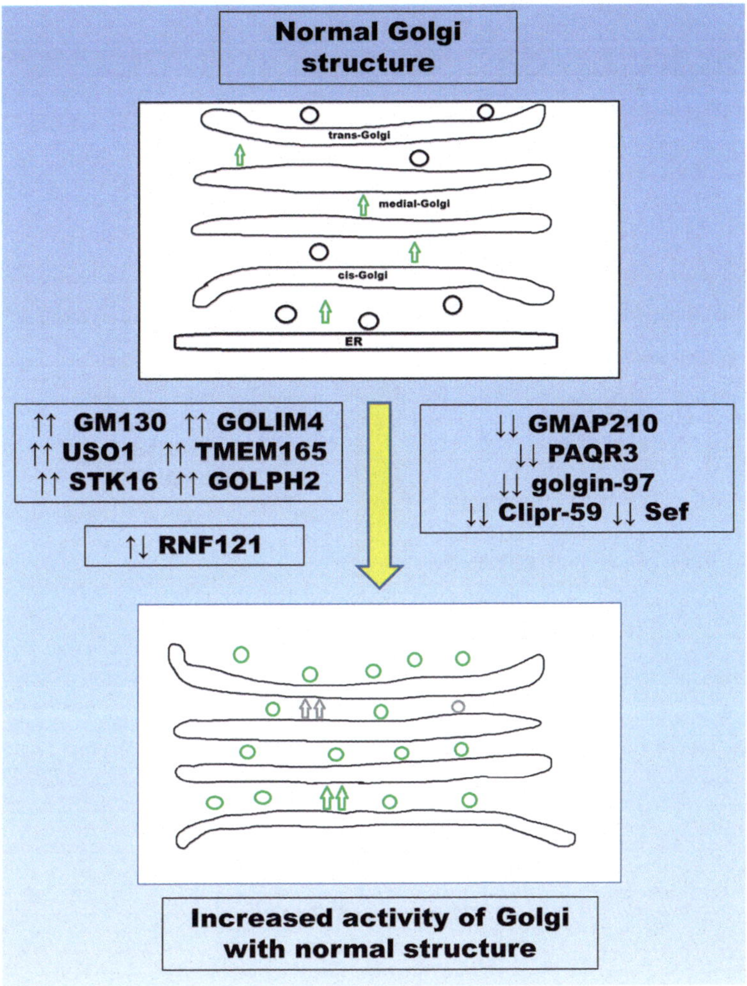

Fig. 20.2 (continued)

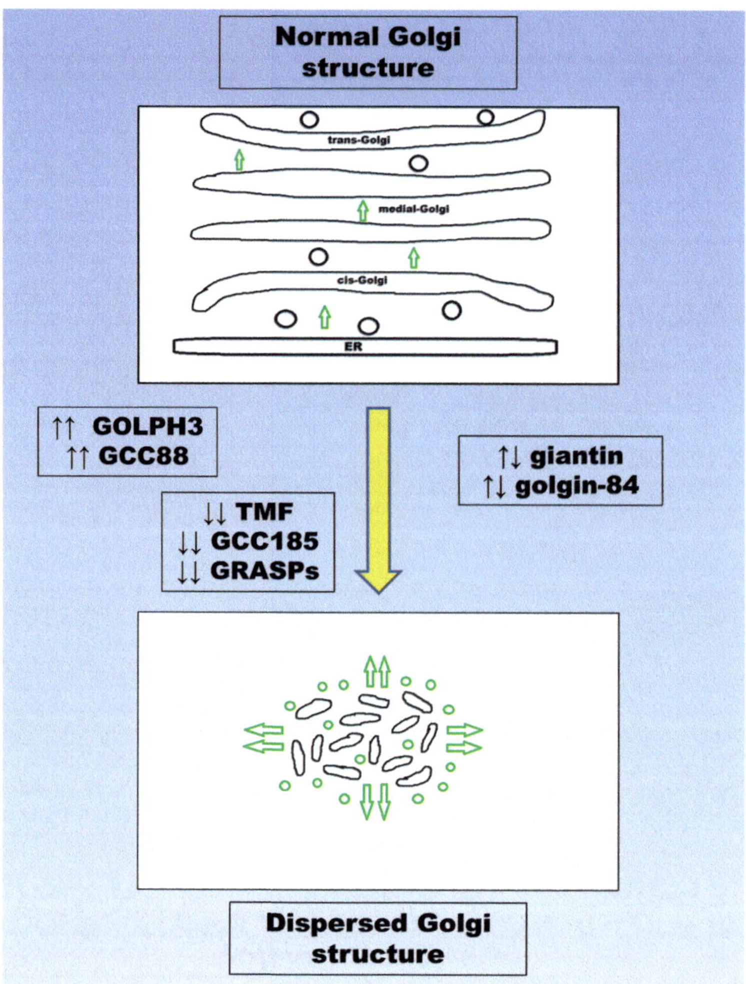

Fig. 20.2 (continued)

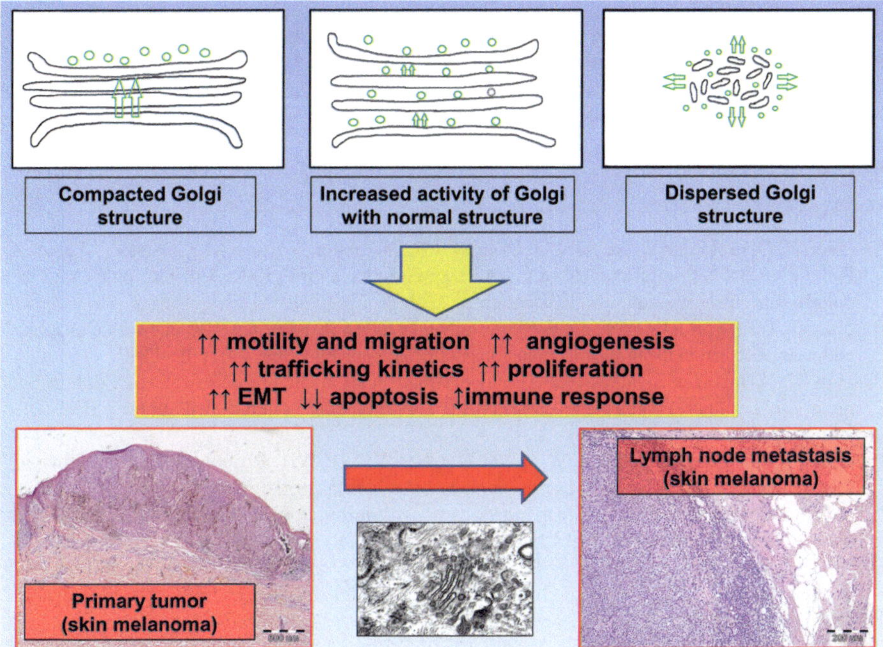

Fig. 20.3 Structural and functional Golgi dysregulation and its role in carcinogenesis. Regardless of the assumed model, the increased activity of oncogenesis initiators and promoters with inhibition of suppressor proteins results in an increased cell motility and migration, increased angiogenesis, significantly activated trafficking kinetics, proliferation, EMT induction, decreased susceptibility to apoptosis-inducing factors, and modulating immune response to tumor cell antigens. Eventually, this will lead to the increased metastatic potential of cancer cells and an increased risk of lymph node and distant metastases

Acknowledgments The authors thank Prof. Marzena Podhorska-Okołów and Dr. Katarzyna Haczkiewicz (Department of Human Morphology and Embryology, Wroclaw Medical University) for preparing the electron microphotographs and Ms. Agnieszka Janczak for her editorial support.

Source of Funding A statutory subsidy by the Polish Ministry of Science and Higher Education as part of grant ST.B130.18.030 (record numbers in the Simple system).

References

Abrham G, Volpe M, Shpungin S, Nir U (2009) TMF/ARA160 downregulates proangiogenic genes and attenuates the progression of PC3 xenografts. Int J Cancer 125:43–53

Bai G, Chu J, Eli M, Bao Y, Wen J (2017) PAQR3 overexpression suppresses the aggressive phenotype of esophageal squamous cell carcinoma cells via inhibition of ERK signaling. Biomed Pharmacother 94:813–819

Bai Y, Cui X, Gao D, Wang Y, Wang B, Wang W (2018a) Golgi integral membrane protein 4 manipulates cellular proliferation, apoptosis, and cell cycle in human head and neck cancer. Biosci Rep 38:BSR20180454

Bai G, Yang M, Zheng C, Zhang L, Eli M (2018b) Suppressor PAQR3 associated with the clinical significance and prognosis in esophageal squamous cell carcinoma. Oncol Lett 15:5703–5711

Bammens R, Mehta N, Race V, Foulquier F, Jaeken J, Tiemeyer M, Steet R, Matthijs G, Flanagan-Steet H (2015) Abnormal cartilage development and altered N-glycosylation in Tmem165-deficient zebrafish mirrors the phenotypes associated with TMEM165-CDG. Glycobiology 25:669–682

Barroso M, Nelson DS, Sztul E (1995) Transcytosis-associated protein (TAP)/p115 is a general fusion factor required for binding of vesicles to acceptor membranes. Proc Natl Acad Sci USA 92:527–531

Baschieri F, Confalonieri S, Bertalot G, Di Fiore PP, Dietmaier W, Leist M, Crespo P, Macara IG, Farhan H (2014) Spatial control of Cdc42 signalling by a GM130-RasGRF complex regulates polarity and tumorigenesis. Nat Commun 5:4839

Baschieri F, Uetz-von Allmen E, Legler DF, Farhan H (2015) Loss of GM130 in breast cancer cells and its effects on cell migration, invasion and polarity. Cell Cycle 14:1139–1147

Bascom RA, Srinivasan S, Nussbaum RL (1999) Identification and characterization of golgin-84, a novel Golgi integral membrane protein with a cytoplasmic coiled-coil domain. J Biol Chem 274:2953–2962

Beischlag TV, Taylor RT, Rose DW, Yoon D, Chen Y, Lee WH, Rosenfeld MG, Hankinson O (2004) Recruitment of thyroid hormone receptor/retinoblastoma-interacting protein 230 by the aryl hydrocarbon receptor nuclear translocator is required for the transcriptional response to both dioxin and hypoxia. J Biol Chem 279:54620–54628

Bekier ME 2nd, Wang L, Li J, Huang H, Tang D, Zhang X, Wang Y (2017) Knockout of the Golgi stacking proteins GRASP55 and GRASP65 impairs Golgi structure and function. Mol Biol Cell 28(21):2833–2842

Bergeron JJM, Au CE, Thomas DY, Hermo L (2017) Proteomics identifies Golgi phosphoprotein 3 (GOLPH3) with a link between Golgi structure, cancer, DNA damage and protection from cell death. Mol Cell Proteomics 16:2048–2054

Bhat G, Hothpet VR, Lin MF, Cheng PW (2017) Shifted Golgi targeting of glycosyltransferases and α-mannosidase IA from giantin to GM130-GRASP65 results in formation of high mannose N-glycans in aggressive prostate cancer cells. Biochim Biophys Acta Gen Subj 1861:2891–2901

Blank B, von Blume J (2017) Cab45-Unraveling key features of a novel secretory cargo sorter at the trans-Golgi network. Eur J Cell Biol 96:383–390

Brémond A, Meynet O, Mahiddine K, Coito S, Tichet M, Scotlandi K, Breittmayer JP, Gounon P, Gleeson PA, Bernard A, Bernard G (2009) Regulation of HLA class I surface expression requires CD99 and p230/golgin-245 interaction. Blood 113:347–357

Burguete AS, Fenn TD, Brunger AT, Pfeffer SR (2008) Rab and Arl GTPase family members cooperate in the localization of the golgin GCC185. Cell 132:286–298

Buschman MD, Rahajeng J, Field SJ (2015a) GOLPH3 links the Golgi, DNA damage, and cancer. Cancer Res 75:624–627

Buschman MD, Xing M, Field SJ (2015b) The GOLPH3 pathway regulates Golgi shape and function and is activated by DNA damage. Front Neurosci 9:362

Chan WL, Steiner M, Witkos T, Egerer J, Busse B, Mizumoto S, Pestka JM, Zhang H, Hausser I, Khayal LA, Ott CE, Kolanczyk M, Willie B, Schinke T, Paganini C, Rossi A, Sugahara K, Amling M, Knaus P, Chan D, Lowe M, Mundlos S, Kornak U (2018) Impaired proteoglycan glycosylation, elevated TGF-β signaling, and abnormal osteoblast differentiation as the basis for bone fragility in a mouse model for gerodermia osteodysplastica. PLoS Genet 14:e1007242

Chang SH, Hong SH, Jiang HL, Minai-Tehrani A, Yu KN, Lee JH, Kim JE, Shin JY, Kang B, Park S, Han K, Chae C, Cho MH (2012) GOLGA2/GM130, cis-Golgi matrix protein, is a novel target of anticancer gene therapy. Mol Ther 20:2052–2063

Chen Y, Chen PL, Chen CF, Sharp ZD, Lee WH (1999) Thyroid hormone, T3-dependent phosphorylation and translocation of Trip230 from the Golgi complex to the nucleus. Proc Natl Acad Sci USA 96:4443–4448

Chen J, Wang F, Xu J, He Z, Lu Y, Wang Z (2016) The role of PAQR3 gene promoter hypermethylation in breast cancer and prognosis. Oncol Rep 36:1612–1618

Choi Y, Kwon CH, Lee SJ, Park J, Shin JY, Park DY (2018) Integrative analysis of oncogenic fusion genes and their functional impact in colorectal cancer. Br J Cancer 119:230–240

Darby S, Sahadevan K, Khan MM, Robson CN, Leung HY, Gnanapragasam VJ (2006) Loss of Sef (similar expression to FGF) expression is associated with high grade and metastatic prostate cancer. Oncogene 25:4122–4127

Darby S, Murphy T, Thomas H, Robson CN, Leung HY, Mathers ME, Gnanapragasam VJ (2009) Similar expression to FGF (Sef) inhibits fibroblast growth factor-induced tumourigenic behaviour in prostate cancer cells and is downregulated in aggressive clinical disease. Br J Cancer 101:1891–1899

Davis FM, Parsonage MT, Cabot PJ, Parat MO, Thompson EW, Roberts-Thomson SJ, Monteith GR (2013) Assessment of gene expression of intracellular calcium channels, pumps and exchangers with epidermal growth factor-induced epithelial-mesenchymal transition in a breast cancer cell line. Cancer Cell Int 13:76

Derby MC, van Vliet C, Brown D, Luke MR, Lu L, Hong W, Stow JL, Gleeson PA (2004) Mammalian GRIP domain proteins differ in their membrane binding properties and are recruited to distinct domains of the TGN. J Cell Sci 117:5865–5874

Derby MC, Lieu ZZ, Brown D, Stow JL, Goud B, Gleeson PA (2007) The trans-Golgi network golgin, GCC185, is required for endosome-to-Golgi transport and maintenance of Golgi structure. Traffic 8:758–773

Diao A, Rahman D, Pappin DJ, Lucocq J, Lowe M (2003) The coiled-coil membrane protein golgin-84 is a novel Rab effector required for Golgi ribbon formation. J Cell Biol 160:201–212

Ding J, Du K (2009) ClipR-59 interacts with Akt and regulates Akt cellular compartmentalization. Mol Cell Biol 29:1459–1471

Ding Z, Liu Y, Yao L, Wang D, Zhang J, Cui G, Yang X, Huang X, Liu F, Shen A (2015) Spy1 induces de-ubiquitinating of RIP1 arrest and confers glioblastoma's resistance to tumor necrosis factor (TNF-α)-induced apoptosis through suppressing the association of CLIPR-59 and CYLD. Cell Cycle 14:2149–2159

Donizy P, Kaczorowski M, Biecek P, Halon A, Szkudlarek T, Matkowski R (2016) Golgi-related proteins GOLPH2 (GP73/GOLM1) and GOLPH3 (GOPP1/MIDAS) in cutaneous melanoma: patterns of expression and prognostic significance. Int J Mol Sci 17:e1619

Duan J, Li X, Huang S, Zeng Y, He Y, Liu H, Lin D, Lu D, Zheng M (2018) GOLPH2, a gene downstream of ras signaling, promotes the progression of pancreatic ductal adenocarcinoma. Mol Med Rep 17:4187–4194

Duhamel S, Hébert J, Gaboury L, Bouchard A, Simon R, Sauter G, Basik M, Meloche S (2012) Sef downregulation by Ras causes MEK1/2 to become aberrantly nuclear localized leading to polyploidy and neoplastic transformation. Cancer Res 72:626–635

Efimov A, Kharitonov A, Efimova N, Loncarek J, Miller PM, Andreyeva N, Gleeson P, Galjart N, Maia AR, McLeod IX, Yates JR 3rd, Maiato H, Khodjakov A, Akhmanova A, Kaverina I (2007) Asymmetric CLASP-dependent nucleation of noncentrosomal microtubules at the trans-Golgi network. Dev Cell 12:917–930

Egerer J, Emmerich D, Fischer-Zirnsak B, Chan WL, Meierhofer D, Tuysuz B, Marschner K, Sauer S, Barr FA, Mundlos S, Kornak U (2015) GORAB missense mutations disrupt RAB6 and ARF5 binding and Golgi targeting. J Invest Dermatol 135:2368–2376

Fan F, Feng L, He J, Wang X, Jiang X, Zhang Y, Wang Z, Chen Y (2008) RKTG sequesters B-Raf to the Golgi apparatus and inhibits the proliferation and tumorigenicity of human malignant melanoma cells. Carcinogenesis 29:1157–1163

Feng L, Xie X, Ding Q, Luo X, He J, Fan F, Liu W, Wang Z, Chen Y (2007) Spatial regulation of Raf kinase signaling by RKTG. Proc Natl Acad Sci U S A 104:14348–14353

Foulquier F, Amyere M, Jaeken J, Zeevaert R, Schollen E, Race V, Bammens R, Morelle W, Rosnoblet C, Legrand D, Demaegd D, Buist N, Cheillan D, Guffon N, Morsomme P,

Annaert W, Freeze HH, Van Schaftingen E, Vikkula M, Matthijs G (2012) TMEM165 deficiency causes a congenital disorder of glycosylation. Am J Hum Genet 91:15–26

Fridmann-Sirkis Y, Siniossoglou S, Pelham HR (2004) TMF is a golgin that binds Rab6 and influences Golgi morphology. BMC Cell Biol 5:18

Fritzler MJ, Lung CC, Hamel JC, Griffith KJ, Chan EK (1995) Molecular characterization of Golgin-245, a novel Golgi complex protein containing a granin signature. J Biol Chem 270:31262–31268

Fuchs Y, Brunwasser M, Haif S, Haddad J, Shneyer B, Goldshmidt-Tran O, Korsensky L, Abed M, Zisman-Rozen S, Koren L, Carmi Y, Apte R, Yang RB, Orian A, Bejar J, Ron D (2012) Sef is an inhibitor of proinflammatory cytokine signaling, acting by cytoplasmic sequestration of NF-κB. Dev Cell 23:611–623

Fujikura D, Ito M, Chiba S, Harada T, Perez F, Reed JC, Uede T, Miyazaki T (2012) CLIPR-59 regulates TNF-α-induced apoptosis by controlling ubiquitination of RIP1. Cell Death Dis 3: e264

García-Mata R, Sztul E (2003) The membrane-tethering protein p115 interacts with GBF1, an ARF guanine-nucleotide-exchange factor. EMBO Rep 4:320–325

Gillingham AK, Pfeifer AC, Munro S (2002) CASP, the alternatively spliced product of the gene encoding the CCAAT-displacement protein transcription factor, is a Golgi membrane protein related to giantin. Mol Biol Cell 13:3761–3774

Gosavi P, Houghton FJ, McMillan PJ, Hanssen E, Gleeson PA (2018) The Golgi ribbon in mammalian cells negatively regulates autophagy by modulating mTOR activity. J Cell Sci 131:jcs211987

Grice DM, Vetter I, Faddy HM, Kenny PA, Roberts-Thomson SJ, Monteith GR (2010) Golgi calcium pump secretory pathway calcium ATPase 1 (SPCA1) is a key regulator of insulin-like growth factor receptor (IGF1R) processing in the basal-like breast cancer cell line MDA-MB-231. J Biol Chem 285:37458–37466

Guinea B, Ligos JM, Laín de Lera T, Martín-Caballero J, Flores J, Gonzalez de la Peña M, García-Castro J, Bernad A (2006) Nucleocytoplasmic shuttling of STK16 (PKL12), a Golgi-resident serine/threonine kinase involved in VEGF expression regulation. Exp Cell Res 312:135–144

Guo W, You X, Xu D, Zhang Y, Wang Z, Man K, Wang Z, Chen Y (2016) PAQR3 enhances Twist1 degradation to suppress epithelial-mesenchymal transition and metastasis of gastric cancer cells. Carcinogenesis 37:397–407

Hayes GL, Brown FC, Haas AK, Nottingham RM, Barr FA, Pfeffer SR (2009) Multiple Rab GTPase binding sites in GCC185 suggest a model for vesicle tethering at the trans-Golgi. Mol Biol Cell 20:209–217

He Q, Gong Y, Gower L, Yang X, Friesel RE (2016) Sef regulates epithelial-mesenchymal transition in breast cancer cells. J Cell Biochem 117:2346–2356

Hennies HC, Kornak U, Zhang H, Egerer J, Zhang X, Seifert W, Kühnisch J, Budde B, Nätebus M, Brancati F, Wilcox WR, Müller D, Kaplan PB, Rajab A, Zampino G, Fodale V, Dallapiccola B, Newman W, Metcalfe K, Clayton-Smith J, Tassabehji M, Steinmann B, Barr FA, Nürnberg P, Wieacker P, Mundlos S (2008) Gerodermia osteodysplastica is caused by mutations in SCYL1BP1, a Rab-6 interacting golgin. Nat Genet 40:1410–1412

Heuer D, Rejman Lipinski A, Machuy N, Karlas A, Wehrens A, Siedler F, Brinkmann V, Meyer TF (2009) Chlamydia causes fragmentation of the Golgi compartment to ensure reproduction. Nature 457:731–735

Hewavitharana T, Wedegaertner PB (2015) PAQR3 regulates Golgi vesicle fission and transport via the Gβγ-PKD signaling pathway. Cell Signal 27:2444–2451

Hoffmann HH, Schneider WM, Blomen VA, Scull MA, Hovnanian A, Brummelkamp TR, Rice CM (2017) Diverse viruses require the calcium transporter SPCA1 for maturation and spread. Cell Host Microbe 22:460–470.e5

Houghton FJ, Chew PL, Lodeho S, Goud B, Gleeson PA (2009) The localization of the Golgin GCC185 is independent of Rab6A/A' and Arl1. Cell 138:787–794

Howley BV, Howe PH (2018) Metastasis-associated upregulation of ER-Golgi trafficking kinetics: regulation of cancer progression via the Golgi apparatus. Oncoscience 5:142–143

Howley BV, Link LA, Grelet S, El-Sabban M, Howe PH (2018) A CREB3-regulated ER-Golgi trafficking signature promotes metastatic progression in breast cancer. Oncogene 37:1308–1325

Hsu RM, Zhong CY, Wang CL, Liao WC, Yang C, Lin SY, Lin JW, Cheng HY, Li PY, Yu CJ (2018) Golgi tethering factor golgin-97 suppresses breast cancer cell invasiveness by modulating NF-κB activity. Cell Commun Signal 16:19

Huang S, Wang Y (2017) Golgi structure formation, function, and post-translational modifications in mammalian cells. F1000Res 6:2050

Huang W, Guo W, You X, Pan Y, Dong Z, Jia G, Yang C, Chen Y (2016) PAQR3 suppresses the proliferation, migration and tumorigenicity of human prostate cancer cells. Oncotarget 33:53948–53958

Infante C, Ramos-Morales F, Fedriani C, Bornens M, Rios RM (1999) GMAP-210, a cis-Golgi network-associated protein, is a minus end microtubule-binding protein. J Cell Biol 145:83–98

Ise M, Kageyama H, Araki A, Itami M (2018) Identification of a novel GORASP2-ALK fusion in an ALK-positive large B-cell lymphoma. Leuk Lymphoma 6:1–5

Jiang Y, Xie X, Zhang Y, Luo X, Wang X, Fan F, Zheng D, Wang Z, Chen Y (2010) Regulation of G-protein signaling by RKTG via sequestration of the G betagamma subunit to the Golgi apparatus. Mol Cell Biol 30:78–90

Jiang Y, Xie X, Li Z, Wang Z, Zhang Y, Ling ZQ, Pan Y, Wang Z, Chen Y (2011) Functional cooperation of RKTG with p53 in tumorigenesis and epithelial-mesenchymal transition. Cancer Res 71:2959–2968

Jiang Y, Su Y, Zhao Y, Pan C, Chen L (2015) Golgi phosphoprotein3 overexpression is associated with poor survival in patients with solid tumors: a meta-analysis. Int J Clin Exp Pathol 8:10615–10624

Jin Y, Dai Z (2016) USO1 promotes tumor progression via activating Erk pathway in multiple myeloma cells. Biomed Pharmacother 78:264–271

Jin T, Ding Q, Huang H, Xu D, Jiang Y, Zhou B, Li Z, Jiang X, He J, Liu W, Zhang Y, Pan Y, Wang Z, Thomas WG, Chen Y (2012) PAQR10 and PAQR11 mediate Ras signaling in the Golgi apparatus. Cell Res 22:661–676

Jing J, Junutula JR, Wu C, Burden J, Matern H, Peden AA, Prekeris R (2010) FIP1/RCP binding to Golgin-97 regulates retrograde transport from recycling endosomes to the trans-Golgi network. Mol Biol Cell 21:3041–3053

Joachim J, Jefferies HB, Razi M, Frith D, Snijders AP, Chakravarty P, Judith D, Tooze SA (2015) Activation of ULK kinase and autophagy by GABARAP trafficking from the centrosome is regulated by WAC and GM130. Mol Cell 60:899–913

Ju Q, Zhao Y, Liu Y, Zhou G, Li F, Xie P, Li Y, Li GC (2013) Monoclonal antibody preparation of Golgi phosphoprotein 2 and preliminary application in the early diagnosis of hepatocellular carcinoma. Mol Med Rep 8:517–522

Kienzle C, Basnet N, Crevenna AH, Beck G, Habermann B, Mizuno N, von Blume J (2014) Cofilin recruits F-actin to SPCA1 and promotes Ca^{2+}-mediated secretory cargo sorting. J Cell Biol 206:635–654

Kim S, Hill A, Warman ML, Smits P (2012a) Golgi disruption and early embryonic lethality in mice lacking USO1. PLoS One 7:e50530

Kim HJ, Lv D, Zhang Y, Peng T, Ma X (2012b) Golgi phosphoprotein 2 in physiology and in diseases. Cell Biosci 2:31

Kodani A, Kristensen I, Huang L, Sütterlin C (2009) GM130-dependent control of Cdc42 activity at the Golgi regulates centrosome organization. Mol Biol Cell 20:1192–2000

Koreishi M, Gniadek TJ, Yu S, Masuda J, Honjo Y, Satoh A (2013) The golgin tether giantin regulates the secretory pathway by controlling stack organization within Golgi apparatus. PLoS One 8:e59821

Kovacs L, Chao-Chu J, Schneider S, Gottardo M, Tzolovsky G, Dzhindzhev NS, Riparbelli MG, Callaini G, Glover DM (2018) Gorab is a Golgi protein required for structure and duplication of Drosophila centrioles. Nat Genet 50:1021–1031

Kurahashi H, Akagi K, Inazawa J, Ohta T, Niikawa N, Kayatani F, Sano T, Okada S, Nishisho I (1995) Isolation and characterization of a novel gene deleted in DiGeorge syndrome. Hum Mol Genet 4:541–549

Lallemand-Breitenbach V, Quesnoit M, Braun V, El Marjou A, Poüs C, Goud B, Perez F (2004) CLIPR-59 is a lipid raft-associated protein containing a cytoskeleton-associated protein glycine-rich domain (CAP-Gly) that perturbs microtubule dynamics. J Biol Chem 279:41168–41178

Lee JS, Kim MY, Park ER, Shen YN, Jeon JY, Cho EH, Park SH, Han CJ, Choi DW, Jang JJ, Suh KS, Hong J, Kim SB, Lee KH (2018) TMEM165, a Golgi transmembrane protein, is a novel marker for hepatocellular carcinoma and its depletion impairs invasion activity. Oncol Rep 40:1297–1306

Li Z, Ling ZQ, Guo W, Lu XX, Pan Y, Wang Z, Chen Y (2015) PAQR3 expression is downregulated in human breast cancers and correlated with HER2 expression. Oncotarget 6:12357–12368

Li RH, Zhang AM, Li S, Li TY, Wang LJ, Zhang HR, Shi JW, Liu XR, Chen Y, Chen YC, Wei TY, Gao Y, Li W, Tang HY, Tang MY (2016) PAQR3 gene expression and its methylation level in colorectal cancer tissues. Oncol Lett 12:1773–1778

Li X, Li M, Chen D, Shi G, Zhao H (2018a) PAQR3 inhibits proliferation via suppressing PI3K/AKT signaling pathway in non-small cell lung cancer. Arch Med Sci 14:1289–1297

Li H, Yang LL, Xiao Y, Deng WW, Chen L, Wu L, Zhang WF, Sun ZJ (2018b) Overexpression of Golgi phosphoprotein 2 is associated with poor prognosis in oral squamous cell carcinoma. Am J Clin Pathol 150:74–83

Lieu ZZ, Gleeson PA (2010) Identification of different itineraries and retromer components for endosome-to-Golgi transport of TGN38 and Shiga toxin. Eur J Cell Biol 89:379–393

Lieu ZZ, Derby MC, Teasdale RD, Hart C, Gunn P, Gleeson PA (2007) The golgin GCC88 is required for efficient retrograde transport of cargo from the early endosomes to the trans-Golgi network. Mol Biol Cell 18:4979–4991

Lieu ZZ, Lock JG, Hammond LA, La Gruta NL, Stow JL, Gleeson PA (2008) A trans-Golgi network golgin is required for the regulated secretion of TNF in activated macrophages in vivo. Proc Natl Acad Sci U S A 105:3351–3356

Lin YC, Chiang TC, Liu YT, Tsai YT, Jang LT, Lee FJ (2011) ARL4A acts with GCC185 to modulate Golgi complex organization. J Cell Sci 124:4014–4026

Ling ZQ, Guo W, Lu XX, Zhu X, Hong LL, Wang Z, Wang Z, Chen Y (2014) A Golgi-specific protein PAQR3 is closely associated with the progression, metastasis and prognosis of human gastric cancers. Ann Oncol 25:1363–1372

Liu KC, Lin BS, Zhao M, Wang KY, Lan XP (2013) Cutl1: a potential target for cancer therapy. Cell Signal 25:349–354

Liu G, Zhang Y, He F, Li J, Wei X, Li Y, Liao X, Sun J, Yi W, Niu D (2014) Expression of GOLPH2 is associated with the progression of and poor prognosis in gastric cancer. Oncol Rep 32:2077–2085

Liu Y, Snedecor ER, Choi YJ, Yang N, Zhang X, Xu Y, Han Y, Jones EC, Shroyer KR, Clark RA, Zhang L, Qin C, Chen J (2016) Gorab is required for dermal condensate cells to respond to hedgehog signals during hair follicle morphogenesis. J Invest Dermatol 136:378–386

Liu J, Yang X, Li B, Wang J, Wang W, Liu J, Liu Q, Zhang X (2017) STK16 regulates actin dynamics to control Golgi organization and cell cycle. Sci Rep 7:44607

Lock JG, Hammond LA, Houghton F, Gleeson PA, Stow JL (2005) E-cadherin transport from the trans-Golgi network in tubulovesicular carriers is selectively regulated by golgin-97. Traffic 6:1142–1156

Lounglaithong K, Bychkov A, Sampatanukul P (2018) Aberrant promoter methylation of the PAQR3 gene is associated with prostate cancer. Pathol Res Pract 214:126–129

Lu L, Hong W (2003) Interaction of Arl1-GTP with GRIP domains recruits autoantigens Golgin-97 and Golgin-245/p230 onto the Golgi. Mol Biol Cell 14:3767–3781

Lu L, Tai G, Hong W (2004) Autoantigen Golgin-97, an effector of Arl1 GTPase, participates in traffic from the endosome to the trans-golgi network. Mol Biol Cell 15:4426–4443

Lu L, Tai G, Hong W (2005) Interaction of Arl1 GTPase with the GRIP domain of Golgin-245 as assessed by GST (glutathione-S-transferase) pull-down experiments. Methods Enzymol 404:432–441

Lührig S, Kolb S, Mellies N, Nolte J (2013) The novel BTB-kelch protein, KBTBD8, is located in the Golgi apparatus and translocates to the spindle apparatus during mitosis. Cell Div 8:3

Luke MR, Kjer-Nielsen L, Brown DL, Stow JL, Gleeson PA (2003) GRIP domain-mediated targeting of two new coiled-coil proteins, GCC88 and GCC185, to subcompartments of the trans-Golgi network. J Biol Chem 278:4216–4226

Luke MR, Houghton F, Perugini MA, Gleeson PA (2005) The trans-Golgi network GRIP-domain proteins form alpha-helical homodimers. Biochem J 388:835–841

Ma Z, Wang Y, Piao T, Li Z, Zhang H, Liu Z, Liu J (2015) The tumor suppressor role of PAQR3 in osteosarcoma. Tumour Biol 36:3319–3324

MacNeil AJ, Pohajdak B (2009) Getting a GRASP on CASP: properties and role of the cytohesin-associated scaffolding protein in immunity. Immunol Cell Biol 87:72–80

Makowski SL, Tran TT, Field SJ (2017) Emerging themes of regulation at the Golgi. Curr Opin Cell Biol 45:17–23

McGee LJ, Jiang AL, Lan Y (2017) Golga5 is dispensable for mouse embryonic development and postnatal survival. Genesis 55. https://doi.org/10.1002/dvg.23039

Meyer S, Nolte J, Opitz L, Salinas-Riester G, Engel W (2010) Pluripotent embryonic stem cells and multipotent adult germline stem cells reveal similar transcriptomes including pluripotency-related genes. Mol Hum Reprod 16:846–855

Micaroni M, Giacchetti G, Plebani R, Xiao GG, Federici L (2016) ATP2C1 gene mutations in Hailey-Hailey disease and possible roles of SPCA1 isoforms in membrane trafficking. Cell Death Dis 7:e2259

Millarte V, Farhan H (2012) The Golgi in cell migration: regulation by signal transduction and its implications for cancer cell metastasis. Sci World J 2012:498278

Missiaen L, Raeymaekers L, Dode L, Vanoevelen J, Van Baelen K, Parys JB, Callewaert G, De Smedt H, Segaert S, Wuytack F (2004) SPCA1 pumps and Hailey-Hailey disease. Biochem Biophys Res Commun 332:1204–1213

Mo P, Yang S (2018) The store-operated calcium channels in cancer metastasis: from cell migration, invasion to metastatic colonization. Front Biosci (Landmark Ed) 23:1241–1256

Mori K, Kato H (2002) A putative nuclear receptor coactivator (TMF/ARA160) associates with hbrm/hSNF2 alpha and BRG-1/hSNF2 beta and localizes in the Golgi apparatus. FEBS Lett 520:127–132

Murphy T, Darby S, Mathers ME, Gnanapragasam VJ (2010) Evidence for distinct alterations in the FGF axis in prostate cancer progression to an aggressive clinical phenotype. J Pathol 220:452–460

Nacak TG, Leptien K, Fellner D, Augustin HG, Kroll J (2006) The BTB-kelch protein LZTR-1 is a novel Golgi protein that is degraded upon induction of apoptosis. J Biol Chem 281:5065–5671

Nakamura N, Lowe M, Levine TP, Rabouille C, Warren G (1997) The vesicle docking protein p115 binds GM130, a cis-Golgi matrix protein, in a mitotically regulated manner. Cell 89:445–455

Naumann N, Schwaab J, Metzgeroth G, Jawhar M, Haferlach C, Göhring G, Schlegelberger B, Dietz CT, Schnittger S, Lotfi S, Gärtner M, Dang TA, Hofmann WK, Cross NC, Reiter A, Fabarius A (2015) Fusion of PDGFRB to MPRIP, CPSF6, and GOLGB1 in three patients with eosinophilia-associated myeloproliferative neoplasms. Genes Chromosomes Cancer 54:762–770

Navarro Negredo P, Edgar JR, Manna PT, Antrobus R, Robinson MS (2018) The WDR11 complex facilitates the tethering of AP-1-derived vesicles. Nat Commun 9:596

Ogino K, Low SE, Yamada K, Saint-Amant L, Zhou W, Muto A, Asakawa K, Nakai J, Kawakami K, Kuwada JY, Hirata H (2015) RING finger protein 121 facilitates the degradation and membrane localization of voltage-gated sodium channels. Proc Natl Acad Sci USA 112:2859–2864

Oka T, Ungar D, Hughson FM, Krieger M (2004) The COG and COPI complexes interact to control the abundance of GEARs, a subset of Golgi integral membrane proteins. Mol Biol Cell 15:2423–2435

Pakdel M, von Blume J (2018) Exploring new routes for secretory protein export from the trans-Golgi network. Mol Biol Cell 29:235–240

Papadopoulos N, Lennartsson J, Heldin CH (2018) PDGFRβ translocates to the nucleus and regulates chromatin remodeling via TATA element-modifying factor 1. J Cell Biol 217:1701–1717

Peng W, Lei Q, Jiang Z, Hu Z (2014) Characterization of Golgi scaffold proteins and their roles in compartmentalizing cell signaling. J Mol Histol 45:435–445

Perez F, Pernet-Gallay K, Nizak C, Goodson HV, Kreis TE, Goud B (2002) CLIPR-59, a new trans-Golgi/TGN cytoplasmic linker protein belonging to the CLIP-170 family. J Cell Biol 156:61–642

Perry E, Tsruya R, Levitsky P, Pomp O, Taller M, Weisberg S, Parris W, Kulkarni S, Malovani H, Pawson T, Shpungin S, Nir U (2004) TMF/ARA160 is a BC-box-containing protein that mediates the degradation of Stat3. Oncogene 23:8908–8919

Petrosyan A, Holzapfel MS, Muirhead DE, Cheng PW (2014) Restoration of compact Golgi morphology in advanced prostate cancer enhances susceptibility to galectin-1-induced apoptosis by modifying mucin O-glycan synthesis. Mol Cancer Res 12:1704–1716

Philips MR (2004) Sef: a MEK/ERK catcher on the Golgi. Mol Cell 15:168–169

Pizzo P, Lissandron V, Pozzan T (2010) The trans-golgi compartment: a new distinct intracellular Ca store. Commun Integr Biol 3:462–464

Popławski P, Piekiełko-Witkowska A, Nauman A (2017) The significance of TRIP11 and T3 signalling pathway in renal cancer progression and survival of patients. Endokrynol Pol 68:631–641

Potelle S, Dulary E, Climer L, Duvet S, Morelle W, Vicogne D, Lebredonchel E, Houdou M, Spriet C, Krzewinski-Recchi MA, Peanne R, Klein A, de Bettignies G, Morsomme P, Matthijs G, Marquardt T, Lupashin V, Foulquier F (2017) Manganese-induced turnover of TMEM165. Biochem J 474:1481–1493

Pranke IM, Morello V, Bigay J, Gibson K, Verbavatz JM, Antonny B, Jackson CL (2011) α-Synuclein and ALPS motifs are membrane curvature sensors whose contrasting chemistry mediates selective vesicle binding. J Cell Biol 194:89–103

Preisinger C, Short B, De Corte V, Bruyneel E, Haas A, Kopajtich R, Gettemans J, Barr FA (2004) YSK1 is activated by the Golgi matrix protein GM130 and plays a role in cell migration through its substrate 14-3-3zeta. J Cell Biol 164:1009–1020

Qiao S, Guo W, Liao L, Wang L, Wang Z, Zhang R, Xu D, Zhang Y, Pan Y, Wang Z, Chen Y (2015) DDB2 is involved in ubiquitination and degradation of PAQR3 and regulates tumorigenesis of gastric cancer cells. Biochem J 469:469–480

Reddy JV, Burguete AS, Sridevi K, Ganley IG, Nottingham RM, Pfeffer SR (2006) A functional role for the GCC185 golgin in mannose 6-phosphate receptor recycling. Mol Biol Cell 17:4353–4363

Rejman Lipinski A, Heymann J, Meissner C, Karlas A, Brinkmann V, Meyer TF, Heuer D (2009) Rab6 and Rab11 regulate chlamydia trachomatis development and golgin-84-dependent Golgi fragmentation. PLoS Pathog 5:e1000615

Ren Y, Cheng L, Rong Z, Li Z, Li Y, Li H, Wang Z, Chang Z (2006) hSef co-localizes and interacts with Ras in the inhibition of Ras/MAPK signaling pathway. Biochem Biophys Res Commun 347:988–993

Ren Y, Cheng L, Rong Z, Li Z, Li Y, Zhang X, Xiong S, Hu J, Fu XY, Chang Z (2008) hSef potentiates EGF-mediated MAPK signaling through affecting EGFR trafficking and degradation. Cell Signal 20:518–533

Ríos RM, Sanchís A, Tassin AM, Fedriani C, Bornens M (2004) GMAP-210 recruits gamma-tubulin complexes to cis-Golgi membranes and is required for Golgi ribbon formation. Cell 118:323–335

Rivero S, Cardenas J, Bornens M, Rios RM (2009) Microtubule nucleation at the cis-side of the Golgi apparatus requires AKAP450 and GM130. EMBO J 28:1016–1028

Rizzo R, Parashuraman S, D'Angelo G, Luini A (2017) GOLPH3 and oncogenesis: what is the molecular link? Tissue Cell 49:170–174

Rong Z, Wang A, Li Z, Ren Y, Cheng L, Li Y, Wang Y, Ren F, Zhang X, Hu J, Chang Z (2009) IL-17RD (Sef or IL-17RLM) interacts with IL-17 receptor and mediates IL-17 signaling. Cell Res 19:208–215

Sato K, Roboti P, Mironov AA, Lowe M (2015) Coupling of vesicle tethering and Rab binding is required for in vivo functionality of the golgin GMAP-210. Mol Biol Cell 26:537–553

Satoh A, Wang Y, Malsam J, Beard MB, Warren G (2003) Golgin-84 is a rab1 binding partner involved in Golgi structure. Traffic 4:153–161

Sechi S, Frappaolo A, Belloni G, Colotti G, Giansanti MG (2015) The multiple cellular functions of the oncoprotein Golgi phosphoprotein 3. Oncotarget 6:3493–3506

Shin JJH, Gillingham AK, Begum F, Chadwick J, Munro S (2017) TBC1D23 is a bridging factor for endosomal vesicle capture by golgins at the trans-Golgi. Nat Cell Biol 19:1424–1432

Sohda M, Misumi Y, Yamamoto A, Nakamura N, Ogata S, Sakisaka S, Hirose S, Ikehara Y, Oda K (2010) Interaction of Golgin-84 with the COG complex mediates the intra-Golgi retrograde transport. Traffic 11:1552–1566

Sohda M, Misumi Y, Ogata S, Sakisaka S, Hirose S, Ikehara Y, Oda K (2015) Trans-Golgi protein p230/golgin-245 is involved in phagophore formation. Biochem Biophys Res Commun 456:275–281

Sorice M, Matarrese P, Manganelli V, Tinari A, Giammarioli AM, Mattei V, Misasi R, Garofalo T, Malorni W (2010) Role of GD3-CLIPR-59 association in lymphoblastoid T cell apoptosis triggered by CD95/Fas. PLoS One 5:e8567

Stogios PJ, Downs GS, Jauhal JJ, Nandra SK, Privé GG (2005) Sequence and structural analysis of BTB domain proteins. Genome Biol 6:R82

Sui J, Li X, Xing J, Cao F, Wang H, Gong H, Zhang W (2015) Lentivirus-mediated silencing of USO1 inhibits cell proliferation and migration of human colon cancer cells. Med Oncol 32:218

Tan X, Banerjee P, Guo HF, Ireland S, Pankova D, Ahn YH, Nikolaidis IM, Liu X, Zhao Y, Xue Y, Burns AR, Roybal J, Gibbons DL, Zal T, Creighton CJ, Ungar D, Wang Y, Kurie JM (2017) Epithelial-to-mesenchymal transition drives a pro-metastatic Golgi compaction process through scaffolding protein PAQR11. J Clin Invest 127:117–131

Taneja TK, Ma D, Kim BY, Welling PA (2018) Golgin-97 targets ectopically expressed inward rectifying potassium channel, Kir2.1, to the trans-Golgi network in COS-7 cells. Front Physiol 9:1070

Tang SL, Gao YL, Hu WZ (2017) PAQR3 inhibits the proliferation, migration and invasion in human glioma cells. Biomed Pharmacother 92:24–32

Tompkins N, MacKenzie B, Ward C, Salgado D, Leidal A, McCormick C, Pohajdak B (2014a) Cytohesin-associated scaffolding protein (CASP) is involved in migration and IFN-γ secretion in natural killer cells. Biochem Biophys Res Commun 451:165–170

Tompkins N, MacNeil AJ, Pohajdak B (2014b) Cytohesin-associated scaffolding protein (CASP) is a substrate for granzyme B and ubiquitination. Biochem Biophys Res Commun 452:473–478

Troadec E, Dobbelstein S, Bertrand P, Faumont N, Trimoreau F, Touati M, Chauzeix J, Petit B, Bordessoule D, Feuillard J, Bastard C, Gachard N (2017) A novel t(3;13)(q13;q12) translocation fusing FLT3 with GOLGB1: toward myeloid/lymphoid neoplasms with eosinophilia and rearrangement of FLT3? Leukemia 31:514–517

Vanegas S, Sua LF, López-Tenorio J, Ramírez-Montaño D, Pachajoa H (2018) Achondrogenesis type 1A: clinical, histologic, molecular, and prenatal ultrasound diagnosis. Appl Clin Genet 11:69–73

Wang Y, Seemann J, Pypaert M, Shorter J, Warren G (2003) A direct role for GRASP65 as a mitotically regulated Golgi stacking factor. EMBO J 22:3279–3290

Wang X, Li X, Fan F, Jiao S, Wang L, Zhu L, Pan Y, Wu G, Ling ZQ, Fang J, Chen Y (2012) PAQR3 plays a suppressive role in the tumorigenesis of colorectal cancers. Carcinogenesis 33:2228–2235

Wang X, Wang L, Zhu L, Pan Y, Xiao F, Liu W, Wang Z, Guo F, Liu Y, Thomas WG, Chen Y (2013) PAQR3 modulates insulin signaling by shunting phosphoinositide 3-kinase p110α to the Golgi apparatus. Diabetes 62:444–456

Wang XW, Wei W, Wang WQ, Zhao XY, Guo H, Fang DC (2014) RING finger proteins are involved in the progression of Barrett esophagus to esophageal adenocarcinoma: a preliminary study. Gut Liver 8:487–494

Wei JH, Seemann J (2009) Mitotic division of the mammalian Golgi apparatus. Semin Cell Dev Biol 20:810–816

Wei JH, Seemann J (2017) Golgi ribbon disassembly during mitosis, differentiation and disease progression. Curr Opin Cell Biol 47:43–51

Wei JH, Zhang ZC, Wynn RM, Seemann J (2015) GM130 regulates Golgi-derived spindle assembly by activating TPX2 and capturing microtubules. Cell 162:287–299

Witkos TM, Lowe M (2016) The Golgin family of coiled-coil tethering proteins. Front Cell Dev Biol 3:86

Wong M, Gillingham AK, Munro S (2017) The golgin coiled-coil proteins capture different types of transport carriers via distinct N-terminal motifs. BMC Biol 15:3

Wu M, Lu L, Hong W, Song H (2004) Structural basis for recruitment of GRIP domain golgin-245 by small GTPase Arl1. Nat Struct Mol Biol 11:86–94

Wu Q, Zhuang K, Li H (2016) PAQR3 plays a suppressive role in laryngeal squamous cell carcinoma. Tumour Biol 37:561–565

Wu MH, Luo JD, Wang WC, Chang TH, Hwang WL, Lee KH, Liu SY, Yang JW, Chiou CT, Chang CH, Chiang WF (2018) Risk analysis of malignant potential of oral verrucous hyperplasia: a follow-up study of 269 patients and copy number variation analysis. Head Neck 40:1046–1056

Xiang P, Sun Y, Liu Y, Shu Q, Zhu Y (2018) Really interesting new gene finger protein 121 is a tumor suppressor of renal cell carcinoma. Gene 676:322–328

Xiu Y, Liu Z, Xia S, Jin C, Yin H, Zhao W, Wu Q (2014) MicroRNA-137 upregulation increases bladder cancer cell proliferation and invasion by targeting PAQR3. PLoS One 9:e109734

Xu Y, Deng N, Wang X, Chen Y, Li G, Fan H (2017) RKTG overexpression inhibits proliferation and induces apoptosis of human leukemia cells via suppression of the ERK and PI3K/AKT signaling pathways. Oncol Lett 14:965–970

Yamane J, Kubo A, Nakayama K, Yuba-Kubo A, Katsuno T, Tsukita S, Tsukita S (2007) Functional involvement of TMF/ARA160 in Rab6-dependent retrograde membrane traffic. Exp Cell Res 313:3472–3485

Yang X, Kovalenko D, Nadeau RJ, Harkins LK, Mitchell J, Zubanova O, Chen PY, Friesel R (2004) Sef interacts with TAK1 and mediates JNK activation and apoptosis. J Biol Chem 279:38099–38102

Yu L, Zhou L, Cheng Y, Sun L, Fan J, Liang J, Guo M, Liu N, Zhu L (2014) MicroRNA-543 acts as an oncogene by targeting PAQR3 in hepatocellular carcinoma. Am J Cancer Res 4:897–906

Yu N, Signorile L, Basu S, Ottema S, Lebbink JHG, Leslie K, Smal I, Dekkers D, Demmers J, Galjart N (2016) Isolation of functional tubulin dimers and of tubulin-associated proteins from mammalian cells. Curr Biol 26:1728–1736

Zemirli N, Pourcelot M, Dogan N, Vazquez A, Arnoult D (2014) The E3 ubiquitin ligase RNF121 is a positive regulator of NF-κB activation. Cell Commun Signal 12:72

Zhang X, Wang Y (2016) GRASPs in Golgi structure and function. Front Cell Dev Biol 3:84

Zhang Y, Jiang X, Qin X, Ye D, Yi Z, Liu M, Bai O, Liu W, Xie X, Wang Z, Fang J, Chen Y (2010) RKTG inhibits angiogenesis by suppressing MAPK-mediated autocrine VEGF signaling and is downregulated in clear-cell renal cell carcinoma. Oncogene 29:5404–5415

Zhang H, Zhao X, Yan L, Li M (2011) Similar expression to FGF (Sef) reduces endometrial adenocarcinoma cells proliferation via inhibiting fibroblast growth factor 2-mediated MAPK/ERK signaling pathway. Gynecol Oncol 122:669–674

Zhang Y, Hu W, Wang L, Han B, Lin R, Wei N (2017) Association of GOLPH2 expression with survival in non-small-cell lung cancer: clinical implications and biological validation. Biomark Med 11:967–977

Zhao Y, Hongdu B, Ma D, Chen Y (2014) Really interesting new gene finger protein 121 is a novel Golgi-localized membrane protein that regulates apoptosis. Acta Biochim Biophys Sin Shanghai 46:668–674

Zhao J, Yang C, Guo S, Wu Y (2015) GM130 regulates epithelial-to-mesenchymal transition and invasion of gastric cancer cells via snail. Int J Clin Exp Pathol 8:10784–10791

Zhao C, Li Y, Chen G, Wang F, Shen Z, Zhou R (2017) Overexpression of miR-15b-5p promotes gastric cancer metastasis by regulating PAQR3. Oncol Rep 38:352–358

Zhou Z, Sun X, Zou Z, Sun L, Zhang T, Guo S, Wen Y, Liu L, Wang Y, Qin J, Li L, Gong W, Bao S (2010) PRMT5 regulates Golgi apparatus structure through methylation of the golgin GM130. Cell Res 20:1023–1033

Zhou F, Wang S, Wang J (2017) PAQR3 inhibits the proliferation and tumorigenesis in esophageal Cancer cells. Oncol Res 25:663–671

Zisman-Rozen S, Fink D, Ben-Izhak O, Fuchs Y, Brodski A, Kraus MH, Bejar J, Ron D (2007) Downregulation of Sef, an inhibitor of receptor tyrosine kinase signaling, is common to a variety of human carcinomas. Oncogene 26:6093–6098

Printed by Printforce, the Netherlands